Lecture Notes in Computer Science 10366

Commenced Publication in 1973
Founding and Former Series Editors:
Gerhard Goos, Juris Hartmanis, and Jan van Leeuwen

More information about this series at http://www.springer.com/series/7409

Lei Chen · Christian S. Jensen
Cyrus Shahabi · Xiaochun Yang
Xiang Lian (Eds.)

Web and Big Data

First International Joint Conference, APWeb-WAIM 2017
Beijing, China, July 7–9, 2017
Proceedings, Part I

 Springer

Editors
Lei Chen
Computer Science and Engineering
Hong Kong University of Science and
   Technology
Hong Kong
China

Christian S. Jensen
Computer Science
Aarhus University
Aarhus N
Denmark

Cyrus Shahabi
Computer Science
University of Southern California
Los Angeles, CA
USA

Xiaochun Yang
Northeastern University
Shenyang
China

Xiang Lian
Kent State University
Kent, OH
USA

ISSN 0302-9743   ISSN 1611-3349   (electronic)
Lecture Notes in Computer Science
ISBN 978-3-319-63578-1   ISBN 978-3-319-63579-8   (eBook)
DOI 10.1007/978-3-319-63579-8

Library of Congress Control Number: 2017947034

LNCS Sublibrary: SL3 – Information Systems and Applications, incl. Internet/Web, and HCI

Printed on acid-free paper

This Springer imprint is published by Springer Nature
The registered company is Springer International Publishing AG
The registered company address is: Gewerbestrasse 11, 6330 Cham, Switzerland

Preface

This volume (LNCS 10366) and its companion volume (LNCS 10367) contain the proceedings of the first Asia-Pacific Web (APWeb) and Web-Age Information Management (WAIM) Joint Conference on Web and Big Data, called APWeb-WAIM. This new joint conference aims to attract participants from different scientific communities as well as from industry, and not merely from the Asia Pacific region, but also from other continents. The objective is to enable the sharing and exchange of ideas, experiences, and results in the areas of World Wide Web and big data, thus covering Web technologies, database systems, information management, software engineering, and big data. The first APWeb-WAIM conference was held in Beijing during July 7–9, 2017.

As a new Asia-Pacific flagship conference focusing on research, development, and applications in relation to Web information management, APWeb-WAIM builds on the successes of APWeb and WAIM: APWeb was previously held in Beijing (1998), Hong Kong (1999), Xi'an (2000), Changsha (2001), Xi'an (2003), Hangzhou (2004), Shanghai (2005), Harbin (2006), Huangshan (2007), Shenyang (2008), Suzhou (2009), Busan (2010), Beijing (2011), Kunming (2012), Sydney (2013), Changsha (2014), Guangzhou (2015), and Suzhou (2016); and WAIM was held in Shanghai (2000), Xi'an (2001), Beijing (2002), Chengdu (2003), Dalian (2004), Hangzhou (2005), Hong Kong (2006), Huangshan (2007), Zhangjiajie (2008), Suzhou (2009), Jiuzhaigou (2010), Wuhan (2011), Harbin (2012), Beidaihe (2013), Macau (2014), Qingdao (2015), and Nanchang (2016). With the fast development of Web-related technologies, we expect that APWeb-WAIM will become an increasingly popular forum that brings together outstanding researchers and developers in the field of Web and big data from around the world.

The high-quality program documented in these proceedings would not have been possible without the authors who chose APWeb-WAIM for disseminating their findings. Out of 240 submissions to the research track and 19 to the demonstration track, the conference accepted 44 regular (18%), 32 short research papers, and ten demonstrations. The contributed papers address a wide range of topics, such as spatial data processing and data quality, graph data processing, data mining, privacy and semantic analysis, text and log data management, social networks, data streams, query processing and optimization, topic modeling, machine learning, recommender systems, and distributed data processing.

The technical program also included keynotes by Profs. Sihem Amer-Yahia (National Center for Scientific Research, CNRS, France), Masaru Kitsuregawa (National Institute of Informatics, NII, Japan), and Mohamed Mokbel (University of Minnesota, Twin Cities, USA) as well as tutorials by Prof. Reynold Cheng (The University of Hong Kong, SAR China), Prof. Guoliang Li (Tsinghua University, China), Prof. Arijit Khan (Nanyang Technological University, Singapore), and

Prof. Yu Zheng (Microsoft Research Asia, China). We are grateful to these distinguished scientists for their invaluable contributions to the conference program.

As a new joint conference, teamwork is particularly important for the success of APWeb-WAIM. We are deeply thankful to the Program Committee members and the external reviewers for lending their time and expertise to the conference. Special thanks go to the local Organizing Committee led by Jun He, Yongxin Tong, and Shimin Chen. Thanks also go to the workshop co-chairs (Matthias Renz, Shaoxu Song, and Yang-Sae Moon), demo co-chairs (Sebastian Link, Shuo Shang, and Yoshiharu Ishikawa), industry co-chairs (Chen Wang and Weining Qian), tutorial co-chairs (Andreas Züfle and Muhammad Aamir Cheema), sponsorship chair (Junjie Yao), proceedings co-chairs (Xiang Lian and Xiaochun Yang), and publicity co-chairs (Hongzhi Yin, Lei Zou, and Ce Zhang). Their efforts were essential to the success of the conference. Last but not least, we wish to express our gratitude to the Webmaster (Zhao Cao) for all the hard work and to our sponsors who generously supported the smooth running of the conference.

We hope you enjoy the exciting program of APWeb-WAIM 2017 as documented in these proceedings.

June 2017
Xiaoyong Du
Beng Chin Ooi
M. Tamer Özsu
Bin Cui
Lei Chen
Christian S. Jensen
Cyrus Shahabi

Organization

Organizing Committee

General Co-chairs

Xiaoyong Du — Renmin University of China, China
BengChin Ooi — National University of Singapore, Singapore
M. Tamer Özsu — University of Waterloo, Canada

Program Co-chairs

Lei Chen — Hong Kong University of Science and Technology, China
Christian S. Jensen — Aalborg University, Denmark
Cyrus Shahabi — The University of Southern California, USA

Workshop Co-chairs

Matthias Renz — George Mason University, USA
Shaoxu Song — Tsinghua University, China
Yang-Sae Moon — Kangwon National University, South Korea

Demo Co-chairs

Sebastian Link — The University of Auckland, New Zealand
Shuo Shang — King Abdullah University of Science and Technology, Saudi Arabia
Yoshiharu Ishikawa — Nagoya University, Japan

Industrial Co-chairs

Chen Wang — Innovation Center for Beijing Industrial Big Data, China
Weining Qian — East China Normal University, China

Proceedings Co-chairs

Xiang Lian — Kent State University, USA
Xiaochun Yang — Northeast University, China

Tutorial Co-chairs

Andreas Züfle — George Mason University, USA
Muhammad Aamir Cheema — Monash University, Australia

ACM SIGMOD China Lectures Co-chairs

Guoliang Li   Tsinghua University, China
Hongzhi Wang   Harbin Institute of Technology, China

Publicity Co-chairs

Hongzhi Yin   The University of Queensland, Australia
Lei Zou   Peking University, China
Ce Zhang   Eidgenössische Technische Hochschule ETH, Switzerland

Local Organization Co-chairs

Jun He   Renmin University of China, China
Yongxin Tong   Beihang University, China
Shimin Chen   Chinese Academy of Sciences, China

Sponsorship Chair

Junjie Yao   East China Normal University, China

Web Chair

Zhao Cao   Beijing Institute of Technology, China

Steering Committee Liaison

Yanchun Zhang   Victoria University, Australia

Senior Program Committee

Dieter Pfoser   George Mason University, USA
Ilaria Bartolini   University of Bologna, Italy
Jianliang Xu   Hong Kong Baptist University, SAR China
Mario Nascimento   University of Alberta, Canada
Matthias Renz   George Mason University, USA
Mohamed Mokbel   University of Minnesota, USA
Ralf Hartmut Güting   Fernuniversität in Hagen, Germany
Seungwon Hwang   Yongsei University, South Korea
Sourav S. Bhowmick   Nanyang Technological University, Singapore
Tingjian Ge   University of Massachusetts Lowell, USA
Vincent Oria   New Jersey Institute of Technology, USA
Walid Aref   Purdue University, USA
Wook-Shin Han   Pohang University of Science and Technology, Korea
Yoshiharu Ishikawa   Nagoya University, Japan

Program Committee

Alex Delis   University of Athens, Greece
Alex Thomo   University of Victoria, Canada

Aviv Segev	Korea Advanced Institute of Science and Technology, South Korea
Baoning Niu	Taiyuan University of Technology, China
Bin Cui	Peking University, China
Bin Yang	Aalborg University, Denmark
Carson Leung	University of Manitoba, Canada
Chih-Hua Tai	National Taipei University, China
Cuiping Li	Renmin University of China, China
Daniele Riboni	University of Cagliari, Italy
Defu Lian	University of Electronic Science and Technology of China, China
Dejing Dou	University of Oregon, USA
Demetris Zeinalipour	Max Planck Institute for Informatics, Germany and University of Cyprus, Cyprus
Dhaval Patel	Indian Institute of Technology Roorkee, India
Dimitris Sacharidis	Technische Universität Wien, Vienna, Austria
Fei Chiang	McMaster University, Canada
Ganzhao Yuan	South China University of Technology, China
Giovanna Guerrini	Universita di Genova, Italy
Guoliang Li	Tsinghua University, China
Guoqiong Liao	Jiangxi University of Finance and Economics, China
Hailong Sun	Beihang University, China
Han Su	University of Southern California, USA
Hiroaki Ohshima	Kyoto University, Japan
Hong Chen	Renmin University of China, China
Hongyan Liu	Tsinghua University, China
Hongzhi Wang	Harbin Institute of Technology, China
Hongzhi Yin	The University of Queensland, Australia
Hua Li	Aalborg University, Denmark
Hua Lu	Aalborg University, Denmark
Hua Wang	Victoria University, Melbourne, Australia
Hua Yuan	University of Electronic Science and Technology of China, China
Iulian Sandu Popa	Inria and PRiSM Lab, University of Versailles Saint-Quentin, France
James Cheng	Chinese University of Hong Kong, SAR China
Jeffrey Xu Yu	Chinese University of Hong Kong, SAR China
Jiaheng Lu	University of Helsinki, Finland
Jiajun Liu	Renmin University of China, China
Jialong Han	Nanyang Technological University, Singapore
Jian Yin	Zhongshan University, China
Jianliang Xu	Hong Kong Baptist University, SAR China
Jianmin Wang	Tsinghua University, China
Jiannan Wang	Simon Fraser University, Canada
Jianting Zhang	City College of New York, USA
Jianzhong Qi	University of Melbourne, Australia

Jinchuan Chen	Renmin University of China, China
Ju Fan	National University of Singapore, Singapore
Jun Gao	Peking University, China
Junfeng Zhou	Yanshan University, China
Junhu Wang	Griffith University, Australia
Kai Zeng	University of California, Berkeley, USA
Karine Zeitouni	PRISM University of Versailles St-Quentin, Paris, France
Kyuseok Shim	Seoul National University, Korea
Lei Zou	Peking University, China
Lei Chen	Hong Kong University of Science and Technology, SAR China
Leong Hou U.	University of Macau, SAR China
Liang Hong	Wuhan University, China
Lianghuai Yang	Zhejiang University of Technology, China
Long Guo	Peking University, China
Man Lung Yiu	Hong Kong Polytechnic University, SAR China
Markus Endres	University of Augsburg, Germany
Maria Damiani	University of Milano, Italy
Meihui Zhang	Singapore University of Technology and Design, Singapore
Mihai Lupu	Vienna University of Technology, Austria
Mirco Nanni	ISTI-CNR Pisa, Italy
Mizuho Iwaihara	Waseda University, Japan
Mohammed Eunus Ali	Bangladesh University of Engineering and Technology, Bangladesh
Peer Kroger	Ludwig-Maximilians-University of Munich, Germany
Peiquan Jin	Univerisity of Science and Technology of China
Peng Wang	Fudan University, China
Yaokai Feng	Kyushu University, Japan
Wookey Lee	Inha University, Korea
Raymond Chi-Wing Wong	Hong Kong University of Science and Technology, SAR China
Richong Zhang	Beihang University, China
Sanghyun Park	Yonsei University, Korea
Sangkeun Lee	Oak Ridge National Laboratory, USA
Sanjay Madria	Missouri University of Science and Technology, USA
Shengli Wu	Jiangsu University, China
Shi Gao	University of California, Los Angeles, USA
Shimin Chen	Chinese Academy of Sciences, China
Shuai Ma	Beihang University, China
Shuo Shang	King Abdullah University of Science and Technology, Saudi Arabia
Sourav S Bhowmick	Nanyang Technological University, Singapore
Stavros Papadopoulos	Intel Labs and MIT, USA
Takahiro Hara	Osaka University, Japan
Taketoshi Ushiama	Kyushu University, Japan

Tieyun Qian	Wuhan University, China
Ting Deng	Beihang University, China
Tru Cao	Ho Chi Minh City University of Technology, Vietnam
Vicent Zheng	Advanced Digital Sciences Center, Singapore
Vinay Setty	Aalborg University, Denmark
Wee Ng	Institute for Infocomm Research, Singapore
Wei Wang	University of New South Wales, Australia
Weining Qian	East China Normal University, China
Weiwei Sun	Fudan University, China
Wei-Shinn Ku	Auburn University, USA
Wenjia Li	New York Institute of Technology, USA
Wen Zhang	Wuhan University, China
Wolf-Tilo Balke	Braunschweig University of Technology, Germany
Xiang Lian	Kent State University, USA
Xiang Zhao	National University of Defence Technology, China
Xiangliang Zhang	King Abdullah University of Science and Technology, Saudi Arabia
Xiangmin Zhou	RMIT University, Australia
Xiaochun Yang	Northeast University, China
Xiaofeng He	East China Normal University, China
Xiaoyong Du	Renmin University of China, China
Xike Xie	University of Science and Technology of China, China
Xingquan Zhu	Florida Atlantic University, USA
Xuan Zhou	Renmin University of China, China
Yanghua Xiao	Fudan University, China
Yang-Sae Moon	Kangwon National University, South Korea
Yasuhiko Morimoto	Hiroshima University, Japan
Yijie Wang	National University of Defense Technology, China
Yingxia Shao	Peking University, China
Yong Zhang	Tsinghua University, China
Yongxin Tong	Beihang University, China
Yoshiharu Ishikawa	Nagoya University, Japan
Yu Gu	Northeast University, China
Yuan Fang	Institute for Infocomm Research, Singapore
Yueguo Chen	Renmin University of China, China
Yunjun Gao	Zhejiang University, China
Zakaria Maamar	Zayed University, United Arab Emirates
Zhaonian Zou	Harbin Institute of Technology, China
Zhengjia Fu	Advanced Digital Sciences Center, Singapore
Zhiguo Gong	University of Macau, SAR China
Zouhaier Brahmia	University of Sfax, Tunisia

Keynotes

Key notes

A Holistic View of Human Factors in Crowdsourcing

Sihem Amer-Yahia

CNRS, University of Grenoble Alpes, Grenoble, France
sihem.amer-yahia@cnrs.fr

Abstract. For over 40 years, organization studies have examined human factors in physical workplaces and their influence on the ability of an individual to perform a task, or a set of tasks, alone or in collaboration with others. In a virtual marketplace, the crowd is typically volatile, its arrival and departure asynchronous, and its levels of attention and accuracy diverse. This has generated a wealth of new research ranging from studying workers' fatigue in task completion to examining the role of motivation in task assignment. I will review such work and argue that we need a holistic view to take full advantage of human factors such as skills, expected wage and motivation, in improving the performance of a crowdsourcing platform.

Experience on XXX Health such as Earth Health and Human Health Though Big Data

Masaru Kitsuregawa[1,2]

[1] The University of Tokyo, Tokyo, Japan
[2] National Institute of Informatics, Tokyo, Japan
kitsure@tkl.iis.u-tokyo.ac.jp

Abstract. We have been working in the area so called 'Health'. In this talk, our experiences on the problem solving for earth environmental health and human health by big data system technologies are presented. We are wondering what type of platform be suitable for societal health as a whole.

Thinking Spatial

Mohamed Mokbel

Department of Computer Science and Engineering, University of Minnesota
mokbel@umn.edu

Abstract. The need to manage and analyze spatial data is hampered by the lack of specialized systems to support such data. System builders mostly build general-purpose systems that are generic enough to handle any kind of attributes. Whenever there is a pressing need for spatial data support, it is considered as an afterthought problem that can be addressed by adding new data types, extensions, or spatial cartridges to existing systems. This talk advocates for dealing with spatial data as first class citizens, and for always thinking spatially whenever it comes to system design. This is well justified by the proliferation of location-based applications that are mainly relying on spatial data. The talk will go through various system designs and show how they would be different if we have designed them while thinking spatially. Examples of these systems include data base systems, big data systems, recommender systems, social networks, and crowd sourcing.

Contents – Part I

Data Mining, Privacy and Semantic Analysis

Text and Log Data Management

Social Networks

Data Mining and Data Streams

Query Processing

Topic Modeling

Contents – Part II

Distributed Data Processing and Applications

Machine Learning and Optimization

Demo Papers

Tutorials

Meta Paths and Meta Structures: Analysing Large Heterogeneous Information Networks

Reynold Cheng[✉], Zhipeng Huang, Yudian Zheng, Jing Yan, Ka Yu Wong, and Eddie Ng

University of Hong Kong, Pokfulam Road, Pokfulam, Hong Kong
{ckcheng,zphuang,ydzheng2,jyan}@cs.hku.hk, kywong2@connect.hku.hk, ngheii@gmail.com
http://www.cs.hku.hk/~ckcheng/

Abstract. A heterogeneous information network (HIN) is a graph model in which objects and edges are annotated with types. Large and complex databases, such as YAGO and DBLP, can be modeled as HINs. A fundamental problem in HINs is the computation of closeness, or relevance, between two HIN objects. Relevance measures, such as PCRW, PathSim, and HeteSim, can be used in various applications, including information retrieval, entity resolution, and product recommendation. These metrics are based on the use of meta-paths, essentially a sequence of node classes and edge types between two nodes in a HIN. In this tutorial, we will give a detailed review of meta-paths, as well as how they are used to define relevance. In a large and complex HIN, retrieving meta paths manually can be complex, expensive, and error-prone. Hence, we will explore systematic methods for finding meta paths. In particular, we will study a solution based on the Query-by-Example (QBE) paradigm, which allows us to discover meta-paths in an effective and efficient manner.

We further generalise the notion of meta path to "meta structure", which is a directed acyclic graph of object types with edge types connecting them. Meta structure, which is more expressive than the meta path, can describe complex relationship between two HIN objects (e.g., two papers in DBLP share the same authors and topics). We will discuss three relevance measures based on meta structure. Due to the computational complexity of these measures, we also study an algorithm with data structures proposed to support their evaluation. Finally, we will examine solutions for performing query recommendation based on meta-paths. We will also discuss future research directions.

1 Background

Heterogeneous information networks (HINs), such as DBLP [5], YAGO [8], and DBpedia [1], have recently received a lot of attention. These data sources, containing a vast number of inter-related facts, facilitate the discovery of interesting knowledge [4,6,7]. Figure 1(a) illustrates an HIN, which describes the relationship among entities of different types (e.g., author, paper, venue and topic). For

© Springer International Publishing AG 2017
L. Chen et al. (Eds.): APWeb-WAIM 2017, Part I, LNCS 10366, pp. 3–7, 2017.
DOI: 10.1007/978-3-319-63579-8_1

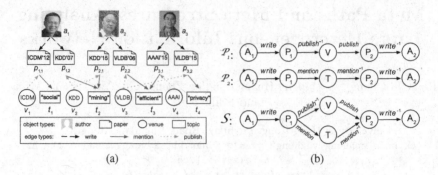

Fig. 1. HIN, meta paths, and meta structures.

example, Jiawei Han (a_2) has *written* a VLDB paper ($p_{2,2}$), which *mentions* the topic "efficient" (t_3).

Given two HIN objects a and b, the evaluation of their *relevance* is of fundamental importance. This quantifies the degree of closeness between a and b. In Fig. 1(a), Jian Pei (a_1) and Jiawei Han (a_2) have a high relevance score, since they have both published papers with keyword "mining" in the same venue (KDD). Relevance finds its applications in information retrieval, recommendation, and clustering [9,10]: a researcher can retrieve papers that have high relevance in terms of topics and venues in DBLP; in YAGO, relevance facilitates the extraction of actors who are close to a given director. As another example, in entity resolution applications, duplicated HIN object pairs having high relevance scores (e.g., two different objects in an HIN referring to the same real-world person) can be identified and removed from the HIN.

Relevance Computation. In this tutorial, we will explore different ways of computing the relevance between two graph objects, for instance, neighborhood-based measures, such as *common neighbors* and *Jaccard's coefficient*; graph-theoretic measures based on random walks, such as *Personalized PageRank* and *SimRank*. These measures do not consider object and edge type information in an HIN. We will discuss the concept of *meta paths* [4,9]. A meta path is a sequence of object types with edge types between them. Figure 1(b) illustrates a meta path $\mathcal{P}_1$, which states that two authors (A_1 and A_2) are related by their publications in the same venue (V). Another meta path $\mathcal{P}_2$ says that two authors have written papers containing the same topic (T). We will discuss several meta-path-based relevance measures, including *PathCount*, *PathSim*, and *Path Constrained Random Walk (PCRW)* [4,9]. These measures have been shown to be better than those that do not consider object and edge type information.

We will further discuss *meta structures*, recently proposed in [3], to depict the relationship of two graph objects. This is essentially a directed acyclic graph of object and edge types. Figure 1(b) illustrates a meta structure $\mathcal{S}$, which depicts that two authors are relevant if they have published papers in the same venue, and have also mentioned the same topic. A meta path (e.g., $\mathcal{P}_1$ or $\mathcal{P}_2$) is a special case of a meta structure. However, a meta path fails to capture such complex relationship that can be conveniently expressed by a meta structure (e.g., $\mathcal{S}$). We will discuss how meta structures can be used to formulate three relevance definitions, as well as their efficient calculation.

Meta Path Discovery. There are often a huge number of meta paths between a pair of HIN objects. This can be very difficult, even for a domain expert, to identify the right meta paths. We will discuss a meta path discovery algorithm, recently proposed by [6], where users provide example instances of source and target objects through a *Query-by-Example* paradigm, to derive meta paths automatically. We will demonstrate a HIN search engine prototype based on this algorithm.

Query Recommendation. We will study the use of meta paths in query recommendation, where queries are suggested to web search users based on their previous query histories. As studied in [2], it is possible to use a knowledge base (a HIN) and its related meta-paths to perform effective query recommendation. The approach is especially useful to *long-tail queries* that rarely appear in query logs.

2   Proposed Schedule

The following is our proposed schedule of the 90-min tutorial.

- **Introduction (15 min).** We will discuss the basic model of HIN, and discuss applications based on it, such as search, relevance computation, query recommendation, and data integration (10 min). We will also introduce meta-paths, a fundamental HIN analysis tool, and give an overview of the tutorial (5 min).
- **Main contents (60 min).** Next, we will introduce meta path, and how it facilitates the computation of various relevance measures (10 min). We then explain the process of discovering meta paths (15 min). We discuss a novel query recommendation framework based on meta paths (15 min). We will also present the meta structures, which is the latest development of meta paths (15 min). We will demonstrate a HIN search engine prototype based on meta paths (5 min).
- **Conclusions (15 min).** We will conclude the tutorial and discuss future directions (5 min). The rest of the time will be dedicated to Q&A (10 min).

3   Intended Audience

The tutorial is designed for researchers interested in latest development in the field of HINs, especially regarding meta-paths for novel applications. The HIN search demonstration will be give insight to software practitioners for developing recommendation facilities for HINs.

4   Biography of Presenters

Reynold Cheng is an Associate Professor of the Department of Computer Science in the University of Hong Kong. He obtained his PhD from Department of Computer Science of Purdue University in 2005. He was granted an Outstanding Young Researcher Award 2011–12 by HKU. He was the recipient of the 2010 Research Output Prize in the Department of Computer Science of HKU. He also received the U21 Fellowship in 2011. He received the Performance Reward in years 2006 and 2007 awarded by the Hong Kong Polytechnic University. He is a member of the IEEE, the ACM, and ACM SIGMOD. He is an editorial board member of TKDE, DAPD and IS, and was a guest editor for TKDE, DAPD, and Geoinformatica. He is an area chair of ICDE 2017, senior PC member of BigData 2017 and DASFAA 2015, PC co-chair of APWeb 2015, area chair for CIKM 2014, and workshop co-chair of ICDE 2014. He received an Outstanding Service Award in CIKM 2009. He has served as PC members and reviewer for top conferences and journals.

Zhipeng Huang is a 2nd year Ph.D. in the CS department of HKU, supervised by Prof. Nikos Mamoulis and Dr. Reynold Cheng. He received his bachelor degree from EECS department of PKU in 2015. His research interests cover data mining, data management and data cleaning.

Yudian Zheng is a 4th year Ph.D. in the CS department of HKU, supervised by Dr. Reynold Cheng. Yudian's research interests cover crowdsourcing, data management and data cleaning. He has published full research papers in well-established database and data mining conferences/journals, including SIGMOD, VLDB, KDD, WWW, ICDE, and TKDE. He has also taken internships in Microsoft Research and Google Research.

Jing Yan is a 1st year MPhil student supervised by Dr. Reynold Cheng in the CS department of HKU. His research interests include data management and data mining, with emphasis on knowledge graphs and data cleaning.

Ka Yu Wong is currently a MSc student of the CS department of HKU.

Eddie Ng is currently a MSc student of the CS department of HKU.

Acknowledgements. Reynold Cheng, Zhipeng Huang, Yudian Zheng, and Jing Yan were supported by the Research Grants Council of Hong Kong (RGC Projects HKU 17229116 and 17205115) and the University of Hong Kong (Projects 102009508 and 104004129).

References

1. Auer, S., Bizer, C., Kobilarov, G., Lehmann, J., Cyganiak, R., Ives, Z.: DBpedia: a nucleus for a web of open data. In: Aberer, K., et al. (eds.) ASWC/ISWC - 2007. LNCS, vol. 4825, pp. 722–735. Springer, Heidelberg (2007). doi:10.1007/978-3-540-76298-0_52
2. Huang, Z., Cautis, B., Cheng, R., Zheng, Y.: KB-enabled query recommendation for long-tail queries. In: CIKM, pp. 2107–2112 (2016)
3. Huang, Z., Zheng, Y., Cheng, R., Sun, Y., Mamoulis, N., Li, X.: Meta structure: computing relevance in large heterogeneous information networks. In: KDD (2016)

4. Lao, N., Cohen, W.W.: Relational retrieval using a combination of path-constrained random walks. Mach. Learn. **81**(1), 53–67 (2010)
5. Ley, M.: DBLP Computer Science Bibliography (2005)
6. Meng, C., Cheng, R., Maniu, S., Senellart, P., Zhang, W.: Discovering meta-paths in large heterogeneous information networks. In: WWW, pp. 754–764 (2015)
7. Mottin, D., Lissandrini, M., Velegrakis, Y., Palpanas, T.: Exemplar queries: give me an example of what you need. PVLDB **7**(5), 365–376 (2014)
8. Suchanek, F.M., Kasneci, G., Weikum, G.: Yago: a core of semantic knowledge. In: WWW, pp. 697–706 (2007)
9. Sun, Y., Han, J., Yan, X., Yu, P.S., Wu, T.: Pathsim: meta path-based top-k similarity search in heterogeneous information networks. In: PVLDB, pp. 992–1003 (2011)
10. Yu, X., Ren, X., Sun, Y., Sturt, B., Khandelwal, U., Gu, Q., Norick, B., Han, J.: Recommendation in heterogeneous information networks with implicit user feedback. In: RecSys, pp. 347–350 (2013)

Spatial Data Processing and Data Quality

TrajSpark: A Scalable and Efficient In-Memory Management System for Big Trajectory Data

Zhigang Zhang, Cheqing Jin$^{(\boxtimes)}$, Jiali Mao, Xiaolin Yang, and Aoying Zhou

School of Data Science and Engineering,
East China Normal University, Shanghai, China
{zgzhang,jlmao1231,xlyang}@stu.ecnu.edu.cn,
{cqjin,ayzhou}@sei.ecnu.edu.cn

Abstract. The widespread application of mobile positioning devices has generated big trajectory data. Existing disk-based trajectory management systems cannot provide scalable and low latency query services any more. In view of that, we present TrajSpark, a distributed in-memory system to consistently offer efficient management of trajectory data. TrajSpark introduces a new abstraction called IndexTRDD to manage trajectory segments, and exploits a global and local indexing mechanism to accelerate trajectory queries. Furthermore, to alleviate the essential partitioning overhead, it adopts the time-decay model to monitor the change of data distribution and updates the data-partition structure adaptively. This model avoids repartitioning existing data when new batch of data arrives. Extensive experiments of three types of trajectory queries on both real and synthetic dataset demonstrate that the performance of TrajSpark outperforms state-of-the-art systems.

Keywords: Big trajectory data · In-memory · Low latency query

1 Introduction

Recently, with the explosive development of positioning techniques and popular use of intelligent electronic devices, trajectory data of MOs (Moving Objects) has been accumulated rapidly in many applications, such as location-based services (LBS) and geographical information systems (GIS). For example, DiDi[1], the largest one-stop consumer transportation platform in China, now has 1.5 million registered active drivers, and provides services for more than 300 million passengers. The total length of all trajectories generated in this platform reaches around 13 billion kilometers in 2015. Moreover, the volume of trajectory data increases in a surging way. In March 2016, the number of trajectories generated in one day has already exceeded 10 million. It is challenging to provide real-time service over such data. However, as almost all of existing trajectory management systems are disk-oriented (e.g., TrajStore [4], Clost [13], and Elite [18]), they cannot support low latency query services upon big trajectory data.

---

[1] http://www.xiaojukeji.com/en/taxi.html.

© Springer International Publishing AG 2017
L. Chen et al. (Eds.): APWeb-WAIM 2017, Part I, LNCS 10366, pp. 11–26, 2017.
DOI: 10.1007/978-3-319-63579-8_2

Recently, in-memory computing systems are a widely used to provide low latency query services. For instance, Spark[2], a distributed in-memory computing system, has been widely used. Spark provides a data abstraction called RDDs (Resilient Distributed Datasets), to maintain a collection of objects that are partitioned across a cluster of machines. Users can manipulate RDDs conveniently through a batch of predefined operations. However, Spark is lack of indexing mechanism upon RDDs and needs to scan the whole dataset for a given query. Recently, some Spark-based system prototypes have been proposed to process big spatial data, including SpatialSpark [19], LocationSpark [14], GeoSpark [20] and Simba [17]. Amongst them, SpatialSpark implements the spatial join query on top of Spark, but it does not index RDDs. GeoSpark provides a new abstraction, called SRDD, to represent spatial objects such as points and polygons. Although it embeds a local index in each SRDD partition, global index is not supported. LocationSpark proposes a solution to solve query skewness. In contrast of them, Simba extends Spark SQL with native support to spatial operations. Meanwhile, it introduces both global and local indexes over RDDs. However, these prototypes view data as a set of spatial points and employ the point-based indexing strategies. Such strategies decrease the trajectory query performance as points of an MO need to be retrieved from different nodes and sorted to form a chronologically ordered sequence [4]. Moreover, they are by nature designed to manage a static dataset and cannot efficiently react to data distribution changes as data increases. To handle a batch of new data, the whole dataset should be repartitioned from scratch, which is quite computation costly.

Inspired by above observations, we design and implement TrajSpark (Trajectory on Spark) system to support low-latency queries over big trajectory data. TrajSpark proposes a new abstraction called IndexTRDD to manage trajectories as a set of trajectory segments. To accelerate query processing, it imports the global and local indexing mechanism which embeds a local hash index in each data partition and builds a global index over these partitions. Furthermore, TrajSpark tracts the change of data distribution by using a time decay model to continuously support efficient management over the daily increasing big trajectory data. Our main contributions can be summarized as follows:

- We first propose TrajSpark to mange the big trajectory data while existing Spark-based systems only support a static big spatial dataset.
- We introduce IndexTRDD, an RDD of trajectory segments, to support efficient data storage and management by incorporating a global and local indexing strategy.
- We monitor the change of data distribution by importing a time decay model which alleviates the repartitioning overhead occurred in existing Spark-based systems and gets a good partition result at the same time.
- We execute three types of trajectory queries on TrajSpark and conduct extensive experiments to evaluate query performance. Experimental results demonstrate the superiority of TrajSpark over other Spark-based systems.

---

[2] http://spark.apache.org/.

The rest of the paper is organized as follows. Section 2 reviews related works. We give an overview of TrajSpark in Sect. 3, and detailed the system in Sects. 4 and 5. In Sect. 6, we introduce the implementation of three typical trajectory queries in TrajSpark. Section 7 provides an experimental study of our system. Finally, we give a brief conclusion in Sect. 8.

2   Related Work

We review the work mostly related to our research in this section.

Centralized Trajectory Management Systems: There are many centralized systems to manage trajectory data. PIST, an off-line system, supports indexes over points. It first partitions data according to a spatial index, and then supports a temporal index in each partition [2]. SETI segments trajectories into sub-trajectories with the guidance of a spatial index, and groups them into a collection of spatial partitions [3]. It shows that supporting index over trajectory segments is more efficient than indexing trajectory points. TrajStore not only uses an I/O cost model to dynamically segment trajectories, but also uses clustering and compressing techniques to reduce storage overhead. But it only supports range query [4]. These systems can not meet the requirement of big data processing as they adopt the centralized architecture.

Disk-Based Distributed Spatial and Spatio-Temporal Data Management Systems: Recently, some distributed disk-based systems have been proposed to manage spatial data by utilizing the Hadoop[3] framework. Spatial-Hadoop [5] pushes spatial data inside Hadoop core by adopting a layered design and supports efficient spatial operations by employing a two-level index structure. AQWA [1] is an improved version of SpatialHadoop by proposing a workload-aware partition strategy which divides those frequently accessed regions into more fine-grained subregions. There are also some systems particularly designed for big spatio-temporal/trajectory data management. PRADASE [10] and Clost [13] are directly built on top of Hadoop and accelerate queries through a global spatio-temporal index. MD-HBase [11], RHBase [6] and GeMesa [7] are built on top of distributed key/value stores, and they use space-fill curves [6,11] and Geohash [7] algorithms to map spatio-temporal points into single-dimension space separately. Different from above works, Elite [18] is built on top of OpenStack[4] for big uncertain trajectorie. Nevertheless, all the above systems are disk-based, and none of them can provide low latency query services.

Memory-Based Spatial Data Management Systems: SharkDB [16] proposes a column-wise storage format to manage trajectory within main memory. However, it is deployed on a big-memory machine and cannot scale out to the distributed environment. Besides, some distributed in-memory spatial data management systems have been proposed. SpatialSpark [19] and GeoSpark [20] are

---

[3] http://hadoop.apache.org/.
[4] http://www.openstack.org/.

two systems built on top of Spark. SpatialSpark is specifically designed for spatial join queries. GeoSpark proposes a new RDD called SRDD to support typical spatial queries. It supports a local index for each partition of SRDD. In comparison of GeoSpark, LocationSpark proposes a solution to solve query skewness by using Bloom Filter [14]. Simba, built on top of Spark SQL, supports multidimensional data queries [17]. Moreover, both a query optimizer which leverages indexes and some spatial-aware optimizations are imported in Simba to improve query efficiency. The above systems process spatial data as independent data points, while trajectory data are usually viewed as a collection of time series. Directly using these systems to process trajectories queries may sacrifice the query efficiency. Moreover, these systems cannot scale well when a new batch of data is imported to the system.

3 System Overview

The architecture of TrajSpark, as shown in Fig. 1, is composed of four layers: (1) *Apache Spark Layer*, where regular operations and fault tolerance mechanisms are supported by Apache Spark, (2) *Trajectory Presentation Layer*, where a new abstraction called IndexTRDD is designed to support indexes over trajectory data. (3) *Assistant Data Layer*, which monitors the change of data distribution and guides the partitioning of forthcoming data. A global index which indexes partitions of IndexTRDD is maintained. (4) *Query Processing Layer*, which processes trajectory queries in an efficient way by utilizing indexes.

- **Apache Spark Layer:** This layer is directly inherited from Apache Spark, and the description of it is omitted in this paper.
- **Trajectory Presentation Layer:** In this layer, the trajectory segments that are spatio-temporally close will be grouped into the same data partition. In each partition, segments belonging to the same MO are depicted in a space-efficient format. A new abstract called IndexTRDD is proposed to organize

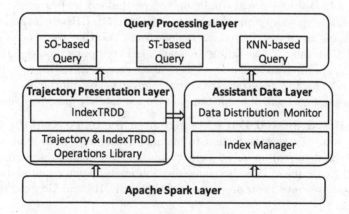

Fig. 1. System overview

all those segments, and a rich operations library is provided to manipulate
trajectories and IndexTRDD. As the core of TrajSpark, we give a detailed
description in Sect. 4.

- **Assistant Data Layer:** A few statistics are maintained in this layer to record
  the change of data distribution. This layer also maintains a global index which
  indexes all partitions of IndexTRDD. This layer is detailed in Sect. 5.
- **Query Processing Layer:** We introduce the implementation of three typical
  trajectory quires in this layer (detailed in Sect. 6).

4   Trajectory Presentation Layer

4.1   Trajectory Segment Presentation

Instead of storing trajectories as a time series directly, the raw trajectories generated from data sources are usually stored as GPS logs and each log record
corresponds to a trajectory point. The schema of such points can be viewed as
a table with the following form: $(MOID, Location, Time, A_1, \cdots, A_n)$, where
$MOID$ is the identification of an MO, $Time$ and $Location$ are the temporal and
spatial information. The rest attributes vary in different data sources. Although
it is simple to represent the trajectories as an RDD of points by directly loading
the raw data into Spark, it leads to a high storage overhead due to the limitation
of row-stores. Moreover, as analyzed in [4], trajectory segment based technologies
can improve the query efficiency more significantly than point based ones.

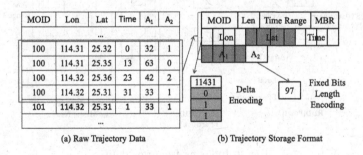

(a) Raw Trajectory Data   (b) Trajectory Storage Format

Fig. 2. Trajectory presentation

To covert raw data into trajectory segments, TrajSpark partitions points that
are spatio-temporally close into the same partition firstly (detailed in Sect. 4.2).
Then, points of the same MO are sorted to form a trajectory segment and the
segment is packed into a space-efficient format as shown in Fig. 2(a). In this
format, values of the same attribute are stored and compressed continuously. For
numberic attributes (such as time and location attribute), data are compressed
by *delta encoding* [4]. For an *enum* attribute (A_2 in Fig. 2(b)), fixed bits length
encoding is used. For other attributes, such as the *string*, we simply use the *gzip*

compression. Besides, we maintain other sketch data such as the $MOID$, Len (length) and MBR (Minimum Bounding Rectangle) of the segment to achieve a quick pruning for queries. Finally, data in each partition are changed to a set of compressed trajectory objects, and the whole dataset is transformed to an RDD of such objects (we call this RDD as TRDD).

4.2   Indexing for Trajectory Data

In the above section, we introduce how to transform the GPS logs into TRDD. While TRDD only supports sequential scan for queries which is very expensive as it needs to access the whole dataset. Hence, we need to support indexing strategy to improve the query efficiency and at the same time without changing the core of Spark. To overcome these challenges, we propose IndexTRDD which changes the storage structure of TRDD by embedding a local hash index in each partition, and get $O(1)$ computation to retain the trajectory of a given MO. Furthermore, we build a global index over data partitions of IndexTRDD to prune irrelevant partitions. Figure 3 details the indexing mechanism which can be divided into three phases: partitioning, local indexing, and global indexing.

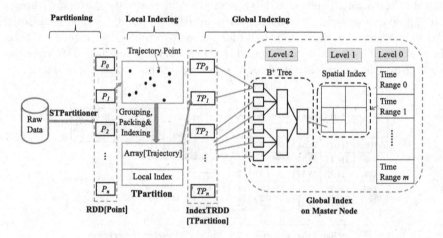

Fig. 3. Indexing from raw data

Partitioning. In this phase, TrajSpark loads the raw dataset from disk into memory as an RDD of trajectory points. This RDD needs to be repartitioned according to the following three constraints: (1) *Data Locality*. Trajectory points that are spatio-temporally close to each other should be assigned to the same partition. (2) *Load Balancing*. All partitions should be roughly of the same size. (3) *Partition Size*. Each partition should have a proper size so as to avoid memory overflow. Spark provides two predefined partitioners for one-dimensional keys, including range and hash partitioner. However, they cannot fit well for multi-dimensional data such as trajectory. To address this problem, TrajSpark defines a new partitioner named **STPartitioner** which contains a spatial Quad tree or k-d tree index. The spatial index can be learned from data distribution and ensures that each leaf node contains the same amount of data. STPartitioner uses the boundaries of those leaf nodes to partition points. Then, trajectory points located in the same boundary are grouped together. Finally, due to the constraint of partition size, TrajSpark splits points belonging to the same boundary into a few data partitions according to $MOID$ (in default) or time attribute, and makes sure each partition satisfy the above constraints.

Local Indexing. After the partitioning phase, the dataset is still an RDD of trajectory points. In this step, we transform the above RDD into TRDD by grouping and packing such points firstly. Then, we add a local index at the head of each partition which maps the MOID of each trajectory to its subscripts. We call the combined data structure of index and trajectory array as TPartition. So the whole dataset is transformed into an RDD of TPartitions, where the RDD is **IndexTRDD**. Finally, we collect the ID (each partition of RDD has a unique ID), the spatial and temporal ranges of each data partition (a TPartition object) to construct the global index.

Global Indexing. The last phase is to build the global index $gIndex$ over all partitions. As shown in Fig. 3, $gIndex$ is a three-level hybrid index. Data is divided by the level-0 coarse time ranges according to its temporal attribute firstly. Each coarse time range corresponds with a level-1 spatial index which is used by STPartitioner. To index partitions that belong to the same spatial boundary, a level-2 B^+-Tree is used. When TrajSpark is initialized with the first batch of data, level-0 index contains only one value (the beginning timestamp of that batch of data), and the level-1 spatial index is the same one used in STPartitioner. The spatial and temporal information collected from all partitions are used to construct the level-2 indexes. Each spatial range in level-1 index corresponds with a level-2 index. TrajSpark keeps $gIndex$ in the memory of master node and updates it when new data partitions arrive. Even for a big trajectory dataset, the number of partitions is not very large (shown in Fig. 4(c)). Thus, the global index can be easily fitted in the memory of master node.

5   Assistant Data Layer

5.1   Data Distribution Monitor

In real applications, new batches of data are appended on an hourly or daily basis [1], and the data distribution changes accordingly. On one hand, a static partitioning strategy results in unbalanced data partitions. On the other hand, if we repartition the whole dataset (required in existing systems [14,17,19,20]) when each batch of data arrives, it leads to an expensive workload. Meanwhile, it is worthless to repartition the old data, because new data are more valuable than those old ones. So, when a new batch of data arrives, TrajSpark tries to only partition this batch of data without touching existing data which differs from the target of AQWA [1] who needs to repartition part of existing data.

Moreover, TrajSpark focuses on long term data distribution changes and also tries to alleviate the influence of temporal changes.

In the light of above considerations, TrajSpark adopts the time decay model to depict the change of data distribution by giving the recent data a higher weight. TrajSpark divides the whole spatial area into $m * m$ fine-grained cells and computes the data distribution by counting the number of points in each cell. When a new batch of data arrives, TrajSpark maintains two matrices: $A_{existing}$ and A_{new}, which separately record the distribution of data that have been loaded in TrajSpark and the new one. After loading the new batch of data, $A_{existing}$ decays weight by dividing γ firstly (γ is the decay factor, which gives *older* data lower weights). Then, A_{new} is added to $A_{existing}$ and set to zero. Observe that after a batch of data is appended, $A_{existing}$ changes accordingly. To better depict the change of $A_{existing}$, we use the notation $A_{existing}^n$ to represent the spatial distribution of data after n batches of data are appended.

To depict the adaptivity of our partitioning strategy, we define a new matrix PA_c (Partition with A) which is initialized with $A_{existing}^0$, and create the spatial index of STPartitioner from PA_c by partitioning the whole spatial area into subregions with equal number of points. After the n-th batch of data is loaded, if the difference between $A_{existing}^n$ and PA_c is larger than a given threshold, it means that the distribution of recently loaded data has greatly changed, TrajSpark updates the value of PA_c with $A_{existing}^n$, and updates STPartitioner using a new spatial index created from the new PA_c to partition the incoming data. In TrajSpark, we use the JSD distance to measure the difference between two data distributions [12] (both the distribution matrices should be normalized before computing). The lazy-update property of time decay model enables TrajSpark to resist abrupt or temporary data distribution changes.

5.2   Index Manager

Index manager mainly supports the update and persistence of the *gIndex*. Two cases will lead to the update of *gIndex*. The first is when the STPartitioner updates its spatial index. At this case a new time range will be added to level-0 index, and the spatial index will be added to level-1 as its children. The second case is when all partitions of the new data have been added to IndexTRDD, the information of these partitions will be added to the level-2 index of *gIndex*. The index manger stores *gIndex* in the memory of the master node. Besides, TrajSpark also chooses to persist it into the file system (after its updating) and has the option of loading it back from the disk. This enables TrajSpark to load indexes back to the system even after system failure. It needs to mention that, TrajSpark supports spatio-temporal operations for *gIndex*, such as *intersect*, *overlap* and so on, to find partitions satisfying the query constraints.

6   Query Processing Layer

Typical trajectory queries include SO (Single Object)-based query [8,10,13], STR (Spatio-Temporal Range)-based [8,15,16] query and KNN (K Nearest

---

Algorithm 1. SO-based query

Input: *moid, tRange*;
Output: one trajectory;
 1: *pids* = gIndex.intersect(*tRange*);
 2: *ts* = IndexTRDD.PartitionPruningRDD(*pids*)
                        .getTraWithID(*moid*).mapValues(sub(*tRange*));
 3: **return** *ts*.reduceByKey(merge).collect();

Neighbor)-based query [8,11,16]. In this section, we introduce how TrajSpark efficiently processes these queries by utilizing indexes and the operation libraries.

6.1 SO-Based Query

An SO-based query retrieves the trajectory of a given MO by receiving two parameters: *moid* and *tRange*, where *moid* denotes the ID of an MO and *tRange* is the temporal constraint. Spark expresses this query as an RDD *filter* action, which requires scanning the whole dataset. TrajSpark can achieve better performance by utilizing indexes. It leverages two observations: (1) The level-0 index is sufficient to prune irrelevant partitions, and the level-2 index can find the partitions whose time ranges are intersected with *tRange*. (2) For each partition, the trajectory of *moid* can be filtered quickly according to the local *hash* index.

Based on the above observations, Algorithm 1 introduces the detailed steps. Firstly, TrajSpark traverses the global index to find data partitions whose time ranges are intersected with the *tRange* (line 1). It needs to mention that, the global index *gIndex* is a spatio-temporal index, and the input parameter for *intersect* operation can also contain a spatial constraint. Next, IndexTRDD calls a Spark API— PartitionPruningRDD, to mark required partitions. Then, TrajSpark randomly accesses the trajectory in each partition according to the given *moid* and finds the sub-trajectory located in the *tRange* (line 2). Finally, all sub-trajectories of the given MO are merged into one. Note that TrajSpark provides the *merge* function to merge two trajectory segments of the same MO.

6.2 STR-Based Query

An STR-based query retrieves trajectories within a spatio-temporal range. It receives two parameters *tRange* and *sRange*. By utilizing the indexes, TrajSpark can also achieve better performance than the *filter* operation of Spark.

Algorithm 2. STR-based query

Input: *tRange, sRange*;
Output: a set of trajectories;
 1: *pids* = gIndex.intersect(*tRange, sRange*);
 2: *ts* = IndexTRDD.PartitionPruningRDD(*pids*)
                        .filter(*tRange, sRange*).mapValues(sub(*tRange, SRange*));
 3: **return** *ts*.reduceByKey(merge).collect();

Algorithm 3. KNN-based query

Input: $IndexTRDD$, $disM$;
Output: the k most similar trajectories to tr;
1: $mbr = tr.MBR$, $tRange = tr.TimeRange$;
2: **repeat**
3:    $pids$ = gIndex.intersect(mbr, $tRange$);
4:    ts = IndexTRDD.PartitionPruningRDD($pids$)
                        .filter(mbr, $tRange$).reduceBykey(merge);
5:    mbr.expand($1 + \alpha$);
6: **until** (ts.size $> k$)
7: $candidate=ts$.collect();
8: **return** $candidate$.map($t \rightarrow$(disM(t, tr),t)).sortByKey.top(k);

Algorithm 2 sketches basic steps to process such queries. At first, TrajSpark traverses the global index to filter partitions that are intersected with the given spatio-temporal range (line 1). Then, for each partition, TrajSpark filters candidates whose spatial bounding box and temporal range are intersected with the given spatio-temporal constraint. Furthermore, it finds a sub-trajectory which is bounded by the spatio-temporal constraint for each candidate (line 2).

6.3 KNN-Based Query

There are many variations of KNN-based query, and we focus on finding top-k trajectories who are most similar to the reference one. This kind of query is very common in trajectory patten analysis, and we represent it with **KNN**(tr, $disM$). Here, tr refers to the query reference, and $disM$ refers to the distance/similarity metric between trajectories (popular metrics such as Euclidean distance, DTW and LCSS are supported in TrajSpark). The processing procedure is shown in Algorithm 3. TrajSpark gets the MBR and time range of tr firstly (line 1). This is because candidate results are spatio-temporally close to the reference, the using of mbr and $tRange$ facilitates the pruning of candidate. Then, TrajSpark filters candidate partitions using the global index (line 3). After that, sub-trajectories are further pruned and merged into complete trajectories (line 4). These trajectories are the candidates of the final result. However, if the number of these candidates is smaller than k, TrajSpark expands the region of mbr (the center of mbr will not change, while the width and length become $1 + \alpha$ ($0 < \alpha < 1$) times) and re-executes the spatio-temporal query until the number of candidates is larger than k. Here, the default value of α is set to 0.2. Finally, TrajSpark measures the similarity for those candidate trajectories and selects k smallest ones as the final result.

7 Experiments

7.1 Experimental Setup

We evaluate the performance of TrajSpark in this section. All experiments are conducted on a 12-node clustering running Spark 1.5.2 over Ubuntu 12.0.4. Each node is equipped with an 8 cores Intel E5335 2.0 GHz processor and 16 GB memory. The Spark cluster is deployed in standalone mode.

Two trajectory datasets of Beijing taxis [9], including a real dataset and a synthetic one, are used to evaluate the performance. The real one, gathered by 13,007 taxis in 3 months (from October to December in 2013), has 2.5 billion records and comprises about 190 GB. Each record contains the following attributes: taxi ID, time, longitude, latitude, speed and many other descriptive information. To better show the scalability of TrajSpark, we generate the synthetic dataset by extending the real one. In the synthetic one, every taxi reports its location every five seconds when it is taken by passengers, and the number of records is 18 billion comprising about 1.4 TB. It needs to mention that the former dataset can be completely loaded into the distributed memory, while the storage overhead of the latter one far exceeds the memory capability of our cluster. Thus, only partial of the synthetic data can be loaded in memory.

We compute the MBR of the spatial range of Beijing and split the rectangle into $1,000 * 1,000$ cells where each cell covers an area of nearly 180 m * 180 m. We compare the performance of TrajSpark with GeoSpark and Simba in terms of query latency and scalability. The latency is represented by the average running time of a few queries, and the scalability is evaluated when different amount of data is loaded into those systems.

7.2 Performance of Data Appending

Firstly, we study the performance of data appending when batches of data are loaded into those systems. Figure 4(a) gives the running time when batches of real dataset are appended (each batch comprises about 32G). In GeoSpark and Simba, the time cost of appending a batch of data increases linearly as the volume of existing data increases, because they should repartition both existing and the new batch of data. While TrajSpark requires less time and the loading time keeps steady with the increase of data volume. This is because TrajSpark only needs to partition the new batch of data and also can reach balanced data partitions. So, in real big data applications where the volume of data grows rapidly, TrajSpark outperforms GeoSpark and Simba significantly.

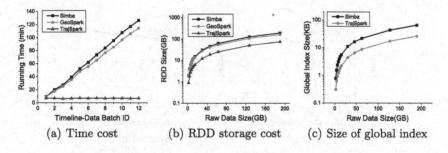

(a) Time cost   (b) RDD storage cost   (c) Size of global index

Fig. 4. Time and storage cost for appending data

Next, we investigate the storage overhead of different systems using the real dataset and show the results with the RDD size of loaded data in Fig. 4(b). TrajSpark has the lowest storage cost due to data compression, while Simba and GeoSpark consume more storage space (about 2–3X), because they should store both the original data and the index tree in each partition. Simba requires more space consumption than GeoSpark since the temporal dimension is also used to index data. We also evaluate the size of global index in these systems. Global index is not supported in GeoSpark, so we only show the results of TrajSpark and Simba. Figure 4(c) indicates that both TrajSpark and Simba have small global index storage overhead (in order of KB). The size of global index in TrajSpark is so small that it can be easily fitted into the memory of the master node. Moreover, there are fewer partitions in TrajSpark due to data compression, so the size of global index in TrajSpark is only about 1/3 that of Simba.

7.3   Query Performance

We first examine the efficiency and scalability of SO-based query. The query latency is represented by the average query time of 100 queries which retrieve the whole history of the given MOs. We increase the volume of dataset by loading the daily generated data. Figure 5 demonstrates that TrajSpark is an order of magnitude faster than Simba, and nearly two orders of magnitude faster than GeoSpark, because GeoSpark needs to scan the whole dataset. Although Simba can prune irrelevant partitions using the global index, it needs to traverse all the content of the selected partitions. In contrast, TrajSpark not only utilizes the global spatio-temporal index to prune partitions but also uses the local hash index to support random access to trajectories. Note that these systems perform better on the real dataset than the synthetic one. This is mainly because the real one can be completely loaded in memory, while only a small part of the synthetic one can be loaded. So queries on the latter dataset require extra I/O cost. Nevertheless, these systems still performs well on the synthetic dataset due to the following reasons: (i) We persist data at the storage level of "MEM_AND_DISK_SER", so hot data can be cached in memory, (ii) By using the global indexes, a huge amount unnecessary I/O costs can be avoided.

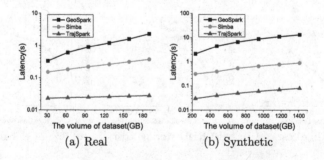

(a) Real   (b) Synthetic

Fig. 5. Performance of SO-based query

Subsequently, we examine the impact of data size for the STR-based query. Since the spatial area is a critical parameter for the query result due to the unbalanced data distribution, we randomly select 100 areas as the spatial constraint for our queries, and each of these areas contains 20*20 cells. These queries only select trajectories generated in the last week. Figure 6(a) and (b) show the performance of these algorithms. We can see that TrajSpark and Simba behave steady, while the query latency of GeoSpark increases linearly. Without a global index, GeoSpark needs to scan the whole dataset. While TrajSpark and Simba utilize the global index to prune data partitions, and the number of data partitions to be scanned does not vary significantly since the query range has not changed greatly. Moreover, both of the latter two systems use a local index to prune trajectories in each partition. Consequently, TrajSpark and Simba are about an order of magnitude faster than GeoSpark. Moreover, TrajSpark is 3–5 times faster than Simba, because it prunes candidates through the MBR and time range of the trajectory. So it can find the result in $O(\log_n)$ (n is the length of a segment) time as the segment is ordered. Differently, Simba needs to sort points of the MO to restore the original segment which costs $O(n \log_n)$.

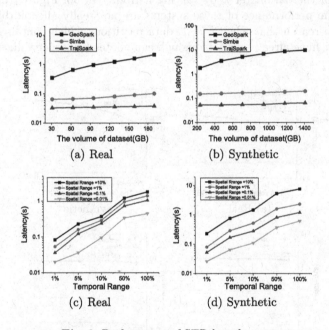

Fig. 6. Performance of STR-based query

Furthermore, we report the performance of our system on STR-based queries under various spatio-temporal ranges. The spatial constraints are 10%, 1%, and 0.1% of the entire region. The temporal constraints are 100%, 50%, 10%, 5% and 1% of the 3 months. As shown in Fig. 6(c) and (d), a large spatial or temporal range usually leads to a longer query latency. But the performance is not essentially linear to the query range, because the number of partitions to be scanned and the amount of data to be accessed in each partition do not grow in a linear way. For example, when the spatial-range is set to 0.01%, and the temporal range increases from 1% to 100%, the query latency grows 30 times. Similarly, when the temporal range is set to 1%, and the spatial range increases from 0.01% to 10%, the query latency grows about 9 times.

Finally, we evaluate the performance of top-k similar trajectory query by using the Euclidean distance as the similarity metric. In this experiment, the trajectories of ten taxis from the same day are selected as the query reference. Figure 7(a) and (b) shows the scalability of TrajSpark when different amount of data is loaded into the system and the value of k is set to 10. TrajSpark is two orders of magnitude faster than GeoSpark, and runs about 4–6x faster than Simba. That is because TrajSpark and Simba prune data partitions with the global index, while GeoSpark has no global index and needs to access all data partitions. In comparison of Simba, TrajSpark does not need to sort the points of each trajectory. This result is similar to that of STR-based query because the core of this query is an iterative spatio-temporal query. Furthermore, we evaluate the impact of the parameter k by varying it from 1 to 50. Figure 7(c) and (d) show that the performance of these systems are not really affected by k. This is due to the reason that when $k = 1$, data partitions which contain the most similar result have already contain enough candidates for larger values of k.

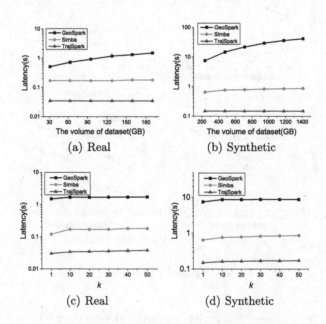

(a) Real   (b) Synthetic

(c) Real   (d) Synthetic

Fig. 7. Performance of KNN-based query

8   Conclusion

To process the massively increasing trajectory data and support near real-time query services, this paper proposes a distributed in-memory system called TrajSpark. This system is built on top of Spark, and proposes IndexTRDD structure that incorporating a global and local indexing mechanism. Additionally, TrajSpark utilizes the time-decaying model to monitor the change of data distribution and enables the data-partition structure to adapt to data changes. We validate the storage overhead, data loading and query latency of TrajSpark by experiments on both real and synthetic datasets. Experimental results show that TrajSpark outperforms existing systems in terms of scalability and efficiency. For future work, we plan to support more complicated operations by utilizing TrajSpark.

Acknowledgement. This paper is supported by the National Key Research and Development Program of China (2016YFB1000905), NSFC (61370101, 61532021, U1501252, U1401256 and 61402180), Shanghai Knowledge Service Platform Project (No. ZF1213).

References

1. Aly, A.M., Mahmood, A.R., Hassan, M.S., Aref, W.G., Ouzzani, M., Elmeleegy, H., Qadah, T.: AQWA: adaptive query-workload-aware partitioning of big spatial data. PVLDB **8**(13), 2062–2073 (2015)
2. Botea, V., Mallett, D., Nascimento, M.A., Sander, J.: PIST: an efficient and practical indexing technique for historical spatio-temporal point data. GeoInformatica **12**(2), 143–168 (2008)
3. Chakka, V.P., Everspaugh, A.C., Patel, J.M.: Indexing large trajectory data sets with seti, vol. 1001, p. 12. Citeseer (2003)
4. Cudré-Mauroux, P., Wu, E., Madden, S.: Trajstore: an adaptive storage system for very large trajectory data sets. In: ICDE, pp. 109–120 (2010)
5. Eldawy, A., Mokbel, M.F.: SpatialHadoop: a MapReduce framework for spatial data. In: ICDE, pp. 1352–1363 (2015)
6. Huang, S., Wang, B., Zhu, J., Wang, G., Yu, G.: R-hbase: a multi-dimensional indexing framework for cloud computing environment. In: ICDM, pp. 569–574 (2014)
7. Hughes, J.N., Annex, A., Eichelberger, C.N., Fox, A., Hulbert, A., Ronquest, M.: Geomesa: a distributed architecture for spatio-temporal fusion. In: SPIE Defense+ Security, p. 94730F (2015)
8. Lange, R., Dürr, F., Rothermel, K.: Scalable processing of trajectory-based queries in space-partitioned moving objects databases. In: SIGSPATIAL, p. 31 (2008)
9. Liu, H., Jin, C., Zhou, A.: Popular route planning with travel cost estimation. In: Navathe, S.B., Wu, W., Shekhar, S., Du, X., Wang, X.S., Xiong, H. (eds.) DASFAA 2016. LNCS, vol. 9643, pp. 403–418. Springer, Cham (2016). doi:10. 1007/978-3-319-32049-6_25
10. Ma, Q., Yang, B., Qian, W., Zhou, A.: Query processing of massive trajectory data based on mapreduce. In: CIKM, pp. 9–16 (2009)

11. Nishimura, S., Das, S., Agrawal, D., El Abbadi, A.: MD-hbase: design and implementation of an elastic data infrastructure for cloud-scale location services. DPD **31**(2), 289–319 (2013)
12. Österreicher, F., Vajda, I.: A new class of metric divergences on probability spaces and its applicability in statistics. AISM **55**(3), 639–653 (2003)
13. Tan, H., Luo, W., Ni, L.M.: Clost: a hadoop-based storage system for big spatiotemporal data analytics. In: CIKM, pp. 2139–2143 (2012)
14. Tang, M., Yu, Y., Malluhi, Q.M., Ouzzani, M., Aref, W.G.: LocationSpark: a distributed in-memory data management system for big spatial data. PVLDB **9**(13), 1565–1568 (2016)
15. Tzoumas, K., Yiu, M.L., Jensen, C.S.: OceanST: a distributed analytic system for large-scale spatiotemporal mobile broadband data. PVLDB **7**, 1561–1564 (2014)
16. Wang, H., Zheng, K., Zhou, X., Sadiq, S.W.: SharkDB: an in-memory storage system for massive trajectory data. In: SIGMOD, pp. 1099–1104 (2015)
17. Xie, D., Li, F., Yao, B., Li, G., Zhou, L., Guo, M.: Simba: efficient in-memory spatial analytics. In: SIGMOD, pp. 1071–1085 (2016)
18. Xie, X., Mei, B., Chen, J., Du, X., Jensen, C.S.: Elite: an elastic infrastructure for big spatiotemporal trajectories. VLDB J. **25**(4), 473–493 (2016)
19. You, S., Zhang, J., Gruenwald, L.: Large-scale spatial join query processing in cloud. In: ICDE Workshops, pp. 34–41 (2015)
20. Yu, J., Wu, J., Sarwat, M.: Geospark: a cluster computing framework for processing large-scale spatial data. In: SIGSPATIAL, pp. 70:1–70:4 (2015)

A Local-Global LDA Model for Discovering Geographical Topics from Social Media

Siwei Qiang, Yongkun Wang$^{(\boxtimes)}$, and Yaohui Jin

Network and Information Center, Shanghai Jiao Tong University, Shanghai, China
{qiangsiwei,ykw,jinyh}@sjtu.edu.cn

Abstract. Micro-blogging services can track users' geo-locations when users check-in their places or use geo-tagging which implicitly reveals locations. This "geo tracking" can help to find topics triggered by certain events in certain regions. However, discovering such topics is very challenging because of the large amount of noisy messages (e.g. daily conversations). This paper proposes a method to model geographical topics, which can filter out irrelevant words by different weights in the local and global contexts. Our method is based on the Latent Dirichlet Allocation (LDA) model but each word is generated from either a local or a global topic distribution by its generation probabilities. We evaluated our model with data collected from Weibo, which is currently the most popular micro-blogging service for Chinese. The evaluation results demonstrate that our method outperforms other baseline methods in several metrics such as model perplexity, two kinds of entropies and KL-divergence of discovered topics.

Keywords: Geolocation · Geographical topics · Topic modeling · Latent Dirichlet Allocation

1 Introduction

Micro-blogging services including Twitter and Weibo have emerged as a medium in spotlight for online users to share breaking news or interesting stories in their lives and update their status anywhere and anytime in their daily lives. With the advancement of positioning technology, the popularity of low-cost GPS chips and wide availability of smart phones, large-scale crowd-generated social media data with geographical records have become prevalent on the web and can also be easily collected. Such textual data with geo-coordinates or geo-tagged locations usually contain landmark information (e.g., scenic spots or famous restaurants) or information on local events (e.g., movies, vocal concerts, exhibitions or sports games), and hence, can provide us rich and interpretable semantics on different locations. It is also possible to infer inherent geographic variability of topics across various locations.

In recent years, a significant amount of research have been conducted on addressing the questions of how the information is created and shared in different

© Springer International Publishing AG 2017
L. Chen et al. (Eds.): APWeb-WAIM 2017, Part I, LNCS 10366, pp. 27–40, 2017.
DOI: 10.1007/978-3-319-63579-8_3

geographic locations and how the spatial and linguistic characteristics of people vary across regions. Among them, a considerable amount of studies have been conducted on GPS-associated documents including organizing geo-tagged documents or photos and studying user movement for POI recommendation, user prediction and time prediction. They try to address the following two needs. The first is to discover different topics of interests those are coherent in geographical regions. For example, a city is usually formed by different functional sub-regions, such as business area, residential area and entertainment area. The second is to comparing the topics discovered across different geographical locations. For example, people would like to know where is the landmarks of the city for tourists or which places they could go when they plan to go shopping or have fun during the weekends.

However, the challenge is that the messages with geo-locations are mixed with overwhelming noisy messages of daily chats or expressions of personal emotions, which have little or no relations to the location context. For example, in the city of Shanghai, *Waitan* is a famous waterfront and one of the most popular scenic spots for tourists. However, even in such a spot, Weibo are still full of daily conversations and greetings such as 'Good night' or 'Have a nice weekend', which has no local semantics. When taking all local posts into account, the meaningfulness or concentrativeness of the discovered topics can be compromised. Therefore, it is very difficult to discover meaningful geo-location topics by existing methods such as inferring occurrences of words from local posts.

This paper proposes an effective method to handle noisy messages and model geographical topics of different locations. The proposed method is based on the Latent Dirichlet Allocation (LDA) [2] topic model. The intuitive idea is that, for a noise specific location, the words used by users are different between (a) daily conversations and (b) the description of the landmark or local events. The former is relatively consistent across different locations, and denoted as *global context*, while the latter, which is essentially helpful to identify the true characteristics of the region or area, varies by sites and are denoted as *local context*. Our method takes the local and global contexts into consideration, and different from all existing models to reveal spatial topics, it models each word to be generated from either its local or the global topic distribution by its estimated probabilities. The proposed strategy is able to distinguish locally featured words from noise and improve the quality of discovered topics.

Our evaluation based on two typical social media datasets. One is from Weibo, which is a Chinese micro-blogging website. Akin to a hybrid of Twitter and Facebook, it is one of the most popular sites in China, in use by well over 30% of Internet users, with a market penetration similar to the United States' Twitter. The other is from Yelp, which publish crowd-sourced reviews about local businesses, as well as the online reservation service and online food-delivery service. Our model is evaluated with several metrics widely used in assessing topic models, such as perplexity and KL-divergence, together two kinds of entropies, topic entropy and location entropy to assess the concentrativeness of the discovered topics. The evaluation results demonstrate that our method outperforms other baseline methods and show its superiority in information filtering and geographical topic discovery.

2   Related Work

In this section we discuss some work related to our study, including geo-tagged social media mining, topic modeling and local word detection.

Mobility or posting pattern mining with geo-tagged social media data become a hot topic with the development of GPS technology. The activities of mobile users are typically represented as follows: a user appears at a certain location (with a pair of latitude and longitude coordinates), and leaves a post (e.g., Weibo or review), which is likely semantically related to the user and/or the location [18]. The mining problem is usually formulated as finding various mobility or posting patterns from user activities, such as frequent patterns, periodic behaviors, representative behaviors and activity recognition [3,7,13]. In literature, numerous methods have been proposed to extract such patterns from the social media data. Representative works include stop and move detection, significant place extraction, frequent regular pattern discovery, transportation mode recognition. However, these works mainly focus on the trajectory or posting patterns, and seldom explore the contextual semantics of user-generated contents.

Topic modeling is a classic task to enable text analysis at a semantic level and to discover hidden semantic structures in a text body. The most representative and widely used topic models are probabilistic latent semantic analysis (pLSA) [1] and LDA [2]. Both are generative statistical models, and assume that in a given dataset each document is associated with a topic distribution, and each topic with a word distribution, the difference is that in LDA, the topic distribution is assumed to have a Dirichlet prior, and in practice, this results in more reasonable mixtures of topics in a document. Recently, in order to support location-aware information retrieval or to compare topics across geographical locations, there are many works in the area of geographical topic modeling [4,5,8,9,12,14–17,19,21–23]. For example, Yin et al. proposes and compares three ways of modeling geographical topics, including a location-driven model, a text-driven model, and a joint model called LGTA [14], which combines geographical clustering and topic modeling into one framework. In this model, the coordinates in each document are drawn from a 2D Gaussian distribution and the region is drawn from a Multinomial distribution over all regions. Hong et al. models diversity in tweets based on topical diversity, geographical diversity, and an interest distribution of the user [16]. Further, it takes the Markovian nature of users' locations into account and identifies topics based on location and language. The spatial Topic (ST) Model for location recommendation has been proposed by Hu and Ester recently to capture the correlation between users' movements and between user interests and the function of locations [18]. A hierarchical topic model which models regional variations of topics has been presented by Ahmed et al., which combines distributions over locations, topics, and over user characteristics, both in terms of location and in terms of their content preferences [20]. Unlike previous work, it automatically infers both the hierarchical structure over content and over the size and position of geographical locations, and gains higher accuracy on location estimation from Tweets. Although all the above works discover regions and geographical topics, they do not consider the overwhelming noisy messages in user-generated contents, which can have a major impact on the results.

Another line relevant to our research is local word detection. The general idea is that when a location specific event of interest takes place, there can be a surge in the volume of documents related to the event, and as a result, such a surge of information can be utilized to identify location-characterized topics. Based on the premise that local words should have concentrated spatial distributions around their location centers, Backstrom et al. proposes a spatial variation model for analyzing geographic distribution of terms in search engine query logs [6], and this method has been used by Cheng et al. to decide whether a word is local or not [12]. Meanwhile, Mathioudakis et al. uses spatial discrepancy to detect spatial bursts, which identifies geographically focused information bursts, attribute them to demographic factors and identify sets of descriptive keywords [10]. In these work, whether the word is local or not is determined by a assigned locality score. However, it is demonstrated by Wu et al. that this method can be erroneous since it assumes one peak density distribution while many local words can have multiple peaks [24]. Our work is different from the existing works in that, word locality is not generated directly but evaluated by the generation probability and is not associated with a static locality score, which means that a word (e.g. car) can be both *non-local* for a majority of locations and also *local* for a few particular locations (e.g. automobile 4 S shops).

3   Method

3.1   Local-Global LDA Model

In this section, we propose a novel topic model for geo-tagged social media texts called LGLDA (Local-Global LDA Model), which combines noise filtering and topic modeling into one framework. To begin with, we define the notations used in this paper as listed in Table 1.

To discover geographical topics, the spatial structure of words should be encoded. The words that are close in space are likely to be clustered into the same geographical topic. However, in our dataset, the geographical distance of two words cannot be calculated due to the loss of the exact geo coordinations, but each peace of text is associated with a location tag, therefore if two words come from texts with the same location tag, they are close, otherwise they are distant. Furthermore, if the exact geo coordinations are assessable, the closeness of any two words can be calculated by Euclidean distance, and our model can be modified by assuming that geographical distribution of each region follows a Gaussian distribution. Hence, the words that are close in space are more likely to belong to the same region, so they are more likely to be clustered into the same topic.

Table 1. Notations used in the paper.

Notation	Description
θ_g	Global topic distribution
θ_l	Local topic distribution for location l
ϕ	Word distribution for topic k
ω	Location relevence
z_e	Latent topic
w	Observed word
α_l	Multinomial distribution prior for θ_l
α_g	Multinomial distribution prior for θ_g
β	Multinomial distribution prior for ϕ
γ	Binomial distribution prior for ω
L	Number of locations
D_l	Number of documents in location l
N_d	Number of words in document d

In our scenario, each geo-located document d is tagged with a location l, and contains a set of words w_d. A geographical topic z is a meaningful theme shared by similar locations, and each location is associated with a topic distribution $p(z|l)$.

We formalize our model based on the following intuitions. Firstly, words close in space are likely to be clustered into the same geographical topic. Therefore, topics are generated from locations instead of individual documents. Secondly, locally featured words have a more compact geographical scope. For example, 'bravo' is a word for a performer, so that it is more possible to be used at a theater, a concert or a stadium rather than other places. On the contrary, noisy words (e.g., happy, love, city) can have a much wider spatial range. However, some words could be local for certain locations although these words are commonly used at many places. In our method, the role (local or non-local) of a word is determined by its generation probabilities of its local and global semantic contexts. Therefore, our model is named as Local-Global LDA model (or LGLDA).

The graphical representation of our model is shown in Fig. 1. Shaded nodes indicate observed variables or priors, while light ones represent latent variables. In order to keep a small set of parameters for simplification, in our model there are one shared set of topics with two different distributions θ_l and θ_g for the local topics and the global topics, respectively. It might be interesting and reasonable to utilize two kinds of ϕ for words' local and global distributions corresponding to the two topic distributions, and we would like to study it in our future work.

For a collection of L locations, geo-tagged by D documents, each contains N words, the topic of each word can either be drawn from θ_l or from θ_g. Topic

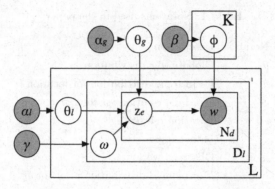

Fig. 1. Graphical representation of the proposed local-global LDA model.

assignment is denoted by z_e, and $e(= l/g)$ indicates whether it is drawn from the local or the global. Since micro-blog is length limited, it is likely to have focused concept. Therefore, if a document is location relevant, each word in it is more likely to be relevant. This relevance of each document is indicated by ω, with binomial distribution prior γ. Finally, the word distribution in K topics is denoted by ϕ. α_l, α_g, β are priors for θ_l, θ_g and ϕ, respectively.

In order to weight between local and global distributions, we add an additional parameter, named local-global weight ratio. Assume the local and global topic distributions are $\theta_l = [p_{l,1}, ..., p_{l,K}]$ and $\theta_g = [p_{g,1}, ..., p_{g,K}]$ respectively, topic assignment is drawn from a concatenated distribution θ in Eq. (1).

$$\theta = \frac{\lambda}{\lambda+1}\theta_l p(e=l|w) \oplus \frac{1}{\lambda+1}\theta_g p(e=g|w)$$
$$= [\frac{\lambda p_{l,1} p_w^{(l)}}{\lambda+1}, ..., \frac{\lambda p_{l,K} p_w^{(l)}}{\lambda+1}, \frac{p_{g,1} p_w^{(g)}}{\lambda+1}, ..., \frac{p_{g,K} p_w^{(g)}}{\lambda+1}]$$

(1)

When λ is too large, the global word set is narrowed and ineffective for noise filtering, while when λ is too small, the size of local words is sparse and it fails to discover meaningful topics. Therefore, an appropriate λ is crucial. In our experiment, it is optimized by estimating the model's perplexity (as illustrated in Fig. 2 in Sect. 4.4).

The generative process of our model is summarized in Algorithm 1.

3.2   Model Inference

Like most Bayesian models, collapsed Gibbs sampling was used for model inference. We present the conditional probability of its latent variables $z_e, \theta_l, \theta_g, \phi$ and w for sampling. Details are omitted for limited space. It is assumed that topic distributions θ_l, θ_g and word distribution ϕ of each topic k are drawn from dirichlet distributions of their respective priors α_l, α_g and β, while locality relevance ω are drawn from a binomial distribution with prior γ.

Algorithm 1. Generative process of LGLDA model.

1: **for each** l-th location **do**
2:      Draw a Dirichlet distribution over all latent topics $\theta_l \sim Dirichlet(\alpha_l)$.
3: **end for**
4: Draw a Dirichlet distribution over all latent topics $\theta_g \sim Dirichlet(\alpha_g)$.
5: **for each** k-th topic **do**
6:      Draw a Dirichlet distribution over all words $\phi \sim Dirichlet(\beta)$.
7: **end for**
8: **for each** l-th location **do**
9:      **for each** d-th Document **do**
10:         Draw a Bernoulli distribution $\omega \sim Dirichlet(\gamma)$.
11:         **for each** w-th word position **do**
12:             Draw a topic from Multinominal distribution $z_e \sim Dirichlet(\theta_{l|g})$.
13:             Draw a word from Multinominal distribution $w \sim Dirichlet(\phi)$.
14:         **end for**
15:      **end for**
16: **end for**

The conditional probability for sampling the topic assignment z_e of each word is computed in Eq. (2), where $z_{e,i} = k$ and $e_i = \kappa$ represent the assignments of the ith word to topic k, and mark the word as local if $\kappa = 1$ otherwise non-local. $\lambda_\kappa = \frac{\lambda}{\lambda+1}(\kappa = l)$ or $\frac{1}{\lambda+1}(\kappa = g)$. $n_{-i,\kappa,k}$ represents the word count with locality assignment κ and topic assignment k, and $-i$ means not including the ith word.

$$
p(z_{e,i} = k, e_i = \kappa | z_{-i}, e_{-i}) \propto \lambda_\kappa \cdot \frac{n_{-i,\kappa,k}^{(d_i)} + \gamma_\kappa}{n_{-i,\cdot,k}^{(d_i)} + \gamma_l + \gamma_g}
$$
$$
\cdot \frac{n_{-i,\kappa,k}^{(l_i)} + \alpha_\kappa}{n_{-i,\kappa,\cdot}^{(l_i)} + K\alpha_\kappa} \cdot \frac{n_{-i,\kappa,k}^{(w_i)} + \beta}{n_{-i,\kappa,k}^{(\cdot)} + W\beta}, \kappa \in \{l, g\}
\tag{2}
$$

Consequently, the topic distribution of each location can be computed in Eq. (3).

$$
p(z_i = k | l_i) = \frac{n_{1,k}^{(l_i)} + \alpha_l}{n_{1,\cdot}^{(l_i)} + K\alpha_l}
\tag{3}
$$

4   Evaluation

4.1   Dataset

In this section, we experimentally evaluate the effectiveness of the proposed method. We report our experimental results on the following two real datasets.

The first come from Weibo [25] (all written in Chinese), which is a Chinese micro-blogging website. Akin to a hybrid of Twitter and Facebook, it is one of the most popular sites in China. Without loss of generality, we only focus on messages in Shanghai (the largest city in China and one of the largest cities in the world by population) in 2015. Since most of the locations have been tagged for only a few times, we only selected those locations with a considerable number of posts within a pre-defined spatial range. The original data was preprocessed by filtering out stop words, then nouns and verbs were extracted as valid words with a Chinese POS (part-of-speech) tagger. Messages with less than three valid words were eliminated.

Our second dataset is from Yelp, which publish crowd-sourced reviews about local businesses, as well as the online reservation service and online food-delivery service. This dataset is publicly available [26], which is collected from Phoenix, which is a US city. In the Yelp dataset, each review has a location that is associated with a unique pair of latitude and longitude coordinates and a business name which is usually correspond to a restaurant, a hotel or an entertainment area. In order to share the same time span with the Weibo dataset, only reviews posted in 2015 is selected.

Some statistics about these two datasets are presented in Table 2.

Table 2. Dataset details

Description	Weibo	Yelp
Total number of message	252173	661833
Total number of location	1088	43990
Average length of message	110.32	33.30

4.2   Comparison Methods

We compare the proposed model LGLDA with the following other methods.

– **TF-IDF with K-means clustering (or TF-IDF)**
  In this method, Weibos are firstly preprocessed and presented as tf-idf weighted vectors and then aggregated by locations. Therefore, the feature vector of each location is summed by all documents tagged with that location. Finally, the feature vectors of each location are clustered by K-means. The center of each cluster represents a topic, and the weight of each element in the vector denotes the importance of the according keyword.
– **LDA model with location aggregation (or LDA)**
  In this method, topic and word distributions are firstly calculated by standard LDA algorithm with all documents, without considering the geo-tagged locations. After the global topics and word distribution in each topic are calculated, they are then aggregated by location. The topic distribution of each location is calculated as the average over all documents geo-tagged with that location.
– **Local LDA model (or LocalLDA)**
  This method is similar to those in previous works. The topics are generated from locations instead of documents. If two words are from the same region,

they are more likely to be clustered into the same topic. However, it is a simplified scenario, usually, if two words are close to each other in space, they are more likely to belong to the same region. In our dataset, the geographical distance of two words cannot be calculated due to the loss of the exact geo coordinations, but become a Boolean variable of whether they are tagged with the same location. Different from LGLDA model, in LocalLDA, there is only a local topic distribution, and all settings are kept the same with the LGLDA model.

4.3 Quantitative Measures

In order to make a comparison between different methods, several quantitative measures are used.

- **Perplexity**
  Perplexity is used to evaluate the performance of topic modeling. Perplexity is the standard metric to evaluate the predictive power and generalizability of a topic model, and is monotonically decreasing with increasing likelihood of the test data set. Hence, a lower perplexity score indicates stronger predictive power.

$$perplexity(D) = exp\{-\frac{\sum_{d\in D}\log p(w_d)}{\sum_{d\in D}N_d}\} \tag{4}$$

  where D is the test collection and N_d is the length of document d.
- **Location and Topic entropies**
  The average topic entropy of each location and the average location entropy of each topic are used to measure the concentrativeness of discovered topics. Each location should have a compact distribution on topics, while each topic should concentrate on a small set of locations.

$$entropy_{topic} = \frac{1}{L}\sum_{L}\sum_{K}p_k^{(l)}\log p_k^{(l)} \tag{5}$$

$$entropy_{location} = \frac{1}{K}\sum_{K}\sum_{L}p_l^{(k)}\log p_l^{(k)} \tag{6}$$

  where $p_k^{(l)}$ and $p_l^{(k)}$ are the estimated probabilities of topic k for location l and location l for topic k, respectively.
- **KL-divergence**
  KL-divergence is used to measure the average distance of word distributions of all pairs of topics. The larger the average KL-divergence is, the more distinct the topics are.

$$D_{KL}(p_i||p_j) = \sum_{L}p_i^{(k)}log\frac{p_i^{(k)}}{p_j^{(k)}} \tag{7}$$

  where $p_i^{(k)}$ and $p_j^{(k)}$ are the estimated probabilities of topic k for location i and location j respectively.

4.4   Settings

In our experiments, the number of topics was set at K to 20 for all models (including K-means), α_l and α_g to 0.1, β to 0.1, γ_l and γ_g to 0.5 for all LDA-based models empirically and were run for 500 iterations.

For the LGLDA model, the local-global weight ratio was determined by model's perplexity as shown in Fig. 2. As can be seen, in the Weibo dataset, with the gradually increasing value of λ from 0.1 to 20, the perplexity descended first and then ascended, and reached minimum at 0.6. Hence, 0.6 is the best value for λ for the Weibo dataset, and was used in our experiments. Performance on the other two metrics also confirmed this selection.

In the Yelp dataset, λ was determined with the same manner, and was finally set at 0.8. The optimal value of λ for the two dataset is quite close. In our future work, we would experiment against more different kinds of dataset to investigate whether the chosen value of λ is a coincidence or it is decided by the intrinsic properties of all user generated social media data.

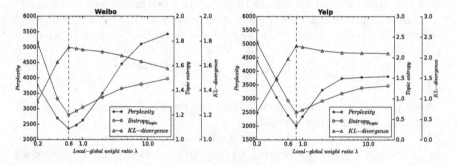

Fig. 2. The impact of the value of local-global weight ratio λ on model's performance.

4.5   Results

In this section, we experimentally evaluate the effectiveness of the LGLDA model, and compare it against the baseline methods.

Locality Score

In order to validate the effectiveness of our LGLDA model to distinguish between local words and noise, we defined the locality score. The locality score is defined as the ratio of the average probability of words generated from the local and the global topic distribution, and it is calculated as in Eq. (8).

$$Locality(d) = \frac{\sum_{w \in d} p(z_{e,w}, e_w = l)}{\sum_{w \in d} p(z_{e,w}, e_w = g)} \tag{8}$$

The locality score is a measurement of the relatedness of the messages to its respective local context or semantics. The higher the score is, the more representative the message is of its tagged location. In order to illustrate the usefulness of this measurement, we sorted all Weibos according to the locality score within

Table 3. Examples of Weibo text with locality score sampled from location Waitan

Locality score	Weibo text
7.18	Cruising along Waitan while taking a close-up view of the Oriental Pearl Tower is a worthwhile trip
0.91	Breakfast at Xitang, lunch at Hangzhou, dinner at Waitan. What an incredible day on the run
0.02	To deal with a difficult customer in the afternoon, I'd have to get up early to do data analysis

each location, and three Weibos were sampled from the collection with distinguishable locality score with location tagged 'Waitan' and shown in Table 3.

As we can see, the result is quite in accordance with our expectation. Weibo with the highest score which includes several landmark names and location specific words is highly relevant, while the other two messages containing only few or no location featured words are weakly related or irrelevant. Therefore, our model has the ability to rank texts according to their location relevence.

Comparative Results with Baselines

In this section, we use the quantitative measures described in Sect. 4.3 to evaluate the performances and show the superiority of our LGLDA model. The used quantitative measures include perplexity, KL-divergence and two kinds of entropies: topic entropy and location entropy.

Table 4. Results of the comparative experiments

Dataset	Method	Perplexity	Topic entropy	Location entropy	KL-divergence
Weibo	LDA	6904.11	2.9680	59.5100	**2.2944**
	LocalLDA	5679.95	1.5156	31.2292	1.5694
	LGLDA	**2357.94**	**1.1998**	**24.2575**	**1.7494**
Yelp	LDA	6320.41	2.3669	44.9711	2.1769
	LocalLDA	4570.38	1.0965	20.8351	1.2330
	LGLDA	**2021.72**	**0.5608**	**10.6569**	**2.2892**

Table 4 gives the comparative results of our LGLDA model with other baselines. As we can see, both in the Weibo dataset and in the Yelp dataset, our LGLDA model outperforms other baselines in almost all quantitative measures. Although Weibo and Yelp have distinctive business purposes and target users in different countries, while their datasets yield different statistic characteristics, our model is more preferable in both scenarios.

The LGLDA model achieves much lower perplexity for the reason that it can separate local and non-local words. Hence the words or the documents are better classified and organized by topics. With noisy words filtered out and only location related words kept, the discovered topics are more distinct and representative of the local semantics, which is reflected by the topic and location entropies.

In Table 4, in the Weibo dataset, the KL-divergence of our model is lower than that of the original LDA model, and the reason may due to the location relevance constraint. Since in the LDA model, all messages are unseparated according to locations, the discovered topics can be formed by messages with different geo tags. Further more, because all messages are deemed as useful, the structure of the clusters discovered is different from the other LDA-based models.

Topic Comparison of Different Methods

In this section, we show the topics discovered by different methods with Weibo and Yelp datasets.

In all the models, we set the number of topics at 20, and since to list all of the keywords in each topic would have taken a lot of space while has little help to gain an useful insight, we only showed the top three topics discovered in each dataset here. The result is shown in Table 5.

Table 5. TOP3 topics discovered by Weibo and Yelp dataset

Weibo				Yelp			
TF-IDF	LDA	LocalLDA	LGLDA	TF-IDF	LDA	LocalLDA	LGLDA
Work, company, mood, women, teacher	Love, feeling, mate, inside, teacher	Work, mood, love, children, phone	Work, company, mood, phone, city	Time, food, service, people, location	Food, table, minutes, server, service	Food, time, minutes, people, nice	Food, service, restaurant, delicious, menu
University, teacher, effort, mood, paper	Teacher, English, school, exam, culture	University, teacher, library, school, birthday	University, school, library, teacher, student	Food, chicken, service, menu, love	Steak, restaurant, dessert, bread, meal	Steak, dinner, table, restaurant, server	Steak, dessert, bread, cheese, salad
Waitan, city, night, Oriental Pearl Tower, restaurant	City, Waitan, Shanghai, Oriental Pearl Tower, international	Waitan, hotel, center, Oriental Pearl Tower, financial	Waitan, Oriental Pearl Tower, Chenghuang Temple, Nanjing Road, Huangpu River	Hotel, stay, pool, desk, vegas	Hotel, vegas, casino, desk, night	Hotel, stay, time, vegas, night	Hotel, vegas, casino, pool, check

In the Weibo dataset, the first topic (2nd row) is composed of words with broader meanings, which can be viewed as noises, while the other two topics (3rd row and 4th row) contain the semantics of education and tourist attractions. As can be seen, keywords discovered by our LGLDA model achieve the best relevance while keywords by other methods include more or less noise (e.g. mood, city, etc.). To further drill down the result, in our LGLDA model, Topic1 together with Topic4 accounts for 98.8% of global topic distribution and contains a large amount of words, which is not related to any specific semantic locations. With these noisy words filtered out, locations can be covered by fewer topics. For example, for Oriental Pearl Tower, the weight of Topic3 in LGLDA is 0.933 compared with a mixed topic constitution discovered by LocalLDA (0.425 for Topic1, 0.414 for Topic3 and 0.161 for all others).

The result of the Yelp dataset is in accordance with the Weibo dataset. Firstly, the topics discovered by our LGLDA model are more distinct. The first topic and the second topic discovered is separable, since the first concentrates on the general aspect of meals or restaurants, while the second concentrates more on the detailed kinds of foods or dishes. However, the topics discovered by other models are more mixed up. Secondly, the keywords of topics discovered by our LGLDA model contain less noises. In contrast, keywords in the topics of other methods in more noisy (universal words may exist in different topics), therefore have impaired the semantic distinctness of the topics discovered.

5 Conclusion

This paper proposes a method, which combines local word filtering and geographical topic modeling into one framework. The proposed LDA-based model LGLDA can effectively distinguish between location related words and a variety of noisy daily interests by properly choosing the local-global weight ratio parameter in the Bayesian model. Results on Weibo collection show the effectiveness of our method over other baselines.

This initial work shows the potential for location-sensitive information retrieval and opens up several interesting future directions. Firstly, we would like to apply our models on other interesting data sources. For example, we can mine interesting geographical topics from the tweets associated with user locations in Twitter. Second, we would like to compare the topics discovered as local and global topics, and investigate the correlation between the topics discovered and human mobility pattern disclosed by other datasets such as cellular signaling and traffic sensor data.

References

1. Hofmann, T.: Probabilistic latent semantic indexing. In: SIGIR, pp. 50–57. ACM (1999)
2. Blei, D.M., Ng, A.Y., Jordan, M.I.: Latent dirichlet allocation. J. Mach. Learn. Res. **3**, 993–1022 (2003)
3. Ashbrook, D., Starner, T.: Using GPS to learn significant locations and predict movement across multiple users. In: UbiComp, pp. 275–286 (2003)
4. Mei, Q., Liu, C., Su, H., Zhai, C.: A probabilistic approach to spatiotemporal theme pattern mining on weblogs. In: WWW, pp. 533–542. ACM (2006)

5. Wang, C., Wang, J., Xie, X., Ma, W.Y.: Mining geographic knowledge using location aware topic model. In: GIR, pp. 65–70. ACM (2007)
6. Backstrom, L., Kleinberg, J., Kumar, R., Novak, J.: Spatial variation in search engine queries. In: WWW, pp. 357–366. ACM (2008)
7. Palma, A.T., Bogorny, V., Kuijpers, B., Alvares, L.O.: A clustering-based approach for discovering interesting places in trajectories. In: SAC (2008)
8. Li, H., Li, Z., Lee, W.C., Lee, D.L.: A probabilistic topic-based ranking framework for location-sensitive domain information retrieval. In: SIGIR, pp. 331–338. ACM (2009)
9. Sizov, S.: Geofolk: latent spatial semantics in web 2.0 social media. In: WSDM, pp. 281–290. ACM (2010)
10. Mathioudakis, M., Koudas, N.: Identifying, attributing and describing spatial bursts. In: Proceedings of the VLDB Endowment, pp. 1091–1102. ACM (2010)
11. Eisenstein, J., Connor, B.O., Smith, N.A., Xing, E.P.: A latent variable model for geographic lexical variation. In: EMNLP, pp. 1277–1287. ACM (2010)
12. Cheng, Z., Caverlee, J., Lee, K.: You are where you tweet: a content-based approach to geo-locating twitter users. In: CIKM, pp. 759–768. ACM (2010)
13. Li, Z., Ding, B., Han, J., Kays, R., Nye, P.: Mining periodic behaviors for moving objects. In: SIGKDD, pp. 1099–1108. ACM (2010)
14. Yin, Z., Cao, L., Han, J., Zhai, C., Huang, T.: Geographical topic discovery and comparison. In: WWW, pp. 247–256. ACM (2011)
15. Ye, M., Yin, P., Lee, W.C., Lee, D.L.: Exploiting geographical influence for collaborative point-of-interest recommendation. In: SIGIR, pp. 325–334. ACM (2011)
16. Hong, L., Ahmed, A., Gurumurthy, S., Smola, A.J., Tsioutsiouliklis, K.: Discovering geographical topics in the Twitter stream. In: WWW, pp. 769–778. ACM (2012)
17. Bauer, S., Noulas, A., Seaghdha, D.O., Clark, S., Mascolo, C.: Talking places: modelling and analyzing linguistic content in foursquare. In: SocialCom/PASSAT, pp. 348–357. IEEE (2012)
18. Hu, B., Ester, M.: Spatial topic modeling in online social media for location recommendation. In: RecSys, pp. 25–32. ACM (2013)
19. Hu, B., Jamali, M., Ester, M.: Spatio-temporal topic modeling in mobile social media for location recommendation. In: ICDM, pp. 1073–1078. ACM (2013)
20. Ahmed, A., Hong, L., Smola, A.J.: Hierarchical geographical modeling of user locations from social media posts. In: WWW, pp. 25–36. ACM (2013)
21. Yuan, Q., Cong, G., Ma, Z., Sun, A., Thalmann, N.M.: Who, where, when and what: discover spatio-temporal topics for twitter users. In: SIGKDD, pp. 605–613. ACM (2013)
22. Kim, Y., Han, H., Yuan, C.: TOPTRAC: topical trajectory pattern mining. In: SIGKDD, pp. 587–596. ACM (2015)
23. Liu, Y., Ester, M., Hu, B., Cheung, D.W.: Spatio-temporal topic models for check-in data. In: ICDM, pp. 889–894. IEEE (2015)
24. Wu, F., Li, Z., Lee, W.C., Wang, H., Huang, Z.: Semantic annotaion of mobility data using social media. In: WWW, pp. 1253–1263. ACM (2015)
25. https://en.wikipedia.org/wiki/Sina_Weibo
26. https://www.yelp.com/dataset_challenge

Team-Oriented Task Planning in Spatial Crowdsourcing

Dawei Gao[1], Yongxin Tong[1(✉)], Yudian Ji[2], and Ke Xu[1]

[1] SKLSDE Lab and IRC, Beihang University, Beijing, China
{david_gao,yxtong,kexu}@buaa.edu.cn
[2] The Hong Kong University of Science and Technology,
Sai Kung, Hong Kong SAR, China
yjiab@connect.ust.hk

Abstract. The rapid development of mobile devices has stimulated the popularity of spatial crowdsourcing. Various spatial crowdsourcing platforms, such as Uber, gMission and Gigwalk, are becoming increasingly important in our daily life. A core functionality of spatial crowdsourcing platforms is to allocate tasks or make plans for workers to efficiently finish the published tasks. However, existing studies usually ignore the fact that tasks may impose different skill requirements on workers, which may lead to decreased numbers of accomplished tasks in real-world applications. In this work, we propose a practical problem called TOTP, *T*eam-*O*riented *T*ask *P*lanning, which not only makes feasible plans for workers but also satisfies the skill requirements of different tasks on workers. We prove the NP-hardness of TOTP, and propose two greedy-based heuristic algorithms to solve the TOTP problem. Evaluations on both synthetic and real-world datasets verify the effectiveness and the efficiency of the proposed algorithms.

Keywords: Spatial crowdsourcing · Task plan · Team formation

1 Introduction

With the rapid development of mobile and intelligent devices, spatial crowdsourcing platforms, such as Uber, gMission [3] and Gigwalk, are gaining increasing popularity. Different from traditional crowdsourcing platforms, tasks published on spatial crowdsourcing platforms require workers to travel to specific locations to accomplish the tasks.

A fundamental issue in spatial crowdsourcing is the planning problem [12,14], which refers to making traveling plans for workers to efficiently finish the published tasks under constraints such as travel budgets and completion time. We argue that such a problem formulation is impractical, because the tasks on real-life spatial crowdsourcing platforms often come with various requirements. Consequently, only workers with the desired skills are able to accomplish the corresponding tasks. Imagine the following scenario. There is a spatial crowdsourcing

© Springer International Publishing AG 2017
L. Chen et al. (Eds.): APWeb-WAIM 2017, Part I, LNCS 10366, pp. 41–56, 2017.
DOI: 10.1007/978-3-319-63579-8_4

platform which provides domestic services. Currently it has three tasks: the first one needs cleaning and tutoring from 3:00 p.m. to 5:00 p.m.; the second requires babysitting and cleaning from 4:00 p.m. to 6:00 p.m.; and the third needs cooking from 6:00 p.m. to 7:00 p.m. There are also some workers on the platform: Paul is skilled at cleaning and babysitting, David is good at cooking and Lucy is skilled at tutoring. Note that it is non-trivial for a single worker, such as Paul or David, to accomplish the requirement of the above tasks. It is also difficult for the platform to make plans for the workers under the spatial and time constraints.

Existing solutions to the planning problems in spatial crowdsourcing do not consider the skill requirements, spatial and time constrains simultaneously. In the above scenario, existing studies will assign the first task to Paul, which will not be completed. To jointly account for the skill requirements of tasks and the spatial and time constraints, our key insight it to assign *a team of workers* to fulfil all the requirements. We propose TOTP, a Team-Oriented Task Planning problem to maximum the total satisfaction of the workers. Note that we use skills to represent the specific requirements of tasks on workers. We illustrate the motivation of TOTP with the following example. In this example, the skills are denoted as $\{e_1, \cdots, e_4\}$.

Table 1. Basic information of tasks and workers in Example 1

Workers			Tasks			
No	Owning skills	Travel budget	No	Required skills	Capacity	Time period
w_1	$\{e_2, e_4\}$	24	t_1	$\{e_2, e_3\}$	2	[5,6]
w_2	$\{e_3\}$	20	t_2	$\{e_2\}$	2	[1,3]
w_3	$\{e_3, e_4\}$	19	t_3	$\{e_2, e_4\}$	1	[7,8]
w_4	$\{e_1, e_2\}$	21	t_4	$\{e_1, e_2, e_3\}$	2	[2,4]
w_5	$\{e_1, e_4\}$	23				

Table 2. Satisfaction between tasks and workers in Example 1

	w_1	w_2	w_3	w_4	w_5
t_1	1	3	2	1	5
t_2	2	3	2	1	4
t_3	5	2	3	4	1
t_4	4	4	2	1	6

Example 1. Suppose we have five workers w_1–w_5 and four tasks t_1–t_4 on a spatial crowdsourcing platform. The locations of the workers and the tasks are shown in the 2D space in Fig. 1a. We use Euclidean distance in this example. Table 1 shows the attributes of the workers and the tasks. The skills of the workers and the distances that he/she would like to travel are shown in the second and third columns. The skill requirements of tasks on workers are shown in

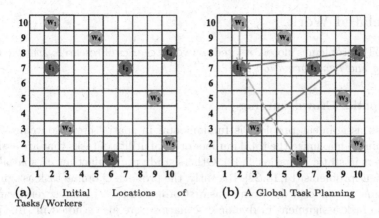

(a) Initial Locations of Tasks/Workers

(b) A Global Task Planning

Fig. 1. Example 1

the fifth column. Capacity shows the maximum number of workers that can participate in the corresponding task. The last column in Table 1 shows the completion time, which is the duration that the assigned worker needs to stay at the task's location. Table 2 shows the satisfaction of workers, which represents the workers' preferences on the tasks. The spatial crowdsourcing platform has to make assignments between the workers and tasks such that the skill requirements of tasks are satisfied and the total satisfaction is maximized. Figure 1b shows a global task planing, i.e. $\{t_1, t_3\}$ for w_1 and $\{t_4, t_1\}$ for w_2, respectively. Notice that t_1 can be accomplished by a team of w_1 and w_2.

Contributions. We propose a more realistic planning problem in spatial crowdsourcing called the *T*eam-*O*riented *T*ask *P*lanning problem (TOTP). As the example indicates, the TOTP problem not only makes plans for each worker but also attempts to satisfy the skill requirements of different tasks. To summarize, our contributions are as follows.

- We identify TOTP, a new spatial crowdsourcing planning problem that accounts for the skill requirements of tasks on workers.
- We prove that the TOTP problem is NP-hard.
- We propose two greedy algorithms to solve the TOTP problem, and analyze the complexity of both algorithms.
- We verify the effectiveness and efficiency of the proposed algortihms through extensive experiments on synthetic and real-world datasets.

In the rest of the paper, we review related work in Sect. 2, formulate the TOTP problem and prove its NP-hardness in Sect. 3. Section 4 presents our algorithms on TOTP problem and Sect. 5 show the experimental evaluations. Finally we conclude this paper in Sect. 6.

2   Related Work

Our TOTP problem is closely related to two categories of research: *spatial crowdsourcing* and *team formation*.

2.1   Spatial Crowdsourcing

The task assignment problem is fundamental in spatial crowdsourcing. Many efforts aim to maximize the total number or total utility of tasks that are assigned to workers in static scenarios [9,19]. Others study the conflict-aware spatial task assignment problems [14,15,22]. Recently, the problem of online task assignment in dynamic spatial crowdsourcing was first proposed by [21], and several variants of online task assignment in dynamic scenarios were also studied in [16,17,20]. Other practical issues such as location privacy protection of workers have also been explored [18]. In addition, [6] introduced the route planning problem for a worker and attempt to maximize the number of complete tasks, while the corresponding online version of [6] is investigated in [11]. [7] studies to assign workers under the spatial and time constraints. [3] summarizes the challenges and opportunities in spatial crowdsourcing. Despite the extensive research efforts on task assignment in spatial crowdsourcing, they all assume simple and homogenous tasks without considering the situations where the tasks are complex and require a team of workers to finish.

2.2   Team Formation

The team formation problem is first proposed by Lappas et al. [10], which aims to find the minimum cost team of experts according to the skills and relationships of users in social networks. Notice that the team formation problem can be reduced from typical NP-complete problems, indicating that the team formation problem is NP-hard. [1] focuses on minimizing the maximum workload when forming teams to cover the skills, and studies both the off-line and on-line settings. [13] studies the team formation problem with capacity constraints on a social network, and presents approximation algorithms with provable guarantees. [2] studies the online team formation problem called the Balanced Social Task Assignment problem, and proposes an online algorithm with provable guarantee. In [8], based the skills of crowd workers, the authors study how to recommend k teams for spatial crowdsourcing tasks. Furthermore, Cheng et al. also proposed the issue of team-oriented task assignment in spatial crowdsourcing recently [5]. The above studies only focus on satisfying the skill requirements. They neither consider the location information and travel budgets, nor address the problem of how to make feasible plans for workers.

3   Problem Statement

In this section we formally define the Team-Oriented Task Planning (TOTP) problem and prove that the problem is NP-hard. We assume $E = <e_1, \cdots, e_m>$

as a universe of m skills throughout this paper. Let T be a set of tasks. Each task t has its own location l_t that requires the assigned workers to travel to. The task t's skill requirement is represented by a subset of E: $E_t = \{e_1, \cdots, e_{|E_t|}\}$. We also define a time interval $[s_1^t, s_2^t]$, which is the starting time and the ending time of the corresponding task t. Note that the assigned workers need to stay at l_t during this time interval. In real-life applications, the number of workers to finish a task is normally limited. Hence we define the task's capacity c_t as the maximum number of workers allowed for the task. We define W as a set of workers. Each worker w has his/her starting location l_w, a set of skills $E_w = \{e_1, \cdots, e_{|E_w|}\}$ and a travel budget B_w, which represents the total travel cost that the worker w would like to spend to accomplish the tasks, which can be money, distance or time.

Definition 1 (Crowd Worker). *A crowd worker ("worker" for short) is denoted as $w = <l_w, E_w, B_w>$, where l_w is the starting location of worker w in the 2D space, E_w is a set of skills that the worker is good at, and B_w is the travel budget that the worker w would like to spend to accomplish the tasks.*

Definition 2 (Crowdsourced Task). *A crowdsourced task ("task" for short) is denoted as $t = <l_t, E_t, Int, c_t>$, where l_t is the location in the 2D space that workers have to travel to, E_t is the task's required skills, $Int = [s_1^t, s_2^t]$ represents the task's time interval during which the task should be accomplished and c_t is the capacity of the task that limits the number of the workers assigned to the task.*

For each worker w, we define $P_w = \{t_1^w, t_2^w, \cdots, t_{|P_w|}^w\}$ as the plan of the arranged tasks in time order. Suppose that t_i and t_{i+1} are two tasks in P_w, a plan is feasible if there is no time conflict among the arranged tasks, and the workers can perform t_{i+1} in time after finishing t_i. To evaluate the travel cost between any two tasks, such as t_i and t_j, we use $cost(l_{t_i}, l_{t_j})$ to represent the travel cost between t_i and t_j. If a worker cannot perform the next task in time, the cost between these two tasks will be ∞. Meanwhile, we define $u(w, t)$ as the satisfaction between task t and worker w, and $\mathcal{U}(w) = \sum_{t_i \in P_w} u(w, t_i)$ as worker w's satisfaction on P_w. Finally we give the definition of feasible plan and the Team-Oriented Task Planning (TOTP) problem.

Definition 3 (Feasible Plan). *A plan P_w is feasible if and only if: $s_2^{t_i} \leq s_1^{t_{i+1}}, \forall 1 \leq i \leq |P_w| - 1$.*

Definition 4 (Team-Oriented Task Planning (TOTP) Problem). *Given a set of tasks $T = \{t_1, t_2, \cdots, t_{|T|}\}$ and a set of workers $W = \{w_1, w_2, \cdots, w_{|W|}\}$ with their associated attributes, the Team-Oriented Task Planning (TOTP) problem is to find feasible plans $A = \cup_w\{P_w\}$ for workers with the maximum utility cost: $Utility(A) = \sum_{w_i \in W} \mathcal{U}(w_i)$, such that the following constraints are satisfied:*

- *Skill constraint: each required skill of the tasks is covered by the assigned workers.*

– *Travel budget constraint: a worker's total travel cost is under his/her travel budget.*
– *Capacity constraint: the number of workers assigned to a task is lower than the task's capacity.*

Theorem 1. *The Team-Oriented Task Planning problem is NP-hard.*

Proof. We prove that the Team-Oriented Task Planning problem is NP-hard by reducing the knapsack problem, a well known NP-complete problem, to the TOTP problem. An instance of the knapsack problem consists of a set of n items $\{x_1, \cdots, x_n\}$ where each item x_i has its value $v_i > 0$, weight $m_i > 0$, and the maximum weight M that the bag can carry. The decision version of the knapsack problem asks whether there is a collection of items $C = \{x_{s_1}, x_{s_2}, \cdots, x_{s_k}\}$ such that $\sum_{i=1}^{k} v_{s_i} = K$ and $\sum_{i=1}^{k} m_{s_i} \leq M$. We construct an instance of the knapsack problem using an instance of the TOTP problem as follows (Table 3).

Table 3. Summary of symbol notations

Notation	Description
w	Worker
t	Task
c_t	Capacity of t
$l_w (l_t)$	Location of w(or t)
T	Set of tasks
W	Set of workers
B_w	Budget of w
$E_w (E_t)$	Skill set of w(or t)
P_w	Plan of w
$A = \cup_w \{S_w\}$	The total plan
$u(t, w)$	Satification between t and w
$\mathcal{U}(w) = \sum_{t \in P_w} u(w, t)$	w's satification on P_w
$Utility(A) = \sum_{w \in W} \mathcal{U}(w)$	The total satification of A

– Let $|W| = 1, W = \{w\}$, and $B_w = M$.
– Let $E_t = E_w, \forall w \in W$ and $\forall t \in T$.
– Each item corresponds to a task in TOTP problem. $u(w, t_i) = \frac{v_i}{\max v_i}, \forall 1 \leq i \leq n$ and the capacities of all the tasks equal to 1.
– Let $s_2^{t_i} < s_1^{t_{i+1}}, \forall 1 \leq i < n$.
– The travel cost of worker w and task t_i is set as: $cost(w, t_i) = \frac{m_i}{2}$.
– The travel cost between two events is constructed as:

$$cost(l_{t_i}, l_{t_j}) = \begin{cases} \frac{m_i + m_j}{2} & 1 \leq i < j \leq n \\ +\infty & otherwise \end{cases}$$

Thus, the problem is to decide if there is a feasible plan P_w for w such that $\sum_{t_i \in P_w} u(w, t_i) = \frac{K}{\max v_i}$ satisfying all the constraints. We can see that if the collection exists, then the plan P_w is feasible, and it satisfies that $\sum_{t_i \in P_w} u(w, t_i) = \frac{K}{\max v_i}$ and the total travel cost is less than M. That is, if the plan P_w exists, then there is a collection C satisfying the constraints on the sum of values and weights. □

4   Algorithms of TOTP Problem

In this section, we present two greedy-based algorithms to solve the TOTP problem.

4.1   Rarest Skill Priority Algorithm

We first present a greedy algorithm called the Rarest Skill Priority Algorithm. The Rarest Skill Priority Algorithm recursively selects such skills that are required by numbers of tasks but only few workers have such skills. We call such skills *rarest skill*. The reason why we choose rarest skills in priority is to avoid the case where the workers possessing the rare skills have been assigned to other tasks and the tasks requiring these skills would never be accomplished. The rarest skills are calculated by $\arg\max_{e \in E} \frac{|\{t|e \in E_t\}|}{|\{w|e \in E_w\}|}$. Then, if a number of workers or tasks own/require the rarest skill, we greedily make an assignment of a pair of $(worker, task)$ such that the utility gain is the largest. The utility gain is defined in Eq. 1.

$$ratio(w, t) = \frac{u(w, t)}{inc\_cost(w, t)} \tag{1}$$

where $inc\_cost(w, t) =$

$$
\begin{cases}
cost(l_w, l_t) & P_w = \emptyset \\[2ex]
\begin{aligned}&cost(l_w, l_t) + cost(l_t, l_{t_1^w})\\ &-cost(l_w, l_{t_1^w})\end{aligned} & s_2^t < s_1^{t_1^w} \\[2ex]
\begin{aligned}&cost(l_{t_i^w}, l_t) + cost(l_t, l_{t_{i+1}^w})\\ &-cost(l_{t_i^w}, l_{t_{i+1}^w})\end{aligned} & s_2^{t_i^w} < s_1^t, s_2^t < s_1^{t_{i+1}^w} \\[2ex]
cost(l_{t_{|P_w|}^w}, l_t) & s_2^{t_{|P_w|}^w} < s_1^t \\[2ex]
\infty & otherwise
\end{cases}
\tag{2}
$$

Equation 1 defines the ratio between the satisfaction and the additional travel cost. With a larger $ratio(w, t)$, the task t is more suitable for w since he/she has a larger satisfaction and less travel cost. In Eq. 2, $inc\_cost(w, t)$ is the additional

---

Algorithm 1. Rarest Skill Priority Algorithm

---

    input : A set of workers W, a set of tasks T and their associated attributes
    output: $A = \cup_w \{P_w\}$
  1: $H \leftarrow \emptyset$
  2: **for** $e_{rare} = \arg\max_{e \in E} \frac{|\{t|e \in E_t\}|}{|\{w|e \in E_w\}|}$ **do**
  3:     $W_{rare} = \{w|e_{rare} \in E_w\}$
  4:     $T_{rare} = \{t|e_{rare} \in E_t\}$
  5:     **for** $t_i \in T_{rare}$ **do**
  6:         $w = \arg\max_{w \in W_{rare}} ratio(w, t_i)$
  7:         $H \leftarrow (w, t_i)$
  8:     **end for**
  9:     **while** $H \neq \emptyset$ **do**
10:         Pop (w, t) from H with the largest ratio
11:         Add t to P_w if $\{t\} \bigcup P_w$ is feasible and t is not full of capacity
12:         $E_t = E_t - E_w$
13:         Update (w, t) in H for each t
14:     **end while**
15: **end for**

---

travel distance when inserting task t to plan $P_w = \{t_1^w, t_2^w, \cdots, t_{|P_w|}^w\}$. The details of $inc\_cost(w, t)$ are shown in Eq. 2. If P_w is an empty set, the worker only needs to travel to l_t to perform the task. Otherwise, when the starting time of t is earlier than that of the first task of P_w, we can insert t into P_w as the first task. Hence the additional travel cost is $cost(l_w, l_t) + cost(l_t, l_{t_1}^w) - cost(l_w, l_{t_1}^w)$. When t's time interval is between task t_i^w and t_{i+1}^w in P_w ($\forall 1 \leq i \leq |P_w| - 1$), we insert the task t in-between and the worker needs to travel to l_t after finishing task t_i^w. Thus the additional cost is $cost(l_{t_i^w}, l_t) + cost(l_t, l_{t_{i+1}^w}) - cost(l_{t_i^w}, l_{t_{i+1}^w})$. Finally, if task t's starting time is after the last task's ending time, we define the additional cost as $cost(l_{t_{|P_w|}^w}, l_t)$.

Algorithm 1 shows the pseudo-code of the Rarest Skill Priority Algorithm. Specifically, when making plans for the workers possessing the rarest skills, we use a heap H to store the worker-task pair (w, t) with the largest ratio for each task in a decreasing order. Then we can pop the pair on the top of H and add it into the worker's plan. In Algorithm 1, we first initialize the heap H (Line 1). Then we find the rarest skill, and the set of workers W_{rare} and tasks T_{rare} that require/own the rarest skill (Lines 2–4). We traverse the set T_{rare} and add the pair with the largest ratio (calculated by Eq. 1) into the heap H for each task (Lines 5–8), pop the pair (w, t) with the largest ratio in H, and add task t to w's plan if it is feasible (Line 10). Afterwards, we update the skill requirement of t to compute the next rarest skill. Because the additional travel cost of the pair associated with w has changed, we need to update the pairs in H (Line 13) for the next iteration. We pop the top pair with the largest ratio until H is empty, and we continue the loop for finding the rarest skill and making assignments (Lines 2–15).

Example 2. Back to Example 1, we first traverse the tasks $t_1 - t_4$ and workers $w_1 - w_5$, and find that e_2 is the rarest skill. Then we make assignment between $W_{rare} = \{w_1, w_4\}$ and $T_{rare} = \{t_1, t_2, t_3, t_4\}$. For each task t in T_{rare}, we find the worker who has the largest ratio with him/her in W_{rare} and push the worker into the heap. In the first iteration, the ratios are shown in Table 4. We choose the pair with the largest ratio for each task and build the heap $H = \{(w_1, t_1), (w_4, t_2), (w_1, t_3), (w_1, t_4)\}$. We pop the largest pair (w_1, t_3) with the ratio 0.51 and add t_3 to w_1's plan. Then we update the pair for t_1, t_2, t_4 in H and finally in this iteration we get $\{(w_1,t_3), (w_1,t_1), (w_4,t_2), (w_4,t_4)\}$. In the following iterations we repeat the process and finally we get $\{t_1, t_3, t_4\}$ for w_1, $\{t_1, t_2\}$ for w_4 and $\{t_4\}$ for w_5.

Table 4. Ratios in the first iteration

	t_1	t_2	t_3	t_4
w_1	0.33	0.34	0.51	0.49
w_4	0.28	0.35	0.50	0.20

Complexity Analysis. In the worst case, Algorithm 1 has to traverse all the skills to cover the requirement. During each iteration, we traverse the rest of skills to find the rarest skill whose time cost is $O(|E|(|W| + |T|))$. Then for each task in W_{rare} we spend $O(|W||T|)$ to find the $(worker, task)$ pair with the largest ratio. Searching through and updating the heap take about $O(|W||T|)$. Thus the time complexity of Algorithm 1 is $O(|E|^2(|W| + |T|) + |E||W||T|)$ in the worst case, where $|E|$ is the number of all skills. The memory cost of Algorithm 1 is mainly to store the heap and the plans for workers, which is $O(|W||T|)$.

4.2   Skill Cover and Utility Priority Algorithm

In this subsection, we present another solution to the TOTP problem, which is called the Skill Cover and Utility Priority Algorithm. In the Skill Cover and Utility Priority Algorithm, we first attempt to cover all the skills of a task like the team formation problem. Specifically, for each task we attempt to form a team to satisfy the skill requirement with a minimal team size. If the team size is smaller than the capacity, in the second step we greedily assign the worker with the largest satisfaction to the tasks. Finally we can obtain a team for the task and satisfy the skill requirements of the tasks.

Algorithm 2 presents the details of the Skill Cover and Utility Priority Algorithm. In lines 1–3, we first sort tasks in an increasing order of staring time and initialize $time_w$ for each worker, which records the available time of the worker. Then we traverse the set of tasks. In line 6 we use function $Dis(.)$ to compute the total travel cost of the plan and pick up the set of workers W' who can participate in task t. In lines 7–11 we initialize the team g and choose the worker

---

Algorithm 2. Skill Cover and Utility Priority Algorithm

input : A set of workers W, a set of tasks T and their associated attributes
output: $A = \cup_w \{P_w\}$
1: $sort(T)$
2: **for** w in W **do**
3:     $time_w = 0$
4: **end for**
5: **for** t in T **do**
6:     $W' = \{w | w \in W$ and $Dis(P_w \cup \{t\}) \leq B_w$ and $time_w \leq s_1^t\}$
7:     $g = \emptyset$
8:     **while** g cannot cover E_t **do**
9:         $w = \text{argmax}_{w \in W'} \{|E_t \cap E_w|\}$
10:        $g = g + \{w\}$
11:    **end while**
12:    **while** $|g| < c_t$ **do**
13:        $w = \arg\max_{w \in W' - g} u(w, t)$
14:        $g = g + \{w\}$
15:    **end while**
16:    **for** w in g **do**
17:        Add t to P_w
18:        $B_w = B_w - cost(w, t)$
19:        $l_w = l_t$
20:        $time_w = s_2^t$
21:    **end for**
22: **end for**
23: **return** $A = \cup_w \{P_w\}$

---

who has the most required skills and add him/her to team g. Then if the size of g is smaller than c_t, we add workers with the largest satisfaction to the team until the task's capacity is fully occupied. Finally, we update the workers' plans, budgets, current locations and $time_w$ in lines 16–21 and continue the iteration for the next task. Specifically, when the size of team g is larger than c_t, we abandon the task and go on to the next task.

Example 3. Back to our Example 1. In the first step the tasks are sorted as $\{t_2, t_4, t_1, t_3\}$. For task t_2, w_3 has all the required skills. We add w_3 to team g and at this time the team size is smaller than 2. Thus, w_5 is added to the team for having the largest satisfaction for t_2. Then for t_4, we find worker w_4 who has the most required skills and add w_4 into the team. Because w_3 has participated in t_2, which conflicts with t_4, we choose w_2 to cover the last skills. At last the team size equals to t_4's capacity. For t_1 and t_3, we run Algorithm 2 in a similar way, and finally we get a team $\{w_1, w_2\}$ for t_1 and $\{w_1\}$ for t_3. The final result is presented in Table 5 and the total satisfaction is 20.

Complexity Analysis. For the two-step greedy algorithm, the time cost to sort tasks and initialize the worker set is $O(|T|ln(|T|) + |W|)$. Then we traverse each task to find a feasible team. Forming team g in lines 8–15 takes $O(|E||W|)$ in

Table 5. Result of Example 1

Worker	w_1	w_2	w_3	w_4	w_5
Plan	$\{t_1, t_3\}$	$\{t_4, t_1\}$	$\{t_2\}$	$\{t_4\}$	$\{t_2\}$

the worst case. Finally updating the workers' attributes takes $O(|W|)$. Therefore the overall time cost of the algorithm is $O(|T||E||W| + |T||W|)$. The major space cost of Algorithm 2 is from storing the set W', T and their associated attributes, whose overall space cost is $O(|W| + |T|)$.

Table 6. Synthetic datasets

Notation	Value
$|W|$	1000,1500,**2000**,2500,3000
$|T|$	250,500,**750**,1000,1250
mean	4,6,**8**,10,12
factor	0.5,1,**2**,4,8
$|E_w|$	3,**6**,9,12,15
$|E_t|$	5,10,**15**,20,25

5   Evaluation

In this section we conduct experiments on both synthetic and real-world datasets. We use the dataset from gMission [4], a spatial crowdsourcing platform, as the real-world dataset. In this dataset, we extract 1000 tasks, each of which is associated with some descriptions introducing the details. Thus, the required skills of the tasks can be extracted from the descriptions, and the worker's skills are learned from his/her history tasks. Table 6 shows the parameters of the synthetic dataset, and the default values are shown in bold. In the synthetic dataset, the numbers of workers $|W|$ and tasks $|T|$ are set between 1000–3000 and 250–1250, respectively. According to the real-world dataset, we let the capacities of workers follow the normal distribution, with the mean between 4–12. In terms of travel budget B_w, we define a parameter $factor$ to vary the travel budget. Then we have $B_w = \min_{t \in T} cost(l_w, l_t) + \frac{\min_{t \in T} cost(l_w, l_t) + \max_{t \in T} cost(l_w, l_t)}{2} * factor$, and we vary $factor$ from 0.5 to 8. Note that the moving distance of a worker is computed in Euclidean distance and it can be easily extended to the road network distance or other distance metrics. Both the numbers of workers' skills $|E_w|$ and tasks' required skills $|E_t|$ follow the normal distribution and the means are between 3–15 and 5–20. In the real-world dataset, we still use $B_w = \min_{t \in T} cost(l_w, l_t) + \frac{\min_{t \in T} cost(l_w, l_t) + \max_{t \in T} cost(l_w, l_t)}{2} * factor$ as the budget of workers, because there are no such parameters in the real-world dataset.

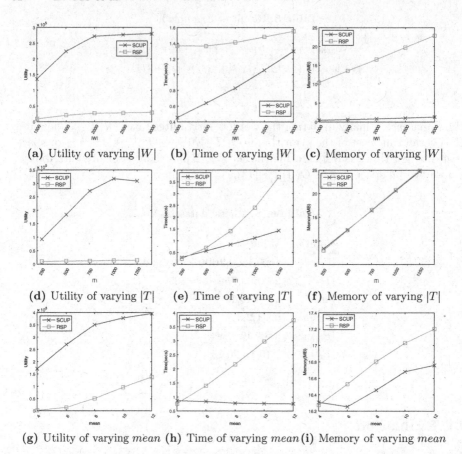

(a) Utility of varying $|W|$ **(b)** Time of varying $|W|$ **(c)** Memory of varying $|W|$

(d) Utility of varying $|T|$ **(e)** Time of varying $|T|$ **(f)** Memory of varying $|T|$

(g) Utility of varying *mean* **(h)** Time of varying *mean* **(i)** Memory of varying *mean*

Fig. 2. Results on varying $|W|$, $|T|$, and *mean*.

We evaluate both the Skill Cover and Utility Priority Algorithm (Algorithm 2), denoted as SCUP and the Rarest Skill Priority Algorithm (Algorithm 1), denoted as RSP. We compare these algorithms in terms of utility, time and memory. When comparing utility, we only compute the satisfaction of the completed tasks, whose skills are completely satisfied.

Effect of $|W|$. Figure 2a to c present the results of varying $|W|$ in the synthetic dataset. The total utility obtained from SCUP is much larger than that from RSP. The running time and the memory cost of both SCUP and RSP are small but increase with $|W|$. The memory of SCUP is smaller than that of RSP, because RSP needs to store the worker-task pairs in the heap.

Effect of $|T|$. Figure 2d to f show the results of varying $|T|$ in the synthetic dataset. The utility of SCUP increases with $|T|$, and stables when $|T|$ reaches 1000. This is because the number of workers and the travel budget B_w are limited, so the workers cannot complete more tasks. The time cost of SCUP is

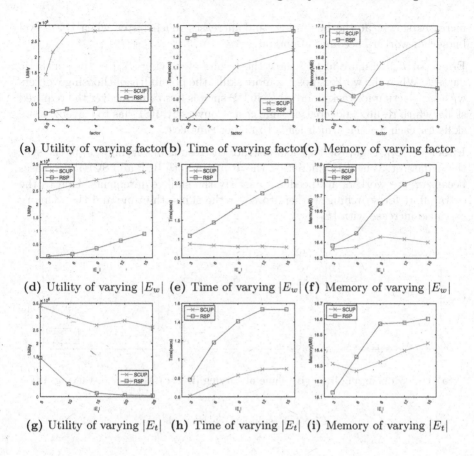

(a) Utility of varying factor **(b)** Time of varying factor **(c)** Memory of varying factor

(d) Utility of varying $|E_w|$ **(e)** Time of varying $|E_w|$ **(f)** Memory of varying $|E_w|$

(g) Utility of varying $|E_t|$ **(h)** Time of varying $|E_t|$ **(i)** Memory of varying $|E_t|$

Fig. 3. Results on varying $factor$, $|E_w|$, and $|E_t|$.

smaller than that of RSP, and the memory cost of SCUP is approximately equal to that of RSP. Tt is because SCUP spends more time to update the heap and traverse the tasks and workers to find the rarest skill.

Effect of $mean$. Figure 2g to i depict the results of varying $mean$ in the synthetic dataset. SCUP performs better than RSP not only in total utility but also in time and memory cost. This is because increasing $mean$ of capacity enables the tasks to accept more workers. Later the number of workers and the workers' travel budget become the bottlenecks for total utility in SCUP. As for the utility of RSP, the increasing $mean$ of capacities increases the possibility of tasks to be accomplished.

Effect of $factor$. Figure 3a to c present the results of varying $factor$ in the synthetic dataset. The influence of $factor$ on SCUP is stronger than that of RSP (see Fig. 3c). It might be because RSP attempts to cover the rarest skills first, which disperses the workers to different tasks. Thus the utility of RSP increases slowly at the beginning, and increases much faster when the travel budget of workers becomes abundant.

Effect of $|E_w|$. Figure 3d to f show the results of varying $|E_w|$ in the synthetic dataset. When the workers possess more skills, the possibility of choosing workers with satisfaction for tasks increases. SCUP spends less time to cover the required skills, which results in decreased time cost. Conversely, RSP has to traverse more skills for each worker, which leads to higher time cost.

Effect of $|E_t|$. Figure 3g to i demonstrate the results of varying $|E_t|$ in the synthetic dataset. When $|E_t|$ increases, the utility of both RSP and SCUP decreases because more workers are needed to satisfy the skill requirements. Due to the extra effort for searching workers to cover the skills, the time and the memory cost also increase with $|E_t|$.

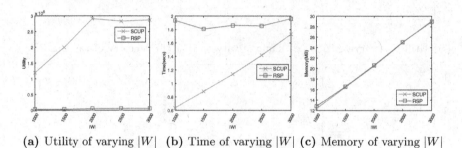

(a) Utility of varying $|W|$ (b) Time of varying $|W|$ (c) Memory of varying $|W|$

Fig. 4. Results on real datasets.

Effect of $|W|$ **in Real Dataset.** Finally Fig. 4a to c show the results of varying $|W|$ in the real-world dataset. When $|W|$ increases, the utility of SCUP increases fast at the beginning. When $|W|$ reaches 2000, the utility stabilizes, since the tasks are consumed. For RSP, the utility exhibits a similar but less dynamic trend to SCUP. The memory costs of the two algorithms are approximately the same. The time costs of both algorithms are small.

Summary. SCUP outperforms RSP in utility in various scenarios. One reason is that when making an assignment between workers and tasks owning/requiring the rarest skill, the tasks' capacities are consumed but only few skills are satisfied. Therefore some tasks' skill requirements cannot be completely satisfied. In contrast, SCUP attempts to cover the skills with a small team of workers, which ensures that the task are actually accomplished. Furthermore, SCUP also outperforms RSP in time and memory.

6   Conclusion

In this paper, we introduce Team-Oriented Task Planning (TOTP) problem, a realistic planning problem in spatial crowdsourcing which attempts to assign workers to suitable tasks. Different from previous research, it also takes into account the skill requirements of tasks on workers. We prove that TOTP is NP-hard, and propose two heuristic algorithms to solve the TOTP problem. Finally we conduct experiments on both synthetic and real-world datasets and verify the effectiveness and the efficiency of the proposed algorithms.

Acknowledgment. This work is supported in part by the National Science Foundation of China (NSFC) under Grant Nos. 61502021, 61328202, and 61532004, National Grand Fundamental Research 973 Program of China under Grant 2012CB316200.

References

1. Anagnostopoulos, A., Becchetti, L., Castillo, C., Gionis, A., Leonardi, S.: Power in unity: forming teams in large-scale community systems. In: CIKM, pp. 599–608 (2010)
2. Anagnostopoulos, A., Becchetti, L., Castillo, C., Gionis, A., Leonardi, S.: Online team formation in social networks. In: WWW, pp. 839–848 (2012)
3. Chen, L., Shahabi, C.: Spatial crowdsourcing: challenges and opportunities. IEEE Data Eng. Bull. **39**(4), 14–25 (2016)
4. Chen, Z., Fu, R., Zhao, Z., Liu, Z., Xia, L., Chen, L., Cheng, P., Cao, C.C., Tong, Y., Zhang, C.J.: gMission: a general spatial crowdsourcing platform. Proc. VLDB Endow. **7**(13), 1629–1632 (2014)
5. Cheng, P., Lian, X., Chen, L., Han, J., Zhao, J.: Task assignment on multi-skill oriented spatial crowdsourcing. IEEE Trans. Knowl. Data Eng. **28**(8), 2201–2215 (2016)
6. Deng, D., Shahabi, C., Demiryurek, U.: Maximizing the number of worker's self-selected tasks in spatial crowdsourcing. In: GIS, pp. 314–323 (2013)
7. Deng, D., Shahabi, C., Zhu, L.: Task matching and scheduling for multiple workers in spatial crowdsourcing. In: GIS, pp. 21:1–21:10 (2015)
8. Gao, D., Tong, Y., She, J., Song, T., Chen, L., Xu, K.: Top-k team recommendation in spatial crowdsourcing. In: Cui, B., Zhang, N., Xu, J., Lian, X., Liu, D. (eds.) WAIM 2016. LNCS, vol. 9658, pp. 191–204. Springer, Cham (2016). doi:10.1007/978-3-319-39937-9_15
9. Kazemi, L., Shahabi, C.: Geocrowd: enabling query answering with spatial crowdsourcing. In: GIS, pp. 189–198 (2012)
10. Lappas, T., Liu, K., Terzi, E.: Finding a team of experts in social networks. In: SIGKDD, pp. 467–476 (2009)
11. Li, Y., Yiu, M.L., Xu, W.: Oriented online route recommendation for spatial crowdsourcing task workers. In: Claramunt, C., Schneider, M., Wong, R.C.-W., Xiong, L., Loh, W.-K., Shahabi, C., Li, K.-J. (eds.) SSTD 2015. LNCS, vol. 9239, pp. 137–156. Springer, Cham (2015). doi:10.1007/978-3-319-22363-6_8
12. Lu, E.H., Chen, C., Tseng, V.S.: Personalized trip recommendation with multiple constraints by mining user check-in behaviors. In: GIS, pp. 209–218 (2012)
13. Majumder, A., Datta, S., Naidu, K.: Capacitated team formation problem on social networks. In: SIGKDD, pp. 1005–1013 (2012)
14. She, J., Tong, Y., Chen, L.: Utility-aware social event-participant planning. In: SIGMOD, pp. 1629–1643 (2015)

15. She, J., Tong, Y., Chen, L., Cao, C.C.: Conflict-aware event-participant arrangement. In: ICDE, pp. 735–746 (2015)
16. She, J., Tong, Y., Chen, L., Cao, C.C.: Conflict-aware event-participant arrangement and its variant for online setting. IEEE Trans. Knowl. Data Eng. **28**(9), 2281–2295 (2016)
17. Song, T., Tong, Y., Wang, L., She, J., Yao, B., Chen, L., Xu, K.: Trichromatic online matching in real-time spatial crowdsourcing. In: ICDE, pp. 1009–1020 (2017)
18. To, H., Ghinita, G., Shahabi, C.: A framework for protecting worker location privacy in spatial crowdsourcing. Proc. VLDB Endow. **7**(10), 919–930 (2014)
19. To, H., Shahabi, C., Kazemi, L.: A server-assigned spatial crowdsourcing framework. ACM Trans. Spat. Algorithms Syst. **1**(1), 2 (2015)
20. Tong, Y., She, J., Ding, B., Chen, L., Wo, T., Xu, K.: Online minimum matching in real-time spatial data: experiments and analysis. Proc. VLDB Endow. **9**(12), 1053–1064 (2016)
21. Tong, Y., She, J., Ding, B., Wang, L., Chen, L.: Online mobile micro-task allocation in spatial crowdsourcing. In: ICDE, pp. 49–60 (2016)
22. Tong, Y., She, J., Meng, R.: Bottleneck-aware arrangement over event-based social networks: the max-min approach. World Wide Web: Internet Web Inf. Syst. **19**(6), 1151–1177 (2016)

Negative Survey with Manual Selection:
A Case Study in Chinese Universities

Jianguo Wu[1], Jianwen Xiang[1], Dongdong Zhao[1(✉)], Huanhuan Li[2],
Qing Xie[1], and Xiaoyi Hu[1]

[1] School of Computer Science and Technology,
Wuhan University of Technology, Wuhan, China
{jgwu,jwxiang,zdd,felixxq,huxiaoyi}@whut.edu.cn
[2] School of Computer Science, China University of Geosciences, Wuhan, China
julylhh@gmail.com

Abstract. Negative survey is a promising method which can protect personal privacy while collecting sensitive data. Most of previous works focus on negative survey models with specific hypothesis, e.g., the probability of selecting negative categories follows the uniform distribution or Gaussian distribution. Moreover, as far as we know, negative survey is never conducted with manual selection in real world. In this paper, we carry out such a negative survey and find that the survey may not follow the previous hypothesis. And existing reconstruction methods like NStoPS and NStoPS-I perform poorly on the survey data. Therefore, we propose a method called NStoPS-MLE, which is based on the maximum likelihood estimation, for reconstructing useful information from the collected data. This method also uses background knowledge to enhance its performance. Experimental results show that our method can get more accurate aggregated results than previous methods.

Keywords: Privacy protection · Negative survey · Reconstruction method

1 Introduction

Nowadays, the rapid development of computer network and big data technologies brings great convenience to people, but it also increases the risk of disclosing sensitive data and personal privacy. Negative Survey [1, 2] is a promising privacy protection technique. In a negative survey, participants are asked to answer a question by selecting a category that they do NOT belong to (this kind of category is called negative category). When the number of categories in a question is larger than 2, the privacy of the participants can be protected because attackers cannot determine the real answer of a participant. After collecting negative survey results, statistical results about population distribution over different categories could be reconstructed by several methods.

Previous works about negative survey mainly focus on models with specific hypotheses, e.g., the probability that participants select negative categories follows the uniform distribution or Gaussian distribution. These models could be reasonable when negative categories are selected by electronic devices instead of humans. However, in some applications with high security requirements, participants need/want to manually

© Springer International Publishing AG 2017
L. Chen et al. (Eds.): APWeb-WAIM 2017, Part I, LNCS 10366, pp. 57–65, 2017.
DOI: 10.1007/978-3-319-63579-8_5

select a negative category as the answer, and they do not want to use electronic devices because the security cannot be guaranteed. Therefore, we conduct a negative survey in Wuhan University of Technology and China University of Geosciences, and the answers are manually selected by participants. Based on the survey results, we have several findings (as shown in Sect. 3.3). Moreover, we propose a method called NStoPS-MLE to reconstruct useful aggregated results. To enhance the performance, the proposed method uses background knowledge about the overall probabilities of selecting negative categories. Experimental results show that NStoPS-MLE performs better than NStoPS [1, 2] and NStoPS-I [5] on most of the questions.

2 Related Work

Negative survey is first proposed by Esponda [1, 2] in 2006. For example, in a positive survey, the question is designed as follow:

What is the rank of your score in your class:

$$\text{A. } 1-5 \quad \text{B. } 6-15 \quad \text{C. } 16-25 \quad \text{D.} \geq 26?$$

In negative survey, this question is designed as follow:
Which is **NOT** the rank of your score in your class:

$$\text{A. } 1-5 \quad \text{B. } 6-15 \quad \text{C. } 16-25 \quad \text{D.} \geq 26?$$

If the rank of Alice's score is 3, in positive survey, she should select A. But in negative survey, she should select one answer among B, C and D at random.

Generally, assume that the number of categories in a question is c, the number of participants is n, and Q is the reconstructed matrix composed by q_{ij}, where q_{ij} denotes the probability that a participant, who actually belongs to the ith category, selects the jth category as negative category. The statistical results collected from negative survey is $r = (r_1 \dots r_c)$, where r_i is the number of participants that select the ith category as negative category. Our goal is to reconstruct aggregated results $t = (t_1 \dots t_c)$ from negative survey results, where t_i denotes the number of participants that actually belong to the ith category. A theoretical model called NStoPS for reconstructing t is: $t = rQ^{-1}$ [1, 2].

Presently, there are some researches about negative survey. Typically, Bao et al. pointed out in [5] that NStoPS would produce unreasonable negative values, and they proposed two algorithms called NStoPS-I and NStoPS-II to handle negative values. Xie et al. [4] proposed Gaussian negative survey, in which the probability that participants select negative categories follows Gaussian distribution. Zhao et al. [6] suggested to use background knowledge in reconstructing useful information from negative survey results. Recently, Esponda et al. [3] proposed a personalized negative survey model, which could meet different privacy requirements from users. Negative survey has been applied to several scenarios. For example, in [7], Horey et al. employed negative survey for collecting anonymous data in sensor networks. In 2012, Horey et al. [8] used negative survey in collecting the location information of users. In [9], Liu et al. applied negative survey to the privacy protection of cloud data. Overall, previous researches (e.g., [1, 2, 4–6]) mainly focus on uniform negative survey or Gaussian negative survey, in which the probability that participants select negative categories is assumed to follow the uniform distribution or Gaussian distribution. In their experiments, they used electronic devices to simulate the negative selection.

3 Overview of the Survey

3.1 Survey Goal and Questionnaire Design

The main goal of our work is carrying out a realistic negative survey and finding its characteristics. We conducted a survey in two universities. Our questionnaire has three parts. The first part is an anonymous positive survey, we assume the statistical results from this part is close to the truth, and we evaluate the reconstructed results from negative survey based on the results from this part. The second part is a real-name negative survey, and the third part is a real-name positive survey. The third part is simply used to construct some background knowledge about the probabilities of negative selection, and the background knowledge will be used in reconstructing results from negative survey. The respondents can answer all surveys or part of them.

Our questions are sensitive issues about university students. Each part has 15 questions, but the questions in the negative survey are designed in a different form. Several examples of questions are listed as follows: "how often do you skip class", "what's the rank of your scores in your class", "how often do you watch xanthic films". In our questionnaire, four questions have 3 categories, six questions have 4 categories, and five questions have 5 categories. To avoid that the order of categories would affect the choice of the respondents, we rearrange the order of categories in the second survey. Because we finally analyze the collected data based on the content of each category, for a convenience, we use "category A, B, C, D, E" in part 2 the same as in part 1 and part 3 in the rest of this paper.

3.2 Data Statistics

We collect data by surveys online and offline. For online surveys, we program the survey website, and participants are guided to the anonymous positive survey, real-name negative survey and real-name positive survey in turn. The answers of participants are automatically stored to the server database. For offline surveys, first, we conduct the anonymous positive survey, and then, we conduct the two real-name surveys. In the end, we collect 811 valid records (corresponds to 811 respondents) from the anonymous positive survey, 550 valid records from the real-name negative survey, and 528 valid records from the real-name positive survey.

The statistical results of each category for anonymous positive survey and real-name negative survey are shown in Table 1. All these results are rounded to one decimal place. The statistical results of the real-name positive survey are not presented because we just use it to get Q in reconstruction.

Table 1. Statistical results of the anonymous positive survey and real-name negative survey.

	Anonymous positive survey					Real-name negative survey				
	A	B	C	D	E	A	B	C	D	E
1	45.5	49.3	3.2	2.0		12.9	8.5	40.4	38.2	
2	80.6	13.6	3.8	2.0		10.5	14.0	15.3	60.2	
3	23.1	33.3	31.6	12.1		35.5	8.5	15.5	40.5	
4	76.6	19.6	2.5	1.4		9.1	10.2	17.5	63.3	
5	13.4	37.1	39.8	5.3	4.3	18.7	10.5	5.3	25.8	39.6
6	32.3	42.8	18.7	3.1	3.1	17.1	3.6	12.5	18.9	47.8
7	29.6	55.0	11.0	4.4		16.9	5.8	13.8	63.5	
8	40.0	54.6	5.4			24.2	15.5	60.4		
9	91.9	6.3	1.8			9.8	25.6	64.5		
10	15.4	37.5	33.3	8.5	5.3	25.1	8.2	9.8	16.0	40.9
11	94.8	4.1	1.1			8.1	22.0	69.8		
12	94.8	3.8	1.3			6.9	48.5	44.5		
13	8.9	56.5	28.2	3.5	3.0	20.5	7.3	8.7	30.4	33.1
14	6.0	18.4	29.6	38.1	7.9	43.8	5.6	7.1	22.0	21.3
15	5.8	4.6	40.8	48.6		38.2	38.0	6.7	17.1	

3.3 Survey Findings

Based on the collected survey data, we have the following findings:

(1) The probability that participants select negative categories might not follow the uniform distribution or Gaussian distribution. We extract all valid records that have the same identity (i.e., name) in the real-name negative survey and positive survey, and we compare the answers of each participant in the two surveys. Finally, we can obtain a matrix Q for each question, and $Q(i, j)$ denotes the percentage of participants who select the jth category in real-name negative survey among those participants who select the ith category in real-name positive survey. The $Q(i, j)$ could represent (at least approximate to) the probability that participants, who actually belong to the ith category, select the jth category in negative survey. We find that $Q(i, j)$ might not follow the uniform or Gaussian distribution, for example, as shown in Fig. 1, the distributions of $Q(1, 1) \sim Q(1, 4)$ and $Q(2, 1) \sim Q(2, 4)$ are neither uniform nor Gaussian distribution.

Moreover, we find participants prefer to select negative categories that have extreme values. For example, among the participants who select category A in the 3rd question (see the example in Sect. 2) in real-name positive survey, about 57% of them select D as negative category in negative survey while D has an extreme value. The percentages of selecting B and C are about 14% and 18%, respectively. Note that there are usually about 35 students in the classes where we conduct surveys.

(2) Typical reconstruction methods (i.e., NStoPS and NStoPS-I) perform poorly on the collected data. As shown in Tables 3, 4 and 5, the accuracy of NStoPS and NStoPS-I on several questions is low. For example, the errors of the results reconstructed by NStoPS on the 2–7th questions are larger than 0.60. The errors of the results

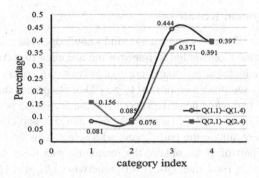

Fig. 1. The distributions of Q(1, 1) ~ Q(1, 4) and Q(2, 1) ~ Q(2, 4).

reconstructed by NStoPS-I on the 2nd, 3rd, 6th, 14th questions are larger than 0.40, and especially, the error is about 0.78 for the 14th question. Those results are almost useless.

(3) Reconstructed results might not be better on the questions with less categories. For example, the result on the 13th question (with 5 categories) is better than several results on the questions with 3 categories or 4 categories (as shown in Table 6).

(4) There might be more unreasonable answers in negative surveys when participants manually select negative categories. For example, we find some participants select the same option for all questions and some participants write fake names in real-name surveys. We regard these records as dirty data and remove them when reconstructing. Moreover, there are some non-negative values in Q matrix for most of the questions (e.g., the 1st, 2nd, 3rd questions, see Table 2). The method of getting Q is showed in Sect. 4.1. It indicates that some participants might have not followed the rule of negative selection, i.e., they have selected a category they really belong to in negative survey, therefore $Q(i, i)$ in some matrices are not 0. Furthermore, we find that the results reconstructed by the theoretical model (i.e., NStoPS) contain unreasonable negative values for most of the questions.

Table 2. Matrices from the samples from the real-name negative and positive survey.

	1st question				2nd question				3rd question			
Q	0.13	0.09	0.41	0.37	0.07	0.12	0.17	0.64	0.11	0.12	0.18	0.59
	0.11	0.1	0.3	0.49	0.24	0.08	0.04	0.64	0.25	0.07	0.2	0.48
	1/3	1/3	0	1/3	1/3	1/3	0	1/3	0.53	0.11	0.06	0.3
	1/3	1/3	1/3	0	1/3	1/3	1/3	0	0.7	0.06	0.12	0.12

4   Reconstruction Algorithm

In this section, a method called NStoPS-MLE is proposed for reconstructing useful aggregated results from negative data.

4.1   Using Background Knowledge

In real world, we usually have some background knowledge. Therefore, we try to use part of Q as background knowledge to improve the accuracy of reconstructed results.

As we have conducted a real-name positive survey, we collect part information about Q by randomly sampling a number of (e.g., 100 or 50) participants and comparing their answers in the real-name negative survey and the real-name positive survey. For example, for the 1st question, $Q(1, 2)$ is set to the percentage of participants who select B in the negative survey among the participants who select A in the real-name positive survey. Except the part of Q we obtain, the remaining part is set as that in uniform negative surveys, i.e., if $i = j$, then $Q(i, j) = 0$; otherwise, $Q(i, j) = 1/(c-1)$.

4.2   NStoPS-MLE

The probability that a participant actually belongs to the ith category and selects the jth category in negative survey is $\frac{t_i}{n} \times q_{ij}$. Consequently, the probability that a participant selects the jth category in negative survey is:

$$p_j = \sum_{i=1}^{c} \frac{t_i}{n} \times q_{ij}. \tag{1}$$

Let $p = (p_1...p_c)$, and in an event of the negative selection on the question in negative survey: the probabilities that the $1^{st}...c^{th}$ category is selected as a negative category are $p_1...p_c$, respectively. The negative selection event happens n times, and the probability that the $1^{st}...c^{th}$ categories are selected as negative categories $r_1...r_c$ times respectively, can be calculated as:

$$\Pr(r|p) = \frac{n!}{r_1! \times ... \times r_c!} p_1^{r_1} \times ... \times p_c^{r_c}. \tag{2}$$

It subjects to multinomial distribution. When reconstructing, we have the observed results $r = (r_1...r_c)$ but $p = (p_1...p_c)$ remains unknown because t is unknown. The reconstruction can be formalized as:

$$\widehat{p}_{mle} = \arg\max_{p \in P} \left\{ \frac{n!}{r_1! \times ... \times r_c!} p_1^{r_1} \times ... \times p_c^{r_c} \right\}. \tag{3}$$

Where P contains all feasible values of p. Because when t is known, p can be calculated from t, we have $\Pr(r|t) = \Pr(r|p)$ and (3) can be converted to:

$$\begin{aligned} \widehat{t}_{mle} &= \arg\max_{t \in T} \left\{ \frac{n!}{r_1! \times ... \times r_c!} \prod_{i=1}^{c} \left(\sum_{j=1}^{c} \frac{t_j}{n} \times q_{ji} \right)^{r_i} \right\} \\ &= \arg\max_{t \in T} \left\{ \sum_{i=1}^{c} r_i \times \log\left(\sum_{j=1}^{c} t_j \times q_{ji} \right) \right\}. \end{aligned} \tag{4}$$

Where T contains all feasible values of t. According to the definition of t, it has the following constrains: $\sum_{i=1}^{c} t_i = n$ and $0 \leq t_i \leq n$. By the constrains, unreasonable negative values can be avoided in reconstruction.

The steps of NStoPS-MLE are shown as follows. Firstly, we counts the total number of participants by $n = r_1 + r_2 + \ldots + r_c$. Next, we revise the unreasonable values of q_{ii} in Q to 0, and scale the other values at the same row by $q_{ij} = q_{ij} / \sum_{j=1\ldots c, j \neq i} q_{ij}$. Then we solves (4) with constrains $\sum_{i=1}^{c} t_i = n$ and $0 \leq t_i \leq n$. Finally, we can get $\widehat{t}$ out. Note that, there are many methods can efficiently solve (4) with constrains, like the interior point algorithm. When $c < 4$ and $n < 1000$, it is feasible in practice to enumerate every possible assignments of t to find the best one according to (4). In our experiments, we solve (4) by the built-in function called fmincon in Matlab.

5 Experimental Results

In this section, we carry out several experiments on reconstructing aggregated results by NStoPS-MLE, and compare it with NStoPS and NStoPS-I.

For each question, we make a list of categories, for which we will collect background knowledge. For each listed category, we randomly select 100 or 50 participants from those who finished the real-name negative survey and positive survey. Table 3 shows the number of the sampled participants for each category in each question. Next, we collect the values in Q from the records of the sampled participants. The remaining part of Q is set according to that of uniform negative surveys. Finally, using the Q, we reconstruct aggregated results $\widehat{t}$ from negative survey results, and we evaluate $\widehat{t}$ by the error formula $\frac{1}{n} \sqrt{\sum_{i=1}^{c} \left(\widehat{t_i} - t_i\right)^2}$ [5].

We carry out the above experiment 30 times for each question, and the average value of error is presented in Tables 4, 5 and 6. Note that, NStoPS and NStoPS-I are executed with the matrix Q for uniform negative surveys. Tables 4, 5 and 6 shows the results for the questions which have three, four and five categories respectively.

As shown in Tables 4, 5 and 6, NStoPS-MLE performs better than NStoPS over all questions, and it performs better than NStoPS-I over most questions. NStoPS-MLE performs worse than NStoPS-I only in the 1st, 8th and 12th questions, because NStoPS-I has already obtained very good results. Note that the effectiveness of the background knowledge about Q is related to the accuracy of the results in the real-name positive survey. However, the "real-name" rule may induce inexact background knowledge.

Table 3. The number of different categories we sample.

	1	2	3	4	5	6	7	8	9	10	11	12	13	14	15
A	100	100	100	100	100	100	100	100	100	100	100	100	0	0	0
B	100	50	100	100	100	100	100	100	0	100	0	0	100	100	0
C	0	0	100	0	100	50	0	0	0	100		0	100	100	100
D	0	0	50	0	0	0	0			0			0	100	100
E					0	0				0			0	0	

Table 4. *Errors* for the questions which have 3 categories.

	8	9	11	12
NStoPS	0.32089732	0.537624429	0.668872531	0.13349797
NStoPS-I	**0.083677434**	0.290930698	0.308991642	**0.129740317**
NStoPS-MLE	0.117258544	**0.163318938**	**0.155915267**	0.131531038

Table 5. *Errors* for the questions which have 4 categories.

	1	2	3	4	7	15
NStoPS	0.417023068	1.071013116	0.646269694	1.133704834	1.113415734	0.477458566
NStoPS-I	**0.126676882**	0.421710766	0.40721515	0.362863376	0.292960686	0.37690809
NStoPS-MLE	0.164332104	**0.261843698**	**0.238704758**	**0.185308482**	**0.241277358**	**0.23388729**

Table 6. *Errors* for the questions which have 5 categories.

	5	6	10	13	14
NStoPS	0.782076105	1.101897433	0.860045928	0.592824026	1.122504328
NStoPS-I	0.309181876	0.477232203	0.236083239	0.19853143	0.782297744
NStoPS-MLE	**0.229283031**	**0.275135786**	**0.180742101**	**0.155891588**	**0.231061221**

The results of NStoPS-MLE seem more stable than that of NStoPS-I, and all *errors* for NStoPS-MLE are less than 0.276, but the *errors* of NStoPS-I in the 2nd, 3rd, 6th and 14th questions are larger than 0.40. Specifically, NStoPS-I has an *error* larger than 0.78 on the 14th question, and that makes its result almost useless.

6 Conclusion and Future Work

In this paper, we present and analyze a real-world negative survey and obtain several findings. Existing reconstruction methods like NStoPS and NStoPS-I perform poorly on the data for several questions. Thus, we propose a method called NStoPS-MLE to effectively reconstruct aggregated results from negative survey results. Experimental results show that NStoPS-MLE using background knowledge performs better than NStoPS and NStoPS-I over most of questions. The method in this paper could be used in real-world negative survey.

In future work, we will try to carry out several surveys to investigate the influence of the privacy degree of different categories on the utility of the collected data. And we will try to use other background knowledge to enhance NStoPS-MLE.

Acknowledgments. This work was partially supported by the National Natural Science Foundation of China (No. 61672398), the Key Natural Science Foundation of Hubei Province of China (No. 2015CFA069), the Applied Fundamental Research of Wuhan (No. 20160101010004), and the Fundamental Research Funds for the Central Universities (No. 173110002).

References

1. Esponda, F.: Negative surveys. arXiv:math/0608176 (2006)
2. Esponda, F., Guerrero, V.M.: Surveys with negative questions for sensitive items. Stat. Probab. Lett. **79**, 2456–2461 (2009)
3. Esponda, F., Kael, H., Victor, M.G.: A statistical approach to provide individualized privacy for surveys. PLoS ONE **11**(1), e0147314 (2016)
4. Xie, H., Kulik, L., Tanin, E.: Privacy-aware collection of aggregate spatial data. Data Knowl. Eng. **70**, 576–595 (2011)
5. Bao, Y., Luo, W., Zhang, X.: Estimating positive surveys from negative surveys. Stat. Probab. Lett. **83**, 551–558 (2013)
6. Zhao, D., Luo, W., Yue, L.: Reconstructing positive surveys from negative surveys with background knowledge. In: The 2016 International Conference on Data Mining and Big Data, pp. 86–99 (2016)
7. Horey, J., Groat, M., Forrest, S., Esponda, F.: Anonymous data collection in sensor networks. In: The Fourth Annual International Conference on Mobile and Ubiquitous Systems: Computing, Networking and Services, pp. 1–8 (2007)
8. Horey, J., Forrest, S., Groat, M.M.: Reconstructing spatial distributions from anonymized locations. In: The 28th International Workshop on Data Engineering, pp. 243–250 (2012)
9. Liu, R., Tang, S.: Negative survey-based privacy protection of cloud data. In: The 2015 International Conference in Swarm Intelligence, pp. 151–159 (2015)

Element-Oriented Method of Assessing Landscape of Sightseeing Spots by Using Social Images

Yizhu Shen$^{(\boxtimes)}$, Chenyi Zhuang, and Qiang Ma

Department of Social Informatics, Graduate School of Informatics,
Kyoto University, Kyoto, Japan
{shen,zhuang}@db.soc.i.kyoto-u.ac.jp, qiang@i.kyoto-u.ac.jp

Abstract. Assessing the quality of sightseeing spots is a key challenge to satisfy the diverse needs of tourists and discover new sightseeing resources (spots). In this paper, we propose an element-oriented method of landscape assessment that analyzes images available on image-sharing web sites. The experimental results demonstrate that our method is superior to the existing ones based on low-level visual features and user behavior analysis.

Keywords: Point of interests · Sightseeing value · Image processing

1 Introduction

The growing importance of tourism to the global economy has highlighted the need for a reliable means of estimating the sightseeing value in order to exploit and distribute resources to sightseeing spots in a rational manner [1–3,15,16].

As a user behavior based method, Zhuang et al. [1,4,5] discover obscure spots with high sightseeing quality but low popularity. However, since their methods rely on the analysis of users' behaviors, the lack of social information for obscure spots makes their solution not so flexible. The content-based method [13] tries to find relationships between low-level visual features and sightseeing value. However, considering the wide variation in photographic skills and techniques of tourists, the performance of this method is limited by its strong dependency on photo quality.

According to a theory proposed in environmental psychology, when people experience a landscape, information is derived through senses, organized, and interpreted by human perception [6]. In this way, a mental model [7] has been devised in which human perception is affected by three aspects: biological factors, cultural factors and individual factors. On the basis of [8], we have proposed three environmental psychology based criteria for sightseeing quality estimation by analyzing social images [15,16]. These criteria for landscape assessment are based on low-level visual features of images shot around the spot. The assessment method [15,16] is still depends on the image quality. However, the quality of

© Springer International Publishing AG 2017
L. Chen et al. (Eds.): APWeb-WAIM 2017, Part I, LNCS 10366, pp. 66–73, 2017.
DOI: 10.1007/978-3-319-63579-8_6

'social images' varies, and many do not satisfy such conditions. In addition, because obscure spots are not well known, there are not enough high-quality images of them that could be used for analysis.

To find a relationship between sightseeing value and landscape elements, in this paper, we propose an element-oriented method that assesses the landscape of a given spot by considering its individual elements. To improve our previous method [15, 16], we extract the landscape elements contained in sightseeing spots and assess the landscape by analyzing them instead of low-level image features. In addition, we also propose a richness-based spot assessment method by taking the richness of the landscape into consideration.

2 Landscape Assessment

2.1 Overview

The input data of our method are sightseeing spots with social images, and the output data are corresponding landscape assessment scores and rankings. As a preliminary process, the data preprocessing obtains representative landscapes image groups and extracts the landscape elements. After that, we estimate the coherence and the visual scale per spot from the images. Then we integrated these two criteria based on the concept of richness in order to rank the spots.

2.2 Data Pre-processing

Filtering. At first, it is needed to filter out the noise social images. Another goal of filtering is to identify the corresponding image groups for scenes at a sightseeing spot that would be interesting to tourists.

Landscape images of scenes that would be of the same interest to tourists tend to be very similar. We use DBCSAN [17] to cluster the images per each spot. The output of this process consists of several image groups, which are supposed to correspond to the scenes of a spot.

Element Extraction. To calculate criteria on landscape elements, we extract elements in the data preparation process after filtering the data. By utilizing the method proposed in [14], our implementation considers only eight of these object classes: building, grass, tree, sky, mountain, water, flower, and road. Connected regions having the same label are treated as an element in the landscape. The extracted landscape elements are used in the element-oriented landscape assessment.

2.3 Assessment Criteria

As an improvement of our previous work, we assess the landscape of a spot with two criteria: (1) coherence and (2) visual scale.

Coherence. According to the research [8], coherence relates to the unity of a scene; it is enhanced by the degree of repetition of colors and textures. On the basis of this definition of unity and coherence, we use patterns as an indicator of the sightseeing value. The patterns are recurrences of a certain composition of color and texture. The color-based pattern calculation is performed in both an intra-element and inter-element way.

(A) Intra-element Coherence of Color is to find the repeated color compositions contained in elements. The details are as follows.

Step 1 (Griding and Clustering): Each element is divided into blocks of 15×15 pixels, and an HSV space-based color histogram representing each block is calculated. Here, a color-based pattern is a repetition of color histogram features contained in an element. The mean-shift method [9] is used to cluster all similar color histograms into groups. In that case, each element is represented by rows of numbers corresponding to the clusters of the color histogram.

Step 2 (Sequence Compression and Comparison): The sequence pattern mining method is adapted on the basis of the definition of coherence so that it can find the patterns inside each element. Since the repetition of a single color feature does not convey much information as far as human perception goes, continuous regions of one color should be excluded from consideration. Therefore, the color-based pattern calculation uses sequence compression wherein adjacent groups having the same numbers are compressed into one group; this reduces redundancy and time complexity.

Step 3 (Finding matched parts): Suppose the element e_i is represented by rows $t_i (i = 1, \ldots m)$ and its corresponding compressed sequence is c_i. We discover the frequent pattern $s_k (k = 1, \ldots, n.)$ in the compressed sequences. Then, we decompress and compare these patterns to refine them. For two patterns $s_{i,k}$ $s_{j,k}$ respectively appearing in rows c_i and c_j (also t_i and t_j), their similarity is defined as follows.

$$Sim(s_{i,k}, s_{j,k}) = 1 - \frac{\|len_t(s_{i,k}) - len_t(s_{j,k})\|}{\|len(t_i) - len(t_j)\|} \tag{1}$$

where, $len_t(s_{i,k})$ denotes the length of $s_{i,k}$ in the original (uncompress) row, and $len(t_i)$ is the length of t_i. In this procedure, only the original pairs with high similarity for each matched pattern can be marked as repetitions patterns $s_i, i = 1, \ldots, n$.

We calculate the color intra-element repetition score as follows.

$$Rep(s_i) = \sigma_{i=1}^{m} \frac{len_t(s_{i,1})}{len(t_i)} \quad Rep_{intra}(e_i) = \frac{\sigma_{k=1}^{n} Rep(s_k)}{Size(e_i)} \tag{2}$$

In this formula, $Rep(s_i, s_j)$ produces a repetition score pattern s_k. The repetition score for the an element e_i is calculated by summing the repetition scores of patterns and normalizing it by $Size(e_i)$, which is the number of sequence numbers contained in element e_i.

(B) Inter-element Coherence of Color. For the inter-element pattern calculation for color, inspired by the research [15], we calculate the color coherence at the level of the overall image $Rep_{inter\_color}(img)$. Considering the potentially large number of elements contained in an image and the sequence numbers contained in each element, comparisons between elements tend to have high time complexity. Therefore, we take the image as a whole and do the calculation in a similar way as the element-based calculation.

Considering both the inter-element and intra-element repetition put important emphasis on the coherence, we combine these two scores as follows to calculate the color coherence score.

$$Coherence_{color}(img) = \sum Rep_{intra\_color}(e_i) \times Rep_{inter\_color}(img) \quad (3)$$

Coherence of Texture. We consider the repeated textures as the repeated or similar patterns in an element. We represent each pattern by using the local binary pattern (LBP) features [10], which is a traditional feature for texture, and perform the same steps as in the color-based pattern calculation.

$$Coherence_{texture}(img) = \sum Rep_{intra\_texture}(e_i) \times Rep_{inter\_texture}(img) \quad (4)$$

Considering that both color and texture are important for the coherence calculation, we define $Co(img)$ as the coherence for an image img as follows.

$$Co(img) = Coherence_{texture}(img) \times Coherence_{color}(img) \quad (5)$$

Visual Scale. The visual scale is defined as a perceptual unit that reflects the openness, depth, and roominess apparent in a landscape [8]. Since both openness and depth are important determiners of the visual scale of a landscape, we calculate values, $op(i)$ and $dp(i)$, for them and use the harmonic value in the visual-scale score $Vi(img)$ for image img. As the way of quantifying openness and depth, we use the GIST [11] based method.

$$Vi(img) = \frac{2}{\frac{1}{op(img)} + \frac{1}{dp(img)}} \quad (6)$$

2.4 Scene Assessment

The criteria mentioned above apply to one image. A method to combine all the scores calculated for the whole image set of the scene I ($I = \{img_1, \ldots, img_n\}$; the image clusters obtained in the data pre-processing) is thus used.

Because of the individual differences in quality between images, for each criteria $c \in \{coherence, visual\text{-}scale\}$, we assume that the scores calculated by the criteria function $Cf(img) \in \{Co(img), Vi(img)\}$ from the image set I are normally distributed and take the highest likelihood value in a 95% confidence interval. The value of a spot I_i is denoted as $LV_c(I_i)$. Besides the criteria score for each image, the diversity of photographers is an important consideration when combining scores because the more people who take photos whose values are close to the most likely value, the more likely the highest likelihood is reliable. Therefore, we calculate the overall value $OS_c(I_i)$ for spot I_i by multiplying the $LV_c(I_i)$ and the fraction of photographers in the 95% confidence interval:

$$OS_c(I_i) = LV_c(I_i) \times \left(1 + \frac{Num_{in}}{Num_{all}}\right), i \in \{1..|I|\}, c \in \{coherence, visual\text{-}scale\}$$
(7)

where Num_{in} is the number of photographers in the 95% confidence interval and Num_{all} denotes the total number of photographers.

We assume that these two criteria have a large impact on the sightseeing value and calculate the combined sightseeing estimation score $Score(I_i)$ as the product of overall values for each criteria score.

$$Score(I_i) = \prod_c OS_c(I_i), c \in \{coherence, visual\text{-}scale\}$$
(8)

2.5 Richness-Based Spot Assessment

On the basis of the research [8], richness is defined as the volume of landscape in sightseeing spots. The image set for a certain sightseeing spot is clustered into image groups, which are treated as the corresponding images for the landscape.

For each image groups g_j, we calculate the proportion of groups of size n_j in the whole image set g_j for sightseeing spot I_i and calculate richness-based scores as follows. In this sense, a landscape scene which has a large number of photos is indicative of the whole sightseeing spot.

$$Score_{richness}(I_i) = \sum_j \frac{n_i}{\sum n} \times Score(g_j)$$
(9)

3 Experimental Evaluation

We compared our method with three baseline methods and the evaluation metric is nDCG [12].

3.1 Dataset

As the experimental data, we collected images of 16 spots from Flickr by keyword based search. The images contained a balance between high-quality spots abundant in natural and cultural elements and low-quality spots that mainly consist of modern architecture so that the experimental data would be unbiased.

To obtain the ground truth, we recruited eight subjects to label each candidate spot in terms of its coherence and visual scale. A five-point scale ranging from 1 for very low value to 5 for very high value was used, and we regarded the average of all the subjects' labels as the ground truth for a spot. Table 1 shows the labeling results and details of our data set.

Table 1. Ground truth: average assessment scores by subjects

Avg. score	Tennryuji Temple	Ninnaji Temple	Kinkakuji Temple	Shisendo	Fushimiji Temple	Hanami Street	Kyoto Station	Daigoji Temple
# of photos	1k	1k	1k	0.4k	1k	1k	1k	1k
# of groups	2	1	2	1	3	2	2	1
Coherence	2.875	2.875	2.75	3.125	3.375	2.5	2.25	2.5
Visual scale	3.375	3	3.25	2.625	2.375	2.75	2.375	3.5
Sightseeing value	4.33	3.17	4	4.17	4.5	3.17	2.33	4
Avg. score	Tai Lake	Jinji Lake	Tiger Hill	Suzhou Museum	Humble Garden	Shantang Street	Guanqian Street	Sekizan Temple
# of photos	1k	0.2k	1k	0.4k	1k	1k	0.3k	0.3k
# of groups	1	1	1	1	1	2	1	1
Coherence	2.625	4.25	3.375	2.25	3.125	2.5	2	3.5
Visual scale	3.75	4.75	3.875	2.375	3.125	2.75	2	2
Sightseeing value	2.83	3.83	3.67	4.17	4.33	3.67	1.67	3.67

3.2   Evaluation of Coherence and Visual Scale Criteria

Figure 1 compares the two kinds of criteria scores with the corresponding ground truth, respectively. Despite the low popularity and small number of images, three obscure spots (Sekizan Temple, Daigoji Temple and Shisen-do) are scored quite accurately, as accurately as the other spots.

Coherence. As shown in Fig. 1, Jinji Lake obtains the highest ground truth coherence score and highest score from our element-oriented method. It seems that the existence of color and texture repetitions for the same kinds of element (e.g., the orderly arrangements of rock pitons) leads to the element-oriented method giving a higher coherence score to Jinji Lake.

The correlation coefficient for Fig. 1 is 0.8193, and the nDCG score for the element-oriented coherence calculation is 0.9722. The overall ranking result, which takes the intra-element factor into consideration, indicates that this method closely matches the human perception of coherence.

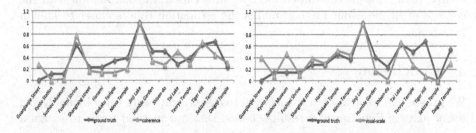

Fig. 1. Comparison with ground truth for coherence and visual scale.

Visual Scale. In the results shown in Fig. 1, the calculated visual-scale scores for Jinji Lake and Tai Lake are clearly different from those of the other spots. The common feature of Jinji Lake and Tai Lake is that most images of them show a wide open vista. Compared with the spots that occupy a small amount of space, spots that have large amounts of space provide a stronger experience of wide-open vistas, in line with the definition of visual scale [8] in environmental psychology. The correlation coefficient for Fig. 1 is 0.5919, and the nDCG score for the visual-scale calculation is 0.9407.

3.3 Evaluation of Spot Ranking

Three baseline methods were implemented in this comparison.

- User-rating method: Here, we calculated the average scores of users' ratings on TripAdvisor (http://www.tripadvisor.com/) as assessment scores.
- NSED method: Proposed in [13], NSED is used to examine the relationship between low-level visual features and perceived naturalness.
- Low-feature based method: Our previous method estimates the sightseeing value by utilizing low-level visual features such as overall tone and photo quality.

The nDCG scores for Low-feature based method, Element-oriented method, User-rating method and NSED method are 0.9277, 0.9381, 0.8376 and 0.9195.

According to the result, the element-oriented method was more accurate than the three baseline methods. The user-rating method, a classic method of the user-behavior approach, is considered to be too inflexible for sightseeing value estimations because scores given by users are affected by both the sightseeing quality and the popularity of a spot. Based on the idea that naturalness tends to have an irregular shape, the NSED method also achieves quite high accuracy in terms of naturalness. However, not only naturalness but also the composition of artificial and natural elements affects the sightseeing value judged by humans. Compared with the NSED method, our method considers factors involved in a sightseeing estimation more comprehensively, in order to satisfy the sightseeing needs of tourists.

The element-oriented assessment outperformed our previous low-feature-based assessments. Note that human perception of coherence is affected by not only the overall color composition but also the relative relationship between element features, which means that the element-oriented method is more sophisticated than a low-feature based method. In addition, in combination with the concept of richness, our method makes the estimation results closer to human perception.

4 Conclusion

We propose an element-oriented method to assess sightseeing value by analyzing social images. Experimental results showed that our method tends to assign a high score to spots with beautiful scenery, wide fields of vision and obvious color tendencies, which are all typical features of good sightseeing spots. In future work, more criteria will be incorporated in our method.

Acknowledgement. This work is partly supported by JSPS KAKENHI (16K12532, 15J01402) and MIC SCOPE (172307001).

References

1. Chen, W.C., Battestini, A., Gelfand, N., Setlur, V.: Visual summaries of popular landmarks from community photo collections. In: ACM Multimedia, pp. 789–792 (2009)
2. Zhuang, C., Ma, Q., Liang, X., Yoshikawa, M.: Anaba: an obscure sightseeing spots discovering system. In: ICME, pp. 1–6 (2014)
3. Hasegawa, K., Ma, Q., Yoshikawa, M.: Trip tweets search by considering spatio-temporal continuity of user behavior. In: Liddle, S.W., Schewe, K.-D., Tjoa, A.M., Zhou, X. (eds.) DEXA 2012. LNCS, vol. 7447, pp. 141–155. Springer, Heidelberg (2012). doi:10.1007/978-3-642-32597-7_13
4. Zheng, Y., Zha, Z., Chua, T.: Research and applications on georeferenced multimedia: a survey. MTAP **51**(1), 77–98 (2011)
5. Zhuang, C., Ma, Q., Liang, X., Yoshikawa, M.: Discovering obscure sightseeing spots by analysis of geo-tagged social images. In: ASONAM, pp. 590–595 (2015)
6. Kaplan, S., Kaplan, R.: Humanscape: Environments for People, pp. 84–90. Duxbury Press, North Scituate (1978)
7. Bourassa, S.C.: The Aesthetics of Landscape. Behaven Press, New York (1991)
8. Tveit, M., Ode, A., Fry, G.: Key concepts in a framework for analyzing visual landscape character. Landsc. Res. **31**(3), 229–255 (2006)
9. Comaniciu, D., Meer, P.: Mean shift: a robust approach toward feature space analysis. IEEE TPAMI **24**(5), 603–619 (2002)
10. Ojala, T., Pietikainen, M., Harwood, D.: A comparative study of texture measures with classification based on featured distributions. Pattern Recogn. **29**(1), 51–59 (1996)
11. Oliva, A., Torralba, A.: Modeling the shape of the scene: a holistic representation of the spatial envelope. IJCV **42**(3), 145–175 (2001)
12. Jarvelin, K., Kekalainen, J.: Cumulated gain-based evaluation of IR techniques. ACM TOIS **20**, 422–446 (2002)
13. Berman, M.G.B., Hout, M.C., Kardan, O.: The perception of naturalness correlates with low-level visual features of environmental scenes. PLoS ONE **9**(12), e114572 (2014)
14. Gould, S.: Multiclass pixel labeling with non-local matching constraints. IEEE CVPR **157**, 2783–2790 (2012)
15. Shen, Y., Ge, M., Zhuang, C., Ma, Q.: Sightseeing value estimation by analyzing geosocial images. IJBDI (2016)
16. Shen, Y., Ge, M., Zhuang, C., Ma, Q.: Sightseeing value estimation by analyzing geosocial images. In: BigMM, pp. 117–124 (2016)
17. Ester, M., Kriegel, H.P., Sander, J.: A density-based algorithm for discovering clusters in large spatial databases with noise. In: KDD, vol. 96, pp. 226–231 (1996)

Sifting Truths from Multiple Low-Quality Data Sources

Zizhe Xie[1]([✉]), Qizhi Liu[1], and Zhifeng Bao[2]

[1] State Key Laboratory for Novel Software Technology,
Nanjing University, Nanjing, China
`forrest0402@smail.nju.edu.cn`, `lqz@nju.edu.cn`
[2] School of CSIT, RMIT University, Melbourne, Australia

Abstract. In this paper, we study the problem of assessing the quality of co-reference tuples extracted from multiple low-quality data sources and finding true values from them. It is a critical part of an effective data integration solution. In order to solve this problem, we first propose a model to specify the tuple quality. Then we present a framework to infer the tuple quality based on the concept of *quality predicates*. In particular, we propose an algorithm underlying the framework to find true values for each attribute. Last, we have conducted extensive experiments on real-life data to verify the effectiveness and efficiency of our methods.

Keywords: Data fusion · Data cleaning · Predicates

1 Introduction

Web data grow at an unprecedented pace following the increasing number of data sources, and people get opportunities to access a wide variety of information and viewpoints from multiple individual sources. Although where to find answers is not troublesome anymore, it remains a big challenge on how to sift true answers from multiple low-quality data sources [7].

Consider a person named Mary shown in Table 1. We collected four tuples for Mary from four sources. Tuple t_i is collected from D_i. Apparently, there are conflicts among the four tuples and we need to resolve conflicts among different salaries and affiliations.

Existing data fusion methods [2,5,6,8] more or less take a voting approach, i.e., accumulating votes from various sources for each value on the same object and selecting the value with the highest vote. However, an inherent limitation of this model is that it is not able to find multiple true values, and it has to depend on a certain relationship among sources to work. Unlike existing methods which make the assumption that only one true value exists, in this paper, we aim to solve the same problem under a more relaxed assumption: multiple true values may exist, and propose a framework to find such true values. In particular, we have made the following contributions:

© Springer International Publishing AG 2017
L. Chen et al. (Eds.): APWeb-WAIM 2017, Part I, LNCS 10366, pp. 74–81, 2017.
DOI: 10.1007/978-3-319-63579-8_7

- We propose a novel data quality model which integrates several data quality criteria and we drop out the common assumption in previous work.
- We propose the concept of *quality predicates* to differentiate true values from false values, which is able to work when there exist multiple true values from data sources.
- For a certain tuple, we propose an algorithm to infer its quality vectors based on its quality predicate(s). The time complexity of finding top-k true values in our method outperforms other rule-based methods.
- We experimentally verify the effectiveness and efficiency of our methods using two real datasets.

The remainder of this paper is organized as follows. Section 2 formally defines the quality model. Section 3 presents the algorithm to infer the *quality vectors* and find true values. Section 4 reports the experiments and we conclude at Sect. 5.

Table 1. Entity instance person for Mary

	Name	Salary	Research area	Affiliation	Publication
t_1:	Mary	142k	Data integration, data cleaning	Amazon	Data integration
t_2:	Mary	120k	Data cleaning	Google	null
t_3:	Mary	88k	Knowledge management	AT&T Labs	A diagnostic tool for data errors
t_4:	Mary	88k	Information retrieve	null	null

2   Model for Tuple Quality

In this section, we will introduce a model to specify the tuple quality.

2.1   Quality Predicates

There have been much work in assessing the tuple quality from plenty of semantic facets such as consistency [4], currency [3] and accuracy [1]. In this paper, we propose three types of *quality predicates* to evaluate tuple quality.

Priority. A *Priority relationship* denotes the relationship defined on attributes. If a set of values of one attribute is messy or impure, it is hard to find true values. However, if attribute values are pure i.e. most sources provide the same value, it is easy to find true values. In our model, we use Shannon entropy to calculate purity. For an attribute A_i: $H(A_i) = -\sum_{x \in X} p(x) \log_2 p(x)$, where X is the set of the classes of values in A_i (null is not a class), and $p(x)$ is the proportion of the number of x to the number of values from A_i. Besides, for the values of the same attribute, the higher the proportion of null values is, the harder the values are to acquire. We use $p_n(A_i)$ to represent the proportion of null values for A_i.

Definition 1 (*Priority predicate*). *For attributes A_i and A_j, if $\frac{H(A_i)}{1-p_n(A_i)} < \frac{H(A_j)}{1-p_n(A_j)}$, we define a priority predicate* $\text{Prior}(A_i, A_j)$.

By Definition 1, a priority relationship is a total order on $A = (A_1, ..., A_l)$. Let $P_{score}(A_i) = \frac{H(A_i)}{1-p_n(A_i)}$. Assuming $P_{score}(A_1) < ... < P_{score}(A_l)$, for any two adjacent $P_{score}(A_i) < P_{score}(A_j)$, we can define a *priority predicate* $\text{Prior}(A_i, A_j)$.

Example 1. In Table 1, we can define three *priority predicates*: Prior(Salary, Research Area), Prior(Salary, Publication) and Prior(Publication, Affiliation).

Status. A *status relationship* is the relationship among the values of the same attribute. The quality of an attribute value changes along with that attribute value.

Definition 2 *(Status predicate). A status predicate* $\text{Stat}(A_i)$ *is in the form* $(\forall t_i, \forall t_j)(P(t_i^{(A_k)}, t_j^{(A_k)}) \wedge \phi(t_i, t_j))$ *representing that the values of* A_i *satisfying a certain condition are worse than the others.*

A status predicate φ is defines on two tuples (it is less likely that a predicate that involves more tuples in real life), and it checks whether $t_i^{(A_k)}$ and $t_j^{(A_k)}$ satisfy the condition defined by P. We then introduce the predicates we use.

For numberic values, we define $P_1(v_1, v_2)$ $(P_2(v_1, v_2))$ to denote v_1 is bigger (less) than v_2. For string typed values, we define $P_3(v_1, v_2)$ $(P_4(v_1, v_2))$ to denote v_1 is longer (shorter) than v_2. We also define $P_5(v_1, v_2)$ $(P_6(v_1, v_2))$ to represent v_1 is more (less) detailed than v_2 in term of their information entropy.

Note that ϕ in *status predicates* is a condition that t_i and t_j must comply. We define $\phi(t_i, t_j) = f_1(t_i^{(A_m)}, t_j^{(A_m)}) \wedge ... \wedge f_l(t_i^{(A_n)}, t_j^{(A_n)})$ where $f_i(v_1, v_2)$ is either $v_1 = v_2$ or $v_1 \neq v_2$.

Example 2. In Table 1, a tuple with a higher salary is of better quality. Hence, we can define $\varphi_4 : \text{Stat}(\text{Salary}) = (\forall t_i, \forall t_j)(P_{lt}(t_i^{(\text{Salary})}, t_j^{(\text{Salary})}))$ to represent a tuple with lower salary is of lower quality.

Interaction. *Interaction* is the relationship among values of different attributes in one tuple. For example, in Table 1, some values are null, which indicates they are of low quality.

Definition 3 *(Interaction predicate). An interaction predicate* $\text{Inter}_\delta(A_1, ..., A_l)$ *represents that when a tuple satisfies condition* δ, *its values of attribute* $A_1, ..., A_l$ *are of low quality.*

For an *interaction predicate*, $\delta = (\forall t_i)(P_1'(t_i^{A_m}, c_1) \wedge ... \wedge P_l'(t_i^{A_n}, c_l))$ where $P_i'(v, c)$ can be any predefined predicate. For *interaction predicate*, we add four more predicates.

For the values of string type, we define $P_7(v, c)$ and $P_8(v, c)$ to represent v_1 contains or not contains c. For both numberic values, we define $P_9(v, c)$ and $P_10(v, c)$ to represent v_1 is equal or not equal to c.

The second argument c is in the range of the dataset. c usually represents a single word or a punctuation. In addition, no matter what schema a instance has, we can define a universal *interaction predciate* $\varphi_5 : \text{Inter}_{(\forall t_i)(P_7(t_i^{(A_i)}, null))}(A_i)$ for every attribute.

3 Deducing Tuple Quality

In this section, we present an algorithm to deduce the *quality vectors* of tuples by *quality predicates*. *Quality vectors* are a n-ary vector representing the quality of a tuple, where the real number in each dimension represents the quality of corresponding attribute value in the tuple.

Our method to find true values is based on the voting approach and assumes that multiple true values exist. For each attribute, if the attribute is labeled as "multi-valued", we will return all the attribute values with non-negative values in *quality vectors*. If the attribute is labeled as "time-sensitive", we will return the attribute value with the highest value in *quality vectors*.

The runtime of our algorithm has the square relation with the size of the tuples. Finding top-k candidates is a polynomial problem after deducing *quality vectors*, while it is NP-hard in a chase-like algorithm [1].

Applying Quality Predicates. *Priority predicates* are used for distinguishing the quality of two similar attribute values by the quality of another attribute value. Hence we apply a priority predicate $Prior(A_i, A_j)$ for two tuples t_1 and t_2 when $q_{t_1}^{(A_j)} = q_{t_2}^{(A_j)}$ and $q_{t_1}^{(A_i)} > q_{t_2}^{(A_i)}$. In this case, if only $t_1^{(A_j)} \neq t_2^{(A_j)}$, we consider $t_1^{(A_j)}$ is of better quality than $t_2^{(A_j)}$. Because priority relationship is a total order relationship on all attributes, we can sort *priority predicates* by their P_{score} of the first attribute in descending order and traverse them in sequence.

Status predicates are used for distinguishing the values in different status. We apply them in the comparison between two tuples i.e. we extract all distinct pairs of tuples and apply *status predicates* on them. During the comparison between t_1 and t_2, if they satisfy a *status predicate* $Stat(A_k) = (\forall t_1, \forall t_2)(P(t_1^{(A_k)}, t_2^{(A_k)}) \wedge \phi(t_1, t_2))$, we decrease η from $q_{t_1}^{(A_k)}$.

The *interaction predicate* is defined on w.r.t. a single tuple. For a *interaction predicate* $Inter_\delta(A_1, ..., A_l)$, if a tuple t_1 satiesfies δ, we decrease η from $q_{t_1}^{(A)}$ where $A = (A_1, ..., A_l)$.

Influence Factor. When applying *quality predicates*, we need to change the values in the corresponding *quality vectors*. We use the word "influence factor" to represent the degree that the *quality vectors* being changed. For simplicity, we set all influence factors when applying *status predicates* and *interaction predicate* to the same value, and set influence factors to 1 when applying *priority predicates* because that is enough to distinguish two different facts of equal quality. We will discuss the influence of different η in Sect. 4.

Execution Order. When applying *quality predicates*, is different applying order leads to a different result? The answer to this question is affirmative. *Status predicates* and *interaction predicates* act on the attribute values that are immutable. Therefore, they can be applied without disturbing each other. However, *priority predicates* act on the variable value of *quality vectors*. Therefore, *priority predicates* should be applied in the last after we inferred *quality vectors*.

Algorithm 1. Deducing the quality vectors of tuples

---

1 function DTQ (M);
   Input : a specification $M = (\mathbf{D}, e, I_s, P_P, P_S, P_I)$
   Output: the *quality vectors* Q of I_s
2 $\mathbf{I}_e \leftarrow \text{PARTITION}(I_s, e)$;
3 $\mathbf{Q} \leftarrow \emptyset$;
4 **for** $I_{e_i} \in \mathbf{I}_e$ **do**
5     $\mathbf{Q}_{e_i} \leftarrow \text{INIT}(I_{e_i})$;
6   **for** $\varphi_s \in P_S$ **do**
7   **for** $(t_i, t_j) \in I_{e_i}$ **do**
8 | $\text{APPLYSTAT}(t_i, t_j, \varphi_s, \mathbf{Q}_{e_i})$;
9   **end**
10   **end**
11   **for** $\varphi_i \in P_I$ **do**
12   **for** $t_i \in I_{e_i}$ **do**
13 | $\text{APPLYINTER}(t_i, \varphi_i, \mathbf{Q}_{e_i})$;
14   **end**
15   **end**
16   **for** $\varphi_c \in \text{TOPSORT}(P_P)$ **do**
17   **for** $I \in \text{SORTBYCLASS}(I_{e_i})$ **do**
18   **for** $(t_i, t_j) \in I$ **do**
19 | $\text{APPLYPRIOR}(t_i, t_j, \varphi_c, \mathbf{Q}_{e_i})$;
20   **end**
21   **end**
22   **end**
23    $\mathbf{Q} \leftarrow \mathbf{Q} \cup \mathbf{Q}_{e_i}$;
24 **end**
25 **return** Q;

Algorithm. We now present the main driver of DTQ. Given M, it first partitions I_s into $\mathbf{I}_e = (I_{e_1}, ..., I_{e_q})$ by entity set $e = (e_1, ...e_q)$. For each I_{e_i}, it initializes an empty $|I_{e_i}| \times n$ *quality matrix* $\mathbf{Q}_{e_i} = (\boldsymbol{q}_{t_1}, ..., \boldsymbol{q}_{t_{|I_{e_i}|}})$, where n presents the number of attributes and each row in $\mathbf{Q}_{e_i}$ represents a *quality vector* of a tuple. Then it starts to apply *status predicates* to deduce *quality vectors*. Assuming $\varphi_s = \text{Stat}(A_k)$, APPLYSTAT will subtract η from $\boldsymbol{q}_{t_i}^{(A_k)}$ if $t_i^{(A_k)}$ and $t_j^{(A_k)}$ satisfy φ_s. Next, DTQ applies *interaction predicates* to further update $\mathbf{Q}_{e_i}$. It traverses each tuple of I_{e_i}. Assuming $\varphi_i = \text{Inter}_\delta(A_1, ..., A_k)$, If t_i satisfies condition δ, then APPLYINTER will substracts η from $\boldsymbol{q}_{t_i}^{(A)}$, where $A = (A_1, ..., A_k)$. Finally, it applies *priority predicates*. Assuming $\varphi_c = \text{Prior}(A_1, A_2)$, if $t_i^{(A_2)} = t_j^{(A_2)}$, APPLYPRIOR will make the comparison between $t_i^{(A_1)}$ and $t_j^{(A_1)}$ and plus 1 to $\mathbf{Q}_{e_i}^{(A_2)}$ of the tuple with greater value on A_1. Notice that before extracting pairs of tuples, it first sorts I_{e_i} in ascending order by the quality of A_2 and groups I_{e_i} by the quality of A_2. We use I to denote a group of tuples that have the same value on A_2.

4   Experimental Study

Using both real-life data and synthetic data, we conducted three sets of experiments to evaluate: (1) the effectiveness of algorithm DTQ on the real dataset, compared with the algorithms of [2,8]; (2) the relationship between the trustworthiness of sources and recall; and (3) the influence of η;

Experimental Setting. We used two real-life datasets[1] (Book and Flight). In our experiment, we used FOIL to discover the *quality predicates*. We manually filtered the predicates of low quality. For all datasets, we set $\eta = 15$ and $\epsilon = 0.1$.

We implemented the following, all in C#: (1) DTQ; (2) TRUTHFINDER [8]; and (3) a truth discovery algorithm SIM [2]. All experiments were conducted on a 64 bit Windows Intel Core(TM) i5-3470 CPU with 16 GB of memory and 1000 GB of storage. Each experiment was repeated 5 times and the average is reported.

Exp-1: Effectiveness of DTQ. Using real life data Book and Flight we evaluated the effectiveness of DTQ compared with SIM [2] and TRUTHFINDER [8].

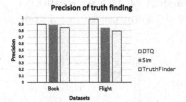

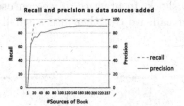

Fig. 1. Results of truth finding.   **Fig. 2.** Applying DTQ on Book.

We tested how many true values were correctly derived by DTQ. As shown in Fig. 1, according to the gold standard, DTQ achieved 90% precision while SIM achieved 89% precision and TRUTHFINDER achieved 85% of precision on Book.

We also ran DTQ on Flight to further verify its effectiveness. In Fig. 1, DTQ get the highest precision on all datasets. In [6], 16 methods including 1 baseline methods, 4 web-link based methods, 3 IR based methods, 7 bayesian based methods and ACCUCOPY were used to find true values on Flight. Among them, ACCUCOPY get the highest precision of 96.0%. However, DTQ get a higher precision than ACCUCOPY. It also can be seen from Fig. 1 that DTQ worked well on Book and Flight.

Exp-2: Trustworthiness of Sources. Using real life data Book and Flight, we applied DTQ to calculate the trustworthiness of each data source and sort data sources by their quality in descending order. For a data source D_i, we calculate

---

[1] All datasets can be download from http://lunadong.com/fusionDataSets.htm.

the trustworthiness of it as $\lambda_i = \sum_{t \in D_i} d(\boldsymbol{q}_t)/|D_i|$, where t is the tuple pertaining to D_i and $|D_i|$ is the size of D_i. By the rank of data sources, we evaluated (a) the change of recall and precision as data sources added; and (b) the distribution of the trustworthiness of data sources.

Recall. As shown in Figs. 2 and 3, we tested the recall as data sources added. In the experiments, we used preference model $d_h(\mathbf{u}) = \mathbf{u}^{(1)} + ... + \mathbf{u}^{(n)}$ to convert a vector to a real number. From the figures, the recall is definitely monotonically increasing as data sources added. All datasets show that true values lie in a few datasets. The curve in Fig. 3 is not as steep and smooth as that in Fig. 2 because the number of data sources is small and we only used a few *quality predicates* to distinguish the trustworthiness of data sources. In conclusion, by scoring data sources, we can use a small amount of data sources to get a same or even better result. Last, an effective rank of data sources verifies the effectiveness of DTQ.

Precision. Figures 2 and 3 also report the precision of DTQ as data sources added. The change of precision is not definitely monotonically increasing. Before the recall reaches the peak, adding more data sources means a larger precision, and the figure of precision during this phase is similar to that of recall. After recall remains stable, more data sources mean more noises, thus the precision may start to shock. In practice, if we can use a small amount of data sources to get a high recall, we can then use a small amount of data sources to get a good precision.

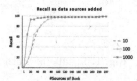

Fig. 3. Applying DTQ on Flight.

Fig. 4. η may affect the precision of DTQ.

Fig. 5. η affects the convergence rate of recall.

Exp-3: Influence Factor η. Intuitively, we should not choose an extremely small value because we use it to filter tuples and a tiny value will not have an effect. We evaluated integer η from two aspects and we set η for all *quality predicates* the same value for simplicity.

Influence on Precision. We first tested the influence of different size of η on the precision of DTQ using Book and Flight. As shown in Fig. 4, we ran DTQ when η in the range of 1 to 100. For Book, when $\eta = 1$, the precision is only 59%. When $1 < \eta < 15$, the precision has a slight wobble. When $\eta \geq 15$, the precision remains 91%. For Flight, the size of η does not influence the result of DTQ. From observation, we found that whether η has an influence depends on *quality predicates*. In practice, this situation is common and there are always imperfect *quality predicates*. Hence, we need to avoid a very large η as well as an extraordinary small η. We set $\eta = 15$ for all our experiments.

Influence on Recall. We then tested the influence of different η in *interaction predicates* on the recall of DTQ on Book. As shown in Fig. 5, we executed DTQ and recorded the recall when $c = 10, 30, 100, 200, 1000$ separately. We found that with the increase in η, the convergence rate of recall is getting slow, which means we need more data sources to achieve an equal result.

5   Conclusion

This paper studied how to the estimate tuple quality and find true values among multiple low-quality data sources. We measured tuple quality by three types of *quality predicates* including *priority predicates*, *status predicates* and *interaction predicates*. These *quality predicates* are in a simple form and can be found automatically by existing methods. By the quality model, we can assess the quality of attribute values and the quality of tuple, and thereby find the true values via *quality vectors*.

In future, we would like to explore how the trustworthiness of data sources affect the accuracy of the true values.

Acknowledgements. This work is supported by the National Natural Science Foundation of China (Grant No. 61572247,61373130).

References

1. Cao, Y., Fan, Y., Yu, W.: Determining the relative accuracy of attributes. In: Proceedings of the 2013 ACM SIGMOD International Conference on Management of Data, pp. 565–576. ACM (2013)
2. Dong, X.L., Berti-Equille, L., Srivastava, D.: Integrating conflicting data: the role of source dependence. Proc. VLDB Endow. **2**(1), 550–561 (2009)
3. Fan, W., Geerts, F., Wijsen, J.: Determining the currency of data. ACM Trans. Database Syst. (TODS) **37**(4), 25 (2012)
4. Fan, W., Li, J., Ma, S., Tang, N., Yu, W.: Towards certain fixes with editing rules and master data. Proc. VLDB Endow. **3**(1–2), 173–184 (2010)
5. Galland, A., Abiteboul, S., Marian, A., Senellart, P.: Corroborating information from disagreeing views. In: Proceedings of the Third ACM International Conference on Web Search and Data Mining, pp. 131–140. ACM (2010)
6. Li, X., Dong, X.L., Lyons, K., Meng, W., Srivastava, D.: Truth finding on the deep web: is the problem solved? Proc. VLDB Endow. **6**, 97–108 (2012). VLDB Endowment
7. Li, Y., Gao, J., Meng, C., Li, Q., Su, L., Zhao, B., Fan, W., Han, J.: A survey on truth discovery. SIGKDD Explor. Newsl. **17**(2), 1–16 (2016)
8. Yin, X., Han, J., Yu, P.S.: Truth discovery with multiple conflicting information providers on the web. IEEE Trans. Knowl. Data Eng. **20**(6), 796–808 (2008)

Graph Data Processing

A Community-Aware Approach to Minimizing Dissemination in Graphs

Chuxu Zhang[1,2], Lu Yu[3], Chuang Liu[1], Zi-Ke Zhang[1(✉)], and Tao Zhou[4(✉)]

[1] Alibaba Research Center for Complexity Sciences,
Hangzhou Normal University, Hangzhou, China
zhangzike@gmail.com
[2] Department of Computer Science and Engineering,
University of Notre Dame, Notre Dame, USA
[3] King Abdullah University of Science and Technology, Jeddah, Saudi Arabia
[4] Big Data Research Center,
University of Electronic Science and Technology of China,
Chengdu, China
zhutou@ustc.edu

Abstract. Given a graph, can we minimize the spread of an entity (such as a meme or a virus) while maintaining the graph's community structure (defined as groups of nodes with denser intra-connectivity than inter-connectivity)? At first glance, these two objectives seem at odds with each other. To minimize dissemination, nodes or links are often deleted to reduce the graph's connectivity. These deletions can (and often do) destroy the graph's community structure, which is an important construct in real-world settings (e.g., communities promote trust among their members). We utilize rewiring of links to achieve both objectives. Examples of rewiring in real life are prevalent, such as purchasing products from a new farm since the local farm has signs of mad cow disease; getting information from a new source after a disaster since your usual source is no longer available, etc. Our community-aware approach, called *constrCRlink* (short for Constraint Community Relink), preserves (on average) 98.6% of the efficacy of the best community-agnostic link-deletion approach (namely, *NetMelt*+), but changes the original community structure of the graph by only 4.5%. In contrast, *NetMelt*+ changes 13.6% of the original community structure.

Keywords: Dissemination control in graph · Community structure · Graph mining

1 Introduction

We address the following problem: given a graph G,[1] can the dissemination of an entity (such as a meme or a virus) be minimized on G while maintaining G's

---

[1] We use the following similar terms in this paper: graph and network, vertex and node, edge and link.

© Springer International Publishing AG 2017
L. Chen et al. (Eds.): APWeb-WAIM 2017, Part I, LNCS 10366, pp. 85–99, 2017.
DOI: 10.1007/978-3-319-63579-8_8

community structure (where nodes within a community have dense connectivity amongst each other, but they have sparse connectivity with others outside their community)? The problem of controlling an entity's spread on a graph has been studied extensively recently [3,5,13,17,19–21], but (to the best of our knowledge) no one has investigated this problem under the constraint of maintaining the graph's community structure as much as possible. Preserving communities in a graph is an important problem in many real-world applications, e.g., individuals trust members of their communities more than non-members because their interactions are more embedded (due to higher link density between members of a community than to outsiders) [2].

The *epidemic tipping point* (i.e., whether a dissemination will die out or not) depends on two factors: (a) the entity's strength and (b) the graph's path capacity [3,17]. We assume that we cannot modify the entity's strength and focus on manipulating the graph's path capacity. However, instead of deleting nodes or links (which affect the graph's community structure), we investigate algorithms that rewire links in order to minimize dissemination *and* minimize change to the community structure of the original (i.e., unperturbed) graph. We quantify minimizing dissemination by the *drop in the largest (in module) eigenvalue of the adjacency matrix*; and measure the amount of change to the community structure of the original (i.e., unperturbed) graph by the *variation of information*, an entropy-based distance function. Thus we focus on solving a realizable problem - namely, how can we efficiently rewire a set of K edges that effectively contain dissemination and maintain community structure.

To solve the aforementioned problem, we present the *CRlink* algorithm (short for Community Relink), which rewires edges in the graph that lead to the largest drop in the leading eigenvalue of the adjacency matrix by choosing the *relink-to* edge with the smallest *eigenscore* within a given community. Furthermore, we present the *constrCRlink* algorithm (short for Constraint Community Relink), which is based on *CRlink* but the rewiring of the edges is based on node-degree constraints. Experiments on a range of different graphs demonstrate the efficiency and effectiveness of *CRlink* and *constrCRlink*. The main contributions of the paper are summarized as: (1) We introduce the problem of minimizing dissemination while preserving community structure on graphs. (2) We propose two efficient and effective algorithms for the aforementioned problem - namely, *CRlink* and *constrCRlink*. (3) Experimental results on various real graphs show that *CRlink* and *constrCRlink* algorithms efficaciously solve in the aforementioned problem.

The rest of paper is organized as follows. Section 2 formally defines the edge rewire manipulation and the new problem of minimizing dissemination on a graph while maintaining the graph's community structure. Section 3 proposes algorithms to solve the problem. Section 4 presents our experiments. Section 5 reviews related works. The paper concludes in Sect. 6.

2   Problem Definition

Table 1 lists the symbols used throughout the paper. We represent an undirected unweighted graph by its adjacency matrix, which is denoted by bold upper-case letter $\mathbf{A}$. Bold lower-case letter $\mathbf{c}$ stands for the community-assignment vector of nodes. The greek letters Φ and Ψ are the sets of deleted and added edges in the rewiring process, respectively. The leading eigenvalue of $\mathbf{A}$ is λ. The bold lower-case letters $\mathbf{u}$ and $\mathbf{v}$ denote the left and right eigenvectors corresponding to λ, respectively.

Table 1. Symbols used in the paper.

Symbol	Definition and description
$\mathbf{A}$	The adjacency matrix of a graph
$\mathbf{A}(i,j)$	The $(i,j)^{\text{th}}$ element of $\mathbf{A}$
$\mathbf{c}$	Community-assignment vector of nodes
$\mathbf{c}(i)$	Community assignment of node i
Φ	Set of deleted edges in rewiring process
Ψ	Set of added edges in rewiring process
λ	The leading eigenvalue of $\mathbf{A}$
$\mathbf{u}, \mathbf{v}$	The left eigenvector and right eigenvector corresponding to λ
n	The number of nodes in graph
m	The number of edges in graph
K	The edge budget

Definition 1 *(Edge Rewiring)*. *Given an undirected edge $e:\langle src, end\rangle$, an edge rewiring on e is a two-step operation: (1) delete e and (2) add a new edge $\hat{e}$ where $\hat{e}$ is either $\langle src, des\rangle$ or $\langle end, des\rangle$, where $des \neq src \neq end$.*

Given the above definition, it is useful to further define two types of edges and three kinds of nodes that participate in the edge rewiring operation. They are: the `rewire-from` edge (denoted by `rf`) is the deleted edge in the rewiring operation, as in edge $e:\langle src, end\rangle$ in Fig. 1. The `rewire-to` edge (denoted by `rt`) is the newly added edge in the rewiring operation, as in edge $\hat{e}:\langle src, des\rangle$ in Fig. 1. The *source* node (denoted by *src*) is the node which is an endpoint in both the `rf` and `rt` edges, as in the node *src* in Fig. 1. The *end* node (denoted by *end*) is the `rewire-from` node, which appears only in the `rf` edge, as in the node *end* in Fig. 1. The *destination* node (denoted by *des*) is the `rewire-to` node, which appears only in the `rt` edge, as in the node *des* in Fig. 1.

In order to design an algorithm for minimizing dissemination while preserving community structure, we need to quantify how we measure the decrease in dissemination and the preservation of community structure. For the former,

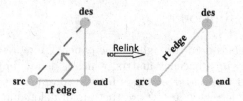

Fig. 1. Example of an edge relink. Edge $\langle src, end \rangle$ is deleted and node src is relinked to node des. The edge between src and des is a new edge in the graph.

Chakrabarti *et al.* [3] and Prakash *et al.* [17] show that the dissemination process disappears in a graph if the strength of the entity (measured by the ratio of its birth rate α over its death rate β) is less than one over the leading eigenvalue λ of $\mathbf{A}$–i.e., $\alpha/\beta < 1/\lambda$. In other words, λ is the only graph-based parameter that determines the tipping point of the dissemination process. The larger the λ, the smaller the dissemination threshold for the entity to spread out. Thus, an ideal strategy for minimizing dissemination on a graph is to *minimize the leading eigenvalue* λ; or alternatively *maximize the drop in the leading eigenvalue* λ. Tong *et al.* [20] estimate the effects of edge removal on λ via an *eigenscore* function. Specifically, they define the eigenscore of an edge $e{:}\langle i, j \rangle$ as the product of the i-th and j-th elements of the left and right eigenvectors corresponding to λ. We use the eigenscore function to select the rf-type edges to be deleted and rt-type edges to be added. An rf-type edge has the largest eigenscore in the graph. An rt-type edge has the smallest eigenscore in the graph. Together these two identify the edge whose rewiring will result in the largest decrease in λ. In addition, we need a way of quantifying how much the community structure of a graph changes as its edges are manipulated. Among the many ways of measuring this quantity, we select the *variation of information* $V(\mathbf{c}, \hat{\mathbf{c}})$ [9]. $V(\mathbf{c}, \hat{\mathbf{c}})$ is a symmetric entropy-based distance function. It measures the "robustness" of a community structure to perturbations in the adjacency matrix. The formal definition of $V(\mathbf{c}, \hat{\mathbf{c}})$ is given in Sect. 4.1. The value of $\frac{V(\mathbf{c}, \hat{\mathbf{c}})}{\log n}$ is between 0 (no change) to 1 (complete change), inclusive.

Finding the set of K edges whose deletion maximizes the drop in λ is an NP-hard problem [20]. The most effective approximate edge-deletion algorithm to maximize the drop in λ recomputes the eigenscores of edges after each edge deletion. This approximate algorithm, called *NetMelt$^+$*, is an improved version of *NetMelt* [20]. Edge rewiring, is a combination of edge deletion and edge addition. Under the same budget K, the best case for edge rewiring is to choose K edges of type rf to delete as in *NetMelt$^+$*, which leads to maximizing the drop in λ. However, with edge rewiring, there are also K edges of type rt that need to be added. These edge additions lead to an increase in λ. Hence, the drop in λ under edge rewiring is always less than the drop under edge deletion. That is, it is impossible to maximize the drop in λ with edge relinkage, whose edge additions are required to minimize $V(\mathbf{c}, \hat{\mathbf{c}})$.

With above analysis, we look for edges that produce a large drop in the λ and a small value of $V(\mathbf{c}, \hat{\mathbf{c}})$. Thus, the problem is formally defined as follows:

Problem 1. *Given a graph* **A** *and an integer (budget)* K, **output** *a set of* K_d
edges of type rf *to be deleted from* **A** *and a set of* K_a *new edges of type* rt *to be
added to* **A**, *which produce a large drop in* λ *and a small value of* $V(\mathbf{c}, \hat{\mathbf{c}})$. *The
budget* K *is equal to* K_d *and* $K_a \leqslant K_d$.

Note that there may be no associated rt-type edge added for a given rf-
type edge deleted (i.e., $K_a \leqslant K_d$). In the following section, we introduce two
algorithms to solve Problem 1.

3 Proposed Algorithms

3.1 Proposed Algorithm: Community Relink (CRlink)

To get the largest drop in λ with edge rewiring (see Definition 1), one can delete
K_d edges of type rf with the highest eigenscores and add K_a previously non-
existent edges of type rt with the lowest eigenscores. We name this simple strategy *GRlink* (short for <u>G</u>lobal <u>Re</u>link). Thus, *GRlink* repeatedly deletes the edge
with the highest eigenscore in the graph and adds the edge with the lowest
eigenscore from one of the endpoints of the deleted edge to any node in the
graph. However, the motivation for edge rewiring (i.e., deletion of an existing
edge followed addition of a new edge) is to maintain the graph's community
structure. The key issue is which previously non-existent edges of type rt are
suitable for addition. *GRlink* chooses the rt edge with the smallest eigenscore
in the whole graph. From the community structure perspective, edge rewiring
among all nodes in the graph may completely change the community structure
because it may decrease the connections among nodes within a community while
increasing the connections across communities, which can lead to different out-
comes for community assignments. Thus, we implement edge rewiring within a
community based on the following considerations:

- Both endpoints of most rf edges are in the same community (i.e., most rf
  edges are "non-bridges"). The average non-bridge edges ratio is over 80% in
  the datasets used in this paper.
- Edge rewiring in the same community is more effective for maintaining community structure than edge rewiring throughout the whole graph.
- In real applications, it is more realizable for an individual to connect to
  another individual who is in the same community (due to higher trust between
  community members).

Algorithm 1 describes the Community Relink (*CRlink*) algorithm. In each
loop of *CRlink*, it first chooses the rf edge with the highest *eigenscore* to delete
and then finds the suitable rt candidate edges whose des node is in the same
community with src node. Finally, it selects the best rt edge with the lowest
eigenscore among these candidates to add. In some loops of *CRlink*, there may
be no associated rt edge for an rf edge due to the within community constraint.
Nonetheless, *CRlink* deletes the rf edges in these loops. Thus, $K_a \leqslant K_d$ in the

---

Algorithm 1. Community Relink (a.k.a. *CRlink*)

---

1 **Input:**Adjacency matrix $\mathbf{A}$, budget K, community vector $\mathbf{c}$;
2 **Output:**K_d deleted edges of type rf indexed by set Φ, and a corresponding K_a added edges of type rt indexed by set Ψ ($K_a \leqslant K_d = K$);
3 Initialize Φ and Ψ to the empty set;
4 **for** $t = 0$ *to* K **do**
5 $\quad$ compute the leading eigenvalue λ of $\mathbf{A}$;
6 $\quad$ compute the corresponding eigenvectors: $\mathbf{u}$ and $\mathbf{v}$;
7 $\quad$ $score(e_{ij})$=$\mathbf{u}(i)\mathbf{v}(j)$ for $i, j = 1, 2, ..., n$;
8 $\quad$ find $e_{del} = e_{\hat{i}\hat{j}} = argmax_{e_{ij}} score(e_{ij})$, where $e_{ij} \notin \Phi$ and $e_{ij} \notin \Psi$;
9 $\quad$ add the edge e_{del} into Φ;
10 $\quad$ **for** $k = 1$ *to* n **do**
11 $\quad\quad$ **if** $\mathbf{c}(\hat{i}) == \mathbf{c}(k)$&&$\mathbf{A}[\hat{i}, k] == 0$ **then**
12 $\quad\quad\quad$ $score(\hat{e}_{\hat{i}k}) = \mathbf{u}(\hat{i})\mathbf{v}(k)$;

13 $\quad\quad$ **if** $\mathbf{c}(\hat{j}) == \mathbf{c}(k)$&&$\mathbf{A}[\hat{j}, k] == 0$ **then**
14 $\quad\quad\quad$ $score(\hat{e}_{\hat{j}k}) = \mathbf{u}(\hat{j})\mathbf{v}(k)$;

15 $\quad$ **if** $\hat{e}_{\hat{i}k} \cup \hat{e}_{\hat{j}k} == \emptyset$ **then**
16 $\quad\quad$ $\hat{e}_{add} = null$ and do not update $\mathbf{A}$

17 $\quad$ **else**
18 $\quad\quad$ find $\hat{e}_{add} = argmin_{(\hat{e}_{\hat{i}k} \cup \hat{e}_{\hat{j}k})} score(\hat{e}_{\hat{i}k} \cup \hat{e}_{\hat{j}k})$,
19 $\quad\quad$ where $\hat{e}_{\hat{i}k} \cup \hat{e}_{\hat{j}k} \notin \Psi$;
20 $\quad\quad$ add the new edge $\hat{e}_{add}$ to Ψ;
21 $\quad\quad$ update added (rt) edges in $\mathbf{A}$;

CRlink algorithm. Note that newly added edges in former steps can not be re-deleted in later steps, as well as newly deleted edges can not be re-added. Thus in Step 8 of Algorithm 1, $e_{ij} \notin \Psi$ avoids the re-deletion of newly added edges. Step 16 and 21 do not update the deleted (rf) edges in $\mathbf{A}$, which guarantees that newly deleted edges will not be re-added.

3.2 Proposed Algorithm: Constraint Community Relink (constrCRlink)

CRlink rewires edges by deleting the edges of type rf with the largest eigenscores and adding the within community edges of type rt with the smallest eigenscores. In this work, we further consider the node degree of *des* when choosing the rt edge to add. An intuitive way is to constrain the degree of *des* node in edge rewiring within community. Adding an edge to a node with small degree impacts the community structure more than adding an edge to a node with large degree. With such consideration, we present the Constraint Community Relink (or *constrCRlink*) algorithm based on *CRlink*. In each iteration of *constrCRlink*, it chooses the rf edge with the highest eigenscore to delete; and rewires one of the endpoints to the corresponding lowest eigenscore rt edge with a small degree *des* node. Similar to *CRlink*, $K_a \leqslant K_d$ in *constrCRlink*. We only need to change Step 11 and Step 13 in Algorithm 1 to get the algorithm for *constrCRlink*:

- Step 11: **if** $\mathbf{c}(\hat{i}) == \mathbf{c}(k)$ && $\mathbf{A}[\hat{i}, k] == 0$ && $d_k \leq \rho$ **then** ...
- Step 13: **if** $\mathbf{c}(\hat{j}) == \mathbf{c}(k)$ && $\mathbf{A}[\hat{j}, k] == 0$ && $d_k \leq \rho$ **then** ...

ρ is a small value parameter for degree constraint and d_k denotes the degree of node k. *constrCRlink* does not consider the special case where the two endpoints are in different communities. This decision is due to two reasons. First, most of the edges in a given graph are non-bridge edges (i.e., with the two endpoints in the same community). Thus, the case where the two endpoints are in different communities has little influence and so *constrCRlink* ignores it. Second, there are a few deleted edges with two endpoints in different communities. This leads to a more stable community structure since edges across communities are deleted while edges within communities are added.

3.3   Algorithm Complexity Analysis

Lemma 1. *The time complexities of CRlink and constrCRlink are $\mathcal{O}(K(m+n))$. The space costs of CRlink and constrCRlink are $\mathcal{O}(n^2)$.*

Proof. In *CRlink* and *constrCRlink*, Steps 5 and 6 take $\mathcal{O}(m+n)$ by Lanczos algorithm [12]. Steps 7 and 8 cost $\mathcal{O}(m)$. The loop from Steps 10 to 14 takes $\mathcal{O}(n)$. Step 18 costs $\mathcal{O}(n)$. Over K iterations, the algorithm takes $\mathcal{O}(K(m+n+m+n+n))$ time. Thus, the time complexities of *CRlink* and *constrCRlink* are $\mathcal{O}(K(m+n))$. In many real graphs, $m \sim n \log n$.

In terms of space, we first need $\mathcal{O}(m)$ to store the original graph **A**. It costs $\mathcal{O}(1)$ and $\mathcal{O}(2n)$ to store the largest eigenvalue and its associated eigenvectors, respectively. In Step 7, it costs $\mathcal{O}(m)$ to store the eigenscores of all edges. Moreover, in the worst case, we need an additional $\mathcal{O}(\binom{n}{2} - m)$ to store the eigenscores of non-existing edges. The storage of deleted edges and added edges take $\mathcal{O}(K)$. Therefore, the total space cost of *CRlink* and *constrCRlink* are $\mathcal{O}(m+1+2n+m+\binom{n}{2}-m+K) \sim \mathcal{O}(m+n+n^2+K)$. Since $n^2 \gg m > n > K$, the total space cost of *CRlink* and *constrCRlink* is $\mathcal{O}(n^2)$ in the worst case.   □

4   Experiments

4.1   Experimental Setup

Datasets. Table 2 lists the graphs used in our experiment. All of them are transformed to undirected and unweighted graphs. We use the following six different types of graphs to evaluate our algorithms:[2] **Facebook user-postings (FB):** We use two graphs of this type. Each node represents a Facebook user. An edge between two users means a "posting" event between them. **Twitter re-tweet (TT):** We use two graphs of this type. A node is a Twitter account. There is an edge between two accounts if a re-tweet event happens between them. **Yahoo! Instant Messenger (YIM):** A node is a Yahoo! IM user. An edge indicates a communication between two users. **Oregon Autonomous System (OG):** A node represents an autonomous system. An edge is a connection inferred from

---

[2] Most of our datasets are available at https://snap.stanford.edu/data/.

the Oregon route-views. **Weibo re-tweet (Weibo):** A node denotes a Sina-Weibo user. There is an edge between two users if a re-tweet event happens between them. **Collaboration Network of ArXiv (CA):** Nodes represent scientists, edges represent collaborations (i.e., co-authoring a paper).

Table 2. Datasets used in our experiments. We use the *Louvain method* [1] to find communities. The number of communities is computed automatically by the method.

Dataset	# of nodes (n)	# of edges (m)	# of communities
FB-1	27,168	26,231	2,154
FB-2	29,557	29,497	1,865
TT-1	25,843	28,124	2,983
TT-2	39,546	45,149	3,920
YIM	50,576	79,219	2,867
OG	7,352	15,665	38
Weibo	34,866	37,849	4,786
CA	5,243	14,484	392

Evaluation Metrics. We consider performances on both the decrease in the leading eigenvalue λ and the change in the community structure $V(\mathbf{c}, \hat{\mathbf{c}})$. Given the original graph $\mathbf{A}$ and the perturbed graph $\hat{\mathbf{A}}$, we have the two evaluation measures: **(a) Drop in the leading eigenvalue:** We define the percent drop in the leading eigenvalue λ as:

$$\Delta\lambda\% = \frac{100 \times (\lambda - \hat{\lambda})}{\lambda},$$

where $\hat{\lambda}$ is the leading eigenvalue of $\hat{\mathbf{A}}$. The higher the $\Delta\lambda\%$, the better the performance. **(b) Change in the community structure:** We use the *variation of information* $V(\mathbf{c}, \hat{\mathbf{c}})$ [9] between the community structures of $\mathbf{A}$ and $\hat{\mathbf{A}}$ since it has all the properties of a proper distance measure. $V(X, Y)$ is defined as:

$$V(X,Y) = H(X|Y) + H(Y|X) = -\sum_{xy} P(x,y) \log \frac{P(x,y)}{P(y)} - \sum_{xy} P(x,y) \log \frac{P(x,y)}{P(x)},$$

where $H(X|Y)$ and $H(Y|X)$ are conditional entropies. $P(x,y) = n_{xy}/n$, $P(x) = n_x/n$ and $P(y) = n_y/n$, where x and y are the community assignments in $\mathbf{c}$ and $\hat{\mathbf{c}}$, respectively. n_{xy} is the number of nodes which belong to community x in $\mathbf{c}$ and community y in $\hat{\mathbf{c}}$. In addition, we normalized $V(\mathbf{c}, \hat{\mathbf{c}})$ by $1/\log n$ since $\log n$ is the maximum value of $V(\mathbf{c}, \hat{\mathbf{c}})$. The lower the $V(\mathbf{c}, \hat{\mathbf{c}})$, the better the performance (i.e., the more the original community structure is preserved). To find communities, we use the Louvain method [1] due to its good performance in both efficacy and efficiency. The choice of community discovery algorithm is orthogonal to our work.

Comparison Methods. We compare the results of six methods: (1) *GRlink*: edge rewiring, rt edges selected from the whole graph based on eigenscore computation. (2) *CRlink*: edge rewiring, rt edges chosen from within a community based on eigenscore computation. (3) *constrCRlink*: edge rewiring, rt edges selected from within a community based on eigenscore computation and degree constraint $\rho = 1$. (4) *NetMelt*: edge deletion, deleted edges selected based on eigenscore computation. (5) *NetMelt$^+$*: edge deletion, an improved version of *NetMelt*, which re-computes the eigenscore after each edge deletion. (6) *RandMelt*: edge deletion, deleted edges are chosen randomly. We run *RandMelt* 100 times and report the average results.

4.2   Experimental Results and Dicussions

Performance w.r.t. $\Delta\lambda\%$ and $V(\mathbf{c}, \hat{\mathbf{c}})$. First, we evaluate the effectiveness of the different methods, in terms of $\Delta\lambda\%$ and $V(\mathbf{c}, \hat{\mathbf{c}})$, across various edge budgets K. Figure 2 shows that *constrCRlink* performs well in terms of $\Delta\lambda\%$ (it is close to *NetMelt$^+$*, which solely optimizes for $\Delta\lambda\%$) and has the smallest $V(\mathbf{c}, \hat{\mathbf{c}})$. *CRlink* also has good performances in both $\Delta\lambda\%$ and $V(\mathbf{c}, \hat{\mathbf{c}})$. So, our algorithms not only have strong impact in containing dissemination but also maintaining community structure. In addition, as discussed in Sect. 3.1, *GRlink* has a large value in $V(\mathbf{c}, \hat{\mathbf{c}})$, i.e., it performs badly in preserving community structure.

Table 3. Results of $\Delta\lambda\%$ and $V(\mathbf{c}, \hat{\mathbf{c}})$ with a fixed budget $P = 100 \times \frac{K}{m} \approx 8\%$. *constrCRlink* preserves on average 98.6% of *NetMelt$^+$*'s efficacy in $\Delta\lambda\%$; and it performs much better in $V(\mathbf{c}, \hat{\mathbf{c}})$.

Dataset	Metric	RandMelt	NetMelt	NetMelt$^+$	GRlink	CRlink	constrCRlink
FB-1	$\Delta\lambda\%$	2.5228	42.118	**64.842**	63.024	63.132	**63.324**
	$V(\mathbf{c}, \hat{\mathbf{c}})$	0.1239	0.1319	**0.1511**	0.2682	0.0510	**0.0483**
FB-2	$\Delta\lambda\%$	4.7317	28.568	**60.312**	58.521	58.798	**58.902**
	$V(\mathbf{c}, \hat{\mathbf{c}})$	0.1530	0.1390	**0.1741**	0.2974	0.0587	**0.0552**
TT-1	$\Delta\lambda\%$	17.803	43.277	**68.820**	66.946	67.257	**67.592**
	$V(\mathbf{c}, \hat{\mathbf{c}})$	0.1648	0.1519	**0.1780**	0.2694	0.0461	**0.0463**
TT-2	$\Delta\lambda\%$	10.161	42.744	**75.985**	74.396	74.428	**74.549**
	$V(\mathbf{c}, \hat{\mathbf{c}})$	0.2073	0.1856	**0.2242**	0.3425	0.0538	**0.0515**
YIM	$\Delta\lambda\%$	14.022	27.282	**68.413**	66.765	57.914	**67.636**
	$V(\mathbf{c}, \hat{\mathbf{c}})$	0.1148	0.0612	**0.0649**	0.2288	0.0454	**0.0452**
OG	$\Delta\lambda\%$	12.001	32.288	**39.227**	38.979	38.875	**38.476**
	$V(\mathbf{c}, \hat{\mathbf{c}})$	0.1609	0.1649	**0.1370**	0.2701	0.0582	**0.0544**
Weibo	$\Delta\lambda\%$	12.489	25.037	**43.639**	43.193	43.382	**43.417**
	$V(\mathbf{c}, \hat{\mathbf{c}})$	0.1446	0.1056	**0.1091**	0.2659	0.0422	**0.0381**
CA	$\Delta\lambda\%$	8.3240	16.429	**47.832**	47.031	35.081	**47.808**
	$V(\mathbf{c}, \hat{\mathbf{c}})$	0.1114	0.0541	**0.0518**	0.1033	0.0274	**0.0274**

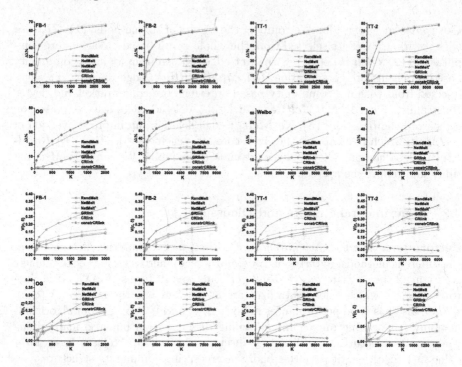

Fig. 2. (Best viewed in color.) $\Delta\lambda\%$ and $V(\mathbf{c}, \hat{\mathbf{c}})$ vs. budget K across different graphs. *constrCRlink*'s $\Delta\lambda\%$ closely shadows that of *NetMelt⁺* across various graphs (the first and second rows); but its $V(\mathbf{c}, \hat{\mathbf{c}})$ is the smallest (the third and fourth rows). (Note: the x-axes are in different scales due to different graph sizes.) (Color figure online)

Besides, Table 3 lists the $\Delta\lambda\%$ and $V(\mathbf{c}, \hat{\mathbf{c}})$ results of different methods with a fixed budget $P = 100 \times \frac{K}{m} \approx 8\%$ across various graphs. The take-away points from this table are: (1) On average, *constrCRlink* preserves 98.6% of *NetMelt⁺*'s efficacy in term of $\Delta\lambda\%$. (2) On average, *constrCRlink* changes the community structure by 4.5%, while *NetMelt⁺* changes it by 13.6%. In other words, *Net-Melt⁺* changes the graph's community structure on average about 3 times more than *constrCRlink*. (3) As expected, $V(\mathbf{c}, \hat{\mathbf{c}})$ of *GRlink* is the largest among all methods because it is agnostic of a node's community structure when it performs edge rewiring.

There are two seemingly counter-intuitive phenomena in Fig. 2 and Table 3. One is that *CRlink* seems to not be as good as *constrCRlink* in $\Delta\lambda\%$, even though *CRlink* has more choices for adding edges. These two methods use different strategies to manipulate the network structure, which result in very different eigenscores. The smallest eigenscore of an edge after *constrCRlink* can be less than the smallest eigenscore of an edge after *CRlink*. The same argument holds when contrasting *GRlink* with *constrCRlink*. The other counter-intuitive phenomenon is that when more edges are modified, $V(\mathbf{c}, \hat{\mathbf{c}})$ of most methods increase as expected, but those of *CRlink* and *constrCRlink* keep decreasing.

The reason for this is because edge-deletion methods and *GRlink* tend to change the community structure more as the edge budget increases, which lead to an increase in the community variation of information. However, *CRlink* and *constrCRlink* rewire edges within communities. This (often) makes the community structure more stable as the edge budget increases, which leads to a decrease in the community variation of information.

Greatest Community Component Visualization. To clearly show the differences in the community structure change across different methods, we extract the Greatest Community Component (GCC, which is the community with the maximum number of nodes among all communities) of the original FB-1 graph and the perturbed FB-1 graphs. For better visualization, we use $K = 1300$ (i.e., $P \approx 5\%$). Figure 3 shows that after applying *CRlink* and *constrCRlink*, the GCCs of their (respective) perturbed graphs are similar to the original GCC. After applying *GRlink* and *NetMelt$^+$*, the GCCs of their (respective) perturbed graphs are different from the original GCC (with many nodes having been assigned to other communities). Therefore, from the visualization perspective, *CRlink* and *constrCRlink* perform well in maintaining the community structure.

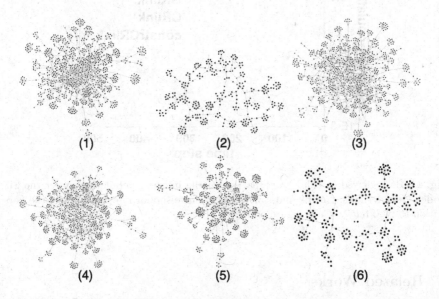

Fig. 3. (Best viewed in color.) GCC visualizations of the original/unperturbed graph and the perturbed graphs. (1), (2), (3), (4), (5) and (6) represent the GCCs of the original FB-1 graph, the graph after *GRlink*, the graph after *CRlink*, the graph after *constrCRlink*, the graph after *NetMelt*, and the graph after *NetMelt$^+$*, respectively. GCCs of the graph after *CRlink* (3) and the graph after *constrCRlink* (4) are the most similar to the original FB-1 graph's GCC (1). (Color figure online)

Simulation of Virus Propagation. We evaluate the effectiveness of our algorithms in terms of minimizing the infected population. Specifically, we simulate the SIS (Susceptible-Infected-Susceptible) model [16] of virus propagation. Due to space limitation, we only report the results on the FB-1 graph. The results on the other graphs are similar. In this experiment, we set the budget K to 2000 and the virus strength s to 0.25. Figure 4 reports the relationship between the rate of infected population and the time step. All results are average values of 100 runs. Obviously, the lower the rate, the better the performance in minimizing dissemination. It can be seen that the infected rate of *constrCRlink* is close to *NetMelt*$^+$'s infected rate. This means that *constrCRlink*, as desired, has similar performance to *NetMelt*$^+$ in dissemination minimization. As shown in the previous sections, *constrCRlink* maintains the community structure of the original/unperturbed graph while *NetMelt*$^+$ does not.

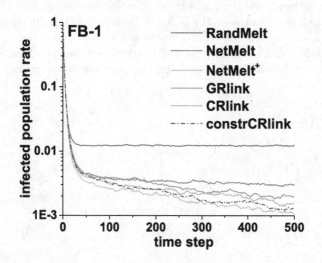

Fig. 4. (Best viewed in color.) Comparison of the infected population under the SIS model. Our methods, *CRlink* and *constrCRlink*, have similar infected rates in the population to *NetMelt*$^+$. (Color figure online)

5   Related Works

The relevant literature for our work can be categorized into two parts: controlling entity dissemination and analyzing community structure.

Controlling Entity Dissemination. The dynamic processes on large graphs like blogs and propagations [8,11] are closely related to entity propagation. For the entity dissemination control, Chakrabarti *et al.* [3] and Prakash *et al.* [17] prove that the only graph-based parameter determining the epidemic threshold is the leading eigenvalue of the adjacency matrix of graph. Tong *et al.* [20] introduce the *NetMelt* algorithm, which minimizes the dissemination on a graph by deleting edges with the largest eigenscore associated with the leading eigenvalue (see Sect. 2 for the definition of eigenscore). Le *et al.* [13] show that *NetMelt* performs poorly on graphs with small eigen-gaps (like many social graphs) and introduce *MET* (short for Multiple Eigenvalues Tracking) to overcome the small eigen-gap problem. *Chan et al.* [4] track multiple eigenvalues for the purpose of measuring graph robustness. Kuhlman *et al.* [10] study contagion blocking in graphs via edge deletion. Saha *et al.* [19] developed *GreedyWalk* approximation algorithms for reducing the spectral radius by removing the minimum cost set of edges or nodes. To the best of our knowledge, no previous work has investigated edge relinkage in order to minimize dissemination while maintaining community structure.

Analyzing Community Structure. Besides entity dissemination control, we try to minimize the change in the graph's community structure after perturbation. Many efforts have been devoted to community structure detection and analysis. The past literatures [1,6,7,18] propose several effective methods to detect communities in real-world graphs. Leskovec *et al.* [14] investigate a range of community detection methods in order to understand the difference in their performances. Nematzadeh *et al.* [15] investigate the impact of community structure on information diffusion with the linear threshold model. Karrer *et al.* [9] study the significance of community structure by measuring its robustness to small perturbations in graph structure. Motivated by this last work, we use the difference in community assignment of each node to quantify the abilities of the different algorithms in preserving the graph's community structure.

6   Conclusion

We present the problem of minimizing dissemination in a population (that is represented as a complex network) while maintaining its community structure (where community is defined as a group of individuals with more links between them than to outside members). Due to the poor performance of edge deletions in preserving community structure, we introduce the edge-rewiring framework and two algorithms: *CRlink* and *constrCRlink*. *CRlink* tends to rewire edges within a community; *constrCRlink* improves *CRlink*'s performance by adding node-degree constraint to rewired edges. Our experimental results on several real-world graphs show that *CRlink* and *constrCRlink* preserve most of the efficacy (more than 98.6%) of *NetMelt*$^+$ in dissemination minimization. Besides, *CRlink* and *constrCRlink* perform much better in preserving community structure (only 4.5% change) than other methods like *NetMelt*$^+$ (with 13.6% change). Furthermore, we investigate the reasons for the different performances of our algorithms.

Acknowledgements. This work was partially supported by Natural Science Foundation of China (Grant Nos. 61673151, 61503110 and 61433014), Zhejiang Provincial Natural Science Foundation of China (Grant Nos. LY14A050001 and LQ16F030006).

References

1. Blondel, V.D., Guillaume, J.-L., Lambiotte, R., Lefebvre, E.: Fast unfolding of communities in large networks. J. Stat. Mech. Theory Exp. **2008**(10), P10008 (2008)
2. Burt, R.S.: Structural Holes: The Social Structure of Competition. Harvard University Press, Cambridge (1992)
3. Chakrabarti, D., Wang, Y., Wang, C., Leskovec, J., Faloutsos, C.: Epidemic thresholds in real networks. ACM Trans. Inf. Syst. Secur. **10**(4), 1 (2008)
4. Chan, H., Akoglu, L., Tong, H.: Make it or break it: manipulating robustness in large networks. In: Proceedings of the 2014 SIAM International Conference on Data Mining, pp. 325–333. SIAM (2014)
5. Chen, C., Tong, H., Prakash, B.A., Eliassi-Rad, T., Faloutsos, M., Faloutsos, C.: Eigen-optimization on large graphs by edge manipulation. ACM Trans. Knowl. Discov. Data **10**(4), 49 (2016)
6. Fortunato, S.: Community detection in graphs. Phys. Rep. **486**(3), 75–174 (2010)
7. Girvan, M., Newman, M.E.: Community structure in social and biological networks. Proc. Natl. Acad. Sci. **99**(12), 7821–7826 (2002)
8. Gruhl, D., Guha, R., Liben-Nowell, D., Tomkins, A.: Information diffusion through blogspace. In: Proceedings of the 13th International Conference on World Wide Web, pp. 491–501. ACM (2004)
9. Karrer, B., Levina, E., Newman, M.E.: Robustness of community structure in networks. Phys. Rev. E **77**(4), 046119 (2008)
10. Kuhlman, C.J., Tuli, G., Swarup, S., Marathe, M.V., Ravi, S.: Blocking simple and complex contagion by edge removal. In: 2013 IEEE 13th International Conference on Data Mining, pp. 399–408. IEEE (2013)
11. Kumar, R., Novak, J., Raghavan, P., Tomkins, A.: On the bursty evolution of blogspace. World Wide Web **8**(2), 159–178 (2005)
12. Lanczos, C.: An iteration method for the solution of the eigenvalue problem of linear differential and integral operators. United States Government Press Office (1950)
13. Le, L.T., Eliassi-Rad, T., Tong, H.: MET: a fast algorithm for minimizing propagation in large graphs with small eigen-gaps. In: Proceedings of the 2015 SIAM International Conference on Data Mining, pp. 694–702. SIAM (2015)
14. Leskovec, J., Lang, K.J., Mahoney, M.: Empirical comparison of algorithms for network community detection. In: Proceedings of the 19th International Conference on World Wide Web, pp. 631–640. ACM (2010)
15. Nematzadeh, A., Ferrara, E., Flammini, A., Ahn, Y.-Y.: Optimal network modularity for information diffusion. Phys. Rev. Lett. **113**(8), 088701 (2014)
16. Pastor-Satorras, R., Vespignani, A.: Epidemic spreading in scale-free networks. Phys. Rev. Lett. **86**(14), 3200 (2001)
17. Prakash, B.A., Chakrabarti, D., Faloutsos, M., Valler, N., Faloutsos, C.: Threshold conditions for arbitrary cascade models on arbitrary networks. In: 2011 IEEE 11th International Conference on Data Mining, pp. 537–546 (2011)
18. Rosvall, M., Bergstrom, C.T.: Maps of random walks on complex networks reveal community structure. Proc. Natl. Acad. Sci. **105**(4), 1118–1123 (2008)
19. Saha, S., Adiga, A., Prakash, B.A., Vullikanti, A.K.S.: Approximation algorithms for reducing the spectral radius to control epidemic spread. In: Proceedings of the 2015 SIAM International Conference on Data Mining, pp. 568–576. SIAM (2015)

20. Tong, H., Prakash, B.A., Eliassi-Rad, T., Faloutsos, M., Faloutsos, C.: Gelling, and melting, large graphs by edge manipulation. In: Proceedings of the 21st ACM International Conference on Information and Knowledge Management, pp. 245–254. ACM (2012)
21. Zhang, Y., Adiga, A., Saha, S., Vullikanti, A., Prakash, B.A.: Near-optimal algorithms for controlling propagation at group scale on networks. IEEE Trans. Knowl. Data Eng. **28**(12), 3339–3352 (2016)

Time-Constrained Graph Pattern Matching in a Large Temporal Graph

Yanxia Xu[1], Jinjing Huang[1], An Liu[1], Zhixu Li[1], Hongzhi Yin[2], and Lei Zhao[1(✉)]

[1] School of Computer Science and Technology, Soochow University, Suzhou, China
xyx.edu@gmail.com, huangjj@siit.edu.cn, {anliu,zhixuli,zhaol}@suda.edu.cn
[2] School of ITEE, The University of Queensland, Brisbane, QLD, Australia
h.yin1@uq.edu.au

Abstract. Graph pattern matching (GPM) is an important operation on graph computation. Most existing work assumes that query graph or data graph is static, which is contrary to the fact that graphs in real life are intrinsically dynamic. Therefore, in this paper, we propose a new problem of Time-Constrained Graph Pattern Matching (TCGPM) in a large temporal graph. Different from traditional work, our work deals with temporal graphs rather than a series of snapshots. Besides, the query graph in TCGPM contains two types of time constraints which are helpful for finding more useful subgraphs. To address the problem of TCGPM, a baseline method and an improved method are proposed. Besides, to further improve the efficiency, two pruning rules are proposed. The improved method runs several orders of magnitude faster than the baseline method. The effectiveness of TCGPM is several orders of magnitude better than that of GPM. Extensive experiments on three real and semi-real datasets demonstrate high performance of our proposed methods.

1 Introduction

Graph pattern matching (GPM) plays a significant role in many fields, such as information retrieval [1], community detecting [2] and biology [3]. The development of GPM undergoes three periods. Firstly, a lot of researchers investigate the problem of querying static graph pattern in static graphs [4–9]. However, graphs in real life are inherently dynamic. Therefore, a lot of efforts have been made in querying static graph pattern in dynamic graphs [10,11]. Recently, to discover more interesting patterns, there exists a few work about querying dynamic graph pattern in dynamic graphs [12].

However, there exist following issues in the traditional work: (1) The dynamic graph is often modeled as a series of static snapshots. However, many complex events in real life cannot be treated as a series of points, such as spread of epidemics, information diffusion, and so on [13]. A series of snapshots cannot illustrate all temporal information and a part of information is missing. (2) The traditional solutions often offer users exponential number of matched

© Springer International Publishing AG 2017
L. Chen et al. (Eds.): APWeb-WAIM 2017, Part I, LNCS 10366, pp. 100–115, 2017.
DOI: 10.1007/978-3-319-63579-8_9

subgraphs [14]. It is daunting for users to inspect all matched subgraphs and find what they really want.

To solve the first issue, we focus on studying temporal graphs. Edges and vertices in a temporal graph exist during a period of time. Temporal graphs can show more information than a series of snapshots. For example, we can clearly see the phenomenon in a temporal graph that event "fever" is during the event "flue". The phenomenon is important for therapy. However, it cannot be illustrated by a series of snapshots.

In order to settle the second issue, we design two types of time constraints in the query graph to reduce the number of matched subgraphs drastically. The first type of time constraint is directed against a single object. For example, we want to find an old classmate who worked at Google for a few years around 2006–2008. The second type of time constraint is directed against the temporal relation among several objects. ALLEN [15] divides the temporal relation into 13 categories which are shown in Table 1.

Table 1. Presentation of ALLEN's 13 temporal relations

Relation Code	Relation	Figure Illustration	Relation Code	Relation	Figure Illustration
1	t_1 before t_2	t_1 t_2	8	t_2 before t_1	t_2 t_1
2	t_1 overlaps t_2	t_1 t_2	9	t_2 overlaps t_1	t_2 t_1
3	t_1 starts t_2	t_1 t_2	10	t_2 starts t_1	t_2 t_1
4	t_1 finishes t_2	t_1 t_2	11	t_2 finishes t_1	t_2 t_1
5	t_1 meets t_2	t_1 t_2	12	t_2 meets t_1	t_2 t_1
6	t_1 contains t_2	t_1 t_2	13	t_2 contains t_1	t_2 t_1
7	t_1 equals t_2	t_1 t_2			

In this paper, we propose a new problem of querying time-constrained graph pattern in a temporal graph. We utilize an example in Fig. 1 to illustrate the problem.

Example 1. Figure 1(b) shows a biology network. A vertex represents a protein and an edge denotes protein-protein interaction. Each edge is associated with a list of time intervals denoted as $(s_1, f_1), (s_2, f_2), \ldots, (s_n, f_n)$ where (s_i, f_i) denotes that interaction starts at time s_i and finishes at time f_i. If a biologist wants to know if there exists a dynamic module [16] in Fig. 1(b) and the dynamic module is described as follows: (1) The module consists of three proteins, i.e.,

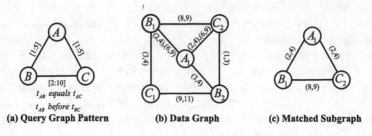

Fig. 1. An example of querying time-constrained graph pattern in a temporal graph

A, B and C. (2) The interaction between A and B exists during time 1 to 5. The interaction between A and C exists during time 1 to 5. The interaction between B and C exists during time 2 to 10. (3) The time of AB equals that of AC and the time of AB should be earlier than that of BC. Figure 1(a) shows the dynamic module. The matched subgraph of Fig. 1(a) is shown in Fig. 1(c) (computing process is discussed in Sect. 4).

Querying time-constrained graph pattern in a temporal graph is not an easy work. As we all know, graph pattern matching is typically defined in terms of subgraph isomorphism which is a NP-hard problem [17]. Besides, taking time constraints into consideration slow down the process drastically. For example, there exists a graph consisting of m nodes and each edge contains n time intervals on average. There may exist $\frac{m \times (m-1)}{2}$ edges. Therefore, there can be $n^{\frac{m \times (m-1)}{2}}$ temporal subgraphs to be checked (the definition of the temporal subgraph is shown in Sect. 3).

In all, the contributions of this paper can be summarized as follows:

- We propose a new problem, i.e., time-constrained graph pattern matching in a large temporal graph. To the best of our knowledge, we are the first to study the problem.
- We propose a baseline algorithm to solve the problem. To improve the efficiency, an improved algorithm is proposed. Besides, two pruning rules are proposed to improve efficiency. The improved algorithm runs several orders of magnitude faster than the baseline algorithm.
- We have carried out a large number of experiments based on three real and semi-real datasets to evaluate the effectiveness and efficiency of our solutions. Experimental results show the high performance of proposed algorithms.

Organization: The rest of the paper is organized as follows. Section 2 reviews related work. In Sect. 3, we introduce several basic concepts and define the problem formally. Details of algorithms are described in Sect. 4. In Sect. 5, we conduct a series of experiments to evaluate the effectiveness and efficiency of proposed methods. Finally, Sect. 6 concludes the paper in brief.

2   Related Work

In this section, we discuss related work on temporal graphs and graph pattern matching.

Temporal Graphs: Recently, a lot of researchers investigate temporal graphs deeply. Huang et al. find minimum spanning trees in a temporal graph [18]. Yang et al. study how to find k dense temporal subgraphs in a temporal graph [19]. Various types of temporal paths are defined to study temporal graphs [20–22]. However, no work queries time-constrained subgraphs in a temporal graph.

Graph Pattern Matching: The problem of querying static graph in static graphs has two lines. One line is subgraph isomorphism which is a NP-hard problem [17]. Recently, Lee et al. [23] re-implement five state-of-the-art subgraph isomorphism algorithms and compare them based on real-life data. However, isomorphism-based graph pattern matching is too strict to find useful patterns. Besides, it is prohibitively expensive when it is applied in large scale graphs. Another line of graph pattern matching is graph simulation [24,25]. Currently, Fan et al. [2] propose a revision of the notion of graph pattern matching, i.e., bounded graph simulation, which overcomes issues in subgraph isomorphism. However, all of work mentioned above cannot solve our problem because both query graph and data graph are static in their work.

In order to solve more practical problems, a lot of researchers investigate deeply on querying static graph in dynamic graphs. These work can be divided into two categories: transactional query and single graph query. In transactional query [26,27], there exist a graph dataset where old graphs can be deleted and new graphs can be added. Yuan et al. [28] propose a one-pass algorithm to update graph indices. In single graph query, nodes and edges in the data graph can be deleted or added. Fan et al. [11] develop incremental algorithms to quickly find results by revising old results. However, none of these work can solve our problem because the query graphs in these work are static.

Recently, a few work focuses on querying dynamic patterns in a dynamic graph. Song et al. [12] aim to find an event pattern over graph stream. However, their work models the dynamic graph as a series of snapshots, which causes the loss of temporal information. Besides, the query pattern only considers partial order constraint on the time of edges [12]. Hence, the algorithms in their work cannot solve our problem.

3   Preliminaries

Temporal Data Graph: Given a directed labeled graph $G = (V, E, L)$, V is a set of vertices, E is a set of edges, L is a label function that assigns labels to vertices and edges. Each edge in E is in the form of (u, v, T) where $u, v \in V$ and T denotes a set of time intervals such that $(s_1, f_1), (s_2, f_2), \ldots, (s_n, f_n)$ where s_i

is starting time of a time interval and f_i is finishing time of a time interval (All algorithms can be used for undirected graphs and each edge in an undirected graph can be seen as a bidirectional edge).

Time-Constrained Query Pattern: A time-constrained query graph $Q = (V_q, E_q, L_q, T_q, TS_q)$ is also a directed graph, V_q is a set of vertices, $E_q \subseteq V_q \times V_q$, L_q is a label function, T_q and TS_q denote two types of time constraints. The first type of time constraint T_q aims at limiting time for a single edge. If there exists a time constraint (e_i, l, h) in T_q, it means that starting time of e_i cannot be earlier than l and finishing time of e_i cannot be later than h. For example, a user wants to find an old friend who worked at a company around 2003–2008 years where 2003 year is a lower bound and 2008 year is an upper bound. Another time constraint TS_q is the constraint on temporal relations between edge e_i and edge e_j. For example, a user wants to go to v_3 from v_1 through v_2 by flight. There exist hidden time constraints on temporal relationship in this query. The finishing time of the flight $v_1 v_2$ must be earlier than the starting time of the flight $v_2 v_3$. ALLEN [15] divides temporal relationship into thirteen types. Table 1 illustrates these thirteen widely used relationships [29]. TS_q is a matrix where $TS_q[i][j]$ is a relation code (Table 1) limited for the temporal relationship between i-th edge and j-th edge.

In the following context, several basic concepts are introduced.

Definition 1. Subgraph Isomorphism [23]: *Given a query graph $Q = (V, E, L)$, a subgraph $g = (V', E', L')$ in the data graph G, a subgraph isomorphism is a bijective function f: $V -> V'$ such that:*

(1) $\forall u \in V, L(u) \subseteq L'(f(u))$.
(2) $\forall (u_i, u_j) \in E$ if and only if $(f(u_i), f(u_j)) \in E'$ and $L(u_i, u_j) = L'(f(u_i), f(u_j))$.

Definition 2. Temporal Subgraph: *Given a temporal data graph $G = (V, E, L)$, a temporal subgraph is a directed labeled graph $g = (V_t, E_t, L_t)$ where $V_t \subseteq V$, L_t is a label function and E_t is a set of edges. Each edge in E_t is in the form of (u, v, t) where $u, v \in V_t$ and t is a time interval. $\forall (u, v, t) \in E_t, \exists (u, v, T) \in E$ and $t \in T$.*

Definition 3. Temporal Subgraph Isomorphism: *Given a query graph $Q = (V_q, E_q, L_q, T_q, TS_q)$, a temporal subgraph $g = (V_t, E_t, L_t)$, a temporal subgraph isomorphism is a bijective function M: $V_q -> V_t$ such that:*

(1) $\forall u \in V_q, L_q(u) \subseteq L_t(M(u))$.
(2) $\forall (u_i, v_i) \in E_q$ if and only if $(M(u_i), M(v_i), t_i) \in E_t$ and $L_q(u_i, v_i) = L_t(M(u_i), M(v_i), t_i)$.
(3) $\forall (e_i, l_i, h_i) \in T_q (e_i = (u_i, v_i)), \exists (M(u_i), M(v_i), t_i) \in E_t$ and $t_i.s \geq l_i$ and $t_i.f \leq h_i$ where s is starting time and f is finishing time.
(4) $\forall (M(u_i), M(v_i), t_i), (M(u_j), M(v_j), t_j) \in E_t$, the temporal relationship between t_i and t_j is $TS[i][j]$.

4   Algorithms

The definitions of formal problem have been proposed in previous section. This section shows details of solutions. We firstly introduce a baseline method, i.e., Time-Constrained Graph Pattern Matching based on Vertex mapping (TCGPM-V). Then, another method, i.e., Time-Constrained Graph Pattern Matching based on Edge mapping (TCGPM-E) is proposed to improve the efficiency. Besides, two pruning rules are proposed to further improve the efficiency.

4.1   *TCGPM-V* Algorithm

Before discussing details of *TCGPM-V*, we firstly define a complete vertex mapping set formally.

Definition 4. A Complete Vertex Mapping Set: *Given a query graph* $Q = (V_q, E_q, L_q, T_q, TS_q)$ *and a data graph* $G = (V, E, L)$, *a complete vertex mapping set* $S_v \subseteq V_q \times V$ *such that:*

(1) For any two elements $(v_i, v_i'), (v_j, v_j') \in S_v$, *if* $v_i \neq v_j$, $v_i' \neq v_j'$.
(2) $\forall v_i \in V_q$ *if and only if* $\exists (v_i, v_i') \in S_v$.

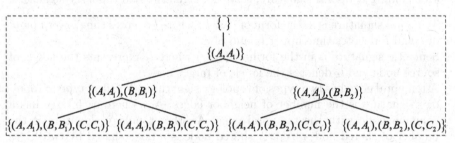

(a) Process of Generating Complete Vertex Mapping Sets

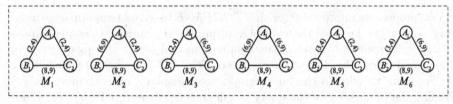

(b) Possible Temporal Subgraphs

Fig. 2. Process of graph pattern matching based on vertex mapping

Algorithm *TCGPM-V* can be divided into two steps. The first step, using depth-first tree search, aims to find all complete vertex mapping sets. The second step enumerates all possible temporal subgraphs and finds correct temporal subgraphs. We utilize Fig. 2 to illustrate the process of *TCGPM-V*. Given a query graph (Fig. 1(a)) and a data graph (Fig. 1(b)), it firstly enumerates all complete vertex mapping sets, i.e., $\{(A, A_1), (B, B_1), (C, C_1)\}, \{(A, A_1), (B, B_1), (C, C_2)\}, \{(A, A_1), (B, B_2), (C, C_1)\}$ and $\{(A, A_1), (B, B_2), (C, C_2)\}$ (shown in Fig. 2(a)). Then, it generates all possible temporal subgraphs shown in Fig. 2(b). The complete vertex mapping sets $\{(A, A_1), (B, B_1), (C, C_1)\}$ and $\{(A, A_1), (B, B_2), (C, C_1)\}$ cannot generate correct temporal subgraphs because there exists no edge between A_1 and C_1. $M_1 - M_4$ are generated by $\{(A, A_1), (B, B_1), (C, C_2)\}$. M_5 and M_6 are generated by $\{(A, A_1), (B, B_2), (C, C_2)\}$. Only M_1 meets the time constraints in Fig. 1(a) and finally M_1 is returned shown in Fig. 1(c). M_2, M_3, M_5 and M_6 cannot meet the constraint that t_{AB} equals t_{AC}. M_4 cannot meet the constraint that t_{AB} is before t_{BC}.

4.2   *TCGPM-E* Algorithm

In previous years, most of studies about subgraph pattern matching were based on vertex mapping. Previous work tried to reduce the number of recursive calls to improve the efficiency. According to in-depth comparison of subgraph isomorphism algorithms [23], we know that signature-pruning is very important for subgraph isomorphism. A temporal edge in a query graph has more signatures than a vertex. The signatures of a temporal edge e can be divided into three types including temporal signature, semantic signature and topology signature:

- Temporal signature is in the form of (e, l, h) where l denotes time lower bound of e and h denotes time upper bound of e.
- Semantic signature is in the form of (l_s, l_t) where l_s represents the labels of source node and l_t denotes the labels of target node.
- A temporal edge has two types of topology signatures. The first type of topology signature is the number of neighbor edges $|N_n|$ where each edge has a common node with the temporal edge. Another topology signature is the number of shared nodes $|N_s|$. A shared node is a node shared by two edges in N_n.

We propose an improved algorithm $TCGPM\text{-}E$ based on temporal edge mapping. To further improve the efficiency of improved algorithms, we decompose a large data graph into a set of small subgraphs and search the query graph in these small subgraphs. The whole process of $TCGPM\text{-}E$ is illustrated as follows: (1) Selecting an edge in the query graph according to a ranking function. (2) Finding the diameter r of the query graph centered at selected edge. (3) Finding all mapping edges M_e of selected edge e. (4) Decomposing the data graph into $|M_e|$ subgraphs. Each subgraph is centered at an edge in M_e and diameter of each subgraph is r. (5) Performing subgraph isomorphism based on edge mapping in each subgraph. (6) Enumerating matched temporal subgraphs.

(1) Ranking Function: To improve the efficiency of the algorithm, it is necessary to minimize the number of components $|M_e|$ which is associated with the selected edge in the query graph. The more mappings the selected edge has, the

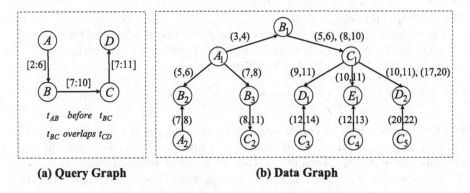

(a) Query Graph   (b) Data Graph

Fig. 3. Example of query and data graphs

more components can be obtained. Hence, it is necessary to find the edge which has less mappings. The ranking function of an edge e is defined as follows:

$$f(e) = \frac{\frac{h-l}{h_{max} - l_{min}} \times \frac{freq(l_s, l_t)}{|E_G|}}{|N_n|} \qquad (1)$$

where h is the time upper bound of e, l is the time lower bound of e, $h_{max} = max(\sum_{i=1}^{|E_G|} h_i)$, $l_{min} = min(\sum_{i=1}^{|E_G|} l_i)$, $freq(l_s, l_t)$ is the number of edges in the data graph which have the same semantic signature with e, $|E_G|$ is the number of edges in data graph and $|N_n|$ is the number of adjacent edges of e. An edge with lowest value is selected. Intuitively, if the interval between l and h is narrower, then less edges may meet the temporal requirements. Besides, e has less mappings if less edges have the same semantic features with e. Finally, the more adjacent edges e has, the less mappings e may have [23].

Let's take Fig. 3 as an example. The value of e_{AB} is $f(e_{AB}) = \frac{\frac{6-2}{11-2} \times \frac{4}{12}}{1} = \frac{4}{27}$. The value of e_{BC} is $f(e_{BC}) = \frac{\frac{10-7}{11-2} \times \frac{2}{12}}{2} = \frac{1}{36}$. The value of e_{CD} is $f(e_{CD}) = \frac{\frac{11-7}{11-2} \times \frac{4}{12}}{1} = \frac{4}{27}$. Finally, e_{BC} is selected.

(2) Graph Diameter: A line of a graph refers a sequence of edges where adjacent edges must have a common vertex. Given an edge e_0 as the center of a graph, a longest line starting from e_0 is $e_0, e_1, e_2, \ldots, e_{i-1}, e_i$ (if $n \neq m$, then $e_n \neq e_m$), then the diameter of a graph centered at e_0 is i.

Let's look at the query graph in Fig. 3. If e_{BC} is set as the center edge, then the longest line is e_{BC}, e_{CD} or e_{BC}, e_{AB}. Hence, the diameter of the query graph is 1.

(3) Edge Mappings: Given an edge e_i in the query graph, if an edge e_j in the data graph is the mapping of e_i, it must meet all following requirements:

- $e_i.l_s = e_j.l_s$ and $e_i.l_t = e_j.l_t$.
- There exists at least one time interval (s, f) in time list of e_j where s is later than time lower bound of e_i and f is earlier than time upper bound of e_i.
- $|e_j.N_n| \geqslant |e_i.N_n|$ and $|e_j.N_s| \geqslant |e_i.N_s|$.

In Fig. 3, only $e_{B_1C_1}$ is a mapping of e_{BC}.

(4) Decomposition: A component can be obtained by the following process: Given a mapping edge e_i and diameter r of a query graph, e_i is firstly put into an empty edge set S_i. Secondly, putting all adjacent edges of edges in S_i into S_i and repeating it r times. Then, a component S_i is obtained. Repeating the process $|M_e|$ times and $|M_e|$ components are obtained.

For example, given a mapping edge $e_{B_1C_1}$ and diameter 1, we finally obtain the component $\{e_{B_1C_1}, e_{A_1B_1}, e_{C_1D_1}, e_{C_1E_1}, e_{C_1D_2}\}$.

(5) Subgraph Isomorphism Based on Edge Mapping: Edge mapping is used in the improved algorithm. We firstly define a complete edge mapping set.

Definition 5. A Complete Edge Mapping Set: *Given a query graph* $Q = (V_q, E_q, L_q, T_q, TS_q)$ *and a data graph* $G = (V, E, L)$, *a complete edge mapping set* $S_e \subseteq E_q \times E$ *must meet the following requirements:*

(1) For any two elements $(e_i, e_i'), (e_j, e_j') \in S_e$, *if* $e_i \neq e_j$, *then* $e_i' \neq e_j'$.
(2) $\forall e_i \in E_q$ *if and only if* $\exists (e_i, e_i') \in S_e$.

Algorithm 1. *RecursiveMapE*

Data: a query patten $Q = (V_q, E_q, L_q, T_q, TS_q)$, a data graph $G = (V, E, L)$ and a partial edge mapping set S_e

Result: a set of complete edge mapping sets

1 **if** S_e *is a complete edge mapping set* **then**
2 | $Output(S_e)$;
3 **else**
4 | $e = NextUnmatchedEdge(Q, S_e)$;
5 | compute a set of mappings E for e;
6 | **for** *each edge* e' *in* E **do**
7 | | put (e, e') into S_e;
8 | | **if** $IsFeasible(S_e, e', e)$ **then**
9 | | | $RecursiveMapE(Q, G, S_e)$;
10 | | remove (e, e') from S_e;

Algorithm 1 shows the pseudo-code of subgraph isomorphism based on edge mapping. Firstly, it checks if S_e is a complete edge mapping set (line 1). If S_e is a complete edge mapping set, S_e is returned (line 2). Otherwise, it needs to extend the partial edge mapping set (lines 4–10). Function $NextUnmatchedEdge$ finds a query edge which is not in S_e (line 4). $NextUnmatchedEdge$ has two rules for finding next edge: (1) The next edge must be an edge connected with a mapped edge which is proved to be efficient [4]. (2) It picks up an edge in query graph according to the ranking function. In line 5, E is a set of mapping edges for e. It extends S_e by adding each possible mapping pairs (e, e') (line 7). After adding a mapping pair, it is necessary to check if a partial edge mapping set is feasible (line 8). If S_e is feasible, it invokes $RecursiveMapE$ to extend S_e until S_e becomes a complete edge mapping set (line 9). The function $IsFeasible$ contains four rules for each added edge mapping pair (e, e') such that: (1) $\forall (e_j, e'_j) \in S_e, e_j.v_t = e.v_s \iff e'_j.v_t = e'.v_s$. (2) $\forall (e_j, e'_j) \in S_e, e_j.v_t = e.v_t \iff e'_j.v_t = e'.v_t$. (3) $\forall (e_j, e'_j) \in S_e, e_j.v_s = e.v_s \iff e'_j.v_s = e'.v_s$. (4) $\forall (e_j, e'_j) \in S_e, e_j.v_s = e.v_t \iff e'_j.v_s = e'.v_t$ where v_s is source node and v_t is target node.

(6) Pruning Rules: After obtaining all complete edge mapping sets, it needs to enumerate all possible temporal subgraphs. The time complexity of enumeration is $|T_a|^{|E_q|} * |R|$ where $|E_q|$ is the number of edges in the query pattern, $|T_a|$ is the average number of time intervals of an edge and $|R|$ is the number of complete edge mapping sets. There are two ways of improving the efficiency. The first way is reducing $|R|$ and the second way is reducing $|T_a|$ ($|E_q|$ is decided by users).

Pruning 1: Given a query pattern $Q = (V_q, E_q, L_q, T_q, TS_q)$ and a complete edge mapping set S_e, $\forall (e_i, e'_i), (e_j, e'_j) \in S_e$, if there are not two time intervals $t_{e'_i}$ and $t_{e'_j}$ that the temporal relation between $t_{e'_i}$ and $t_{e'_j}$ is $TS_q[i][j]$, then S_e can be eliminated. For example, in Fig. 3, $\{(e_{AB}, e_{A_1 B_1}), (e_{BC}, e_{B_1 C_1}), (e_{CD}, e_{C_1 D_2})\}$ is a complete edge mapping set. However, there are not two time intervals $t_{B_1 C_1}$ and $t_{C_1 D_2}$ that $t_{B_1 C_1}$ overlaps $t_{C_1 D_2}$. Hence, this complete edge mapping set can be eliminated.

Pruning 2: Given a query pattern $Q = (V_q, E_q, L_q, T_q, TS_q)$ and a complete edge mapping set S_e, $\forall (e_i, e'_i), (e_j, e'_j) \in S_e$, if there exists no time interval $t_{e'_j}$ that $t_{e'_i}$ has $TS_q[i][j]$ temporal relation with $t_{e'_j}$, then $t_{e'_i}$ can be eliminated. For example, in Fig. 3, $\{(e_{AB}, e_{A_1 B_1}), (e_{BC}, e_{B_1 C_1}), (e_{CD}, e_{C_1 D_1})\}$ is a complete edge mapping set. The time interval $(5, 6)$ of $e_{B_1 C_1}$ can be eliminated because there exists no time interval $t_{C_1 D_1}$ that $(5, 6)$ overlaps $t_{C_1 D_1}$.

Algorithm 2 shows the pseudo-code of $TCGPM\text{-}E$. At first, it selects an edge e in query graph according to the ranking function (line 1). Then, it computes the diameter r of query graph centered at selected edge e by BFS (breadth-first-search) and mappings of e (line 2). For each mapping edge e_i of e, it extracts a component g from the data graph G by BFS (line 4). Then, the first mapping pair (e, e_i) is put into the partial edge mapping set S_e (lines 6–7). All complete edge mapping sets R_g in the subgraph g are computed by invoking function $RecursiveMapE$ (line 8). $Pruning1$ is used to reduce the number of complete edge mapping sets (line 9). Lines 10–17 show the process of generating all correct

---

Algorithm 2. *TCGPM-E*

Data: a query patten $Q = (V_q, E_q, L_q, T_q, TS_q)$ and a data graph $G = (V, E, L)$
Result: a set of temporal subgraphs

1 select a query edge e with lowest ranking value;
2 compute the query graph diameter r centered at e and the mapping set M_e of e;
3 **for** *each $e_i \in M_e$* **do**
4 | $g = GenComponent(e_i, r, G)$;
5 | **if** *the size of g is larger than Q* **then**
6 | $S_e = \emptyset$;
7 | put (e, e_i) into S_e;
8 | $R_g = RecursiveMapE(Q, g, S_e)$;
9 | reduce the number of complete edge mapping sets $|R_g|$ by pruning 1;
10 | **for** *each complete edge mapping set $S_e \in R_g$* **do**
11 | reduce average time intervals T_a by pruning 2;
12 | **while** *exist new temporal subgraphs E_t* **do**
13 | **for** *each edge $e_q \in E_q$, $(e_q, e'_q) \in S_e$* **do**
14 | select a time interval $t_{e'_q}$ from the time intervals list of e'_q;
15 | put $(e'_q, t_{e'_q})$ into E_t;
16 | **if** *E_t satisfies all time constraints* **then**
17 | $Output(E_t)$;

---

temporal subgraphs. Lines 13–15 enumerate all possible temporal subgraphs w.r.t. a complete edge mapping set S_e. If a temporal subgraph E_t meets all time constraints, E_t is returned (lines 16–17).

5   Experiments

In this section, we experimentally evaluate effectiveness and efficiency of proposed algorithms. All algorithms are implemented using java in a Linux machine with 96 CPU, 2.60 GHz, 1T RAM. We use two real datasets named Contact[1], SEQ[2] and a semi-real dataset named Patents[3]. Patents dataset does not have temporal information. In order to keep the real temporal distribution, we transfer the temporal information of dataset SEQ to dataset Patents. All results are average value based on three runs (Table 2).

5.1   Effectiveness

At first, we evaluate the effectiveness of Time-Constrained Graph Pattern Matching (TCGPM). Let's look at the Fig. 4. If a lady works at a company in zone one

---

[1] http://www.sociopatterns.org/datasets.
[2] https://gtfsrt.api.translink.com.au.
[3] https://snap.stanford.edu/data/cit-Patents.html.

Table 2. Datasets

Dataset	Contact	SEQ	Patents
Num. of vertices	327	12,654	3,774,768
Num. of static edges	5,818	15,522	16,518,948
Num. of temporal edges	188,508	701,585	746,556,059
Avg. time intervals	32	45	45
Num. of labels	9	23	421

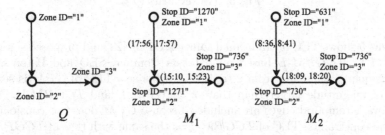

Fig. 4. Illustration of useless matched temporal subgraphs

and she needs to have dinner with a friend in zone two after work. After dinner, she has to go home in zone three. Hence, she queries a traffic patten shown in Q. Both M_1 and M_2 are returned by traditional solutions in GPM (Graph Pattern Matching) [23]. M_1 is not a correct result because the time of edge (1270, 1271) is later than that of edge (1271, 736). When the lady arrives at stop 1271, she is too late to go to stop 736. M_2 is not a good choice for the lady because she could not be off duty at 8:36 a.m. When constraints on temporal relation between two edges are considered, M_1 is not returned by $TCGPM$. When constraints on the time of a single edge can be considered, M_2 is not returned by $TCGPM$. Hence, $TCGPM$ can filter out a lot of useless subgraph so that users can inspect results and find what they really want quickly. To quantify effectiveness of solutions, we define the following performance metric:

Temporal Edge Coverage (TEC): A temporal edge is an edge attached with one time interval. Temporal edge coverage not only considers the number of temporal edges in returned temporal subgraphs, it also considers frequency of each temporal edge in returned subgraphs. Given a data graph $G = \{V_G, E_G, L_G\}$ and a set of returned temporal subgraphs $\{M_1, M_2, \ldots, M_n\}$, TEC is defined as:

$$TEC = \sum_{j=1}^{n} |E_{M_j}| \left/ \sum_{i=1}^{|E_G|} |e_i.T| \right. \tag{2}$$

where $|E_{M_j}|$ is the number of temporal edges in M_j and $|e_i.T|$ is the number of time intervals of edge e_i in the data graph. The challenge is to keep TEC as low as possible.

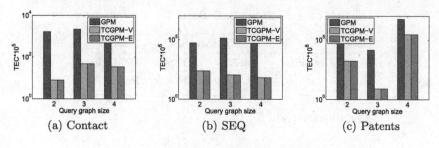

Fig. 5. TEC evaluation

Figure 5 shows TEC of traditional solution *GPM* [23] and proposed solutions *TCGPM-V*, *TCGPM-E* based on datasets Contact, SEQ and Patents. We vary the query graph size from 2 to 4. As we can see, TEC of *GPM* is several orders of magnitude larger than that of *TCGPM-E* and *TCGPM-V*. A lot of irrelevant temporal edges are included because *GPM* does not consider the temporal constraints. TEC of *TCGPM-E* is the same with that of *TCGPM-V* because *TCGPM-E* offer users the same results with *TCGPM-V*.

5.2 Efficiency

In this subsection, we mainly evaluate (1) efficiency of proposed algorithm by comparing it with previous algorithms; (2) the effect of parameters on efficiency; (3) efficiency of proposed pruning rules.

A lot of traditional algorithms, i.e., Ullman [30], VF2 [4] and GraphQL [5], can be used to discover all complete vertex mapping sets. Hence, we re-implement three traditional algorithms in the baseline method and compare them with the improved method *TCGPM-E*. VF2 is used in *TCGPM-F*, GraphQL is used in *TCGPM-G* and Ullman is used in *TCGPM-U*.

Figure 6(a)–(c) show the performance in Contact, SEQ, Patents datasets respectively. The proposed method *TCGPM-E* runs several orders of magnitude faster than other algorithms. Then, we analyze the effect of several parameters, i.e., average time intervals of an edge, number of labels and data graph size (the number of edges in the data graph). Figure 6(d) shows that when average time intervals of an edge becomes larger, the runtime of algorithms increases because the time complexity of generating all temporal subgraphs is $|T_a|^{|E_q|} * |R|$ where $|T_a|$ is average time intervals. Figure 6(e) depicts that increasing number of labels leads to decreasing of runtime because when the data graph size is fixed, the number of labels is larger, there exist less mappings. Figure 6(f) demonstrates that the larger the data graph size is, the longer the runtime is.

Figure 6(g)–(i) illustrate the efficiency of pruning rules. *Pruning** denotes that both *Pruning*1 and *Pruning*2 are used. We can see clearly that both *Pruning*1 and *Pruning*2 can save much runtime. Besides, when two pruning rules are used simultaneously, the performance is better. In Fig. 6(i),

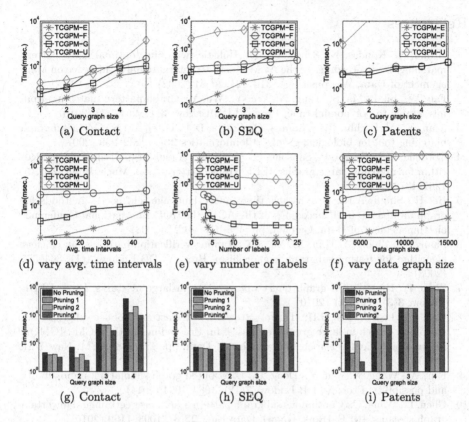

Fig. 6. Efficiency evaluation of proposed algorithms

the performance of pruning rules is not obvious due to semi-log coordinates system. When query graph size is four, *Pruning∗* can save about 30% of run-time.

6   Conclusion

In this paper, we propose the new problem, time-constrained graph pattern matching in a temporal graph. We propose two algorithms, *TCGPM-V* and *TCGPM-E*. Besides, we design two pruning rules. Extensive experiments demonstrate the high performance of proposed solutions. Due to large volumes of temporal graph data, we will study disk representation in the future work.

Acknowledgments. This work was supported by the National Natural Science Foundation of China under Grant Nos. 61572335 and 61572336, the Natural Science Foundation of Jiangsu Province of China under Grant No. BK20151223, the Natural Science Foundation of Jiangsu Provincial Department of Education of China under Grant No. 12KJB520017, and Collaborative Innovation Center of Novel Software Technology and Industrialization, Jiangsu, China.

References

1. Bruno, N., Koudas, N., Srivastava, D.: Holistic twig joins: optimal XML pattern matching. In: Proceedings of the ACM SIGMOD International Conference on Management of Data, Wisconsin, pp. 310–321. ACM (2002)
2. Fan, W., Li, J., Ma, S., Tang, N., Wu, Y., Wu, Y.: Graph pattern matching: from intractable to polynomial time. Proc. VLDB Endow. **3**(1), 264–275 (2010)
3. Tian, Y., McEachin, R.C., Santos, C., States, D.J., Patel, M.: SAGA: a subgraph matching tool for biological graphs. Bioinformatics **23**(2), 232–239 (2007)
4. Cordella, L.P., Foggia, P., Sansone, C., Vento, M.: A (sub)graph isomorphism algorithm for matching large graphs. IEEE Trans. Pattern Anal. Mach. Intell. **26**(10), 1367–1372 (2004)
5. He, H., Singh, A.K.: Graphs-at-a-time: query language and access methods for graph databases. In: Proceedings of the ACM SIGMOD International Conference on Management of Data, Canada, pp. 405–418. ACM (2008)
6. Shang, H., Zhang, Y., Lin, X., Yu, J.X.: Taming verification hardness: an efficient algorithm for testing subgraph isomorphism. Proc. VLDB Endow. **1**(1), 364–375 (2008)
7. Zhao, P., Han, J.: On graph query optimization in large networks. Proc. VLDB Endow. **3**(3), 340–351 (2010)
8. Han, W., Lee, J., Lee, J.H.: Turbo$_{iso}$: towards ultrafast and robust subgraph isomorphism search in large graph databases. In: Proceedings of the ACM SIGMOD International Conference on Management of Data, pp. 337–348. ACM, New York (2013)
9. Fan, W., Wang, X., Wu, Y., Deng, D.: Distributed graph simulation: impossibility and possibility. Proc. VLDB Endow. **7**(12), 1083–1094 (2014)
10. Chen, L., Wang, C.: Continuous subgraph pattern search over certain and uncertain graph streams. IEEE Trans. Knowl. Data Eng. **22**(8), 1093–1109 (2010)
11. Fan, W., Wang, X., Wu, Y.: Incremental graph pattern matching. ACM Trans. Database Syst. **38**(3), 18–118 (2013)
12. Song, C., Ge, T., Chen, C.X., Wang, J.: Event pattern matching over graph streams. Proc. VLDB Endow. **8**(4), 413–424 (2014)
13. Khurana, U., Deshpande, A.: Storing and analyzing historical graph data at scale. In: Proceedings of the 19th International Conference on Extending Database Technology, France, pp. 65–76 (2016)
14. Fan, W., Wang, X., Wu, Y.: Diversified top-k graph pattern matching. Proc. VLDB Endow. **6**(13), 1510–1521 (2013)
15. Allen, J., Allen, J.: Maintaining knowledge about temporal intervals. Commun. ACM **26**(11), 832–843 (1983)
16. Jin, R., McCallen, S., Liu, C., Xiang, Y., Almaas, E., Zhou, X.J.: Identifying dynamic network modules with temporal and spatial constraints. In: Proceedings of the Pacific Symposium, USA, pp. 203–214 (2009)
17. Shamir, R., Tsur, D.: Faster subtree isomorphism. J. Algorithms **33**, 267–280 (1999)
18. Huang, S., Fu, A.W., Liu, R.: Minimum spanning trees in temporal graphs. In: Proceedings of the 2015 ACM SIGMOD International Conference on Management of Data, Melbourne, pp. 419–430 (2015)
19. Yang, Y., Yan, D., Wu, H., Cheng, J., Zhou, S., Lui, J.C.S.: Diversified temporal subgraph pattern mining. In: Proceedings of the 22nd International Conference on Knowledge Discovery and Data Mining, San Francisco, pp. 1965–1974 (2016)

20. Wu, H., Huang, Y., Cheng, J., Li, J., Ke, Y.: Reachability and time-based path queries in temporal graphs. In: 32nd IEEE International Conference on Data Engineering, Helsinki, pp. 145–156 (2016)
21. Wu, H., Cheng, J., Huang, S., Ke, Y., Lu, Y., Xu, Y.: Path problems in temporal graphs. Proc. VLDB Endow. **7**(9), 721–732 (2014)
22. Kossinets, G., Kleinberg, J.M., Watts, D.J.: The structure of information pathways in a social communication network. In: Proceedings of the 14th International Conference on Knowledge Discovery and Data Mining, Las Vegas, pp. 435–443 (2008)
23. Lee, J., Han, W., Kasperovics, R., Lee, J.: An in-depth comparison of subgraph isomorphism algorithms in graph databases. Proc. VLDB Endow. **6**(2), 133–144 (2012)
24. Henzinger, M.R., Henzinger, T.A., Kopke, P.W.: Computing simulations on finite and infinite graphs. In: 36th Annual Symposium on Foundations of Computer Science, Milwaukee, pp. 453–462 (1995)
25. Ma, S., Cao, Y., Fan, W., Huai, J., Wo, T.: Strong simulation: capturing topology in graph pattern matching. Trans. Database Syst. **39**(1), 4 (2014)
26. Yuan, D., Mitra, P., Yu, H., Giles, C.L.: Iterative graph feature mining for graph indexing. In: 28th International Conference on Data Engineering, USA, pp. 198–209 (2012)
27. Chen, C., Yan, X., Yu, P.S., Han, J., Zhang, D., Gu, X.: Towards graph containment search and indexing. In: Proceedings of the 33rd International Conference on Very Large Data Bases, Austria, pp. 926–937 (2007)
28. Yuan, D., Mitra, P., Yu, H., Giles, C.L.: Updating graph indices with a one-pass algorithm. In: Proceedings of the 2015 ACM SIGMOD International Conference on Management of Data, Melbourne, pp. 1903–1916 (2015)
29. Chen, Y., Peng, W., Lee, S.: Mining temporal patterns in time interval-based data. IEEE Trans. Knowl. Data Eng. **27**(12), 3318–3331 (2015)
30. Ullmann, J.R.: An algorithm for subgraph isomorphism. J. ACM. **23**, 31–42 (1976)

Efficient Compression on Real World Directed Graphs

Guohua Li[1], Weixiong Rao[1($\boxtimes$)], and Zhongxiao Jin[2]

[1] School of Software Engineering, Tongji University, Shanghai, China
{03melon,wxrao}@tongji.edu.cn
[2] SAIC Motor AI Lab, Shanghai, China
jinzhongxiao@saicmotor.com

Abstract. Real world graphs typically exhibit power law degrees, and many of them are directed graphs. The growing scale of such graphs has made efficient execution of graph computation very challenging. Reducing graph size to fit in memory, for example by using the technique of lossless compression, is crucial in cutting the cost of large scale graph computation. Unfortunately, literature work on graph compression still suffers from issues including low compression ration and high decompression overhead. To address the above issue, in this paper, we propose a novel compression approach. The basic idea of our graph compression is to first cluster graph adjacency matrix via graph structure information, and then represent the clustered matrix by lists of encoded numbers. Our extensive evaluation on real data sets validates the tradeoff between space cost saving and graph computation time, and verifies the advantages of our work over state of art *SlashBurn* [1,2] and *Ligra+* [3].

Keywords: Compression · Graph clustering · Graph computation algorithms

1 Introduction

Graphs are becoming increasingly important for numerous applications, ranging across the domains of World Wide Web, social networks, bioinformatics, computer security and many others. Such real world graphs typically exhibit power law degrees, and many graphs are directed, such as Web graph and Twitter social graph. Given the real world directed graphs, the growing scale of such graphs has made efficient execution of graph computation very challenging. Some previous work (e.g., PowerGraph [4], Ligra [5], Trinity [6]) fit large graphs in distributed shared memory. However, such systems require a terabyte of memory and expensive cost to maintain distributed computers. Reducing graph size to fit in memory, e.g., by lossless compression, is crucial in cutting the cost of large-scale graph computation and thus motivates our work in this paper.

Literature work typically adopted two following techniques. (*1*) Compression of adjacency lists when graphs are represented as adjacency lists. For example,

© Springer International Publishing AG 2017
L. Chen et al. (Eds.): APWeb-WAIM 2017, Part I, LNCS 10366, pp. 116–131, 2017.
DOI: 10.1007/978-3-319-63579-8_10

recent work *Ligra+* [3] adopts the widely used encoding schemes (such as byte coding, run-length byte coding and nibble coding) over adjacency lists for smaller graph size. (*2*) Compression of adjacency matrix when graphs are represented as adjacency matrix. This technique (e.g., adopted by *SlashBurn* [1,2]) frequently exploit graph clustering algorithms to compress graphs.

Unfortunately, the two approaches above still suffers from issues to compress real work directed graphs. For example, the technique of compressing adjacency lists including *Ligra+* is frequently with a low compression ratio, when compared with the compression of adjacency matrix. Alternatively, the clustering techniques, frequently used by compression of adjacency matrix, are known to perform not very well on real world directed graphs due to no good cut of such graphs. Finally, many compression techniques including both *Ligra+* and *SlashBurn* could compromise graph computation efficiency, due to nontrivial decompression overhead.

To address the above issue, in this paper, we propose a novel compression algorithm on real world directed graphs. The proposed approach can be intuitively treated as a hybrid of the two approaches above: we first perform an effective clustering algorithm and then represent the resulting adjacency matrix by lists of encoded numbers. In this way, our approach has the chance to greatly reduce the graph size.

We make the following contributions to enable our work. (*1*) The proposed clustering algorithm leverages real world graph structure information (indegrees, outdegrees and sibling neighbors) to permute graph vertices for high space cost saving. (*2*) Our scheme to encode the clustered graphs can greatly optimize graph computation (e.g., Breadth-First-Traversal: BFS) efficiency with no or trivial decompression overhead. Thus, the proposed graph compression does not compromise graph computation time. (*3*) Our extensive experiment on real data sets studies the tradeoff between space cost saving and graph computation time, and verifies the advantages of our work over state of arts *SlashBurn* and *Ligra+*. For example, when compared with *SlashBurn* on a real data set Amazon graph, our work uses 35.77% fewer nonempty blocks (when both *SlashBurn* and our work uses the same approach to encode nonempty blocks for fairness), and 4.79× faster time of running BFS. When compared with *Ligra+*, our scheme can achieve 20.24% less space cost saving and 8.24× higher speedup ratio to run the BFS algorithm on an example graph data set.

The rest of the paper is organized as follows. We first give the problem definition and encoding scheme in Sect. 2, and next present the clustering algorithm in Sect. 3. After that, Sect. 4 evaluates our approach, and Sect. 5 reviews literature work. Finally Sect. 6 concludes the paper.

2 Problem Definition and Encoding Scheme

In this section, we first give the problem definition (Sect. 2.1), and present the scheme to encode adjacency matrices (Sect. 2.2).

2.1   Problem Definition

Given a directed graph $\mathcal{G}$, we model it as an asymmetric binary matrix $\mathcal{M}$. The binary element $e_{i,j}$ in the i-th row and j-th column indicates whether or not there exists a directed edge from the i-th vertex to j-th vertex (where i and j denote vertex IDs). When dividing the matrix $\mathcal{M}$ to blocks of $b \times b$ elements, we determine whether a block is empty or not. Specifically, if the $b \times b$ elements in a block are all zeros, we say that the block is *empty* and otherwise *non-empty*. Now we can implicitly measure graph space cost by counting those nonempty blocks, denoted by $B(\mathcal{M})$.

Our basic idea is to permute matrix rows and columns (i.e., re-ordering vertex IDs), such that the 1-elements in the permuted matrix is co-clustered. Thus, we have chance to minimize $B(\mathcal{M})$ and reduce graph space cost.

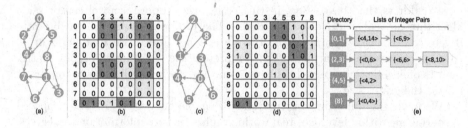

Fig. 1. Permutation of graph vertices: (a-b) directed graph and adjacency matrix before permutation, (c-d) directed graph and adjacency matrix after permutation, (e) representation of the permuted graphs by lists of encoded numbers.

Example 1. Figure 1(a) gives a directed graph. Figure 1(b) shows the associated binary matrix with $B(\mathcal{M}) = 10$ non-empty blocks (we divide the matrix into 2×2 elements). After permuting the graph we will have the resulting graph and associated matrix in Fig. 1(c-d). Now we have $B(\mathcal{M}) = 7$ non-empty blocks in Fig. 1(d).

Problem 1. Given a directed graph $\mathcal{G}$ and associated matrix $\mathcal{M}$, we want to find a permutation of vertex IDs $\pi : V \to [n]$ with the objective to minimize the count $B(\mathcal{M})$ after the permutation.

Note that Problem 1 is NP-hard (see [7]) and no efficient solution can solve it with polynomial time. Consider that real world graphs follow power-law degree distributions with few 'hub' nodes having very high degrees and majority of the nodes having low degrees. Traditional clustering approaches, such as spectral clustering [8], co-clustering [9], cross-associations [10], and shingle-ordering [11], do not work well on such real work graphs due to no good cuts.

2.2   Graph Representation by Encoding Non-empty Blocks

When a directed graph is represented as a binary matrix above, we encode those nonempty blocks (consisting of $b \times b$ elements) into lists of integers. Specially, we first use a directory to maintain the matrix row IDs. Each element in the directory (i.e., an associated matrix row ID) refers to a list of integers to encode such non-empty blocks. For example, in Fig. 1(b), since we divide the matrix into 2×2 blocks, each element in the directory contains two row IDs when the each of such IDs is with at least one non-empty block. Since the row IDs 6 and 7 are with the all empty blocks, the directory does not contain the row IDs 6 and 7. Thus, among the 9 row IDs (from 0 to 8), we have totally four elements in the directory: $\{0, 1\}$, $\{2, 3\}$, $\{4, 5\}$ and $\{8\}$.

Next, to encode a non-empty block, we use two numbers. The 1st number is the left-most column ID of such a block. Next by treating the binary elements inside the block as the binary form of an integer, we can use the integer number to represent the whole block. For example, for the left nonempty block in the row IDs 0 and 1, its left-most column ID is 4; the $2 \times 2 = 4$ binary elements 1110 are encoded to be an integer 14. Thus, we use an integer pair $\{4, 14\}$ to encode the block. Similar situation holds for other non-empty blocks. Figure 1(d) gives the directory and lists of integer pairs to encode the graph in Fig. 1(c).

Note that we might adopt an alternative approach to compress nonempty blocks. For example, *SlashBurn* adopted gZip to compress nonempty blocks. However, our encoding scheme above offers the obvious benefit: the decoding overhead from the encoded numbers to original binary elements is trivial, when compared with the gZip decompression. Our experiments will verify the benefit when compared with other encoding and compression approaches.

It is not hard to find that the space cost of such encoded lists also depends upon $B(\mathcal{M})$. Thus, the next section we will present the permutation algorithm.

3   Graph Clustering

We first highlight the solution overview (Sect. 3.1), and next give the algorithm details (Sect. 3.2).

3.1   Solution Overview

In order to understand the proposed clustering algorithm of the permutation of graph vertices, we first give two observations of real world graph structure. First let us consider the hub vertices with high in-degrees in real world directed graphs. Due to the power-law in-degree distributions in such graphs, few *hub* vertices are with a large amount of in-coming edges (e.g., a lot of *spoke* fans in Twitter social network), indicating very high in-degrees; yet the majority of vertices have low in-degrees. Given the hub and spoke vertices, we give the following observation.

Observation 1. *For two hub vertices pointed by a large amount of spoke neighbors, it is not rare that such hub vertices share many common spoke neighbors.*

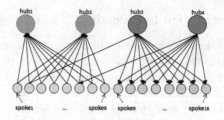

Fig. 2. Hubs and spokes

Intuitively, if the similarity of such spokes is high, we would like to permute the hub vertices together in the matrix columns. Meanwhile, if two spoke neighbors share common hubs (for example, due to the similar interest in social networks). The similarity of such spokes is high and we also place such spokes together in the matrix rows. In Fig. 2, hub_1 and hub_2 share 8 incoming neighbors $spoke_1...spoke_8$, and $spoke_1...spoke_8$ share the outgoing neighbors hub_1 and hub_2.

We next consider the hubs with high out-degrees. Real world directed graphs also follow power-law out-degree distribution, i.e., few hub vertices are with a very large amount of spoke neighbors (a.k.a very high out-degrees) and the majority of vertices are with low out-degrees. We again give the following observation.

Observation 2. *For two hub vertices with high out-degrees, if they share many spoke neighbors, we would like to place the hub vertices together in the matrix rows. Meanwhile, if two spoke vertices share many incoming hub neighbors, we similarly place the spoke vertices together in the matrix columns. In Fig. 2, hub_3 and hub_4 share 8 outgoing neighbors $spoke_9...spoke_{16}$, and $spoke_9...spoke_{16}$ share the incoming neighbors hub_3 and hub_4.*

For illustration, Fig. 3(a-b) visualize the matrices $\mathcal{M}$ before and after permutation (we will soon present the permutation algorithm in the next section). In the matrices, each row (resp. column) indicates a source vertex (resp. destination vertex). If there is a directed edge from source i to destination j (where $1 \leq i, j \leq V$), we plot a point in the coordination (i, j).

In Fig. 3(a), the points are randomly distributed in the matrix before permutation. Instead after permutation, we follow the above observations to purposely co-cluster the points in the following areas, as shown in Fig. 3(b). Specially, (i) for the hubs with high in-degrees (i.e., a large amount of in-coming spoke neighbors), we purposely permute such hubs and spokes together in the matrix to make sure the points w.r.t such hubs and spokes to be co-clustered in the vertical zones; (ii) for the hubs with high out-degrees (i.e., a large amount of out-going spoke neighbors), we again co-cluster the points w.r.t such hubs and spokes in the horizonal zone; and (iii) for those vertices with both high in-degrees and high out-degrees, we co-cluster the associated points in the top-left zone. Since

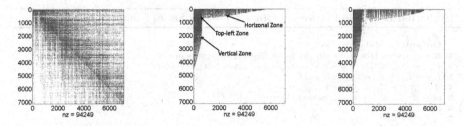

Fig. 3. Visualization of Wiki-Vote graph (from left to right): (a-b) the matrices before and after permutation, (c) the matrix after permutation by setting a different parameter value.

all points are co-clustered in the three zones above, no other points are in the rest of the matrix. Thus, we have chance to minimize the count of non-empty blocks B($\mathcal{M}$).

3.2 Algorithm Detail

In this section, we give the algorithm detail to permute graph vertex IDs, such that the points are co-clustered in the associated zones. Our algorithm will output a list P of vertices, such that the i-th vertex in the list P (with $0 \leq i \leq |V| - 1$) will be re-assigned with the new vertex ID i. It means that the i-th vertex in P, though with the original ID i', will be re-ordered from the old ID i' to a new ID i.

In the algorithm, suppose that the list P has already been added by i items (i.e., old vertex IDs) then the key question is which vertex should be selected from the remaining $(|V| - i + 1)$ IDs. To this end, we leverage the two observations in Sect. 2.1: (i) the power-law distribution of vertex degrees and (ii) the similarity of vertices (measured by common neighbors), to design the permutation algorithm. On the overall, the algorithm first selects a set of candidate vertices based on vertex degrees (either in-degrees or out-degrees), and next use common neighbor information to finalize the vertex selection among the candidate vertices.

Candidate Selection: Algorithm 1 leverages degree information to select k candidate vertices from the vertex set S. First, we choose the top-k highest in-degree and out-degree vertices, respectively (lines 1–2). Such vertices are in-degree and out-degree hubs, i.e., L_{id} and L_{od}. After that, we perform the union and intersection on L_{id} and L_{od}, and have two corresponding sets $L_{\sqcup}$ and $L_{\sqcap}$.

If the cardinality $|L_{\sqcup}|$ is smaller than k, we have at most k vertices which are still unprocessed. We then simply return such vertices as candidates (line 4). Otherwise, if we have $|L_{\sqcup}| == k$, such vertices are those with the top-k in-degrees and out-degrees, and line 5 also returns such k vertices (line 5). Finally given $|L_{\sqcup}| > k$, we have some or none of vertices in L_{id} which appear in L_{od}. Thus, we first choose $L_{\sqcap}$ (line 7) and next select the top-k' highest in-degree

---

Algorithm 1. Select (Directed Graph G, Vertex Set S, Number k)

Input: Vertex set S, parameter k
Output: Candidate vertices set
1 $L_{id} \leftarrow$ vertex IDs in S with the top-k highest in-degrees in G;
2 $L_{od} \leftarrow$ vertex IDs in S with the top-k highest out-degrees in G;
3 $L_\cap \leftarrow L_{id} \cap L_{od}$, $L_\sqcup \leftarrow L_{id} \sqcup L_{od}$;
4 **if** $|L_\sqcup| < k$ **then return** $L_\sqcup$;
5 **else if** $|L_\sqcup| == k$ **then return** $L_\sqcup$;
6 **else**
7 $\quad$ $k' \leftarrow k - |L_\cap|$;
8 $\quad$ select the nodes $T \sqsubseteq (L_\sqcup - L_\cap)$ with top-k' highest in-degrees or out-degrees;
9 $\quad$ **return** $L_\cap \sqcup T$;

---

or out-degree vertices among the vertices $(L_\sqcup - L_\cap)$ (line 8) together as the candidates (line 9).

Common Neighbors: Beside the power law degree distribution, common neighbor information is also important to our algorithm. The basic idea is to find those vertices sharing with high number of common neighbors.

Algorithm 2 gives the detail to unify the selection of both candidate vertices and common neighbors. First line 3 selects the at most k candidate vertices C (see Algorithm 1); line 4 adds all in-neighbors and out-neighbors of C to a new set N. Next, for each pairwise members $u \neq v \in N$, we find their common neighbors $Nbr(u, v)$, and add the number $|Nbr(u, v)|$ of such common neighbors (as a value) and the pairs (u, v) (as a key) to a key-value pair list L (line 6). By sorting the pair list L (line 7), we then select the key (u, v) with the highest $|Nbr(u, v)|$ and add k vertices to P, if such vertices are not inside P (lines 8–11). After that, in case that $cnt'(<k)$ vertices are selected by the steps above, we can still select the remaining $(k - cnt')$ vertices from the previously selected vertices C to make sure that k vertices are added to P (lines 12–14). We repeat the above steps until all vertices have been selected to P (lines 2–15).

In the algorithm above, the running time depends upon the number of members in N (see line 6). Thus, we can limit the number of members $m \in N$, for example by requiring that m at least has two (in- and out-) neighbors. In addition, the number k is an important parameter in the algorithm above. The tuning of k involves the tradeoff between the compression quality and running time. Figure 3(b-c) visualizes two matrices when setting $k = 1$ and $k = 0.005n$, respectively. It is not hard to find that Fig. 3(b) is with a higher point density than Fig. 3(c). A larger k could speedup the running time of Algorithm 2, but at cost of higher compression ratio.

4   Evaluation

4.1   Experimental Setting

Table 1 lists five real world directed graphs used in our experiment. The data sets are collected by Stanford Large Network Dataset Collection (SNAP) [12].

Algorithm 2. Permute (Graph $G = \{V, E\}$, Number k)

Input: Graph $G = (V, E)$, parameter k
Output: Permutation of vertex list
1 initiate an empty list P and empty set S; $S \leftarrow V$; $cnt \leftarrow 0$;
2 **while** $cnt < |V|$ **do**
3 $\quad$ $C \leftarrow select(G, S, k)$; $N \leftarrow C$; $cnt' = 0$;
4 $\quad$ **foreach** $c \in C$ **do** $\{N \leftarrow (Nbr_{in}(c) \sqcup Nbr_{out}(c))\}$;
5 $\quad$ initiate an empty key-val pair list L;
6 $\quad$ find the common neighbors $Nbr(u, v)$ of pairwise members $u \neq v \in N$, and add the $\langle key \leftarrow (u, v), val \leftarrow |Nbr(u, v)| \rangle$ pairs to list L;
7 $\quad$ sort the key-val pair list L by descending order of values;
8 $\quad$ **for** $(i = 0; i < k \&\& i < -L-;)$ **do**
9 $\quad\quad$ select the key (u, v) from the i-th pair from L;
10 $\quad\quad$ **if** $u \notin P$ **then** $\{$add u to P; remove u from S; $i++$; $cnt'++\}$;
11 $\quad\quad$ **if** $v \notin P$ **then** $\{$ add v to P; remove v from S; $i++$; $cnt'++ \}$;
12 $\quad$ **if** $cnt' < k$ **then**
13 $\quad\quad$ select $(k - cnt')$ vertices from C, add them to P, remove them to S;
14 $\quad\quad$ $cnt' \leftarrow k$;
15 $\quad$ $cnt \leftarrow cnt + cnt'$;

Table 1. Statistics of the real data sets

Name	Node	Edge	Description
Amazon0302	262, 111	1, 234, 877	Amazon product co-purchasing network from March 2 2003
soc-Slashdot0811	77, 360	905, 468	Slashdot social network from November 2008
soc-Epinions1	75, 879	508, 837	Who-trusts-whom network of Epinions.com
Wiki-Vote	7, 115	103, 689	Wikipedia who-votes-on-whom network
p2p-Guntella05	8, 846	31, 839	Gnutella peer to peer network from August 5 2002

Given the data sets, we evaluate the performance of our scheme to answer the following questions:

- Q_1: How well does our scheme perform when compared with other graph re-ordering approaches?
- Q_2: How well is our scheme comparable to other graph compression approaches?
- Q_3: How fast our scheme can perform graph computation with no or partial graph decompression?

In order to answer the questions above, we use the following performance metrics.

(1) Given a graph's adjacency matrix, we divide it into blocks of $b \times b$ elements. Each block consists of at most $b \times b$ elements. Thus, the number of *non-empty blocks* measures the density of the adjacency matrix. A smaller number of

non-empty blocks means a better re-ordering algorithm and thus leads to less space cost.

(2) We follow the previous work [1,2] and encode the graph adjacency matrix into binary bits, and define a cost function by assuming that a compression method achieves the *information theoretic* lower bound. After that, we compute the *bits per edge*, which is computed by (i) the totally required bits using information-theoretic coding methods against (ii) the number of edges. This metric measures how many bits are used to store a single edge on average. A smaller amount of bits per edge indicates less space cost used to encode a graph.

(3) We measure the used *space cost* of a compressed graph when maintained on disk. In addition, the *compression ratio*, measured by the ratio of used space cost before and after a graph compression approach is adopted, is also used to measure the goodness of a graph compression approach.

(4) Finally we are interested in the *speedup ratio* of a graph computation algorithm which is computed by the running time of performing graph computation over original graphs against the time over compressed graphs. The speedup ratio is helpful to understand (i) the benefit of reduced I/O cost (due to smaller graph size by compression) to speedup the graph computation and (ii) the introduced decompression overhead if any. We use the Breadth-first search (BFS)-based graph traversal algorithm and Dijkstra-based shortest paths as the two examples of the graph computation. In case that decompression is required, the running time needs to include both decompression time and graph computation time.

Based on the above metrics, we mainly compare our scheme with other re-ordering algorithms, plus another graph compression approach *Ligra+*.

- *Natural*. Natural reordering of graph vertices means the original adjacency matrix. For some graphs, the natural reordering provides high locality among consecutive vertices.
- *Degree Sort*. The reordering is based on the decreasing degrees of graph vertices.
- *SlashBurn* [1,2] is the most similar to our work. Note that *SlashBurn* is originally designed for undirected graphs. In order to make sure that *SlashBurn* works for directed graphs, we have to adapt the asymmetric matrices associated with directed graphs by redundantly adding extra rows (or columns) with respect to source vertices (or destination vertices) in the directed graphs. Then the elements associated with such extra rows (or columns) are all 0-elements. Now with the adapted matrices, we next adopt *SlashBurn* to reorder such matrices.
- *Ligra* [5] and *Ligra+* [3]. *Ligra+* also adopts a graph reordering approach. However, differing from *SlashBurn* and our work, it improves locality by assigning neighbor IDs close to vertex ID [13]. After that, it sorts edges and encodes the difference. It is useful to understand how our work and *SlashBurn* is comparable to other graph compression approaches such as *Ligra+*. Note

that *Ligra* and *Ligra+* leverage multicore parallel functionality to greatly speedup graph computation and compression/decompression.

We implement our approach by GCC, and evaluate all experiments on a Linux workstation with 6 cores of Xeon 2.60 GHz CPU, 64 GB RAM and 1 TB SATA disk. Our algorithm expects to select the top-k degree vertices with similar sibling relationship during each iteration. Following the poke game, we name our algorithm Ace_Up.

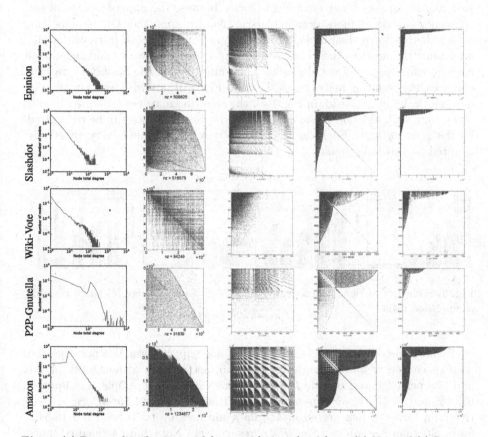

Fig. 4. (a) Degree distribution, and four reordering algorithms: (b) Natural (c) Degree sort (d) SlashBurn (e) Ace Up

4.2   Comparison of Reordering Algorithms

In this section, we compare the proposed algorithm with three other re-ordering approaches including *Nature, Degree sort* and *SlashBurn.*

To capture the overall degree distribution, Fig. 4(a) plots the degrees of five used data sets. The x-axis represents vertex degrees (here a vertex degree means the sum of in-degree and out-degree of a vertex), and the y-axis is the frequency of those vertices with the associated degree. In general, all of the directed graphs follow very skewed degree distribution.

Given such directed graphs, we plot the adjacency matrices after the vertex IDs of such graphs are re-ordered by four algorithms as shown in Fig. 4(b-e). In all five data sets, our scheme is always with the least area to plot the points associated with the graphs and thus the least amount of nonempty blocks. *Natural* reordering instead requires the maximum number of nonempty blocks. *Degree sort* reordering uses fewer amount of blocks, because the upper-left area of the adjacency matrix is more dense than the Nature approach. Our scheme can achieve better result than *SlashBurn*. It is mainly because *SlashBurn* works only on symmetric matrices (i.e., undirected graphs) and we have to add redundant rows (or columns) for directed graphs. Thus, after performing *SlashBurn* on such matrices, the resulting matrices, as shown in Fig. 4(d), are with equal number of rows and columns. Instead, in Fig. 4(e), the resulting matrices generated by our scheme are with different number of rows and columns (which can be recognized by the x and y axis). Such result indicates that our scheme is very effectively adapted to directed graphs.

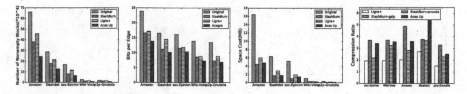

Fig. 5. From left to right: (a) nonempty blocks (b) bits per edge (c) space cost (d) compression ratio

Figure 5 directly measures the number of nonempty blocks, bits per edge and used space cost of four schemes. For the two metrics such as nonempty blocks and bits per edge, our scheme Ace_Up performs best. For example, Ace_Up uses 63.07% and 35.77% fewer non-empty blocks than *Nature* and *SlashBurn*, respectively; in terms of bits per edge, Ace_Up again uses up to 41.96% and 16.54% fewer bits than *Natural* and *SlashBurn*, respectively. Note that for the used space cost, *SlashBurn* uses gZip to compress each non-empty blocks and thus leads to the least space cost. Instead, we use integer pairs to encode nonempty blocks (see Sect. 2.2). Our scheme can achieve less space cost than *Ligra+* but sightedly higher space cost than *SlashBurn*. In case that our scheme Ace_Up and *Slash-Burn* both adopt the same compression approach (either gZip or the encoding approach), Ace_Up can use less space cost than SlashBurn. Nevertheless, we will show soon that the gZip-based SlashBurn will incur much higher running time for BFS traversal than our scheme.

Finally, Fig. 5(d) measures the compression ratio of four schemes. In this figure, after we use *SlashBurn* on directed graphs, we next use gZip and the encoding scheme in Sect. 2.2 to compress non-empty blocks. Thus, we have two version of *SlashBurn*, namely *SlashBurn-gZip* and *SlashBurn-encode*. Now it is clear that Ace_Up leads to a higher compression ratio than *SlashBurn-encode*, though lower than *SlashBurn-gZip*. Among these algorithms, *Ligra+* shows the smallest compression ratio.

4.3 Graph Computation Speedup Ratio

In this section, we measure the speedup ratio of two graph computation algorithms (BFS traversal and Dijkstra-based shortest paths). The ratio is computed by the running time used by the computation on original graphs and the time on compressed graphs. In terms of *Ligra+*, we measure the ratio by the running time of *Ligra* (i.e., the graph computation engine without adopting graph compression technique) and the time of *Ligra+*. A larger ratio indicates faster computation on compressed graphs and otherwise slower computation.

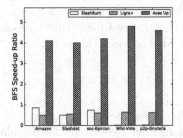 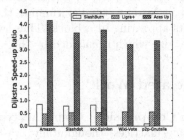

Fig. 6. (a) Speed up ration of BFS (left); (b) speed up ratio of Dijkstra

In Fig. 6, Ace_Up is with much higher speedup ratios than both *SlashBurn* and *Ligra+*. For example, to perform BFS on Amazon data set, Ace_Up is with 4.42× and 8.24× higher speedup ratio than *SlashBurn* and *Ligra+*, respectively. It is because of the non-trivial decompression overhead caused by *SlashBurn* and *Ligra+*. In particular, both *SlashBurn* and *Ligra+* are with the speedup ratios smaller than 1.0, indicating that the decompression overhead compromises the reduced I/O overhead (due to smaller space cost of compressed graphs). Instead, our scheme Ace_Up is based on the encoded integers and the decoding time from encoded integer numbers to binary bits is trivial.

Note that though with a speedup ratio smaller than 1.0, the absolute running time of *Ligra+* is much smaller than our scheme Ace_Up. It is mainly caused by the parallel computation engine and parallel decompression offered by the excellent system design of *Ligra* and *Ligra+*.

4.4 Effect of Parameter *b*

Finally we are interested in how the parameter *b* affects the performance of our scheme in terms of space cost and graph computation. In Table 2, we vary *b*

Table 2. Space cost and computation time of various b

	Nonempty blocks			Compression ratio			BFS traversal time		
	$b=16$	$b=32$	$b=64$	$b=16$	$b=32$	$b=64$	$b=16$	$b=32$	$b=64$
Amazon	290,265	241,936	200,442	2.3080	3.4179	3.3471	0.4136	0.1909	0.2168
Slashdot	189,085	125,058	67,491	3.2482	3.5617	3.8467	0.035883	0.02215	0.02302
soc-Epinions	101,036	62,596	33,446	2.9418	3.5995	3.1143	0.030643	0.02015	0.02277
Wiki-Vote	21,169	8,406	2,573	6.09144	5.3768	3.3831	3.280E−05	2.642E−05	2.59E−05
p2p-Gnutella	12,859	8,154	3,398	2.06967	2.5190	2.1411	2.918E−05	2.542E−05	2.47E−05

by 16, 32 and 64, and divide graph adjacency matrices into $b \times b$ blocks. It is obvious that the parameter b will affect the number of nonempty blocks and bits per edge. On the overall, a bigger b leads to fewer amount of nonempty blocks and bits per edge. However, in terms of graph computation time (in this table, we give BFS traversal time only. Due to space limit of the table, we do not give the running time of Dijkstra algorithm), a larger b, for example $b = 64$, does not necessarily leads to smaller running time. It is because a larger b means more elements appearing in a block. Thus, we have to spend more time to lookup a specific edge which is encoded by the block (due to more elements).

5   Related Work

Graph Compression: Graphs are typically represented by adjacency lists and adjacency matrices, though they are implemented by various file formats to store graphs. When a graph is maintained by adjacency lists of sorted vertex IDs, some previous work [14,15] compressed the lists by difference of vertex IDs. The WebGraph [15] compresses the graph by representing adjacent nodes by the small difference to previous value, instead of original large IDs. Since the large IDs require more space (i.e., word length), the maintenance of small difference of continuous IDs can save space cost. [16] improve the framework of WebGraph by proposing a different ordering [16]. Still following the similar idea to maintain the differences between consecutive IDs, $Ligra+$ optimizes the previous works by much more complex coding schemes such as the run-length encoded byte coding scheme.

Some work represents a graph by a matrix instead of adjacency lists. GBase [17] and $SlashBurn$ [1,2] are based on graph adjacency matrixes. They divide the matrices into blocks of matrix elements. Consider that the adjacency matrix of a real work graph is usually very sparse. $SlashBurn$ can co-cluster the elements together more densely inside the adjacency matrix. The basic idea of $SlashBurn$ [1] is to remove the vertices of highest degree and next to find a giant connected component (GCC) in rest of the graph to repeat this process. After clustering the matrix and dividing the matrix into element of blocks, $SlashBurn$ adopts gZip to *compress* such blocks for less space cost. $SlashBurn$ works well on power-law graphs. However, SlashBurn suffers from some limitation. First, it works only on undirected graphs and cannot be directly used to compress directed graphs.

Moreover, gZip is known as a heavy-level compression approach, and the decompression overhead of gZip is non-trivial. Since we cannot directly perform graph computation on compressed graphs, we have to spend non-trivial decompression overhead before graph computation works on originally uncompressed graphs.

Some work proposes graph storage format to maintain sparse matrices, such as Compressed Sparse Blocks (CSB) [18]. It is essentially the same as Compressed Sparse Row (CSR) [19] and allows vector multiplication in parallel. It uses simple tuple to representation to store a matrix as row index, column index and the nonzero value. In addition, when maintaining only those graphs without edge weights, the CSR in PrefEdge [20] maintains two components (row index and column index) but with no element values. In addition, some works such as [21] consider graph compression as a matrix factorization problem. Graph computation can be directly performed on such factorized matrices. However, the factorization incurs information loss.

Graph Computation: There are two storage styles when storing a data set, one is to store in main memory and another is in external memory. So the optimization of computation can lay on reduce the querying time on graph or optimize on I/O-efficiency.

When graphs are compressed, we frequently have to recover original graphs by decompression. Non-trivial decompression overhead could compromise the I/O reduction achieved by the reduced space cost when compressed graphs are on disk. Few work in literature has studied the optimization of computation efficiency on compressed graphs. For example, by translating many graph computation by matrix operations, Pegasus [22] can benefit from the less I/O overhead caused by the smaller size of graphs on disk when the graph is compressed (as shown in GBase [17] and SlashBurn [1,2]). However, after the graph is loaded from disk to main memory, decompression is still needed before graph computation. As shown in Pegasus, sparse matrix-vector (spMV) multiplication that is frequently used by many graph computation. Some previous work [23,24] has explored the problem of running algorithms on compressed inputs in spMV context. Though with some promising results, such previous work only studied the specific spMV computation and it cannot be comfortably extended to other matrix computation (for example matrix eigenvalue decomposition), which could be used by some graph computation.

Fan et al. [25] gives a framework for query-preserving graph compression. It proposes two compression methods preserving reachability queries and pattern matching queries, based on bounded simulation. Though such queries can be performed directly on compressed graphs, due to a lossy compression, it is unknown whether or not other queries expect the reachability can performed well on the compressed graphs.

6 Conclusion and Future Work

Given real world directed graphs, we study the problem to permute graph vertices in order to minimize the amount of nonempty blocks. The proposed algorithm leverages graph structure information (e.g., in-degrees, out-degrees, and sibling neighbors) to re-order vertex IDs for the permutation. We perform extensive experiments on real world directed graphs and show that our work can outperform the state of art *SlashBurn* in terms of space cost saving and graph computation time. As future work, we continue to study the performance of graph compression. For example, we are interested in how modern hardware (such as multicore computers and clustered machines) can speedup graph compression and decompression.

Acknowledgement. This work is partially supported by National Natural Science Foundation of China (Grant Nos. 61572365, 61503286), Science and Technology Commission of Shanghai Municipality (Grant Nos. 14DZ1118700, 15ZR1443000, 15YF1412600), and Huawei Innovation Research Program (HIRP).

References

1. Kang, U., Faloutsos, C.: Beyond 'caveman communities': hubs and spokes for graph compression and mining. In: 11th IEEE International Conference on Data Mining, ICDM 2011, Vancouver, BC, Canada, 11–14 December 2011, pp. 300–309 (2011)
2. Lim, Y., Kang, U., Faloutsos, C.: SlashBurn: graph compression and mining beyond caveman communities. IEEE Trans. Knowl. Data Eng. **26**(12), 3077–3089 (2014)
3. Shun, J., Dhulipala, L., Blelloch, G.E.: Smaller and faster: parallel processing of compressed graphs with ligra+. In: 2015 Data Compression Conference, DCC 2015, Snowbird, UT, USA, 7–9 April 2015, pp. 403–412 (2015)
4. Low, Y., Gonzalez, J., Kyrola, A., Bickson, D., Guestrin, C., Hellerstein, J.M.: Distributed graphlab: a framework for machine learning in the cloud. PVLDB **5**(8), 716–727 (2012)
5. Shun, J., Blelloch, G.E.: Ligra: a lightweight graph processing framework for shared memory. In: ACM SIGPLAN Symposium on Principles and Practice of Parallel Programming, PPoPP 2013, Shenzhen, China, 23–27 February 2013, pp. 135–146 (2013)
6. Shao, B., Wang, H., Li, Y.: Trinity: a distributed graph engine on a memory cloud. In: Proceedings of the ACM SIGMOD International Conference on Management of Data, SIGMOD 2013, New York, NY, USA, 22–27 June 2013, pp. 505–516 (2013)
7. Johnson, D.S., Krishnan, S., Chhugani, J., Kumar, S., Venkatasubramanian, S.: Compressing large Boolean matrices using reordering techniques. In: (e)Proceedings of the Thirtieth International Conference on Very Large Data Bases, Toronto, Canada, 31 August–3 September 2004, pp. 13–23 (2004)
8. Ng, A.Y., Jordan, M.I., Weiss, Y.: On spectral clustering: analysis and an algorithm. In: Advances in Neural Information Processing Systems. 14th International Conference on Neural Information Processing Systems: Natural and Synthetic, NIPS 2001, Vancouver, British Columbia, Canada, 3–8 December 2001, pp. 849–856 (2001)
9. Davis, J.V., Kulis, B., Jain, P., Sra, S., Dhillon, I.S.: Information-theoretic metric learning. In: Machine Learning, Proceedings of the Twenty-Fourth International Conference (ICML 2007), Corvallis, Oregon, USA, 20–24 June 2007, pp. 209–216 (2007)

10. Chakrabarti, D., Papadimitriou, S., Modha, D.S., Faloutsos, C.: Fully automatic cross-associations. In: Proceedings of the Tenth ACM SIGKDD International Conference on Knowledge Discovery and Data Mining, Seattle, Washington, USA, 22–25 August 2004, pp. 79–88 (2004)

11. Chierichetti, F., Kumar, R., Lattanzi, S., Mitzenmacher, M., Panconesi, A., Raghavan, P.: On compressing social networks. In: Proceedings of the 15th ACM SIGKDD International Conference on Knowledge Discovery and Data Mining, Paris, France, 28 June–1 July 2009, pp. 219–228 (2009)

12. Leskovec, J., Krevl, A.: SNAP datasets: Stanford large network dataset collection, June 2014. http://snap.stanford.edu/data

13. Shun, J., Dhulipala, L., Blelloch, G.: Tutorial: large-scale graph processing in shared memory. https://github.com/jshun/ligra/blob/master/tutorial/tutorial.pdf

14. Adler, M., Mitzenmacher, M.: Towards compressing web graphs. In: Data Compression Conference, DCC 2001, Snowbird, Utah, USA, 27–29 March 2001, pp. 203–212 (2001)

15. Boldi, P., Vigna, S.: The webgraph framework II: codes for the world-wide web. In: 2004 Data Compression Conference (DCC 2004), 23–25 March 2004, Snowbird, UT, USA, p. 528 (2004)

16. Boldi, P., Santini, M., Vigna, S.: Permuting web and social graphs. Internet Math. 6(3), 257–283 (2009)

17. Kang, U., Tong, H., Sun, J., Lin, C., Faloutsos, C.: GBASE: an efficient analysis platform for large graphs. VLDB J. 21(5), 637–650 (2012)

18. Buluç, A., Fineman, J.T., Frigo, M., Gilbert, J.R., Leiserson, C.E.: Parallel sparse matrix-vector and matrix-transpose-vector multiplication using compressed sparse blocks. In: Proceedings of the 21st Annual ACM Symposium on Parallelism in Algorithms and Architectures, SPAA 2009, Calgary, Alberta, Canada, 11–13 August 2009, pp. 233–244 (2009)

19. https://en.wikipedia.org/wiki/Sparse_matrix

20. Nilakant, K., Dalibard, V., Roy, A., Yoneki, E.: PrefEdge: SSD prefetcher for large-scale graph traversal. In: International Conference on Systems and Storage, SYSTOR 2014, Haifa, Israel, 30 June–02 July 2014, pp. 4:1–4:12 (2014)

21. Nourbakhsh, F., Bulò, S.R., Pelillo, M.: A matrix factorization approach to graph compression with partial information. Int. J. Mach. Learn. Cybern. 6(4), 523–536 (2015)

22. Kang, U., Tsourakakis, C.E., Faloutsos, C.: PEGASUS: a peta-scale graph mining system. In: The Ninth IEEE International Conference on Data Mining, ICDM 2009, Miami, Florida, USA, 6–9 December 2009, pp. 229–238 (2009)

23. Buluç, A., Williams, S., Oliker, L., Demmel, J.: Reduced-bandwidth multithreaded algorithms for sparse matrix-vector multiplication. In: 25th IEEE International Symposium on Parallel and Distributed Processing, IPDPS 2011, Anchorage, Alaska, USA, 16–20 May 2011 - Conference Proceedings, pp. 721–733 (2011)

24. Karakasis, V., Gkountouvas, T., Kourtis, K., Goumas, G.I., Koziris, N.: An extended compression format for the optimization of sparse matrix-vector multiplication. IEEE Trans. Parallel Distrib. Syst. 24(10), 1930–1940 (2013)

25. Fan, W., Li, J., Wang, X., Wu, Y.: Query preserving graph compression. In: Proceedings of the ACM SIGMOD International Conference on Management of Data, SIGMOD 2012, Scottsdale, AZ, USA, 20–24 May 2012, pp. 157–168 (2012)

Keyphrase Extraction Using Knowledge Graphs

Wei Shi[1], Weiguo Zheng[1(✉)], Jeffrey Xu Yu[1], Hong Cheng[1], and Lei Zou[2]

[1] The Chinese University of Hong Kong, Sha Tin, Hong Kong
{shiw,wgzheng,yu,hcheng}@se.cuhk.edu.hk
[2] Peking University, Beijing, China
zoulei@pku.edu.cn

Abstract. Extracting keyphrases from documents automatically is an important and interesting task since keyphrases provide a quick summarization for documents. Although lots of efforts have been made on keyphrase extraction, most of the existing methods (the co-occurrence based methods and the statistic-based methods) do not take semantics into full consideration. The co-occurrence based methods heavily depend on the co-occurrence relations between two words in the input document, which may ignore many semantic relations. The statistic-based methods exploit the external text corpus to enrich the document, which introduces more unrelated relations inevitably. In this paper, we propose a novel approach to extract keyphrases using knowledge graphs, based on which we could detect the latent relations of two keyterms (i.e., noun words and named entities) without introducing many noises. Extensive experiments over real data show that our method outperforms the state-of-art methods including the graph-based co-occurrence methods and statistic-based clustering methods.

1 Introduction

The continuously increasing text data, such as news, articles, and papers, make it urgent to design an effective and efficient technique to extract high-quality keyphrases automatically since keyphrases help us to have a quick knowledge of the text. Keyphrase extraction also provides useful resources for text clustering [9], text classification [30], and document summarization [28]. There are two main types of studies on keyphrase extraction, i.e., supervised and unsupervised. Majority of the supervised methods regard keyphrase extraction as a binary classification task [12,13,15,26,29], and take some features, such as term frequency-inverse document frequency (*tf-idf*) and the position of the first occurrence of a phrase, as the inputs of a Naive Bayes classifier [24]. However, a binary classifier fails to rank the phrases since each candidate phrase is classified independently with others. More importantly, supervised methods demand a lot of training data, i.e., manually labeled keyphrases. This is extremely expensive and time-consuming in domain-specific scenarios. In order to reduce manpower, investigating comparative unsupervised methods is highly desired. Thus, we focus on studying unsupervised methods to extract keyphrases from a single input document (e.g., news and article).

© Springer International Publishing AG 2017
L. Chen et al. (Eds.): APWeb-WAIM 2017, Part I, LNCS 10366, pp. 132–148, 2017.
DOI: 10.1007/978-3-319-63579-8_11

The existing unsupervised approaches can be divided into two categories, i.e., co-occurrence based methods and statistic-based methods, as shown in Table 1. The co-occurrence based methods, e.g., *SW* [11], *TextRank* [21], *ExpandRank* [27], *CM* [5], build a word co-occurrence graph exploiting the word co-occurrence relations that are obtained from the input document, and then apply some ranking algorithms such as PageRank [23] and betweenness [7] on the graph to get the ranking score of each word. Based on the ranking score, the *top-k* phrases are returned as keyphrases. The statistic-based methods, e.g., *Wan'07* [28], *SC* [18], and *TPR* [17], explore some external text corpus to assist keyphrase extraction. For example, *SC* [18] uses the Wikipedia [1] articles, based on which several statistical distances can be computed, such as cosine similarity, Euclidean distance, Pointwise Mutual Information (PMI), and Normalized Google Similarity Distance (NGD) [6]. It first clusters candidate words based on the statistical distance, and then selects the exemplar terms. Finally, the candidate phrases that contain exemplar terms are selected as keyphrases.

Table 1. Overview of the approaches

Category	Methods	Information sources
Co-occurrence based	*CM* [5], *SW* [11], *TextRank* [21], *ExpandRank* [27]	*Input documents*
Statistic-based	*Wan'07* [28], *SC* [18], *TPR* [17]	*External text corpus, e.g., Wikipedia article*
Semantics-based	*Our method*	*Knowledge graphs, e.g., DBpedia*

However, the existing methods suffer from two drawbacks: *information loss* and *information overload*.

- Information Loss. The co-occurrence based methods heavily depend on the co-occurrence relations between two words. If two words never occur together within a predefined window size in an input document, there will be no edges to connect them in the built co-occurrence graph even though they are semantically related. Furthermore, as reported in [11], the performance improves little by increasing the window size w if w exceeds a small threshold (w ranges from 3 to 9).
- Information Overload. The statistic-based methods exploit external text corpus to enrich the input document. Nevertheless, the real meanings of words in the document may be overwhelmed by the large amount of introduced external texts. Furthermore, they can only acquire very limited useful knowledge about the words in the input document since they just use the statistical information of two words in the external texts actually.

Running Example. *Figure 1 shows a document from the dataset DUC2001, where the keyphrases are labeled in red font.*

In the co-occurrence based methods, the words adjacent to more other words tend to rank higher. Let us consider two candidate phrases "disaster" and "Hurricane Andrew". They do not occur together within a window if the window size is 10. Thus no edges will be induced to connect them in the co-occurrence graph, which indicates they are not highly related while "Hurricane Andrew" is an instance of "disaster". Hence, "disaster" is not delivered as a keyphrase using the co-occurrence based methods, e.g., TextRank and ExpandRank.

The statistic-based methods directly map a word to the words with the same surface form in the external text resources. The real meaning of a word in the input document may be overwhelmed by the extended corpus. For instance, the most common meaning of "house" in Wikipedia is "House Music", which is different from the meaning "United States House of Representatives" in the example document. Hence, statistic-based methods fail to exclude the wrong senses and result in bad similarity relations.

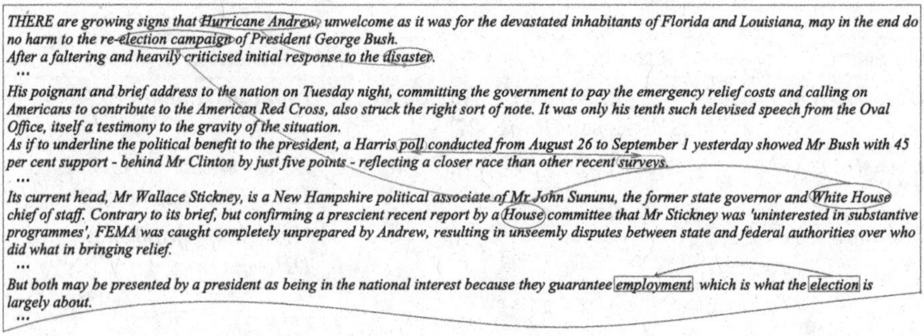

THERE are growing signs that Hurricane Andrew, unwelcome as it was for the devastated inhabitants of Florida and Louisiana, may in the end do no harm to the re-election campaign of President George Bush.
After a faltering and heavily criticised initial response to the disaster.

His poignant and brief address to the nation on Tuesday night, committing the government to pay the emergency relief costs and calling on Americans to contribute to the American Red Cross, also struck the right sort of note. It was only his tenth such televised speech from the Oval Office, itself a testimony to the gravity of the situation.
As if to underline the political benefit to the president, a Harris poll conducted from August 26 to September 1 yesterday showed Mr Bush with 45 per cent support - behind Mr Clinton by just five points - reflecting a closer race than other recent surveys.

Its current head, Mr Wallace Stickney, is a New Hampshire political associate of Mr John Sununu, the former state governor and White House chief of staff. Contrary to its brief, but confirming a prescient recent report by a House committee that Mr Stickney was 'uninterested in substantive programmes', FEMA was caught completely unprepared by Andrew, resulting in unseemly disputes between state and federal authorities over who did what in bringing relief.
...

But both may be presented by a president as being in the national interest because they guarantee employment, which is what the election is largely about.
...

Fig. 1. An example of input document. (Color figure online)

In this paper, we propose a novel method to extract keyphrases by considering the underlying semantics. Relying on semantics, we can address the problems above. (1) Using semantics, we can find the latent relations between two keyterms (see Sect. 2.1). As shown in the example document in Fig. 1, although the distance between "disaster" and "Hurricane Andrew" in the document is large, it is easy to add an edge between them if we know "Hurricane Andrew" is an instance of "disaster" according to their semantic relation. (2) In order to avoid introducing many noisy data, we attempt to include the useful information by incorporating semantics. For example, through entity linking, our method could select "United States House of Representatives" as the proper meaning of "house" in the example document in Fig. 1 and only bring in the corresponding semantic relations, thus excluding the noisy relations caused by "House Music".

In order to incorporate the semantics for keyphrase extraction, we resort to knowledge graphs in this paper. A knowledge graph, e.g., DBpedia [3], captures lots of entities, and describes the relationships between the entities. Note that two earlier work communityRank [8] and SemanticRank [25] utilize the Wikipedia page linkage as the external knowledge. However, they just use the link information in a coarse-grained statistical way. Different from the work above, we propose to incorporate semantics into keyphrase extraction by adopting the structure of knowledge graph (e.g., DBpedia).

The major contributions of this paper are summarized as follows:

- We are the first to use the structure of knowledge graphs to provide semantic relations among keyterms in keyphrase extraction task.
- We propose a systematic framework that integrates topic clustering and graph ranking for keyphrase extraction.
- By considering the real observations, we give a novel and reasonable definition of keyphrases, i.e., the important noun phrases of a document.
- Extensive experiments over real data show that our method achieves better performance compared with the classic co-occurrence based methods and statistic-based methods.

The rest of this paper is organized as follows. Section 2 formulates the task of the work and gives an overview of our method. To cover more topics of a document, we perform the topic clustering in Sect. 3. Section 4 studies how to incorporate knowledge graphs to model the keyterms. Followed by the process of generating keyphrases in Sect. 5. Section 6 reports experiments on real data. In Sect. 7, we review related work in recent years. Finally, we conclude this work in Sect. 8.

2 Problem Definition and Framework

In this section, we first formulate our problem and then give the overview of our methods. Table 2 lists the notations used in the paper.

2.1 Problem Formulation

A *named entity* is a real-world object such as location, organization, person and so on. To name a few, "George Bush", "Florida", "Red Cross", "Gulf of Mexico". A *noun phrase* is composed of continuous noun words or named entities. For example, "president George Bush", "Hurricane Andrew", "New York City" are all noun phrases.

Noun words and *named entities* in the document are called **Keyterms**. For example, "hurricane", "Florida", "George Bush" are all keyterms.

Definition 1 (Keyphrase). *A keyphrase consists of keyterms and is highly relevant to the topics of a document.*

For instance, "president george bush", "hurricane andrew", "florida", "louisiana", "election campaign", "emergency relief" and "disaster" are the keyphrases of the example document in Fig. 1.

Table 2. Notations

Notation	Definition
d	Document
$G = (V, E, LA)$	Knowledge graph
$P = \{p_1, p_2, ..., p_n\}$	A group of noun phrases
$KT = \{kt_1, kt_2, ..., kt_m\}$	A group of keyterms
$C = \{c_1, c_2, ..., c_r\}$	Clusters of KT
$AN = \{v_{a_1}, v_{a_2}, ..., v_{a_t}\}$	A group of anchor nodes
$EV \subset V$	A group of expanded nodes
L	The path between two anchor nodes
$EV.L$	The expanded nodes on the path L
$KG_h(AN)$	h-hop keyterm graph induced by AN
$\mu(EV.L, AN)$	The semantic relatedness between $EV.L$ and AN
τ	The semantic relatedness threshold

Remark. In this paper, we specify that a keyphrase should be a noun phrase. The reasons are listed as follows: (1) Noun phrases are more substantive, while adjectives always serve as the descriptive function. Table 3 presents some phrases with and without adjectives, while the meanings of phrases do not change much. Moreover, the rate of noun phrases in the manually labeled keyphrases by Wan and Xiao is 64% [27]. (2) In general, different annotators may label different groups of keyphrases according to their understanding. But they always reach high agreement on noun phrases, which is verified through our experiments (see Sect. 6.3).

Table 3. Phrases with and without adjectives

With adjectives	W/o adjectives
Widespread destruction	Destruction
Severe damage	Damage
Serious injury	Injury
Massive aircraft	Aircraft
Large immigrant population	Immigrant population
Small twister	Twister

Problem 1 (Keyphrase Extraction). Keyphrase extraction is the task of extracting a group of keyphrases from a document with good coverage of the topics.

2.2 Framework of Our Method

Figure 2 shows our framework. Given a document, we first select the keyterms. Then, we apply clustering algorithms on keyterms based on the semantic similarity which aims at covering more topics of the document. The keyterms in each cluster are linked to entities in the knowledge graph. For each cluster, taking the mapped entities as anchor nodes, we can detect the latent keyterm-keyterm relations in the input document through extracting the h-hop keyterm graph (please refer to Definition 2) from the knowledge graph. Then, we apply Personalized PageRank (PPR) [10] on each keyterm graph and obtain the ranking score of each keyterm. To distinguish the importance of different clusters, we regard the centers of all the clusters as keyterms and form a new cluster and then rank these centers using the similar method as keyterms. Finally, the keyphrases are generated based on the ranking scores of keyterms.

Topic clustering. In general, a document consists of multiple topics. Thus we propose a clustering method to cover more topics. Specifically, the keyterms are clustered according to the semantic similarity. Then we try to generate keyphrases from each cluster. Different from the existing methods that employ single word or n-gram as the clustering element, we use keyterms (i.e., the noun words and named entities) in this paper.

Keyterm graph construction. For the keyterms in each cluster, we build a keyterm graph by exploiting the structure of the knowledge graph. The keyterm graph describes the semantic relations among keyterms.

Graph-based ranking and phrase generation. We can adopt the ranking algorithms, such as PPR and SimRank [14], to compute the importance of each keyterm. Then we can get the ranking score of each candidate phrase. The k candidate phrases with the largest scores are delivered as keyphrases.

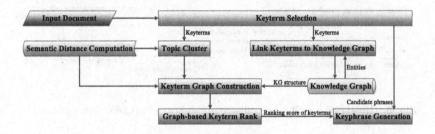

Fig. 2. System framework.

3 Topic Clustering

Since a document always has more than one topics, keyphrases should represent all these topics. Hence we propose to cluster the keyterms according to semantics, and then generate the keyphrases based on the clusters.

3.1 Keyterm Selection

The existing method *SC* [18] also proposes a clustering approach to cover the topics of a document. However, it clusters all words of the document without distinguishing the parts of speech, which may introduce many noisy data. Mostly, noun words and named entities describe the key points of a document. Hence, we propose to take noun words and named entities as keyterms. Since the knowledge graph consists of entities and relations, it is easy to integrate the knowledge graph into keyphrase extraction by mapping keyterms to entities. To extract the keyterms, we resort to the existing NLP tools, e.g., Stanford CoreNLP [20].

3.2 Keyterm Clustering

In this paper, we apply k-means clustering algorithm [19] to divide the keyterms into r clusters. K-means is one of the simplest unsupervised learning algorithms to solve the well-known clustering problem. It partitions m objects into r clusters such that each object belongs to the closest cluster in terms of mean distance.

For each cluster, we select the keyterm which is nearest to the corresponding centroid and has candidate entities in the knowledge graph G as the final centroid. In order to convey the semantics, we use Google Word2vec [22] to compute the semantic distance of two keyterms. If a keyterm kt is a named entity that consists of multiple words, we compute the vector representation of kt according to the words that are contained in kt as the work [16] does.

Example 1. Figure 3 shows the clustering result on the example document, where r is set to be 7. As the color deepened, the corresponding cluster is more important and relevant to the main topics of the example document. Keyterms in blue are the cluster centroids.

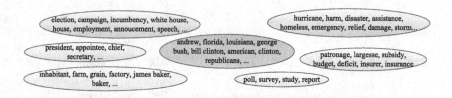

Fig. 3. Clustering result when cluster number $r = 7$. (Color figure online)

4 Incorporating Knowledge Graphs into Keyphrase Extraction

As discussed above, considering the semantics, we can detect more relations, which can improve the ranking results. In this paper, we propose to adopt knowledge graph to assist keyphrase extraction. However, it is unclear how to integrate knowledge graph in a better way since there is no such related study. We try to give a solution in this work.

4.1   Keyterm Linking

Given a knowledge graph G, the keyterm linking task is to link keyterms to G, i.e., find the mapping entity in G for each keyterm. A keyterm, regarded as a surface form, usually has multiple mapping entities. For example, the keyterm "Philadelphia" may refer to a city or a film. There have been many studies that focus on sense disambiguation. Most of the existing methods demand that each entity in G has a page p that describes the entity. Then the semantic relatedness between p and the input document d could be computed, e.g., computing the cosine similarity between them. The entity that corresponds to the page with the largest similarity score is selected as the mapping entity.

Since the surface forms are keyterms, we build a mapping dictionary for keyterms, which materializes the possible mappings for a keyterm kt. Each possible mapping is assigned a prior probability. To compute the prior probability, we resort to Wikipedia articles. For each linkage (i.e., the underlined phrase that is linked to a specific page) in Wikipedia, we can get a pair of surface form and entity. Then we can get the probability of each candidate entity e_i by calculating the proportion of its co-occurrence number with kt over the total occurrence number of kt as shown in (1).

$$Pr\{(kt, e_i) \,|kt\} = \frac{T\,(kt, e_i)}{T\,(kt)}, \tag{1}$$

where $T\,(kt, e_i)$ is the co-occurrence number of kt and e_i, and $T\,(kt)$ is the occurrence number of kt. Given a keyterm kt, we explore the dictionary to find its candidate mappings. For an entity in Wikipedia, we use the corresponding Wikipedia page as its context. For entities in DBpedia, we use the dataset "long_abstracts_en.ttl" as the context. If the knowledge graph has page descriptions, we can integrate the semantic relatedness (we adopt cosine similarity between the context of an entity and the input document in our experiments) and the prior probability to decide the correct mapping.

Example 2. Consider the keyterm "house" in the example document, there are several candidate entities in Wikipedia, including "House music", "United States House of Representatives", "House", "House system". Our method successfully selects "United States House of Representatives" as the appropriate entity.

4.2   Keyterm Graph Construction

After the keyterm linking, we obtain the mapping entities of keyterms, which are also called *anchor nodes*. Next we build a *h-hop* keyterm graph to capture the semantic relations among keyterms for each cluster. Consider a knowledge graph $G = (V, E, LA)$, where V is the set of nodes, E is the set of edges, and LA is the set of node labels. Let AN denote the set of anchor nodes, i.e., $AN = \{v_{a_1}, v_{a_2}, ..., v_{a_t}\} \subset V$. Definition 2 describes the *h-hop* keyterm graph.

Definition 2 (H-hop Keyterm Graph). *Given the set of anchor nodes AN in the knowledge graph G, the h-hop keyterm graph, denoted by $KG_h(AN)$, is the subgraph of G that includes all paths of length no longer than h between v_{a_i} and v_{a_j}, where $v_{a_i}, v_{a_j} \in AN$ and $i \neq j$.*

Each path in the *h-hop keyterm graph* describes the relation between two anchor nodes. Since the anchor nodes are the mapping entities for keyterms, the *h-hop keyterm graph* models the semantic relations among the keyterms. The newly introduced nodes in $KG_h(AN)$ excluding the anchor nodes are the *expanded nodes*, denoted by EV. In order to eliminate the noisy nodes, we need to refine the *h-hop keyterm graph* by considering the semantics of the whole set of anchors. Algorithm 1 shows the process of constructing the *h-hop keyterm graph*. For $\forall v_{a_i} \in AN$, we apply Breadth-first search (BFS) to extract the paths no longer than h between v_{a_i} and v_{a_j} ($\forall v_{a_j} \in AN$ and $i \neq j$). Next we remove the paths which are less related with AN. Let us consider the path L between two anchor nodes in $KG_h(AN)$. The set of expanded nodes in L is denoted as $EV.L$. Then we compute the semantic relatedness $\mu(EV.L, AN)$ between $EV.L$ and AN. If $\mu(EV.L, AN)$ is less than a threshold τ, L is removed from the *h-hop keyterm graph*. To compute the semantic relatedness $\mu(EV.L, AN)$, we utilize Word2vec and compute cosine similarity between the vector representations of $EV.L$ and AN. The complexity of Algorithm 1 is $\mathcal{O}(N \cdot (\frac{|E|}{|V|})^h)$, where N is the number of anchor nodes.

Example 3. Figure 4 shows part of the h-hop keyterm graph for the cluster in the middle of Fig. 3, where ellipses are anchor nodes and rectangles are expanded nodes. As the dotted bordered rectangle shows, although "History of the United States Republican Party" connects with "republicans" and "george bush", it is less associated with other anchor nodes. Hence, it is removed in the expansion process.

After incorporating the knowledge graph, we can add some relations that are derived from the input document to the keyterm graph. Consider two anchor

---

Algorithm 1. *H-hop Keyterm Graph* Construction
___
Input: Knowledge graph $G = (V, E, LA)$, anchor nodes AN, semantic relatedness threshold τ
Output: *H-hop keyterm graph*
1: **for** $v_{a_i} \in AN$ **do**
2:    **for** $v_{a_j} \in AN$ and $i \neq j$ **do**
3:       $Path(v_{a_i}, v_{a_j}) \leftarrow$ extract all the simple paths no longer than h between v_{a_i} and v_{a_j}
4:       **for** $L \in Path(v_{a_i}, v_{a_j})$ **do**
5:          $\mu(EV.L, AN) \leftarrow$ compute the semantic relatedness between $EV.L$ and AN
6:          **if** $\mu(EV.L, AN) < \tau$ **then**
7:             remove L from $Path(v_{a_i}, v_{a_j})$
8: **return** the subgraph consisting of $Path(v_{a_i}, v_{a_j})$

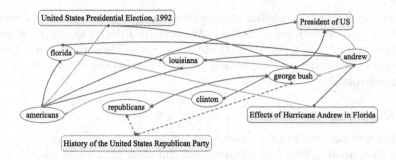

Fig. 4. H-hop keyterm graph.

nodes v_{a_i} and v_{a_j}. If their corresponding keyterms kt_i and kt_j occur in the same window, an edge is added between v_{a_i} and v_{a_j}. For each cluster, we build a keyterm graph. To measure the importance of each cluster, we can use the centroids of all clusters as the input keyterms, and build a corresponding keyterm graph for all the centroids using the same method.

5   Keyphrase Generation

5.1   Keyterm Ranking

After constructing keyterm graphs, the next focus is how to measure the importance of different keyterms. Note that our keyterm graph is a semantic graph, where edges reflect semantic relations between keyterms. But the existing methods show that term frequency is a very important feature of keyphrase. Hence, we need a graph centrality method that ranks keyterms by considering both the semantic structure of the keyterm graph and the frequency of keyterms. Although PageRank ranks nodes using the graph structure, the jump probability of PageRank is normalized, which indicates that it fails to reflect the difference of node features. In contrast, Personalized PageRank (PPR), which is a variant of PageRank, assigns a biased jump probability to each node and ranks graph nodes based on the graph structure. Therefore, it is able to reflect the keyterm frequency by assigning the jump probability according to the keyterm frequency. We formulate the PPR based on keyterm frequency as follows:

$$\mathrm{PPR}\left(v_i\right) = \lambda \times \sum_{v_j \to v_i} \frac{\mathrm{PPR}\left(v_j\right)}{\mathrm{outdegree}\left(v_j\right)} + (1 - \lambda) \times p\left(v_i\right), \qquad (2)$$

where $\mathrm{PPR}\left(v_i\right)$ is the PPR score of v_i, $\mathrm{outdegree}\left(v_j\right)$ is the outdegree of v_j, $p\left(v_i\right)$ is the jump probability of v_i and λ is a damping factor which is usually set to 0.85 [23]. The jump probability $p\left(v_i\right)$ is computed according to (3), where N is the number of nodes.

$$p\left(v_i\right) = \frac{\mathrm{tf}\left(v_i\right)}{\sum\limits_{i=1}^{N} \mathrm{tf}\left(v_i\right)}, i = 1, ..., N. \qquad (3)$$

Since the expanded nodes do not have matching keyterms in the document, their jump probabilities are set to be 0. We perform PPR on the keyterm graph that corresponds to each cluster and cluster centroids. Then we obtain the ranking scores of the corresponding anchor nodes, including the PPR score of each cluster centroid.

5.2   Keyphrase Generation

With the ranking scores of keyterms, next we extract the keyphrases. The main idea is to generate candidate phrases first and then rank them based on the scores of keyterms.

The existing phrase generation methods [18,27] adopt the syntactic rule $(JJ) * (NN|NNS|NNP) +$, where "JJ" is an adjective, "NN", "NNS", and "NNP" are nouns, "*" and "+" mean zero or more adjectives and at least one noun word should be contained in the candidate phrase. For instance, given a sentence "*Federal/NNP Emergency/NNP Management/NNP Agency/NNP has/VBZ become/VB the/DT ultimate/JJ patronage/NN backwater/NN*", using this rule, candidate phrases "federal emergency management agency" and "ultimate patronage backwater" are generated.

We propose to extract important noun phrases as keyphrases in Sect. 2.1. Hence, we design a rule by removing the adjectives, i.e., $(NN|NNS|NNP) +$. Following this rule, except named entities, all the candidate phrases are a chain of continuous noun words. For the example sentence, candidate phrases "federal emergency management agency" and "patronage backwater" are generated.

After generating the candidate noun phrases, we propose a method to compute the ranking score of a candidate phrase by combining the scores of keyterms contained in it. Since keyterms in different clusters cannot be compared directly, we put forward a measurement to compare keyterms from different clusters. In the semantic subgraph construction step, we build a keyterm graph for cluster centroids. Similar to the keyterms in the same cluster, different cluster centroids can be compared as well. It makes sense that if a cluster contains more keyterms, i.e., tf is larger, the cluster is more important. So we combine the ranking score of a cluster centroid and the total tf of the cluster to form a comprehensive score for the cluster, as shown in (4). Next we combine the PPR score of a keyterm and the ranking score of its cluster to form a global utility score for the keyterm, as shown in (5).

$$SC(c_i) = \alpha \times PPR(c_i) + (1 - \alpha) \times \frac{\sum_{j=1}^{t} tf(kt_{ij})}{\log_2 t}, \forall c_i \in C, kt_{ij} \in c_i. \tag{4}$$

$$Score(kt_{ij}) = PPR(kt_{ij}) \times SC(c_i), \quad \forall kt_{ij} \in c_i, \forall c_i \in C. \tag{5}$$

Now, each keyterm is comparable in the whole document. The score of a candidate noun phrase $p_i \in P$ is the sum of scores of keyterms contained in it, as (6) shows.

$$Score(p_i) = \sum_{kt_{ij} \in p_i} Score(kt_{ij}). \tag{6}$$

After we get the scores of all candidate noun phrases, we select the top-K phrases with the largest ranking scores as the keyphrases of the input document.

6   Experimental Evaluation

For ease of presentation, our proposed _knowledge graph_ based method to _rank_ keyterms for keyphrase extraction is denoted as _KGRank_.

6.1   Datasets and Evaluation Metrics

We use the dataset DUC2001[1] to evaluate the performance of our method. The manually labeled keyphrases on this dataset are created in the work [27]. The dataset contains 308 news articles. The average length of the documents is about 700 words. And each document is manually assigned about 10 keyphrases. Since our task is to extract noun phrases as defined in Sect. 2.1, to construct the golden standard, we drop adjectives from the manual phrases.

In the experiments, we adopt precision, recall, and F-measure to evaluate the performance of keyphrase extraction. Their formal definitions are given in (7), (8), and (9) respectively, where $count_{correct}$ is the number of correct keyphrases extracted by automatic method, $count_{output}$ is the number of all keyphrases extracted automatically, and $count_{manual}$ is the number of manual keyphrases. We use Wikipedia and DBpedia as knowledge graphs respectively to assist keyphrase extraction.

The codes to preprocess data are implemented in Python. The other algorithms are implemented in C++. All experiments are conducted on a Windows Server with 2.4 GHz Intel Xeon E5-4610 CPU and 384 GB memory.

$$precision = \frac{count_{correct}}{count_{output}}. \tag{7}$$

$$recall = \frac{count_{correct}}{count_{manual}}. \tag{8}$$

$$\text{F-measure} = \frac{2 \times precision \times recall}{precision + recall}. \tag{9}$$

6.2   Comparison with Other Algorithms

We compare our method _KGRank_ (Wikipedia as the knowledge graph) and _DBRank_ (DBpedia as the knowledge graph) with two co-occurrence graph based methods SingleRank [27] and ExpandRank [27], one cluster-based method SCCooccurrence [18], one statistic-based clustering method SCWiki [18], and _tf-idf_ based methods. Our method which adopts Wikipedia as the knowledge graph achieves better performance than using DBpedia. The reason we think is due to the better coverage ratio of keyterms in Wikipedia, which is about 0.8, while

---

[1] http://www-nlpir.nist.gov/projects/duc/past_duc/duc2001/data.html.

the coverage ratio of DBpedia is about 0.6. For *KGRank*, the best results are achieved at $r = 3$ and $\tau = 3.0$ (Euclidean distance threshold). For *DBRank*, the best results are achieved at $r = 5$ and $\tau = 0.4$ (cosine similarity threshold). The windowsize W for SingleRank is varied from 2 to 20. The best results are achieved at $W = 17$. For ExpandRank, we set the expanded document number $k = 5, 10$, and vary windowsize W from 2 to 20. The best results are achieved at $k = 10$ and $W = 17$.

Figure 5 shows the performances of the seven methods on precision, recall, and F-measure respectively, where the keyphrase number $K = 5, 10, 15, 20$. It is clear that our method almost always performs the best. With the growth of K, the precision decreases and recall increases. F-measure achieves the best performance when K is 10, which is the average number of manual keyphrases. Note that SC selects keyphrases according to the exemplars, it is not able to control the number of keyphrases. For example, it cannot return enough phrases when $K = 20$.

Table 4 shows the keyphrases extracted by six methods above from the example document when $K = 10$. The keyphrases in red and bold are correct keyphrases.

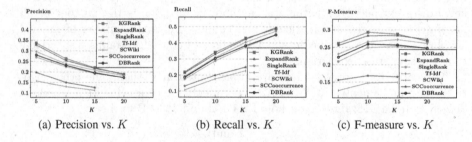

(a) Precision vs. K (b) Recall vs. K (c) F-measure vs. K

Fig. 5. Comparison with other algorithms

Table 4. Case study: keyphrase extraction results for the example document

Method	Keyphrases
KGRank	**florida**, **hurricane andrew**, emergency relief cost, **disaster**, relief, persident carter, **president george bush**,president, dollars, **election campagin**
ExpandRank	**hurricane andrew**, american insurance services group, **president george bush**, andrew card, **florida**, us insurer, president carter, federal emergency management agency, emergency relief cost, president
SingleRank	**president george bush**, president carter, **hurricane andrew**, american insurance services group, andrew card, president, white house chief, emergency relief cost, grain export programme, federal emergency management agency
SCWiki	sign, **election campaign**,**president george bush**, initial response, administration, assistance, business, farm, presidential rival
SCCooccurrence	sign, **hurricane andrew**, inhabitant, **florida**, **louisiana**, **election campaign**, **president george bush**, initial response, president, administration
Tf-idf	**president george bush**, andrew card, **hurricane andrew**, american insurance services group, president carter, emergency relief cost, wallace stickney, federal emergency management agency, grain export programme, james baker

6.3 Effect of Noun Phrase

To validate our opinion that *different annotators reach high agreement on noun phrases*, we randomly select 20 articles and annotate keyphrases that may contain adjectives for each article, denoted by S. Then we compare our manually labeled keyphrases with the benchmark that is used in the work [27], denoted by T. We find that the proportion of common general phrases is $|S \cap T|/|S \cup T| = 44\%$. Meanwhile, the proportion of common noun phrases is $|S' \cap T'|/|S \cup T| = 70\%$, where S' and T' are the keyphrases by removing adjectives from S and T, respectively. Thus, *this observation confirms our conclusion that different annotators reach high agreement on noun phrases.*

The phenomenon above also can be convinced through automatic keyphrase extraction methods. We compare the common keyphrases extracted by different methods, SingleRank [27], ExpandRank [27], and SCCooccurrence [18]. The rate of noun phrases in the common general keyphrases that are extracted by the three methods is 72%.

To further study the two types of candidate phrases, we use phrases containing adjectives as keyterms and noun phrases as keyterms respectively to extract keyphrases. The results are shown in Table 5. It is clear that noun phrases perform better than adjective noun phrases.

Table 5. Adjective noun phrase vs. noun phrase

Phrase type	Precision	Recall	F-Measure
Adjective noun phrase	0.16461	0.215415	0.184411
Noun phrase	0.205	0.30502	0.243754

6.4 Effect of Clustering Parameter

There are a lot of clustering algorithms (e.g., hierarchical clustering, k-means and spectral clustering) and similarity measures (e.g., cosine similarity, Wikipedia-based similarity, and Euclidean distance) can be used in the task of clustering keyterms. Since Google Word2vec performs well in measuring semantic similarity between words and k-means is a simple and effective method to clustering high-dimensional points, we adopt Word2vec vector and k-means in this paper.

We study the impacts of cluster number r in k-means on performance. Figure 6(a) shows the effects of cluster number r when semantic relatedness threshold τ is 3.0 and the keyphrase number K is 10 using Wikipedia as the knowledge graph. As shown in Fig. 6(a), the topic based clustering could improve the performance of keyphrase extraction. Keyterms in the same cluster tend to have more semantic relations, so the h-hop keyterm graph would contain less noises.

6.5   Effect of Keyterm Graph

Due to the large vertex degree of Wikipedia and DBpedia, we fix $h = 2$ in our experiments. If we do not screen out the expanded nodes, the extracted keyterm graph would inevitable contain a lot of nodes that are not related to the topics of the input document. So we use the semantic distance between expanded nodes and anchor nodes to filter the noisy expanded nodes. In this subsection, we study how the semantic relatedness threshold τ affects the performance of keyphrase extraction.

Figure 6(b) and (c) show the effects of relatedness threshold τ using Wikipedia and DBpedia respectively, where the cluster number r is 10 and the keyphrase number K is 10. From the two figures, we can see a suitable semantic relatedness threshold could improve the performance since it filters some noisy expanded nodes. A large semantic relatedness threshold would decrease the improvement of performance since it filters too much semantic relations including meaningful relations and expanded nodes.

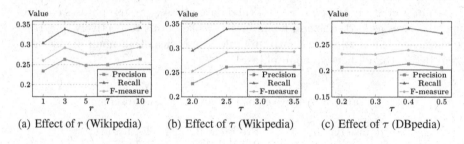

(a) Effect of r (Wikipedia)   (b) Effect of τ (Wikipedia)   (c) Effect of τ (DBpedia)

Fig. 6. Effect of cluster number r and similarity threshold τ

7   Related Work

The co-occurrence based methods, e.g., *TextRank* [21], *SingleRank* [27], *ExpandRank* [27], build a word co-occurrence graph and apply PageRank or its variation to rank the words. *TextRank* [21] is the first to solve the task by building a word co-occurrence graph and applying PageRank algorithm on the graph to rank words. If top-k ranked words are adjacency in the document, then they combine a phrase. *ExpandRank* [27] explores neighborhood information to help keyphrase extraction. Given a document, it first adopts cosine similarity to find the top-k similar documents and combines the $k + 1$ documents as a set. Then it builds a word co-occurrence graph on the document set and gets the ranking score of each word by PageRank. Finally, the score of a candidate phrase is the sum of the scores of the words contained in it. *SW* [11] adopts connectedness centrality, betweenness centrality, and a combination centrality algorithm to rank candidate phrases. Experiments show that there is no obvious difference in the three measures. And they achieve comparable performances with supervised methods KEA [29]. *CM* [5] compares several centrality measures: degree, closeness [2], betweenness, and eigenvector centrality. The experiments on three datasets show that degree is the best measure in the undirected graph, which indicates that tf is a very important feature for keyphrase extraction. So in our method, we choose to use PPR algorithm to rank the keyterms, where the jump probability is set to be proportional to tf.

Instead of just using the input document, the statistic-based methods, e.g., $Wan'07$ [28], SC [18], TPR [17], exploit some external resources. TPR [17] builds a word co-occurrence graph. The difference is that it trains Latent Dirichlet Allocation (LDA) model [4] on Wikipedia articles and gets some topics on the words. Then it runs Topical PageRank [10] under each topic, where the jump probability of each word is related with its occurrence probability under the current topic. The final ranking score of each word is the sum of its ranking score under each topic multiply by its occurrence probability given the topic. However, for a single document, the number of topics is much smaller than the topical number used in the experiments.

8   Conclusions

In this paper, we study on automatic single-document keyphrase extraction task. We propose a novel method that combines semantic similarity clustering algorithms with knowledge graph structure to help discover semantic relations hidden in the input document and cover more topics. We also design a measure to compare different clusters by constructing a semantic graph for the centroids of all clusters. Experiments show that our method achieves better performances than the state-of-art methods.

Acknowledgements. This work is supported by Research Grant Council of Hong Kong SAR No. 14221716 and The Chinese University of Hong Kong Direct Grant No. 4055048 and NSFC under grant Nos. 61622201, 61532010, 61370055, 61402020 and Ph.D. Programs Foundation of Ministry of Education of China No. 20130001120021.

References

1. Wikipedia. http://en.wikipedia.org/
2. Bavelas, A.: Communication patterns in task-oriented groups. J. Acoust. Soc. Am. **22**(6), 725–730 (1950)
3. Bizer, C., Lehmann, J., Kobilarov, G., Auer, S., Becker, C., Cyganiak, R., Hellmann, S.: DBpedia - a crystallization point for the web of data. J. Web Sem. **7**(3), 154–165 (2009)
4. Blei, D.M., Ng, A.Y., Jordan, M.I.: Latent Dirichlet allocation. J. Mach. Learn. Res. **3**, 993–1022 (2003)
5. Boudin, F.: A comparison of centrality measures for graph-based keyphrase extraction. In: IJCNLP 2013, pp. 834–838 (2013)
6. Cilibrasi, R., Vitányi, P.M.B.: The Google similarity distance (2004). CoRR, abs/cs/0412098
7. Freeman, L.C.: A set of measures of centrality based on betweenness. Sociometry **40**, 35–41 (1977)

8. Grineva, M.P., Grinev, M.N., Lizorkin, D.: Extracting key terms from noisy and multitheme documents. In: WWW 2009, pp. 661–670 (2009)
9. Hammouda, K.M., Matute, D.N., Kamel, M.S.: CorePhrase: keyphrase extraction for document clustering. In: Perner, P., Imiya, A. (eds.) MLDM 2005. LNCS, vol. 3587, pp. 265–274. Springer, Heidelberg (2005). doi:10.1007/11510888_26
10. Haveliwala, T.H.: Topic-sensitive PageRank. In: WWW 2002, pp. 517–526 (2002)
11. Huang, C., Tian, Y., Zhou, Z., Ling, C.X., Huang, T.: Keyphrase extraction using semantic networks structure analysis. In: ICDM 2006, pp. 275–284 (2006)
12. Hulth, A.: Improved automatic keyword extraction given more linguistic knowledge. In: EMNLP 2003, pp. 216–223 (2003)
13. Hulth, A.: Reducing false positives by expert combination in automatic keyword indexing. In: RANLP 2003, pp. 367–376 (2003)
14. Jeh, G., Widom, J.: SimRank: a measure of structural-context similarity. In: Proceedings of the Eighth ACM SIGKDD International Conference on Knowledge Discovery and Data Mining, Edmonton, Alberta, Canada, 23–26 July 2002, pp. 538–543 (2002)
15. Jiang, X., Hu, Y., Li, H.: A ranking approach to keyphrase extraction. In: SIGIR 2009, pp. 756–757 (2009)
16. Kusner, M.J., Sun, Y., Kolkin, N.I., Weinberger, K.Q.: From word embeddings to document distances. In: ICML 2015, pp. 957–966 (2015)
17. Liu, Z., Huang, W., Zheng, Y., Sun, M.: Automatic keyphrase extraction via topic decomposition. In: EMNLP, pp. 366–376 (2010)
18. Liu, Z., Li, P., Zheng, Y., Sun, M.: Clustering to find exemplar terms for keyphrase extraction. In: EMNLP 2009, pp. 257–266 (2009)
19. Lloyd, S.P.: Least squares quantization in PCM. IEEE Trans. Inf. Theory $28(2)$, 129–136 (1982)
20. Manning, C.D., Surdeanu, M., Bauer, J., Finkel, J.R., Bethard, S., McClosky, D.: The Stanford CoreNLP natural language processing toolkit. In: ACL 2014, pp. 55–60 (2014)
21. Mihalcea, R., Tarau, P.: TextRank: bringing order into text. In: EMNLP 2004, pp. 404–411 (2004)
22. Mikolov, T., Chen, K., Corrado, G., Dean, J.: Efficient estimation of word representations in vector space. CoRR (2013)
23. Page, L., Brin, S., Motwani, R., Winograd, T.: The PageRank citation ranking: bringing order to the web (1999)
24. Russell, S.J., Norvig, P.: Artificial Intelligence - A Modern Approach: the Intelligent Agent Book. Prentice Hall Series in Artificial Intelligence. Prentice Hall, Englewood Cliffs (1995)
25. Tsatsaronis, G., Varlamis, I., Nørvåg, K.: SemanticRank: ranking keywords and sentences using semantic graphs. In: COLING 2010, pp. 1074–1082 (2010)
26. Turney, P.D.: Learning to extract keyphrases from text (2002). CoRR, cs.LG/0212013
27. Wan, X., Xiao, J.: Exploiting neighborhood knowledge for single document summarization and keyphrase extraction. ACM Trans. Inf. Syst. $28(2)$, 8 (2010)
28. Wan, X., Yang, J., Xiao, J.: Towards an iterative reinforcement approach for simultaneous document summarization and keyword extraction. In: ACL 2007, vol. 7, pp. 552–559 (2007)
29. Witten, I.H., Paynter, G.W., Frank, E., Gutwin, C., Nevill-Manning, C.G.: KEA: practical automatic keyphrase extraction. In: Proceedings of the Fourth ACM Conference on Digital Libraries, pp. 254–255 (1999)
30. Youn, E., Jeong, M.K.: Class dependent feature scaling method using naive Bayes classifier for text datamining. Pattern Recogn. Lett. $30(5)$, 477–485 (2009)

Semantic-Aware Partitioning on RDF Graphs

Qiang Xu[1], Xin Wang[1,4,5]($\boxtimes$), Junhu Wang[2], Yajun Yang[1,5],
and Zhiyong Feng[3,5]

[1] School of Computer Science and Technology, Tianjin University, Tianjin, China
{xuqiang3,wangx,yjyang}@tju.edu.cn
[2] School of Information and Communication Technology, Griffith University,
Gold Coast, Australia
j.wang@griffith.edu.au
[3] School of Computer Software, Tianjin University, Tianjin, China
zyfeng@tju.edu.cn
[4] State Key Laboratory of Novel Software Technology, Nanjing University,
Nanjing, China
[5] Tianjin Key Laboratory of Cognitive Computing and Application, Tianjin, China

Abstract. With the development of the Semantic Web, an increasingly large number of organizations represent their data in RDF format. A single machine cannot efficiently process complex queries on RDF graphs. It becomes necessary to use a distributed cluster to store and process large-scale RDF datasets that are required to be partitioned. In this paper, we propose a *semantic-aware partitioning* method for RDF graphs. Inspired by the PageRank algorithm, classes in the RDF *schema graphs* are ranked. A novel partitioning algorithm is proposed, which leverages the semantic information of RDF and reduces crossing edges between different fragments. The extensive experiments on both synthetic and real-world datasets show that our semantic-aware RDF graph partitioning outperforms the state-of-the-art methods by a large margin.

Keywords: Graph partitioning · Semantic-aware · RDF graph · Schema

1 Introduction

The *Resource Description Framework* (RDF) is a W3C recommendation for describing and organizing resources on the Semantic Web. The RDF data model is a type of graph model, which consists of a set of triples (s, p, o), where s is the *subject*, p the *predicate*, and o the *object*. Each triple (s, p, o) represents a statement that s has the relationship p with o, thus being a directed edge in the RDF graph. With the Linked Data initiative, large amounts of RDF graphs have been publicly released, which often contain millions or even billions of triples.

How to partition a big RDF graph while reducing its crossing edges between different machines and improving the performance of RDF query processing has been recognized as a challenging issue. There exists several methods to partition

L. Chen et al. (Eds.): APWeb-WAIM 2017, Part I, LNCS 10366, pp. 149–157, 2017.
DOI: 10.1007/978-3-319-63579-8_12

RDF graphs in terms of vertices [2,3], edges [4], and paths [5]. However, these methods are all based on the traditional graph partitioning, which merely consider the structural information of graphs. In this paper, we propose a *semantic-aware* RDF partitioning algorithm that leverages the type information specified by the rdf:type predicate to achieve high-quality partitioning of RDF graphs.

Our main contributions include: (1) we propose an efficient algorithm, inspired by PageRank, for ranking classes in RDF schema graphs; (2) a novel RDF partitioning method is proposed, which can use the semantic information embedded RDF graphs to reduce crossing edges between different fragments; (3) the extensive experiments on the synthetic and real-world RDF graphs have been conducted to verify the effectiveness and efficiency of our approach. The experimental results show that our approach outperforms the state-of-the-art methods by a large margin.

2   Related Work

The existing partitioning methods on general graphs can be classified as follows:

Vertex-Based Approach. Thus far, METIS [3] is widely considered to be the most popular graph partitioning algorithm. In [2], Huang et al. propose an RDF partitioning approach based on METIS, which duplicates the vertices that are within n-hop distance from the boundary vertices and assigns them into the same partition. Although this approach can make more queries being executed locally, it may result in a large amount of data redundancy.

Edge-Based Approach. Margo and Seltzer [4] propose a scalable distributed graph partitioner (Sheep) which can produce edge partitions by reducing the graph to an elimination tree without sacrificing the partition quality. However, the cost of edge cut in this approach is unbounded and may be prohibitively high.

Path-Based Approach. Wu et al. [5] present a method to partition RDF graphs with paths as the basic partition unit, which partitions paths sharing merged vertex into the same fragment. By using the end-to-end path, this method makes use of the structural information of RDF graphs to minimize the number of crossing edges. However, the sizes of paths may differ significantly, so using the number of paths as the balance metrics may lead to data skewness.

RDF graphs are such kind of graphs that have inherent semantics. Thus far, the existing works of partitioning RDF graphs mainly include two types. The first type is to adapt the METIS method to some extent for RDF graphs. In fact, it is also a vertex-based approach with the feature of METIS, which cannot leverage the semantics embedded in RDF graphs. The second type is a naive hash partitioning approach, which generates a key according to each subject (or object) and assigns an RDF triple to a fragment simply based on its key. Thus, the number of crossing edges is large and the cost of RDF query processing may be expensive. In contrast, our partitioning method is able to take full advantages of the semantics inherently embedded in RDF graphs.

3   Preliminaries

An RDF dataset is a set of triples which can be represented as a directed labeled graph, i.e., RDF graph, which is defined as follows.

Definition 1 (RDF graph). *Let U and L be the disjoint infinite sets of URIs and literals, respectively. A tuple $(s, p, o) \in U \times U \times (U \cup L)$ is called an* RDF triple, *where s is the* subject, *p is the* predicate *(a.k.a.* property*), and o is the* object. *A finite set of RDF triples is called an* RDF graph.

Given an RDF graph T, we use $S(T)$, $P(T)$, and $O(T)$ to denote the set of subjects, predicates, and objects in T, respectively. For a certain subject $s_i \in S(T)$, we refer to the triples with the same subject s_i collectively as the *entity* s_i, denoted by $Ent(s_i) = \{t \in T \mid \exists p, o \text{ s.t. } t = (s_i, p, o)\}$.

We can use RDF Schema (RDFS) to define classes of entities and the relationships between these classes. For example, $(s, \texttt{rdf:type}, C)$ declares that the entity s is an instance of the class C. Given an RDF graph T, we assume that for each subject $s \in S(T)$ there exists at least a triple $(s, \texttt{rdf:type}, C) \in Ent(s)$, denoted by $s \in C$. We believe that this assumption is reasonable since every entity should belong to at least one type in the real world.

Given RDF graph T and a subject $s \in S(T)$, the *class set* of s is $\mathcal{C}(s) = \{C \mid s \in C \wedge \nexists C' \text{ s.t. } (C', \texttt{rdfs:subClassOf}, C) \in T\}$. The *direct class* of s is defined as $C(s) = \bigcap_{C_i \in \mathcal{C}(s)} C_i$. The *class system* of an RDF graph T is defined as $\mathcal{C}(T) = \{C(s) \mid s \in S(T)\}$.

Definition 2 (Schema graph). *Given an RDF graph T, the* schema graph *of T is an undirected labeled graph, denoted by $G_S(T) = (V_S, E_S, l_S)$, where (1) $V_S = \mathcal{C}(T)$ is the finite set of vertices; (2) $E_S \subseteq V_S \times V_S$ is the finite set of edges; (3) $l_S : E_S \rightarrow P(T)$ is a function that assigns a predicate as the label to an edge; (4) for an edge $e = (C_1, C_2) \in E_S$, a triple exists $(s, p, o) \in T$ such that $C_1 = C(s)$, $C_2 = C(o)$, and $l_S(e) = p$.*

Let T be an RDF graph and $C_i, C_j \in V_S$. We use $T(C_i, C_j)$ to denote the set of all triples whose subjects are of type C_i and objects are of type C_j. $T(C_i, C_j, p_k)$ is used to denote the subset of $T(C_i, C_j)$ where the predicates are p_k, and the set of predicates in $T(C_i, C_j)$ is denoted as $P(C_i, C_j)$.

Definition 3 (Predicate ratio). *Given an RDF graph T and its schema graph $G_S(T)$, the* predicate ratio *of a predicate p_k between two classes $C_i, C_j \in V_S$ is defined as* $\text{pr}(C_i, C_j, p_k) = \frac{|T(C_i, C_j, p_k)|}{|T(C_i, C_j)|}$.

Given an RDF graph T and its schema graph $G_S(T)$, the cardinality of a predicate p_k between two classes $C_i, C_j \in V_S$ is denoted by $\text{card}(C_i, C_j, p_k) = (m, n)$ which means that for each entity s in C_i there are on average n entities in C_j that are connected to s via triples, and vice versa, for each entity s in C_j there are on average m entities in C_i that are connected to s via triples.

Definition 4 (Cardinality factor). *The cardinality factor of a predicate p_k between two classes $C_i, C_j \in V_S$ is defined as*

$$cf(C_i, C_j, p_k) = \begin{cases} 1 & \text{if } n \geq m \wedge \mathsf{card}(C_i, C_j, p_k) = (1, n) \\ -1 & \text{if } n < m \wedge \mathsf{card}(C_i, C_j, p_k) = (m, 1) \\ \frac{n-m}{\max(m,n)} & \text{otherwise} \end{cases} \quad (1)$$

4   Ranking RDF Classes

The algorithm classRank, inspired by PageRank, is shown in Algorithm 1. Intuitively, we consider the k classes with top-k highest *rank scores* (denoted as rs) as the *significant* classes. The differences between classRank and PageRank are: (1) instead of distributing rank score evenly in PageRank, in classRank, the rank score of each class is divided and distributed to its adjacent classes in proportion to $\mathsf{pr} \cdot |\mathsf{cf}|$; (2) a fragment of rank score is distributed to an adjacent class only if the corresponding $\mathsf{cf} < 0$.

Algorithm 1: classRank // Rank classes in an RDF schema graph

Input : A schema graph G_S and the number of iterations k.
Output: The schema graph G_S with rs on each vertex.

1 **foreach** *class* $C_i \in V_S$ **do** $C_i.rs \leftarrow 1/|V_S|$;   // Initialise rs
2 **while** $k > 0$ **do**
3   **foreach** *class* $C_i \in V_S$ **do** $C_i.\delta \leftarrow 0$; // δ will record the change of rs
4   **foreach** *class* $C_i \in V_S$ **do**
5        $P_{in} \leftarrow \{(C_j, p_k) \mid \mathsf{cf}(C_i, C_j, p_k) < 0\}$;
6   **foreach** $(C_j, p_k) \in P_{in}$ **do**
7           $C_j.\delta \leftarrow C_j.\delta + C_i.rs \cdot \mathsf{pr}(C_i, C_j, p_k) \cdot |\mathsf{cf}(C_i, C_j, p_k)|$;
8           $C_i.\delta \leftarrow C_i.\delta - C_i.rs \cdot \mathsf{pr}(C_i, C_j, p_k) \cdot |\mathsf{cf}(C_i, C_j, p_k)|$;
9   **foreach** *class* $C_i \in V_S$ **do** $C_i.rs \leftarrow C_i.rs + C_i.\delta$;
10    $k \leftarrow k - 1$;
11 sort($C.rs$);   // In descending order
12 **return** G_S

Algorithm 1 assigns each class in the RDF schema graph an initial rank score rs (line 1). The rank score rs of C_i is distributed to each of C_i's neighbouring classes C_j, and the amount of the rank score frament $rs \cdot \mathsf{pr}(C_i, C_j, p_k) \cdot |\mathsf{cf}(C_i, C_j, p_k)|$ is added to $C_j.rs$. This process (lines 2–10) will be iteratively executed k times. During each iteration, for each class C_i, the variable δ is used to record the changed value of rs. If the cf value of two adjacent classes C_i and C_j is less than zero (line 5), then C_i will distribute a fragment of rs to C_j (lines 6–8). At the end of each iteration, we update the rs of each class by adding its δ to rs (line 9).

It can be observed that the time complexity of classRank is bounded by $O(|V_S|^2|AE_m|)$, where $|V_S|$ is the size of the RDF schema graph and $|AE_m|$ is the maximum number of edges between adjacent vertices in the schema graph. Note that the size of the RDF schema graph G_S is much less than its corresponding RDF graph.

5   Semantic-Aware Partitioning

We can obtain the k classes with top-k highest rs values from the result of Algorithm 1. Based on these top-k ranked classes, we propose a novel RDF partitioning algorithm semPartition, which is able to leverage the inherent semantics embedded in RDF graphs as heuristic information to make a better RDF partitioning. Note that *entity s* is the basic partitioning unit in our algorithm.

Algorithm 2: semPartition // Partition an RDF graph

Input : An RDF graph T, the schema graph G_S with top-k classes returned by Algorithm 1, the number of fragments n, and the *threshold*.

Output: A fragmentation $\mathcal{F} = \{F_1, \ldots, F_n\}$ of T.

```
 1 for i ← 1 to k do                                    // Partitioning phase
 2 |   foreach s ∈ C do                                 // Class C with the highest rs
 3 |   |   if s has not been processed then
 4 |   |   |   d ← hash(s) mod n; F_d ← F_d ∪ Ent(s);
 5 |   foreach C_i ∈{C's neighbouring classes} do
 6 |   |   foreach p ∈ {edges between C and C_i} do
 7 |   |   |   if cf(C, C_i, p) > threshold ∧ T(C, C_i, p) has not been processed then
 8 |   |   |   |   foreach o ∈ {o | ∃ (s,p,o) ∈ T ∧ s ∈ C ∧ o ∈ C_i} do
 9 |   |   |   |   |   if frag(o) = null then
10 |   |   |   |   |   |   d ← frag(s); F_d ← F_d ∪ Ent(o);

11 foreach T(C_i, C_j, p) that has not been processed do  // Postprocessing phase
12 |   switch (s,p,o) ∈ T(C_i, C_j, p) do
13 |   |   case frag(s) ≠ null ∧ frag(o) = null
14 |   |   |   d ← frag(s); F_d ← F_d ∪ Ent(o);
15 |   |   case frag(s) = null ∧ frag(o) ≠ null
16 |   |   |   d ← frag(o); F_d ← F_d ∪ Ent(s);
17 |   |   case frag(s) = null ∧ frag(o) = null
18 |   |   |   if cf(C_i, C_j, p) > threshold then
19 |   |   |   |   d ← hash(s) mod n; F_d ← F_d ∪ Ent(s);
20 |   |   |   else
21 |   |   |   |   d ← hash(o) mod n; F_d ← F_d ∪ Ent(o);

22 return F = {F_1, ..., F_n};
```

Algorithm 2 consists of two phases. During the partitioning phase (lines 1–10), the algorithm starts with the class C with the highest rs. For each subject s in the class C, $Ent(s)$ is assigned to a certain fragment F_d randomly by a hash function hash(s) on subjects and a modulo of the number of fragments n (lines 2–4). Then, for each neighbouring class C_i of C, the factor $cf(C, C_i, p)$ and the *threshold* are used to decide whether an object o in class C_i need to be distributed into the same fragment as the subject s (i.e., frag(s)) (lines 5–10). If $cf(C, C_i, p) > threshold$ and $T(C, C_i, p)$ has not been processed yet, then objects in class C_i will be partitioned (lines 8–10), otherwise the partitioning of C_i will be ignored and the algorithm will (go to line 2 to) start a new iteration to process the class with the next highest rs in the top-k ranked classes.

In the postprocessing phase (lines 11–21), the remaining RDF triples are partitioned in three cases: (1) if the subject s has been assigned to a fragment but not the object o, then $Ent(o)$ is assigned to the same fragment as s (lines 13–14); (2) if the object o has been assigned to a fragment but not the subject s, then $Ent(s)$ is assigned to the same fragment as o (lines 15–16); (3) if neither the subject s nor object o has been assigned to a fragment, then the assignment of s or o is determined by the comparison of $cf(C_i, C_j, p)$ and *threshold*.

We have proved that Algorithm 2 can get a fragmentation $\mathcal{F} = \{F_1, \ldots, F_n\}$ of an RDF graph T, and every $(s, p, o) \in T$ satisfies $(s, p, o) \in F_i \wedge F_i \in \mathcal{F}$.

It can be observed that the time complexity of semPartition is bounded by $O(|deg_m||AE_m||C_m| + |E_S|)$, where $|deg_m|$ is the largest degree in G_S, $|AE_m|$ is the maximum number of edges between adjacent vertices in G_S, $|C_m|$ is the largest number of subjects in a class, and $|E_S|$ is the number of edges in G_S.

6   Experiments

We have conducted the experiments on both synthetic and real-world RDF graphs to verify the effectiveness and efficiency of our method. The prototype program, which is implemented in Python, is deployed on a desktop computer that has a Intel i5-6500 CPU with 4 cores of 3.2 GHz, 8 GB memory, 500 GB disk, and 64-bit Ubuntu 14.04 as the OS. We generated 3 synthetic datasets using the LUBM benchmark, i.e., LUBM3, LUBM50, and LUBM100, which have 337K, 6.9M, and 13.9M RDF triples, respectively. We also run our algorithms on the real-world RDF graph DBpedia (version 2015-10) with 23M RDF triples. We compared our method with hash partitioning and METIS, and we did not consider the method in [5] because its sophisticated partitioning heuristics suffer from high preprocessing cost and high replication that is verified by [1].

Experiment 1: Performance with Different Parameters. Let n denotes the number of fragment. We carried out experiments on LUBM100 and DBpedia, in which k varies from 2 to 14 and 1 to 400, respectively. As shown in Fig. 1(a) and (c), where $n = 5$, different values of k have significant impact on the partitioning results. The results are much better with $k = 2, 4, 6$ on LUBM100 and $k = 1$ on DBpedia. After k is over a certain value, the larger k is, the more subjects whose classes in top-k are partitioned randomly rather than according to the

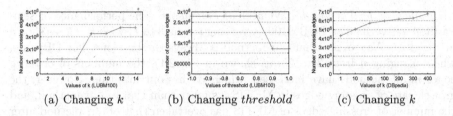

Fig. 1. The result of changing parameters

cf factor between classes, which results in the increasing of crossing edges. As shown in Fig. 1(b), semPartition was executed over LUBM100 with $k = 6$ and $n = 5$. When $threshold \geq 0.9$, the results are the best. The reason is that the $threshold$ determines the strength of the cf factor between classes.

Experiment 2: Comparison Between Algorithms. We compared semPartition with hash partitioning and METIS. The comparison experiments were conducted on LUBM and DBpedia whose results are shown in Fig. 2.

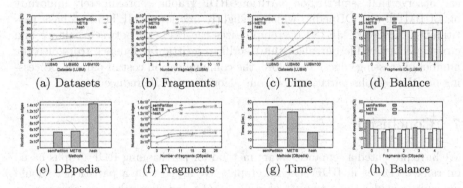

Fig. 2. Comparison between algorithms

(1) Different Size of Datasets. We conducted experiments on three LUBM datasets with $n = 5$. When $k = 6$ and $threshold = 0.9$, our algorithm showed a much better result than hash partitioning and METIS, as shown in Fig. 2(a). We can observe that semPartition demonstrates a better fragmentation in which the numbers of crossing edges are 9.5k, 604.6k, and 1212.7k on LUBM3, 50, and 100, respectively. Our algorithm can generate the best partitioning on LUBM. For DBpedia, as shown in Fig. 2(e), when $k = 1$, $threshold = 0$, and $n = 20$, the number of the crossing edges of METIS and hash partitioning is more than that of semPartition. We believe that the reason why our method outperforms METIS includes: (1) semPartition fully leverages semantics in RDF graphs, while MEITS only considers the structure of graphs; (2) semPartition uses RDF triples as the units of partitioning, while METIS uses vertices as the units of partitioning.

(2) Scalability by Varying the Number of Fragments. The performance of sem-Partiton was verified on LUBM100 by varying n from 3 to 11 with $k = 2$, as shown in Fig. 2(b). We can see that our method is always the best one. For DBpedia, when changing n from 3 to 25, as shown in Fig. 2(f), the numbers of crossing edges in all methods have increased, which verifies our intuition. However, the growth rate of crossing edges in METIS is higher than that of our method, and the number of crossing edges of METIS has overtaken that of our method after the number of fragments becomes larger than 20.

(3) Efficiency on Different Datasets. We verified the time efficiency of our method on LUBM and DBpedia. As shown in Fig. 2(c) and (g), hash partitioning has the least execution time on both types of datasets, while the execution times of our algorithm are competitive to that of METIS, since they are of the same order of magnitude though semPartition is just slower than METIS by a constant factor 1.14 on DBpedia. Not surprisingly, as hash partitioning does not need to consider any structural or semantic information, it is the fastest one.

(4) Balance on Different Datasets. As shown in Fig. 2(d) and (h), when $n = 5$, we can observe that semPartition partitions RDF graphs approximately uniformly on LUBM100 and DBpedia, which is slightly inferior to METIS and hash partitioning. However, METIS overemphasizes the trade-off between balance and crossing edges, while our algorithm sacrifices a little bit balance to reduce the number of crossing edges. Therefore, the communication cost of parallel processing on RDF graphs partitioned by our algorithm can be reduced.

7   Conclusion

We have presented a semantic-aware method for partitioning RDF graphs based on ranked classes in RDF schema graphs. We argue that a partitioner should not only consider the structure of RDF graphs, but also take into account the inherent RDF semantics. Our experimental results on both synthetic and real-word data have verified the effectiveness and efficiency of our method, which outperforms the state-of-the-art RDF graph partitioning methods by a large margin.

Acknowledgments. This work is supported by the National Natural Science Foundation of China (61572353), the Natural Science Foundation of Tianjin (17JCY-BJC15400), the Open Fund Project of State Key Lab. for Novel Software Technology (Nanjing University) (KFKT2015B20), and the Australian Research Council (ARC) Discovery Project (DP130103051).

References

1. Harbi, R., Abdelaziz, I., Kalnis, P., Mamoulis, N.: Evaluating SPARQL queries on massive RDF datasets. Proc. VLDB Endowment **8**(12), 1848–1851 (2015)

2. Huang, J., Abadi, D.J., Ren, K.: Scalable SPARQL querying of large RDF graphs. Proc. VLDB Endowment **4**, 1123–1134 (2011)
3. Karypis, G., Kumar, V.: A fast and high quality multilevel scheme for partitioning irregular graphs. SIAM J. Sci. Comput. **20**(1), 359–392 (1998)
4. Margo, D., Seltzer, M.: A scalable distributed graph partitioner. Proc. VLDB Endowment **8**(12), 1478–1489 (2015)
5. Wu, B., Zhou, Y., Yuan, P., Liu, L., Jin, H.: Scalable SPARQL querying using path partitioning. In: IEEE International Conference on Data Engineering, ICDE 2015, Seoul, South Korea, April, pp. 795–806 (2015)

An Incremental Algorithm for Estimating Average Clustering Coefficient Based on Random Walk

Qun Liao, Lei Sun, He Du, and Yulu Yang[(⊠)]

College of Computer and Control Engineering,
Nankai University, Tianjin, China
{liaoqun, sunleier, duhe}@mail.nankai.edu.cn,
yangyl@nankai.edu.cn

Abstract. Clustering coefficient is an important measure in social network analysis, community detection and many other applications. However, it is expensive to compute clustering coefficient for the real-world networks, because many networks, such as Facebook and Twitter, are usually large and evolving continuously. Aiming to improve the performance of clustering coefficient computation for the large and evolving networks, we propose an incremental algorithm based on random walk model. The proposed algorithm stores previous random walk path and updates the the average clustering coefficient estimation through reconstructing partial path in an incremental approach, instead of recomputing clustering coefficient from scratch as long as graph changes. Theoretical analysis suggests that the proposed algorithm improves the performance of clustering coefficient estimation for dynamic graphs effectively without sacrificing in accuracy. Extensive experiments on some real-world graphs also demonstrate that the proposed algorithm reduces the running time significantly comparing with a state-of-art algorithm based on random walk.

Keywords: Clustering coefficient · Graph analysis · Incremental algorithm · Random walk

1 Introduction

Clustering coefficient [1] plays an important role in mining useful knowledge from real-world networks, such as the World Wide Web and online social networks. Clustering coefficient describes the homophily (people become friends with those similar to themselves) and transitivity (friends of friends become friends) of a network, which is widely used in community detection [2], spam detection [3], protein-protein interaction network analysis [4] and many other applications.

In many real-world applications the networks are large and evolving all the time. For example, Google crawls more than 600 K new webpages every second [5] and the number of users of Facebook is over 1.6 billions and increases rapidly. Though many researches have optimized the performance of computing clustering coefficient based on static graph model, it is still expensive to compute clustering coefficient for large and evolving networks.

© Springer International Publishing AG 2017
L. Chen et al. (Eds.): APWeb-WAIM 2017, Part I, LNCS 10366, pp. 158–165, 2017.
DOI: 10.1007/978-3-319-63579-8_13

Aiming to improve the performance of clustering coefficient computation for dynamic graphs, we propose an incremental algorithm in this paper. It computes clustering coefficient via random walk initially. As the graph changes, it reuses previous random walk and gets result updated via reconstructing random walk path incrementally, instead of recomputing from scratch. Theoretical analysis and experimental evaluations both demonstrate that the proposed algorithm improves the performance effectively.

2 Preliminary and Prior Works

The clustering coefficient was first introduced in [1] to describe the degree of how nodes are closed to their neighbors. In this paper, we focus on the average clustering coefficient [6], a widely used version of clustering coefficient.

Let $G = (V, E)$ be an undirected connected graph with n nodes and m edges. We donate by v_i a node in G and by d_i the degree of node v_i. We define $D = \sum_{i=1}^{n} d_i$. The adjacency matrix, donated by A, is a $n \times n$ symmetric matrix, $A_{ij} = A_{ji} = 1$ if and only if there is an edge between v_i and v_j, and $A_{ij} = A_{ji} = 0$ otherwise. For any node v_i, $A_{ii} = 0$.

If $A_{ij} = A_{ik} = 1$, and $j < k$, the triplet (v_j, v_i, v_k) is defined as a wedge. For any wedge (v_j, v_i, v_k), if $A_{jk} = 1$, we define it a triangle, denoted as $<v_j, v_i, v_k>$. For any node v_i, the number of triangles $<v_j, v_i, v_k>$ is donated by l_i. It is obvious that l_i is equal to the number of edges between v_i's neighbors.

The local clustering coefficient of any node v_i is donated by c_i. For node v_i where $d_i > 1$, we define c_i as $2l_i/d_i(d_i-1)$. For node v_i where $d_i \leq 1$, c_i is defined as 0. The average clustering coefficient, donated by c_l, is the average of local clustering coefficient over the set of nodes, which is defined as Eq. (1).

$$c_l = \frac{1}{n} \sum_{i=1}^{n} c_i \tag{1}$$

It is expensive to compute c_l for a large graph due to counting the number of triangles is a challenging task. The best known algorithm for exact triangles counting requires $O(m^{2\omega/(\omega+1)})$ time [7], where $\omega < 2.376$ is the exponent of matrix multiplication [8]. However, it is not practical for large graphs due to its considerable cost of memory.

Random sampling based algorithms [9, 10] are preferable because an accurate approximation is sufficient in a large amount of real-world applications. Among these algorithms, the random walk based algorithm [9] performs well in both accuracy and efficiency. Its another advantage is that it only relies on external access to the graph and requires no prior knowledge of the graph (e.g. the number of nodes). It makes the algorithm adapt to the real-world social network analysis well, because for many online social networks it is not realistic to access the entire graph or another extra information for security and performance reasons.

A main limitation of the work in [9] is that it deals with graphs statically. So it is still expensive to deal with dynamic graphs due to the global recomputation as graph changes. Though streaming algorithms, such as Buriol et al. [11] and Becchetti et al. [12], estimate the clustering coefficient in a streaming model, these algorithms require to access each edge once at least, which limits their performance and make them inefficient comparing to the algorithms based on random walk, which only access a small number of edges among the whole graph.

3 Incremental Algorithm

3.1 Method

To improve the performance of computing clustering coefficient for dynamic graphs, we proposed an incremental algorithm based on the idea of reusing as much previous results as possible. The proposed algorithm stores previous estimated clustering coefficient and random walk path. It gets the estimation updated as graph evolves via reconstructing partial of the stored random walk path. We suppose that the average clustering coefficient is computed initially by the algorithm in [9], and we call the algorithm in [9] as baseline algorithm in the rest of this paper. The baseline algorithm generates a random walk with r steps, $R = (x_1, x_2, \ldots, x_r)$ representing the random walk path, and estimates c_l according to $\hat{c}_l = \Phi_l / \Psi_l$, where $\hat{c}_l$ is the estimation, Φ_l and Ψ_l are defined as Eq. (2) and Eq. (3). And the variable φ_k is defined as Eq. (4).

$$\Phi_l = \frac{1}{r-2} \sum_{k=2}^{r-1} \phi_k (d_{x_k} - 1)^{-1} \tag{2}$$

$$\Psi_l = \frac{1}{r} \sum_{k=1}^{r} (d_{x_k})^{-1} \tag{3}$$

$$\varphi_k = A_{x_{k-1}, x_{k+1}}, 2 \le k \le r - 1 \tag{4}$$

As an arbitrary edge e_{uv} added or removed, the proposed algorithm finds the set of positions where u and v emerging in the random walk path. We donate by X the set of positions of u in the random walk path, $X = \{x \mid$ the xth entry of R is $u\}$. Similarly, we define the set of positions of v, donated by Y. Then we compute the position where we start to replace the random walk path. We define two nonnegative integers z_1 and z_2. z_1 is the minimum entry in both X and Y and z_2 is the maximum number in these two sets. $z_1 = \min\{z \mid z \in X \cup Y\}$ and $z_2 = \max\{z \mid z \in X \cup Y\}$.

If z_1 equals to z_2, we compare z_1 and r, the length of the stored random walk path. If z_1 is greater than $r/2$, we start replacing the random walk path from z_1 to the end of the random walk path. Otherwise, we start from z_1 and replace the random walk path backward to its beginning. If z_1 and z_2 are not equal, we compute the summation of them. If their summation is smaller than r, we replace the random walk path from z_2 to the end. Otherwise, we start our replacement from z_1 backward to the beginning. Figure 1 sketches how the stored random walk updated in different situations.

If X and Y are both empty, the proposed algorithm returns previous result directly without any recomputation. As the steps in the random walk path replaced, we also

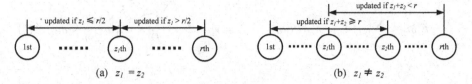

Fig. 1. Random walk updated according to the relations between r, z_1 and z_2

update φ_k and d_k correspondingly and recompute Φ_l and Ψ_l according to Eqs. (2) and (3) respectively. So we get the estimation of clustering coefficient updated eventually.

3.2   Correctness

It is instinctive to explain why the proposed incremental algorithm works. So we only give a simple version of explanation due to limitations on space. Supposing an arbitrary edge changes, there are only two situations: (1) the stored random walk path passes the endpoints of the changed edge; (2) the stored random walk path misses the endpoints.

In the first situation, the proposed algorithm updates random walk around the newly changed graph, thus it is equivalent to computing the clustering coefficient via the baseline algorithm on the newly changed graph.

In the second situation, the random walk doesn't get involved in the changing of graph, which is equivalent to computing the clustering coefficient via the baseline algorithm on the newly changed graph without accessing the newly changed edges. It is supposed there is only an edge changed at a time point, which takes a very tiny percentages of the millions of edges. The length of random walk is always relatively short, only takes 1 or 2 percentages of the number of nodes. Therefore, it is in a quite large probability that the baseline algorithm gets estimations updated for the newly changed graph without accessing the endpoints of the changed edge.

Consequently, the proposed incremental algorithm is correct as long as the baseline algorithm is correct. A detailed proof of the correctness of the baseline algorithm is provided in [9]. So we could confirm that the proposed algorithm is correct.

3.3   Computing Complexity

The main computation of the proposed algorithm is reconstructing the stored random walk path and updating Φ_l and Ψ_l. Thus the amount of computation of the proposed algorithm is linear to the number of steps of the stored random walk path rerouted.

It is intuitive that most edges in the graph are not accessed in the random walk. As e_{uv} arriving at time t, a random walk is needed to be updated only if it passes through either u or v. The probability that node u is accessed by a random walk, donated by π_u, is equal to d_u/D, which is proved in [9]. Define M to be the number of random walk path needed to be updated as an arbitrary e_{uv} changes, we get its expectation $E[M]$ as Eq. (5). Because the upper bound of amount of steps needed to be replaced is r. Hence, the amount of expected work as a randomly picked edge changes is $O(mr^2)$ at most.

$$E[M] \leq \sum_{e_{uv} \in E} r(\pi_{u_t} + \pi_{v_t}) \Pr[u_t = u, v_t = v] \approx 2mr \frac{\sum_{u \in V} d_u}{D} = 2mr \frac{n\bar{d}}{2m} = mr \qquad (5)$$

4 Evaluation

4.1 Experiment Setup

We evaluate the accuracy and performance of our algorithm via comparing with the algorithm in [9] as baseline. We implement both the algorithms in Java and run our experiments on a machine with Intel(R) Core(TM) i7-2600 CPU@3.80 GHz and 8 GB of RAM. We use some real-world graphs downloaded from SNAP [13]. The key parameters of these graphs are listed in Table 1.

Table 1. Main parameters of data sets

Graph	Background	Nodes	Edges	Average clustering coefficient
com-Amazon	Product network	335 K	926 K	0.3967
com-DBLP	Collaboration network	317 K	1050 K	0.6324
web-Stanford	Web graph	282 K	2312 K	0.5976

We use graphs with edges added to simulate evolving graphs in our experiments. We randomly remove a certain number (e.g. 1000, 2000, ..., 10000) of edges in each graph and use the rest part as the initial graph. We compute the average clustering coefficient on the initial graph via the baseline algorithm. Then we add the removed edges one by one and update the estimations via the proposed algorithm and its competitor. For both algorithms, we set $r = 0.02n$ by default, which is enough for accurate estimations [9].

4.2 Accuracy

We use RMSE, defined as Eq. (6), to evaluate the accuracy. We run each experiment for 1000 times and compute the average RMSE in our evaluation.

$$RMSE = \sqrt{E\left[(\hat{c}_l/c_l - 1)^2\right]} \qquad (6)$$

The comparison of RMSE between our algorithm and the baseline algorithm is depicted as Fig. 2, where the dotted lines show the average RMSE among experiments with different number of added edges. Smaller RMSE means performing better in accuracy.

We find that for all experiments, our algorithm achieves close (or even smaller) RMSE as its non-incremental competitor. The average RMSE of the proposed

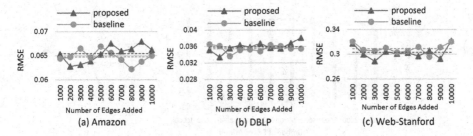

Fig. 2. Comparison of RMSE

algorithm is larger about 1% (on Amazon and DBLP) and smaller about 2% (on Web-Stanford) than the baseline algorithm. The results demonstrate that the accuracy of our algorithm is in the same order of the baseline.

We also evaluate the effects of r, the length of the random walk path, on the accuracy of our algorithm. We run experiments with $r = 0.01n$ and $r = 0.02n$ respectively. A distinct comparison of the accuray of experiments with different r is depicted as Fig. 3. It is clear that as the rise of r, RMSE decreases distinctly. More precisely, as r increases as much as 200% and RMSE decreases by over 28% at most.

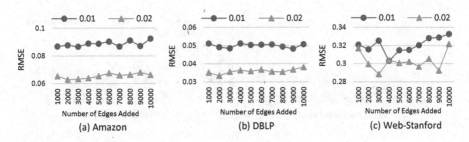

Fig. 3. Comparison of RMSE with different length of random walk path

4.3 Performance

To evaluate performance of our algorithm, we compare the running time on each graph. We find that our incremental algorithm runs much faster than the baseline algorithm. To depict the performance improvement directly, we present the speedup ratio of the incremental algorithm to the baseline algorithm in Fig. 4, where the dotted lines present the trends of speedup ratio as the number of added edges increasing.

From the results we find that the proposed algorithm improves the performance of estimating clustering coefficient for dynamic graphs significantly. Comparing to the baseline algorithm, it speeds up the computation over 20 times around all experiments. Moreover, it speeds up the computation more than 140 times at the best cases (on Amazon). The trends of the speedup ratios among all experiments also provide an obvious rise as the amount of added edges increasing. It implies that the proposed algorithm performs better for graph with a larger number of changed edges.

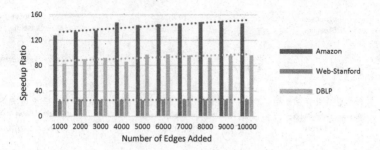

Fig. 4. Speedup ratio of the proposed algorithm

The comparison of running time of experiments on different graphs is shown as Fig. 5(a). It depicts the running time of the proposed algorithm on each graphs for 1000 times respectively. We find that the running time of the proposed algorithm is almost linear with the number of edges. Moreover, the running time of experiments is roughly proportional to the number of edges of the graph. The running time on Web-Stanford dataset is about 3.2 times that of the runing time on DBLP, and almost 4.9 times that of the runing time on Amazon. We also count the number of times of the endpoints of the changed edges emerging in the stored path. As Fig. 5(b) depicted, the counted number is linear with $O(rm)$, which supports our analysis above.

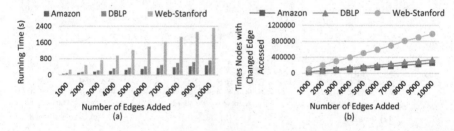

Fig. 5. Performance of the proposed algorithm

We run experiments with $r = 0.01n$ and $r = 0.02n$ respectively, to evaluate the effects of r on the performance of the proposed algorithm. We find that the running time is about linear to r^2 on each graph, depicted as Fig. 6. It meets our analysis in theory. As there are 10000 edges inserted, the running time of the proposed algorithm with

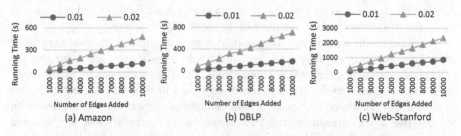

Fig. 6. Comparison of running time with different length of random walk path

$r = 0.02n$ is about 4 times that of experiment with $r = 0.01n$. The result is intuitive, a longer random walk means a higher probability of the changed edges accessed by the random walk and a bigger amount of computation as a random walk updated.

5 Conclusion

In this paper, we propose an incremental algorithm for estimating the average clustering coefficient for dynamic graphs based on random walk. As an arbitrary edge is added and/or removed, the proposed algorithm replaces partial random walk path around the evolving part and updates the estimation based on previous result. The performance improvement of the proposed algorithm is verified through analysis in theory and experiments on real-world graphs. Comparing with a state-of-art algorithm based on random walk, the proposed algorithm improves the performance effectively without sacrifices in accuracy.

References

1. Watts, D.J., Strogatz, S.H.: Collective dynamics of 'small-world' networks. Nature **393** (6684), 440–442 (1998). doi:10.1038/30918
2. Zhang, P.: Wang. J., Li. X.: Clustering coefficient and community structure of bipartite networks. Phys. A: Stat. Mech. Appl. **387**(27), 6869–6875 (2008). doi:10.1016/j.physa. 2008.09.006
3. Xu, Q., Xiang, E.W., Yang, Q.: SMS spam detection using noncontent features. IEEE Intell. Syst. **27**(6), 44–51 (2012). doi:10.1109/MIS.2012.3
4. Stelzl, U., Worm, U., Lalowski, M.: A human protein-protein interaction network: a resource for annotating the proteome. Cell **122**(6), 957–968 (2005)
5. Google Inside Search. https://www.google.com/intl/ta/insidesearch/howsearchworks/thestory/index.html
6. Costa, L.F., Rodrigues, F.A., Travieso, G.: Characterization of complex networks: a survey of measurements. Adv. Phys. **56**(1), 167–242 (2007). doi:10.1080/00018730601170527
7. Alon, N., Yuster, R., Zwick, U.: Finding and counting given length cycles. Algorithmica **17** (3), 209–223 (1997)
8. Coppersmith, D., Winograd, S.: Matrix multiplication via arithmetic progressions. J. symbolic Comput. **9**(3), 251–280 (1990)
9. Katzir, L., Hardiman, S.J.: Estimating clustering coefficients and size of social networks via random walk. ACM Trans. Web (TWEB) **9**(4), 19 (2015). doi:10.1145/2790304
10. Schank, T., Wagner, D.: Approximating clustering-coefficient and transitivity. Universität Karlsruhe, Fakultät für Informatik (2004)
11. Buriol, L.S., Frahling, G., Leonardi, S.: Counting triangles in data streams. In: 25th ACM SIGMOD-SIGACT-SIGART Symposium on Principles of Database Systems, pp. 253–262. ACM, New York (2006). doi:10.1145/2902251.2902283
12. Becchetti, L., Boldi, P., Castillo, C.: Efficient semi-streaming algorithms for local triangle counting in massive graphs. In: 14th ACM SIGKDD International Conference on Knowledge Discovery and Data Mining, pp. 16–24. ACM, New York (2008). doi:10.1145/1401890.1401898
13. Stanford Large Network Dataset Collection. http://snap.stanford.edu/data/index.html

Data Mining, Privacy and Semantic Analysis

Deep Multi-label Hashing for Large-Scale Visual Search Based on Semantic Graph

Chunlin Zhong[1(✉)], Yi Yu[2], Suhua Tang[3], Shin'ichi Satoh[2], and Kai Xing[1]

[1] University of Science and Technology of China, Hefei, China
chlzhong@mail.ustc.edu.cn, kxing@ustc.edu.cn
[2] National Institute of Informatics, Tokyo, Japan
{yiyu,satoh}@nii.ac.jp
[3] The University of Electro-Communications, Tokyo, Japan
shtang@uec.ac.jp

Abstract. Huge volumes of images are aggregated over time because many people upload their favorite images to various social websites such as Flickr and share them with their friends. Accordingly, visual search from large scale image databases is getting more and more important. Hashing is an efficient technique to large-scale visual content search, and learning-based hashing approaches have achieved great success due to recent advancements of deep learning. However, most existing deep hashing methods focus on single label images, where hash codes cannot well preserve semantic similarity of images. In this paper, we propose a novel framework, deep multi-label hashing (DMLH) based on a semantic graph, which consists of three key components: (i) Image labels, semantically similar in terms of co-occurrence relationship, are classified in such a way that similar labels are in the same cluster. This helps to provide accurate ground truth for hash learning. (ii) A deep model is trained to simultaneously generate hash code and feature vector of images, based on which multi-label image databases are organized by hash tables. This model has excellent capability in improving retrieval speed meanwhile preserving semantic similarity among images. (iii) A combination of hash code based coarse search and feature vector based fine image ranking is used to provide an efficient and accurate retrieval. Extensive experiments over several large image datasets confirm that the proposed DMLH method outperforms state-of-the-art supervised and unsupervised image retrieval approaches, with a gain ranging from 6.25% to 38.9% in terms of mean average precision.

Keywords: Learning based hashing · Deep hashing · Image retrieval · Convolutional neural networks

1 Introduction

With the explosive growth of image contents aggregated in the Internet, it is getting more and more important to search images efficiently over massive databases. Latest research shows that learning based hashing is a promising solution

© Springer International Publishing AG 2017
L. Chen et al. (Eds.): APWeb-WAIM 2017, Part I, LNCS 10366, pp. 169–184, 2017.
DOI: 10.1007/978-3-319-63579-8_14

to content-based visual retrieval and search over large-scale databases [1–3] due to their high efficiency in computation and storage. The idea of image hashing is to map high-dimensional image features to low-dimensional binary hash codes while still preserving semantic similarity between images. In this way, calculating the similarity is easier and quicker based on compact hash codes of images than exhaustive search with high-dimensional features. For example, Hamming distance is usually used to describe image similarity in binary hashing space [4,5], the smaller the distance is, the more similar two images are.

Existing content-based image hashing methods for visual search mainly can be divided into two categories: data-independent and data-dependent [6,7]. Data-independent methods generate hash functions randomly without considering any prior information, whose performance is unstable and heavily relies on the qualities of hashing functions. In contrast, data-dependent methods [8] are attracting more attentions due to their better performance, which learn hash functions by exploring the attributes of prior information. Early data-dependent approaches [9,10] learn hash codes from hand-crafted features of images. Features like GIST and Scale-Invariant Feature Transform (SIFT) are extracted from images and then mapped to compact hash codes. However, the distances in hand-crafted features sometimes cannot well capture the semantic similarity from human perception [11], which limit their performance. Recently, deep learning algorithms have become more and more popular because of their powerful abilities in various multimedia applications such as image retrieval and recognition. Inspired by the advancement of deep learning, researchers have proposed many image hashing algorithms based on convolutional neural networks (CNNs) [12–14]. While the majority of existing deep hashing approaches aim to learn models from single label images, which has limited semantic representation of images.

In this paper, we introduce a novel Deep Multi-Label Hashing (DMLH) framework to realize visual content search on large-scale image databases. The main characteristic of our DMLH is to mine the abundant semantic information hidden in multi-label images and preserve it in hash code and feature vector. This method includes the following three main components: (i) Pre-processing via label clustering. The same concept may be annotated with different labels(such as sea and ocean) by different users. Label clustering is to classify similar labels to the same cluster and provide accurate ground truth for the training step. (ii) Deep learning-based hashing. A CNNs based deep learning framework is constructed to learn feature vectors and compact hash codes simultaneously. (iii) Visual content search based on hashing and refined ranking. In the online stage, the feature vector and hash code of a query image are predicted by the trained model, using which we can retrieve most similar images from large-scale dataset rapidly in two steps.

Our deep hashing architecture for visual content search over large-scale datasets contributes in the following aspects:

(1) A semantic graph is proposed to cluster labels to reduce data sparsity meanwhile preserving inner information of images based on the co-occurrence

relation among labels. This helps maintain semantic information more sufficiently and generate more accurate ground truth for the training step compared with other descriptors.

(2) A novel deep CNNs framework is presented to learn visual contents and compact description of images with multi-label semantic vector as input directly, by fine tuning the pre-trained Caffe ImageNet model, which has powerful ability in learning the semantic similarities between multi-label images.

(3) We proposed to rank with high-dimensional feature vector the candidate images found by hash code of the query image, which helps to improve retrieval performance significantly with little computational consumption.

Extensive evaluations on several benchmark datasets confirm that our method achieves high performance in terms of several evaluation metrics compared with other state-of-the-art image retrieval approaches, with a gain ranging from 6.25% to 38.9% in terms of mean average precision.

The rest of this paper is organized as follows: The related work about image retrieval is briefly discussed in Sect. 2. Section 3 introduces our deep hashing-based visual search framework in detail. Experimental setting and results are presented in Sect. 4. Finally, we conclude our paper in Sect. 5.

2 Related Work

In recent years, researchers have proposed many methods for hashing based image retrieval [1,9,15,16]. Here we focus on data-dependent methods. These methods can be divided into three categories: unsupervised hashing, supervised hashing and semi-supervised hashing algorithms.

Unsupervised hashing algorithms, such as K-Means Hashing (KMH) [10], Spectral Hashing (SH) [9] and ITerative Quantization (ITQ) [6], do not consider any label information when learning hash functions. SH is a pioneering and classic image retrieval algorithm, which generates hash codes from query images by non-linear functions and Principal Component Analysis (PCA). KMH performs k-means clustering on data firstly and then uses Hamming distances between cluster indices to indicate the similarity among images.

Considering pre-labeled information when training a model, supervised hashing algorithms [2,16,17] can further boost image retrieval performance compared to unsupervised methods. Liu et al. [2] proposed a kernel-based supervised hashing model based on limited amounts of information by minimizing the distances of similar image pairs and maximizing the distances of dissimilar pairs. Zhang et al. [16] learned image hash functions based on a latent factor model. Lin et al. [18] achieved fast supervised hashing and high precision by training boosted decision trees.

Sometimes only little information is available for object retrieval. In this case, semi-supervised hashing approaches have remarkable performance. One of them proposed by Wang et al. [4] is designed to learn hash functions through minimizing hash code loss on the labeled data while maximizing variance over the labeled and unlabeled data. Semi-supervised hashing algorithm is useful when labeled data is limited.

Compared to shallow models, deep learning models have shown significant performance in image classification [19,20], retrieval [21], and feature extraction [22,23]. Encouraged by the powerful learning capability of deep learning like CNNs, many frameworks have been proposed by researchers to learn compact hash codes for images to achieve scalable image retrieval [12,14,21,24]. Oquab *et al.* [25] proved that rich image semantic information can be extracted from the mid-levels of CNNs. Erin *et al.* [13] learned a supervised deep hashing (SDH) framework by dividing a training set into positive and negative sample pairs. Kevin *et al.* [26] extracted a hidden layer of CNNs model to generate hash codes for images. This approach is constructed on the successful ImageNet architecture described in [19] and uses single label as supervisal information. However, due to the semantic limitation, single label supervisal information cannot describe the contents of images well and results in semantic information loss, as shown in Fig. 1. Lai *et al.* [12] presented an image hashing approach by incorporating a triplet ranking loss function in deep CNNs model to preserve relative similarities between images. Zhao *et al.* [24] constructed a deep learning framework based on semantic ranking with multi-label images. Xia *et al.* [27] embedded the pair-wised image similarity matrix into a deep convolutional model to learn hash functions. These image comparison based hashing algorithms achieve good performance in image retrieval. However, the pre-processing similarity matrixes limit their availabilities in practice. For example, suppose 100k images (a small dataset in large-scale image retrieval) are used in the training phase, the size of pair-wise similarity matrix will be 10 billion, consuming considerable computation and storage resources.

Deep learning algorithms especially CNNs have significant performance in image processing. Inspired by their powerful learning abilities, in this paper, we propose a novel image hashing learning framework DMLH by combining multi-label supervision information and deep CNNs model. Not only semantic information of images is adequately taken into account in our algorithm, but also the practicability and scalability can be ensured in large-scale data environment. The detailed strategy is discussed in the next section.

Single-label:
Sunset

Multi-label:
Sun, Cloud, Sea,
Tree, Sky, Dark

Single-label:
Nature

Multi-label:
Mountain, Flower,
Cloud, Grass, Sky

Fig. 1. Example of image label information.

3 Deep Multi-label Hashing Based Visual Search

As introduced in the previous section, to obtain the discriminating capability of features, the CNNs were used to extract visual features and a hashing layer was combined together to learn binary codes through single-label supervision. However, an image usually contains a diverse content with different aspects. Obviously, single-label deep hashing limits the semantic descriptions of images. For the purpose of improving semantic representations of images and narrowing semantic gap, the task of multi-label based deep hashing is attracting more attentions. In this work, we mainly focus on developing multi-label deep hashing architecture for efficient and accurate visual search of images on large databases.

Figure 2 shows our framework, which consists of two main stages: offline training and online processing. In the offline stage, a training dataset with multi-label images is used to train a deep multi-label hashing model that predicts hash code and feature vector for an image. To provide accurate ground truth, a pre-processing step, label clustering, is performed. Particularly, the set of diverse image labels is classified to semantic clusters according to label co-occurrences in the annotations of images. Similar labels are classified to the same cluster as far as possible. Then, cluster labels are taken as ground truth to train a multi-label hashing model by aid of CNNs. With the trained model, images in the database are organized in hash tables according to their hash codes. In the online stage, a query image without any label is taken as an input. With the trained deep model, its hash code and feature vector are predicted. The predicted hash code helps to find a small portion of candidates from the database, based on which a ranked list is generated by computing the similarity with feature vector. In the following, we describe the main components of the DMLH framework separately.

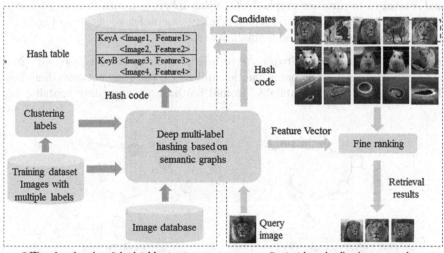

Offline deep learning & hash table structure   Content-based online image search

Fig. 2. The framework of DMLH.

3.1   Label Clustering

Multi-label image hashing methods always have superior performance compared with single-label hashing approaches, because multiple labels contain more accurate and abundant semantic information compared with single label, such as images shown in Fig. 1. However, the same objects often are described by different words due to the diversity of language habits. For example, we can use "people", "human" and "person" to represent "people". What's more, different objects often co-occur in the same image, such as sky and cloud, tree and mountain, which represent another kind of semantic similarity. But each image usually only contains a small number of objects, and is annotated with a small set of labels from a large corpus. Thus, multi-label image hashing approaches are faced with the diversity and sparsity of labels, and computational complexity if a model such as bag-of-words is used to represent the annotation of an image.

For our DMLH framework, in order to reduce computational cost and generate more reliable ground truth for the training step, we propose a novel graph partition algorithm to cluster labels, which contains three parts: *Building Label Co-occurrence Graph*, *Partitioning Semantic Graph*, and *Extracting Image Semantic Vector*.

Building Label Co-occurrence Graph. Collaborative Filtering (CF) [28] is one of most efficient algorithms in recommendation systems, and is widely used in industry nowadays. Take item-based CF recommendation as an example. One shopping cart contains a subset of goods, and the similarity between goods is represented by their co-occurrence frequency in the same carts. Two goods are highly similar if they often occur in the same carts. Motivated by the idea of item-based CF, we take labels as goods and images as carts, and use label co-occurrence frequency to represent label semantic similarity. On this basis, we build a label co-occurrence graph (LCG) using labels as nodes and the co-occur frequency of labels as the weight of edges connecting nodes, as shown in Fig. 3(a), in which there are 8 labels and 9 similarity values.

Partitioning Semantic Graph. According to the idea of item-based CF [28], in LCG, two labels have higher semantic similarity if the edge between them has a larger weight. To extract reliable ground truth and reduce data sparsity, we

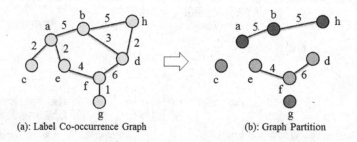

(a): Label Co-occurrence Graph   (b): Graph Partition

Fig. 3. Label clustering to preserve semantic similarity.

propose a graph partition algorithm to cluster labels. We hope that labels with semantic similarity can be classified into the same cluster. Our graph partition algorithm is shown in Alg 1. Given a LCG graph (line 2) and K, the number of connected sub-graph (CSG) (line 3), our algorithm removes the edge with the smallest weight in each iteration until the LCG is divided into K CSGs (line 6–10). Here, a CSG is a sub-graph where from each node there is at least one path to all other nodes. Therefore, a CSG can be built starting from any of its nodes, by iteratively add the neighbors of nodes already found in the CSG. For example, if we set $K = 4$, the LCG in Fig. 3(a) will be divided into 4 CSGs after graph partitioning, as shown in Fig. 3(b). Removing the edge with the smallest weight, this algorithm keeps the edges with large weights in CSGs, preserving the semantic similarity information as far as possible.

Alg 1: Graph Partition

```
program
1    Output: Sg;      # K clusters of labels
2    Input : LCG;     # Label Co-occurrence Graph
3            K;       # Number of target CSGs
4    var
5        NumOfCSG = 1;
6    While NumOfCSG < K do:
7        edge = FindEdgeWithSmallestWeight(LCG);
8        Remove edge from LCG;
9        NumOfCSG = FindNumberOfConnectedSubGraph(LCG);
10   end while
11   Regard all labels in each CSG as a cluster and add them in Sg;
12   return Sg with K clusters of labels;
end.
```

Extracting Image Semantic Vector. The whole label set is clustered into K subsets by graph partitioning Alg 1. In each label subset, there are the highest semantic similarity between each other. Having generating K CSGs, we extract semantic vectors for each multi-label image. Each image is represented by a K-dimensional image semantic vector $I \in \{0, 1\}^K$, each dimension of which corresponds to one cluster. Given an image with multiple labels, a bit of the vector will be set to 1 if any of its labels appears in the corresponding cluster; Otherwise, this bit is 0. In this way, two images will be represented by similar semantic vectors if there is high similarity between their labels.

The label clustering algorithm not only preserves the semantic similarities between images, but also reduces the sparsity of labels, which help to extract compact and reliable semantic feature vectors from images for the training step and improve the performance of the DMLH model.

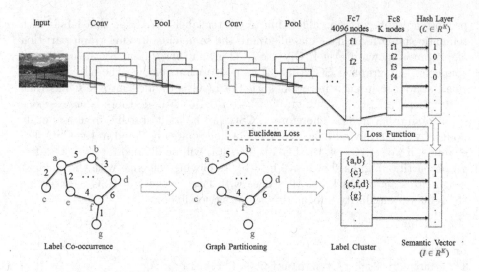

Fig. 4. CNNs hashing structure to learn image hash code and feature vector.

3.2 Deep CNNs Based Hashing Learning

Suppose we have an image dataset $D = \{x_i\}_{i=1}^{N}$ and a class label set $L = \{y_j\}_{j=1}^{M}$. Each image $x_i \in D$ is associated with $S_i \subseteq L$, a small subset of all labels. $|S_i|$, the size of S_i, is 1 in a single-label task and greater than 1 in our multi-label task. The target of image hashing is to learn a mapping as:

$$F : D \to \{0,1\}^K \tag{1}$$

where an input image x_i is encoded into a K-dimensional hash code $H^i = F(x_i)$, and $K \ll M$. The hash codes should preserve the semantic similarities between images after mapping. In specific, if image x_i is similar to x_j, the hash codes H^i and H^j should have a small Hamming or Euclidean distance. Otherwise, the distance between H^i and H^j should be large.

Here, we construct a deep learning structure, based on the pre-trained ImageNet model proposed by Krizhevsky *et al.* [19]. The ImageNet model is trained on the 1.2 million images in the benchmark dataset ImageNet, which achieves high accuracy in classifying images into 1,000 object classes. Specifically, the ImageNet model contains five convolutional layers, three pooling layers and two full-connected (Fc) layers. Except for the Fc8 layer, all layers are followed by a ReLU activation layer. The detailed parameter settings can be found in [19].

The structure of our deep multi-label hashing model is shown in Fig. 4, which contains two main steps: (1) deep CNNs used to extract semantic representations of images and (2) deep hashing mapping semantic information to compact hash codes. The deep CNNs step contains eight layers, the first seven layers of which have the same structure as ImageNet framework [19]. The number of nodes in the last layer, Fc8, is equal to hash code length K. The deep CNNs step is followed by a nonlinear hashing layer, which is used to convert the output of Fc8 layer to hash codes. Here we select $f(\boldsymbol{x}) = Sigmoid(\boldsymbol{x}) \in \boldsymbol{F}$ as the hash function and all input nodes from Fc8 are mapped to $\{0,1\}$ approximately. $\boldsymbol{H}_i$, the ith bit in the hash code $\boldsymbol{H}$, is generated as follows:

$$H_i = sgn(f(\boldsymbol{x}) - 0.5) = \begin{cases} 1 \ f(\boldsymbol{x}) > 0.5, \\ 0 \ otherwise. \end{cases} \tag{2}$$

In the training phase, the initial weights of the first seven layers of our model are set the same as in the pre-trained ImageNet model provided by Caffe group, and the connection parameters between Fc7 and Fc8 layers are initialized randomly. Then, we fine-tune the DMLH model on different image datasets to adapt the model to different retrieval requirements. The training goal of our DMLH is to keep the hash codes consistent with the image semantic vectors extracted from images in Sect. 3.1. The Euclidean distance between hash code $\boldsymbol{H}$ and image semantic vector $\boldsymbol{I}$ is selected as the loss function for back-propagating:

$$Loss = \frac{1}{2K}\|\boldsymbol{H} - \boldsymbol{I}\|_2^2 = \frac{1}{2K}\sum_{i=1}^{K}(\boldsymbol{H}_i - \boldsymbol{I}_i)_2^2. \tag{3}$$

where $\boldsymbol{H}, \boldsymbol{I} \in \boldsymbol{R}^K$. Most deep multi-label image hashing methods train models with a pairwise or triple-wise similarity between images, such as [24,27]. Except for causing computational explosion, these methods usually need to define complex similarity descriptions between images in the training step. In comparison, our DMLH method uses the image semantic vectors directly, which help our model to extract and preserve visual contents of images sufficiently, achieving high performance and scalability in large-scale visual content search.

3.3 Visual Search by a Combination of Hashing and Fine Ranking

The target of visual content search in this paper is to find top N images most similar to the query image from the database. Our method is performed in two steps for rapid and accurate image retrieval: (1) Deep multi-label hashing is employed for a "coarse" search, narrowing the search region to images having a non-negligible similarity to the query. (2) Ranking of the coarse search is refined by further using feature vector, returning images with a much higher similarity.

Coarse Search. Given a query image $\boldsymbol{Q}$, we firstly extract the output of the hash layer from the trained DLMH model and convert it to hash code $\boldsymbol{H}_Q$ with Eq. 2. Then, we use $\boldsymbol{H}_Q$ as a key to build a candidate pool $\boldsymbol{P}$ from the hash table of the image database. An image with hash code $\boldsymbol{H}_i$ is added in $\boldsymbol{P}$ if the Hamming distance between $\boldsymbol{H}_Q$ and $\boldsymbol{H}_i$ is less than a pre-defined threshold.

Fine Ranking. Studies in [19,30] show that the Fc6–8 layers of the ImageNet model can preserve sufficient visual information of an input image. The output of Fc7 layer is a 4,096 dimensional feature vector, which preserves more abundant visual information of images compared with hash code. Given one query image $\boldsymbol{Q}$

and each image i its candidate pool P, we extract high-dimensional feature vector v_Q and v_i for Q and i from Fc7 layer, respectively. Then, we rank image $i \in P$ based on the Euclidean distance between v_Q and v_i to improve the performance of the proposed DMLH method. The smaller the distance is, the more similar two images are. The top N similar images in the candidate pool P are the final retrieval results for the query image Q.

4   Experiment

The performance of DMLH is verified in this section. Our experiments are conducted on a Centos7.2 server, which contains CUDA7.0, Caffe0.14.5 and python2.7. Furthermore, it is configured with E5-1650v4 CPU (3.6 GHz), DDR4-2400 Memory (64 G) and GeForce GTX TITAN GPU (6144 MB).

4.1   Datasets

Two benchmark datasets are chosen to evaluate the performance of our strategy.

NUS-WIDE Dataset. This dataset [29] contains 269,648 images collected from the social media sharing website Flickr. Each image is associated with one or several labels in 81 concepts, such as sky, people, and ocean. More specifically, images are marked by a small subset of 5,018 unique semantic tags. For a fair comparison, we follow the works in [7,12,27] to use the images associated with the 21 most frequent concepts, where each concept has more than 5,000 images. Our experimental image datasets contain 4,509 tags out of 5,018 unique tags, with approximately all semantic labels. We randomly select 100 images from the top 21 concepts to form query set and 10,500 images are used in the training step, 500 samples for each concept.

MIRFLICKR-25K Dataset. This dataset [30] consists of 25,000 images also collected from the Flickr. All images are associated to 24 concepts, such as animals, car and night. 14 stricter concepts are used to label images if a concept is salient in one image. After confirmed, the stricter labels are ignored in our experiments since all images with stricter labels have the concepts before intensified. Further, each image is annotated by a 1,384 dimensional semantic tag vector, in which each tag has appeared more than 20 times in the image dataset. 2,400 images, 100 per concept, are extracted to build the query set and the 9,600 images, 400 images per concept, are used to train the deep CNNs model.

4.2   Evaluation Methods

DMLH algorithm is a two-step visual content search method in which candidate images generated by hash codes are ranked with high-dimensional feature vectors. We evaluate and compare DMLH with several state-of-the-art image hashing methods, including unsupervised methods SH [9], LSH [31], ITQ [6], and supervised methods SDH [32], KSH [2]. For DMLH, we use the raw images as input. For the other baseline methods, one image is represented by a 512-dimensional GIST feature vector. For a fair comparison, we have modified the parameters of baseline methods provided by the original authors to fit with experimental datasets. All approaches use the same ground-truth: a retrieved image is irrelevant if it shares no common concepts with the query image.

4.3 Evaluation Metrics

Three metrics are selected to measure the retrieval performance of different methods: (1) Precision of top N retrieved images, which is calculated as the ratio of the number of correctly retrieved images to that of all retrieved images. (2) Recall, which measures the proportion of similar samples that can be retrieved from the image database successfully. (3) Mean Average Precision (MAP), which provides a single figure measure of quality across recall levels.

4.4 Hashing and Ranking

The verification results (Precision and Total Searching Time) of the visual content search method proposed in Sect. 3.3 are shown in Fig. 5. Directly ranking images via exhaustive search (Ranking) using high-dimensional feature victors has the best precision because semantic information of images is better preserved. However, it consumes much more time than the other two methods. Hash codes as hashing key (Hashing) help to search images with $O(1)$ time, which achieves fast image search with the time consumption being almost negligible. However, its precision is poor. In our DMLH method, images in the database are assigned to buckets based on the hash codes. In the retrieval stage, candidate images are found by hash codes firstly. Then, high-dimensional feature vectors are used to refine the retrieval results. In this way, its retrieval time is reduced to about 3/10 compared with Ranking while the degradation in precision is suppressed. Therefore, DMLH achieves a better tradeoff than the two extremes (Hashing and Ranking) in terms of retrieval performance (precision) and retrieval time.

4.5 Results on NUS-WIDE

In the offline stage, the semantic graph is constructed on the co-occurrence relations between 4,509 image tags. Then, 128, 96 or 64 subgraphs are generated by $Alg1$. The proposed DMLH is constructed on the opensource Caffe framework. Fig. 6 shows the Precision and Recall results of different hashing methods on the NUS-WIDE dataset. As can be seen, compared with unsupervised approaches, supervised approaches have better performance by training models under the guidance of a supervised information. And the proposed DMLH significantly outperforms that of other state-of-the-art image retrieval algorithms, regardless of the number of hash bits, 64, 96, 128, used in the hashing. Learning compact image description by a deep CNNs model, our method has better capability to utilize semantic supervised information to capture the visual features of images.

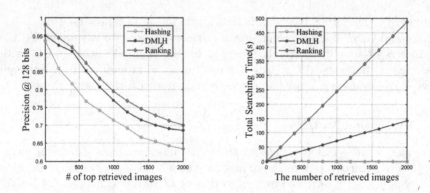

Fig. 5. The equilibrium between image ranking and efficiencies. Hashing means retrieving images only with hash codes. And ranking means that images are retrieved only by high-dimensional feature vectors.

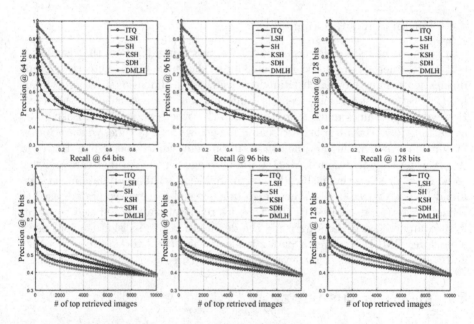

Fig. 6. Comparison of hashing performance of our approach and other hashing methods based on multi-label feature dataset: NUS-WIDE.

Table 1. Comparison of MAP with different methods on NUS-WIDE. We calculate the MAP values within the top 5,000 returned samples.

Method	ITQ [6]	LSH [31]	SH [9]	KSH [2]	SDH [32]	**DMLH**
64 bits	0.452	0.426	0.391	0.493	0.502	**0.543**
96 bits	0.465	0.456	0.414	0.501	0.514	**0.553**
128 bits	0.479	0.466	0.427	0.507	0.528	**0.561**

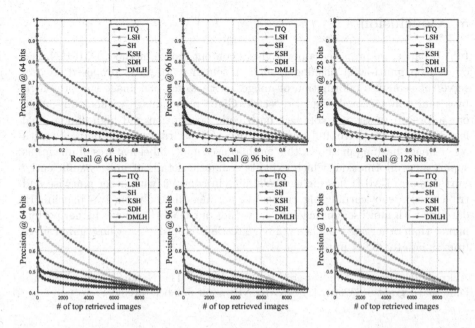

Fig. 7. Comparison of hashing performance of our approach and other hashing methods based on multi-label feature dataset: MIRFLICKR-25K.

Further, we calculate the MAP values within the top 5,000 retrieved images. Table 1 shows the evaluation results of MAP of different retrieval methods, which indicate that our DMLH has the best ability in retrieving similar images. Its gain over other methods ranges from 6.25% to 38.9% according to Table 1.

4.6 Results on MIRFLICKR-25K

For the MIRFLICKR-25K, we follow the same experimental setup as NUS-WIDE, where the semantic graph is built on the co-occurrence relations among 1,384 semantic tags. Figure 7 and Table 2 show the experimental results of different image retrieval approaches. As can be seen, our DMLH achieves the best performance and is much superior than other hashing methods.

Table 2. Comparison of MAP with different methods on MIRFLICKR-25K. We calculate the MAP values within the top 5,000 returned samples.

Method	ITQ [6]	LSH [31]	SH [9]	KSH [2]	SDH [32]	**DMLH**
64 bits	0.426	0.410	0.398	0.439	0.464	**0.501**
96 bits	0.430	0.425	0.410	0.454	0.477	**0.509**
128 bits	0.440	0.431	0.420	0.459	0.489	**0.530**

5 Conclusion

In this paper, we proposed a novel visual content search method, DMLH, based on a deep multi-label hashing model. A semantic graph is proposed to preserve the semantic information of images, with which one image is represented by a multi-dimensional semantic vector. Given a raw image without any label, our approach predicts a compact hash code description and high-dimensional feature vector via a carefully designed deep CNNs framework. Compared with other image hashing methods, this method uses multi-label features of images as supervising information, which exploits semantic details of images sufficiently. Furthermore, DMLH has higher scalability in contrast with other pairwise and triplet-wise deep learning algorithms. Experimental results on two benchmark datasets with multi semantic labels show that our DMLH achieves higher performance than other state-of-the-art image retrieval approaches, whose gain ranges from 6.25% to 38.9% in terms of MAP.

Acknowledgments. This research is supported by The National Natural Science Fund under grant 61332004, JSPS KAKENHI grant number 16K16058.

References

1. Norouzi, M., Blei, D.M.: Minimal loss hashing for compact binary codes. In: The 28th International Conference on Machine Learning (ICML 2011), pp. 353–360 (2011)
2. Liu, W., Wang, J., Ji, R., Jiang, Y.-G., Chang, S.-F.: Supervised hashing with kernels. In: 2012 IEEE Conference on Computer Vision and Pattern Recognition (CVPR), pp. 2074–2081. IEEE (2012)
3. Sun, Y., Wang, X., Tang, X.: Deep learning face representation from predicting 10,000 classes. In: Proceedings of the IEEE Conference on Computer Vision and Pattern Recognition, pp. 1891–1898 (2014)
4. Wang, J., Kumar, S., Chang, S.-F.: Semi-supervised hashing for scalable image retrieval. In: 2010 IEEE Conference on CVPR, pp. 3424–3431. IEEE (2010)
5. Liu, W., Wang, J., Kumar, S., Chang, S.-F.: Hashing with graphs. In: Proceedings of the 28th International Conference on Machine Learning (ICML 2011) (2011)
6. Gong, Y., Lazebnik, S.: Iterative quantization: a procrustean approach to learning binary codes. In: 2011 IEEE Conference on Computer Vision and Pattern Recognition (CVPR), pp. 817–824. IEEE (2011)
7. Li, W.-J., Wang, S., Kang, W.-C.: Feature learning based deep supervised hashing with pairwise labels. arXiv preprint arXiv:1511.03855 (2015)
8. Zheng, Y., Guo, Q., Tung, A.K., Wu, S.: LazyLSH: approximate nearest neighbor search for multiple distance functions with a single index. In: Proceedings of the 2016 International Conference on Management of Data. ACM (2016)
9. Weiss, Y., Torralba, A., Fergus, R.: Spectral hashing. In: Advances in Neural Information Processing Systems, pp. 1753–1760 (2009)
10. He, K., Wen, F., Sun, J.: K-means hashing: an affinity-preserving quantization method for learning binary compact codes. In: Proceedings of the IEEE Conference on Computer Vision and Pattern Recognition, pp. 2938–2945 (2013)

11. Wan, J., Wang, D., Hoi, S.C.H., Wu, P., Zhu, J., Zhang, Y., Li, J.: Deep learning for content-based image retrieval: a comprehensive study. In: Proceedings of the 22nd ACM international conference on Multimedia, pp. 157–166. ACM (2014)
12. Lai, H., Pan, Y., Liu, Y., Yan, S.: Simultaneous feature learning and hash coding with deep neural networks. in: Proceedings of the IEEE Conference on Computer Vision and Pattern Recognition, pp. 3270–3278 (2015)
13. Erin Liong, V., Lu, J., Wang, G., Moulin, P., Zhou, J.: Deep hashing for compact binary codes learning. In: Proceedings of the IEEE Conference on Computer Vision and Pattern Recognition, pp. 2475–2483 (2015)
14. Wang, J., Song, Y., Leung, T.: Learning fine-grained image similarity with deep ranking. In: Proceedings of the IEEE Conference on CVPR, pp. 1386–1393 (2014)
15. Norouzi, M., Fleet, D.J., Salakhutdinov, R.R.: Hamming distance metric learning. In: Advances in NIPS, pp. 1061–1069 (2012)
16. Zhang, P., Zhang, W., Li, W.-J., Guo, M.: Supervised hashing with latent factor models. In: Proceedings of the 37th International ACM SIGIR Conference on Research & Development in Information Retrieval, pp. 173–182. ACM (2014)
17. Norouzi, M., Punjani, A., Fleet, D.J.: Fast search in hamming space with multi-index hashing. In: 2012 IEEE Conference on Computer Vision and Pattern Recognition (CVPR). IEEE, pp. 3108–3115 (2012)
18. Lin, G., Shen, C., Shi, Q., van den Hengel, A., Suter, D.: Fast supervised hashing with decision trees for high-dimensional data. In: Proceedings of the IEEE Conference on Computer Vision and Pattern Recognition, pp. 1963–1970 (2014)
19. Krizhevsky, A., Sutskever, I., Hinton, G.E.: ImageNet classification with deep convolutional neural networks. In: Advances in Neural Information Processing Systems, pp. 1097–1105 (2012)
20. Gao, J., Jagadish, H.V., Lu, W., Ooi, B.C.: DSH: data sensitive hashing for high-dimensional k-NNsearch. In: Proceedings of the 2014 ACM SIGMOD International Conference on Management of Data. ACM, pp. 1127–1138 (2014)
21. Babenko, A., Slesarev, A., Chigorin, A., Lempitsky, V.: Neural codes for image retrieval. In: European Conference on Computer Vision, pp. 584–599 (2014)
22. Sermanet, P., Eigen, D., Zhang, X., Mathieu, M., Fergus, R., LeCun, Y.: OverFeat: integrated recognition, localization and detection using convolutional networks. arXiv preprint arXiv:1312.6229 (2013)
23. Szegedy, C., Liu, W., Jia, Y., Sermanet, P., Reed, S., Anguelov, D., Erhan, D., Vanhoucke, V., Rabinovich, A.: Going deeper with convolutions. In: Proceedings of the IEEE Conference on Computer Vision and Pattern Recognition, pp. 1–9 (2015)
24. Zhao, F., Huang, Y., Wang, L., Tan, T.: Deep semantic ranking based hashing for multi-label image retrieval. In: Proceedings of the IEEE Conference on Computer Vision and Pattern Recognition, pp. 1556–1564 (2015)
25. Oquab, M., Bottou, L., Laptev, I., Sivic, J.: Learning and transferring mid-level image representations using convolutional neural networks. In: Proceedings of the IEEE Conference on Computer Vision and Pattern Recognition, pp. 1717–1724 (2014)
26. Lin, K., Yang, H.-F., Hsiao, J.-H., Chen, C.-S.: Deep learning of binary hash codes for fast image retrieval. In: Proceedings of the IEEE Conference on Computer Vision and Pattern Recognition Workshops, pp. 27–35 (2015)
27. Xia, R., Pan, Y., Lai, H., Liu, C., Yan, S.: Supervised hashing for image retrieval via image representation learning. In: AAAI, vol. 1, p. 2 (2014)

28. Pazzani, M.J., Billsus, D.: Content-based recommendation systems. In: Brusilovsky, P., Kobsa, A., Nejdl, W. (eds.) The Adaptive Web. LNCS, vol. 4321, pp. 325–341. Springer, Heidelberg (2007). doi:10.1007/978-3-540-72079-9_10
29. Chua, T.-S., Tang, J., Hong, R., Li, H., Luo, Z., Zheng, Y.-T.: NUS-WIDE: a real-world web image database from national University of Singapore. In: Proceedings of ACM Conference on Image and Video Retrieval (CIVR 2009), 8–10 July 2009
30. Mark, B.T., Huiskes, J., Lew, M.S.: New trends and ideas in visual concept detection: The MIR flickr retrieval evaluation initiative. In: Proceedings of the 2010 ACM International Conference on Multimedia Information Retrieval, MIR 2010. New York, NY, USA, pp. 527–536. ACM (2010)
31. Gionis, A., Indyk, P., Motwani, R., et al.: Similarity search in high dimensions via hashing. VLDB **99**(6), 518–529 (1999)
32. Shen, F., Shen, C., Liu, W., Tao Shen, H.: Supervised discrete hashing. In: Proceedings of the IEEE Conference on CVPR, pp. 37–45 (2015)

An Ontology-Based Latent Semantic Indexing Approach Using Long Short-Term Memory Networks

Ningning Ma, Hai-Tao Zheng[✉], and Xi Xiao

Tsinghua-Southampton Web Science Laboratory, Graduate School at Shenzhen, Tsinghua University, Beijing, China
mnn15@mails.tsinghua.edu.cn, {zheng.haitao,xiaox}@sz.tsinghua.edu.cn

Abstract. Nowadays, online data shows an astonishing increase and the issue of semantic indexing remains an open question. Ontologies and knowledge bases have been widely used to optimize performance. However, researchers are placing increased emphasis on internal relations of ontologies but neglect latent semantic relations between ontologies and documents. They generally annotate instances mentioned in documents, which are related to concepts in ontologies. In this paper, we propose an Ontology-based Latent Semantic Indexing approach utilizing Long Short-Term Memory networks (LSTM-OLSI). We utilize an importance-aware topic model to extract document-level semantic features and leverage ontologies to extract word-level contextual features. Then we encode the above two levels of features and match their embedding vectors utilizing LSTM networks. Finally, the experimental results reveal that LSTM-OLSI outperforms existing techniques and demonstrates deep comprehension of instances and articles.

1 Introduction

With the rapid development of Internet, the information is growing explosively on the web and document indexing technology is becoming increasingly more crucial. The core task of existing search engine is to understand the real intention of user and semantic meanings of web content, aiming to obtain more semantically expressive resources. Generally, the semantic indexing methods in the literature can be distinguished according to the following two categories: (1) ontology-based approaches utilizing ontology and knowledge base as background knowledge and (2) statistical approaches, to identify groups of words that commonly appear together and therefore may jointly describe a particular reality [16].

The first category which utilizes ontologies and knowledge bases [3,11,14] to understand semantic context has recently been an area of considerable interest in semantic indexing. An *ontology* is a formal knowledge description of concepts and their relationships, an ontology together with a set of individual *instances* of classes constitutes a *knowledge base*. There are two main challenges: utilizing relations among concepts in ontologies and mapping information in documents

© Springer International Publishing AG 2017
L. Chen et al. (Eds.): APWeb-WAIM 2017, Part I, LNCS 10366, pp. 185–199, 2017.
DOI: 10.1007/978-3-319-63579-8_15

into knowledge bases. However, researchers are placing increased emphasis on internal relations of ontologies but neglect latent semantic relations between ontologies and documents. They generally annotate all instances mentioned in documents, which are related to concepts in ontologies.

Meanwhile, as for the second category, probabilistic topic models view each document as a mixture of various topics and each topic as a mixture of words [4,13], which possess fully generative semantics of documents. The sense of a word is a hidden random variable that is inferred from data. However, they do not carry the explicit notion of sense that is necessary for word sense disambiguation. Fortunately, background ontologies and knowledge bases are generally exploited to determine the meaning and the contextual information of an ambiguous word.

Thus, we propose an Ontology-based Latent Semantic Indexing model utilizing topic models and Long Short-Term Memory networks (LSTM-OLSI). An instance's contextual information is extracted utilizing semantic relations in ontology knowledge base; a document's *general topic* (a sequence of words) is extracted by an importance-aware topic model. The similarity between an instance and a document is measured by the distance between their corresponding sequences (an instance's contextual information and a general topic information) embedding vectors computed by the LSTM networks. The results indexed by the instances with profound meanings intuitively depict that LSTM-OLSI outperforms existing techniques and demonstrates deep comprehension of instances and articles.

Our main contribution is outlined as follows:

- We take into account both word-level contextual information to resolve word sense ambiguity and document-level semantics to clarify the semantics of the documents.
- We propose an importance-aware topic model to generate a general topic, which is a comprehensive topic composed of all the subtopics.
- Further, we present a strategy for explicitly encoding semantic relationships between the documents and the knowledge base using LSTM networks.

In Sect. 2, we discuss related work in ontology-based and machine learning based indexing approaches. We present the LSTM-OLSI model in Sect. 3. Further, we discuss the experimental methods and compare the LSTM-OLSI with some state-of-the-art methods in Sect. 4. Finally, Sect. 5 concludes our work.

2   Related Work

Several research approaches for indexing documents have been proposed. We classify them as ontology-based indexing approaches and machine learning based indexing approaches, discuss these two types of related work.

Ontology-Based Indexing Approaches. For a variety of text analysis problems, the knowledge representation is useful, such as document similarity computation, search result re-ranking and semantic indexing [15,17]. There are a variety of semantic search approaches [3,14,19] utilizing ontology knowledge base which span the four main processes of an IR system: indexing, querying, searching and ranking [9]. The ideal indexing is to choose a set of features to represent documents. Posch [19] enriched domain-specific ontologies with ency-clopedic background knowledge, leveraged textual and structural information to implement automatic classification and subject indexing of documents. Gödert [11] proposed an index structure which is based on the concept of ontology. Lee et al. [14] presented effective semantic indexing and search techniques considering the semantic relationships in ontologies and proposed a weighting measure for the semantic relationships. Concretely, they considered the number of meaning-ful semantic relationships, the coverage of the keywords, and the discriminating power of the keywords. Hahm [10] proposed an indexing method dealing with semantic path in ontologies, which computed semantic scores among different instances in ontologies for each term. Those semantic instances with shorter paths had larger score.

Machine Learning Based Indexing Approaches. Several studies have developed indexing techniques leveraging supervised and unsupervised machine learning methods. A standardized supervised learning approach is the *bag-of-words* approach, which represents documents by the words they contain and neglectes the order of the words. Moreover, sequences of words (*n-grams*) are utilized to represent a document and the words are weighted by different schemes such as TF-IDF. However, these approaches cannot depict the semantic meaning of the documents. To address this shortcoming, some novel methods are pro-posed. Latent Semantic Indexing (LSI) [8] maps documents to low-dimensional concept vectors utilizing a document-word matrix called singular value decompo-sition (SVD). The relevance of a document to one or several keywords is assumed to be proportional to the cosine similarity between their concept vectors. A sig-nificant step forward in this regard was Probabilistic Latent Semantic Indexing (PLSI) [13] model, which views documents as mixtures of topics and ranks the documents by the probability of the query given the document distribution over topics. Utilizing topic models is one of the state-of-art unsupervised learning approaches to index documents. Latent Dirichlet Allocation (LDA) [4] extends PLSI and assumes that topic distributions have a Dirichlet prior. Newman et al. [18] exploited an unsupervised Bayesian model and applied the method Dirichlet process segmentation for extracting key phrases from a document. Ma et al. [16] proposed a semantic search method based on LDA and view each theme gen-erated from topic models as the basic message block that waits constantly to be searched. Chebil et al. [6] exploited a possibilistic network that carries out partial matching between documents and a biomedical vocabulary to index the biomedical documents.

3   Ontology-Based Latent Semantic Indexing Model

We design a novel semantic indexing framework by taking into account both the word-level contextual information and the document-level semantics, to resolve word sense ambiguity and to clarify the semantics of the documents, respectively.

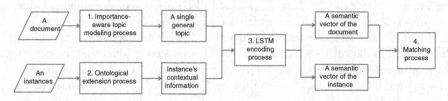

Fig. 1. The architecture of LSTM-OLSI.

The LSTM-OLSI is performed as follows: Firstly, we extract a single *general topic* (a sequence of words) for each document utilizing an importance-aware topic model inspired by Latent Dirichlet allocation (LDA) [4] (Step 1). After that, we extract an instance's contextual information utilizing semantic relations in ontology knowledge base (Step 2). Finally, we encode a semantic vector of a general topic and a semantic vector of an instance using LSTM networks, respectively (Step 3). And we measure the similarity between an instance and a document utilizing the cosine similarity between the semantic vectors of two sequences (Step 4).

In general, the overall LSTM-OLSI model consists of the following three essential components (see Fig. 1), which are described in more detailed in Sects. 3.1, 3.2 and 3.3:

(1) The probabilistic topic modeling process (Step 1);
(2) The ontological extension process (Step 2); and
(3) The LSTM sequence encoding and matching process (Step 3 and Step 4).

3.1   Probabilistic Topic Modeling Process

The importance-aware topic model in LSTM-OLSI is inspired by Latent Dirichlet allocation (LDA) [4]. We start with a brief introduction of LDA. LDA models each of D documents as a mixture over K latent topics, each of which describes a multinomial distribution over a W word vocabulary. The original space of vocabulary is mapped to several topics, which can better depict the semantic meaning of the documents. After conducting the Gibbs sampling process to train the model, given the parameters α and β, the joint distribution of a topic mixture θ, a set of N topics z, and a set of N words w, it is easy to obtain the following useful results related to the words, documents and topics in the documents set:

- Latent topics with the most likely words in each topic and a topic-to-word probability distribution (i.e., $p(w \mid z, \beta)$);
- A document-to-topic probability distribution (i.e., $p(z \mid \theta)$).

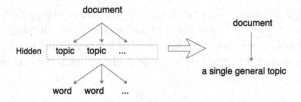

Fig. 2. The generation process of the *general topic*.

The LDA model is somewhat more elaborate than the three-level models often studied in the classical hierarchical Bayesian literature. However, to index the documents with high accuracy, we target to take a sequence of the most semantically expressive words to represent each document. Certainly, it is inappropriate to take one or more topics with higher probabilities to represent a document, for the reason that the words with higher probabilities in the topics with lower probabilities may be ignored. Therefore, we propose an approach to directly compute the semantic relations between the documents and the words. We generate a single *general topic* for each document which differs from conventional topic models (Fig. 2). By marginalizing over the hidden topic variable z, however, we can understand LDA as a two-level model and capture direct semantic relations between the documents and the words. To capture relations between the documents and the words, we derive the document-specific word probability distribution $p(w \mid \theta, \beta)$ as:

$$p(w \mid \theta, \beta) = \sum_z p(w \mid z, \beta) p(z \mid \theta) \tag{1}$$

However, the semantically expressive capacities are different between the two levels of probabilities, namely, the document-specific topic probabilities and the topic-specific word probabilities. In other words, without considering the different importance between the above two levels of probabilities, some words with low first-level probabilities but high second-level probabilities may be computed with a high correlation score. For example, we assume a document has a *health* related topic with a probability of 0.6 and has a *technology* related topic with a probability of 0.2. So the theme of the document is more likely about *health*. We also assume that the *health* related topic has a word *disease* with a probability of 0.2, but the *technology* related topic has a word *tech* with a probability of 0.8. So the relevant score of *tech* is calculated as $0.2 \times 0.8 = 0.16$, and the relevant score of *disease* is computed as $0.6 \times 0.2 = 0.12$, which is the smaller than *tech*. As we described above, the word *disease* is more relevant to this document than *tech*, since the theme of this document is more likely about *health*.

Therefore, we propose a self-adaptive approach utilizing an importance-aware manner applied to various realistic scenarios. Specifically, we define a correlation score $Correl(w_n, d)$ of a word w_n for a document d is computed as:

$$Correl(w_n, d) = \sum_z \left(f(p(w_n \mid z, \beta)) g(p(z \mid \theta_d)) \right) \tag{2}$$

where $f(p(w_n \mid z, \beta))$ represents the importance score of the probability of word w_n occurring in topic z, $g(p(z \mid \theta_d)$ represents the importance score of the probability of topic z occurring in document d for any given word.

The $f(p)$ score function and the $g(p)$ score function are given as follows:

$$f(p) = p^r \tag{3}$$

$$g(p) = p^q \tag{4}$$

where r is the penalty factor acting on the topic specific word probability and q is the penalty factor acting on the document specific topic probability.

The penalty factor q greater than 1 can increase the gap between those topics with larger probabilities and those with smaller probabilities. If q is less than 1, it can reduce the gap. The same applies to the penalty factor r which acts on the topic-to-word probability. As we have described above, in this indexing system, the document-to-topic level is more semantically expressive than the topic-to-word level, so we fix $r = 1$ and tune q. Finally, we select the first several W words in the whole vocabulary with the highest correlation scores for each document, thus deriving the *general topic*, which is a comprehensive topic composed of all the sub-topics.

It is a key issue in the topic model to determine the value of parameter K, the number of latent topics. Inappropriate setting of K customarily imposes an adverse impact on the modeling results. In particular, we computed the perplexity of a held-out test set to evaluate the models. The perplexity, used by convention in language modeling, is monotonically decreasing in the likelihood of the test data, and is algebraically equivalent to the inverse of the geometric mean per-word likelihood [4]. More formally, for a test set of M documents, the perplexity is:

$$perplexity(D_{test}) = exp(-\frac{\sum_{d=1}^{M} log(p(\mathbf{w}_d))}{\sum_{d=1}^{M} N_d}) \tag{5}$$

where $\mathbf{w}_d$ denotes text document d and N_d represents the number of words in document d. Moreover, a lower perplexity score indicates a better generalization performance [4].

3.2 Ontological Extension Process

The DBpedia ontology is leveraged to depict the word-level semantics in the process of LSTM-OLSI, collaborating with the probabilistic topic model which depicts the document-level semantics. The main tasks performed by this ontological extension process (Step 2 in Fig. 1) are discussed in detail below (Fig. 3).

For each instance, it has a verbal description in the ontology and we extend its semantic representations utilizing the ontology and knowledge base to depict its semantic meaning. The semantic representation of an instance is illustrated in its verbal description and its contextual instances' verbal descriptions. Intuitively, instances are stored in RDF statements which are triples of the form (*subject*, *property*, *object*). The form of *subject* or *object* typically indicates an instance in the knowledge base. A semantic path is composed of one or more *properties*. As the length of a semantic path gets longer, the relevance between the source and the destination decreases [14]. Therefore, for the single instance, we regard the set of its adjacent instances directly connected with a 1-length path in the knowledge base as its contextual instances.

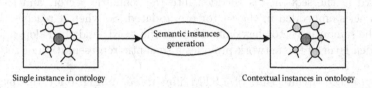

Single instance in ontology   Contextual instances in ontology

Fig. 3. The ontological extension process.

The verbal descriptions of the instance and its contextual instances are regarded as the instance's ontological contextual information, which essentially are a sequence of words and depict the word-level semantics.

3.3   LSTM Sequence Encoding and Matching Process

In Sect. 3.1, a document is encoded into a *general topic* (a sequence of words), and in Sect. 3.2, an instance is encoded with its ontological contextual information (also a sequence of words). We treat the above two levels of information as two sequences of words with internal structures, i.e., word dependencies. And the two sequences are about the same length.

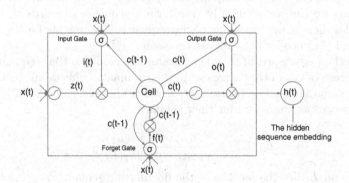

Fig. 4. The LSTM architecture used for sequence embedding.

The recurrent neural networks (RNN), a type of deep neural networks, have been widely used in time sequence modeling. However, it is generally difficult to learn the long term dependency within a sequence due to vanishing gradients problem. One of the effective solutions is using memory cells named Long Short Term Memory (LSTM). Therefore, we use LSTM recurrent networks to sequentially take each word in a sentence, extract its information, and embed it into a semantic vector. Due to its ability to capture long term memory, the LSTM accumulates increasingly richer information as it goes through the sequence. The encoding process is performed word-by-word sequentially. At each time step, a new word in the sequence is encoded into the semantic vector, and the word dependencies embedded in the vector are updated. So when it reaches the last word, the semantic vector has embedded all the words and their dependencies, the hidden layer of the network provides a semantic representation of the whole sequence [12].

We use the architecture of LSTM illustrated in Fig. 4 for the proposed sequence embedding method. LSTM has three gates: input gate ($i(t)$), forget gate ($f(t)$) and output gate ($o(t)$) which slows down the disappearance of past information and makes Backpropagation through Time (BPTT) easier. In this figure, $\sigma()$ is the sigmoid function, $c(t)$ is the cell state vector and $h(t)$ is the hidden activation vector, which can be used as a semantic representation of the t-th word. We utilize this architecture to find an embedded vector for each word, then use the $h(last)$ corresponding to the last word in the sentence as the semantic vector for the entire sequence.

However, the core task is the learning of the embedding vector for a sentence, that is, to train a model that can automatically transform a sequence of words to a vector that encodes the semantic meaning of the sequence. Our approach of sequence encoding is inspired by the work in sequence to sequence learning (seq2seq) [21]. They utilize a recurrent network as an encoder to read in an input sequence into a hidden state, which is the input to a decoder recurrent network that predicts the output sequence. To derive the sequence embedding, we encode a sequence and reconstruct the original sequence to train the seq2seq model. Thus we can automatically transform a sequence of words to a vector that encodes the semantic meaning of the sequence. The weights for the decoder network and the encoder network are the same.

The method developed in this paper trains the model so that sequences that are paraphrase of each other are close in their semantic embedding vectors. We adopt the cosine similarity $C(document, instance)$ between the semantic vectors of two sentences as a measure for their similarity:

$$C(D, I) = \frac{h_D(L_D)^T h_I(L_I)}{\|h_D(L_D)\| \cdot \|h_I(L_I)\|} \tag{6}$$

where L_D and L_I are the lengths of the document's semantic sequence D and the instance's semantic sequence I, respectively, $h_D()$ and $h_I()$ are the hidden activation vectors of D and I, respectively.

4   Experiments

To demonstrate the reliability and stability of our approach, we conduct further experiments. The ontology-based processing part is fully-implemented in Java and the topic modeling part is in Scala. The LSTM sequence encoding and matching process is implemented utilizing TensorFlow [2].

We first describe the datasets, the preprocessing details, the performance measure method and the baseline methods in Sect. 4.1. Then we report the experimental process and the comparison with the other four baseline models in Sect. 4.2. Finally, we illustrate the results indexed by many instances with profound meanings, which intuitively depict that LSTM-OLSI has deeper semantic comprehension of both instances and news articles.

4.1   Experimental Setup

Dataset. We exploit a real world news corpus collection, a movie sentiment dataset and a rich ontology knowledge base to conduct the experiments.

- **MSNNews:** The news articles are extracted from a large news corpus, which are news articles searched from MSN news web pages. We organize volunteers to classify these news articles manually into categories according to its article content, and we select five categories: crime, health, politics, soccer and technology. We select one million news articles and current affairs happened recently, and the average word count of news articles is about 250.
- **IMDB:** The IMDB movie sentiment dataset contains 25,000 labeled and 50,000 unlabeled documents in the training set and 25,000 in the test set [1]. The average length of each document is 241 words and the maximum length of a document is 2,526 words. We utilize this dataset to train the seq2seq model and to derive the LSTM networks' weights.
- **DBpedia:** We exploit the version of DBpedia 2016-04 as instance data base, which involves 9.5 billion pieces of information (RDF triples). The DBpedia ontology currently covers 685 concepts which form a subsumption hierarchy, described by 2,795 different properties and contains about 4,233,000 instances. The knowledge base is big enough to contain most domains of knowledge in our daily life so the knowledge in daily news can be represented.

Data Preprocessing. Before our algorithm is capable to index news documents, we first perform certain preprocessing procedures. First, we extract context information from DBpedia knowledge base to establish our experimental basis for indexing. In the context of this work, we extract 6.0M instances from DBpedia that we eventually utilize in our work. Next, after extracting news corpus involving the five categories, we lowercase all characters, perform word segmentation and remove stop words. In each category, we randomly held out 10% of the data for test purposes and trained the models on the remaining 90%, to conduct 10-fold cross-validation.

Performance Measure. We conduct an evaluable method to estimate accuracy of the indexing results. More concretely, if a document in the corresponding category *crime* is searched by the query keyword *crime*, we consider it as a relevant news document. We compute precision as:

$$precision = \frac{|relevant\_news| \cap |retrieved\_news|}{|retrieved\_news|} \tag{7}$$

and recall as:

$$recall = \frac{|relevant\_news| \cap |retrieved\_news|}{|relevant\_news|} \tag{8}$$

Baseline Methods. We exploit a comparative evaluation with four indexing approaches to estimate our approach: a topic modeling-based approach, an ontology-based approach, a LSTM sentence matching approach and a TF-IDF approach.

LDA Based Approach. The first baseline model in comparison is a probabilistic topic modeling approach from Ma et al. [16] which generates the latent topics with the most likely words in each topic. According to a user query, the first several thematic keywords with the highest probabilities of each possible semantically related topic are recommended to the user. With the user-selected topic or topics of interest, a ranked list of documents is returned.

Ontology-Based Approach. The second baseline approach utilizes ontology-based method. In this method [11], indexing of documents is a statement about the aboutness of documents, that is, an indexing term is only assigned if the corresponding concepts are covered issues within the context of the document.

LSTM Text Matching Approach. Another baseline method utilizes LSTM networks to encode and match the documents and the instances' ontological information, but it does not encode the documents into the *general topics*.

TF-IDF. The last baseline model in comparison is the standardized indexing model Term Frequency Inverse Document Frequency (TF-IDF) [20], a statistical measure used to evaluate how important a word is to a document in a collection or corpus. The importance increases proportionally to the number of times a word appears in the document but is offset by the frequency of the word in the corpus.

4.2 Experimental Results

The experiments conducted in the implementation and evaluation stages are classified as two categories of experiments. The first category aims to tune the parameters and coefficient in order to optimize the performance of the proposed

LSTM-OLSI. The set of experiments belonging to the first category are: initial-izing α, β and the number of iterations in topic model; tuning the number of topics (K) by computing perplexity; and tuning the length of the *general topic* (W) and the penalty factor q utilizing a grid search method. The second cate-gory aims to highlight the effectiveness of LSTM-OLSI. The set of experiments belonging to the second category are: comparing the performance of LSTM-OLSI to some existing approaches, evaluating the deep semantic comprehension capacity of LSTM-OLSI. The next subsections discuss the above two categories in details.

Implementation and Optimization

First, we tune the parameters and coefficient to optimize the performance of LSTM-OLSI. The probabilistic topic model is trained with 5000 iterations of Gibbs sampling using $\alpha = 50/|K|$ and $\beta = 0.01$ (the default values used, e.g., in [5,23]).

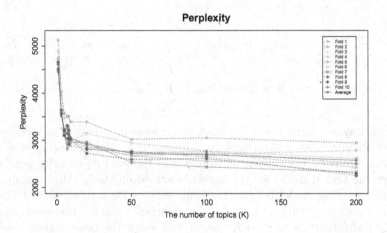

Fig. 5. Perplexity resulting curve using 10-fold cross-validation.

Afterwards, in order to determine the optimal number of topics for the news corpus (parameter K), we present a 10-fold cross-validation process to compute the curves of perplexity (Fig. 5). The ten *Folds* represent the result when the corresponding 10% held out data was treated as the test data and *Average* depicts the average value of the 10 folds. It is depicted that in most cases, the values of perplexity reach relatively low scores when K is greater than 50. In addition, the *Average* curve gets its lowest point in the case that the number of topics is 50. Therefore, the optimal number of topics of the news corpus should be 50.

Finally, we compute the length of the single *general topic* (W) and the penalty factor (q). We utilize the grid search method to tune W in the range from 100 to 300 and $q = \{1, 2, 3\}$. Figure 6 (left) depicts the average precision of the five categories during tuning parameter W and q. The curves reach their maximum values in the case that $W = 120$. Figure 6 (right) depicts when W is greater than 120, the recall curves slightly increase when W is growing. Basically, in our experiments, the setting of $W = 120$ and $q = 3$ consistently provides an optimal performance than other configurations of W. To keep the balance of the precision and recall effectively, we fix $W = 120$ for these data sets. As for the distinct data set, the performance of LSTM-OLSI is also optimal when $q = 3$, that is, LSTM-OLSI is fairly robust to the change of data sets. So we fix $W = 120$ and $q = 3$ for the news corpus.

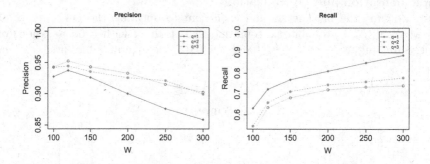

Fig. 6. Grid search of W and q.

In this way, we derive a *general topic* (a sequence of words) of each document and each instance's ontological contextual information (also a sequence of words). We utilize both the word2vec embeddings and the seq2seq encoder weights to initialize the LSTM model for the sequence embedding task. After training the sequence encoding model for roughly 500K steps with a batch size of 128 [7], we obtain the embedding vectors of the news articles and the instances. Finally, the cosine similarity is utilized to match and index the news articles and the instances.

Evaluation

To highlight the effectiveness of our indexing approach, we compare the performance of LSTM-OLSI with other approaches, namely, the LDA based method, ontology-based indexing model (OIM in Fig. 7), the LSTM text matching method and the TF-IDF method (see Fig. 7). For each approach, we computed the precision on the five categories of news corpus respectively and the average precision of them. We consider TF-IDF as the baseline against the other comparing approaches to evaluate the stability of the news corpus since TF-IDF is the standardized indexing model. Figure 7 indicates LSTM-OLSI (95.4%) has statistically significant improvements over LSTM (92.8%), LDA (91.3%), OIM (90.2%) and TF-IDF (82.3%), respectively. The proposed LSTM-OLSI outperforms all the other methods with a significant margin.

To intuitively evaluate the performance of LSTM-OLSI, we show many intuitive examples. Take a semantic instance *Heart and Crime* as an example, the

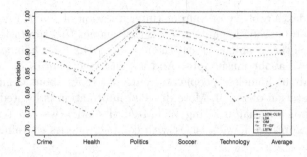

Fig. 7. Comparison with baseline models.

word *heart* is semantically relevant to news articles about heart and love, and the word *crime* is relevant to the news articles about violence and crime. It is interesting to see that the theme of the indexed news articles is an ingenious combination of *heart* related topics and *crime* related topics, which is maternal and parent-child related crime (see Table 1).

Table 1. Intuitive LSTM-OLSI results of profound instances.

Instance	Sample indexed news titles
Heart and crime	Mom who killed kids in reincarnation case gets life Knife with blood found in home where children killed Dad of Ohio girl found dead in crib gets 3 years in prison
Knife play	Pokémon Go: armed robbers use game to lure players into trap Vigil held for teen shot dead after basketball game
Politics of love	Seen and not heard: homeless people absent from election even as ranks grow For Clinton, sisterhood is powerful - and Trump helps Barkley: 'More power to Draymond for slapping the hell out of that kid'
Music technology	IBM is making a music app that can create entirely new songs just for you CloudPlayer now lets you take your playlists everywhere Apple links with NASA to make music from space
Sports nutrition	Top 10 rules you must follow every day to lose 10 pounds Make These 3 changes, burn more calories
Happy nation	Defender Riise announces retirement at 35 The incredible career of Zlatan Ibrahimović

Moreover, we exploit more intuitive examples to estimate the performance of LSTM-OLSI. Table 1 reports some sample news results indexed by several instances with profound meanings. LSTM-OLSI demonstrates deep comprehension besides literal comprehension of the instances. For example, the instance

Knife Play indexes robbery or murder (interpretation of *Knife*) news relevant to the Pokémon Go mobile game or basketball games (interpretation of *Play*). Since music has the meanings of art, the instance *Music Technology* can index technology news about music apps. And the instance *Politics of Love* indexes politic news about homeless people and current events about Clinton's sisterhood (interpretation of *Love*). More detailed information is located in Table 1. The results indicate that utilizing LSTM-OLSI model, we can retrieve more semantically expressive news articles with complex queries containing multiple topics.

5   Conclusions and Future Work

The work presented in this paper develops a novel ontology-based latent semantic indexing approach to extract both word-level contextual features (via ontology and knowledge base) and document-level contextual features (via the importance-aware topic model) from text. Then the LSTM networks make effective use of the extracted two levels of features to encode and match the documents and the instances. We summarized a method of selecting adjustable parameters to achieve higher accuracy according to the complexity and the diversity of the news documents. Moreover, the indexing results indicate that LSTM-OLSI outperforms the state-of-the-art and has deep semantic comprehension of both instances and news articles. Our future work will further extend the methods to include (1) Using the proposed semantic indexing method for other important information retrieval tasks for which semantic indexing has a key role, (2) Developing more general version of the proposed model exploiting more knowledge bases.

Acknowledgments. This research is supported by National Natural Science Foundation of China (Grant No. 61375054), Natural Science Foundation of Guangdong Province Grant No. 2014A030313745 Basic Scientific Research Program of Shenzhen City Grant No. JCYJ20160331184440545), and Cross fund of Graduate School at Shenzhen, Tsinghua University (Grant No. JC20140001).

References

1. Maas, A.L., Daly, R.E., Pham, P.T., Huang, D., Ng, A.Y., Potts, C.: Learning word vectors for sentiment analysis. In: ACL (2011)
2. Abadi, M., Agarwal, A., Barham, P., Brevdo, E., Chen, Z., Citro, C., Corrado, G.S., Davis, A., Dean, J., Devin, M., Ghemawat, S.: Tensorflow: large-scale machine learning on heterogeneous distributed systems (2016). arXiv preprint arXiv:1603.04467
3. Alec, C., Reynaud-Delaître, C., Safar, B.: An ontology-driven approach for semantic annotation of documents with specific concepts. In: ISWC 2016 (2016)
4. Blei, D.M., Ng, A.Y., Jordan, M.I.: Latent Dirichlet allocation. J. Mach. Learn. Res. **3**(January), 993–1022 (2003)

5. Borisov, A., Serdyukov, P., de Rijke, M.: Using metafeatures to increase the effectiveness of latent semantic models in web search. In: WWW, April 2016
6. Chebil, W., Soualmia, L.F., Omri, M.N., Darmoni, S.J.: Indexing biomedical documents with a possibilistic network. J. Assoc. Inf. Sci. Technol. **67**, 928–941 (2015)
7. Dai, A.M., Le, Q.V.: Semi-supervised sequence learning. In: NIPS, pp. 3079–3087 (2015)
8. Deerwester, S., Dumais, S.T., Furnas, G.W., Landauer, T.K., Harshman, R.: Indexing by latent semantic analysis. J. Am. Soc. Inf. Sci. **41**(6), 391 (1990)
9. Fernández, M., Cantador, I., López, V., Vallet, D., Castells, P., Motta, E.: Semantically enhanced information retrieval: an ontology-based approach. Web Semant.: Sci. Serv. Agents World Wide Web **9**(4), 434–452 (2011)
10. Hahm, G.J., Yi, M.Y., Lee, J.H., Suh, H.W.: A personalized query expansion approach for engineering document retrieval. Adv. Eng. Inform. **28**(4), 344–359 (2014)
11. Gödert, W.: An ontology-based model for indexing and retrieval. J. Assoc. Inf. Sci. Technol. **67**(3), 594–609 (2016)
12. Hochreiter, S., Schmidhuber, J.: Long short-term memory. Neural Comput. **9**(8), 1735–1780 (1997)
13. Hofmann, T.: Probabilistic latent semantic indexing. In: Proceedings of SIGIR, pp. 50–57. ACM, August 1999
14. Lee, J., Min, J.K., Oh, A., Chung, C.W.: Effective ranking and search techniques for web resources considering semantic relationships. IPM **50**(1), 132–155 (2014)
15. Lehmann, J., Isele, R., Jakob, M., Jentzsch, A., Kontokostas, D., Mendes, P.N., Bizer, C.: DBpedia-a large-scale, multilingual knowledge base extracted from Wikipedia. Semant. Web **6**(2), 167–195 (2015)
16. Ma, B., Zhang, N., Liu, G., Li, L., Yuan, H.: Semantic search for public opinions on urban affairs: a probabilistic topic modeling-based approach. IPM **52**(3), 430–445 (2016)
17. Mukherjee, S., Ajmera, J., Joshi, S.: Unsupervised approach for shallow domain ontology construction from corpus. In: Proceedings of WWW, pp. 349–350. ACM, April 2014
18. Newman, D., Koilada, N., Lau, J.H., Baldwin, T.: Bayesian text segmentation for index term identification and keyphrase extraction. In: COLING, pp. 2077–2092, December 2012
19. Posch, L.: Enriching ontologies with encyclopedic background knowledge for document indexing. In: Mika, P., et al. (eds.) ISWC 2014. LNCS, vol. 8797, pp. 537–544. Springer, Cham (2014). doi:10.1007/978-3-319-11915-1_36
20. Sparck Jones, K.: A statistical interpretation of term specificity and its application in retrieval. J. Doc. **28**(1), 11–21 (1972)
21. Sutskever, I., Vinyals, O., Le, Q.V.: Sequence to sequence learning with neural networks. In: NIPS, pp. 3104–3112 (2014)
22. Wang, Q., Xu, J., Li, H., Craswell, N.: Regularized latent semantic indexing. In: Proceedings of SIGIR, pp. 685–694. ACM, July 2011
23. Wei, X., Croft, W.B.: LDA-based document models for ad-hoc retrieval. In: Proceedings of SIGIR, pp. 178–185. ACM, August 2006

Privacy-Preserving Collaborative Web Services QoS Prediction via Differential Privacy

Shushu Liu, An Liu[(✉)], Zhixu Li, Guanfeng Liu, Jiajie Xu, Lei Zhao, and Kai Zheng

School of Computer Science and Technology, Soochow University, Suzhou, China
anliu@suda.edu.cn

Abstract. Collaborative Web services QoS prediction has become an important tool for the generation of accurate personalized QoS. While a number of achievements have been attained on the study of improving the accuracy of collaborative QoS prediction, little work has been done for protecting user privacy in this process. In this paper, we propose a privacy-preserving collaborative QoS prediction framework which can protect the private data of users while retaining the ability of generating accurate QoS prediction. We introduce differential privacy, a rigorous and provable privacy preserving technique, into the preprocess of QoS data prediction. We implement the proposed approach based on a general approach named Laplace mechanism and conduct extensive experiments to study its performance on a real world dataset. The experiments evaluate the privacy-accuracy trade-off on different settings and show that under some constraint, our proposed approach can achieve a better performance than baselines.

Keywords: Collaborative QoS prediction · Privacy-preserving · Differential privacy · Data distribution

1 Introduction

Quality of service (QoS) has been widely used for describing nonfunctional characteristics of web services. QoS-based Web services selection, composition and recommendation [1,9,10] have been discussed extensively in the recent literature. A common assumption of these proposed approaches is that accurate QoS values of Web Services are always available. It is, however, still an open problem to obtain accurate QoS values. On one hand, the QoS values advertised by service providers or third-party communities are not accurate to service users, as they are susceptible to the uncertain Internet environment and user context. On the other hand, it is impractical for service users to directly evaluate the QoS of all available services due to the constraints of time, cost and other resources. As an effective solution to this problem, personalized collaborative web services QoS prediction [22,24] which draw from personalized recommendation [17,18] has received much attention recently. The basic idea is that similar users tend

© Springer International Publishing AG 2017
L. Chen et al. (Eds.): APWeb-WAIM 2017, Part I, LNCS 10366, pp. 200–214, 2017.
DOI: 10.1007/978-3-319-63579-8_16

to observe similar QoS for the same service, so it is possible to predict the QoS value of the service observed by a user based on the QoS values of the service observed by the similar users to this particular user. By this kind of computation, different users are typically given different QoS prediction values even for the same service and the final prediction values in fact depends on their specific context. Based on these provided QoS values, a variety of techniques have been employed to improve the quality especially accuracy of prediction [19,21].

Though many achievements have been attainted on the study of improving the accuracy of collaborative QoS prediction, little work has been done for protecting user privacy in this process. In fact, the observed QoS values could be sensitive information, so users may not be willing to share them with others. For example, the observed response time reported by a user typically depends on her location [19], which means that the user's location could be deduced from the QoS information she provided. Consequently, an interesting but challenging question is whether or not a recommender system can make accurately personalized QoS prediction for users while protecting their privacy.

Homomorphic encryption [8] which allows computations to be carried out on ciphertext is a straightforward way to achieve privacy. However, all these operations require not only a large computation cost [7,11], but also sustained communication between parties [3]. Not even to mention the difficulty to apply some complicated computations into the encrypted domain. Hence, it is infeasible to deal with our problem by the usage of Homomorphic encryption.

Another technique, randomized perturbation which is proposed by Polat and Du [15], claimed that accurate recommendation could still be obtained while randomness from a specific distribution are added to the original data to prevent information leakage. The same idea is introduced in a recent work [28] which is also achieved by adding random in a certain range to the original data. However, the range α of randomness was chosen by experience and does not have provable privacy guarantees. What's worse, it is recognized that with the application of the clustering on the perturbed data, adversaries can accurately infer users' private data with accuracy up to 70% [23].

Though the privacy protection of randomized perturbation is insecure, it inspires us to design a lightweight and provable perturbation. Specifically, we develop our privacy preserving QoS prediction for users with the integration of a strong and provable privacy model, differential privacy, which is the state-of-art technique for privacy preserving data publishing. Differential privacy [6] has drawn much research attention literally, as it aims at providing effective means to minimize the noise added to the original data with respect to a specific privacy.

Despite the prosperity of differential privacy, applications of QoS prediction is rather limited. To the best of our knowledge, [12,13] are two differential privacy based privacy preserving recommendation systems which are the most related works to our problem. Machanavajjhala et al. [12] studied the privacy preserving of personalized social recommendation which is solely based on user's social graph. With differential privacy, sensitive links in the social graph can be preserved effectively which means that attackers cannot deduce the existence of a single link in the graph by passively observing a recommendation result. But, it is also found that good recommendations were achievable only under weak privacy parameters, or only for a small fraction of users. McSheery and Mironov [13] applied differential privacy to collaborative filtering, a general solution for recommendation systems. They split the recommendation algorithms into two parts; they are the *learning phase* which can be performed with differential privacy guarantees, and *individual recommendation phase* in which the learned results are used for individual prediction. Different from the work done by [12,13], we focus on the privacy guarantees of data publishing instead of knowledge learning and we explore additional approaches beyond those being investigated in [13], like latent factor models.

To sum up, the main contribution of this work is to formulate a differential privacy based privacy preserving collaborative Web Services QoS prediction. The task is non-trivial and our approach has the following advantage:

- For the approach we consider, privacy-preserving algorithms can be parameterized to essentially match the prediction to their non-private analogues.
- By integrating the privacy guarantees into an application, we can provide user with unfettered access to the raw data.
- Experiments on the real world dataset show that prediction accuracy on our disguised data is very close to that on users' private data.

This paper is organized as follows: Sect. 2 introduces some techniques used to building our privacy-preserving solution. Section 3 presents the system architecture of our privacy-preserving QoS prediction framework and the detail of our approach. Experimental results of proposed framework are presented in Sect. 4. Finally, Sect. 5 concludes the paper.

2   Differential Privacy

It's necessary to distinguish between differential privacy and traditional cryptosystems. Differential privacy gives a rigorous and quantitative definition on privacy leakage under a very strict attack model, and has it proved. Based on the idea of differential privacy, user can get privacy protection at utmost with ensuring the availability of data. The biggest advantage of this method is: although based on data distortion, the noise needed for perturbation is independent of data size. We can achieve high level of privacy protection by adding a very small amount of noise [20]. Despite many privacy preserving methods, like k-anonymity and l-diversity, have been proposed, differential privacy is still recognized as the most rigorous and robust privacy preserving model because of its solid mathematical foundation.

2.1   Security Definition Under Differential Privacy

There are two hypothesizes of differential privacy. On one hand, the output of any computation such as SUM, should not be affected by any operation like inserting or deleting a record. On the other hand, it gives a rigorous and quantitative definition on the privacy leakage under a very strict attack model: an attacker cannot distinguish a record with a probability more than ϵ even she has the knowledge of the entire dataset except the target one. The formal definition is as follows.

Definition 1 (ϵ-Differential Privacy [5]). A randomized function K gives ϵ-differential privacy if for all data sets D_1 and D_2 differing on at most one element, and all $S \subseteq Range(K)$,

$$\frac{Pr[K(D_1 \in S)]}{Pr[K(D_2 \in S)]} \leq exp(\epsilon) \tag{1}$$

D is a database of rows, D_1 is a subset of D_2 and the larger data set D_2 contains exactly one additional row. The probability space $Pr[.]$ in each case is over the coin flips of K. The privacy parameter $\epsilon > 0$ is public, and a smaller ϵ yields a stronger privacy guarantee.

Since differential privacy is defined under probabilistic, any method to achieve this is necessarily random. Some of these, like the Laplace mechanism [6], rely on adding controlled noise. Others, like the exponential mechanism [14] and posterior sampling [4], sample from a problem-dependent distribution instead. We will elaborate the construction in the following part.

2.2 Laplace Mechanism via Global Sensitivity

Apart from the definition of differential privacy, Dwork et al. [6] also claimed that differential privacy can be achieved by adding random noise with distribution like Laplace. A random variable has a Laplace(μ, b) distribution if its probability density function is:

$$f(x \mid \mu, b) = \frac{1}{2b} exp(-\frac{|x - \mu|}{b}) \tag{2}$$

μ and b are the location parameter and scale parameter, respectively. For the sake of simplicity, we set $\mu = 0$, so the distribution can be regarded as the symmetric exponential distribution with the standard deviation of $\sqrt{2}b$.

To add noise with Laplace distribution, b is set to $\Delta f/\epsilon$ and the generation of noise is referred as:

$$laplace(\Delta f/\epsilon) \tag{3}$$

here, Δf is global sensitivity, the definition is given next. ϵ is privacy parameter which used to leverage the privacy. As we can see from the equation, the added noise is proportional to Δf, and is inversely proportional to ϵ.

Definition 2 (Global Sensitivity [5]). For $f: D \rightarrow R^d$, the L_k-sensitivity of f is

$$\Delta f = max_{D_1, D_2} ||f(D_1) - f(D_2)||_k \tag{4}$$

for all D_1, D_2 differing in at most one element and $|| \cdot ||_k$ denotes the L_k-norm.

3   Privacy Preserving Collaborative Web Service Prediction

3.1   System Model

As we have discussed in introduction, [23] has testified that randomized perturbation is not safe as it can be inferred by the technique of clustering, but the
system model proposed by [28] is mature and suitable for many scenarios, so we
adopt and adapt this model here. Specifically, each user disguises her observed
QoS values of the services she has invoked and collected locally, and then sends
to the server, the owner of all disguised QoS values. It's safe to upload QoS
values since the server cannot derive any sensitive information about individual
with disguised data. However, the data disguising scheme should still be able to
allow the server to conduct collaborative filtering (either neighborhood-based or
model-based) from the disguised data. Based on the predicted QoS values, the
server can run a variety of applications such as QoS-based selection, composition
and recommendation (Fig. 1).

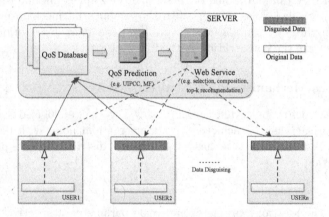

Fig. 1. Privacy preserving collaborative QoS prediction

Data disguising is the key component of privacy preserving collaborative web
service QoS prediction. The basic idea of data disguising is to perturb the raw
data with randomness in such properties: (a) randomness should be able to
guarantee no sensitive information (such as each individual user's QoS value)
can be deduced from perturbed data; (b) though information of individual is
limited, the aggregate information of these users can still be evaluated with
decent accuracy when the number of users is significantly large. Such property
is useful for computations that are based on aggregate information. For those
computation, we can still generate meaningful outcome without knowing the
exact values of individual data items because the needed aggregate information
can be estimated from the perturbed data.

Another focus of our method is the trade-off between accuracy and privacy. The more the number of randomness, the bigger the gap between disguised data and original data, which presents a higher level of privacy. Oppositely, the less the randomness number, the more obvious the data characteristics. For the computation based on context, this means a more accurate outcome. It has been an open problem to deal with the trade-off between the accuracy and privacy. In this paper, the privacy is parameterized by ϵ and given by each user. By taking advantages of differential privacy, the randomness number added in the observed QoS values is the least which preserves a decent accuracy with respect to a specific privacy.

3.2 Privacy Preserving Collaborative Web Service QoS Prediction

Collaborative filtering (CF) is a mature technique adopted by most modern recommender systems. In this section, we adopt two representative CF approaches: Neighborhood-based Collaborative Filtering and Model-based Collaborative Filtering. We will show how to integrate differential privacy into two representative CF approaches for Web services QoS prediction. More details about these two methods can be found in [16,24].

Differential Privacy Based Data Disguising. We begin with the data disguising. We use r_{ui} to denote a QoS value collected by user u for web service i, r_u for the entire vector of QoS values evaluated by user u, and similarly, I_{ui} and I_u denote the binary elements and vectors indicating the presence of QoS values respectively. $c_u = |I_u|$ is the number of QoS values evaluated by user u. In our exposition, differential privacy is the key technique used for data disguising. Laplace mechanism [6] obtains ϵ-differential privacy by adding noise of Laplace distribution.

Definition 3 (Laplace Mechanism [5]). Given a function $g: D \rightarrow R^d$, the following computation maintains ϵ-differential privacy:

$$X = g(x) + Laplace(\Delta f / \epsilon) \tag{5}$$

We distinguish between disguised data and original data with upper case and lower case, respectively. ϵ is privacy parameter used to leverage the privacy and smaller ϵ provides a stronger privacy guarantee. Δf is global sensitivity, the definition is given forehead. Here, we compute Δf with L_1-norm:

$$\Delta f = max_{D_1, D_2} ||g(D_1) - g(D_2)||_1 \tag{6}$$

For simpleness, ϵ-differential privacy of each user u is achieved by the following equation:

$$R_{ui} = r_{ui} + Laplace(\Delta f / \epsilon) \tag{7}$$

where, Δf is defined as the maximum difference between QoS values, which is:

$$\Delta f = max(r_{ui} - r_{uj}) \tag{8}$$

After disguising, all user sends disguised QoS values R_u to server, sensitive information about original data r_{ui} is preserved by randomness. However, the aggregate information of users can still be estimated. Thus, QoS prediction can be performed with direct access to R_{ui} independently.

Collaborative Web Service QoS Prediction. Next, we will show how to extend the two representative collaborative filtering approaches to perform our differential privacy based QoS prediction based on disguised data.

(1) Neighbourhood-Based Collaborative Filtering

Here, we divide all process into three parts: z-score normalization, data disguising and QoS prediction.

Step 1: to eliminate the difference between user data and facilitate better accuracy, the user needs to perform z-score normalization on the observed QoS data. Z-score normalization is performed on the QoS value with the following equation:

$$q_{ui} = (r_{ui} - \bar{r}_u)/\omega_u \tag{9}$$

where $\bar{r}_u$ is the mean and ω_u is the standard deviation of QoS vector r_u. After the normalization, QoS data have a zero mean and unit variance.

Step 2: user perform disguising on the normalized QoS value by:

$$Q_{ui} = q_{ui} + Laplace(\Delta f/\epsilon) \tag{10}$$

where, ϵ, the privacy parameter, is set by user u. Δf is defined according to the distribution of QoS value, which is: $\Delta f = max(r_{ui} - r_{uj})$. After disguising, the user sends their own disguised values Q_u to server, sensitive information about original data q_{ui} is preserved by randomness. Nevertheless, the aggregate information of users can still be estimated. Thus, QoS prediction can be performed with direct access to Q_{ui}.

During the process of QoS prediction, two types of similarity are calculated in order to improve prediction accuracy: *user similarity* and *service similarity*. In particular, the similarity between two users u and v are calculated based on the services they have commonly invoked using the following equations:

$$Sim(u, v) = \frac{\sum_{s_i \in S}(r_{u,i} - \bar{r}_u)(r_{v,i} - \bar{r}_v)}{\sqrt{\sum_{s_i \in S}(r_{u,i} - \bar{r}_u)^2 \sum_{s_i \in S}(r_{v,i} - \bar{r}_v)^2}}, \tag{11}$$

where $S = S_u \cap S_v$ is the set of services that user u and user v have commonly invoked, $r_{u,i}$ is the QoS value of service i observed by user u, $\bar{r}_u$ is the average QoS value of all services observed by user u.

However, due to the disguising of QoS values, at server side we only have the disguised QoS value Q_{ui}, rather than true value q_{ui}. Therefore, we consider to employ Q_{ui} to approximately compute the similarity value as follows.

According to the z-normalization, $\omega_u = \sqrt{\sum_{s_i \in S}(r_{u,i} - \bar{r}_u)^2/c_u}$, and by substituting this formula into computation, the similarity can be calculated as

$$Sim(u,v) = \frac{\sum_{s_i \in S}(r_{u,i} - \bar{r}_u)(r_{v,i} - \bar{r}_v)}{\omega_u \omega_v \sqrt{c_u c_v}}, \tag{12}$$

also, we observe that during the z-normalization, $q_{ui} = (r_{ui} - \bar{r}_u)/\omega_u$. Then, it is easy to get that

$$Sim(u,v) = \frac{\sum_{s_i \in S} q_{u,i} q_{v,i}}{\sqrt{c_u c_v}}, \tag{13}$$

Next, we will prove that though with data disguising, the scalar product property between two vectors remains the same. To make it clear, we denote the two vectors as $a = (a_1, a_2, \cdots, a_n)$ and $b = (b_1, b_2, \cdots, b_n)$ respectively. After disguising, the two vectors are $A = (A_1, A_2, \cdots, A_n)$ and $B = (B_1, B_2, \cdots, B_n)$. We have

$$\begin{aligned}
AB &= \sum_{i=1}^{n} A_i B_i \\
&= \sum_{i=1}^{n}(a_i + Laplace(\Delta f_a/\epsilon_a))(b_i + Laplace(\Delta f_b/\epsilon_b)) \\
&= \sum_{i=1}^{n}(a_i b_i + Laplace(\Delta f_a/\epsilon_a)Laplace(\Delta f_b/\epsilon_b)) \\
&\quad + \sum_{i=1}^{n}(a_i Laplace(\Delta f_b/\epsilon_b) + b_i Laplace(\Delta f_a/\epsilon_a))
\end{aligned}$$

because a_i and $Laplace(\Delta f_b/\epsilon_b)$ are independent vectors and $Laplace(\Delta f_b/\epsilon_b)$ is symmetric exponential distribution with $\mu = 0$, we have $\sum a_i Laplace(\Delta f_b/\epsilon_b) \approx 0$. Likewise, we have

$$\sum b_i Laplace(\Delta f_a/\epsilon_a) \approx 0 \text{ and } \sum Laplace(\Delta f_a/\epsilon_a)Laplace(\Delta f_b/\epsilon_b) \approx 0. \tag{14}$$

Hence, we derive the following equation:

$$AB \approx \sum a_i b_i = ab. \tag{15}$$

Furthermore, we can get

$$Sim(u,v) \approx \frac{\sum_{s_i \in S} Q_{u,i} Q_{v,i}}{\sqrt{c_u c_v}}. \tag{16}$$

Note that $Sim(u,v)$ is ranging from $[-1, 1]$, and a larger value indicates that two users (or services) are more similar [24].

Based on the above similarity values, the QoS value of service i observed by user u can be predicted directly. We make use of the similar users to user u through the following equation:

$$q'_{u,i} = \bar{Q}_u + \sum_{v \in User} \frac{Sim(u,v)(q_{v,i} - \bar{q}_v)}{\sum_{v \in User} Sim(u,v)}, \tag{17}$$

Like the user based QoS prediction, the item based QoS prediction can be computed the same, or as proved in [24], these two ways can be combined together to improve the accuracy of QoS prediction. Due to the limit of space, we omit the description of these approaches here.

(2) Model-Based Collaborative Filtering

Matrix factorization (MF) [25] is a typical solution of model based collaborative filtering which improves the accuracy of prediction effectively by studying latent factor models.

Suppose that the observed QoS values of n users and m services are in a sparse matrix denoted by Q_{n*m} where each element q_{ij} reflects a QoS value of user i for service j. With the input of Q_{n*m}, MF aims to factorize the user-services matrix Q_{n*m} into two latent matrices of a lower dimension d: user-factor matrix U_{n*d} and service-factor matrix V_{m*d}. Then, vacant elements in Q_{n*m} can be approximated as the product of U and V, i.e., unknown QoS value q'_{ij} is evaluated by $q'_{ij} = U_i \cdot V_j^T$.

MF is often transformed into an optimization problem, and the local optimal solution is obtained by iteration. The objective function (or loss function) of MF is defined as:

$$min_{U,V} \sum_{q_{ij} \in Q} [(q_{ij} - U_i V_j^T)^2 + \lambda(||U_i||^2 + ||V_j||^2)] \tag{18}$$

The first part is the squared difference between the existing QoS matrix and the predicted one, but only for elements that have evaluated by users. The latter part is the regularization term, added to deal with overfitting induced by the sparsity of input. By dealing with this optimization, we get user-factor matrix U_{n*d} and service-factor matrix V_{m*d} eventually.

Alternative least squares (ALS) and stochastic gradient descent (SGD) are two commonly used methods for solving this optimization problem. We take SGD as our solution since ALS is more difficult which requires the computation of inverse matrix. Iterative equations of SGD are as follows:

$$U_i \leftarrow U_i + \gamma((q_{ij} - U_i V_j^T)V_j - \lambda' U_i) \tag{19}$$

$$V_j \leftarrow V_j + \gamma((q_{ij} - U_i V_j^T)U_i - \lambda' V_j) \tag{20}$$

γ is the learning rate, λ' is the regularization coefficient. The selection of both parameters affects the result significantly. When the value of γ is big, it results in divergence and the results cannot get into convergence. To get a convergence, we set γ to a small value, 0.001 by experience, though it requires a longer training time. And λ' is set to 0.01 which is also selected by experience.

In the first iteration, U and V are set randomly. But a better selection can help to accelerate the computation effectively. Hence, we initialise U and V near the average of all QoS value that have been observed. The iteration will terminate when the objective function value is less than a certain threshold.

4   Experiments

In this section, we conduct three series of experiments on a real data set to evaluate our privacy-preserving QoS prediction framework.

4.1   Experimental Setup

We first note that a real Web services QoS dataset is introduced in [26,27], which includes QoS values of 5,825 real-world Web services observed by 339 users. This dataset is quite useful when studying the accuracy of QoS prediction. According to the dataset, we focus on two representative QoS attributes: response time (RT) and throughout (TP). Table 1 describes the statistics of the dataset, AVE and STD is the average and standard deviation of data respectively, density means the ratio of observed data to all data. More details of the dataset can be found in [26,27]. During the presentation of our experiment, we set the performance of RT in the left and TP in the right.

We use cross validation to train and evaluate the QoS prediction. The dataset here is collected motivated and complete, but in practice, for limited time and resource, a user usually invokes only a handful of services, and the density of data is under 10% generally. To simulate such sparsity in our experiment, we randomly remove entries from the full dataset and only keep a small density of historical QoS values as our training set. And the removed data is treated as testing set for accuracy evaluation.

Then, we perform algorithms of QoS prediction on training set and evaluate the prediction on testing set. Four algorithms are implemented and evaluated here. UIPCC which is proposed in [24] is a representative implementation of neighbourhood-based collaborative filtering and MF introduced in [25] is an implementation of model-based collaborative filtering. LUIPCC and LMF are two methods intergrading with differential privacy which is achieved by Laplace mechanism.

To quantize the accuracy of QoS prediction, we employ Root Mean Square Error (RMSE) as the metric which has been widely used in related work (e.g., [2,13]):

$$RMSE = \frac{\sqrt{\sum_R (q_{ui} - q'_{ui})^2}}{|R|} \tag{21}$$

R consists of all value needed to be predicted in training set and $|R|$ is the number of set R. q'_{ui} is the predicted value of set R and q_{ui} is the corresponding value in testing set. Generally, a smaller RMSE indicates a better prediction.

Table 1. Statistic of datasets

QoS	# User	# Service	AVE	STD	Density
RT (s)	339	5825	0.90	1.973	94.8%
TP (kpbs)	339	5825	47.56	110.797	92.7%

Noted that, the default setting of parameters follows Table 2. We choose parameters of UIPCC and MF by experiences of [24,25]. Generally, ϵ is set to 0.5 by default which can preserve sufficient privacy.

Table 2. Parameter setting

UIPCC	$k = 20$	$\lambda = 0.1$	-
MF	$d = 20$	$\gamma = 0.001$	$\lambda' = 0.01$
Laplace	$\epsilon = 0.5$	-	-

4.2 Privacy vs Accuracy

Figure 2 is the comparison corresponding to RT and TP between our differential privacy based QoS prediction and original approaches under varying privacy. By introducing differential privacy into QoS prediction, users can achieve privacy. But for users who adopt our approaches, they do need to consider the trade-off between privacy and accuracy. On one hand, a user can attain high level privacy by adding more Laplace noises which definitely decreases the utility of data. On the other extremal hand, a user can get a 100% accuracy without adding any Laplace noises. To study the performance of changing accuracy, we perform algorithms of QoS prediction on testing set and evaluate the prediction on testing set. The privacy parameter, ϵ, changes from 0.5 to 4 stepped by 0.5. We can observe that both LUIPCC and LMF decrease in RMSE when ϵ gets larger. A larger ϵ means a looser privacy constraints and the utility of data is less limited, thus user can get a better accuracy. It is also noticeable that when ϵ gets larger, e.g., larger than 2.0 in Fig. 2, our privacy-preserving approaches, both LUIPCC and LMF, can acquire almost the same or even more accuracy than UIPCC. Especially, when ϵ is as large as 4, the prediction accuracy of LMF is much better than the baseline UIPCC. Additionally, we find that MF outperforms UIPCC. This suggests the superior effectiveness of model-based approaches in capturing the latent structure of the QoS data, which conforms to the results reported in [25].

Another fact which requires our attention is that though a recent work [28] claims a better performance than both original algorithms, UIPCC and MF, the randomness added to prevent the information leakage is not large enough, adversaries can accurately infer users' privacy data with the application of the clustering [23].

To sum up, our differential privacy based algorithms can provide a privacy preserving QoS prediction with a parameterized privacy. And the results show that recommendation on our disguised user data is very close to that on user's private data under a loose constraint.

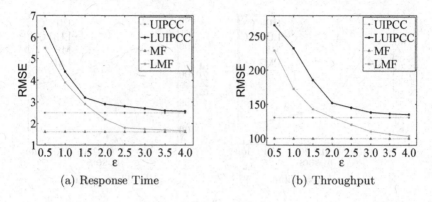

(a) Response Time   (b) Throughput

Fig. 2. Privacy vs accuracy

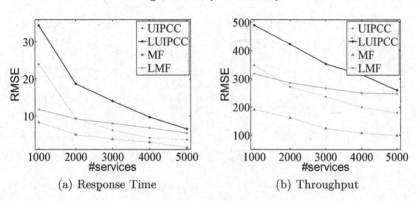

(a) Response Time   (b) Throughput

Fig. 3. Influence of services

4.3 Influence of Data Size

To evaluate the influence of data size, we design the experiments by changing the number of services and users respectively. In Fig. 3, the number of users is set to 339 and the number of service is varying from 1000 to 5000 with a step 1000, where the service is selected randomly from the original dataset. And the other parameters of the experiment are set as Table 2. We do the same experiment setting in Fig. 4 which contains 5825 services.

It is straightforward that both the number of services and the number of users have a positive influence on the accuracy of algorithms which means that the more the data is given, the better the prediction can be. In other words, with more data, we can provide a better accuracy.

Another finding is that though the accuracy differs significantly between different data size, the trend of original algorithms and our differential privacy based algorithms are the same, such as the trend of UIPCC and LUIPCC or the trend of MF and LMF. It infers an dramatically advantage of differential privacy that the noise needed for data disguising is independent of data size, so users can achieve a high level of privacy protection by adding a very small amount of noise.

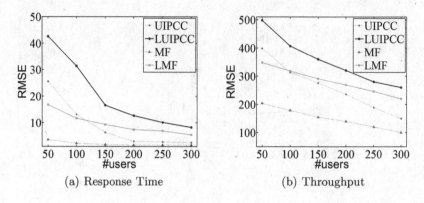

(a) Response Time

(b) Throughput

Fig. 4. Influence of users

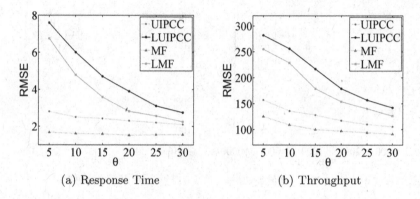

(a) Response Time

(b) Throughput

Fig. 5. Influence of density

4.4 Influence of Density

In addition to data size, density denoted as θ is also a subject to the performance of algorithms. Figure 5 presents the results of the accuracy comparison under different density. Though the influence of density on original algorithms is not obvious, it does have an significant influence of our differential based algorithms. The dataset with a higher density performs better. This result implies that the density is also a crucial factor for determining the performance of our differential privacy based approaches. More importantly, we can observe that when the number of services gets larger, the gap between traditional approaches and our differential privacy based approaches gets smaller. More precisely, in Fig. 5, when the density is set to 5, the gap between LUIPCC and UIPCC is 5. However, when the density increased to 30, the gap between LUIPCC and UIPCC decreases to 1. So, users are suggested to use the dataset with a higher density to preserves a closer prediction to original results.

5   Conclusion

To the best of our knowledge, this is the first piece of work that introduces differential privacy into a collaborative web services QoS prediction framework. Differential privacy gives a rigorous and quantitative definition on privacy leakage under very strict constraints. Based on the idea of differential privacy, users can get privacy protection at utmost with ensuring the availability of data. Empirical results show that our framework provides a secure and accurate collaborative Web services QoS prediction.

Acknowledgment. Research reported in this publication was partially supported Natural Science Foundation of China (Grant Nos. 61572336, 61572335, 61402313) and Natural Science Foundation of Jiangsu Province of China under Grant No. BK20151223.

References

1. An, L., Liu, H., Li, Q., Huang, L., Xiao, M.: Constraints-aware scheduling for transactional services composition. J. Comput. Sci. Technol. **24**(4), 638–651 (2009)
2. Berlioz, A., Friedman, A., Kaafar, M.A., Boreli, R., Berkovsky, S.: Applying differential privacy to matrix factorization. In: The ACM Conference, pp. 107–114 (2015)
3. Canny, J.: Collaborative filtering with privacy via factor analysis. In: International ACM SIGIR Conference on Research and Development in Information Retrieval, pp. 238–245 (2002)
4. Dimitrakakis, C., Nelson, B., Mitrokotsa, A., Rubinstein, B.I.P.: Robust and private Bayesian inference. Arxiv **8776**, 291–305 (2014)
5. Dwork, C.: Differential privacy. In: Bugliesi, M., Preneel, B., Sassone, V., Wegener, I. (eds.) ICALP 2006. LNCS, vol. 4052, pp. 1–12. Springer, Heidelberg (2006). doi:10.1007/11787006_1
6. Dwork, C., Mcsherry, F., Nissim, K.: Calibrating noise to sensitivity in private data analysis. In: VLDB Endowment (2014)
7. Erkin, Z., Veugen, T., Toft, T., Lagendijk, R.L.: Generating private recommendations efficiently using homomorphic encryption and data packing. IEEE Trans. Inf. Forensics Secur. **7**(3), 1053–1066 (2012)
8. Gentry, C.: A fully homomorphic encryption scheme (2009)
9. Liu, A., Li, Q., Huang, L., Xiao, M.: FACTS: a framework for fault-tolerant composition of transactional web services. IEEE Trans. Serv. Comput. **3**(1), 46–59 (2010)
10. Liu, A., Li, Q., Huang, L., Ying, S., Xiao, M.: Coalitional game for community-based autonomous web services cooperation. IEEE Trans. Serv. Comput. **6**(3), 387–399 (2013)
11. Liu, A., Zhengy, K., Liz, L., Liu, G., Zhao, L., Zhou, X.: Efficient secure similarity computation on encrypted trajectory data. In: IEEE International Conference on Data Engineering, pp. 66–77 (2015)
12. Machanavajjhala, A., Korolova, A., Sarma, A.D.: Personalized social recommendations: accurate or private. In: VLDB Endowment (2011)

13. Mcsherry, F., Mironov, I.: Differentially private recommender systems: building privacy into the net. In: ACM SIGKDD International Conference on Knowledge Discovery and Data Mining, pp. 627–636 (2009)
14. Mcsherry, F., Talwar, K.: Mechanism design via differential privacy. In: IEEE Symposium on Foundations of Computer Science, FOCS 2007, pp. 94–103 (2007)
15. Polat, H., Du, W.: Privacy-preserving collaborative filtering using randomized perturbation techniques. In: Proceedings of 3rd IEEE International Conference on Data Mining (ICDM 2003), 19–22 December 2003, Melbourne, Florida, USA, pp. 625–628 (2003)
16. Salakhutdinov, R., Mnih, A.: Probabilistic matrix factorization. In: International Conference on Neural Information Processing Systems, pp. 1257–1264 (2007)
17. Shang, S., Chen, L., Wei, Z., Jensen, C., Wen, J.R., Kalnis, P.: Collective travel planning in spatial networks. IEEE Trans. Knowl. Data Eng. **28**(5), 1132–1146 (2016)
18. Shang, S., Ding, R., Zheng, K., Jensen, C.S., Kalnis, P., Zhou, X.: Personalized trajectory matching in spatial networks. VLDB J. **23**(3), 449–468 (2014)
19. Tang, M., Jiang, Y., Liu, J., Liu, X.: Location-aware collaborative filtering for QoS-based service recommendation. In: IEEE International Conference on Web Services, pp. 202–209 (2012)
20. Yanga, L.I., Wen, W., Xie, G.Q.: Survey of research on differential privacy. Appl. Res. Comput. **29**(9), 3201–3582 (2012)
21. Yu, Q., Zheng, Z., Wang, H.: Trace norm regularized matrix factorization for service recommendation. In: IEEE International Conference on Web Services, pp. 34–41 (2013)
22. Zhang, Q., Ding, C., Chi, C.H.: Collaborative filtering based service ranking using invocation histories. In: IEEE International Conference on Web Services, pp. 195–202 (2011)
23. Zhang, S., Ford, J., Makedon, F.: Deriving private information from randomly perturbed ratings. In: SIAM International Conference on Data Mining, 20–22 April 2006, Bethesda, MD, USA (2006)
24. Zheng, Z., Ma, H., Lyu, M.R., King, I.: WSRec: a collaborative filtering based web service recommender system. In: IEEE International Conference on Web Services, pp. 437–444 (2009)
25. Zheng, Z., Ma, H., Lyu, M.R., King, I.: QoS-aware web service recommendation by collaborative filtering. IEEE Trans. Serv. Comput. **4**(2), 140–152 (2010)
26. Zheng, Z., Zhang, Y., Lyu, M.R.: Distributed QoS evaluation for real-world web services. In: IEEE International Conference on Web Services, pp. 83–90 (2010)
27. Zheng, Z., Zhang, Y., Lyu, M.R.: Investigating QoS of real-world web services. IEEE Trans. Serv. Comput. **7**(1), 32–39 (2014)
28. Zhu, J., He, P., Zheng, Z., Lyu, M.R.: A privacy-preserving QoS prediction framework for web service recommendation. In: IEEE International Conference on Web Services, pp. 241–248 (2015)

High-Utility Sequential Pattern Mining with Multiple Minimum Utility Thresholds

Jerry Chun-Wei Lin[1]([✉]), Jiexiong Zhang[1], and Philippe Fournier-Viger[2]

[1] School of Computer Science and Technology,
Harbin Institute of Technology Shenzhen Graduate School, Shenzhen, China
jerrylin@ieee.org, jiexiong.zhang@foxmail.com
[2] School of Natural Sciences and Humanities,
Harbin Institute of Technology Shenzhen Graduate School, Shenzhen, China
philfv@hitsz.edu.cn

Abstract. High-utility sequential pattern mining is an emerging topic in recent decades and most algorithms were designed to identify the complete set of high-utility sequential patterns under the single minimum utility threshold. In this paper, we first propose a novel framework called high-utility sequential pattern mining with multiple minimum utility thresholds to mine high-utility sequential patterns. A high-utility sequential pattern with multiple minimum utility thresholds algorithm, a lexicographic sequence (LS)-tree, and the utility-linked (UL)-list structure are respectively designed to efficiently mine the high-utility sequential patterns (HUSPs). Three pruning strategies are then introduced to lower the upper-bound values of the candidate sequences, and reduce the search space by early pruning the unpromising candidates. Substantial experiments on real-life datasets show that our proposed algorithms can effectively and efficiently mine the complete set of HUSPs with multiple minimum utility thresholds.

Keywords: Data mining · Sequence · High-utility sequential pattern · Multiple thresholds

1 Introduction

Sequential pattern mining (SPM) [1–4] has been emerging as an interesting and critical topic in recent years. The main target of SPM is to discover the set of frequent sequences measured by a user-specified minimum support threshold. This process may, however, not be informative for decision makers since they cannot discover the patterns with high profit or having great impact. To address this issue, high-utility itemset mining (HUIM) has been introduced [5,6] to consider both the quantity and the unit of profit of itemsets to mine the high-utility itemsets (HUIs). Considering the ordered sequences in real-life situations, high-utility sequential pattern mining (HUSPM) [7–10] was introduced for mining more informative sequential patterns. However, HUSPM meets an important

L. Chen et al. (Eds.): APWeb-WAIM 2017, Part I, LNCS 10366, pp. 215–229, 2017.
DOI: 10.1007/978-3-319-63579-8_17

limitation since it necessitates to measure all patterns with a single minimum utility threshold for finding the complete set of high-utility sequential patterns (HUSPs). Utilizing a single threshold for all itemsets or sequences in databases means that they are treated as the same importance, which is not convincing in real-world situations.

In this paper, we first design a new framework called high-utility sequential pattern mining with multiple minimum utility thresholds for mining the set of HUSPs. It allows to set different thresholds for items instead of a single minimum utility threshold and avoid the "rare item" problem [11]. Besides, a lexicographic-sequence (LS)-tree is introduced as the search space to mine the complete set of HUSPs. A novel compressed utility-linked (UL)-list structure is further designed to store the information of patterns. Three pruning strategies are then developed to reduce the search space and improve mining performance of the designed algorithm, which can be observed in the experiments.

2   Preliminaries and Problem Statement

Let $I = \{i_1, i_2, \ldots, i_m\}$ be a finite set of distinct items. A quantitative itemset, denoted as $v = [(i_1, q_1)(i_2, q_2) \ldots (i_c, q_c)]$, is a subset of I and each item in the quantitative itemset is associated with a quantity (internal utility). An itemset, denoted as $w = [i_1, i_2, \ldots, i_c]$, is a subset of I without quantities. A quantitative sequence is an ordered list of one or more quantitative itemsets, which is denoted as $s = <v_1, v_2, \ldots, v_d>$. A sequence is an ordered list of one or more itemsets without quantities, which is denoted as $t = <w_1, w_2, \ldots, w_d>$. For convenience, we use "q-" as the abbreviation of "quantitative". Thus, the "q-sequence" indicates the sequences with quantities, and "sequence" indicates the sequences without quantities, which can be also defined for the "q-itemset". For example, $<[(a, 2)\ (b, 1)], [(c, 3)]>$ is a q-sequence while $<[ab], [c]>$ is a sequence. $[(a, 2)\ (b, 1)]$ is a q-itemset and $[ab]$ is an itemset. A quantitative sequential database is a set of transactions $D = \{S_1, S_2, \ldots, S_n\}$, where each transaction $S_q \in D$ is a q-sequence, and has a unique identifier q, named its *SID*. In addition, each item in D is associated with a profit (external utility), and denoted as $pr(i_j)$.

Table 1. A quantitative sequential database

SID	Q-sequence
S_1	$<[(a{:}2)(c{:}3)], [(a{:}3)(b{:}1)(c{:}2)], [(a{:}4)(b{:}5)(d{:}4)], [(e{:}3)]>$
S_2	$<[(a{:}1)(e{:}3)], [(a{:}5)(b{:}3)(d{:}2)], [(b{:}2)(c{:}1)(d{:}4)(e{:}3)]>$
S_3	$<[(e{:}2)], [(c{:}2)(d{:}3)], [(a{:}3)(e{:}3)], [(b{:}4)(d{:}5)]>$
S_4	$<[(b{:}2)(c{:}3)], [(a{:}5)(e{:}1)], [(b{:}4)(d{:}3)(e{:}5)]>$
S_5	$<[(a{:}4)(c{:}3)], [(a{:}2)(b{:}5)(c{:}2)(d{:}4)(e{:}3)]>$
S_6	$<[(f{:}4)], [(a{:}5)(b{:}3)], [(a{:}3)(d{:}4)]>$

A running example of the quantitative sequential database is shown in Table 1, which consists of 6 transactions and 6 items. The external utility of each item is defined in *profit-table* as: $\{pr(a):5, pr(b):3, pr(c):4, pr(d):2, pr(e):1, pr(f):6\}$. From the given example, it can be seen that $[(a;2)(c:3)]$ is the first *q*-itemset in a transaction S_1. The quantity of an item (a) in this *q*-itemset is 2, and its utility is calculated as $(2 \times 5)(= 10)$.

Definition 1. *The minimum utility threshold of an item (i_j) in a quantitative sequential database D is denoted as $mu(i_j)$, and the multiple minimum utility threshold table (denoted as MMU-table) consists of the minimum utility threshold of each item in D, which can be defined as:*

$$MMU\text{-}table = \{mu(i_1), mu(i_2), \ldots, mu(i_m)\}. \tag{1}$$

In Table 1, we assume that the *MMU-table* is defined as *MMU-table* $= \{mu(a), mu(b), mu(c), mu(d), mu(e), mu(f)\} = \{500, 500, 500, 200, 500, 70\}$.

Definition 2. *The minimum utility threshold of a sequence t is denoted as $MIU(t)$, which is the least mu value among the items in t and defined as:*

$$MIU(t) = min\{mu(i_j) \mid i_j \in t\}. \tag{2}$$

In Table 1, $MIU(<[a]>) = min\{mu(a)\} = 500$, and $MIU(<[ad]>) = min \{mu(a), mu(d)\} = min\{500, 200\} = 200$.

Definition 3. *The utility of an item (i_j) in a q-itemset v is denoted as $u(i_j, v)$, and defined as:*

$$u(i_j, v) = q(i_j, v) \times pr(i_j), \tag{3}$$

where $q(i_j, v)$ is the quantity of (i_j) in v, and $pr(i_j)$ is the profit of (i_j).

In Table 1, the utility of an item (c) in the first *q*-itemset of S_1 is calculated as: $u(c, [(a:2)(c:3)]) = q(c, [(a:2)(c:3)]) \times pr(c) (= 3 \times 4)(= 12)$.

Definition 4. *The utility of a q-itemset v is denoted as $u(v)$ and defined as:*

$$u(v) = \sum_{i_j \in v} u(i_j, v). \tag{4}$$

In Table 1, $u([(a:2)(c:3)]) = u(a, [(a:2)(c:3)]) + u(c, [(a:2)(c:3)]) (= 2 \times 5 + 3 \times 4)(= 22)$.

Definition 5. *The utility of a q-sequence $s = <v_1, v_2, \ldots, v_d>$ is denoted as $u(s)$ and defined as:*

$$u(s) = \sum_{v \in s} u(v). \tag{5}$$

In Table 1, $u(S_1) = u([(a:2)(c:3)]) + u([(a:3)(b:1)(c:2)]) + u([(a:4)(b:5)(d:4)]) + u([(e:3)])(= 22 + 26 + 43 + 3)(= 94)$.

Definition 6. *The utility of a quantitative sequential database D is denoted as* $u(D)$ *and defined as:*

$$u(D) = \sum_{s \in D} u(s). \tag{6}$$

In Table 1, $u(D) = u(S_1) + u(S_2) + u(S_3) + u(S_4) + u(S_5) + u(S_6)(= 94 + 67 + 56 + 67 + 76 + 81)(= 441)$.

Definition 7. *Given a q-sequence* $s = <v_1, v_2, \ldots, v_d>$ *and a sequence* $t = < w_1, w_2, \ldots, w_{d'}>$, *iff* $d = d'$ *and the items in* v_k *are the same as the items in* w_k *for* $1 \leq k \leq d$, *t matches s, which is denoted as* $t \sim s$.

In Table 1, $<[ac], [abc], [abd], [e]>$ matches S_1. However, the target sequence may have multiple matches in a q-sequence. For example, $<[a],[b]>$ has three matches as $<[a:2],[b:1]>$, $<[a:2],[b:5]>$ and $<[a:3],[b:5]>$ in S_1. This feature brings more challenges to the designed framework.

Definition 8. *Given two itemsets w and w', w is said to be contained in w' as* $w \subseteq w'$ *iff w is a subset of w'. Given two q-itemsets v and v', v is said to be contained in v' as* $v \subseteq v'$ *iff for any item in v, there exists the same item having the same quantity in v'.*

In Table 1, an itemset $[ac]$ is contained in the itemset $[abc]$. The q-itemset $[(a:2)(c:3)]$ is contained in $[(a:2)(b:1)(c:3)]$, but is not contained in $[(a:2)(b:3)(c:1)]$.

Definition 9. *Given two sequences* $t = <w_1, \ldots, w_d>$ *and* $t' = <w'_1, \ldots, w'_{d'}>$, *t is said to be contained in t' as* $t \subseteq t'$ *iff there exists an integer sequence* $1 \leq k_1 \leq k_2 \leq \cdots \leq d'$ *such that* $w_j \subseteq w'_{k_j}$ *for* $1 \leq j \leq d$. *Given two q-sequences* $s = <v_1, \ldots, v_d>$ *and* $s' = <v'_1, \ldots, v'_{d'}>$, *s is said to be contained in s' as* $s \subseteq s'$ *iff there exists an integer sequence* $1 \leq k_1 \leq k_2 \leq \cdots \leq d'$ *such that* $v_j \subseteq v'_{k_j}$ *for* $1 \leq j \leq d$. *For convenience, we use* $t \subseteq s$ *to indicate that* $t \sim s_k \wedge s_k \subseteq s$.

In Table 1, $<[(a:2)],[(e:3)]>$ and $<[(a:4)],[(e:3)]>$ are contained in S_1, but $<[(a:1)],[e:3]>$ and $<[(a:4)],[(e:4)]>$ are not contained in S_1.

Definition 10. *The utility of a sequence t in a q-sequence s is denoted as* $u(t, s)$ *and defined as:*

$$u(t, s) = max\{u(s_k) \mid t \sim s_k \wedge s_k \subseteq s\}. \tag{7}$$

In Table 1, $u(<[a],[b]>, S_1) = max\{u(<[a:2],[b:1]>), u(<[a:2],[b:5]>), u(<[a:3],[b:5]>)\} = max\{13, 25, 30\} = 30$. From this example, it shows that a sequence has multiple utility values in a q-sequence, which is much different from traditional SPM and HUIM.

Definition 11. *The utility of a sequence t in a quantitative sequential database D is denoted as* $u(t)$ *and defined as:*

$$u(t) = \sum_{s \in D} \{u(t, s) \mid t \subseteq s\}. \tag{8}$$

In Table 1, $u(<[a],[b]>) = u(<[a],[b]>, S_1) + u(<[a],[b]>, S_2) + u(<[a],[b]>, S_3) + u(<[a],[b]>, S_4) + u(<[a],[b]>, S_5)$ $(= 30 + 31 + 27 + 37 + 35)(= 160)$.

Definition 12 (High-Utility Sequential Pattern, HUSP). *A sequence t in a quantitative sequential database D is defined as a high-utility sequential pattern (HUSP) iff its utility is no less than the minimum threshold of t as:*

$$HUSP \leftarrow \{t \mid u(t) \geq MIU(t)\}. \tag{9}$$

In Table 1, $u(<[a],[b]>)(= 160)$ and $MIU(<[a],[b]>)(= 500)$; $<[a],[b]>$ is not a HUSP since $u(<[a],[b]>)(= 160) < MIU(<[a],[b]>)$ $(= 500)$.

Problem Statement: Given a quantitative sequential database and a *MMU-table*, the problem of high-utility sequential pattern mining with multiple minimum utility thresholds is to discover the complete set of HUSPs whose utility values are no less than their *MIU* values.

3   Proposed Framework

Based on the above concepts, a baseline high-utility sequential pattern algorithm with multiple minimum utility thresholds is first designed. The proposed algorithm first scans the database to find the 1-sequences for building the lexicographic sequence (LS)-tree. For each node in the LS-tree, a corresponding projected database is built and consisting of the utility-linked (UL)-lists transformed from the transactions in the original database, which is used to calculate the actual utilities and upper-bound values of the generated candidates. Each UL-list can be used to represent each transaction (*q*-sequence). To generate the child nodes (supersets) of the node in the LS-tree, the *I-Concatenation* and *S-Concatenation* operations are used to combine the candidate HUSPs with items, forming new candidate HUSPs. Each new candidate HUSP (child node) will be evaluated to determine whether it is an actual HUSP and whether the algorithm should explore its supersets (child nodes). The above processes are recursive performed until no candidates are required to be determined. After that, the set of HUSPs are then returned.

3.1   Lexicographic Sequence (LS)-tree

In the designed algorithm, a lexicographic sequence (LS)-tree is built to ensure the **completeness** and **correctness** for mining the HUSPs. The database is first scanned to find the satisfied 1-sequences against their *PMIU* value (which will be described below). The LS-tree is then built from 1-sequences by adopting a depth-first search strategy. The child nodes the are combination results of *I-Concatenation* or *S-Concatenation* of the parent node. Figure 1 shows a LS-tree built from the running example of Table 1. In Fig. 1, a circle represents an *I-Concatenation* sequence while the square represents a *S-Concatenation* sequence. Notice that nodes in the LS-tree represent the candidates (search space) of the HUSPs.

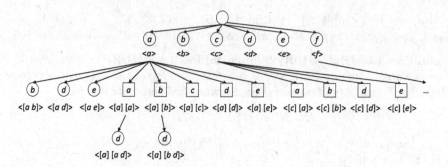

Fig. 1. A lexicographic sequence (LS)-tree

3.2 Utility-Linked (UL)-List Structure

In the designed algorithm, the utility and upper-bound values of candidates are calculated from each transaction. This process has the problem of multiple matches, and requires more computations. To handle this situation, we introduce a novel compact utility-linked (UL)-list structure to store the utility information of each transaction. This structure efficiently helps generate the utility of the *I-Concatenation* and *S-Concatenation* sequences for later mining process. The UL-list structure of S_1 is shown in Table 2.

Table 2. The utility-linked (UL)-list structure of S_1

U&P information	$<[(a, 10, 84, 3)(c, 12, 72, 5)], [(a, 15, 57, 6)(b, 3, 54, 7) (c, 8, 46, -)], [(a, 20, 26, -)(b, 15, 11, -)(d, 8, 3, -)], [e, 3, 0, -]>$
Header_table	$(a, 1) (b, 4) (c, 2) (d, 8) (e, 9)$

For the header_table in the UL-list structure, it represents the set of distinct items with their first occurrence positions in the transformed transaction. In Table 2, the distinct items of S_1 are (a), (b), (c), (d), and (e) and their first occurrence positions in S_1 are respectively 1, 4, 2, 8 and 9. For the U&P (utility and position) information, each element respectively represents the **(1) item name**, the **(2) utility of the item**, the **(3) remaining utility of the item**, and the **(4) next position of the item**. In Table 2, the utility of the item (a) in the first element is calculated as 10 in S_1; the total utility except the item (a) in S_1 (named as remaining utility) is calculated as 84, and the next position of the item (a) in S_1 is found as 3. The utility and the remaining utility of item can be used to calculate the utility values and the upper-bound values of patterns respectively. The next position of item will be used for concatenation and selecting the maximal utility values (based on Definition 10) and the maximal upper-bound values of patterns. For each node in the LS-tree, transactions containing this node (sequence) are transformed into a utility-linked (UL)-list and attached to the projected database of this node. The utilities and upper-bound values of the candidates can be easily calculated from the projected database based on the UL-list structure.

3.3 Concatenations

In the designed algorithm, two operations such as *I-Concatenation* and *S-Concatenation* are used to generate the child nodes (supersets) of the processed nodes.

Definition 13. Given a sequence t and an item i_j, the *I-Concatenation* of t with i_j is to append i_j to the last itemset of t, denoted as $<t \oplus i_j>_{I-Concatenation}$; the *S-Concatenation* of t with i_j is to add i_j as a new itemset to the last of t, denoted as $<t \oplus i_j>_{S-Concatenation}$.

For example, given a sequence $t = <[a], [b]>$ and a new item (c), we can obtain that $<t \oplus c>_{I-Concatenation} = <[a], [bc]>$ and $<t \oplus c>_{S-Concatenation} = <[a], [b], [c]>$. Based on those two operations, the candidates can be thus easily formed in the search space for mining the HUSPs.

As mentioned before, the UL-list structure can be used to find the utilities and the upper-bound values of the candidates for deriving HUSPs. A sequence may, however, find the multiple matches in a q-sequence, which makes a sequence have multiple utilities in a q-sequence. Thus, it necessitates to find the positions of the matches to calculate the utilities and the upper-bound values of the processed node (sequence). For convenience, the position of the last item within each match is defined as the **concatenation point**, and the first concatenation point is named as **start point**. For example in Table 1, assume a sequence $t = <[a], [b]>$, it has three matches in S_1 as $<[a:2], [b:1]>$, $<[a:2], [b:5]>$ and $<[a:3], [b:5]>$. The concatenation points of t in S_1 respectively are 4, 7 and 7, and the start point is 4. The candidate items for *I-Concatenation* are the items appearing in the same itemsets where concatenation points appear. In the above example, the candidate items for *I-Concatenation* are $\{(c:2), (d:4)\}$. The items in the itemsets after the start point are the candidate items for *S-Concatenation*. In the above example, the start point $(= 4)$ appears in the second itemset. Hence, the items after the second itemset are candidate items for *S-Concatenation*, which are $\{(a:4), (b:5), (d:4), (e:3)\}$. Besides, to discover the complete set of HUSPs, the designed algorithm enumerates all candidates by two concatenations in *lexicographic-ascending* order.

3.4 Proposed Algorithm

Since the downward closure property does not hold in the problem of high-utility sequential pattern mining with multiple minimum utility thresholds, new downward closure property is required to reduce the search space for mining the HUSPs. Details are described below.

Definition 14. *Given two q-sequences s and s', if $s \subseteq s'$, the extension of s in s' is the rest of s' after s, which can be denoted as $<s'-s>_{rest}$. Given a sequence t and a q-sequence s, if $t \subseteq s$, the extension of t in s is the rest of s after s_k, which can be denoted as $<s - t>_{rest}$, where s_k is the first match of t in s.*

For example, given two q-sequences $s = <[a:2],[b:5]>$ and S_1 in Table 1, the extension of s in S_1 is $<S_1 - s>_{rest} = <[(d:4)],[(e:3)]>$. Given a sequence $t = <[a],[b]>$, there exists three matches of t in S_1, and the first one is $<[a:2],[b:1]>$. Thus, $<S_1 - t>_{rest} = <[(c:2)],[(a:4)(b:5)(d:4)],[(e:3)]>$.

Definition 15. *The extension items of a sequence t in a quantitative sequential database D is denoted as $I(t)_{rest}$ and defined as:*

$$I(t)_{rest} = \{i_j \mid i_j \in <s - t>_{rest} \wedge t \subseteq s \wedge s \in D\}. \tag{10}$$

In the above example, $I(<[a],[b]>)_{rest} = \{a,b,c,d,e\}$.

To speed up mining process and maintain the downward closure property, a sequence-weighted utilization (SWU) [12] was thus proposed to maintain the sequence-weighted downward closure (SWDC) property for mining the HUSPs. Based on the SWDC property, it can ensure that if the SWU of a sequence t is less than a threshold, the utility of t is less than the threshold; the utilities of the supersets of t are also less than the threshold. Numerous unpromising candidates can be pruned. The SWU of a sequence t is still much larger than the actual utility of t. Thus, the potential utility (PU) model [12] was also introduced to estimate the lower upper-bound values of the candidates. However, in the designed framework, the thresholds for sequences are different; the SWDC and PU properties cannot be directly applied. To address this issue, we propose the following theorems to maintain the downward closure property in the proposed framework.

Definition 16. *The potential minimum utility threshold of a sequence t in a quantitative sequential database D is denoted as $PMIU(t)$ and defined as:*

$$PMIU(t) = min\{mu(i_j) \mid i_j \in t \vee i_j \in I(t)_{rest}\}. \tag{11}$$

In Table 1, $PMIU(<[a],[b]>) = min\{mu(i_j) \mid i_j \in <[a],[b]> \vee i_j \in \{a,b,c,d,e\}\} = min\{500, 500, 500, 200, 500\} = 200$.

Theorem 1. *Given a sequence t, let UB be the upper-bound of t, if $UB < PMIU(t)$, then t and the supersets of t cannot be HUSPs.*

Proof. Let t' be a superset of t, we can obtain that $\{i_j \mid i_j \in t'\} \subseteq \{i_j \mid i_j \in t \vee i_j \in I(t)_{rest}\}$ and $I(t')_{rest} \subseteq I(t)_{rest}$. Based on the definition of upper-bound, $u(t') \leq UB$ and $u(t) \leq UB$. Then, we have $u(t') \leq UB < PMIU(t) = min\{mu(i_j) \mid i_j \in t \vee i_j \in I(t)_{rest}\} \leq min\{mu(i_j) \mid i_j \in t' \vee i_j \in I(t')_{rest}\} = PMIU(t') \leq min\{mu(i_j) \mid i_j \in t'\} = MIU(t'), u(t) \leq UB < PMIU(t) = min\{mu(i_j) \mid i_j \in t \vee i_j \in I(t)_{rest}\} \leq min\{mu(i_j) \mid i_j \in t\} = MIU(t)$. Thus, we can obtain that t and t' cannot be the HUSPs.

From Theorem 1, it can be seen that if the upper-bound of a sequence t is less than the *PMIU* of t, then t and the supersets of t are not HUSPs. The upper-bound (*UB*) is the maximal possible utility, which can be either the *SWU* and *PU* of t, and the *PMIU* is the least minimum utility threshold of the sequence. With the help of the above theorems, we can ensure that the designed algorithm maintains the completeness and correctness of the discovered HUSPs. Based on the proposed framework and the above theorem, the baseline algorithm for HUSPM with multiple minimum utility thresholds and PGrowth mining approach are respectively introduced in Algorithms 1, 2 and 3.

The **PGrowth** function uses the depth-first search strategy to enumerate the possible sequences in our defined *alphabetic-ascending* order. The enumerated sequences are based on the *I-Concatenation* and *S-Concatenation* operations. After the concatenation operations, new sequences are measured by **Judge** function. Details are given in Algorithm 3.

Algorithm 1. Proposed algorithm

Input: D, a quantitative sequential database; *utable*, a utility table containing the unit profit of each item; *MMU-table*, a table containing the minimum utility threshold of each item.

Output: The set of *HUSPs*.

1 scan D to: 1). calculate $u(s)$ for each $s \in D$; 2). build the UL-list for each $s \in D$;
2 $HUSPs \leftarrow \varnothing$;
3 **for** *each* $i_j \in D$ **do**
4     $PD(<i_j>) \leftarrow \{$the UL-list of $s \mid <i_j> \subseteq s \land s \in D\}$;
5   calculate $SWU(<i_j>)$, $u(<i_j>)$, $PMIU(<i_j>)$, and $MIU(<i_j>)$;
6   **if** $SWU(<i_j>) \geq PMIU(<i_j>)$ **then**
7   **if** $u(<i_j>) \geq MIU(<i_j>)$ **then**
8             $HUSPs \leftarrow HUSPs \cup <i_j>$;
9   **PGrowth**($<i_j>$, $PD(<i_j>)$, $HUSPs$);

10 **return** $HUSPs$;

Algorithm 2. PGrowth(*prefix*, $PD(prefix)$, $HUSPs$)

---

1 scan $PD(prefix)$ to get C^I; // measured by SWU
2 **for** *each* $i_j \in C^I$ **do**
3   **Judge**($<prefix \oplus i_j>_{I-Concatenation}$, $PD(prefix)$, $HUSPs$);

4 scan $PD(prefix)$ to get C^S; // measured by SWU
5 **for** *each* $i_j \in C^S$ **do**
6   **Judge**($<prefix \oplus i_j>_{S-Concatenation}$, $PD(prefix)$, $HUSPs$);

Algorithm 3. Judge(*prefix'*, *PD(prefix)*, *HUSPs*)

1 $PD(prefix') \leftarrow \{$the UL-list of $s \mid prefix' \subseteq s \wedge s \in PD(prefix)\}$;
2 calculate $u(prefix')$, $PU(prefix')$, $PMIU(prefix')$, and $MIU(prefix')$;
3 **if** $PU(prefix') \geq PMIU(prefix')$ **then**
4   **if** $u(prefix') \geq MIU(prefix')$ **then**
5          $HUSPs \leftarrow HUSPs \cup prefix'$;
6   **PGrowth**(*prefix'*, *PD(prefix')*, *HUSPs*);

3.5   Designed Pruning Strategies

The designed baseline algorithm can effectively discover the complete set of HUSPs. However, the search space of the algorithm is still large since the SWU and PU are the over-estimated upper-bound values, which are both larger than the actual utility of the pattern. To reduce a large number of candidates, three strategies are designed to improve the performance of the baseline algorithm. First, we will introduce a tighter upper-bound for mining HUSPs. Details are respectively given below.

Definition 17. *The maximal extension utility of a sequence t in a q-sequence s is denoted as $MEU(t, s)$ and defined as:*

$$MEU(t, s) = max\{u(s_k) + u(< s - s_k >_{rest}) \mid t \sim s_k \wedge s_k \subseteq s\}. \qquad (12)$$

In Table 1, given a sequence $t = <[a], [b]>$, t has 3 matches in S_2, which are $<[a:1], [b:3]>, <[a:1], [b:2]>$ and $<[a:5], [b:2]>$. Thus, $u(<S_2 - <[a:1], [b:3]>>_{rest}) = u(<[(d:2)], [(b:2)(c:1)(d:4)(e:3)]>) = 25$; we cal also obtain that $u(<S_2 - <[a:1], [b:2]>>_{rest}) = u(<[(c:1)(d:4)(e:3)]>) = 15$ and $u(<S_2 - <[a:5], [b:2]>>_{rest}) = u(<[(c:1)(d:4)(e:3)]>) = 15$. The utilities of 3 matches respectively are 14, 11 and 31. Thus, $MEU(<[a], [b]>, S_2) = max\{14 + 25, 11 + 15, 31 + 15\} = 46$.

Definition 18. *The maximal extension utility of a sequence t in D is denoted as $MEU(t)$ and defined as:*

$$MEU(t) = \sum_{s \in D} \{MEU(t, s) \mid t \subseteq s\}. \qquad (13)$$

In Table 1, given a sequence $t = <[a], [b]>$, $MEU(t)$ is calculated as $(67 + 46 + 37 + 48 + 54)$ $(= 252)$, which is smaller than $PU(t)$ $(= 279)$.

Theorem 2. *Given a quantitative sequential database D, and two sequences t and t'. If $t \subseteq t'$, we can obtain that*

$$MEU(t') \leq MEU(t). \qquad (14)$$

Proof. Suppose s is a transaction in D and contains t and t'. Let s_q be a q-sequence satisfying $\{u(s_q) + u(<s - s_q>_{rest})\} = MEU(t, s)$, where $t \sim s_q \wedge s_q \subseteq s$. Let $s_{q'}$ be a q-sequence satisfying $\{u(s_{q'}) + u(<s - s_{q'}>_{rest})\} = MEU(t', s)$ where $t' \sim s_{q'} \wedge s_{q'} \subseteq s$. Since $t \subseteq t'$, we can divide t' into two parts as the prefix t and the extension e such that $t + e = t'$. Similarly, the $s_{q'}$ can also be divided into two parts as the prefix $s_{q'_t}$ matching t and the extension $s_{q'_e}$ matching e such that $s_{q'_t} + s_{q'_e} = s_{q'}$. Thus, $MEU(t', s) = \{u(s_{q'}) + u(<s - s_{q'}>_{rest})\} = \{u(s_{q'_t}) + u(s_{q'_e}) + u(<s - s_{q'}>_{rest})\} \le \{u(s_{q'_t}) + u(<s - s_{q'_t}>_{rest})\} \le \{u(s_q) + u(<s - s_q>_{rest})\} = MEU(t, s)$. We can thus obtain that $MEU(t') = \sum_{s \in D}\{MEU(t', s) \,|\, t' \subseteq s\} \le \sum_{s \in D}\{MEU(t, s) \,|\, t' \subseteq s\} \le \sum_{s \in D}\{MEU(t, s) \,|\, t \subseteq s\} = MEU(t)$.

Theorem 2 indicates that if the MEU value of a sequence t is less than the minimum utility threshold, the MEU values of the supersets of t are less than the minimum utility threshold.

Theorem 3 (MEU Strategy, MEUS). *Given a quantitative sequential database D and a sequence t, we can obtain that*

$$MEU(t) \le PU(t) \le SWU(t) \tag{15}$$

Proof. Since $u(s_k) \le max\{u(s_k) \,|\, t \sim s_k \wedge s_k \subseteq s\} = u(t, s)$ and $u(<s - s_k>_{rest}) \le u(<s - t>_{rest})$, $MEU(t, s) = max\{u(s_k) + u(<s - s_k>_{rest}) \,|\, t \sim s_k \wedge s_k \subseteq s\} \le u(t, s) + u(<s - t>_{rest}) \le u(s)$. Thus $MEU(t) = \sum_{s \in D}\{MEU(t, s) \,|\, t \subseteq s\} \le \sum_{s \in D}\{u(t, s) + u(<s - t>_{rest}) \,|\, t \subseteq s\} = PU(t) \le \sum_{s \in D}\{u(s) \,|\, t \subseteq s\} = SWU(t)$.

Theorem 3 indicates that the MEU model is a tighter upper-bound compared to the PU and SWU models. With the MEU model, the designed algorithm can reduce more candidates than that of PU and SWU models. The MEU model can be used to estimate the utility values of the candidate sequences and the supersets of candidate sequences. Based on the definition of HUSP and the above theorems, we can obtain that if the MEU of a sequence t is less than the $PMIU$ of t, t and the supersets of t will not be considered as the HUSPs.

For a candidate sequence t, amounts of candidate items are still required to be processed for *I-Concatenation* and *S-Concatenation*. To reduce the number of candidate sequences in advance, we then propose the look ahead strategy (LAS) to remove the unpromising candidate items.

Theorem 4 (Look Ahead Strategy, LAS). *Given a sequence t and a quantitative sequential database D, two situations can be considered to generate the supersets as:*

(1) if i_j is a I-Concatenation candidate item of t, the maximal utility of $< t \oplus i_j >_{I-Concatenation}$ is no more than $\sum_{s \in D}\{MEU(t, s) \,|\, < t \oplus i_j >_{I-Concatenation} \subseteq s\}$.

(2) if i_j is a S-Concatenation candidate item of t, the maximal utility of $< t \oplus i_j >_{S-Concatenation}$ is no more than $\sum_{s \in D}\{MEU(t, s) \,|\, < t \oplus i_j >_{S-Concatenation} \subseteq s\}$.

Proof. For (1), let $t' = <t \oplus i_j>_{I-Concatenation}$ for convenience. From Definition 17 and Theorem 2, we have $u(t') \leq MEU(t')$ and $MEU(t',s) \leq MEU(t,s)$. Thus $u(t') \leq MEU(t') = \sum_{s \in D}\{MEU(t',s)\,|\,t' \subseteq s\} \leq \sum_{s \in D}\{MEU(t,s)\,|\,t' \subseteq s\}$. In the same way, (2) is true.

Based on the LAS, it can be used to quickly remove the unpromising candidate items for *I-Concatenation* and *S-Concatenation* of sequence t without calculating the *MEU* values of $<t \oplus i_j>_{I-Concatenation}$ and $<t \oplus i_j>_{S-Concatenation}$. Since the upper-bound in LAS can be calculated from the utility linked lists of t, LAS can remove unpromising candidate items in advance. As a result, the algorithm reduces the computation cost by exploring a smaller set of candidate items for concatenation of t.

Theorem 5 (Irrelevant Item Pruning Strategy, IPS). *Given a sequence t and an item $i_j \in I(t)_{rest}$, the maximal utility of $<t \oplus i_j>_{I-Concatenation}$ or $<t \oplus i_j>_{S-Concatenation}$ is no more than $\sum_{s \in D}\{MEU(t,s)\,|\, <t \oplus i_j>_{I-Concatenation} \subseteq s \vee <t \oplus i_j>_{S-Concatenation} \subseteq s\}$.*

Proof. For $<t \oplus i_j>_{I-Concatenation}$, based on Theorem 4, it can be found that $\sum_{s \in D}\{MEU(t,s)\,|\, <t \oplus i_j>_{I-Concatenation} \subseteq s \vee <t \oplus i_j>_{S-Concatenation} \subseteq s\} \geq \sum_{s \in D}\{MEU(t,s)\,|\, <t \oplus i_j>_{I-Concatenation} \subseteq s\} \geq u(<t \oplus i_j>_{I-Concatenation})$. For $<t \oplus i_j>_{S-Concatenation}$, it can be proven in the same way.

With the help of IPS, the remaining utility values of candidate sequences in each transaction decrease since many irrelevant items can be greatly pruned. As a result, the MEU values of candidate sequences can be greatly decreased; more candidates will be removed by the IPS.

4   Experimental Results

For the convenience, the baseline and the baseline algorithm with three pruning strategies are respectively named as **Baseline** and **Baseline_S** in the experiments. All the algorithms were implemented in Java and experiments were carried out on a personal computer equipped with an Intel Core2 i7-4790 CPU and 8 GB of RAM, running the 64-bit Microsoft Windows 7 operating system. To automatically set different minimum utility thresholds for items, we then defined that: $mu(i_j) = max\{\beta \times u(i_j), LMU\}$, where LMU is a user-specified parameter for the least minimum utility threshold, and $u(i_j)$ is the utility of each item i_j in the database. The β is a constant factor to adjust the mu values of items. Four real-life [13] datasets were used in the experiments to evaluate the performance of our proposed approaches. More details of the used datasets can be found in [13].

4.1   Runtime

Figure 2 shows the runtime of the compared approaches under a fixed β with various LMU. From Fig. 2, it can be seen that the Baseline_S algorithm outperforms the Baseline algorithm. The reason is that the Baseline_S algorithm has tighter upper-bound to prune the unpromising candidates early than that of the Baseline algorithm. It can be seen that the runtime of the proposed approaches increases along with the decrease of *LMU*. When *LMU* is set as a low value, the runtime of the Baseline algorithm sharply increases due to a very large candidates, such as in Fig. 2(b)–(d). Notice that there are no results of the Baseline algorithm when *LMU* is set as a very low value, such as in Fig. 2(c) and (d). The reason is that the large dataset with some long sequences causes a very large number of candidates. Thus, the designed Baseline approach cannot return the results within a reasonable time.

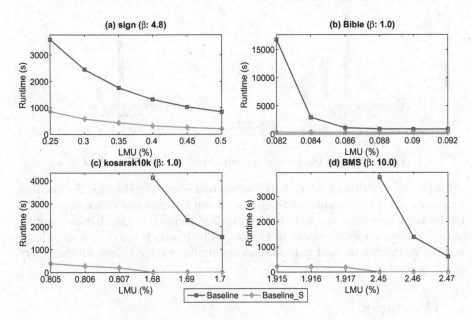

Fig. 2. Runtime under fixed β with various *LMU*

4.2 Number of Candidates

The number of candidates and the number of actual discovered HUSPs of the compared approaches under a fixed β with various *LMU* is evaluated and shown in Fig. 3. From Fig. 3, it can be seen that the number of candidates of the Baseline_S algorithm is much less than that of the Baseline algorithm. This shows that the designed pruning strategies can greatly reduce the unpromising candidates for mining the HUSPs, and the requirements of runtime can be greatly improved. From the results, it also can be seen that the difference of the number of candidates is more obvious than that of the runtime. This is also reasonable since it takes computations to apply the pruning strategies for reducing the number of candidates.

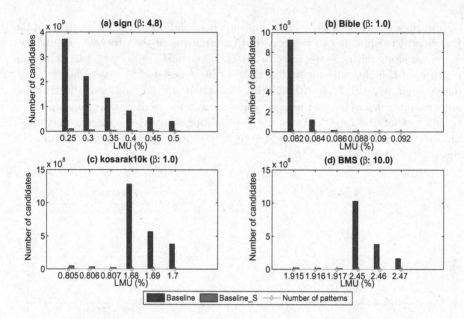

Fig. 3. Number of candidates under fixed β with various LMU

When the LMU is set lower, it is obvious to see that the Baseline_S algorithm generates much less candidates than that of the Baseline algorithm. Especially, the Baseline algorithm has no results in Fig. 3(c) and (d). Notice that the number of candidates sharply decreases in Fig. 3(c). The reason is that there are a large number of candidates and patterns having similar utility values within a very small interval in this dataset.

5   Conclusion

In this paper, we propose a novel high-utility sequential pattern mining with multiple minimum utility thresholds framework for mining HUSPs. Based on the proposed framework, a baseline approach is first proposed. With the help of the designed LS-tree, UL-list structure, and the properties of HUSPM, the proposed algorithm can discover the complete set of HUSPs with multiple minimum utility thresholds. To improve the performance of the baseline algorithm, three pruning strategies are then introduced to lower the upper-bound value of the sequences and reduce the search space to find the HUSPs. Results are then evaluated to show the effectiveness and efficiency for mining HUSPs in terms of runtime and number of candidates.

Acknowledgment. This research was partially supported by the National Natural Science Foundation of China (NSFC) under grant No. 6150309, by the Research on the Technical Platform of Rural Cultural Tourism Planning Basing on Digital Media under grant 2017A020220011, and by the CCF-Tencent Project under grant No. IAGR20160115.

References

1. Agrawal, R., Srikant, R.: Mining sequential patterns. In: The International Conference on Data Engineering, pp. 3–14 (1995)
2. Han, J., Pei, J., Mortazavi-Asl, B., Chen, Q., Dayal, U., Hsu, M.: Freespan: frequent pattern-projected sequential pattern mining. In: ACM SIGKDD International Conference on Knowledge Discovery and Data Mining, pp. 355–359 (2000)
3. Fournier-Viger, P., Lin, J.C.W., Kiran, R.U., Koh, Y.S., Thomas, R.: A survey of sequential pattern mining. Data Sci. Pattern Recogn. 1(1), 54–77 (2017)
4. Pei, J., Han, J., Mortazavi-Asl, B., Wang, J., Pinto, H., Chen, Q., Dayal, U., Hsu, M.: Mining sequential patterns by pattern-growth: the prefixspan approach. IEEE Trans. Knowl. Data Eng. 16(11), 1424–1440 (2004)
5. Chan, R., Yang, Q., Shen, Y.D.: Mining high utility itemsets. In: IEEE International Conference on Data Mining, pp. 19–26 (2003)
6. Liu, Y., Liao, W., Choudhary, A.N.: A two-phase algorithm for fast discovery of high utility itemsets. In: Pacific-Asia Conference on Advances in Knowledge Discovery and Data Mining, pp. 689–695 (2005)
7. Yin, J., Zheng, Z., Cao, L., Song, Y., Wei, W.: Efficiently mining top-k high utility sequential patterns. In: IEEE International Conference on Data Mining, pp. 1259–1264 (2013)
8. Lan, G., Hong, T., Tseng, V.S., Wang, S.: Applying the maximum utility measure in high utility sequential pattern mining. Expert Syst. Appl. 41(11), 5071–5081 (2014)
9. Alkan, O.K., Karagoz, P.: Crom and huspext: improving efficiency of high utility sequential pattern extraction. IEEE Trans. Knowl. Data Eng. 27(10), 2645–2657 (2015)
10. Wang, J., Huang, J., Chen, Y.: On efficiently mining high utility sequential patterns. Knowl. Inf. Syst. 49(2), 597–627 (2016)
11. Liu, B., Hsu, W., Ma, Y.: Mining association rules with multiple minimum supports. In: ACM SIGKDD International Conference on Knowledge Discovery and Data Mining, pp. 337–341 (1999)
12. Yin, J., Zheng, Z., Cao, L.: USpan: an efficient algorithm for mining high utility sequential patterns. In: ACM SIGKDD International Conference on Knowledge Discovery and Data Mining, pp. 660–668 (2012)
13. Fournier-Viger, P., Lin, J.C., Gomariz, A., Gueniche, T., Soltani, A., Deng, Z., Lam, H.T.: The SPMF open-source data mining library version 2. In: The European Conference on Machine Learning and Knowledge Discovery in Databases, pp. 36–40 (2016)

Extracting Various Types of Informative Web Content via Fuzzy Sequential Pattern Mining

Ting Huang[1,2], Ruizhang Huang[1,2,3(✉)], Bowei Liu[1,2], and Yingying Yan[1,2]

[1] College of Computer Science and Technology, Guizhou University,
Guiyang, Guizhou, China
durant.huang@gmail.com, rzhuang@gzu.edu.cn, bwei.liu@gmail.com,
yyingy0921@gmail.com
[2] Guizhou Provincial Key Laboratory of Public Big Data, Guizhou University,
Guiyang, Guizhou, China
[3] State Key Laboratory for Novel Software Technology, Nanjing University,
Nanjing, People's Republic of China

Abstract. In this paper, we present a web content extraction method to extract different types of informative web content for news web pages. A fuzzy sequential pattern mining method, namely FSP, is developed to gradually discover fuzzy sequential patterns for various types of informative web content. To avoid the situation that the usage of HTML tags may be changed with the development of web technology, fuzzy sequential patterns are mined using a stable feature, in particular, the number of tokens in each line of source code. We have conducted extensive experiments and good clustering properties for the discovered sequential patterns are observed. Experimental results demonstrate that the FSP method is effective compared with state-of-the-art content extraction methods. Besides main articles of web pages, it can also find other types interesting web content such as article recommendations and article titles effectively.

Keywords: Content extraction · Fuzzy sequential pattern · Recommendation discovery

1 Introduction

With the increasing usage of the Internet, web pages become one of the most important information sources. It is required by many applications that the content of web pages be collected and analyzed appropriately. Most of traditional web content extraction methods focus on extracting main articles of web pages. However, besides main articles, there are a number other types of informative web content. For example, titles of news articles are of special usage and are usually given additional emphasis in the task of news document analysis. List of news links on web pages are also useful because it often refers to the news article recommendations which are needed for studying document relationships. These informative web content blocks are either grouped with the main article

© Springer International Publishing AG 2017
L. Chen et al. (Eds.): APWeb-WAIM 2017, Part I, LNCS 10366, pp. 230–238, 2017.
DOI: 10.1007/978-3-319-63579-8_18

as a single continuous unit or are discarded as non-informative content for the traditional web content extraction approaches. Therefore, it is useful to develop a web content extraction method which could identify and extract different types of informative web content from web pages for real usage.

The first contribution of this paper is to tackle the task of extracting various types of informative web content. A fuzzy sequential pattern mining method is developed to recognize patterns for different types of informative web content with a small set of sample web pages. The patterns are then used to identify informative web content blocks for new coming web pages. The second contribution of this paper is to use stable features for recognizing patterns of informative web content. Instead of HTML tags which are normally used to guide the display format of a web page and changes along with the developing of web technology, we make use of the information that can be obtained from the content-text of the web page. When discarding HTML tags, we observe that one useful information for identifying informative web content is the number of words in each line of HTML source code, namely, text length. Based on this observation, we make use of the variance of the text length as a key indicator of the informative web content. The fuzzy sequential pattern mining method is developed to discover sequential patterns with different level of fuzziness from web pages.

We have conducted extensive experiments with web pages collected from different websites. Experimental results demonstrate that our proposed method is effective for extracting various types of web content where all web content are discovered with the same process of FSP. The remaining parts of this paper are organized as follows, Sect. 2 introduces our problem more deeply. The proposed FSP method is described in detail in Sect. 3. Experimental results are presented in Sect. 4.

2 Problem Definitions

2.1 Problem Description

There are two interesting observations when going through the source code of each web page. The first observation is that each web block is related to a segment of source code separated by a number of HTML tags. As a result, a web page block can be coded with a sequence of number of tokens in each line, namely text length. The second observation is that web pages from the same website always maintain similar web structures. In Fig. 1 we select four web pages from one single website. For each web page, the length for each line of source code is depicted. It is obvious that the source code of each web page is composed of a number of common web segments depicted by a set of rectangles and circles. Web blocks presented by those dash rectangle web segments are all non-informative structural content in all four web pages, such as navigation bars and title banners. These web blocks are with the same web codes in all web pages. Web segments indicated by dash circles are used to code less structural parts of the web page which are similar but not the same through all web pages. Form the web display, we found that these part of source codes are all used to depict article recommendations. Web segments indicated by regular rectangles

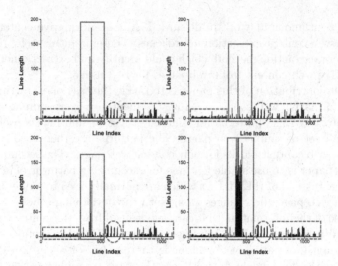

Fig. 1. Sequential text length of source code for four pages from the same web site.

are all used to code main articles of the web page which are nonstructural and are far different for different web pages. Therefore, we can encode web segments by fuzzy sequential patterns of text length. The more structural a web block, the more precise the pattern is. For less structural web blocks, fuzzy patterns can be used to mark the blocks.

2.2   Definitions

Definition 1 (Item). *Given a single line of a web page source code, let* $[l_{min}, l_{max}]$ *be the length range of the line, where* l_{min} *is used to indicate the minimal number of tokens in a line and* l_{max} *is used to indicate the maximal number of tokens in a line. An item* ι *is denoted by the length range* $[l_{min}, l_{max}]$ *of the source line.*

Definition 2 (Sequential Item). *Let* $I = \{\iota_1, \iota_2, \ldots, \iota_n\}$ *be a set of items. A sequential item* s *is an ordered list of item denoted as* $\langle \iota_{o1}, \iota_{o2}, \ldots, \iota_{on} \rangle$ *where* $\iota_{oi} \in I$ *for* $1 \leq i \leq n$.

Definition 3 (Fuzzy Sequential Item). *Let* s *be a sequential item. A fuzzy sequential item* ϵ *is defined as* s *together with the ordered list of fuzzy factors for each item in* s, *denoted as* $(\langle \iota_1, \iota_2, \ldots, \iota_n \rangle, \langle f_1, f_2, \ldots, f_n \rangle)$ *where* f_i *is the fuzzy factor for item* ι_i *respectively.*

Definition 4 (Fuzzy Identical). *Let* $\epsilon_1 = (\langle \iota_{11}, \iota_{12}, \ldots, \iota_{1n} \rangle, \langle f_{11}, f_{12}, \ldots, f_{1n} \rangle)$, *and* $\epsilon_2 = (\langle \iota_{21}, \iota_{22}, \ldots, \iota_{2n} \rangle, \langle f_{21}, f_{22}, \ldots, f_{2n} \rangle)$ *be two fuzzy sequential item.* ϵ_1 *and* ϵ_2 *are regarded to be identical i.e.* $\epsilon_1 = \epsilon_2$, *if and only if* $\iota_{1i} \wedge \iota_{2i} \neq NULL$ *for all* $1 \leq i \leq n$.

In our approach, we use an item to represent one line of HTML source code. l_{min} and l_{max} is first initialized by the exact line length and will be scaled with fuzzy factor in later processes. A fuzzy sequential pattern p is represented by four components. The first component is a representative sequential item s_p which is used to match the web segments. The second component, denoted as f_p is the fuzzy factor of the sequential pattern. f_p is calculated by selecting half of each item with large values and taking their average. The third component, denoted as m_p is the block size of the sequential pattern p which is regarded as the number of items in p. The fourth component, denoted as t_p is the text length of the fuzzy sequential pattern p. t_p is calculated by selecting half of the item with large values of l_{max} and taking their average of l_{max}. Specifically, a fuzzy sequential pattern is represented as $p = (s_p : \langle \iota_1, \iota_2, \ldots, \iota_{mp} \rangle, f_p, m_p, t_p)$.

3   Fuzzy Sequential Pattern Mining (FSP)

The overall design of the fuzzy sequential pattern mining (FSP) method is depicted in Fig. 2. Given a set of sample web pages, we removed all the HTML tags and employed a state-of-the-art web segmentation method proposed by Kohlschutter and Nejdl [2] for discovering web segments. A set of web segments are obtained for each web page, denoted as $\epsilon_w = \{\epsilon_e\}$. As a result, a set of segment set is obtained, denoted as $\{\epsilon_w\}$, for all the sample web pages. $\{\epsilon_w\}$ is used as the input to the FSP method. There are two main processes in the FSP method. The first process is used to discover frequent fuzzy sequential items from the set of web segments. All discovered frequent sequential items are then be passed to the pattern generation process for generating fuzzy sequential patterns and adjust the web segments with fuzzy factor. Adjusted web segments will then be used to mine frequent fuzzy sequential items in the next iteration. Detail explanations on the two main processes are described as follow.

Frequent Fuzzy Sequential Items Mining. Given a set of web segments, we mine the frequent fuzzy sequential items with an Apriori-style algorithm. The only difference between our method to the standard Apriori is that the items used in our method should be fuzzy sequential item as described in Definition 3. Two items are regarded as the same when they are fuzzy identical as described in Definition 4. Besides, there is one parameter in the process of our Apriori-style algorithm, namely $minSup$, which controls the minimal percentage of occurrences of a frequent fuzzy sequential item in the set of web segments.

Pattern Generation. In the process of pattern generation, each frequent fuzzy sequential items ϵ_p is evaluated for generating fuzzy patterns p_{ϵ_p}. In the FSP method, ϵ_p can be used to generate fuzzy sequential patterns only if it is fuzzy identical to a web segments ϵ_e. Note that the pattern p_{ϵ_p} can be generated with ϵ_p as described in Sect. 2.2. ϵ_e will be removed as the web segments has been identified by a certain pattern. Otherwise, ϵ_p is used to recognize web segments

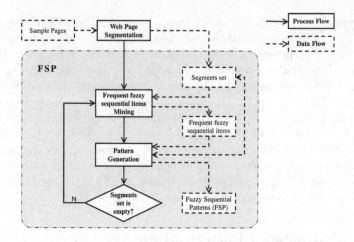

Fig. 2. The overall architecture design of the fuzzy sequential pattern mining.

or parts of web segments that cannot be marked by any frequent fuzzy sequential items. These web segments are scaled according to the current fuzzy factor for the next round. Detail explanation of the pattern generation process is discussed in Algorithm 1.

By combining the features of fuzzy sequential patterns, in particular, the pattern fuzzy factor f_p, the pattern size m_p, and the pattern length t_p, good clustering properties of fuzzy sequential patterns can be observed. We employed the standard multi-class SVM model [3] to identify the types of fuzzy sequential patterns. The SVM model is trained by the labeled fuzzy sequential patterns discovered from the sample web pages and is applied to all patterns discovered from various web sites.

Content Extraction. Given the set of fuzzy sequential patterns discovered, for a new web page, web content blocks are extracted in the process of content extraction. Web segments are first generated with web segmentation method introduced in [2]. Each web segment is then marked by fuzzy sequential patterns. The matched pattern with the smallest fuzzy factor is return.

4   Experiments

4.1   Experiment Setup

We arbitrarily crawled 2000 news web pages each website from 10 popular English news websites. Each web page was manually annotated with article title, main article, and article recommendations.

There is only one parameter to be setup in our proposed FSP method, in particular, we set *minimum support* of the fuzzy sequential item mining algorithm as described in Sect. 3 to 80% for all experiments. For each website, we randomly

---

Algorithm 1. Pattern Generation

Input:
    - Frequent fuzzy sequential items $\{\epsilon'_p\}$;
    - Web segment set $\{\epsilon_w\}$; - Current fuzzy factor f;
Output:
    - Adjusted web segment set $\{\epsilon_w\}$; - Fuzzy sequential pattern set $\{p\}$

 1: **for** each we segment $\epsilon_e \in \{\epsilon_w\}$ **do**
 2:     Scale the range $[l_{min}, l_{max}]$ of each $\iota_i \in \epsilon_e$;
 3: **for** each frequent fuzzy sequential items ϵ_p **do**
 4:     **for** each web page w **do**
 5:         match ϵ_p with it's supported web segments ϵ_e, $\epsilon_e \in \epsilon_w$;
 6:         **if** $\epsilon_p = \epsilon_e$ **then**
 7:            Remove ϵ_e from $\{\epsilon_e\}$;
 8:            **if** $p \notin \{p\}$ **then**
 9:                $\{p\} \leftarrow$ Generate Pattern p_{ϵ_p};
10:         **else**
11:            **for** each $\iota_i \in \epsilon_e$ which be matched by ϵ_p **do**
12:                Unscale ι_i;
13: Empty $\{\epsilon_p\}$;
14: Adjust fuzzy factor for next round: $f{+}{+}$;

---

select 5 pages for discovering fuzzy sequential patterns and the remaining pages are used for testing. Following the discussion in Sect. 3, we trained a multi-class SVM classifiers for identifying the type of web content for each sequential fuzzy pattern.

For comparative study, we investigated 4 state-of-the-art web content extraction approaches with different strategies for extracting informative web contents. The first method is Content Code Blurring, denoted as CCB [1]. The second method is CETR designed by Weninger et al. [6]. The third method is a standard vision-based method implemented by Popela, denoted as jVIPS [4]. The CETD is a state-of-the-art DOM-based content extraction method which involves the usage of text density [5]. Note that all these approaches are designed to extract the main article of web pages. Other informative parts discussed in our paper are regarded as noise and discarded.

4.2 Experiment Results

Experimental Results on Main Article Extraction. Table 1 presents the experimental results of our proposed FSP method on main article extraction. It shows that the FSP method achieves the best performances for 8 out 10 datasets. There is only one dataset, the Mail Online dataset, from which the FSP gets slightly worse results. The reason is that web pages from the Mail Online dataset contain a large number of long user comments which have similar written style with main articles. These long user comments confuse the FSP method. In real practice, we found that the FSP recognized a number of small web segments labeled as main articles. This problem can be greatly improved if we reconstruct the web segments by linking those continuous small main article web segments and only selecting the largest one as the final result.

Table 1. Experimental results on main article content extraction. Winners are bold

Sources	CCB	CETR	jVIPS	CETD	FSP		
	F_1-score				Precision	Recall	F_1-score
YAHOO	93.09%	95.21%	93.51%	95.84%	97.21%	98.56%	**97.88%**
NY Times	89.57%	86.92%	91.26%	**96.95%**	97.23%	96.07%	**96.65%**
BBC	88.71%	90.66%	90.28%	96.10%	98.44%	96.60%	**97.51%**
CNN	86.09%	92.83%	84.58%	**97.17%**	95.86%	97.11%	96.48%
NBC	83.45%	93.17%	81.06%	94.41%	93.78%	96.54%	**95.14%**
Washington Post	83.14%	93.41%	82.96%	95.43%	95.83%	97.81%	**96.81%**
Huffington Post	82.19%	91.32%	88.97%	94.58%	96.28%	99.02%	**97.63%**
Mail Online	91.03%	**96.51%**	89.87%	96.31%	89.68%	91.44%	90.55%
The Guardian	80.80%	90.84%	87.14%	93.21%	94.37%	95.01%	**94.69%**
Reuters	84.10%	91.82%	90.41%	**93.99%**	94.76%	92.99%	**93.87%**
Average	85.01%	91.34%	85.50%	95.24%	95.34%	96.12%	**95.72%**

Experimental Results on Article Recommendation and Article Title Extraction. Besides main articles, the proposed FSP method is also able to detect other interesting web contents like article recommendations and article titles. Experimental result shows that the FSP method obtains good performance for extracting article recommendations. The average F_1-score for the 10 datasets is 86.60%. For some dataset, such as the BBC, the FSP is able to achieve over 90% on the F_1-score.

For the experimental results on article title extraction, Table 2 shown that the FSP performs slightly worse compared with the main article and article recommendation results. The main reason is that the FSP method uses the web segments as the input, which are obtained by a state-of-the-art web segmentation method [2]. However, web segments of article title are not precise. For some news pages, news article titles are merged to the main article segments when there is no obvious HTML tag boundary between them. For some other news pages, main articles and articles recommendations can be split to small pieces due to the inserted advertisements or format tags. Those extremely small segments may be accidentally mislabeled as main articles or article recommendations in our

Table 2. Experimental results on article recommendation and article title extraction.

Sources	Rcommendation			Title		
	Precision	Recall	F_1-score	Precision	Recall	F_1-score
YAHOO	87.57%	74.47%	80.49%	58.09%	84.47%	68.84%
NY Times	80.76%	94.73%	87.19%	67.31%	79.47%	72.89%
BBC	90.12%	95.87%	**92.91%**	65.33%	75.65%	70.11%
CNN	78.61%	99.07%	87.66%	71.09%	79.04%	74.85%
NBC	83.57%	92.04%	87.60%	68.72%	82.04%	74.79%
Washington Post	77.24%	87.68%	82.13%	60.38%	77.68%	67.95%
Huffington Post	76.34%	90.24%	82.71%	62.07%	78.44%	69.30%
Mail Online	79.85%	88.13%	83.79%	64.59%	70.13%	67.25%
The Guardian	89.03%	95.01%	91.92%	57.86%	71.05%	63.78%
Reuters	86.34%	93.17%	89.63%	63.47%	72.17%	67.54%
Average	82.94%	91.04%	**86.60%**	63.58%	77.52%	**69.75%**

experiments. However, given these difficulties, the average results of article title extraction on F_1-score is about 70% which can surely be improved by considering more pattern features such as the text-hyperlink ratios.

5 Conclusions

In this paper, we proposed a novel method, namely FSP, to extract various web content blocks from web pages, by only using text length information of HTML source codes. A fuzzy sequential pattern mining approach is employed to discover sequential patterns which are then be used to mark different web blocks. Experimental results demonstrate that the proposed FSP method is effective. Besides the main article, more interesting web blocks, such as article recommendation blocks and article title blocks, can be discovered.

Acknowledgments. This work is supported by Nation Natural Science Foundation of China (Nos. 61462011, 61202089), Introduced Talents Science Projects of Guizhou University (No. 2016050) and the Graduate Innovated Foundation of Guizhou University Project No. 2011015.

References

1. Gottron, T.: Content code blurring: a new approach to content extraction. In: 19th International Workshop on Database and Expert Systems Application, DEXA 2008, pp. 29–33. IEEE (2008)

2. Kohlschütter, C., Nejdl, W.: A densitometric approach to web page segmentation. In: Proceedings of 17th ACM Conference on Information and Knowledge Management, pp. 1173–1182. ACM (2008)
3. Liu, Y., Zheng, Y.F.: One-against-all multi-class SVM classification using reliability measures. In: Proceedings of 2005 IEEE International Joint Conference on Neural Networks, IJCNN 2005, vol. 2, pp. 849–854. IEEE (2005)
4. Popela, T.: Implementace algoritmu pro vizualni segmentaci www stranek. In: Master's thesis, BRNO University of Technology (2012)
5. Sun, F., Song, D., Liao, L.: DOM based content extraction via text density. In: Proceedings of 34th International ACM SIGIR Conference on Research and Development in Information Retrieval, pp. 245–254. ACM (2011)
6. Weninger, T., Hsu, W.H., Han, J.: CETR: content extraction via tag ratios. In: Proceedings of 19th International Conference on World Wide Web, pp. 971–980. ACM (2010)

Exploiting High Utility Occupancy Patterns

Wensheng Gan[1], Jerry Chun-Wei Lin[1(✉)], Philippe Fournier-Viger[2],
and Han-Chieh Chao[1,3]

[1] School of Computer Science and Technology, Harbin Institute of Technology
Shenzhen Graduate School, Shenzhen, China
wsgan001@gmail.com, jerrylin@ieee.org, hcc@ndhu.edu.tw
[2] School of Natural Sciences and Humanities, Harbin Institute of Technology
Shenzhen Graduate School, Shenzhen, China
philfv@hitsz.edu.cn
[3] Department of Computer Science and Information Engineering, National Dong
Hwa University, Hualien, Taiwan

Abstract. Most studies have considered the frequency as sole interestingness measure for identifying high quality patterns. However, each
object is different in nature, in terms of criteria such as the utility, risk,
or interest. Besides, another limitation of frequent patterns is that they
generally have a low occupancy, and may not be truly representative.
Thus, this paper extends the occupancy measure to assess the utility
of patterns in transaction databases. The High Utility Occupancy Pattern Mining (HUOPM) algorithm considers user preferences in terms
of frequency, utility, and occupancy. Several novel data structures are
designed to discover the complete set of high quality patterns without
candidate generation. Extensive experiments have been conducted on
several datasets to evaluate the effectiveness and efficiency of HUOPM.

Keywords: Frequency · Occupancy · Utility occupancy · Data
structure

1 Introduction

Frequent itemset mining (FIM) and association rule mining (ARM) are some
of the most important and fundamental KDD techniques [1,2,5,6], which are
applied in numerous domains. In recent decades, the task of frequent pattern
mining has been extensively studied by mainly considering the frequency measure for selecting patterns. In real-life applications, the importance of objects
or patterns is often evaluated in terms of implicit factors such as the utility,
interestingness, risk or profit. Hence, the knowledge that actually matters to
user may not be found using traditional FIM and ARM algorithms. Thus, a
utility-based mining framework called high utility pattern mining (HUPM) was
proposed [4,9], which considers the relative importance of items (item utility).
The utility (i.e., importance, interest or risk) of each item can be predefined

© Springer International Publishing AG 2017
L. Chen et al. (Eds.): APWeb-WAIM 2017, Part I, LNCS 10366, pp. 239–247, 2017.
DOI: 10.1007/978-3-319-63579-8_19

based on users' background knowledge or preferences, it has become an emerging research topic in recent years [3,4,7–9,13].

Recently, a study [12] has shown that considering the *occupancy* of patterns is critical for many applications. This allows to find patterns that are more representative, and thus of higher quality. However, implicit factors such as the utility, interestingness, risk or profit of objects or patterns are ignored [12], and it ignores the fact that items/objects may appear more than once in transactions. Besides, HUPM does not assess the occupancy of patterns, the discovered high utility patterns may be irrelevant or even misleading if they are frequent but are not representative of the supporting transactions. Hence, it is desirable to find patterns that are representative of the transactions where they occur. Recently, the OCEAN algorithm was proposed to address the problem of high utility occupancy pattern mining by introducing the utility occupancy measure [11]. However, OCEAN fails to discover the complete set of high utility occupancy patterns and also encounter performance problems.

Therefore, this paper proposes an effective and more efficient algorithm named **H**igh **U**tility **O**ccupancy **P**attern **M**ining in transactional databases (**HUOPM**). The proposed algorithm extracts patterns based on users' interests, pattern frequency and utility occupancy, and considers that each item may have a distinct utility. The major contributions are summarized as follows. (i) A novel and effective HUOPM algorithm is proposed to address the novel research problem of mining high utility occupancy patterns with the utility occupancy measure. To the best of our knowledge, no prior algorithms address this problem successfully and effectively. (ii) Two data structures called utility-occupancy list (UO-list) and frequency-utility table (FU-table), are developed to store the required information about a database, for mining HUOP without repeatedly scanning the database. And the remaining utility occupancy is utilized to calculate an upper bound to reduce the search space. (iii) Extensive experiments have been conducted on real-world datasets to evaluate how effective and efficient the proposed HUOPM algorithm is.

2   Preliminaries and Problem Statement

Let $I = \{i_1, i_2, ..., i_m\}$ be a finite set of m distinct items in a transactional database $D = \{T_1, T_2, ..., T_n\}$, where each quantitative transaction $T_q \in D$ is a subset of I, and has a unique identifier *tid*. The support count of an itemset X, denoted as $sup(X)$, is the number of *supporting transactions* containing X [2]. Transaction T_q supports X if $X \subseteq T_q$, and the set of transactions supporting X is denoted as Γ_X. Thus, $sup(X) = |\Gamma_X|$. An itemset X is a frequent pattern (*FP*) in D if $sup(X) \geq \alpha \times |D|$, where α is minimum support threshold.

Definition 1. Each item i_m in a database D has a unit profit denoted as $pr(i_m)$, which represents its relative importance to the user. Item unit profits are indicated in a profit-table, denoted as $ptable = \{pr(i_1), pr(i_2), ..., pr(i_m)\}$. The utility of an item i_j in T_q is $u(i_j, T_q) = q(i_j, T_q) \times pr(i_j)$, in which $q(i_j)$ is the quantity of i_j in T_q. The utility of an itemset X in T_q is $u(X, T_q) = \sum_{i_j \in X \wedge X \subseteq T_q} u(i_j, T_q)$. Thus, the utility of X in D is $u(X) = \sum_{X \subseteq T_q \wedge T_q \in D} u(X, T_q)$. The transaction utility of T_q is $tu(T_q) = \sum_{i_j \in T_q} u(i_j, T_q)$ [4,9].

Definition 2. The utility occupancy of an itemset X in T_q and D are respectively defined as $uo(X, T_q) = \dfrac{u(X, T_q)}{tu(T_q)}$ and $uo(X) = \dfrac{\sum_{X \subseteq T_q \wedge T_q \in D} uo(X, T_q)}{|\Gamma_X|}$.

Definition 3. Given a minimum support threshold α $(0 < \alpha \leq 1)$ and a minimum utility occupancy threshold β $(0 < \beta \leq 1)$, an itemset X in a database D is said to be a high utility occupancy pattern, denoted as *HUOP*, if it satisfies the following two conditions: $sup(X) \geq \alpha \times |D|$ and $uo(X) \geq \beta$.

3   Proposed Algorithm for Mining HUOPs

The search space of HUOPM can be represented as a Set-enumeration tree [10]. The *downward closure* property does not hold for HUOPs, it is a critical issue to design more suitable data structures and efficient pruning strategies to reduce the search space for mining HUOPs.

Definition 4. A frequency-utility tree (FU-tree) is designed as a sorted Set-enumeration tree using the total order $\prec$ on items. Each child node (called extension node) is generated by extending its prefix (parent) node.

Definition 5. The remaining utility occupancy of an itemset X in T_q is denoted as $ruo(X, T_q)$, and defined as the sum of the utility occupancy values of each item appearing after X in T_q: $ruo(X, T_q) = \dfrac{\sum_{i_j \notin X \wedge X \subseteq T_q \wedge X \prec i_j} u(i_j, T_q)}{tu(T_q)}$.

Definition 6. The utility-occupancy list (UO-list) of an itemset X in a database D is a set of tuples corresponding to transactions where X appears. A tuple contains $<tid, uo, ruo>$ for each T_q containing X, the uo element is the utility occupancy of X in T_q, i.e., $uo(X, T_q)$; the ruo element is defined as the remaining utility occupancy of X in T_q, w.r.t. $ruo(X, T_q)$.

Definition 7. A frequency-utility table (FU-table) of an itemset X contains four informations: the name of the itemset X, the support of X, the sum of the utility occupancies of X in database D (**$uo(X)$**), and the sum of the remaining utility occupancies of X in D (**$ruo(X)$**, and $ruo(X) = \dfrac{\sum_{X \subseteq T_q \wedge T_q \in D} ruo(X, T_q)}{|\Gamma_X|}$).

The construction procedure of the UO-list and FU-table is shown in Algorithm 1, it returns X_{ab} which having the UO-list and FU-table of X_{ab}, denoted as $X_{ab}.UOL$ and $X_{ab}.FUT$.

Lemma 1. Let there be a subtree rooted at X, and Γ_X be the supporting transactions of X. For any possible itemset W in the subtree, we have: $uo(W) \leq \dfrac{\sum_{W \subseteq T_q \wedge T_q \in D}(uo(X, T_q) + ruo(X, T_q))}{|\Gamma_W|}$.

Lemma 2. Given a minimum support threshold α, a subtree rooted at X. For any pattern W in the subtree, an upper bound on the utility occupancy of W is: $\hat{\phi}(W) = \dfrac{\sum_{top\alpha \times |D|, T_q \in \Gamma_X}(uo(X, T_q) + ruo(X, T_q))}{\alpha \times |D|} \geq uo(W)$. Thus, we can directly calculate an upper bound $\hat{\phi}(W)$ on the utility occupancy of a subtree rooted at a processed node X.

Algorithm 1. Construction procedure

Input: X, X_a: extension of X by adding a, X_b: extension of X by adding b.
Output: X_{ab}.

```
1  set X_ab.UOL ← ∅, X_ab.FUT ← ∅;
2  for each tuple E_a ∈ X_a.UOL do
3      if ∃E_a ∈ X_b.UOL ∧ E_a.tid == E_b.tid then
4          if X.UOL ≠ ∅ then
5              Search for element E ∈ X.UOL, E.tid = E_a.tid;
6              E_ab ←< E_a.tid, E_a.uo + E_b.uo − E.uo, E_b.ruo >;
7              X_ab.FUT.uo += E_a.uo + E_b.uo − E.uo;
8              X_ab.FUT.ruo += E_b.ruo;
9          else
10             E_ab ←< E_a.tid, E_a.uo + E_b.uo, E_b.ruo >;
11             X_ab.FUT.uo += E_a.uo + E_b.uo;
12             X_ab.FUT.ruo += E_b.ruo;
13         X_ab.UOL ← X_ab.UOL ∪ E_ab;
14         X_ab.FUT.sup ++;
15     else
16         X_a.FUT.sup - -;
17         if X_a.FUT.sup < α × |D| then
18             return null;
19  return X_ab
```

Theorem 1 (Global downward closure property in the FU-tree). *In the FU-tree, if a tree node is a FP, its parent node is also a FP. Let X^k be a k-itemset (node) and its parent node be denoted as X^{k-1}, which is a (k-1)-itemset. The relationship $sup(X^k) \leq sup(X^{k-1})$ holds.*

Theorem 2 (Partial downward closure property in the FU-tree). *In the FU-tree, the upper bound on utility occupancy of any node in a subtree is no greater than that of its parent node always holds, that is $\hat{\phi}(X^k) \leq \hat{\phi}(X^{k-1})$.*

Strategy 1. *If a tree node X has a support count less than $(\alpha \times |D|)$, then any nodes containing X (i.e. all supersets of X) can be directly pruned.*

Strategy 2. *In the FU-tree, if the upper bound on the utility occupancy of a tree node X is less than β, then any nodes in the subtree rooted at X w.r.t. all extensions of X can be directly pruned and do not need to be explored.*

Strategy 3. *During the construction of the UO-list, if the remaining support of X_a for constructing X_{ab} is less than $\alpha \times |D|$, the support of X_{ab} will also be less than $\alpha \times |D|$, X_{ab} is not a FP, and not a HUOP. Then the construction procedure return null, as shown in Algorithm 1 Lines 16–18.*

Based on the above pruning strategies, Algorithm 2 shows the pseudocode of the designed HUOPM algorithm. Details of the **HUOP-Search** and the **UpperBound** procedures are shown in Algorithms 3 and 4, respectively.

Algorithm 2. HUOPM algorithm

Input: D, *ptable*, α, β.

Output: The complete set of high utility occupancy patterns.

1 scan D to calculate the $sup(i)$ of each item $i \in I$ and and $tu(T_q)$ of each transaction;

2 find $I^* \leftarrow \{i \in I | sup(i) \geq \alpha \times |D|\}$;

3 sort I^* in the designed total order $\prec$;

4 scan D once to build the UO-list and FU-table for each 1-item $i \in I^*$;

5 call **HUOP-Search**$(\phi, I^*, \alpha, \beta)$;

6 return *HUOPs*

Algorithm 3. HUOP-Search procedure

Input: X, *extenOfX*, α, β.

Output: The complete set of *HUOPs*.

1 for *each itemset $X_a \in extenOfX$* **do**

2   obtain $sup(X_a)$ and $uo(X_a)$ from the built $X_a.FUT$;

3   if $sup(X_a) \geq \alpha \times |D|$ **then**

4   if $uo(X_a) \geq \beta$ **then**

5              $HUOPs \leftarrow HUOPs \cup X_a$;

6          $\hat{\phi}(X_a) \leftarrow$ **UpperBound**$(X_a.UOL, \alpha)$;

7   if $\hat{\phi}(X_a) \geq \beta$ **then**

8              $extenOfX_a \leftarrow \emptyset$;

9   for *each $X_b \in extenOfX$ such that $X_a \prec X_b$* **do**

10                  $X_{ab} \leftarrow X_a \cup X_b$;

11   call **Construct**(X, X_a, X_b);

12   call **HUOP-Search**$(X_a, extenOfX_a, \alpha, \beta)$;

13 return *HUOPs*

Algorithm 4. UpperBound procedure

Input: $X_q.UOL$, α.
Output: The upper bound $\hat{\phi}(X_a)$.
1 $sumTopK \leftarrow 0$, $\hat{\phi}(X_a) \leftarrow 0$, $V_{occu} \leftarrow \emptyset$;
2 calculate $(uo(X, T_q) + ruo(X, T_q))$ of each tuple from the built $X_a.UOL$ and put them into the set of V_{occu};
3 sort V_{occu} by descending order as $V_{occu}^{\downarrow}$;
4 **for** $k \leftarrow 1$ to $\alpha \times |D|$ in $V_{occu}^{\downarrow}$ **do**
5 $\quad \lfloor \quad sumTopK \leftarrow sumTopK + V_{occu}^{\downarrow}[k]$;
6 $\hat{\phi}(X_a) = \dfrac{sumTopK}{\alpha \times |D|}$;
7 **return** $\hat{\phi}(X_a)$

---

4   Experimental Results

Only two prior studies, DOFIA [12] and OCEAN [11], are closely related to our work. Major differences are that DOFIA considers both the frequency and occupancy for mining high qualified patterns, but the occupancy is based on the number of items in itemsets/transactions rather than the concept of utility [12]. Thus, OCEAN [11] is implemented to generate high utility occupancy patterns (denoted as HUOPs*), and HUOPs are generated by HUOPM. All algorithms in the experiments are implemented using the Java language and executed on a PC with an Intel Core i5-3460 3.2 GHz processor and 4 GB of memory, running on the 32 bit Microsoft Windows 7 platform. Two real-world datasets, BMSPOS [1] and chess [1], are used in the experiments with a simulation method [13]. Note that there are three versions, $HUOPM_{P12}$ (with strategies 1 and 2), $HUOPM_{P13}$ (with strategies 1 and 3), and $HUOPM_{P123}$ (with strategies 1, 2 and 3), are compared to evaluate the efficiency of HUOPM.

Table 1. Number of patterns under various parameters

BMSPOS (β: 0.3)	α (%)	0.010	0.015	0.020	0.025	0.030	0.035
	HUOPs*	29389	17005	11876	8961	7065	5884
	HUOPs	29444	17048	11899	8987	7094	5905
chess (β: 0.3)	α (%)	60	61	62	63	64	65
	HUOPs*	2423	1776	1374	1108	903	695
	HUOPs	11469	8949	7075	5510	4273	3351
BMSPOS (α: 0.02%)	β (%)	0.24	0.26	0.28	0.30	0.32	0.34
	HUOPs*	287850	106239	34195	11876	5089	2696
	HUOPs	287902	106285	34233	11899	5115	2721
chess (α: 62%)	β (%)	0.26	0.28	0.30	0.32	0.34	0.36
	HUOPs*	9820	4837	1374	514	125	14
	HUOPs	24905	14050	7075	3153	1170	357

Pattern Analysis. To analyze the usefulness of the HUOPM framework, the derived pattern HUOPs* and HUOPs are evaluated. Results for various parameter values are shown in Table 1. It can be clearly observed that the sets of derived HUOPs* is always smaller than that of HUOPs, which means that numerous interesting patterns are effectively discovered by HUOPM, while most of them are missed by OCEAN. Thus, OCEAN fails to discover the complete set of HUOPs. Both HUOPs* and HUOPs decrease when α is increased, less HUOPs* and HUOPs are obtained when β is set higher. Besides, the number of missing patterns (i.e. HUOPs - HUOPs*) sometimes increases when varying β, while it sometimes decreases. It can be concluded that the proposed HUOPM algorithm is acceptable and solve a serious shortcoming of OCEAN.

Efficiency Analysis. A performance comparison of the different strategies used in HUOPM is presented in Fig. 1. It can be clearly observed that the runtime of OCEAN is the worst, compared to the other algorithms in most cases. We also draw the following conclusions. (i) The difference between $HUOPM_{P12}$ and $HUOPM_{P123}$ indicates that pruning strategy 3 always reduces the search space by pruning subtrees. (ii) By comparing $HUOPM_{P13}$ and $HUOPM_{P123}$, it can be concluded that pruning strategy 2, using the upper bound of utility occupancy, provides a trade-off between efficiency and effectiveness. Because it needs to spend additional time to calculate the upper bounds, but sometimes unpromising itemsets can be directly pruned by other pruning strategies. (iii) HUOPM applies pruning strategies to prune unpromising itemsets early, which greatly speed up the mining efficiency, compared to the OCEAN algorithm.

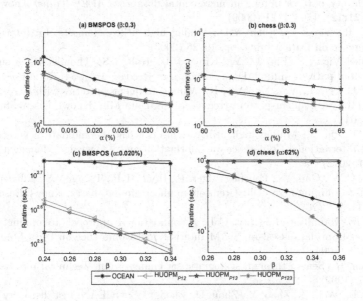

Fig. 1. Runtime under various parameters.

5 Conclusions

In this paper, we proposed an effective HUOPM algorithm to address a new research problem of mining high utility occupancy patterns with utility occupancy. The utility occupancy can lead to useful itemsets that contribute a large portion of total utility for each individual transaction representing user interests or user habit. Based on the two *downward closure* properties and pruning strategies, HUOPM can directly mine HUOPs without candidate generation. Extensive experiments show that HUOPM can efficiently find the complete set of HUOPs and significantly outperforms the state-of-the-art OCEAN algorithm.

Acknowledgments. This research was partially supported by the National Natural Science Foundation of China (NSFC) under grant No. 61503092, by the Research on the Technical Platform of Rural Cultural Tourism Planning Basing on Digital Media under grant 2017A020220011, and by the Tencent Project under grant CCF-Tencent IAGR20160115.

References

1. Frequent itemset mining dataset repository (2012). http://fimi.ua.ac.be/data/
2. Agrawal, R., Srikant, R.: Fast algorithms for mining association rules in large databases. In: International Conference on Very Large Data Bases, pp. 487–499 (1994)
3. Ahmed, C.F., Tanbeer, S.K., Jeong, B.S., Le, Y.K.: Efficient tree structures for high utility pattern mining in incremental databases. IEEE Trans. Knowl. Data Eng. **21**(12), 1708–1721 (2009)
4. Chan, R., Yang, Q., Shen, Y.D.: Mining high utility itemsets. In: International Conference on Data Mining, pp. 19–26 (2003)
5. Fournier-Viger, P., Lin, J.C.W., Kiran, R.U., Koh, Y.S., Thomas, R.: A survey of sequential pattern mining. Data Sci. Pattern Recogn. **1**(1), 54–77 (2017)
6. Han, J., Pei, J., Yin, Y., Mao, R.: Mining frequent patterns without candidate generation: a frequent-pattern tree approach. Data Min. Knowl. Discov. **8**(1), 53–87 (2004)
7. Liu, M., Qu, J.: Mining high utility itemsets without candidate generation. In: ACM International Conference on Information and Knowledge Management, pp. 55–64 (2012)
8. Lin, J.C.W., Gan, W., Fournier-Viger, P., Hong, T.P., Tseng, V.S.: Efficient algorithms for mining high-utility itemsets in uncertain databases. Knowl. Based Syst. **96**, 171–187 (2016)
9. Yao, H., Hamilton, J., Butz, C.J.: A foundational approach to mining itemset utilities from databases. In: SIAM International Conference on Data Mining, pp. 211–225 (2004)
10. Rymon, R.: Search through systematic set enumeration. Technical reports (CIS), p. 297 (1992)
11. Shen, B., Wen, Z., Zhao, Y., Zhou, D., Zheng, W.: OCEAN: fast discovery of high utility occupancy itemsets. In: Pacific-Asia Conference on Knowledge Discovery and Data Mining, pp. 354–365 (2016)

12. Tang, L., Zhang, L., Luo, P., Wang, M.: Incorporating occupancy into frequent pattern mining for high quality pattern recommendation. In: Proceedings of the 21st ACM International Conference on Information and knowledge management, pp. 75–84 (2012)
13. Tseng, V.S., Shie, B.E., Wu, C.W., Yu, P.S.: Efficient algorithms for mining high utility itemsets from transactional databases. IEEE Trans. Knowl. Data Eng. **25**(8), 1772–1786 (2013)

Text and Log Data Management

Translation Language Model Enhancement for Community Question Retrieval Using User Adoption Answer

Ming Chen, Lin Li[(✉)], and Qing Xie

School of Computer Science and Technology,
Wuhan University of Technology, Wuhan, China
{erlangera,cathylilin,felixxq}@whut.edu.cn

Abstract. Community Question Answering (CQA) services on Web provide an important alternative for knowledge acquisition. As an essential component of CQA services, question retrieval can help users save much time by finding relevant questions. However, there is a "gap" between queried questions and candidate questions, which is called lexical chasm or word mismatch problem. In this paper, we improve traditional Topic inference based Translation Language Model (T^2LM) by using the topic information of queries. Moreover, we make use of user information, specifically the number of user adoption answers, for further enhancing our proposed model. In our model, the translation model and the topic model "bridge" the word gap by linking different words. Besides, user information that has no direct relation with semantics is used to help us "bypass" the gap. By combining both of them we obtain a considerable improvement for the performance of question retrieval. Experimental results on a real Chinese CQA data set show that our proposed model improves the retrieval performance over T^2LM baseline by 7.5% in terms of Mean Average Precision (MAP).

Keywords: Question retrieval · Translation model · Topic model · User information

1 Introduction

Community Question Answering (CQA) services have become a popular alternative for online information access, such as Yahoo! Answers[1], Zhihu[2] and Sogouwenwen[3]. Meanwhile, a huge number of User Generated Content (UGC) has been accumulated in the form of Question and Answer (QA) pairs. Their large amount of access historical data makes it possible that users can find the answers of their questions from answered questions. As we know, question

---

[1] http://answers.yahoo.com.
[2] http://www.zhihu.com.
[3] http://wenwen.sogou.com.

© Springer International Publishing AG 2017
L. Chen et al. (Eds.): APWeb-WAIM 2017, Part I, LNCS 10366, pp. 251–265, 2017.
DOI: 10.1007/978-3-319-63579-8_20

retrieval in CQA services returns several relevant questions with possible answers directly. By that way, users do not need to wait for answers from human, which helps users save a lot of time. Therefore, the retrieval of relevant issues and their corresponding answers become an important task for CQA services. Here we define question retrieval as a task where new questions are used as queries to find relevant questions that have already been provided answers. For simplicity and consistency, we use the term "query" to denote a new question raised by a user and "question" to denote the answered question in the CQA archive [16].

Meanwhile, question retrieval can also be considered as a solution of traditional Question Answering problem, by transforming the focus of the Question Answering task from answer extraction, answer matching and answer ranking to searching relevant questions with good ready answers [14]. One of the major challenges for question retrieval is the lexical gap, i.e., the word mismatch between queried questions and candidate questions. For example, "Where can I listen to rock for free online?" and "I need a music sharing website." probably have the same meaning but in different word forms. In addition, the limited length of questions causes the sparsity of word features [17]. Therefore, retrieval models based on word frequency and document frequency statistics are no longer suitable for question retrieval tasks. Jeon et al. [2] proposed language model based question retrieval model. In their model, questions are ranked by their similarities to a query which depends on the exact match of words. However, the model cannot solve the word mismatch problem between the query and their relevant QA pairs in an archive.

To conquer the gap, on the one side, researchers are constantly trying to develop more enhanced models that can bridge the chasm by linking different words. Since the relationship between different words can be modeled through word-to-word translation probabilities, translation-based approaches have obtained some good results [3–7]. To control the noises in translation model, some researchers introduced potential topic information in translation-based model, namely, the topic inference based translation model [11] that is the state-of-the-art translation-based language model. On the other side, researchers are incessantly looking for new ways that can help people bypass the gap. Omari et al. [18] proposed an approach that ranks answers based on their novelty. Other work mainly resorts to query expansion. For example, category Information [19], Key Concept [16,22], Dependency Relation [20] and Topic Information [21] were used to expand queries.

In this paper, we focus on the improvement of translation-based language model by metadata information. We take two major actions to solve the word mismatch problem in question retrieval. Firstly, we improve topic inference based translation model by introducing the topic information of queries. Our improved approach controls the translation noises by leveraging the topic information and balances the impact of each topic by using the topic information of query as weights. Secondly, we add user adoption answer number information into the improved model. Questions from users who have more adoption answers will be ranked higher than others. We do that not only for inspiring users to answer questions, but also because of our assumption that questions from users who have better answer behavior may have higher quality. Specifically detailed analysis will be introduced in Sect. 3.3. By combining both, we further improve the performance of question retrieval in CQA.

The rest of this paper is organized as follows: Sect. 2 introduces the related work about question retrieval. Section 3 describes our improved retrieval model and the strategy of combining the model with user adoption answer number. Experiments and result analysis are reported in Sect. 4. Finally, conclusions and future work are discussed in Sect. 5.

2   Related Work

Ponte and Croft [1] proposed language model based information retrieval model firstly. Then it is widely used in all relevant areas of information retrieval. Jeon et al. [2] took the lead in applying the language model to question retrieval. They exploited Unigram language model to model QA pairs in CQA services and it was applied to find similar questions. However, the above approaches cannot bridge the lexical chasm well. In other words, these methods cannot find relevant questions that have different words from queries.

2.1   Translation-Based Model

To compensate for the lack of traditional retrieval methods, researchers introduced statistical machine translation model into information retrieval model. They used word-to-word translation probabilities to model the relationship between different words.

Berger et al. [3] introduced statistical techniques to bridging the lexical gap in FAQ retrieval. They studied similarity calculation method in question retrieval from the lexical level towards the semantic level firstly. Riezler et al. [4] availed of monolingual translation based retrieval model for answer retrieval. They utilized sentence level paraphrasing approach to capture similarities between questions and answers. Xue et al. [5] presented a question retrieval model that combined a translation-based language model for the question part with a query likelihood method for the answer part. Since they used QA pairs as training data of translation model, the improvements are limited. Bernhanrd and Gurevych [6] relied high quality parallel corpora that has QA pairs collected from the WikiAnswer website, the definitions and glosses of the same term in different lexical semantic resources, and then used this corpora to train a translation model. Finally they obtained a better result. Zhou et al. [7] availed of phrase level translation model to capture similarities between query and question and get a better result than word level.

However, the current translation models in question retrieval only rely on statistics co-occurrence information in parallel corpus to capture word-to-word translation probabilities, which makes the word-to-word translation information have lots of noises. Especially in the present circumstances, monolingual parallel corpus with high quality is few and researchers usually use QA pairs or question title and question description pairs as training data. Figure 1 show an aligned

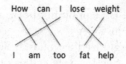

Fig. 1. An aligned example of IBM Model 1

example of IBM Model 1. From Fig. 1 we can see that there is no correlation between word pairs except "weight" and "fat". That makes the word-to-word translation probabilities produced by translation model are biased and affects the performance of retrieval.

2.2   Topic-Based Model

Topic modelling based approaches, such as PLSA [12] and LDA [13], provide an elegant mathematical tool to analyze shallow semantics. Naturally, these techniques have attracted question retrieval researchers attention for a long time.

Wei and Croft [8] proposed an LDA-based document model within the language modeling framework, and evaluated it on several TREC collections. Cai et al. [9] combined the semantic similarity based latent topics with the translation-based language model to improve the retrieval performance. Ji et al. [10] presented Question-Answer Topic Model to model the question-answer relationships, with the assumption that a question and its paired answer share the same topic distribution. However, questions and answers are different in many aspects. They do not share the same topic distribution in many cases. Caused by this, the disadvantage of their model is obvious. Zhang et al. [11] proposed a model that controls translation noise by leveraging the topic information. They focused on similarity of topic distribution between word in query and question. They utilized word distribution information under topic to improve accuracy of word-to-question similarity and further obtained better performances. But their model did not consider the topic information of query that is also valuable for question retrieval.

In this paper, we utilize the topic distribution information of queries to improve the performance of retrieval on the basis of Zhang's model [11]. We add the topic information of queries as weights into the process of word-to-question similarity to balance the impact of each topic.

2.3   Metadata Enhanced Model

Taking into account the social properties of CQA services, QA pairs are accompanied by metadata in real CQA archives usually. The metadata include information of categories, information of askers, information of responders, and information of other users' feedback and so on. It is natural to think that this kind of metadata information is valuable to improve the performance of question retrieval. The category information attracts the attention of researchers immediately because of their relationship with semantics. Cai et al. [9] presented topic model involved category information to discover the latent topics in the content of questions. Cao et al. [14] proposed the category smoothing based and question classification based methods to enhance the performances of existing models. Certainly the methods depend on metadata information that is not always feasible for researchers.

In this paper, we focus on exploiting user information, specifically askers' adoption answer number, to improve the performance of question retrieval. Because questions from user who have better answer behavior could have higher value and they are more acceptable to user.

From the above work in question retrieval, using translation model to capture the similarity of words and using the metadata to improve the performance are two state-of-the-art techniques for question retrieval. In this paper, we take the advantages of the previous approaches and further integrate topic information and user information into a unified probabilistic ranking model to tackle the lexical gap in question retrieval.

3 User Adoption Answer Number for Translation Language Enhancement Model

In this section, we give a brief introduction about the Topic inference based Translation Language Model (T^{2}LM) [11]. Then we present two our main contributions. Firstly, by using the topic information of queries as weights to balance the impact of each topic, we improve the T^{2}LM. Secondly, by introducing the adoption answer information of a question asker into the improved model, we further optimize the ranking list.

3.1 Topic Inference Based Translation Language Model

In the T^{2}LM, given a query $query$ and a QA pair (q, a) consisted of a question q and an answer a, a ranking score $P(query|(q, a))$ is computed as follows:

$$P(query|(q,a)) = \prod_{w \in query} \left(\frac{|(q,a)|}{|(q,a)| + \lambda} P_{t^2lm}(w|(q,a)) + \frac{\lambda}{|(q,a)| + \lambda} P_{ml}(w|C) \right) \tag{1}$$

$$P_{t^2lm}(w|(q,a)) = \mu_1 P_{ml}(w|q) + \mu_2 \left(\sum_{t \in q} p(w|t) P_{ml}(t|q) \right)$$
$$+ \mu_3 \left(\sum_{t \in q} P_{ml}(t|q) \sum_{i=1}^{K} p(w|z_i) p(t|z_i) \right) + \mu_4 P_{ml}(w|a) \tag{2}$$

$$P_{ml}(w|q) = \frac{tf_{w,q}}{|q|}, P_{ml}(w|a) = \frac{tf_{w,a}}{|a|}, P_{ml}(w|C) = \frac{tf_{w,C}}{|C|} \tag{3}$$

The explanations of terms in Eqs. 1 to 3 are showed in Table 1. w is a word in query $query$, and t is a word in question q. $|q|, |a|, |C|$ have similar meanings to

Table 1. Explanation of terms

Term	Explanation
C	The background collection
λ	The smoothing parameter
$|(q,a)|$	The word lengths of (q,a)
$tf_{w,q}$	The frequency of term w in q
$P_{ml}(w|q)$	The maximum likelihood estimate of word w in q
$p(w|t)$	The probability that t is the translation of word w
$p(w|z_i)$	The distribution probabilities of word w under topic z_i
K	The topic number

$|(q,a)|$; $tf_{w,a}$ and $tf_{w,C}$ have similar meanings to $tf_{w,q}$; $P_{ml}(w|a)$ and $P_{ml}(w|C)$ have similar meanings to $P_{ml}(w|q)$; $p(t|z_i)$ has a similar meaning to $p(w|z_i)$; and μ_1, μ_2, μ_3 and μ_4 balance the impact of each component and $\mu_1+\mu_2+\mu_3+\mu_4 = 1$.

3.2 Incorporating the Topic Information of Query

From the Eq. 2 we can see that the significance of each topic is equal in T²LM. And the word topic distribution probabilities from static corpus are used statically for a dynamic query. Although the diverse topic information of various queries is beneficial for question retrieval generally, the information is ignored. In this paper, we propose an approach to exploit this topic information. In our approach, the topic information of queries is used as weights of each topic to improve the process of capture word-to-question similarity. More formally, given a query $query$ and a topic z_i, $P(query|z_i)$ denoting the weight of topic z_i for query $query$ is computed as follows:

$$P(query|z_i) = \frac{\prod_{w \in query} p(w|z_i)}{\sum_{j=1}^{K} \prod_{w \in query} p(w|z_j)} \tag{4}$$

Here w is a word in query $query$; $p(w|z_i)$ and $p(w|z_j)$ are the distribution probability of word w under topic z_i and z_j; and K is topic number. The denominator in Eq. 4 may be zero in some cases. To solve the problem, we make a compromise that we set $P(query|z_i) = 1/K$ for each topic z_i if the problem happens. Then the $P(q|z_i)$ is used as weights of each topic in the process of capturing word-to-question similarity. The specific method is showed as follows:

$$P_{t^2lm^*}(w|(q,a)) = \mu_1 P_{ml}(w|q) + \mu_2 \left(\sum_{t \in q} P(w|t) P_{ml}(t|q) \right)$$
$$+ \mu_3 \left(\sum_{t \in q} P_{ml}(t|q) \cdot K \cdot \sum_{i=1}^{K} P(query|z_i) \cdot p(w|z_i) \cdot p(t|z_i) \right)$$
$$+ \mu_4 P_{ml}(w|a) \tag{5}$$

Here w is a word in query *query*. By multiplying K we control the range of topic part unchanged so that scope of μ_3 is the same as before. We denote the improved model as T^2LM^* in this paper.

3.3 Incorporating the User Information

Thinking about PageRank, a technique for Web search, we see that the importance of web pages relied on web structure building by links have nothing to do with the web content. This technique optimizes the ranking lists of search results significantly. Inspiring by that, we consider a way to measure the importance of questions. In CQA archive, there are no links between QA pairs. What come into our mind is that users' feedback information and user answer number information. For feedback information, there is a problem that users' feedback object are answers usually, users can thumb up or down an answer. It has no direct relation with questions. However, user answer number information can be contacted with questions by question askers. In all kinds of user information, user adoption answer number is most representative for user answer behavior. Therefore, we develop a technique to measure the importance of question by its asker adoption answer number.

Our experiment data set contains metadata of user information including the answer number of each user and their adoption rate. We can get adoption answer number of each question's asker by letting answer number multiply adoption rate. We suspect that the data are related to the retrieval results. So we did some experiments about our conjecture. We selected 100 questions from the data set as queries, and utilized Lucene, TLM and T^2LM to get top 10 candidate questions. Then each question in the result was labelled with "relevant" or "irrelevant" by persons. Table 2 and Fig. 2 show the statistical results of a survey on retrieval result.

Table 2. Distributions of question askers' adoption answer number (Sum denotes the summary of the three models; Background indicates the results of all 578926 questions in data set).

Number	Lucene		TLM		T^2LM		Sum		Background
	Relevant	Irrelevant	Relevant	Irrelevant	Relevant	Irrelevant	Relevant	Irrelevant	
0	0.6307	0.6994	0.6410	0.6311	0.6642	0.6192	0.6453	0.6499	0.6614
1–5	0.1477	0.1600	0.1667	0.2017	0.1418	0.2002	0.1521	0.1873	0.1722
6–10	0.0682	0.0303	0.0513	0.0415	0.0448	0.0359	0.0547	0.0359	0.0412
11–15	0.0284	0.0218	0.0192	0.0237	0.0224	0.0336	0.0233	0.0264	0.0228
16–20	0.0284	0.0145	0.0256	0.0261	0.0299	0.0208	0.0280	0.0205	0.0183
21–30	0.0114	0.0218	0.0128	0.0178	0.0224	0.0220	0.0155	0.0205	0.0191
31–40	0.0227	0.0073	0.0192	0.0095	0.0149	0.0081	0.0190	0.0083	0.0117
41–50	0.0057	0.0048	0.0064	0.0071	0.0000	0.0069	0.0040	0.0063	0.0072
51–100	0.0114	0.0145	0.0256	0.0166	0.0224	0.0255	0.0198	0.0189	0.0174
101–200	0.0114	0.0109	0.0128	0.0083	0.0075	0.0081	0.0105	0.0091	0.0101
201–	0.0341	0.0145	0.0192	0.0166	0.0299	0.0197	0.0277	0.0169	0.0187

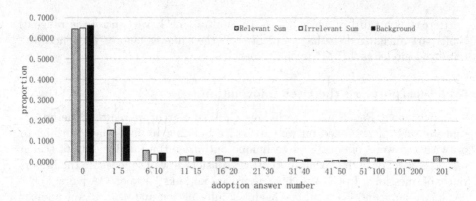

Fig. 2. Distribution of question askers' adoption answer number

Table 3. Average adoption answer number of question asker (Average* indicates that the averages are captured after removing outliers; Sum denotes the summary of the three models; Background denotes the result came from entire data set)

	Lucene		TLM		T²LM		Sum		Background
	Relevant	Irrelevant	Relevant	Irrelevant	Relevant	Irrelevant	Relevant	Irrelevant	
Average	17.381	12.644	14.160	58.676	31.657	53.317	21.066	41.546	39.025
Average*	10.382	3.926	6.183	4.738	7.786	6.162	8.117	4.942	5.035

From Table 2 and Fig. 2, we can find that most askers have no adoption answer. And relevant question and irrelevant question have similar interval distribution on asker's accepted answer number. To find the difference between them, we calculate their average. Unexpectedly, we get that the average number of irrelevant question is bigger than relevant. That is contrary to our conjecture. However, by studying the retrieval result carefully we find the reason is that there are some outliers in our data. The outliers often have more than 20,000 adoption answers which make a huge impact on the results. To eliminate the effects of outliers, we decided to clear it. In our plan, the top 2% users in adoption answer number are removed. After removing outliers, we get Average* which show us askers of relevant question have more accepted answer than irrelevant, on average. The result is showed in Table 3.

By combining Tables 2 and 3, it is appropriate to consider if question from asker have more adoption answers, it may have higher correlation probability. The reason may be that those users who have answered more questions correctly will give higher quality questions. And those questions have higher likelihood of being accepted. So the data of question asker's adoption answer may be beneficial to question retrieval. Based on this we further enhance our proposed T²LM* by adding the information. More formally, given a QA pair (q, a), the $S_{ui}((q, a))$ that denotes user information part score in our model is computed as follows:

$$S_{ui}((q, a)) = \begin{cases} s(u_j) & (s(u_j) \leq 20) \\ 20 & (s(u_j) > 20) \end{cases}, where \ s(u_j) = \sqrt{A_{u_j} \cdot R_{u_j}} \qquad (6)$$

Here user u_j is asker of q; A_{u_j} and R_{u_j} are answer number and the adoption rate of u_j. In order to eliminate the impact of abnormal data and enhance the applicability of the model, instead of clearing the outliers directly, we set a limit for the score. The score will not be more than 20, for the reason that the maximum of adoption answer number near 400 after removing top 2%. After normalizing, we get $P_{ui}((q,a))$. Then we introduce it into our model. More formally, given a query $query$ and a QA pair (q,a), the ranking score $P(query|(q,a))$ is computed as follows:

$$P(query|(q,a)) = (1-\theta)P_{ui}((q,a))$$
$$+ \theta \prod_{w \in query} \left(\frac{|(q,a)|}{|(q,a)|+\lambda} P_{t^2lm^*}(w|(q,a)) + \frac{\lambda}{|(q,a)|+\lambda} P_{ml}(w|C) \right) \quad (7)$$

Here θ controls the impact of user information component. We call our final improved model as User information for Topic inference based Translation Language Model (UT^{2}LM).

4   Experiments

In this section, experiments are conducted on a real CQA archive to demonstrate the effectiveness of our proposed two question retrieval models.

4.1   Experimental Setup

The data set in our experiment comes from NDBC CUP2016, specifically comes from Sogouwenwen that is one of most popular CQA services in China. It is a Chinese data set with two parts. One is QA pairs and the other is user information. For QA pairs, each QA pair consists of three fields: "question" and "answer", as well as metadata "asker ID". For user information, each of user information consists of three fields: "ID", "answer number" and "adoption rate". The data set contains 1,729,263 QA pairs, which have 578608 single questions, and 2,047,669 user information.

For data preprocessing, we segment words and remove the stop words for each QA pair in the data set firstly. Then we select 300 QA pairs from the data set randomly. After removing QA pairs that question lengths are too short and questions are same, we get 253 questions as queries. Among them, 210 questions which are selected by randomly are used as test set to test and other 43 questions are used as development set to adjust the parameters.

We use the GIZA++ toolkit [15] for learning the IBM Translation Model 1 to get the word-to-word translation probabilities. We pool the QA pairs and the answer-question pairs together as the input to this toolkit [5]. We get a word-to-word translation probability list after training. For topic model part, we think of questions as documents and utilize LDA [13] model to model question set. All 578608 single questions are used as data of Gibbs sampling to get the word topic distribution probabilities.

For a practical Question Retrieval System, search results should be presented to the user in a hierarchical way where a list of questions is first presented, and after the user selects a specific question the corresponding answers are presented [5]. Question list is more concise and efficient than QA pair list. Thus, in our experiments, relevance judgments are based on questions. It is easy to transform a QA pair rank into a question rank by taking the highest rank among a group of QA pairs for the same question. In this paper, all the experiments are conducted by getting QA pair ranks firstly and then transforming it into question ranks.

We use three question retrieval models, the Language Model (LM) [2], the Translation-based Language Model (TLM) [5] and the Topic Inference-based Translation language Model (T^2LM) [11], as baseline methods. We conduct experiments to demonstrate the effect of our proposed two models in Sect. 3, T^2LM^* and UT^2LM. For each method, the top 10 retrieval results are kept.

As a common evaluation process used by other researchers in QA field, we recruit three students who their mother tongue is Chinese to label the relevance of the candidate questions regarding to queries. Given a query and its candidate questions list, a student is asked to label the questions in list with "relevant" or "irrelevant". If a candidate question is considered semantically similar to the query, the student will label it as "relevant"; otherwise, the student will label it as "irrelevant". As a result, each candidate question gets three labels and the majority of the labels are taken as the final decision for a query candidate pair.

In order to evaluate the performance of different models, we employ Mean Average Precision (MAP), Mean Reciprocal Rank (MRR) and Precision at 1 (P@1) as evaluation measures [22]. These measures are widely used in the literature for question retrieval in CQA.

By experiments on development set, we adjust the parameters of the baseline systems and our systems to the best. Some results are showed in Figs. 3, 4 and 5. From Fig. 3 we can see that MAP reduces with μ_2 growing overall. Figure 5 shows that with the growth of θ MAP grows first and then decreases and maximizes at $\theta = 0.66$. Finally for T^2LM^* we set $\mu_1 = 0.5$, $\mu_2 = 0.1$, $\mu_3 = 0.3$, $\mu_4 = 0.1$, $K = 70$, and for UT^2LM we set $\mu_1 = 0.5$, $\mu_2 = 0.1$, $\mu_3 = 0.3$, $\mu_4 = 0.1$, $\theta = 0.66$, $K = 70$. We set smoothing parameter $\lambda = 1$ according to the previous work [5].

4.2   Experimental Results

Table 4 shows the compare results of our models and baseline models. We can see that among the three baseline models, T^2LM performs the best and LM performs the worst; TLM between LM and T^2LM. TLM and T^2LM has better performance since they are able to retrieval question that do not share common words with the query, but are semantically similar to the query. T^2LM introduces the topic information to control the translation noise in translation model, it has the best performance. Comparing results of LM and TLM, we find the improvement is very small. We speculate that this is related to our training data of translation model. We use QA pair to train IBM Model 1. That makes word-to-word translation probabilities inaccurate. Showing in the T^2LM^*, the

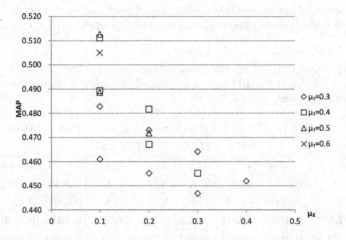

Fig. 3. The influence of μ_1, μ_2 on the performance of $\mathrm{T}^2\mathrm{LM}^*$

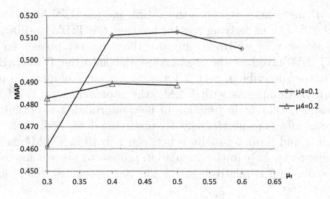

Fig. 4. The influence of μ_1 on the performance of $\mathrm{T}^2\mathrm{LM}^*$ when $\mu_2 = 0.1$

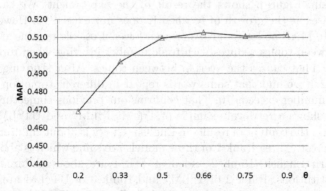

Fig. 5. The influence of θ on the performance of $\mathrm{UT}^2\mathrm{LM}$

μ_3 is very small. That means translation model part has very small weight. The results are consistent with the reported in previous work [11].

Table 4. Question retrieval results (Bold indicates our method and the corresponding results. T^2LM^* and UT^2LM have a statistically significant improvement over the baseline using the t-test, p-value < 0.05)

	MAP	% of MAP improvements over				P@1	MRR
		LM	TLM	T^2LM	T^2LM^*		
LM	0.3583	N/A	N/A	N/A	N/A	0.2516	0.1958
TLM	0.3746	+4.5	N/A	N/A	N/A	0.2846	0.2372
T^2LM	0.4361	+21.7	+16.4	N/A	N/A	0.3208	0.2641
T^2LM^*	**0.4592**	**+28.1**	**+21.3**	**+5.2**	**N/A**	**0.3385**	**0.3195**
UT^2LM	**0.4687**	**+30.8**	**+25.1**	**+7.5**	**+2.1**	**0.3401**	**0.3244**

The T^2LM^* performs better than the T^2LM, and the UT^2LM performs better than the T^2LM^*. The underlying reasons are that the T^2LM^* utilizes the topic information of query as weight to improve the topic component on the basis of T^2LM. UT^2LM introduce the user information into the T^2LM^*. Because of the reason that we describe in Sect. 3.3, the performance of question retrieval is improved further. Comparing with T^2LM^*, the improvement of UT^2LM is small too. The reason is that main function of user information is optimization the ranking list, not affects result strongly. By observing data, we find the difference between Rank 1 and Rank 5 usually is between 1 to 10 in T^2LM^*, here we take the logarithmic technology in the calculation process to prevent underflow. And the optimal value of θ is relatively small 0.5 that illustrates this problem from the side.

The topic number K in the LDA model also has an impact on retrieval result, so we conduct an experiment on development set to find the relationship between them. Figure 6 shows the result of the experiment. We can see that MAP increases as the growth of K when K does not exceed 70. However, when exceeding 70, MAP no longer grows or even lower. Possible reasons are that distinction between topics reduces with topic number growing when topic number is too large. That causes the overlap between topics. After studying the result of LDA model, we find that some words repeat in different topic top words list when topic number exceeds 70. That confirms our previous conjecture.

Figure 7 shows a retrieval example of T^2LM, T^2LM^* and UT^2LM. Here we briefly explain the content of results in English. Query is about children's English learning interest. Three Rank 1 of three models are same which is about helping children learn English. Rank 2 and 3 of T^2LM are about children's English learning institutions. Rank 2 of T^2LM^* and Rank 3 of UT^2LM are same and both are about helping children learn English. Rank 3 of T^2LM^* and Rank 2 of UT^2LM are same and they are about cultivating children's interest in learning

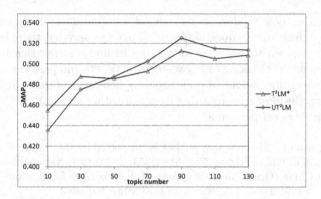

Fig. 6. The influence of topic number on the performance of T^2LM^* and UT^2LM

Query: 我想让孩子学英语，提高一下英语口语水平，但孩子不想学怎么办？

Rank	T^2LM	T^2LM^*	UT^2LM
1	**想提高孩子的英语口语，该如何培养呢？**	**想提高孩子的英语口语，该如何培养呢？**	**想提高孩子的英语口语，该如何培养呢？**
2	暑假了 让孩子去那学点英语呢	**想帮助孩子学好英语口语，该怎么做呢？**	<u>我是想提高小孩对英语的兴趣，掌握学习英语的方法。</u>
3	昆明英语口语培训学校提高英语能力 孩子想上个英语补习班去那儿很有名？	<u>我是想提高小孩对英语的兴趣，掌握学习英语的方法。</u>	**想帮助孩子学好英语口语，该怎么做呢？**

Fig. 7. An example of T^2LM, T^2LM^*, UT^2LM (Bold indicates relevant; Underline indicates the most relevant)

English which are the most relevant result with the query. From Fig. 7 we can see that T^2LM^* performs better than T^2LM. The results all have a relationship with English, but the results of T^2LM^* and UT^2LM focus the topic on the methods and interests in learning English. And we can see that UT^2LM improve ranking list by user information.

5   Conclusion and Future Work

Question retrieval is an important component in Community Question Answering (CQA) services. In this paper, we propose a novel approach by first using topic information of query to improve the quality of the word-to-question similarity estimates. In addition, we exploit user adoption answer number associated with questions in CQA archives for further improving the performance of question retrieval. Experiments on a real CQA data set from Sogouwenwen demonstrate the effectiveness of the proposed retrieval model.

Although the topic model can be used as a potential semantic extension to enhance the performance of question retrieval in CQA services, there is a problem of ambiguity between topics. It is interesting to find an approach to solve the problem. Meanwhile, thinking of keyword extraction techniques, it could be very meaningful to capture weight of each topic by keywords of query instead of all words in query. Besides, it is promising to explore other metadata in CQA archive for improving retrieval performance.

Acknowledgement. This research project is supported by the National Natural Science Foundation of China (Grant Nos: 61602353, 61303029), National Social Science Foundation of China (Grant No: 15BGL048), Hubei Province Science and Technology Support Project (2015BAA072), 863 Program (2015AA015403).

References

1. Ponte, J.M., Croft, W.B.: A language modeling approach to information retrieval. In: Proceedings of the 21st Annual International ACM SIGIR Conference on Research and Development in Information Retrieval, pp. 275–281 (1998)
2. Jeon, J., Croft, W.B., Lee, J.H.: Finding similar questions in large question and answer archives. In: CIKM, pp. 84–90 (2005)
3. Berger, A.L., Caruana, R., Cohn, D., Freitag, D., Mittal, V.O.: Bridging the lexical chasm: statistical approaches to answer-finding. In: Proceedings of the 23rd Annual International ACM SIGIR Conference on Research and Development in Information Retrieval, pp. 192–199 (2000)
4. Riezler, S., Vasserman, A., Tsochantaridis, I., Mittal, V.O., Liu, Y.: Statistical machine translation for query expansion in answer retrieval. In: ACL, pp. 464–471 (2007)
5. Xue, X., Jeon, J., Croft, W.B.: Retrieval models for question and answer archives. In: Proceedings of the 31st Annual International ACM SIGIR Conference on Research and Development in Information Retrieval, pp. 475–482 (2008)
6. Bernhard, D., Gurevych, I.: Combining lexical semantic resources with question & answer archives for translation-based answer finding. In: ACL, pp. 728–736 (2009)
7. Zhou, G., Cai, L., Zhao, J., Liu, K.: Phrase-based translation model for question retrieval in community question answer archives. In: ACL, pp. 653–662 (2011)
8. Wei, W., Croft, W.B.: LDA-based document models for ad-hoc retrieval. In: Proceedings of the 29th Annual International ACM SIGIR Conference on Research and Development in Information Retrieval, pp. 178–185 (2006)
9. Cai, L., Zhou, G., Liu, K., Zhao, J.: Learning the latent topics for question retrieval in community QA. In: IJCNLP, pp. 273–281 (2011)
10. Ji, Z., Xu, F., Wang, B., He, B.: Question-answer topic model for question retrieval in community question answering. In: CIKM, pp. 2471–2474 (2012)
11. Zhang, W.N., Zhang, Y., Liu, T.: A topic inference based translation model for question retrieval in community-based question answering services. Chin. J. Comput. **38**(2), 313–321 (2015)
12. Hofmann, T.: Unsupervised learning by probabilistic latent semantic analysis. Mach. Learn. **45**, 177–196 (2001)
13. Blei, M.D., Ng, Y.A., Jordan, I.M.: Latent dirichlet allocation. J. Mach. Learn. Res. **3**, 993–1022 (2003)

14. Cao, X., Cong, G., Cui, B., Jensen, C.S., Yuan, Q.: Approaches to exploring category information for question retrieval in community question-answer archives. Acm Trans. Inf. Syst. **30**(2), 1–38 (2012)
15. Och, F.J., Ney, H.: Improved statistical alignment models. In: ACL, pp. 440–447 (2000)
16. Zhang, W.N., Ming, Z.Y., Zhang, Y., Liu, T., Chua, T.S.: Capturing the semantics of key phrases using multiple languages for question retrieval. IEEE Trans. Knowl. Data Eng. **28**(4), 888–900 (2016)
17. Chen, L., Jose, J.M., Yu, H., Yuan, F., Zhang, D.: A semantic graph based topic model for question retrieval in community question answering. In: The Ninth ACM International Conference on Web Search and Data Mining, pp. 287–296 (2016)
18. Omari, A., Carmel, D., Rokhlenko, O., Szpektor, I.: Novelty based ranking of human answers for community questions. In: International ACM SIGIR Conference on Research and Development in Information Retrieval, pp. 215–224 (2016)
19. Yuan, Q., Cong, G., Sun, A., Lin, C.Y., Thalmann, N.M.: Category hierarchy maintenance: a data-driven approach. In: The 35th International ACM SIGIR Conference on Research and Development in Information Retrieval, pp. 791–800 (2012)
20. Zhang, W., Ming, Z., Zhang, Y., Nie, L., Liu, T., Chua, T.S.: The use of dependency relation graph to enhance the term weighting in question retrieval. In: COLING, pp. 3105–3120 (2012)
21. Dijk, D.V., Tsagkias, M., Rijke, M.D.: Early detection of topical expertise in community question answering. In: The International ACM SIGIR Conference, pp. 995–998 (2015)
22. Manning, C.D., Raghavan, P., Schütze, H.: Introduction to Information Retrieval, pp. 139–159. Cambridge University Press, Cambridge (2008)

Holographic Lexical Chain and Its Application in Chinese Text Summarization

Shengluan Hou[1,2(✉)], Yu Huang[1,2], Chaoqun Fei[1,2],
Shuhan Zhang[1,2], and Ruqian Lu[1,3]

[1] Key Laboratory of Intelligent Information Processing of Chinese Academy of
Sciences, Institute of Computing Technology,
Chinese Academy of Sciences, Beijing, China
houshengluan1989@163.com
[2] University of Chinese Academy of Sciences, Beijing, China
[3] Key Lab of MADIS, Academy of Mathematics and Systems Science,
Chinese Academy of Sciences, Beijing, China

Abstract. Lexical chain has been widely used in many NLP areas. However, when using it for Web text summarization, especially for domain-specific text summarization, we got low accuracy results. The main reason is that traditional lexical chains only take nouns into consideration while information of other grammatical parts is missing. We introduce lexical chains of predicates and adjectives (adverbs) respectively. These three types of lexical chains together are called holographic lexical chains (HLCs), which capture most of the information included in the text. A specifically designed construction method for HLC is presented. We applied HLC method to Chinese text summarization and used machine learning methods whose features are adapted to the new method. In a comparative study of Chinese foreign trade texts, we got summarization results with accuracy of 86.88%. Our HLC construction method obtained improvements of 7.02% in accuracy than the known best methods in Chinese text summarization.

Keywords: Holographic lexical chain · Text summarization · Machine learning · Lexical cohesion

1 Introduction

Lexical cohesion is a classical tool for analyzing the content of natural language text. Lexical chain is a list of "about the same thing" words, which exploit the lexical cohesion among related words and contributes the continuity of text meaning [13, 19]. Lexical chain has been widely used in many natural language applications, such as text summarization [1–4, 10], machine translation [23], discourse quality measurement [20] and some other areas [5, 17, 22].

The influx of large amount of web documents in the web age results in a great deal of attention on automatic text summarization [14, 15]. Lexical chain is a good tool for text summarization because of its easy computation and high efficiency. There are many efforts of lexical chain based text summarization [2, 4, 10, 19]. Generally, these lexical chain based text summarization methods can be decomposed into two

L. Chen et al. (Eds.): APWeb-WAIM 2017, Part I, LNCS 10366, pp. 266–281, 2017.
DOI: 10.1007/978-3-319-63579-8_21

procedures: (1) Lexical chains construction; (2) Select sentences relating to lexical chains as summary.

In the first procedure, several lexical chains will be computed from each article. Barzilay and Elhadad described the procedure of lexical chain construction as composed of three steps [2]:

1. Select candidate words, including nouns and named entities;
2. For each candidate word, find an appropriate chain to insert, relying on a relatedness criterion between candidate word and members of the lexical chains;
3. If found, insert the word into the chain and update it accordingly; If not, construct a new chain and insert the word into the new chain.

However, when using it for Web text summarization, especially for domain-specific text summarization, we got low accuracy results. The main reason is that traditional approaches for lexical chain construction only take nouns (and named entities, as shown in [2]) into consideration, while information of other grammatical parts, such as predicates or adjectives, which carry important information for text summarization, is missing. On the other hand, these methods adopt the assumption that "one sense per discourse", which had been proved not valid by other researchers [9].

The novelty of our approach is to introduce other two kinds of lexical chains, namely the predicates lexical chains and adjectives (adverbs) lexical chains. In this way we have lexical chains for all three major grammatical parts. These three types of lexical chains together are therefore called holographic lexical chains (HLCs for short), which capture most of the information included in the text. A specifically designed construction method for HLCs is presented. Moreover, we present a scoring criterion to identify strong chains and weak chains.

To illustrate this, let us consider the following text with seven sentences:

Rail freight giant Aurizon has cancelled its $91 million effort to standardize legacy systems onto a single SAP platform after an internal review found the project was at risk of delays and overspend.
In early 2014 the company revealed its plan to consolidate 18 separate legacy systems for logistics, planning, scheduling, ordering, and billing onto SAP HANA and SAP's supply chain execution (SCE) platform 9.1.
The rail operator at the time said the system would improve visibility across the supply chain by "integrating long and short-term planning with resource availability and customer demand".
It had previously implemented SAP for ERP and asset maintenance, and called the HANA and supply chain platform a logical extension.
However, after three years the project will now be cancelled and an impairment charge of $64 million recorded in the company's first quarter FY17 results, Aurizon said today.
Around $27 million of the project's total $91 million cost remains capitalized for software and licenses that are still in use, it said.
Aurizon CEO Andrew Harding said the project was not delivering value for the business.

Fig. 1. An example of Web text

The HLCs can be identified as:

Noun lexical chains:
  rail[1], rail[3];
  Aurizon[1], Aurizon[5], Aurizon[7];
  million[1], million[5], million[6], million[6];
  legacy system[1], SAP platform[1], legacy system[2], SAP HANA[2], SAP[2],
  platform[2], system[3], SAP[4], ERP[4], HANA[4], platform[4], software[6];
  project[1], project[5], project[6], project[7].
Predicate lexical chains:
  cancel[1], cancel[5];
  reveal[2], say[3] say[7].
Adjective(Adverb) chains:
  single[1], separate[2];
  long[3], short[3].

Fig. 2. An example of HLCs

In Fig. 2, the digits in square brackets are sentence numbers.

We further investigate the connections between text summarization and lexical cohesion. We apply HLC to Chinese text summarization. Our Chinese text summarization method can be viewed as two-step process, too. The first step is to build HLC of a text. The second step is selecting summarizing sentences using machine learning methods whose features are adapted to our new HLC construction method.

Our aim is to:

1. Design an accurate and efficient HLC construction algorithm, which can process documents of any length;
2. Design a robust Chinese text summarizer based on HLC approach together with machine learning techniques.

The rest of this paper is organized as follows. In the next section, we survey the related works about lexical chain construction and text summarization. Section 3 defines and illustrates the concept of HLC. A novel HLC construction algorithm and scoring criterion to identify strong chains and weak chains are also presented in this section. Section 4 illustrates HLC's application in Chinese text summarization. The experiments setup and experimental results analysis are shown. Finally, we conclude the paper with hints on future work in Sect. 5.

2   Related Works

Morris and Hirst first introduced the concept of lexical chain and showed it can be an indicator of the whole text structure [13]. They used Roget's thesaurus as the major knowledge base for computing lexical chains. Since there was no computer-readable thesaurus at that time, their algorithm cannot be implemented on the computer.

Hirst and St-Onge presented the first computer-implemented lexical chain construction algorithm, which used WordNet [12] as knowledge base [7]. They defined three kinds of word sense relations. When inserting a candidate word into an existing chain, an extra-strong relation is sought throughout all chains. If a chain is found, the word is inserted with the appropriate sense and the senses of other words are updated; If not, strong relation is preferred to medium-relation to sought. Note that there is no limit in distance for extra-strong relations, seven and three sentences are limited for strong and medium-strong relations respectively.

However, Hirst and St-Onge's "greedy disambiguation" resulted in low word sense disambiguation (WSD) accuracy, since their method disambiguates a word at its first encountering. Barzilay and Elhadad proposed a new method according to all possible alternatives of word senses and chose the best one among them [2]. They also used WordNet as knowledge base. This method significantly improves WSD accuracy but with exponential complexity. It constructs all possible interpretations of candidate words and selects that interpretation with the strongest cohesion, which causes the inefficiency.

Based on Barzilay and Elhadad's work, Silber and McCoy defined a linear time algorithm [19]. They rewrote Wordnet noun database and tools to facilitate efficient access, but still with low accuracy in WSD.

Galley and McKeown further investigated automatic lexical chains construction [6]. They separated WSD and lexical chain construction into two different sub-tasks and adopted the assumption of "one sense per discourse". All word semantic links were linked to a disambiguation graph in time $O(n)$ and all wrong links are removed from this disambiguation graph.

Remus and Biemann presented three knowledge-free methods for lexical chain construction [18]. They used the Latent Dirichlet allocation (LDA) topic model for estimating the semantic closeness of candidate terms. These corpus-driven methods adopt the idea of interpreting lexical chains as clusters and a particular set of lexical chains as a clustering. Other lexical chain applications are mostly based on Galley and McKeown's methods.

The above lexical chain methods only take nouns into consideration. Few efforts focused on verb or adjective lexical chains. Novischi and Moldovan used WordNet to create two types of lexical chains for question answering. Verb lexical chains were used for propagating verb arguments, and noun lexical chainss were used to link semantically related arguments [16]. Jarmasz and Szpakowicz constructed lexical chains using Roget's thesaurus, whose hierarchical structure allow them to build lexical chains using nouns, adjectives, verb, adverbs and interjections [8].

Text summarization is a difficult but important research area in the web age of information explosion. Generally, the methods can be categorized into two approaches: abstract-based approaches and extraction-based approaches. Abstract-based approaches can be considered as content synthesis of source text, which needs deep natural language understanding. Extraction-based approaches extract important sentences to generate a concise and coherent version of original text. Extraction-based approaches are main approaches for text summarization.

Lexical chain based text summarization is a method of extraction-based summarization. Many efforts of lexical chain based text summarization have been developed to date. Barzilay and Elhadad presented a method of scoring chains. The sentences containing these strong chains are chosen as summary sentences [2]. Ercan and Cicekli implemented Galley's lexical chain construction method and represented topics by sets of co-located lexical chains to take advantage of more lexical cohesion clues [4]. They segmented text into topic episodes by means of lexical chain clustering. The final summary was sentences selected from each topic segments.

3 Holographic Lexical Chains

Since Morris and Hirst first proposed the concept of lexical chain, many lexical chain construction methods along with lexical chain based applications have been presented to date. However, these lexical chain construction methods only take nouns (and/or named entities) into consideration, while information of other grammatical parts, such as predicates or adjectives, which often carry important information for text summarization, is neglected.

We introduce other two kinds of lexical chains: predicates lexical chains and adjectives (adverbs) lexical chains in this section. Nouns, predicates and adjectives (adverbs) form the major grammatical parts of a sentence. Their corresponding lexical chains capture most of the information included in the text, which we call holographic lexical chains.

The candidate words of holographic lexical chains contain three parts: noun candidate words, predicate candidate words and adjective (adverb) candidate words. All the words were stemmed first. We selected all nouns and named entities as noun candidate words, which describe the topics of a text. For each sentence, there's only one predicate, which describes the property that a subject has and can be recognized as the root of dependency tree. All predicates from each sentence constitute predicate candidate words. All adjectives and adverbs were selected as adjective (adverb) candidate words, which qualify nouns.

Definition 1 Holographic Lexical Chains (HLCs). Holographic lexical chains of a discourse include three types of lexical chains. They can be summarized with a triple:

$$HLCs = <NChains, PChains, AChains > \tag{1}$$

where

(1) N*Chains* are noun lexical chains, each of which is a list of semantically related nouns, whose elements consist of nouns or named entities;
(2) P*Chains* are predicate lexical chains, each of which is a list of semantically related predicates. The elements in each chain are usually verbs;
(3) A*Chains* are adjective (adverb) lexical chains, each of which is a list of semantically related adjectives and/or adverbs.

Let N*Chains* ∪ P*Chains* ∪ A*Chains* = $\{C_1, C_2, \cdots, C_k\}$, where each $C_i (1 \leq i \leq k)$ is a lexical chain. We have the following properties:

(1) $C_i \cap C_j = \emptyset$, $1 \leq i, j \leq k$ and $i \neq j$;
(2) $C_1 \cup C_2 \cup \cdots \cup C_k = C$, C is the set of all candidate words.

Property (1) means there's no common words between two different lexical chains. In other words, every word occurrence can only belong to one lexical chain, there is no overlapping word occurrences. Property (2) means for all candidate words, each word should belong to one chain.

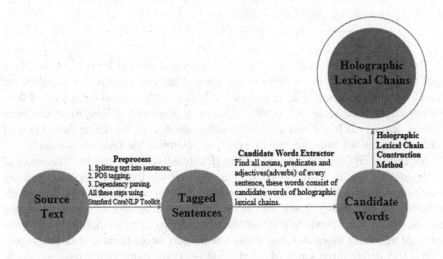

Fig. 3. Process of holographic lexical chains construction

For each source text, the process of HLC construction is shown as follows (Fig. 3):

In the preprocessing procedure, we use Stanford CoreNLP Toolkit [11] for sentence splitting, POS tagging and dependency parsing. Then all nouns, predicates and adjectives (adverbs) are extracted, which form candidate words of HLCs. Finally, we present holographic lexical chain construction method to compute HLCs.

Algorithm 1 Holographic lexical chain construction method

Input: Three types of candidate words, denoted by **cws**.
Output: Holographic lexical chains, denoted by **hlcs**.

```
nounChains =
    getLexicalChains(cws.nouns);
predicateChains =
    getLexicalChains(cws.predicates);
adjectiveChains =
    getLexicalChains(cws.adjectives);
hlcs.setNounChains(nounChains);
hlcs.setPredicateChains(predicateChains);
hlcs.setAdjectiveChains(adjectiveChains);
```

The function "getLexcialChains" above is a general method for constructing three types of lexical chains. The novelty of our method has two aspects: (1) we adopt the idea of "disambiguation graph" in Galley and McKeown's method but get rid of their assumption of "one sense per word", different appearances of one word can have different senses; (2) In the word sense disambiguation step, we define three kinds of word sense relations and each with a weight to determine the final word sense.

We choose HIT IR-Lab Tongyici Cilin (Extended version) (means 'Thesaurus of Synonym Words', TCE for short) [21] as our knowledge base for automatically qualifying semantic similarity between words. TCE groups words into five levels according to their senses. Each line in the fifth level refers to a common semantic concept, which corresponds to a synset that contains words of similar meaning. Since one word may have more than one sense, a word may occur in more than one synset.

We also define three kinds of word sense relations: extra-strong relation, strong relation and medium-strong relation. They are different from the three kinds of relations in Hirst and St-Onge's definition. Empirically, we define extra-strong relation as a word and its literal repetition, strong relation as two synonyms words or words with lowest common ancestor level 4 in TCE, and medium-strong as two words belonging to the same word class, where the lowest common ancestor level is 3 in TCE.

The function "getLexcialChains" has the following three steps:

1. Disambiguation graph construction. Disambiguation graph here is an undirected graph, whose vertex denotes word occurrence, and edge denotes the semantic relationship between two word occurrences.

2. Word sense disambiguation. If a word occurrence has more than one senses, find and preserve the most appropriate sense. Remove other senses and their corresponding edges.
3. Final lexical chains building. Traverse disambiguation graph and the words related by edges form one lexical chain. All building lexical chains consist of the final lexical chains.

Algorithm 2 Disambiguation graph construction

Input: Candidate word list, denoted by **cwl**.
Parameter: The maximum number of sentences between two words who has strong relation, denoted by **msr**; The maximum number of sentences between two words who has medium-strong relation, denoted by **mmsr**.
Output: Disambiguation graph.

Create an empty disambiguation graph, denoted by **dg**.
for each word cw_i in **cwl**
  add cw_i to **dg**;
  for each word gw_j in **dg**
    if cw_i and gw_j has extra-strong relation
      create edge between cw_i and gw_j;
    else if cw_i and gw_j has strong relation and
        distance(cw_i, gw_j)<=**msr**
      create edge between cw_i and gw_j;
    else if cw_i and gw_j has medium-strong relation
        and distance(cw_i, gw_j)<=**mmsr**
      create edge between cw_i and gw_j;
    end if
  end **for**
end **for**

The "distance(cw_i, gw_j)" in Algorithm 2 denotes the number of sentences between the sentence that contains cw_i and sentence contains gw_j. Empirically, we choose the parameters "**msr**" and "**mmsr**" in Algorithm 2 as 7 and 3 respectively.

Algorithm 3 Word sense disambiguation

Input: Disambiguation graph, denoted by *dg*.
Output: Disambiguation graph after word sense disambiguation.

---

for each word cw_i in *dg*
  if cw_i has more than one sense
    for each sense scw_i^j of cw_i
      compute the weight of scw_i^j;
    end **for**
    preserve the sense with highest weight, remove other senses and their corresponding edges;
  end **if**
end **for**

In Algorithm 3, the weight of cw_i's sense scw_i^j means the sum weight of all edges that scw_i^j related to. If scw_i^j connects to n vertexes $\{v_1, v_2, \cdots v_n\}$ and their corresponding edges are $\{e_1, e_2, \cdots e_n\}$, then the weight of scw_i^j is:

$$W_{scw_i^j} = \sum_{k=1}^{n} W_{relationType} * \frac{1}{distance(cw_i, v_k)} \qquad (2)$$

where

$$W_{relationType} = \begin{cases} 1, extra - strong\ relation \\ 0.5, strong\ relation \\ 0.3, medium - strong\ relation \end{cases} \qquad (3)$$

and $distance(cw_i, v_k)$ denotes the number of sentences between the sentence that contains cw_i and sentence contains v_k.

After construction of HLC, there may be some weak chains, such as the chains whose length is 1. One must identify the useful and strong chains. We propose a scoring criterion for chains in HLC. For each chain, the sore can be computed as:

$$Score = ChainStrength * HomogeneityIndex \qquad (4)$$

where

$$ChainStrength = \frac{D_{chainspan}}{N_{all\ sentences}} * L * \sum W_{relationType} \qquad (5)$$

and

$$HomogeneityIndex = 1 - \frac{N_{distinctMembers}}{L} \tag{6}$$

The symbol L in formulas (5) and (6) denotes the number of all words in the chain. In formula (5), $D_{chainspan}$ denotes the number of sentences included in a text block, in which the first and the last sentence contain the first and the last element of the chain respectively; $N_{all\ sentences}$ denotes the number of all sentences of the text; $\sum W_{relationType}$ denotes the sum of all weights of relation between any two words in the chain. $N_{distinctMembers}$ in formula (6) denotes the size of distinct members in the chain. For example, the distinct members in chain "apple, apple, fruit, banana, fruit" are "apple, fruit, banana", whose number is 3.

For each type of lexical chains in HLC, the strong chains satisfy the following formula:

$$Score > AverageScore + \alpha * StandardDeviation \tag{7}$$

The weak chains satisfy the following formula:

$$Weak\ chain = \begin{cases} length < 2, \\ Score < AverageScore - \beta * StandardDeviation. \end{cases} \tag{8}$$

Where $AverageScore$ and $StandardDeviation$ mean the average score and standard deviation of this type of chains belonging to a text, respectively. Here α and β are parameters which can be determined by specific requirements. Empirically, α and β can be determined as 2 by default.

When using HLC to specify applications, weak chains can be removed or only strong chains be selected for better performance and high efficiency.

4 Application in Chinese Text Summarization

In order to verify the validity of our HLC method, we applied it for Chinese text summarization. Since there is no Chinese text summarization benchmark dataset, we chose Chinese foreign trade texts as the domain of our experimental corpus and extracted articles from the Internet. We randomly selected 159 texts as experiment dataset, from which the summary sentences are tagged manually. The average sentence length of these texts is 53.59 and the average number of sentences in each text is 13.53. Excerpts of one of this text is shown in Fig. 4.

In the summary generation step, we used machine learning methods rather than heuristic rules. We treated summary sentence selection as a binary classification problem. Features used in this step are the following (Table 1):

a. Co-appearance of words from different types of lexical chains. Since nouns, predicates and adjective (adverb) constitute all major grammatical parts, we took those sentences as summary sentences, which contain words from two or more types of lexical chains. For example, if a sentence contains three types of lexical chain words "进出口(Import-Export), 增长(Rise), 明显(Obviously)", the sentence may be a summary sentence. There are four vectors of this feature:

对于一季度的外贸形势,商务部长高虎城近日表示,对2月份可能出现的外贸数据波动,各界要有充分的思想准备。

For the foreign trade situation in the first quarter, Minister of Commerce Gao Hucheng said recently that various circles of society should be fully prepared to undertake the probable data fluctuation of foreign trade in February.

高虎城指出,春节因素并不是造成中国开年进出口"双降"的唯一原因。

Gao Hucheng pointed out that the Spring Festival is not the only reason for the "double down" of China's import and export.

由于全球市场需求的疲弱及大宗商品价格的低迷,世界主要经济体的进出口均表现不佳,造成了全球贸易开局不佳的态势。

Due to the weak global market demand and the sluggish bulk commodity prices, the world's major economic entities were the poor performance of imports and exports, resulting in a poor start of global trade trend.

Fig. 4. An example of Chinese foreign trade text (excerpt)

Table 1. Four vectors of word co-appearence feature

Noun + Predicate + Adjective (Adverb)
Noun + Predicate
Noun + Adjective (Adverb)
Predicate + Adjective (Adverb)

b. Appearance of the first word of any lexical chain. It means that this sentence is the begin of a new topic, which should be considered as a summary sentence.

c. Appearance of the representative word of any lexical chain. A representative word is the highest frequency word of a lexical chain. It means that the content of this sentence is closely related to the topic of this lexical chain.

d. Appearance of two or more elements of the same lexical chain. It means that this sentence is more closely related to the lexical chain than the others.

e. Appearance of words from more lexical chains than the others.

f. The first sentence of a text.

g. The long sentence. We define long sentence here as:

$$Length > AverageLength + \lambda * StandardDeviation \tag{9}$$

where λ is a parameter which was taken as 2 in this experiment.

According to these features the summary sentence of example text in Fig. 1 is its first sentence.

In our experiment, after HLC construction, the algorithm filters out weaker lexical chains for better performance and high efficiency (Table 2).

Table 2. HLC filter operation

Chain type	Filter operation
Noun chains	Select strong chains
Predicate chains	Remove weak chains
Adjective (Adverb) chains	Remove weak chains

We used both linear regression and support vector machine methods with different feature combinations for comparative study.

To evaluate our method comprehensively, we used four evaluation criteria: accuracy, precision, recall and F1 value, which measure the performance of our method on different aspects (Table 3).

Table 3. The confusion matrix

All sentences	Predicted	
	Predicted as summary sentences	Predicted as non-summary sentences
Tagged sentences	TP	FN
Non-tagged sentences	FP	TN

Where TP represents the number of tagged sentences that are correctly predicted as summary sentences; FN represents the number of tagged sentences that are falsely predicted as non-summary sentences; FP represents the number of non-tagged sentences that are falsely predicted as summary sentences; TN represents the number of non-tagged sentences that are correctly predicted as non-summary sentences.

The evaluation criteria are defined as follows.

Accuracy: the percentage of correctly predicted sentences out of all sentences:

$$Accuracy = \frac{TP + TN}{TP + TN + FP + FN} \tag{10}$$

Precision: the percentage of correctly predicted as summary sentences out of all tagged sentences:

$$Precision = \frac{TP}{TP + FP} \tag{11}$$

Recall: the percentage of correctly predicted as summary sentences out of all predicted sentences:

$$Recall = \frac{TP}{TP + FN} \tag{12}$$

F1: the harmonic mean of Precision and Recall, which is calculated as:

$$F1 = \frac{2 \times Precision \times Recall}{Precision + Recall} \tag{13}$$

We used ten-fold cross validation and computed the average value of those above four criteria.

The evaluation results using ten-fold cross validation are shown in Table 4, where "a", "b" and "c" denote the first three features respectively. "All" denotes all above features from "a" to "g". "All-a" denotes all above features except "a". "a + b + c" denotes combination of features "a", "b" and "c". The accuracy of 86.88% showed that HLC is a good tool for Chinese text summarization. Furthermore, experimental results confirmed that the feature of word co-appearance plays an important role in Chinese text summarization, which captures most information of a sentence. Support vector machine method outperforms linear regression method.

Table 4. Evaluation results (%) of text summarization

Methods	Features	Average accuracy	Average precision	Average recall	Average F1
Support vector machine	All	86.88	84.40	73.89	76.47
	All-a	83.67	85.61	60.32	67.35
	a + b + c	80.23	73.21	71.38	69.54
Linear regression	All	83.31	78.34	69.78	71.51
	All-a	82.19	79.51	58.11	66.93
	a + b + c	79.05	81.23	63.65	69.62

Table 4 only shows the average values of four criteria, from which we cannot get the data fluctuation information of ten-fold experimental results. We take the results of support vector machine method as an example. Figure 5 shows the data fluctuation of four criteria from ten-fold experiments, from which we can see that the fluctuating range of experimental results is small. Our experimental results of ten-fold cross validation remain stable.

The reasons of failure of identifying the correct summary sentences including two aspects: (1) the tagged sentences were falsely classified as non-summary sentences, because existing features cannot cover all their characteristics; (2) the non-tagged sentences were falsely classified as summary sentences, because these sentences capture common features with summary sentences. According to statistics, the percentage of reason (1) is almost the same as that of reason (2).

For comparison, we re-implemented three main traditional lexical chain construction methods, and used them to construct traditional noun lexical chains on the same Chinese foreign trade corpus. These three methods are listed as follows:

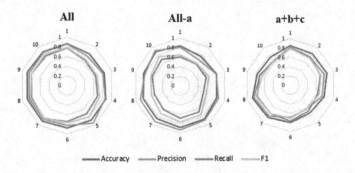

Fig. 5. Data fluctuation of four criteria (support vector machine method)

H&S: Algorithm by Hirst and St-Onge [7], with TCE as its knowledge base.
B&E: Algorithm by Barzilay and Elhadad [2], with TCE as its knowledge base.
G&M: Algorithm by Galley and McKeown [6], with TCE as its knowledge base.

Previous lexical chain based text summarization methods used heuristic rules to select summary sentences, such as appearance of the head word of a lexical chain, appearance of representative word of a lexical chain, co-appearance of two or more lexical chains. We used support vector machine method whose features are adapted to these above rules.

The accuracy results in Table 5 show the advantage of our innovation in HLC construction method and feature selection outperforms other methods with improvement of 7.02% than the best of them(G&M method).

Table 5. Comparison (%) of different lexical chain construction methods for text summarization

Algorithm	Accuracy	Precision	Recall	F1
H&S method	76.39	67.37	60.34	60.82
B&E method	79.25	69.85	65.90	63.75
G&M method	79.86	70.21	64.28	64.03
Our method	**86.88**	**84.40**	**73.89**	**76.47**

Meanwhile, we tested the running efficiency of our HLC method and other three traditional lexical chain methods on the same platform. We used a PC with an I5 CPU (2.50 GHz * 4 Cores) and 16 G memory.

Our method can construct three types of lexical chains at the same time while the other three methods only construct noun lexical chains. Nevertheless, as Fig. 6 shows, the time cost of our method is nearly the same as H&S method and G&M method. These three methods are considerably faster than B&E method.

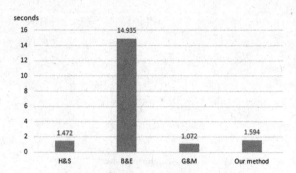

Fig. 6. Time cost of different methods

5 Conclusion

In this paper, we presented our approach of holographic lexical chains (HLCs), which contain three kinds of lexical chains: noun lexical chains, predicate lexical chains and adjective (adverb) lexical chains. HLCs capture most of the information included in the text. A specifically designed HLC construction method and scoring criterion are also presented in this paper. We applied HLC technique to Chinese text summarization. In the summary sentence selection step, we used two machine learning methods: linear regression and support vector machine whose features are adapted to HLC and other text features. Comparative experiments on Chinese foreign trade text summarization demonstrate that our holographic lexical chain construction method outperforms other methods in Chinese text summarization.

One of the future works will be further improving efficiency and accuracy of the HLC technique. A possible way is to integrate HLC with some other features, such as those of discourse structure. Another important task is further applying HLC to other NLP applications.

Acknowledgement. This work was supported by National Key Research and Development Program of China under grant 2016YFB1000902, National Natural Science Foundation of China (No. 61232015, 61472412, 61621003), Beijing Science and Technology Project: Machine Learning based Stomatology and Tsinghua-Tencent-AMSS Joint Project: WWW Knowledge Structure and its Application.

References

1. Alam, H., Kumar, A., Nakamura, M., et al.: Structured and unstructured document summarization: design of a commercial summarizer using lexical chains. In: ICDAR, vol. 3, pp. 1147 (2003)
2. Barzilay, R., Elhadad, M.: Using lexical chains for text summarization. Adv. Autom. Text Summar. 111–121 (1999)
3. Brügmann, S., Bouayad-Agha, N., Burga, A., et al.: Towards content-oriented patent document processing: intelligent patent analysis and summarization. World Patent Inf. **40**, 30–42 (2015)

4. Ercan, G., Cicekli, I.: Lexical cohesion based topic modeling for summarization. In: Gelbukh, A. (ed.) CICLing 2008. LNCS, vol. 4919, pp. 582–592. Springer, Heidelberg (2008). doi:10.1007/978-3-540-78135-6_50

5. Feng, W.L.: Research of theme statement extraction for chinese literature based on lexical chain. Int. J. Multimedia Ubiquitous Eng. **11**(6), 379–388 (2016)

6. Galley, M., McKeown, K.: Improving word sense disambiguation in lexical chaining. In: IJCAI, vol. 3, pp. 1486–1488 (2003)

7. Hirst, G., St-Onge, D.: Lexical chains as representations of context for the detection and correction of malapropisms. WordNet Electr. Lex. Database **305**, 305–332 (1998)

8. Jarmasz, M., Szpakowicz, S.: Not as easy as it seems: automating the construction of lexical chains using *Roget's Thesaurus*. In: Xiang, Y., Chaib-draa, B. (eds.) AI 2003. LNCS, vol. 2671, pp. 544–549. Springer, Heidelberg (2003). doi:10.1007/3-540-44886-1_48

9. Krovetz, R.: More than one sense per discourse. NEC Princeton NJ Labs., Research Memorandum (1998)

10. Li, J., Sun, L., Kit, C., et al.: A query-focused multi-document summarizer based on lexical chains. In: Proceedings of Document Understanding Conference (2007)

11. Manning, C.D., Surdeanu, M., Bauer, J., Finkel, J., Bethard, S.J., McClosky, D.: The stanford CoreNLP natural language processing toolkit. In: Proceedings of the 52nd Annual Meeting of the Association for Computational Linguistics: System Demonstrations, pp. 55–60 (2014)

12. Miller, G.A.: WordNet: a lexical database for English. Commun. ACM **38**(11), 39–41 (1995)

13. Morris, J., Hirst, G.: Lexical cohesion computed by thesaural relations as an indicator of the structure of text. Comput. Linguist. **17**(1), 21–48 (1991)

14. Munot, N., Govilkar, S.S.: Comparative study of text summarization methods. Int. J. Comput. Appl. **102**(12), 33–37 (2014)

15. Nenkova, A., McKeown, K.: A survey of text summarization techniques. In: Aggarwal, C., Zhai, C. (eds.) Mining Text Data, pp. 43–76. Springer, US (2012)

16. Novischi, A., Moldovan, D.: Question answering with lexical chains propagating verb arguments. In: Proceedings of the 21st International Conference on Computational Linguistics and the 44th Annual Meeting of the Association for Computational Linguistics, pp. 897–904. Association for Computational Linguistics (2006)

17. Qian, T., Ji, D., Zhang, M., et al.: Word sense induction using lexical chain based hypergraph model. In: COLING, pp. 1601–1611 (2014)

18. Remus, S., Biemann, C.: Three knowledge-free methods for automatic lexical chain extraction. In: HLT-NAACL, pp. 989–999 (2013)

19. Silber, H.G., McCoy, K.F.: Efficiently computed lexical chains as an intermediate representation for automatic text summarization. Comput. Linguist. **28**(4), 487–496 (2002)

20. Somasundaran, S., Burstein, J., Chodorow, M.: Lexical chaining for measuring discourse coherence quality in test-taker essays. In: COLING, pp. 950–961 (2014)

21. Che, W., Li, Z., Liu, T.: LTP: a chinese language technology platform. In: Proceedings of the Coling 2010: Demonstrations, pp 13–16, Beijing, China, August 2010

22. Wei, T., Lu, Y., Chang, H., et al.: A semantic approach for text clustering using WordNet and lexical chains. Expert Syst. Appl. **42**(4), 2264–2275 (2015)

23. Xiong, D., Ding, Y., Zhang, M., et al.: Lexical chain based cohesion models for document-level statistical machine translation. In: EMNLP, pp. 1563–1573 (2013)

Authorship Identification of Source Codes

Chunxia Zhang[1(✉)], Sen Wang[1], Jiayu Wu[1], and Zhendong Niu[2]

[1] School of Software, Beijing Institute of Technology, Beijing, China
{cxzhang,wangyuwangsen,2220160656}@bit.edu.cn
[2] School of Computer Science, Beijing Institute of Technology, Beijing, China
zniu@bit.edu.cn

Abstract. Source code authorship identification is an issue of authorship identification from documents, and it is to identify authors of source codes or programs based on source code examples of programmers. The main applications of authorship identification of source codes include software intellectual property infringement, malicious code detection and software maintenance and update. This paper proposes an approach of constructing author profiles of programmers based on a logic model of continuous word-level n-gram and discrete word-level n-gram, and a multi-level context model about operations, loops, arrays and methods. Further, we employ the technique of sequential minimal optimization for support vector machine training to identify authorship of source codes. The advantage of author profiles in this paper can discover explicit and implicit personal programming preference patterns of and between keywords, identifiers, operators, statements, methods and classes. Experimental results on programs from two open source websites demonstrate that our approach achieves a high accuracy and outperforms the baseline methods.

Keywords: Authorship identification · Source code · Software forensics · Discrete word-level n-gram · Sequential minimal optimization

1 Introduction

Source code authorship identification is to identify authors of source codes or programs based on code examples of a given set of candidate programmers [1–6]. Authorship identification of source codes is an issue of authorship identification which aims to determine authors of texts such as literatures, articles, essays, emails, blogs, source codes and online forum messages [7,8]. In addition, authorship identification of source codes is an important task in the field of software forensics, and this field is to analyze software source codes or executable codes in order for distinguishing authors of softwares or personalities of those authors [9,10].

The wide applications of authorship identification technique of source codes include software intellectual property infringement, malicious code detection, and software maintenance and update [1,3,10–14]. First, software intellectual

© Springer International Publishing AG 2017
L. Chen et al. (Eds.): APWeb-WAIM 2017, Part I, LNCS 10366, pp. 282–296, 2017.
DOI: 10.1007/978-3-319-63579-8_22

property infringement contains copyright or patent infringement. Authorship identification of source codes can be used to settle ownership disputes of unauthorised source codes [1,10,12–14]. Second, malicious code detection is to detect computer viruses, computer worms, spyware and adware and so on [1,3,12,14]. Authorship identification can help to decide authors or developers of malicious codes. Third, authorship identification of source codes can be utilized to identify authors of previous programs or program modules and track the authorship of program variations in the process of software maintenance and update [1,12].

However, it is infeasible and time-consuming to recognize authors of source codes in a manual way [13,15]. Hence, this paper focuses on how to identify authors of source codes or programs. The differences between authorship identification of source codes and authorship identification of natural language texts are given as follows [14,16–18]: (1) a natural language is a kind of open and complicated language. However, a programming language in which source codes are written is a type of formal and restrictive language. (2) The flexibility of source codes is mainly embodied in the aspects of layout, style, structure and logic of programs. And features about layout, style, structure and logic of source codes usually rely on personal experiences and habits of programmers.

The major challenges of authorship identification of source codes include (1) how to extract features which are independent of functions or purposes of source codes, and specific names of identifiers such as variables and methods; (2) how to extract features which are relatively steady across different programs of the same programmer and in the evolving process of programming characteristics of developers [6,18]; (3) how to extract features which can highlight distinctivenesses among different programmers [13,15].

In this paper, we propose an approach of constructing author profiles of programmers based on a logic source code-oriented model of continuous word-level n-gram and discrete word-level n-gram, and a multi-level context model about operations, loops, arrays and methods. Further, we employ the technique of sequential minimal optimization (SMO) for support vector machine (SVM) training to identify authorship of source codes. Experimental results on programs from two open source websites demonstrate that our approach achieves a high accuracy and outperforms the baseline methods.

The principal contributions of this work are given as follows. (1) An approach of building author profiles of programmers based on a logic source code-oriented model of continuous word-level n-gram and discrete word-level n-gram, and a multi-level context model about operations, loops, arrays and methods is proposed in this paper. The author profiles capture explicit and implicit personal programming characteristics of different programmers on the using of keywords, identifers, operators, statements, methods and classes, and programming collation patterns between those different granularities of components of programming languages. Moreover, extracted features within author profiles are purpose-independent, and are not restricted by specific user-defined names of identifiers such as variables, methods, classes and interfaces. (2) This paper provides a promising technique to fulfill the task of source code authorship identification.

Experimental results on programs from two open source websites show that the identification performance based on our author profiles of programmers by using SMO outperforms two baseline methods, is better than those of present features in related works and those of decision tree and random forest.

The rest of the paper is organized as follows. Section 2 introduces related works about authorship identification of source codes. Section 3 presents our source code authorship identification algorithm. Experimental results are given in Sect. 4. Section 5 concludes the paper and discusses future works.

2   Related Works

The authorship identification techniques of source codes include ranking approaches and machine classifier approaches [14,17]. Further, the latter kind of approaches can be classified into three types of methods: statistical analysis, machine learning, and similarity calculation methods [14,17].

The representative works which used ranking approaches consist of works of Burrows et al. [13,14,19–22] and Frantzeskou et al. [16]. Burrows et al. [13, 14,19–22] adopted an information retrieval technique to deal with the task of authorship identification of source codes. First, they converted programs into token sequences with operator tokens, keyword tokens, and white space tokens. Second, transformed token sequences into a token-level n-gram representation. Third, built an index of all programs and queried the index for a test program with unknown authors. Fourth, selected the authors of top-ranked programs as the authors of the test program. Frantzeskou et al. [16,23] applied a method of Source Code Author Profiles with byte-level n-gram features to solve the problem of source code authorship identification. Here, an author profile is composed of the most frequent n-grams in training data of that author. To a test program with unknown authorship, they first computed the number of common n-grams between the test program and each author profile, and then chose the author with the most common n-grams as the author of that test program.

Krsul and Spafford [24], Ding and Samadzadeh [2] and MacDonell et al. [12] utilized statistical analysis methods to recognize authorship of source codes. Krsul and Spafford [14,24] extracted metrics which contain programming layout metrics, style metrics and structure metrics, and used a multiple discriminant analysis method to identify authorship of source codes. MacDonell et al. [12] calculated metrics about whitespace features, some specific character-level features and some keyword features, and employed a multiple discriminant analysis approach to address the issue of source code authorship identification. Furthermore, Ding and Samadzadeh [2] adopted an canonical discriminant analysis technique to judge authorship of source codes.

The main works of machine learning approaches include the works of Mac-Donell et al. [12], Kothari et al. [1] and Elenbogen and Seliya [25]. MacDonell et al. [12] used feed-forward neural networks and case-based reasoning to identify authorship of source codes. Kothari et al. [1] extracted programming style metrics and character-level n-gram features, and utilized the Bayes classifier and the voting feature intervals to determine authorship of source codes. In addition, Elenbogen and Seliya [25] computed style metrics including number of lines of code, number of comments, average length of variable names, number of variables, and adopted an decision tree model to decide authorship of source codes.

Lange and Mancoridis [3] and Shevertalov et al. [26] employed the similarity calculation method to identify authorship of source codes. The two works computed distances between a program with unknown authors and programs of known authors, and used the nearest-neighbor method to judge authorship of codes.

3 An Authorship Identification Approach of Source Codes

We first give the definition of source code authorship identification.

Definition 1. Given a set P of programmers and their source code examples, the task of *source code authorship identification* is to determine which programmer in the set P wrote a program with unknown authors.

3.1 Overview of Our Approach

Our approach of authorship identification of source codes includes two phases: author profiles construction and author identification, as shown in Fig. 1. The phase of author profiles construction is composed of four steps: (1) programming layout feature extraction, (2) programming style feature extraction, (3) programming structure feature extraction, and (4) programming logic feature extraction based on a continuous word-level n-gram model, a discrete word-level n-gram model and a character-level n-gram model. Furthermore, a method based on the sequential minimal optimization for SVM training is used to identify authorship of source codes or programs.

3.2 Feature Extraction

Author profiles of programmers play an important role in the issue of source code authorship identification. Author profiles should have relative stability within a programmer and highlight distinctivenesses among programmers.

The features constituting author profiles can be divided into four categories: programming layout features, programming style features, programming structure features and programming logic features. Table 1 gives the sets of programming features which are used in our approach and other related works, while Table 2 demonstrates the sets of programming features which are proposed in our approach. Tables 1 and 2 show the category, ID, feature focus and feature description. Here, the feature focus means the programming topic or entity which the feature is concerned with. In this paper, lines of a program are classified into four kinds of lines: code lines, comment lines, blank lines, and hybrid lines of

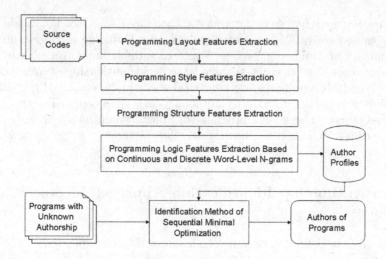

Fig. 1. The process of our authorship identification approach of source codes.

code and comment. A hybrid line of code and comment means a line which contains a piece of code and a piece of comment. Moreover, non-comment lines of a program contain code lines and blank lines.

Programming Layout Feature Extraction. We will present how to extract programming layout features in this subsection.

Definition 2. *Programming layout features are features which reflect the layout of a program or the arrangement of source codes and comments of a program.*

The feature focuses of the programming layout features in Tables 1 and 2 constitute a set {import, if statement, format of operator, format of loop, leading whitespaces of lines, percentage of blank lines}. For example, the feature focus of the feature fl_1 in Table 2 is "import" which denotes the action of importing package. And fl_1 means whether there is at least a blank line between the last line includes the keyword "import" and the next line which contains codes.

The reason that features fl_1, fl_2 and fl_3 in Table 2 are proposed in our approach is given as follows. The goal of fl_1 is to describe the layout characteristic between lines of importing packages and other code lines. Furthermore, fl_2 and fl_3 are to represent the arrangement of whitespaces within for loop statements. Actually, those three features are independent of purposes of programs and exhibit preferences of programmers to the layout about import and for loops. The set of programming layout features in our experiments includes layout features in Tables 1 and 2, i.e., $\{fl_1, fl_2, fl_3, tl_1, tl_2, tl_3, tl_4\}$.

Programming Style Feature Extraction. This subsection will discuss programming style feature extraction from source codes.

Table 1. Programming features used in our approach and other works

Category	ID	Feature focus	Feature description
Layout	tl_1	Format of operator	Average number of whitespaces adjacent to operators
	tl_2	If statement	Is the frequency of left braces on the end of lines including "if(condition)" is greater than that of left braces on the next lines of lines including "if(condition)"
	tl_3	Percentage of blank lines	Percentage of blank lines to all lines of a program
	tl_4	Leading whitespaces of lines	Average number of leading whitespaces per line
Style	ts_1	Number of comment lines	Is the number of hybrid lines is greater than that of comment lines
	ts_2	Percentage of comments	Percentage of comment lines and hybrid lines to non-comment lines
	ts_3	Variable length	Average number of characters per variable
	ts_4	Frequency of for/while loop	Comparing the frequency of the keyword "for" with that of "while"
	ts_5	Frequency of if/switch	Comparing the frequency of the keyword "if" with that of "switch"
	ts_6	Percentage of static global variables	Percentage of the number of static global variables to non-comment lines of a program
	ts_7–ts_9	Public, private, protected	Respective percentages of the keywords "public", "private" and "protected"
	ts_{10}	Kinds of operators	The kinds of occurred operators
	ts_{11}	Percentage of operators	Percentage of the number of occurred operators to non-comment lines
	ts_{12}–ts_{15}	Percentage of methods	Respective percentages of four kinds of methods including int, char, void and String methods
	ts_{16}	Percentage of methods	Percentage of methods to non-comment lines
	ts_{17}–ts_{19}	Variable names	Do variable names include underscores, numbers and uppercase letters, respectively
	ts_{20}	Percentage of variables	Percentage of the number of variables to non-comment lines
	ts_{21}	Number of variables	Sum of the numbers of all variables
	ts_{22}	Go to	Is the statement "go to" used in a program
Structure	tr_1	Average length of lines	Average number of characters per line
Logic	tg_1	Character-level n-gram	Frequency of character-level n-grams

Table 2. Programming features proposed in our approach

Category	ID	Feature focus	Feature focus proposed in our work	Metric proposed in our work	Feature description
Layout	fl_1	Import	✓	✓	Is there at least a blank line between the last line includes the keyword "import" and the next line which contains codes
	fl_2	Format of for loop	✓	✓	Percentage of whitespaces between left and right parentheses of the form "for(...)" to all "for" loops
	fl_3	Format of for loop	✓	✓	Is there at least a whitespace between left parentheses and right parentheses of the form "for(...)"
Style	fs_1	Percentage of comments	×	✓	Percentage of comment lines and hybrid lines to all lines of a program
	fs_2	Percentage of keywords	×	✓	Percentage of the number of keywords to non-comment lines of a program
	fs_3	Percentage of loops	×	✓	Percentage of loops of "for", "while" and "do-while" to non-comment lines
	fs_4	Definition of loop variable	✓	✓	Percentage of loop variables in lines which include the keyword "for" and definitions of those variables to all "for" loops
	fs_5	Two-dimensional array	✓	✓	Is a two-dimensional array used in a program
	fs_6	Array subscript	✓	✓	Is a computational formula used within array subscripts in a program
	fs_7	Addition operation	✓	✓	Percentage of the form like "i+=j" to total of the forms like "i=i+j" and "i+=j"
	fs_8	Return statement	✓	✓	Does "return 0;" occur in a program
	fs_9	Percentage of import	✓	✓	Percentage of lines which include the keyword "import" to all lines of a program
	fs_{10}	Frequency of class and interface	×	✓	Is the frequency of interfaces greater than that of classes
Structure	fr_1	Percentage of defined methods	×	✓	Percentage of defined methods to non-comment lines
	fr_2	Average length of comments	×	✓	Average length of comment lines and hybrid lines
	fr_3	Average length of methods	×	✓	Average number of lines per method
Logic	fg_1	Word-level n-gram	✓	✓	Frequency of word-level continuous n-grams of a program
	fg_2	Word-level n-gram	✓	✓	Frequency of word-level discrete n-grams of a program

Definition 3. *Programming style features* are ones which embody preferences of programmers to stylistic characteristics of programming such as variable names and variable lengths.

We will explain why the style features $fs_1, fs_2, \ldots, fs_{10}$ in Table 2 are designed in our approach. (1) fs_1, fs_2, fs_3 and fs_7 are to reflect personal traits of programmers about the using of comments, keywords, loop statements, and statements of add operation. In particular, the feature fs_4 means that whether a developer frequently defines loop variables in for loop statements. (2) The essence of features fs_6 and fs_5 is to reflect the facts about how to define array subscripts for programmers and whether a two-dimensional array is used in their programs. (3) The feature fs_8 is to depict the habit that a developer write a statement "return 0;" on the end of a void method. (4) The aim of the features fs_9 and fs_{10} is to capture the characteristics of using frequencies of import, interface and class for programmers. In our experiments, we used a set of programming style features which is composed of style features in Tables 1 and 2, i.e., $\{fs_1, fs_2, \ldots, fs_9, fs_{10}, ts_1, ts_2, \ldots, ts_{21}, ts_{22}\}$.

Programming Structure Feature Extraction. We will address how to extract programming structure features in this subsection.

Definition 4. *Programming structure features* are features which reflect the structure characteristics of programs such as average length of methods, which are usually associated with programmers' experiences.

The cause that the features fr_1, fr_2 and fr_3 are proposed in our method lie in that those three features embody the preferences of developers to the application of methods and comments. The set of programming structure features in our experiments contains fr_1, fr_2, fr_3 and tr_1 in Tables 1 and 2.

Logic Feature Extraction Based on Continuous and Discrete Word-Level n-Gram Models. Figure 2 demonstrates a continuous word-level n-gram model of source codes and a discrete word-level n-gram model of source codes [27].

Definition 5. *A continuous word-level n-gram model of source codes* is a contiguous sequence of n items from a program. Here, items of programs are defined as any strings which are separated by whitespaces. Items can be keywords, operators, use-defined identifiers, punctuations and statements and so on. For a sequence $t_1\ t_2\ \ldots\ldots\ t_{p-1}\ t_p$ of items of a program, we can obtain sequences of items shown in Fig. 2 according to that model.

For instance, to the following program (1), we can get sequences of 3 items, as shown in (2), based on the continuous word-level 3-gram model.

$$public\ \sqcup\ void\ \sqcup\ setUseGradient(\ \sqcup\ boolean\ \sqcup\ useGradientValue\ \sqcup\ )\quad (1)$$

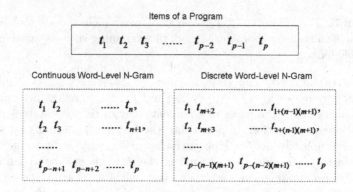

Fig. 2. Continuous and discrete word-level n-gram models of source codes.

(1) *public   void   setUseGradient(*
(2) *void   setUseGradient(boolean*
(3) *setUseGradient(boolean   useGradientValue*   (2)
(4) *boolean   useGradientValue   )*

Definition 6. For a sequence $t_1 t_2 \ldots\ldots t_{p-1} t_p$ of items of a program, *a discrete word-level n-gram model with m skip items* means a sequence of n items, like $t_k t_{k+m+1} \ldots\ldots t_p$ $(1 \leq k \leq p - (n-1)(m+1))$, shown in Fig. 2.

As an illustration, to the program (1), we can acquire sequences of two items with one skip item, as shown in (3), according to the discrete word-level 2-gram model with one skip item.

(1) *public   setUseGradient(*
(2) *void   boolean*
(3) *setUseGradient(useGradientValue*   (3)
(4) *boolean   )*

We first extract a set of the top-k most frequent continuous word-level n-gram sequences, a set of the top-k discrete word-level n-gram sequences, and a set of character-level n-gram sequences from programs in the corpus. The feature fg_1, fg_2 and tg_1 of a program are occurring frequencies of sequences in those three sets, respectively.

The reason that we proposed features based on the continuous and discrete word-level n-gram models are analyzed as follows. (1) Extracted features based on the continuous and discrete word-level n-gram models reflect the using preferences of keywords, identifiers, operators and statements. (2) The continuous word-level n-gram model capture implicit programming patterns of programmers, that is, the collocation patterns between keywords and user-defined identifiers, between operators and identifiers. Here, user-defined identifers include variable names, methods names and class names and so on. However, it is difficult for the character-level n-gram model to achieve this goal. (3) Features

founded on the discrete word-level n-gram model discover underlying collocation patterns of developers between keywords themselves, between user-defined
identifies themselves, between keywords and operators. (4) Features based on
the continuous and discrete word-level n-gram models are easy to be extracted
from source codes, and they do not need any preprocessing. Therefore, those
features based on those two models are irrelevant to the specific purposes of
programs, can embody unconscious, relatively stable and personal programming
characteristics of programmers.

3.3 An Authorship Identification Algorithm

Programs in the corpus have been represented as feature vectors after the process
of feature extraction. Further, the problem of source code author identification
was transformed into a multi-class classification problem in this paper. The classifier of sequential minimal optimization (SMO) for SVM training [28–30] was
employed to determine authorship of source codes.

The sequential minimal optimization is designed to solve the quadratic programming problem, that is, the Lagrangian dual problem of SVM, as shown
in (4) [28–30]:

$$\max_{\alpha} \sum_{i=1}^{n} \alpha_i - \frac{1}{2} \sum_{i=1}^{n} \sum_{j=1}^{n} y_i y_j K(x_i, x_j) \alpha_i \alpha_j,$$
$$s.t. \quad 0 \le \alpha \le C, \quad for \ i = 1, 2, ..., n, \tag{4}$$
$$\sum_{i=1}^{n} y_i \alpha_i = 0,$$

where x_i is the input vector of a training source code example, y_i is a class label
for x_i, n is the number of training examples, α_i are Lagrange multipliers, C is
an hyperparameter, and $K(x_i, x_j)$ is the kernel function.

The advantages of the SMO approach are that it reduces the computation
time and obtains a better scaling characteristic than the SVM training method
[28–30]. Now we give our identification algorithm in Algorithm 1.

Algorithm 1. *Authorship identification from source codes*

Input: The source codes C of anonymous authors.
Output: the author of each program in C.
1: **for** $c_i \in C$, i=1, 2, ..., n **do**
2:     Build the layout feature set $\{fl_1, fl_2, fl_3, tl_1, tl_2, tl_3, tl_4\}$.
3:     Extract the style feature set $\{fs_1, fs_2, ..., fs_{10}, ts_1, ts_2, ..., ts_{22}\}$.
4:     Construct the structure feature set $\{fr_1, fr_2, fr_3, tr_1\}$.
5:     Extract the logic feature set $\{fg_1, fg_2, tg_1\}$.
6:     Build the feature vector v of c_i.
7: **end for**
8: Employ the sequential minimal optimization (SMO) to classify the authorship of
   each program.

---

4 Experiments

4.1 Experimental Results

The two datasets of source codes in the java programming language are used in our experiments, which come from two source code websites (i.e., github.com, planet-source-code.com), respectively. The first dataset include 8000 programs of eight programmers (1000 programs per person), while the second dataset is composed of 502 programs of 53 programmers, and is a imblanced dataset. We utilize the 10-fold cross-validation to evaluate the performance of our approach.

Two baseline methods were implemented on the same two datasets for performance comparison. Those two method are ones which apply the decision tree based on the feature set B_1 and B to identify authorship of source codes, respectively. The formulas in (5) define seven sets $B_1, B_2, B, F_1, F_2, F_3, F$.

$$
\begin{aligned}
B_1 &= \{tg_1\}, \ B_2 = \{tl_1, ..., tl_4, ts_1, ..., ts_{22}, tr_1\} \\
B &= B_1 \cup B_2 \\
F_1 &= \{fl_1, fl_2, fl_3, fs_1, ..., fs_{10}, fr_1, fr_2, fr_3\} \\
F_2 &= \{fg_1\}, \ F_3 = \{fg_2\} \\
F &= F_1 \cup F_2 \cup F_3
\end{aligned}
\tag{5}
$$

Table 3. The identification accuracy with DT, RF and SMO on dataset 1

	B_1	B	F	$F \cup B_1$	$F \cup B$
	Features of the 1st baseline	Features of the 2nd baseline	Our features	Features in B_1 and our features	Features in B and our features
Decision tree (%)	95.95	96.03	96.45	97.61	97.54
Random forest (%)	97.74	97.66	98.08	98.00	98.05
SMO (%)	97.68	97.73	98.40	99.03	99.08

Table 4. The identification accuracy with DT, RF and SMO on dataset 2

	B_1	B	F	$F \cup B_1$	$F \cup B$
Decision tree (%)	71.31	70.52	72.71	73.31	72.51
Random forest (%)	72.91	74.30	81.27	75.70	74.70
SMO (%)	72.31	75.50	80.28	82.47	83.47

Tables 3 and 4 show the identification accuracy of decision tree (DT), random forest (RF) and sequential minimal optimization (SMO) by using the features B_1, B, F, $F \cup B_1$ and $F \cup B$ on dataset 1 and dataset 2, respectively. B_1 is the feature set of character-level 6-grams. F_2 and F_3 are the feature set of continuous word-level 3-grams and the feature set of discrete word-level 2-grams with 1-skip item. The dimensions of B_1, F_2 and F_3 are all 2000.

The following facts can be seen from Tables 3 and 4. (1) With DT, RF and SMO on dataset 1 and dataset 2, the accuracy of F, $F \cup B_1$ and $F \cup B$ is higher than those of B_1 and B. Hence, our proposed features which either are or are not integrated with feature sets B_1 and B of two baseline methods improve the accuracy of B_1 and B. (2) The feature set $F \cup B$ including our proposed features and B achieves the highest accuracy (i.e., 99.08% and 83.47%) on dataset 1 and dataset 2 by utilizing the SMO technique. (3) SMO obtains higher accuracy than those of DT and RF on both dataset 1 and dataset 2 on three feature sets B, $F \cup B_1$ and $F \cup B$.

Table 5. The identification accuracy on dataset 1 by using single feature set

	B_1	B	B_2	F_1	F_2	F_3
	Features of the 1st baseline	Features of the 2nd baseline	Present features in Table 1	Our features in F_1	Our features in F_2	Our features in F_3
Decision tree (%)	95.95	96.03	62.44	54.00	96.60	96.58
Random forest (%)	97.74	97.66	76.23	65.59	98.15	97.83
SMO (%)	97.68	97.73	54.90	48.04	98.13	98.06

Table 6. The identification accuracy on dataset 2 by using single feature set

	B_1	B	B_2	F_1	F_2	F_3
Decision tree (%)	71.31	70.52	51.20	45.82	71.12	69.32
Random forest (%)	72.91	74.30	71.51	57.57	77.89	77.49
SMO (%)	72.31	75.50	51.79	24.30	74.30	80.28

Tables 5 and 6 give the the identification accuracy of DT, RF and SMO on dataset 1 and dataset 2 by using feature sets B_1, B_2, B, F_1, F_2 and F_3, respectively. Experimental results in Tables 5 and 6 show that the accuracy of F_2 and F_3 is higher than those of B_1, B_2 and B on dataset 1 by using DT, RF and SMO, while the performance of F_3 is higher than those of B_1, B_2 and B on dataset 2 by applying RF and SMO. Therefore, the fact demonstrate the validity of our proposed feature sets F_2 and F_3. Moreover, A combination of the feature set F_2 with SMO on dataset 1 gets the highest performance in Table 5, while a combination of the feature set F_3 with SMO on dataset 2 obtains the highest performance in Table 6.

4.2   Parameters Analysis

In addition, we evaluate the influence of different dimensions of F_2 and F_3. The dimensions of F_2 and F_3 are set as 1000, 2000, 3000, 4000 and 5000. Figures 3(a), (b) and 4(a), (b) illustrate the accuracy curves of $F \cup B_1$ and $F \cup B$ by using DT, RF and SMO on dataset 1 and dataset 2, respectively. The curves in Figs. 3 and 4 indicate that the accuracy of SMO is higher than that of DT and RF by using $F \cup B_1$ or $F \cup B$ within all five cases on both dataset 1 and dataset 2. As a whole, the performance with B and our proposed feature set F by using SMO on dataset 1 and dataset 2 achieves the highest accuracy.

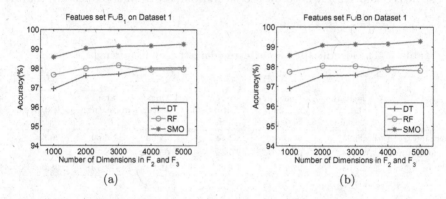

Fig. 3. The identification accuracy of different feature dimensions on dataset 1

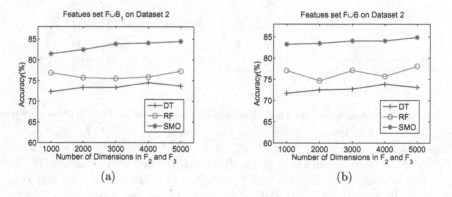

Fig. 4. The identification accuracy of different feature dimensions on dataset 2

5   Conclusion

The task of author identification of source codes is an significant issue in the field of author identification from on-line or off-line texts. Author identification has been widely used in many areas including computer forensics, network public opinion monitoring, intellectual property protection and information security [2,6]. In this paper, an approach of building author profiles of programmers is proposed, which is based on a logic model of continuous word-level n-grams and discrete word-level n-grams, and a multi-level context model about operations, loops, arrays and methods. Moreover, the classifier method of sequential minimal optimization for training SVM is applied to identify authors of source codes. The distinguish characteristic of author profiles not only represent personal preferences to using components of programming languages, i.e., keywords, identifers, operators, statements, methods and classes; but also express explicit and implicit personal combinative patterns among those components. Features within author profiles are independent of purposes of programs, are not constrained by specific user-defined names of variables, methods, classes and parameters and so on. Comparative experimental results on programs from two open source websites indicate that our constructed author profiles with SMO significantly increase the performance of source code author identification. In the future, we will introduce feature selection algorithms to the task of author identification of source code.

Acknowledgments. This work was supported by the National Natural Science Foundation of China (NO. 61672098, NO. 61272361).

References

1. Kothari, J., Shevertalov, M., Stehle, E., et al.: A probabilistic approach to source code authorship identification. In: 4th International Conference on Information Technology, pp. 243–248 (2007)
2. Ding, H., Samadzadeh, M.H.: Extraction of java program fingerprints for software authorship identification. J. Syst. Softw. **72**(1), 49–57 (2004)
3. Lange, R., Mancoridis, S.: Using code metric histograms and genetic algorithms to perform author identification for software forensics. In: 9th Annual Conference on Genetic and Evolutionary Computation, pp. 2082–2089 (2007)
4. Tennyson, M.F.: On improving authorship attribution of source code. In: International Conference on Digital Forensics and Cyber Crime, pp. 58–65 (2012)
5. Gray, A., Sallis, P., MacDonell, S.: Identified: a dictionary-based system for extracting source code metrics for software forensics. In: International Conference on Software Engineering: Education and Practice, pp. 252–259 (1998)
6. Zhang, C., Wu, X., Niu, Z., et al.: Authorship identification from unstructured texts. Knowl.-Based Syst. **66**, 99–111 (2014)
7. Tennyson, M.F., Mitropoulos, F.J.: A bayesian ensemble classifier for source code authorship attribution. In: International Conference on Similarity Search and Applications, pp. 265–276 (2014)
8. Stamatatos, E.: A survey of modern authorship attribution methods. J. Am. Soc. Inf. Sci. Technol. **60**(3), 538–556 (2009)
9. Spafford, E.H., Weeber, S.A.: Software forensics: tracking code to its authors. Comput. Secur. **12**, 585–595 (1993)

10. Software Forensics. http://en.wikipedia.org/wiki/Software_forensics
11. Bandara, U., Wijayarathna, G.: Source code author identification with unsupervised feature learning. Pattern Recogn. Lett. **34**(3), 330–334 (2013)
12. MacDonell, S., Gray, M., MacLennan, G., Sallis, P.: Software forensics for discriminating between program authors using case-based reasoning, feed-forward neural networks and multiple discriminant analysis. In: 6th International Conference on Neural Information Processing, pp. 66–71 (1999)
13. Burrows, S., Tahaghoghi, S.M.M.: Source code authorship attribution using n-grams. In: 12th Australasian Document Computing Symposium, pp. 32–39 (2007)
14. Burrows, S., Uitdenbogerd, A.L., Turpin, A.: Comparing techniques for authorship attribution of source code. Softw.: Pract. Exp. **44**(1), 1–32 (2014)
15. Bandara, U., Wijayarathna, G.: Deep neural networks for source code author identification. In: International Conference on Neural Information Processing, pp. 368–375 (2013)
16. Frantzeskou, G., Stamatatos, E., Gritzalis, S.: Supporting the cybercrime investigation process: effective discrimination of source code authors based on byte-level information. In: 2nd International Conference on E-business and Telecommunication Networks, pp. 163–173 (2005)
17. Frantzeskou, G., Gritzalis, S., MacDonell, S.G.: Source code authorship analysis for supporting the cybercrime investigation process. In: 1st International Conference on E-business and Telecommunication Networks, pp. 85–92 (2004)
18. Krsul, I.: Authorship analysis: identifying the author of a program. Technical report TR-94-030, Purdue University (1994)
19. Burrows, S., Tahaghoghi, S.M.M., Zobel, J.: Efficient plagiarism detection for large code repositories. Softw.-Pract. Exp. **37**(2), 151–175 (2007)
20. Burrows, S., Uitdenbogerd, A.L., Turpin, A.: Temporally robust software features for authorship attribution. In: 33rd Annual International Computer Software and Applications Conference, pp. 599–606 (2009)
21. Burrows, S.: Source code authorship attribution. Ph.D. thesis. RMIT University, Melbourne, Australia (2010)
22. Burrows, S., Uitdenbogerd, A.L., Turpin, A.: Application of information retrieval techniques for source code authorship attribution. In: 14th International Conference on Database Systems for Advanced Applications, pp. 699–713 (2009)
23. Frantzeskou, G., Stamatatos, E., Gritzalis, S., et al.: Identifying authorship by byte-level n-grams: the source code author profile (SCAP) method. Int. J. Digit. Evid. **6**(1), 1–18 (2007)
24. Krsul, I., Spafford, E.H.: Authorship analysis: identifying the author of a program. Comput. Secur. **16**(3), 233–257 (1997)
25. Elenbogen, B.S., Seliya, N.: Detecting outsourced student programming assignments. J. Comput. Sci. Coll. **23**(3), 50–57 (2008)
26. Shevertalov, M., Kothari, J., Stehle, E., Mancoridis, S.: On the use of discretised source code metrics for author identification. In: 1st International Symposium on Search Based Software Engineering, pp. 69–78 (2009)
27. N-gram. http://en.wikipedia.org/wiki/N-gram
28. Platt, J.: Sequential minimal optimization: a fast algorithm for training support vector machines (1998). http://citeseerx.ist.psu.edu/viewdoc/summary?doi=10.1.1.55.560
29. Sequential minimal optimization. http://en.wikipedia.org/wiki/Sequential_minimal_optimization
30. Sequential minimal optimization. http://blog.csdn.net/yclzh0522/article/details/6900707

DFDS: A Domain-Independent Framework for Document-Level Sentiment Analysis Based on RST

Zhenyu Zhao[1,3], Guozheng Rao[1,3(✉)], and Zhiyong Feng[2,3]

[1] School of Computer Science and Technology,
Tianjin University, Tianjin, China
{zhaozhenyu_tx, rgz}@tju.edu.cn
[2] School of Computer Software, Tianjin University, Tianjin, China
zyfeng@tju.edu.cn
[3] Tianjin Key Laboratory of Cognitive Computing and Application,
Tianjin, China

Abstract. Document-level sentiment analysis is among the most popular research fields of nature language processing in recent years, in which one of major challenges is that discourse structural information can be hardly captured by existing approaches. In this paper, a domain-independent framework for document-level sentiment classification with weighting rules based on Rhetorical Structure Theory is proposed. First, original textual documents are parsed into rhetorical structure trees through a preprocessing pipeline. Next, the sentiment score of elementary discourse units is computed via sentence-level sentiment classification method. Finally, according to the rhetorical relation between neighbor discourse units, we define weighting schema and composing rules based on which scores of elementary discourse units are summed recursively to the whole document. Experiment results show that our approach has better performance on datasets in different domains, compared with state-of-art document-level sentiment analysis systems based on RST, and the best result is 15% higher than baseline.

Keywords: Sentiment analysis · Rhetorical Structure Theory · Domain independent

1 Introduction

With internet integrated everywhere in daily life, more and more people share their opinions such as movie reviews, service feedback and so on, on social media or online communities. Researches has proved that authors' sentiments implicated by such data can be of great value to business activities as well as public security [1–4].

Although sentiment analysis and opinion mining has been attractive in recent years and yielded a great number of excellent results in different respects [5, 6], there are still lots of challenges to be settled. This work mainly focuses on the classification of sentiment polarities of textual documents, especially long texts.

Compared with sentiment classification at sentence-level, that of document-level has its own special challenges. For example: there is a common phenomenon that

© Springer International Publishing AG 2017
L. Chen et al. (Eds.): APWeb-WAIM 2017, Part I, LNCS 10366, pp. 297–310, 2017.
DOI: 10.1007/978-3-319-63579-8_23

several opposite sentiments may inhere in the same article, where the opposite opinion may serve as a foil to the main opinion or the author just wants to make his review more comprehensive, which are illustrated as follows:

> "[It could have been a great movie.]A[it does have beautiful scenery,]B[some of the best since Lord of the Rings.]C[The acting is well done,]D[and I really liked the son of the leader of the Samurai.]E[He was a likeable chap,]F[and I hate to see him die.]G[But, other than all that, this movie is nothing more than hidden rip-offs.]H" [7].

Although the review has a great length of praise, from sentence B to G, the author's main idea is to disparage the movie, which can be generated just by the sentences A and H. However it is easy for human to make such a judgement but hard for a computer.

The main reason is that the current approaches can hardly comprehend the structure and rhetorical manners of the article. To address this problem, an intuitive idea is to compose sentences' scores to the whole text based on Rhetorical Structure Theory (RST; Mann and Thompson [8]), where relations between neighbor sentences are defined and several connected discourse units are separated into Nucleus (the more important part) and Satellite (the less important part). Specifically, the example showed in last paragraph can be parsed as Fig. 1 according to RST.

The leaf nodes are Elementary Discourse Units (EDUs) and the nodes with a line below the label are nucleus. From Fig. 1, although the discourse has a lot of positive words, we still know the discourse unit H is the main opinion of the author because H is the Nucleus towards the composition of discourse units from A to G, which have a great meaning for document-level sentiment analysis.

As automatic RST parser has improved considerably, for example, the state-of-the-art automatic RST parsing system DPLP (also employed in this paper) proposed by Ji [9], it is time to think about the possibility and practicability of combining discourse parsing with sentence-level sentiment analysis approaches to deal with document-level sentiment classification.

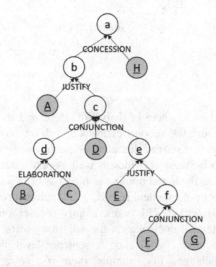

Fig. 1. Representation of the example after RST parsing

In this paper, we proposed a Domain-independent Framework for Document-level Sentiment analysis based on RST (DFDS), which has been proved to have more competitive results on datasets in different domains. The main contribution of this work is listed as follows:

- A novel domain-independent document-level sentiment analysis framework is presented, which combines Rhetorical Structure Theory on discourse parsing and recursive neural network on sentiment classification of elementary discourse units;
- Effective weighting schema and composing rules of discourse units are proposed;
- We explore the possibility and performance of composing different sentence-level sentiment analysis methods with RST parser. And we believe that with improvement of automatic discourse parsing, the contribution of this work will have referential significance.

2   Preliminaries

2.1   Rhetorical Structure Theory

Rhetorical Structure Theory is a compositional framework of discourse parsing that defines 24 relations of discourse structure, such as condition, concession and so on. Elementary discourse units (EDUs) are combined into larger discourse units according to RST, recursively covering the whole text in the end. In each relation, there may be a nucleus and a satellite which is called "Nucleus-Satellite" relation or several nucleus which is called "Multinuclear". The nucleus discourse unit plays a more important role in the larger discourse unit, while the satellite is less important.

RST has been widely use in discourse structure parsing. In the early work on discourse parsing, hand-crafted rules and heuristics were applied to build discourse parsing trees according to Rhetorical Structure Theory [10]. Soricut and Marcu introduced probabilistic models to identify elementary discourse units and built discourse parsing trees, which resultsin the birth of automatic parser SPADE [11]. But SPADE can only parse discourse within sentence-level. Feng and Hirst developed an RST discourse parser, based on the HILDA discourse parser and rich linguistic features [12]. By combining the machinery of large-margin transition based structure prediction, the automatic discourse parser named DPLP has reached a satisfactory accuracy about 62%, while the accuracy of human-annotated relation judgement is just 65.8% [9]. Li et al. first proposed to use dependency structure to represent the relations between EDUs and got competitive performance [13].

2.2   Document-Level Sentiment Analysis

There has been plenty of sentiment analysis approaches proposed for document-level sentiment polarity classification. Pang and Lee proposed a novel machine-learning method with text-categorization techniques to classify the subjective portions of the document [14]. Sharma et al. used three popular sentiment lexicons to extract sentiment representing features and used BPANN to do classification [15]. Tang et al. proposed

Gated Recurrent Neural Networks to overcome the challenge of encoding the intrinsic relations between sentences in the semantic meaning of a document [16]. Xu et al. presented Cached Long Short-Term Memory neural networks to capture the overall semantic information and get the state-of-the-art results on three publicly available document-level sentiment analysis dataset [17]. Approaches based on supervised learning seem to have more competitive accuracy than unsupervised methods, but they always have strong dependence on domain and scale of datasets, which makes most document-level sentiment analysis methods inefficient and domain-dependent.

Contrary to consistent improvements on discourse parsing, there are few efforts to incorporate RST into sentiment analysis. Voll and Taboada proved that RST could improve the results of lexicon-based sentiment analysis with manually-annotated RST parse trees [18]. However, it's very time-consuming and expensive to annotate relation between spans. Heerschop et al. proposed Pathos, a framework based on a document's discourse structure [19]. Wang et al. incorporated hierarchical discourse structure into an unsupervised sentiment analysis framework, but they used manually-annotated discourse parses and a small dataset (604 reviews) [20]. Li et al. focused on the automation of recognizing intra-sentence level discourage relations for polarity classification with unsupervised method [21].

The researches summarized above specialized in document-level classification, but just applied RST on intra-sentence level. Hogenboom et al. employed a weighting scheme as well as HILDA on just one dataset [22], which is similar to our approach, but our schema and rules do not only differentiate satellite and nucleus weights by types of relations in RST, but also the sentiment polarities of satellite and nucleus. Moreover we conduct experiments on four datasets in different domains and get better results. Bhatia et al. showed that RST can improve text-level sentiment analysis and proposed three methods, which employed the dependency-based discourse tree. But their work based on lexicon just considered dependency of satellite and nucleus instead of the rhetorical relation they belong to, and machine learning based methods they proposed had little improvement compared with their baseline [7].

3   Overall Framework

The overall framework is shown in Fig. 2. Firstly, original texts are preprocessed through a NLP pipeline, during which tokenization, part-of-speech tagging and sentence splitting have been done. We apply StanfordCoreNLP[1] in this step in view of its good performance. Secondly, texts are parsed into rhetorical structure trees and structural information in a text can be extracted. At the same time, the sentiment scores of sentences are computed via a sentence-level analysis method. Finally, according to rhetorical structure trees and relations, weighting rules are proposed to sum up the scores of sentences recursively to the whole document and then the sentiment polarities will be classified.

---

[1] Available at http://stanfordnlp.github.io/CoreNLP/.

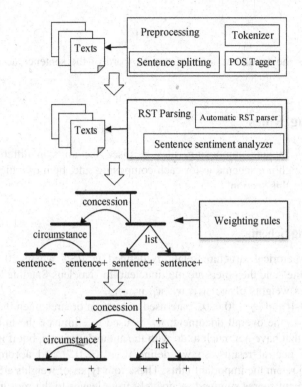

Fig. 2. The overall framework of our method

In a RST parsing tree, sentiment scores of nodes except leaves, which correspond to each elementary discourse unit, are computed via the following formula:

$$S_i = \sum_{j=1}^{k} \varphi(s_{ij}) \qquad (1)$$

Where s_{ij} is the jth child node of the node n_i, which have k child nodes. The sentiment scores of child nodes functioned in weighting rules φ sum up to get the sentiment score S_i. If $S_i > 0$, we believe the node n_i is positive, and if $S_{root} > 0$, we believe the document is positive, otherwise n_i is negative except the case of $S_i = 0$, which is called neutral. Weighting rules φ of each relation will be defined and introduced in detail in Sect. 4.

For leaf nodes, sentence-level sentiment analysis approaches are introduced in to get the sentiment scores of elementary discourse units. To select a more effective approach, we compare two methods based on lexicon with Stanford-Sentiment Annotator[2]. For lexicon based methods, the score of a sentence is computed via the following formula:

---

[2] http://nlp.stanford.edu/sentiment.

$$S = \sum_{i=1}^{n} w_i \qquad (2)$$

Where w_i is the sentiment score of the ith word of the sentence according to the lexicon.

4   Weighting Rules

Weighting rules include composing rules of discourse units in different rhetorical relations and weighting schema w for each composing rule, both of which are defined and explained in this section.

4.1   Weighting Schema

There are 24 rhetorical structure relations defined in RST where 20 relations are "Nucleus-Satellite" and the others are multinuclear. In "Nucleus-Satellite" relations, we define five types weights of w : w_{vs}, w_s, w_l, w_{vl}, w_e.

$w_s \in [0.3, 0.4]$ and $w_l \in [0.6, 0.7]$ are used to weaken or strengthen the influence of discourse units to the overall document; w_{vs} is used to eliminate the influence of the discourse units that have not much to do with the author's attitude, but if $w_{vs} = 0$, there will be lots of neutral results, so we define $w_{vs} \in (0, 0.2]$ and accordingly define $w_{vl} \in [0.8, 1)$ to retain the important units. These four types of weights are all below 1, thus with sentiment scores summed recursively from leaves to the root in a RST tree, sentences more far away from the root, which means they are less important, will have less influence on the final sentiment score of the overall document. Particularly, w_e is defined more than 1 to outstand kiscourse unit, such as for conclusion sentences.

4.2   Composing Rules of Rhetorical Structure Relations

In "Nucleus-Satellite" relations, given a node n_i and sentiment scores of its two child nodes, $s_i(nucleus)$ and $s_i(satellite)$, we separate 20 RST N-S relations into 7 categories functioned by different composing rules:

Categories 1.
Relations: *antithesis, concession*
Composing rules:

If $s_i(nucleus) * s_i(satellite) < = 0$

$$S_i = w_l * s_i(nucleus) - w_s * s_i(satellite)$$

else

$$S_i = w_l * s_i(nucleus) + w_s * s_i(satellite)$$

Explanation: The nucleus span is thought to be the point of the authors, while the satellite often has the opposite polarity like" Even though the picture is fine, I don't like the film without the plot". In other cases, the score of nucleus part can be 0, but we can take use of the opposite number of the satellite part to get the sentiment polarity of larger discourse unit. If the satellite and nucleus have the same polarity, we just add them up but put the nucleus at a prominent position via a larger weight.

Categories 2.
Relation: *circumstance, background*
Composing rules:

$$S_i = w_{vl} * s_i(nucleus) + w_{vs} * s_i(satellite)$$

Explanation: As for the relations of *circumstance* and *background*, the satellite always indicate the time or the place in which the nucleus part takes place, which has not much to do with the opinion of authors, so we set w_{vs} as the weight of satellite part to eliminate its influence, and w_{vl} as the weight of nucleus part.

Categories 3.
Relation: *condition*
Composing rules:

$$S_i = -w_{vl} * s_{ia}(nucleus) + w_{vs} * s_i(satellite)$$

Explanation: It seems hard to tell whether the polarity of nucleus in relation of *condition* can represent author' true opinion, but in the domain of reviews, the satellite part always states the case that actually does not happen, so the nucleus part most likely stands in the opposite position with the author.

Categories 4.
Relation: *motivation, purpose*
Composing rules:

If $s_i(nucleus) * s_i(satellite) < = 0$

$$S_i = w_{vl} * s_i(nucleus) + w_{vs} * s_i(satellite)$$

else

$$S_i = w_l * s_i(nucleus) + w_s * s_i(satellite)$$

Explanation: In most cases, the satellite of relation *motivation* and *purpose* has the same polarity with that of the nucleus, and we just sum them up multiplying different weights. If their polarities are opposite, we only retain the scores of the nucleus in order to eliminate the uncertainty of the satellite.

Categories 5.
Relation: *evidence, justification, restatement, reason, result, enablement*
Composing rules:

If $s_i(nucleus) * s_i(satellite) < = 0$

$$S_i = w_{vs} * s_i(nucleus) + w_{vl} * s_i(satellite)$$

else

$$S_i = \max\{s_i(nucleus), s_i(satellite)\}$$

Explanation: As for these relations, in most cases, the satellite should have the same polarity with the nucleus, and we choose the maximum as the score of the parent span. Since the accuracy of sentence-level sentiment analysis method still doesn't reach 100%, if the polarities of the satellite and the nucleus are different, the satellite are the detailed description of the author's attitude.

Categories 6.
Relation: *evaluation, conclusion*
Composing rules:

$$S_i = w_e * s_i(nucleus) + w_{vs} * s_i(satellite)$$

Explanation: The nucleus part in relations of evaluation and conclusion is always the summary of previous article, thus it is reasonable to enlarge the proportion of the nucleus.

Categories 7.
Relation: *other N-S relations in RST*
Composing rules:

$$S_i = w_l * s_i(nucleus) + w_s * s_i(satellite)$$

Explanation: With regard of the others N-S relations in RST, we just add the scores of the nucleus and the satellite up, although there may be several relations which may make some difference to the sentiment classification of whole texts.

What's more, for multinuclear relations, we follow the fomula:

$$S_i = \sum_{j=1}^{k} S_{ij} \tag{3}$$

Where S_{ij} is the sentiment score of jth child node among k child nodes of node n_i.

5   Experiments

We conduct experiments for document-level sentiment classification, based on sentence-level method and weighting rules. We describe the details and results of the experiments in this section.

5.1   Datasets and Setup

Experiments are conducted on four datasets[3] in different domains, which were collected by Blitzer et al. The datasets contain book reviews denoted as "books", DVD reviews denoted as "dvd", electronics reviews denoted as "elec" and houseware reviews denoted as "houseware". All four datasets have both 1000 positive documents and 1000 negative documents.

Before weighting rules are used, original texts have to be preprocessed into RST trees. Firstly, Stanford CoreNLP is applied to do tokenization, POS tagging and splitting the texts into sentences. Secondly, we use an automatic parser named DPLP[4] proposed by Ji et al. [13] to parse the texts to RST trees. In order to use the weighting rules we defined, we restructure the RST trees with java and for leaf nodes, different sentence-level sentiment classification methods are introduced to compute the sentiment scores of EDUs.

Since there are few related work on document-level sentiment analysis with RST, we employed the state-of-the-art document-level sentiment classification system based on RST, Discourse Depth Reweighting (DDR) proposed by Bhatia et al. [7], as our baseline. It should be noted that there are three methods proposed in Bhatia's work, but only DDR based on lexicon, which is the baseline we choose in this paper, has great improvement then their baselines while the other two methods seem to make little contribution. What's more, most machine learning methods do not have comparability with our work since all weighting schema, composing rules and sentence-level sentiment in our work do not need to be trained on the datasets used in experiments, which makes our method domain-independent and efficient. However, most document-level sentiment analysis methods based on machine learning, such as neural network, need large scale of data to train a model and the test performance have strong dependence on the domain of train datasets.

5.2   Determination of Weights w

We have defined 5 types of weights (w_{vs}, w_s, w_l, w_{vl}, w_e) in Sect. 4 to adjust the influence of discourse units in different categories, but specific value will be determined in this Sect.

8 weighting schemas denoted as I, II, III,…are presented and tested on the same dataset (book reviews). We apply two lexicons to get sentiment scores of sentences and

---

[3] Available at http://www.cs.jhu.edu/~mdredze/datasets/sentiment/.
[4] Available at https://github.com/jiyfeng/DPLP.

evaluate the performance of each schema. The first lexicon[5], which is proposed by Wilson et al. and has also been applied in our baseline, contains 2721 positive words and 4914 negative words. The second lexicon denoted as SWN[6] (sentiWordNet) has 21714 positive words and 25173 negative words. Distribution of Part of speech of words contained in both lexicon is shown in Table 1.

Table 1. Distribution of part-of-speech of words.

Lexicon	Noun		Adjective		Verb		Adverb	
	Pos	Neg	Pos	Neg	Pos	Neg	Pos	Neg
Wilson et al.	1030	2002	1541	2462	473	1061	0	0
SentiWordNet	10082	13171	6330	7960	2928	3415	2674	627

The 8 schemas are listed in detail in the Table 2. It should be noted that we don't want to train a schema by methods of machine learning, which may perform better accuracy with a sentence-level sentiment analysis method on this dataset, because this schema will have strong dependence on that dataset and can hardly adapt datasets in different domains.

Table 2. Proposed weighting schemas

Number	w_{vs}	w_s	w_l	w_{vl}	w_e
I	0.2	0.4	0.6	0.8	4
II	0.1	0.4	0.6	0.9	4
III	0.1	0.3	0.7	0.9	4
IV	0.2	0.35	0.65	0.8	4
V	0.1	0.35	0.65	0.9	2
VI	0.2	0.3	0.7	0.8	2
VII	0.15	0.35	0.65	0.85	2
VIII	0.15	0.4	0.6	0.85	2

Two lexicon based methods are employed to evaluate the 8 schemas and choose the best one among them as the final schema to be used in DFDS. We also compare the weighting schemas we proposed with the weighting strategy that our baseline used. The comparison of accuracy among 9 weighting methods are illuminated as Fig. 3.

Figure 3 shows the performance of each schema with two lexicon based methods. We can see that schema III has better performance, which motivated us to apply it as final weighting schema of our framework. In addition, with the same lexicon the schema III has better results than our baseline DDR, which means the schema we present on RST is more reasonable.

---

[5] Available at http://mpqa.cs.pitt.edu/lexicons/subj_lexicon/.
[6] Available at http://sentiwordnet.isti.cnr.it/.

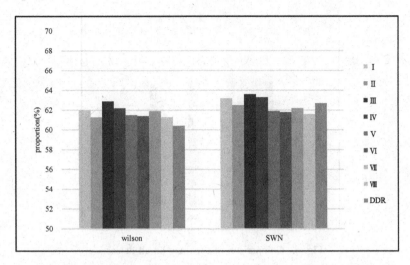

Fig. 3. Comparison of accuracy of weighting schemas with two lexicon based methods

What's more, we also find that all results of 9 weighting methods are improved after applying SWN lexicon. Investigation in results shows that lots of sentences with sentiments are judged to neutral. This is because some sentiment words are not collected in the lexicon collected by Wilson, which results in large amount of neutral classification results, especially in short documents. As shown in Table 1, SWN lexicon contains far more sentiment words than Wilson lexicon, thus gets better accuracy using the same weighting schema. This result also means better lexicon or sentence-level method may also result in better accuracy, so we apply Stanford-Sentiment Annotator as the sentence-level sentiment analysis method of our framework because of its good performance on sentence-level sentiment polarity classification. It must be noted that this annotator does not have the ability to classify the document-level textual corpus.

5.3 Experiments Results and Analysis

Experiments of document-level sentiment classification are conducted on four different domain datasets. We evaluate and compared our method with the baseline on F-score and accuracy. The results have been shown in Figs. 4 and 5.

From Fig. 4 we can see that our method outperforms better accuracy on all four datasets than our baseline. The best result got on DVD reviews dataset has reached more than 75% and has about 15% better than DDR. Even on the dataset of houseware reviews, where DFDS gets the lowest accuracy, it still has about 7% better than DDR. Although four datasets are from different domains, the accuracy of our method maintains at a level about 70%, which indicates good adaptability for different domains.

Figure 5 shows the comparison of F-score on all datasets of both methods, and results show that DFDS outperforms better than DDR. On dvd review dataset, DFDS reach about 77% with about 15% better than the baseline. On both book review and electronic datasets, F-score of DFDS maintains about 70%, while DDR is just about 60%.

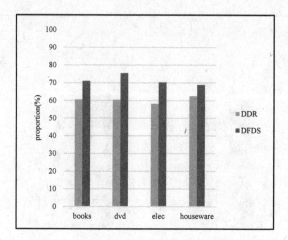

Fig. 4. Comparison of accuracy on all datasets

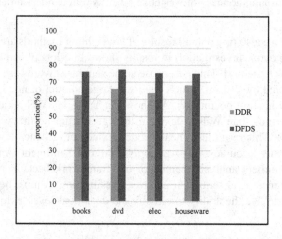

Fig. 5. Comparison of F-score on all datasets

The results above show that our method perform better than our baseline DDR, but the method of sentence-level sentiment classification is based on Wilson lexicon, which has been shown to have much fewer sentiment words. On the contrary, we proposed novel weighting schemas and composing rules and employed better sentence-level analysis method in DFDS, so it is necessary to prove whether the weighting schemas and composing rules we used make sense. Therefore, we extended the work of DDR, and use Stanford-Sentiment Annotator to do sentence-level sentiment classification, denoted as DDRS. We evaluate the accuracy of DDRS on all four datasets with comparison to DDR and DFDS, which is shown in Table 3.

We also compare the proportion of neutral results that the three methods have judge out and the detail statistics are also listed in Table 3. The results showed that our method have better results on all four datasets than DDRS, which have the same

Table 3. Comparison of accuracy for DDR, DDRS and DFDS (%).

Datasets	DDR		DDRS		DFDS	
	Accuracy	Neutral	Accuracy	Neutral	Accuracy	Neutral
books	60.5	10.4	68.9	9.2	71.0	3.0
dvd	60.4	12.2	73.7	6.9	75.4	1.3
elec	58.2	17.8	63.9	9.4	70.3	4.8
hourseware	62.5	17.4	66.0	10.2	68.8	4.8

sentence-level sentiment analysis approach. That means the weighting schemas and composing rules proposed in this paper performs better on semantic relations capture and document-level sentiment classification. In addition, the proportion of neutral of DFDS is much less than our baseline and DDRS, which also definitely results in the improvement of our method. What's more, DDRS performs better than DDR on all datasets, which means the accuracy of overall document classification based on RST has positive correlation with performance of sentence-level sentiment analysis method.

6 Conclusion

To capture the semantic relation between sentences is still a great challenge in document-level sentiment classification. Improvement of discourse structure parsing in recent years motivated us to present a document-level sentiment analysis framework DFDS. We also proposed novel weighting schema and composing rules of discourse units and apply sentence-level sentiment classification method in our framework. Experiments has shown that our method gets competitive results on datasets in different domains.

Acknowledgement. This work is supported bythe National Natural Science Foundation of China (NSFC) (61373165, 61373035 and 61672377).

References

1. O'Connor, B., Balasubramanyan, R., Routledge, B.R., et al.: From tweets to polls: linking text sentiment to public opinion time series. ICWSM **11**, 122–129 (2010)
2. Musto, C., Semeraro, G., Lops, P., et al.: CrowdPulse: a framework for real-time semantic analysis of social streams. Inf. Syst. **54**, 127–146 (2015)
3. Bollen, J., Mao, H., Zeng, X.: Twitter mood predicts the stock market. J. Comput. Sci. **2**(1), 1–8 (2011)
4. Smailović, J., Grčar, M., Lavrač, N., et al.: Stream-based active learning for sentiment analysis in the financial domain. Inf. Sci. **285**(1), 181–203 (2014)
5. Liu, B., Zhang, L.: A survey of opinion mining and sentiment analysis. In: Aggarwal C., Zhai, C. (eds.) Mining Text Data, pp. 415–463. Springer, US (2012)
6. Pang, B., Lee, L.: Opinion mining and sentiment analysis. Found. Trends Inf. Retrieval **2** (1-2), 1–135 (2008)

7. Bhatia, P., Ji, Y., Eisenstein, J.: Better document-level sentiment analysis from RST discourse parsing. arXiv preprint arXiv:1509.01599 (2015)
8. Mann, W.C., Thompson, S.A.: Rhetorical structure theory: description and construction of text structures. In: Kempen, G. (ed.) Natural Language Generation, pp. 85–95. Springer, Netherlands (1987)
9. Ji, Y., Eisenstein, J.: Representation learning for text-level discourse parsing. In: Meeting of the Association for Computational Linguistics, pp. 13–24, USA (2014)
10. Corston-Oliver, S.H.: Beyond string matching and cue phrases: improving efficiency and coverage in discourse analysis. In: The AAAI Spring Symposium on Intelligent Text Summarization, pp. 9–15 (1970)
11. Soricut, R., Marcu, D.: Sentence level discourse parsing using syntactic and lexical information. In: Conference of the North American Chapter of the Association for Computational Linguistics on Human Language Technology, vol. 1, pp. 149–156. Association for Computational Linguistics (2004)
12. Feng, V.W., Hirst, G.: Text-level discourse parsing with rich linguistic features. In: Proceedings of the 50th Annual Meeting of the Association for Computational Linguistics: Long Papers, vol. 1, pp. 60–68. Association for Computational Linguistics (2012)
13. Li, S., Wang, L., Cao, Z., et al.: Text-level discourse dependency parsing. Meet. Assoc. Comput. Linguist. 1, 25–35 (2014)
14. Pang, B., Lee, L.: A sentimental education: sentiment analysis using subjectivity summarization based on minimum cuts. In: Meeting on Association for Computational Linguistics, p. 271. Association for Computational Linguistics (2004)
15. Sharma, A., Dey, S.: A document-level sentiment analysis approach using artificial neural network and sentiment lexicons. ACM SIGAPP Appl. Comput. Rev. 12(4), 67–75 (2012)
16. Tang, D., Qin, B., Liu, T.: Document modeling with gated recurrent neural network for sentiment classification. In: Conference on Empirical Methods in Natural Language Processing, pp. 1422–1432, Portugal (2015)
17. Xu, J., Chen, D., Qiu, X., et al.: Cached Long Short-Term Memory Neural Networks for Document-Level Sentiment Classification. arXiv preprint arXiv:1610.04989 (2016)
18. Voll, K., Taboada, M.: Not all words are created equal: extracting semantic orientation as a function of adjective relevance. In: Orgun, M.A., Thornton, J. (eds.) AI 2007. LNCS, vol. 4830, pp. 337–346. Springer, Heidelberg (2007). doi:10.1007/978-3-540-76928-6_35
19. Heerschop, B., Goossen, F., Hogenboom, A., et al.: Polarity analysis of texts using discourse structure. In: ACM Conference on Information and Knowledge Management. DBLP, pp. 1061–1070, Glasgow, United Kingdom (2011)
20. Wang, F., Wu, Y., Qiu, L.: Exploiting discourse relations for sentiment analysis. In: COLING: Posters, pp. 1311–1320 (2012)
21. Li, J., Zhou, Y., Liu, C., et al.: Sentiment classification of Chinese contrast sentences. In: Zong, C., Nie, JY., Zhao, D., Feng, Y. (eds.) Natural Language Processing and Chinese Computing, vol. 496, pp. 205–216. Springer, Heidelberg (2014)
22. Hogenboom, A., Frasincar, F., De Jong, F., et al.: Using rhetorical structure in sentiment analysis. Commun. ACM 58(7), 69–77 (2015)

Fast Follower Recovery for State Machine Replication

Jinwei Guo[1], Jiahao Wang[1], Peng Cai[1(✉)], Weining Qian[1], Aoying Zhou[1], and Xiaohang Zhu[2]

[1] Institute for Data Science and Engineering, East China Normal University, Shanghai 200062, People's Republic of China
{guojinwei,jiahaowang}@stu.ecnu.edu.cn,
{pcai,wnqian,ayzhou}@sei.ecnu.edu.cn
[2] Software Development Center, Bank of Communications, Shanghai 201201, People's Republic of China
zhu_xiaohang@bankcomm.com

Abstract. The method of state machine replication, adopting a single strong Leader, has been widely used in the modern cluster-based database systems. In practical applications, the recovery speed has a significant impact on the availability of the systems. However, in order to guarantee the data consistency, the existing Follower recovery protocols in Paxos replication (e.g., Raft) need multiple network trips or extra data transmission, which may increase the recovery time. In this paper, we propose the **F**ollower **R**ecovery using **S**pecial mark log entry (FRS) algorithm. FRS is more robust and resilient to Follower failure and it only needs **one** network round trip to fetch the **least** number of log entries. This approach is implemented in the open source database system OceanBase. We experimentally show that the system adopting FRS has a good performance in terms of recovery time.

Keywords: Raft · State machine replication · Follower recovery

1 Introduction

Paxos replication is a popular choice for building a scalable, consistent and highly available database systems. In some Paxos variants, such as Raft [1], when a replica recovers as a Follower, it is ensured that its state machine is consistent with the Leader's. Therefore, the recovering Follower has to discard the invalid log entries (e.g., they are not consistent with the committed ones) and get the valid log records from the Leader. Figure 1 shows an example of log state of 3-replica system at a particular point in time. R1 is the Leader at that moment, and R3 which used to be the Leader crashes.

The *cmt* (*commit point*) indicates that the log before this point can be applied to local state machine safely, which is persistent in local. The *div* (*divergent point*) is the log sequence number (LSN) of the first log entry which are invalid in log order. The *lst* is the last LSN in the log. Note that the *cmt*, *div* and *lst*

© Springer International Publishing AG 2017
L. Chen et al. (Eds.): APWeb-WAIM 2017, Part I, LNCS 10366, pp. 311–319, 2017.
DOI: 10.1007/978-3-319-63579-8_24

of R3 are 2, 5 and 6 respectively. Therefore, when R3 restarts and recovers, it can reach a consistent state only if it discards the log in the range [5, 6], gets remaining log from the Leader (R1) and applies the committed log entries.

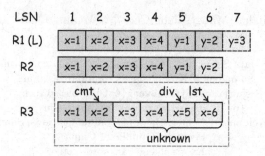

Fig. 1. An example of log state of 3-replica at one point in time.

Unfortunately, R3 does not know the state of the log after the LSN point 3. The existing approaches of Follower recovery do not pay attention to locating the *divergent point* directly. In Raft [1], the recovering Follower discards the last entry in local by executing AppendEntries remote procedure call (RPC) repeatedly until it finds the *divergent point*[1]. This approach need numerous network interactions. The Spinnaker [2] truncates the log after *cmt* (which may be *zero* if it is damaged) logically and gets the write data after *cmt* from the Leader, which increases the number of transmitted log entries and need a complicated mechanism to guarantee the data consistency of log replacement.

In this work, we present the **F**ollower **R**ecovery using **S**pecial mark log entry (FRS) algorithm, which does not depend on *commit point* strongly and requires only one network round trip for fetching the minimum log entries from the Leader when a Follower is recovering. The following is the list of our main contributions.

- We give the notion of the special mark log entry, which is the delimiter at the start of a term.
- We introduce the FRS algorithm, explain why this mechanism works and analyze it together with other approaches.
- We have implemented the FRS algorithm in the open source database system OceanBase[2]. The performance analysis demonstrates the effectiveness of our method in terms of recovery time.

2 Preliminaries

2.1 The Overview of Paxos Replication

Paxos replication, adopting the *strong leadership* and *log coherency* features of Raft, is introduced first. During normal processing, only the Leader can accept

---

[1] There is a optimization in Raft for reducing the number of network interactions, but the optimized approach does not find the *divergent point* directly yet.

[2] https://github.com/alibaba/oceanbase/.

the write requests from clients. When receiving a write, the Leader generates a log record which contains the monotonically increasing LSN, its own *term_id* and updated data. The Leader replicates the log record to all Followers. When the entry is persisted on a majority of nodes, the Leader can apply the write to local state machine and update the *cmt* piggybacked by the next log message. The Followers only store successive commit logs, and replay the local log before the point (*cmt* + 1) in order of log sequence.

The Leader sends heartbeats to all followers periodically in order to maintain its authority. If a Follower finds that there is no Leader, it becomes a Candidate and increases the current *term_id*, and then launches a Leader election. According to the newest log entry, a new Leader with the highest *term_id* and LSN can provide services only when it receives a majority of votes.

2.2   Handling Log Entries in Unknown State

Recall that a replica node flushes a log entry to local storage first. Then the log entries, whose LSN is $\leq cmt$, can be safely applied to local state machine. However, the state of the log after *cmt* is unknown. In other words, a log entry whose LSN is greater than *cmt* may be committed or invalid.

Fortunately, a recovering Follower f can get the *cmt* from local—which is denoted by *f.cmt*—and ensure that the log entries whose LSN is equal to or less than this point are committed. Next, it needs to handle the local log after *f.cmt* carefully. There are two ways to handle the log entries in unknown state:

- **Checking (CHK)**: The recovering Follower f gets the next log entry from the Leader l, and then checks whether it is continuous with local last log record. If not, it discards the log entry and repeats the above process; otherwise, it is back to normal.
- **Truncating (TC)**: The recovering Follower f gets the log after the local *cmt* from the Leader l. Then it replaces the local log after the *cmt* with received log, all of these operations should be done atomically.

It is clear that the CHK is simple and the TC is complicated, because the replacement operation of TC can be interrupted and more steps are required to handle each exceptions, e.g., if a recovering Follower fails again and the appending operation is not finished, it has to do the appending when it restarts. Both of approaches can not locate the *div* directly, which leads to more network round trips or more transmitted log entries in the Follower recovery. We will propose a new approach in the next sections, which needs only one network round trip for getting the minimum number of log entries.

3   The Special Mark Log Entry

In order to reduce the overhead of Follower recovery, we must provide additional mechanism, which can record necessary information used in the recovery of a

Follower. In this section, we introduce the special mark log entry and how a new Leader utilizes the special entry to take over the requests form the clients.

A special mark log entry S is the delimiter at the start of a term. Let S_i and S denote the special mark log entry of the term i and the set of all existing special entries respectively. And we use the notation $l.par$ to access the parameter named par of a log entry l (e.g., $S_i.lsn$ represents the LSN of special log entry S_i). In order to distinguish the other log entries from the special ones, we call them the normal operation log entries, which are the members of the set $\mathcal{N}$. And a mark flag is embedded in each log entry. An S, differing from the normal operation log entries, does not contain any operation data except the mark flag which is set to **true**.

When a replica is elected as a new Leader, a new term gets started. The new Leader must take some actions—which guarantee that the local log entries from previous term are committed—before it provides normal services. Therefore, the Leader using special mark log entry has to take following steps to take over:

(1) According to the new term id t, the Leader generates the special mark log entry S_t. More specifically, it produces a log record with a greater LSN and the new term id t, and sets its $mark\_flag$ to true. Then the Leader sends S_t to all other replicas.
(2) The Leader gets local cmt, and replays local log entries to this point in the background. It obtains the Followers' information and refreshes the cmt periodically until this point is equal to or greater than $S_t.lsn$. Note that the Leader can not service the requests from the clients in this phase.
(3) The Leader can safely apply the whole local log entries to local memory table. After that it can provide the clients with the normal services.

4   Follower Recovery

When a replica node recovers as a Follower from a failure, it has to take some measures to ensure that its own state is consistent with the Leader's. In other words, the recovering Follower has to discard the inconsistent log entries in local storage and get the missing committed log from the Leader. Then it can apply the log records to local memtable, so that the Follower can reach a consistent state with the Leader. The procedure of Follower recovery using special mark log entry (FRS) is shown as follows:

(1) The recovering Follower gets the cmt information and the last LSN $last\_point$ from local first. Then it obtains the last committed special log entry s (i.e., there exists a log entry l where $l \in \mathcal{N}$ and $l.lsn > s.lsn$). Two variables $start$ and end are set to $\max(commit\_point, s.lsn)$ and $last\_point$ respectively. The $start$ indicates the last committed log entry's LSN, which can be figured out by the Follower itself. After that, the Follower sends a confirm request to the Leader with these information.

(2) When the Leader receives the confirm request from the Follower, it gets the embedded parameters *start* and *end* first. Then the Leader obtains the first special log entry *s* in the log range $(start, end]$. If no one satisfies the requirement, the Leader replies a result value *zero* to the Follower; otherwise, it returns the LSN of the record *s*.

(3) When the Follower receives the response of the confirm request, it checks the result value *v*. If *v* is not *zero*, the Follower will first discard the entries from the local log after the index which is equal to $v - 1$. Then it gets the rest log from the Leader, and appends these entries to the local. If *v* is *zero*, the Follower will not copy with the local log.

When a Follower recovers from a failure, it first the above procedure to eliminate invalid log entries in local storage and to acquire the coherent ones from the Leader. Then it processes the log replication using the normal approaches described in Sect. 2.1. Note that the Follower can apply the log entries—whose LSNs are ≤ the local *cmt*—to local state machine in parallel with the recovery.

The FRS algorithm is applied not only to the restarting from a failure but also to the scenario where a Follower finds that a new Leader is elected. Generally, if the term is changed, the lease of previous Leader is expired, and the Follower will turn to Candidate and then convert to Follower again when it knows that the new Leader is not itself. In this case, the Follower executes the FRS algorithm actively. There is another case that a Follower receives a special mark log entry or a log entry containing a newer term value. In this scenario, the Follower dose not change its role, and it only discards the buffered log and executes the FRS algorithm as well.

5 Performance Evaluation

5.1 Experimental Setup

We conducted an experimental study to evaluate the performance of the proposed FRS algorithm, which is implemented in OceanBase 0.4.2.

Cluster Platform: We ran the experiments on a cluster of 12 machines, and each machine is equipped with a 2-socket Intel Xeon E5606 @2.13 GHz (a total of 8 physical cores), 96 GB RAM and 100 GB SSD while running CentOS version 6.5. All machines are connected by a gigabit Ethernet switch.

Database Deployment: The Paxos group (RootServer and UpdateServer as a member) is configured with 3-way replication and each of them is deployed on a singe machine in the cluster. Each pair of MergeServer and ChunkServer is deployed on one of the other 9 servers.

Competitors: We compare the FRS algorithm to other approaches described in the Sect. 2.2. In order to increase efficiency, the CHK approach is responsible for locating the *divergent point*. When finding the log index, the recovering Follower requests the log after *div* (including) as a group to the Leader. Therefore, the number of requests of CHK is about half of the results described in Raft. The Follower adopting the TC approach gets the log after *cmt* which is contained by one message package from the Leader.

Benchmark: We adopted YCSB [3]—a popular benchmark with key-value workloads from Yahoo—to evaluate our implementation. Since we pay attention to the Follower recovery relying on log replication, we modify the workload to have a read/write ratio of 0/100. The clients, which request the writes to the database system, are deployed on the MergeServer/ChunkServer nodes. The size of each write is about 100 bytes value.

5.2   Experimental Results

To measure the Follower recovery, we need to kill a replica node and then restart it. Therefore, each experimental case is conducted as follows first. We ensure that the three replicas work normally and about one million records are inserted into the system. Next, we make the Leader r disconnect from the other replicas before its lease expires in 2 s. When r loses the leadership and a new Leader is elected, we kill and restart r. Then r recovers as a Follower.

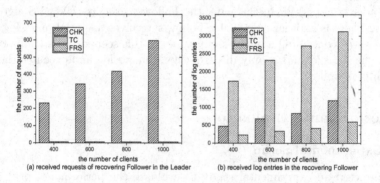

Fig. 2. The statistics of a recovering Follower, which used to be a Leader and failed in different workloads measured by the number of clients.

Follower Recovery Statistics: We first measure the statistic of the Follower recovery in terms of received requests in the Leader and received log entries in the recovering Follower. Therefore, we get the results by adding code to cumulate the corresponding values in the program.

Figure 2 shows the statistics of a recovering Follower, which fails in different workloads denoted by different numbers of clients. As the number of clients increases, we find that the size of the log after local *cmt* or *div* becomes larger, which can lead to more requests and received log entries in the Follower recovery. Figure 2(a) shows that CHK approach needs hundreds of requests to locate the *div* point, and the other approaches need only few network interactions in the recovery phase. Figure 2(b) shows that the TC transmits the most log entries containing the unnecessary ones. Since the log transmitted in locating *div* is not preserved, the number of log received are about twice as the FRS's. All of these results conform to the analysis described above.

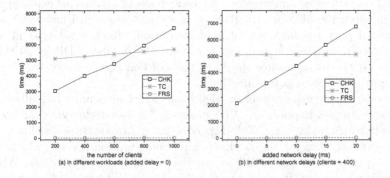

Fig. 3. Follower recovery time with invalid *cmt*.

Follower Recovery Time: Note that the recovery of a Follower has two phases: handling log in unknown state and applying the log to local machine state. Since the applying phases of all the Follower recovery approaches are the same, we only measure the time of handing the uncertain log entries. In this experiment, the recovering Follower fails when the number of clients (workload) is 400.

Figure 3 shows the Follower recovery time with invalid *cmt* when the system is in different workloads or network delays. We use Linux tool tc to add network delay in the specified ethernet interface of the recovering Follower. We find that the recovery time of FRS is shortest and is not impacted by workloads and network delays at all. As the workload increases in Fig. 3(a), the result of CHK was linearly correlated with the number of clients, which indicates that the higher workload can lead to more time of handling a request in the Leader. A recovering Follower adopting TC needs to get the whole log from the Leader. Due to the long time of transforming the whole log, the recovery time of TC—which is between 5 s and 6 s—is not desired. The trend in Fig. 3(b) is similar to 3(a). Therefore, we can conclude that the FRS has the best performance in terms of recovery time, no matter what the workload or the network delay is, and no matter whether the *cmt* is valid.

6 Related Work

State machine replication (SMR) [4], a fundamental approach to designing fault-tolerant services, can ensure that the replica is consistent with each other only if the operations are executed in the same order on all replicas. In reality, database systems utilize lazy master replication protocol [5] to realize SMR.

Paxos replication is widely used to implement a highly available system. [6] describes some algorithmic and engineering challenges encountered in moving Paxos from theory to practice and gives the corresponding solutions. To further increase understandability and practicability, some multi-Paxos variants adopting strong leadership are proposed. In Spinnaker [2], the replication protocol is based on this idea. Nevertheless, its Leader election relies an external coordination service ZooKeeper [7]. Raft [1] is a consensus algorithm designed for educational purposes and ease of implementation, which is adopted in many open source database systems, e.g., CockroachDB [8] and TiDB [9]. Although these protocols can guarantee the correctness of Follower recovery, they neglect to locate the *divergent point* directly.

There are many other consensus algorithms which are similar to multi-Paxos. The Viewstamped Replication (VR) protocol [10] is a technique that handles failures in which nodes crash. Zab [11] (ZooKeeper Atomic Broadcast) is a replication protocol used for the ZooKeeper [7] configuration service, which is an Apache open source software implemented in Java. In these protocols, When a new Leader is elected, it replicates its entire log to Followers. [12] presents the differences and the similarities between Paxos, VR and Zab.

7   Conclusion

Follower recovery is not non-trivial in Paxos replication systems, which not only guarantees the correctness of data, but also should make the recovering replica back to normal fast. In this work, we introduced the Follower recovery using special mark log entry (FRS) algorithm, which does not rely on the *commit point* and transmits the least number of log entries by using only one network round trip. The experimental results demonstrate the effectiveness of our method in terms of recovery time.

Acknowledgments. This work is partially supported by National High-tech R&D Program (863 Program) under grant number 2015AA015307, National Science Foundation of China under grant numbers 61432006 and 61672232, and Guangxi Key Laboratory of Trusted Software (kx201602).

References

1. Ongaro, D., Ousterhout, J.: In search of an understandable consensus algorithm. In: ATC, pp. 305–320 (2014)
2. Rao, J., Shekita, E.J., Tata, S.: Using Paxos to build a scalable, consistent, and highly available datastore. In: VLDB, pp. 243–254 (2011)
3. Cooper, B.F., Silberstein, A., Tam, E., et al.: Benchmarking cloud serving systems with YCSB. In: Socc, pp. 143–154 (2010)
4. Schneider, F.B.: Implementing fault-tolerant services using the state machine approach: a tutorial. ACM Comput. Surv. **22**(4), 299–319 (1990)
5. Gray, J., Helland, P., O'Neil, P., et al.: The dangers of replication and a solution. SIGMOD Rec. **25**(2), 173–182 (1996)

6. Chandra, T.D., Griesemer, R., Redstone, J.: Paxos made live: an engineering perspective. In: PODC, pp. 398–407 (2007)
7. ZooKeeper website. http://zookeeper.apache.org/
8. CockroachDB website. https://www.cockroachlabs.com/
9. TiDB website. https://github.com/pingcap/tidb
10. Oki, B.M., Liskov, B.H.: Viewstamped replication: a new primary copy method to support highly-available distributed systems. In: PODC, pp. 8–17 (1988)
11. Junqueira, F.P., Reed, B.C., Serafini, M.: Zab: high-performance broadcast for primary-backup systems. In: DSN, pp. 245–256 (2011)
12. Van Renesse, R., Schiper, N., Schneider, F.B.: IEEE TDSC 12(4), 472–484 (2015)

Laser: Load-Adaptive Group Commit in Lock-Free Transaction Logging

Huan Zhou[1], Huiqi Hu[1(✉)], Tao Zhu[1], Weining Qian[1], Aoying Zhou[1], and Yukun He[2]

[1] School of Data Science and Engineering, East China Normal University, Shanghai, China
{zhouhuan,zhutao}@stu.ecnu.edu.cn, {hqhu,wnqian,ayzhou}@dase.ecnu.edu.cn
[2] Bank of Communications, Shanghai, China
he.yk@bankcomm.com

Abstract. Log manager is a key component of DBMS and is considered as the most prominent bottleneck in the modern in-memory OLTP system. In this paper, by addressing two existing performance hurdles in the current procedure, we propose a high-performance transaction logging engine Laser and integrate it into OceanBase, an in-memory OLTP system. First, we present a lock-free transaction logging framework to eliminate the lock contention. Then we make theoretical analysis and propose a judicious grouping strategy to determine an optimized group time for different workloads. Experiment results show that it improves 1.4X–2.4X throughput and reduces more than 60% latency compared with current methods.

Keywords: Logging · Group commit · Lock-free · Load-adaptive

1 Introduction

Recent years have seen a shift in the design of high throughput OLTP systems: from the conventional transaction engine to the widespread adoption of multi-core memory system. To improve concurrency, many transaction engines focus on eliminating fundamental bottlenecks, such as lock-based shared data structure, concurrency control, centralized log manager etc. Among them, the log manager is considered as the most prominent bottleneck due to centralized design and dependence on I/O [2]. The state of art method [4] integrates three most widely used techniques, i.e. parallel buffering, flush pipelining and group commit, to form an efficient logging procedure and results show that the procedure can significantly achieve better performance than the traditional logging approaches.

In this paper, we propose a high-performance transaction logging engine called Laser. In particular, we observe two defects that limit the performance of the existing procedure: (1) current approach depends on a lock-based method to manage transaction log records, it involves many lock contentions and reduces the CPU utilization as load increases [6]. (2) existing method uses a fixed group

© Springer International Publishing AG 2017
L. Chen et al. (Eds.): APWeb-WAIM 2017, Part I, LNCS 10366, pp. 320–328, 2017.
DOI: 10.1007/978-3-319-63579-8_25

commit strategy which cannot achieve a good performance when the workload changes. To obtain better latency and throughput, We present a new lock-free transaction logging framework with the help of a well-designed multi-group structure and CAS operation and propose an adaptive group commit where we make theoretical analysis and propose a judicious grouping strategy to determine an optimized grouping time when the workload varies. Then we implement these methods and integrate them into the in-memory OceanBase OLTP system. Results show that it achieves 1.4X–2.4X better performance over the compared methods in throughput as well as reduces more than 60% latency.

2   Related Work

The write-ahead logging (WAL) [5] is widely employed in database systems to provide data durability and recovery. Compared to traditional system, the latest OLTP systems have demonstrated significant performance improvement, however, the log manager is still prone to bottlenecks due to its centralized structure [2].

Many technologies are explored to reduce the overhead of logging. Johnson et al. [4] identify four bottlenecks of the write-ahead logging named I/O-related delay, log-induced lock contention, context switching and log buffer contention. Parallel buffering [4] is used to reduce the log buffer contention. Group commit [1] reduces the I/O-related delay by aggregating multiple log records into one I/O operation. Helland et al. [3] turns out that the database can set group timer to minimize average response time, however it assumes the system load is unchangeable and only examines the effect of grouping time on CPU response time based on the traditional single thread system. Aether [4] utilizes flush pipelining to reduce the overhead of context switching and integrates it with parallel buffering and group commit to form an efficient logging procedure.

3   Preliminary

Transaction logging generally consists of two distinct steps: *pre-logging* and *commit logging* . In the pre-logging step, each transaction fills its log records into an in-memory log buffer. It first acquires a unique log sequence number(LSN) using a lock to indicate its allocated space within the buffer, then copies the log records into the log buffer. The usage of lock is to keep transactions acquiring their LSN in a monotonous serial order. In the second step, log records are physically flushed into disk following the LSN order through some I/O operations.

Three mature techniques such as parallel buffering, flushing pipelining and group commit have been adopted to optimize the procedure as illustrated in Fig. 1. It is non-trivial to combine the three techniques together in a logging procedure. As log records are filled in parallel and must be written to disk in LSN order, log records of large LSN orders cannot be flushed until the front transactions (transactions with small LSN orders) completed buffering. To this end, the flushing thread has to identify a "safe" region (of offset) in the log buffer

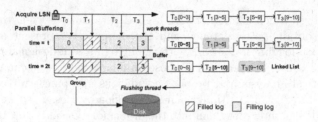

Fig. 1. Transaction commit processing in memory transaction engine

for group commit. In the example, before notifying its region to flushing thread, the work thread of T_1 must wait until T_0 releases its buffer space. To solve the problem, the state of art [4] forms a linked list to release the buffer region in LSN order. For log records, work threads acquire their LSN and enqueue themselves into the list which protected by a lock. Each node in the list contains a "safe" region which indicates the range of log records starting from it and ending at the first successor that has not yet finish buffering. The "safe" region is figured out in a delegated way. Once a work thread finishes filling log records, it first abandons its node in the linked list and merges its offset into the range of its predecessor. When a flushing thread triggers log flushes using group commit policies, it first figures out buffer region of finished log records from the head of linked list. Log records within the region can be flushed into disks.

4   Lock-Free Transaction Logging Framework

Data Structure. We use a buffer (denoted by $\mathcal{B}$) with a constant size $|\mathcal{B}|$ to store log records. The buffer is used in a round-robin manner. We rely on a multi-group and a global offset of $\mathcal{B}$ (denoted by o_f) to manage the logging.

The multi-group structure is formed by a sufficiently large array denoted by $\{G_0, G_1 \cdots G_n\}$ (n is large, e.g. $n = 10000$) and each group G_i is consisted of a sextuple $\langle \texttt{group\_state}_i, \text{LSN}_i, s_i, e_i, n_i, f_i \rangle$, where $\texttt{group\_state}_i$ is the current state of the group, LSN_i is used by work threads to acquire LSN, s_i and e_i are the *logical* start and end offset of the group in $\mathcal{B}$, n_i is used to record the number of active transactions which do not complete filling in the group, f_i is used to indicate whether G_i is frozen. Details of their usages are introduced subsequently. Each group has three possible states `Available`, `Ready` and `Durable`. `Available` means the group is empty and the values (LSN_i, s_i, e_i, n_i, f_i,) of the group can be set. `Ready` indicates the work threads that join into the group can start to acquire LSN and their log records are allowed to be filled into $\mathcal{B}$. `Durable` means that all the work threads in the group have completed buffering their log records and can flush log records into disk. It is worth notice that the multi-group is also used in a round-robin manner.

We also maintain a logical offset o_f to mark the start offset of transaction log records which will be flushed (i.e. the start offset for the next flush). Notice that we all use logical offset here $(o_f/s_i/e_i)$ and their physical address can be easily corresponded as $o_f\%|\mathcal{B}|$, $s_i\%|\mathcal{B}|$ and $e_i\%|\mathcal{B}|$ respectively.

Transaction Logging Procedure. As shown in Fig. 2, we adopt a lock-free mechanism which maintains a global 128 bits structure $\mathcal{Q} = \langle G_i, r_i, o_i \rangle$, where G_i is the group used for acquiring LSN, r_i and o_i are used to record the relative log serial number and offset in the G_i. When a work thread comes to acquire a LSN, it first retrieves $\mathcal{Q}$ and generates a new $\mathcal{Q}$ by increasing $r_i = r_i + 1$, $o_i = o_i + |T|$ (where $|T|$ is the size of log records) with a CAS operation. After acquiring $\mathcal{Q}$, it sets the $n_i = n_i + 1$. When G_i is ready, the work thread detects if its log records can be buffered by comparing $s_i + o_i - o_f <= |\mathcal{B}|$. If the $\mathcal{B}$ has enough space, the work thread assigns the LSN of its transaction as $\text{LSN}_i + r_i - 1$ and fills its log records with its offset. After completing buffering, it decreases the value of n_i by one, then it can turn to serve other transactions. When the last work thread completes buffering, it changes the state of group from **Ready** to **Durable**. A work thread confirms itself as the last thread if it finds $n_i = 0$ and its frozen indicator $f_i = $ true assigned by a grouping thread.

Note that LSN_i and s_i is required for the work thread. Both of them are computed in a *grouping thread* which is used to construct groups. When the condition of group commit is satisfied, the grouping thread generates a new $\mathcal{Q}$ as $\langle G_{i+1}, r_{i+1} = 0, o_{i+1} = 0 \rangle$ with a CAS operation. Then it sets the end offset of the previous group G_i as $e_i = s_i + o_i$, $f_i = true$. Next if the state of G_{i+1} is **Available**, it assigns $s_{i+1} = e_i$, $\text{LSN}_{i+1} = \text{LSN}_i + r_i$ and makes the state into **Ready** so that the next coming transaction can acquire LSN on G_{i+1}.

The flushing thread maintains a position indicator (denoted by seq_L) to determine which group will be written into disk with an I/O operation. When $seq_L = i$, if the state of G_i becomes **Durable**, the flushing thread flushes its contained log records from $(s_i\%|\mathcal{B}|, e_i\%|\mathcal{B}|)$ (or $(s_i\%|\mathcal{B}|, |\mathcal{B}|)$ and $(0, e_i\%|\mathcal{B}|)$) into disk, then it increases $o_f = e_i$, $group_state_i = $ **Available** and increases seq_L to directing the next group for the next flush.

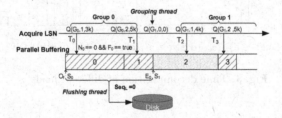

Fig. 2. Lock-free transaction logging framework

5   Load-Adaptive Group Committing

Observation. In this section, we investigate the influence of different grouping strategy on logging performance. We observe that the best grouping time (group timer) is distinguishing for different loads as shown in Fig. 3. We exploit the observation by decomposing the executing time of transaction logging into five main components (1) W_a: the average time that a transaction waits for a group changing to `Ready`. (2) W_b: the average time that a transaction waits for its group becoming `Durable`. (3) W_c: the average time that a transaction waits for its log records to be flushed after its group becomes `Durable`. (4) W_d: the average time that the flushing thread writes the log records into disk. (5) W_e: the rest time spent for logging.

We breakdown the executing time when load throughput and log record size are different in Fig. 3. Consider load throughput, given a small one (320 k tps), group timer $= 0.9$ ms have the largest latency where W_b takes the most time cost, because the procedure provides a large time window of 0.9 ms to make threads acquire their LSNs and buffer their log records. In fact, Fig. 3(c) proves 0.15 ms is a sufficient gap when the throughput is 320 k. On the contrary, 0.15 ms is not sufficient to sever when the throughput is high due to the smaller timer generates more groups to be flushed in every seconds. Many transactions in groups have to wait for the flushing thread, which causes a largest overhead of W_c. Timer $= 0.9$ ms significantly reduces the overhead as it lowers the generation rate of group. Similarly, increasing the record size extends the time to flush a group (W_d) in Fig. 3(d). Therefore there is a largest overhead of W_c when group timer is 0.15 ms and record size is 160 byte. On the contrary, the larger timer does not have the overhead of W_c.

In summary, *when the load throughput or the record size grows, it involves increasing overhead of W_c. Expand the grouping time increases cost of transactions waiting for the group becoming durable (W_b), but it can reduce the overhead of transactions in a group waiting for flushing (W_c).*

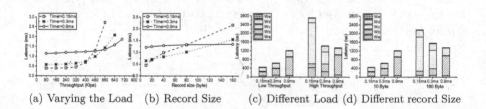

(a) Varying the Load   (b) Record Size   (c) Different Load   (d) Different record Size

Fig. 3. Time decomposition of different loads

Theoretical Analysis. We further discuss the above conclusion through some theoretical analysis. First, we demonstrate an I/O property between the log size and its respect flushing time. Note that it is a general property for I/O operation that well recognized by many data accessing test. Figure 4(a) demonstrates the property: *the amortized time for writing a larger size of log is less than a smaller one.* For example, when it writes 100 kb log, it spends about 0.4 ms, however, if it only costs about 0.7 ms to write 500 kb (5 times large than 100 kb) log. Based on the property, we analyze the reason why increasing the grouping time can reduce the overhead of W_c. Let λ and $|\bar{T}|$ denote the load throughput and log record size. Let $\mathcal{D}$ denotes the group timer and $\mathcal{F}(|\bar{G}|)$ be the total time to write log records where $|\bar{G}|$ is the *size of grouping log* that contained in one group for flushing. Obviously, if $\mathcal{F}(|\bar{G}|) > \mathcal{D}$, the procedure produces overhead of W_c since the group requires to wait its prior group. $|\bar{G}|$ can be computed by the following equation:

$$|\bar{G}| = \begin{cases} |\bar{T}| * \mathcal{D} * \lambda & \lambda < 1/t_{CAS} \\ |\bar{T}| * \mathcal{D} * 1/t_{CAS} & \lambda \geq 1/t_{CAS}. \end{cases}$$

where t_{CAS} is the constant time for a transaction to acquire LSN with the atomic CAS operation and $\mathcal{D} * 1/t_{CAS}$ is the largest number of contained log records. Therefore, when $|\bar{T}|$ is constant, we adjust the value of $\mathcal{D}$ according to the variation of λ to make sure $\mathcal{F}(|\bar{G}|) \leq \mathcal{D}$. Similarly, we can change the $\mathcal{D}$ based on the variation of $|\bar{T}|$ when λ is fixed. As shown in Fig. 4(b), if $\mathcal{D} = 0.15$ ms and $|\bar{T}| = 10$ byte when $\lambda = 320k$ tps, $\mathcal{F}(|\bar{G}|) = 0.158$ ms. This is close to its group timer, thus there is almost not overhead of W_c. If the record size increases, e.g. $|\bar{T}| = 160$ byte, the $|\bar{G}|$ enlarges and $\mathcal{F}(|\bar{G}|)$ becomes 0.3 ms where $\mathcal{F}(|\bar{G}|) > \mathcal{D}$ and causes large overhead of W_c. Now consider using the larger group timer (e.g. $\mathcal{D} = 0.9$ ms), though $|\bar{G}|$ increases several times as larger $\mathcal{D}$, but only smaller growth is generated on flushing time due to the I/O property. When $\mathcal{D} = 0.9$ ms and $|\bar{T}| = 160$ bytes, $\mathcal{F}(|\bar{G}|)$ is 0.35 ms where $\mathcal{F}(|\bar{G}|) << \mathcal{D}$ and avoids the overhead of W_c.

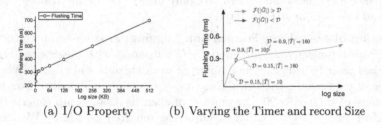

(a) I/O Property   (b) Varying the Timer and record Size

Fig. 4. Exploiting the Flushing Time

To conclude, we find *(1) increasing the grouping time will increase the time of group being durable (W_b), but it reduces the overhead of W_c. (2) When the flushing time is less than the group timer ($\mathcal{F}(|\bar{G}|) < \mathcal{D}$), it eliminates or significantly reduces the overhead of W_c. (3) When $\mathcal{D}$ is small, $\mathcal{D}$ is smaller than $\mathcal{F}(|\bar{G}|)$, but when $\mathcal{D}$ is large enough, $\mathcal{D}$ will be larger than $\mathcal{F}(|\bar{G}|)$ due to the I/O property.*

Grouping Strategy. Based on the above conclusions, we propose our group strategy as *choosing the group timer which generates smallest cost of W_b by eliminating the overhead of W_c*. Formally, it is the minimum group timer $\mathcal{D}^*$ that avoids W_c by satisfying the following equation:

$$\mathcal{D}^* \geq \mathcal{F}(|\bar{G}|_{\mathcal{D}^*}). \tag{1}$$

where $|\bar{G}|_{D*}$ is the size of grouping log by utilizing D^* as group timer.

To implement the grouping strategy, we monitor the flushing time W_d (where W_d is closed to $\mathcal{F}(|\bar{G}|_{D*})$). We turn the group timer based on the previous value and the current flushing time in case of it changing too heavily at a time. In particular, it is tuned as $D = 1/2 * D + 1/2 * \mathcal{F}(|\bar{G}|_{D*})$. D is tuned by the grouping thread and after several times, D will become close to $\mathcal{F}(|\bar{G}|_D)$. If we observe W_a increase heavily over a threshold τ (e.g. 0.1 ms), it means the load throughput has increased and the group is not available, we also increase D as $D = D + \delta$ where δ is a constant time (e.g. $\delta = 2\tau$).

6 Experiment

Experimental Setup. The experiments are conducted on a linux server with 268 GB main memory and *two Interl Xeon E5-2630@2.20 GHZ processors*, each with *10* physical cores. We use RAID5 with flash-based write cache (FBWC) which has high performance on I/O accesses. We implement Laser and the comparing methods into OceanBase [7]. We compare Laser with (1) **Baseline**: the origin transaction logging manager that Oceanbase adopts, it uses a single logging thread to acquire LSN order and fills logs in sequence, and flushes the logs with group commit. (2) **Aether** [4]: it utilizes parallel buffering, flush pipeline and group commit to form a logging procedure as described in Sect. 3. We test the methods on YCSB workload which is popular in evaluating the read/write performance for a database system. We only utilize the update transaction and the record size is from 10 B to 160 B.

Evaluation of Lock-Free Transaction Logging. First we compare the *scalability, peak throughput, latency* and *CPU utilization* with Baseline, Aether and Laser+Fixed timer by varying the number of work threads and clients. When we vary one of parameters, the rest parameters are setting by default values where the number of threads is 20, the number of client is 3200, group timer = 0.3 ms and record size is 10 byte. The results are reported in Fig. 5. We can see Laser+Fixed timer always achieves the best throughput under all the situations, which improves 1.2X–2X better performance.

Scalability. As shown in Fig. 5(a), the performance of Laser+Fixed timer increases nearly linearly when the number of work threads grows. The peak throughput of Aether increases slowly when the number of work threads is bigger than 12 due to acquiring LSN based-on lock which limits the scalability. Baseline quickly becomes saturated when thread count is 8 since the single logging thread becomes the critical bottleneck.

Client-Side Throughput. Figure 5(b) shows the performance of Laser+Fixed timer increases when the number of client varies however Aether increases very slowly when client = 1600 and Baseline becomes saturated when client = 800.

Latency. Figure 5(c) shows Laser+Fixed timer always takes the lowest time when the number of client grows, which improves the latency of Aether and Baseline more than 45%.

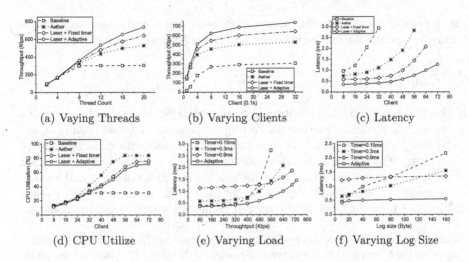

(a) Vaying Threads   (b) Varying Clients   (c) Latency

(d) CPU Utilize   (e) Varying Load   (f) Varying Log Size

Fig. 5. Evaluation of optimized transaction logging

CPU Utilization. Figure 5(d) shows the utilization of Laser+Fixed timer increases gently until the number of client equals 64. However, the cpu utilization of Aether increases fleetly until the client count is 56 due to lock contention appears. The CPU utilization of Baseline does not increase when the number of client equals 32 due to its single thread is saturate.

Evaluation of Adaptive Group Commit. Next we compare the adaptive grouping strategy with three fixed group timers in YCSB. We vary the number of client to increase load throughput (each client generates $10k$ transactions per second) and grow the record size where the default value of load is $320k$. Results are shown in Fig. 5(e) and (f). Adaptive group commit always has lowest latency compared to fixed group timer. For instance, when the record size is 10 bytes, the latency of timer 0.15 ms, 0.3 ms and 0.9 ms is roughly 0.44 ms, 0.64 ms and 1.21 ms respectively, and the latency of Adaptive is 0.4 ms. Meanwhile, the adaptive group commit improves the peak throughput. Adaptive can serve for 730 k load throughput, and the timer 0.15 ms, 0.3 ms, 0.9 ms only serve for 560 k, 640 k and 700 k respectively.

Overall Performance. Finally, we integrate the lock-free logging framework and the adaptive group commit as Laser+Adaptive and evaluate the overall performance. Results are reported in Fig. 5. We can see (1) Laser+Adaptive offers the **best throughput**, which improves **1.4X–2.4X** better peak throughput than Baseline and Aether. For example, in Fig. 5(b), when the number of client equals 3200, the peak throughput of Laser+Adaptive is about 730 k, where the peak throughput of Baseline, Aether and Laser+Fixed timer are 300 k, 520 k and 640 k respectively. (2) Laser+Adaptive has the **lowest latency** when varying load throughput which improves the latency by **60%**. For example, in Fig. 5(c), when the number of client is 32, the latency of Baseline, Aether and Laser+Fixed timer is 2.9 ms, 1.1 ms and 0.6 ms respectively, and Laser+Adaptive is nearly 0.4 ms.

7   Conclusion

In this paper, we propose an optimized transaction logging engine by proposing a new lock-free transaction logging to improve scalability based on a designed multi-group structure and CAS operation, and presenting a judicious grouping strategy which economizes the running time of logging for varied workload through some theoretical analysis. Implementation in Oceanbase and experiment results show the new logging engine can reduce more than 60% latency and achieve 1.4X–2.4X better throughput compared with existing methods.

Acknowledgement. This work is partially supported by National Hightech *R&D* Pro- gram (863 Program) under grant number 2015AA015307, National Science Foundation of China under grant numbers 61332006, 61432006 and 61672232, and the Youth Science and Technology - Yang Fan Program of Shanghai (17YF1427800).

References

1. Hagmann, R.: Reimplementing the Cedar file system using logging and group commit. In: SOSP, pp. 155–162. ACM (1987)
2. Harizopoulos, S., Abadi, D.J., Madden, S., Stonebraker, M.: OLTP through the looking glass, and what we found there. In: SIGMOD, pp. 981–992. ACM (2008)
3. Helland, P., Sammer, H., Lyon, J., Carr, R., Garrett, P., Reuter, A.: Group commit timers and high volume transaction systems. In: Gawlick, D., Haynie, M., Reuter, A. (eds.) HPTS 1987. LNCS, vol. 359, pp. 301–329. Springer, Heidelberg (1989). doi:10.1007/3-540-51085-0_52
4. Johnson, R., Pandis, I., Stoica, R., Manos, A.: Aether: a scalable approach to logging. PVLDB **3**(1–2), 681–692 (2010)
5. Mohan, C., Haderle, D., Lindsay, B., Pirahesh, H., Schwarz, P.: ARIES: a transaction recovery method supporting fine-granularity locking and partial rollbacks using write-ahead logging. TODS **17**(1), 94–162 (1992)
6. Wang, T.Z., Johnson, R., Stoica, R., Manos, A.: Scalable logging through emerging non-volatile memory. PVLDB **7**(10), 865–876 (2014)
7. OceanBase website. https://github.com/alibaba/oceanbase/

Social Networks

Detecting User Occupations on Microblogging Platforms: An Experimental Study

Xia Lv[1], Peiquan Jin[1,2(✉)], Lin Mu[1], Shouhong Wan[1,2],
and Lihua Yue[1,2]

[1] School of Computer Science and Technology,
University of Science and Technology of China, Hefei 230027, China
jpq@ustc.edu.cn
[2] Key Laboratory of Electromagnetic Space Information,
Chinese Academy of Sciences, Hefei 230027, China

Abstract. User occupation refers to the professional position of a user in real world. It is very helpful for a number of applications, e.g., personalized recommendation and targeted advertising. However, because of the risk of privacy leaks, many users do not provide their occupation information on microblogging platforms. This makes it hard to detect user occupations on microblogging platforms. In this paper, we conduct an experimental study on this issue. Particularly, we propose an experimental framework of detecting user occupations on microblogging platforms. We first implement a number of classification models and devise various sets of features for user occupation detection. Then, we propose to construct an occupation-oriented lexicon, which is collected by an iterative extension algorithm considering semantic similarity and importance between words. We combine the lexicon with the word embedding approach to detect user occupations. We conduct comprehensive experiments and present a set of experimental results. The results show that the lexicon-based word embedding method achieves higher accuracy compared with traditional feature-base classification models.

Keywords: Occupation detection · Feature extraction · Word embedding

1 Introduction

Microblog platforms have been an importance source for information extraction [1]. Users' information on microblog platforms is very helpful for many applications such as personalized recommendation and targeted advertising, due to the great number of microblog users. There are many aspects regarding user information, among which user occupation information is of particular business values for commercial applications [2]. However, because of privacy considerations, users usually do not provide their occupations on microblogging platforms. Thus, it is a challenging issue to automatically detect user occupation on microblogging platforms.

There are already some efforts concentrating on mining users' profile information [3–6, 12, 14, 15, 18, 22], such as detection of gender, age, and political orientation. A multi-source integration framework concentrates more on building feature set of

L. Chen et al. (Eds.): APWeb-WAIM 2017, Part I, LNCS 10366, pp. 331–345, 2017.
DOI: 10.1007/978-3-319-63579-8_26

content model and network information, which conducts extensive empirical studies for user occupation industry inference [24], and latent feature representation such as word clusters and embedding is used to classify occupations, but they only generate simple semantic features without comparison to classical models [25]. However, predicting user occupations based on microblog platforms is still a new problem. So far, only a few research works are focused on it, and most of them are related to "*occupation field*" [10, 23–25]. An occupation field indicates an area, which is less specific than an occupation. Thus, it is not sufficient for real applications. For example, "*entertainment*" is an "*occupation field*", but we may wonder whether a user is an actor/actress or a singer.

In this paper, we focus on detecting user occupation on microblogging platforms like Sina Weibo. We address it by leveraging available information such as observable digit contents, user behavior, custom tags, and linguistic messages of users. We experimentally compare several classical models over different feature sets to measure the performance on user occupation detection. Further, we propose a lexicon-based feature selection method and combine it with the word embedding method for user occupation detection. Briefly, we make the following contributions in this paper:

(1) We build a framework for detecting microblog user occupations. It takes advantages of two kinds of information resources, namely the observable digit information and the linguistic messages of users.
(2) We extract features from message contents according to three linguistic models, i.e., *BOW (bag-of-words)*, *n-gram*, and *topic model*. Then, we apply these feature sets into a number of classification models including *the logistic regression model* (LR), SVM, and *the random forest model*.
(3) We propose to construct an occupation-oriented lexicon that adapts semantic similarity and word importance to refine the feature set. The occupation-oriented lexicon is used to simplify the feature selection work and reduce the dimension of features. We integrate the lexicon with the word embedding method to detect user occupations and the experimental results suggest the effectiveness of this design.
(4) We conduct experiments on a real data set from Sina Weibo. The performance of different classification models over different feature sets is compared, and the lexicon-based method is also evaluated in terms of different settings.

2   The Proposed Framework for User Occupation Detection

Figure 1 shows the framework of detecting user occupations on microblogging platforms. It consists of occupation identification, feature extraction, occupation-oriented lexicon integration and classification. In the first stage of occupation identification, we screen the signed "*V*" users from huge and mixed unlabeled datasets, and then choose 8 occupations to label the users via manual and rule-based methods. We collect microblogs and custom tags by a web crawler according to selected users IDs, finally forming the whole occupation labeled dataset. In the feature-extraction stage, we divide the dataset into digit contents and message contents. The digit contents consist of basic user profiles, social influence, and user behavior features, while the message contents contain features described by three linguistic models including *BOW (bag-of-words)*,

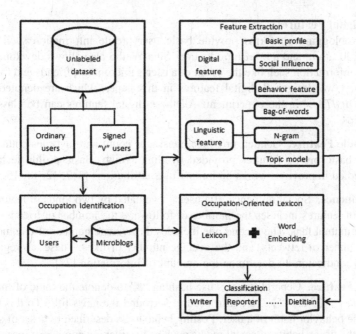

Fig. 1. Framework of microblog user occupation detection

n-gram, and *topic model*. Next, we utilize message contents to generate an occupation-oriented lexicon, and remove irrelevant words. The remaining words are represented by the word embedding method. Finally, we perform feature-based extraction for user occupations through a number of classification models.

2.1   Occupation Identification

Sina Weibo offers APIs for developers to collect data. However, due to the limitation of the APIs, we can only obtain a set of *"verified users"*, whose occupations are mandatorily labeled. Then, we extract the most popular occupations from the set of verified users to form the occupation candidates for the experiments. In this paper, we finally prepare eight occupations which are *writer, reporter, lawyer, photographer, actor, singer, doctor, and dietitian*. These occupations are used as the targets to be detected.

   We randomly choose 8000 user IDs in the verified user set and use these user IDs to fetch their microblogs from Sina Weibo. For each user, we finally collect a set of microblogs, which contains 100 to 500 text messages. A user containing less than 100 microblogs are not considered.

2.2   Feature Extraction

In this part we describe in detail two types of information, namely digital information and linguistic messages, which can help characterize a user.

2.2.1 Digital Features

Most microblogging platforms provide basic user profile information such as user nickname, location, and a brief introduction. Sina Weibo also allows developer to get basic user information such like the count of a user's microblogs, friends and followers. In summary, we design 38 digital features in this paper, which are denoted by the symbol "*DIGITAL*" in the experiment. All these digital features can be classified as three groups.

Basic Profile Feature. Gender, province, messages count and favorites count are the only four basic profile features provided by Sina Weibo. Basic profile features are widely studied in previous works on mining user attributions [3–7, 12, 14, 15, 18, 22].

Social Influence. Social influence of a user is evaluated by two kinds of features. The first kind of features includes the number of followers, the number of friends, and the number of mutual fans. The second one includes the average number of comments, the average number of retweets, and the average number of likes. These are regarded as quantitative indicators to determine the amount of information [21].

Behavior Feature. Generally, users use hashtag "#" to denote the topic of messages. The hashtags " 【" and "】 " are also used to surround the news title. To this end, we can find the behavior habit of a user. Posting behavior is described by a set of statistics capturing the usage habits of social media such as the average number of messages per day [6]. Such information is useful for constructing a model of a user intuitively [4]. In this paper, we consider the following behavior features: the hashtag count per message, the average number of topics and news, the average number of messages per day, and the average number of messages per hour within one day.

2.2.2 Linguistic Features

Linguistic content information contains user name, description, custom tag and messages posted by users. We concatenate the name, description and custom tag together as a short introduction for each user. In addition, we explore various linguistic content features based on three linguistic models, as detailed below.

Bag-of-Words (BOW). The bag-of-words model is a simplifying representation of a text. We always label one person as "*writer*" according to some keywords captured like "*new book*", "*publishing house*" and many book title marks when scanning one's home page, as in life, we also classify different types by typical keywords in classification task, so a bag of words could represents a text in general. Rao and colleagues manually built a list of words to characterize sociolinguistic behaviors, but it's difficult to translate into strong class-indicative because of much manual effort. Instead, we use *chi-square (CHI)* to select feature words, which measures the degree of the independence between the feature and categories. In addition, we use *term frequency–inverse document frequency (tfidf)* which reflects how important a word is to a document in a collection or corpus to represent words [9]. Three bags of words are extracted, we name them "*EMOTION*", "*ENGLISH*" and "*WORDS*", while emotion and English characters are all replaced by specific characters in "*WORDS*".

N-gram. N-gram is a contiguous sequence of n items from a given sequence of text or speech. An n-gram of size 1 is referred to as a "*unigram*"; size 2 is a "*bigram*". Rao et al. uses a mixture of sociolinguistic features and n-gram models to represent twitters [4]. In this paper we will also utilized this as an approach for our task by deriving the unigram and bigrams of the text. We use "*UNIGRAM*" and "*BIGRAM*" to express this feature set.

Topic Feature. Because of the short-text property of microblogs, the method of representing a microblog by single words cannot reflect the topics of the microblog. For this reason, we use topic models to extract the topic features of microblogs. As the *Latent Dirichlet Allocation (LDA)* model [6, 13] and the *Biterm Topic Model (BTM)* model [17] are commonly used as topic models, we consider these two models in our work. Specially, for the LDS model, we concatenate all users' messages as the input and each user is represented as a multinomial distribution over different number of topics. For the BTM model, we concatenate user name, description and custom tag into a short text introduction, and use the BTM model to enhance the topic learning on the new combined short introduction. The combined feature set of LDA and BTM is named "*T-DISTRIBUTION*" in Table 1. In addition to topic distribution, topic related words are also indispensable when judging user occupations. Thus, we use *word2vec* to obtain topic related words and to produce word embedding. Through word2vec, a relationship lexicon responding to preset occupations can be obtained, which is named "*T-WORDS*" in Table 1.

Table 1. Feature sets

Feature		Dimension	Description
DIGITAL		38	Basic profile, social influence, behavior feature
BOW	*EMOTION*	1501	Emotion feature. E.g. [cool], [orz]
	ENGLISH	1746	Capital English character. E.g. TFBOYS
	WORDS	9030	Ordinary words
N-gram	*UNIGRAM*	2276	Unigram model. E.g. a, happy, ending
	BIGRAM	7132	Bigram model. E.g. a happy, happy ending
Topic	*T-WORDS*	2333	Topic related words
	T-DISTRIBTION	$2*k$	Topic distribution(BTM + LDA), k is topic number

2.3 Classification Models

There are a number of classification models proposed by the machine learning area. In this paper, we compare three classification models: a linear SVC classifier with L2 regularized logistic regression [23], a Support Vector Machine (SVM) classifier with linear kernel [4], and an ensemble classification of Random Forest.

3 Occupation-Oriented Lexicon Integration

Traditional classification focuses much on feature extraction. Most of the existing classification works on social network platforms have limitation because of data sparseness and noise. Classical models may produce many features, but semantic features are likely to be ignored. To solve this problem, we propose to build an occupation-oriented lexicon to improve the performance of user occupation detection. This is motivated by the fact that people usually conjecture their occupations by some typically words; thus we can combine occupation-oriented lexicon with word embedding to represent user features.

As shown in Fig. 2, we first use Word2vec to train the corpus. Then, we get a model of word embedding, which expresses similarity between words. We use TextRank and CHI to generate the keyword set, which contains important words in users' microblogs. Moreover, we invite several persons to conduct questionnaires, during which each involved person is asked to give the words related with different occupations. These related words are used as an input of our lexicon extension algorithm. Finally, we combine lexicon with word embedding to get occupation-oriented word embedding features.

There are three tools used in Fig. 2, namely word2vec, CHI, and TextRank. As described in the above section, word2vec can learn the vector representations of words in the high dimensional vector space; it can find the semantic relationships by computing the cosine distances between words. Thus, we apply its word embedding to compute similarity and express words. CHI aims to select feature words, which measures the degree of the independence between the feature and categories. In addition, TextRank was proposed to solve keyword extraction and it is tasked with the automatic identification of terms that best describe the subject of a document. Therefore, we use these two methods to generate keyword set.

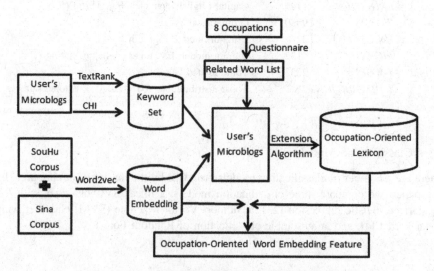

Fig. 2. Flow of occupation-oriented lexicon integration

The lexicon extension algorithm in Fig. 2 is an iterative algorithm that integrates important and occupation-oriented lexicon. Algorithm 1 shows its detailed process.

Algorithm 1 Occupation-Oriented Lexicon Extension Algorithm

Input: OList: occupation list; W_{w2v}: the similar word dictionary of word2vec training model;

K_{dic}: the keyword dictionary of TextRank and CHI generating; R: the related words list of different occupations from questionnaire survey; m: number of words extract in

Output: OOL: occupation-oriented lexicon

---

1. $t = 0$;
2. **While**(1):
3. $\quad$ $t = t + 1$;
4. $\quad$ **For** each occupation occ **in** OL **do**
5. $\quad\quad$ **If** $OOL_{occ} = \emptyset$ **then**
6. $\quad\quad\quad$ **For** each word w **in** R_{occ} **do**
7. $\quad\quad\quad\quad$ **If** w **in** K_{dic} **and** w **in** W_{w2c} **then**
8. $\quad\quad\quad\quad\quad$ $OOL_{occ}.\,\mathrm{add(w)}$;
9. $\quad\quad$ Candidate Words List : CWList = {};
10. $\quad\quad$ **For** each word w **in** OOL_{occ} **do**
11. $\quad\quad\quad$ **If** w **in** W_{w2c} **then**
12. $\quad\quad\quad\quad$ Similarity Map: SMap = {};
13. $\quad\quad\quad\quad$ **For** W_{sim} **in** SimilarityList(w) **do**
14. $\quad\quad\quad\quad\quad$ **If** W_{sim} **in** K_{dic} **and** W_{sim} **in** W_{w2c} **then**
15. $\quad\quad\quad\quad\quad\quad$ $V_{sim} = \mathrm{Sum}(Distance(W_{sim}, each\ word\ of\ R_{occ}))$;
16. $\quad\quad\quad\quad$ SMap. put(W_{sim}, V_{sim});
17. $\quad\quad\quad\quad$ Sort(SMap);
18. $\quad\quad\quad\quad$ **If** SMap. size $> m$ **then**
19. $\quad\quad\quad\quad\quad$ CWList. add(SMap. get(top m));
20. $\quad\quad\quad\quad$ **else** CWList. add(SMap);
21. $\quad\quad$ $OOL_{occ}.\,add(CWList)$;
22. $\quad\quad$ Distinct(OOL_{occ});
23. $\quad$ $Lexicon_t = \mathrm{Sum}(OOL_{occ})$;
24. $\quad$ **If** Acc($Lexicon_t$) < Acc($Lexicon_{t-1}$) **then**
25. $\quad\quad$ **break**;
26. **Return** $Lexicon_{t-1}$

As shown in Algorithm 1, we utilize keywords and related words to filter out noise to identify occupation. The functions *SimilarityList* and *Distance* are computed by word2vec. We combine word embedding and lexicon to predict, and the accuracy whether improved or not guides to stop iterating. Stated another way, we execute the cyclical function continue until the experiment accuracy of lexicon integrated this round is lower compared to last round.

4 Experimental Results

4.1 Experimental Settings

We use the eight occupations discussed in Sect. 2.1 to label the 8000 verified users. Each 1000 users are randomly divided into a training set (70% users) and a testing set (30% users). Before implementing linguistic models, we use ICTCLAS to segment the text words. Classification is performed using *Scikit-learn* [8, 26], which is a Python module integrating a wide range of state-of-the-art machine learning algorithms. We compare three machine learning methods: a linear SVC classifier with L2 regularized logistic regression, an SVM classifier with linear kernel and an ensemble classification of Random Forest on 3-fold cross validation of the training set. The SVM model has 4 parameters of "*C*" and the Random Forest has 23 candidate parameters of tree number (*n_estimators*) which are between 40 and 500.

The corpus of word2vec is the combination of SouHu news and microblogs, which contains 69.3 billion effective words. We adapt the *Skip-gram* model and set the window size to 8 to train the corpus with different word representation sizes. The TextRank and CHI are used on the microblogs of the labeled 8000 users, in which the window size in extracting keywords is set to 5. Finally, we get 22.75 thousand keywords and 9030 words selected by CHI.

We use *accuracy, macro-averaging precision, recall*, and *F-measure* as the metrics [6]. Let U be the user set, N be the number of occupations, if we detect $U_{correct}$ users correctly from U_{test} users, the accuracy is computed as Formula 1. The macro-averaging precision and recall for each occupation are expressed by Formulas 2 and 3, respectively, where U_k is the user number of occupation k, and $U_{k,predict}$ is the number of users that are detected to have occupation k.

$$\text{accuracy} = \frac{|U_{correct}|}{|U_{test}|} \tag{1}$$

$$\text{precision} = \text{avg}\left(\frac{|U_{k,correct}|}{|U_{k,predict}|}\right) k = 1, 2, \ldots, N \tag{2}$$

$$\text{recall} = \text{avg}\left(\frac{|U_{k,correct}|}{|U_k|}\right) k = 1, 2, \ldots, N \tag{3}$$

4.2 Performance of Topic Model

First of all, the prediction performances for various numbers of topics are compared based on the topic model. We use unified Random Forest with the same parameter to predict. The results are shown in Table 2.

As shown in Table 2, with the increasing of the topic size, the precision of prediction keeps growing until the topic size is 30, and then has a slight decreasing trend. Thus we use 30 topics for the next experiments.

Table 2. Performance of *T-DISTRIBTION* with different topic sizes (%)

Metric	Topic size							
	10	15	20	25	30	35	40	45
Precision	63.71	68.79	70.54	71.92	75.54	74.62	73.67	73.92
Recall	64.40	69.53	71.26	72.54	75.09	74.97	74.29	74.33
F-measure	64.05	69.16	70.90	72.23	75.31	74.80	73.98	74.12

4.3 Performance on Different Feature Sets

In this section, we compare different feature sets by 3 methods. Table 3 shows the results on occupation prediction with different features. In this table, we use "*B-ALL*" to express all bag-of-words features including "*EMOTION*", "*ENGLISH*" and "*WORDS*", and we use *tfidf* to represent word feature. Similarly, we use "*N-ALL*" to express the combination of unigram and bigram, and "*T-ALL*" indicates the topic feature involving topic distribution and topic related words. We choose the best performance with various parameter candidates. Then, we list the accuracy of each occupation among the best global performance in Table 4. In the Table 3 we use "*P*" represents precision, "*R*" represents recall, and "*F*" represents F-measure).

Table 3. Performance on different feature sets (%)

Features		Classifier								
		LR			SVM			Random forest		
		P	R	F	P	R	F	P	R	F
DIGITAL		34.56	34.58	34.57	34.08	33.66	33.87	45.52	45.62	45.57
BOW	26.49	26.54	26.52	37.04	36.31	36.67	40.36	40.33	40.34	40.34
	35.67	35.58	35.63	42.92	44.79	43.83	45.58	45.62	45.60	45.60
	73.26	72.25	72.75	74.71	75.45	75.08	77.18	76.79	76.98	76.98
	72.66	72.08	72.37	74.29	75.03	74.66	77.45	77.08	77.27	77.27
N-gram	62.95	62.29	62.62	69.71	70.38	70.04	73.08	69.71	70.04	70.04
	72.52	71.62	72.07	74.96	75.66	75.31	77.43	77.04	77.24	77.24
	73.17	72.50	72.83	74.83	75.58	75.21	77.18	76.58	76.88	76.88
Topic	65.15	64.50	64.82	71.58	72.71	72.14	77.00	76.46	76.73	76.73
	72.81	71.88	72.34	68.04	69.47	68.75	75.77	75.29	75.53	75.53
	67.05	66.46	66.75	74.42	75.46	74.94	77.56	77.08	77.32	77.32
ALL		75.42	74.67	75.04	75.95	75.21	75.58	78.42	77.92	78.17

According to Table 3, we can see that the topic feature on the fewer feature dimensions reaches higher accuracy except some high dimensional feature set like "*WORDS*" and "*BIGRAM*". The features reflecting the main theme of the text, for instance, "*EMOTION*" and "*ENGLISH*", can only reflect small crowd's behavior. The Random Forest model performs best among the three classifiers when a fit parameter is used.

4.4 Performance on Different Feature Units

There is a feasible method provided by *Scikit-learn*, which can compute the contribution of each feature unit in the detection process. Therefore, we use the parameter *"feature_importances"* of the model trained by *Random Forest*, and then get the contribution of each feature. We list top 10 features of *"DIGITAL"* in Table 4, and show several greatest contributive topic words of *"T-WORDS"* in Table 5.

Table 4. Top 10 important features of *DIGITAL* (%)

Feature unit	Contribution value
avgForward	0.0482728
avgBookmark	0.0472507
favouritesCount	0.0421770
avgTopic	0.0412609
time-1	0.0357934
avgNews	0.0353702
avgZan	0.0352574
followerCount	0.0314369
statusesCount	0.0299582
biFollowerCount	0.0294038

Table 5. Most contributive topic words of *T-WORDS* (%)

Occupation	Most Contributive Topic Words
writer	波德莱尔(Baudelaire) 小说家(novelist) 文学史(history of literature) 本书(this book) 文坛(the literary world) 副刊(supplement) 主人公(leading character in a novel) 作品集(Portfolio)
reporter	新闻办(Information Office) 采访时(in the/an interview) 郭伟(Wei Guo) 本刊(this newspaper)
lawyer	律师事务所(law office) 许兰亭(LanTing Xu) 案件(case) 法援(legal aid) 案子(case) 世联(GlobalLink) 事务所(office) 诉讼费(legal fare)
photographer	摄像(camera shooting) 摄像机(vidicon) 拍照(take photos) 相片(photograph) 杂志(magazine) 照相(take pictures)
actor	演出(performance) 演员(actor) 丁军(Jun Ding) 童谣(Yao Tong) 新人(a new people) 米学东(XueDong Mi) 嘉宾(guest) 应采儿(CaiEr Ying) 星运(luck of star) 饰(portray)
singer	演唱(sing) 歌手集(sings) 民谣(balled) 作词(write lyrics) 歌词(lyric) 演唱会(vocal concert) 个人专辑(personal albums) 二胡(Erhu)
doctor	医疗(medical) 治疗学(acology) 患儿(Children patient) 医学部(Department of Medicine) 手术刀(scalpel) 医生(doctor) 医疗保险(medical insurance) 门诊部(out-patient department)
dietitian	营养学(nutriology) 食品(food) 吃富(eat something) 一杯(one cup) 餐单(menu) 食用(edible) 瘦体(thin body) 一罐(one tin)

Table 4 shows that popularity and speech recognition play an important role in the detection, which are indicated by the feature "*avgForward*" and "*avgZan*". We can also identify a reporter due to the high rate of reprinting, reflected by the feature "*avgTopic*" and "*avgNews*". In addition, "*time-1*" is a special feature that means the number of the microblogs posted between twelve and one o'clock, indicating that the user sleeps much later. The results in Table 5 show the most contributive topic words of "*T-WORDS*". With this mechanism, we can build thesaurus for various occupations and select useful topic-related words.

4.5   Impact of the Lexicon Size

To verify the impact of the lexicon extension, we apply 300 dimensions of word embedding to represent remaining words, and integrate all vector representations by computing the sum of each dimension. What's more, in order to utilize more information such as *tf* (term-frequency) and *tfidf* (term frequency and inverse document frequency), we test three methods (as shown in Table 6): word-embedding, *tf* * word-embedding, and *tfidf* * word-embedding. In this experiment, we use the Logistic Regression model as the classifier. The results are shown in Table 6.

Table 6. Accuracy of different lexicon size (%)

#Words	Method	Accuracy
1537	word-embedding	77.54
	tf * word-embedding	75.08
	tfidf * word-embedding	73.46
3530	word-embedding	82.50
	tf * word-embedding	78.92
	tfidf * word-embedding	78.21
6480	word-embedding	80.62
	tf * word-embedding	78.04
	tfidf * word-embedding	75.83

As shown in Table 6, *tf* * word-embedding, and *tfidf* * word-embedding both lead to the decreasing of accuracy. We give two possible explanations. First, *tfidf* is helpful to improve the weight of rare words, because we have removed unimportant words previously. Second, combing the word-embedding feature with *tf* or *tfidf* is likely to result in the loss of some semantics.

The results in Table also show that the accuracy of classification increases with the increasing of the number of words. Thus, we can know that only two rounds of extension steps can extract appropriate lexicon words. Further, when we use the lexicon in user occupation detection, it is helpful to filter noise and improve the effectiveness of detection.

4.6 Lexicon-Based Word Embedding

In this section, we show the performance of word embedding with different dimensions using the lexicon consisting of 3530 words (see Table 6). We compare three classifiers and show the best result of each in Table 7.

Table 7. Performance on different word dimension (%)

Dimension	Classifier		
	LR	SVM	Random forest
100	76.79	76.46	64.04
200	80.21	79.96	66.21
300	82.50	82.33	67.04
400	83.62	83.58	68.21
500	84.25	84.62	67.50
600	84.58	85.46	67.29
700	85.42	86.29	68.67
800	85.42	86.21	68.62
900	85.12	86.54	68.08
1000	86.08	87.12	67.46

As shown in Table 7, with the increasing of the word dimension, we get better performance of classification. It shows that combining the occupation-oriented lexicon with the word embedding method is able to achieve higher accuracy compared with traditional classification models. *LR* and *SVM* produce similar accuracy. Specially, *SVM* achieves the best accuracy when it is set to *"linear"*. To this end, the linear model is more suitable for user occupation classification.

The best result for different user occupations are shown in Table 8. The *"lawyer"* gets highest accurate prediction, while the *"reporter"* is the hardest one to identify. It indicates that the lawyer and the doctor group incline to post messages related to their occupation, while others make less mention of occupation. Specially, reporters usually reprint various kind of news involving other field, thus making much noise of messages and increasing the difficulty to identify.

Table 8. Accuracy of each occupation among best performance (%)

Metric	Occup.							
	writer	reporter	lawyer	photographer	actor	singer	doctor	dietitian
Precision	78.33	77.33	92.67	82.00	79.67	83.00	86.00	90.00
Recall	78.20	78.24	92.21	82.14	81.02	82.04	87.46	87.52
F-measure	78.31	77.78	92.44	82.07	80.34	82.52	86.72	88.74

5   Related Work

Many previous work has been done to mine users' attributions on social media such as Twitter and Sina Weibo, including age [14, 19, 22], language [14], gender [18], location of origin [4], and political orientation [4]. Such previous work used a mixture of sociolinguistic features and n-gram models. Most attributes are inferred from the messages posted by users. In [24], the authors proposed a multi-source integration framework concentrating on building a feature set of contents and network information. Latent feature representation such as word clusters and embedding was used in [25] to classify occupations. In [10], the researchers proposed a classification method for detecting user occupations in microblogs. It uses the domain-specific features from the text, user behavior and social network. The work in [23] considered personal information, community structure and unlabeled data together to identify professions.

Topic model (e.g. LDA and PLSA) is very helpful in microblog analysis [20]. Some researchers adopted the Single-pass Clustering technique by using LDA to extract the hidden microblog topics information [11], and other researchers proposed a model named TweetLDA for Twitter classification tasks [13]. A bi-term topic model (BTM) was proposed for modeling topics in short texts [17]. This model is demonstrated to be helpful for Twitter analysis. Based on topic modeling, topic clustering was also studied in [16].

The work of this paper focuses on microblog user occupation detection and differs from previous researches on microblog user information extraction. We conduct an experimental study to evaluate the traditional feature-based classification models on user occupation detection. We fuse aggregated features according to user basic profile information and user messages, and adopt three classical linguistic models to extract features. Furthermore, we propose to build an occupation lexicon to improve the effectiveness of user occupation detection.

6   Conclusion

Detecting user occupation from microblog platform is an important issue for information extraction on short texts. This paper presents a framework for classifying eight occupations based on Sina Weibo, and addresses the task through three classifiers and the word embedding method. The result indicates that topic features perform well and we get the best results when using the Random Forest model. We also propose an occupation-oriented lexicon and integrate it with word embedding. The experimental results show that the lexicon-based features with lower dimension achieve higher accuracy compared with traditional models.

In future work, we will investigate deep learning for user occupation extraction on microblogging platforms. As deep learning and multi-layered neutral networks have been proven to be very effective in many applications such as image classification and retrieval [27], they are likely to have high performance on user occupation detection.

Acknowledgements. This work is supported by the National Science Foundation of China (61379037 and 71273010).

References

1. Zheng, L., Jin, P., Zhao, J., Yue, L.: A fine-grained approach for extracting events on microblogs. In: Decker, H., Lhotská, L., et al. (eds.) DEXA 2014, LNCS, vol. 8644, pp. 275–283. Springer, Heidelberg (2014)
2. Lv, X., Jin, P., Yue, L.: User occupation prediction on microblogs. In: Li, F., Shim, K., et al. (eds.) APWeb 2016, LNCS, vol. 9932, pp. 497–501. Springer, Heidelberg (2014)
3. Mislove, A., Viswanath, B., Gummadi, K., Druschel, P.: You are who you know: inferring user profiles in online social networks. In: Third International Conference on Web Search and Web Data Mining (WSDM), pp. 251–260. ACM, New York (2010)
4. Rao, D., Yarowsky, D., Shreevats, A., Gupta, M.: Classifying latent user attributes in Twitter. In: 2nd International Workshop on Search and Mining User-Generated Contents, pp. 37–44. ACM, Toronto (2010)
5. Burger, J., Henderson, J., Kim, G., et al.: Discriminating gender on Twitter. In: 2011 Conference on Empirical Methods in Natural Language Processing, pp. 1301–1309. ACL, Stroudsburg, PA, USA (2011)
6. Pennacchiotti, M., Popescu, A.: A machine learning approach to Twitter user classification. In: Fifth International Conference on Weblogs and Social Media, pp. 281–288. The AAAI Press, Barcelona, Catalonia, Spain (2011)
7. Golbeck, J., Robles, C., Turner, K.: Predicting personality with social media. In: CHI 2011 Extended Abstracts on Human Factors in Computing Systems, pp. 253–262. ACM, Vancouver (2011)
8. Pedregosa, F., Varoquaux, G., Gramfort, A., et al.: Scikit-learn: machine learning in Python. J. Mach. Learn. Res. 12, 2825–2830 (2011)
9. Rajaraman, A., Ullman, J.: Mining of Massive Datasets. Cambridge University Press, Cambridge (2012)
10. Zhou, M., Xu, Y., Zhao, X.: Study of feature extract on microblog user occupation classification. In: Fourth International Symposium on Information Science and Engineering, pp. 20–23. IEEE CS, Shanghai (2012)
11. Huang, B., Yang, Y., Mahmood, A., et al.: Microblog topic detection based on LDA model and single-pass clustering. In: Yao, J., et al. (eds.) RSCTC 2012. LNCS, vol. 7413, pp. 166–171. Springer, Heidelberg (2012). doi:10.1007/978-3-642-32115-3_19
12. Tinati, R., Carr, L., Hall, W., et al.: Identifying communicator roles in Twitter. In: 21st International Conference on World Wide Web, pp. 1161–1168. ACM, Lyon (2012)
13. Quercia, D., Askham, H., Crowcroft, J.: TweetLDA: supervised topic classification and link prediction in Twitter. In: 4th Annual ACM Web Science Conference, pp. 247–250. ACM, Evanston (2012)
14. Nguyen, D., Gravel, R., Trieschnigg, D., et al.: How old do you think i am? A study of language and age in Twitter. In: Seventh International AAAI Conference on Weblogs and Social Media, pp. 439–448. The AAAI Press, Cambridge (2013)
15. Schwartz, H., Eichstaedt, J., Kern, M., et al.: Personality, gender, and age in the language of social media: the open-vocabulary approach. PLoS ONE 8(9), e73791 (2013)
16. Zhao, J., Li, X., Jin, P.: A time-enhanced topic clustering approach for news web search. Int. J. Database Theory Appl. 5(4), 1–10 (2012)
17. Yan, X., Guo, J., Lan, Y., et al.: A biterm topic model for short texts. In: 22nd International Conference on World Wide Web, pp. 1445–1456. ACM, Rio de Janeiro (2013)
18. Huang, F., Li, C., Lin, L.: Identifying gender of microblog users based on message mining. In: Li, F., Li, G., et al. (eds.) WAIM 2014. LNCS, vol. 8485, pp. 488–493. Springer, Cham (2014). doi:10.1007/978-3-319-08010-9_54

19. Li, Y., Liu, T., Liu, H., et al.: Predicting microblog user's age based on text information. In: Lin, X., Manolopoulos, Y., et al. (eds.) WISE 2013. LNCS, vol. 8180, pp. 510–515. Springer, Berlin, Heidelberg (2014). doi:10.1007/978-3-642-41230-1_45

20. Yang, S., Kolcz, A., Schlaikjer, A., et al.: Large-scale high-precision topic modeling on Twitter. In: 20th ACM SIGKDD International Conference on Knowledge Discovery and Data Mining, pp. 1907–1916. ACM, New York (2014)

21. Wu, X., Wang, J.: Micro-blog in China: identify influential users and automatically classify posts on Sina micro-blog. J. Ambient Intell. Humanized Comput. 5(1), 51–63 (2014)

22. Chen, L., Qian, T., Wang, F., et al.: Age detection for Chinese users in Weibo. In: Dong, X., Yu, X., et al. (eds.) WAIM 2015. LNCS, vol. 9098, pp. 83–95. Springer, Cham (2015). doi:10.1007/978-3-319-21042-1_7

23. Tu, C., Liu, Z., Sun, M.: PRISM: profession identification in social media with personal information and community structure. In: Zhang, X., Sun, M., et al. (eds.) CNCSMP 2015. CCIS, vol. 568, pp. 15–27. Springer, Singapore (2015)

24. Huang, Y., Yu, L., Wang, X., et al.: A multi-source integration framework for user occupation inference in social media systems. World Wide Web 18(5), 1247–1267 (2015)

25. Preoţiuc-Pietro, D., Lampos, V., Aletras, N.: An analysis of the user occupational class through Twitter content. In: 53rd Annual Meeting of the Association for Computational Linguistics, pp. 1754–1764. ACL, Beijing (2015)

26. Scikit-Learn. http://scikit-learn.org/stable/. Accessed 21 Apr 2017

27. Wan, S., Jin, P., Yue, L.: An approach for image retrieval based on visual saliency. In: 2009 International Conference on Image Analysis and Signal Processing, pp. 172–175. IEEE CS, Linhai (2009)

Counting Edges and Triangles in Online Social Networks via Random Walk

Yang Wu[1(⊠)], Cheng Long[2], Ada Wai-Chee Fu[1], and Zitong Chen[1]

[1] The Chinese University of Hong Kong, Shatin, Hong Kong
{yangwu,adafu,ztchen}@cse.cuhk.edu.hk
[2] Queen's University Belfast, Belfast, Northern Ireland
cheng.long@qub.ac.uk

Abstract. Online social network (OSN) analysis has attracted much attention in recent years. Edge and triangle counts are both fundamental properties in OSNs. However, for many OSNs, one can only access parts of the network using the application programming interfaces (APIs). In such cases, conventional algorithms become infeasible. In this paper, we introduce efficient algorithms for estimating the number of edges and triangles in OSNs based on random walk sampling on OSNs. We also derive a theoretical bound on the sample size and number of APIs calls needed in our algorithms for a probabilistic accuracy guarantee. We ran experiments on several publicly available real-world networks and the results demonstrate the effectiveness of our algorithms.

Keywords: Graph sampling · Random walk · Triangle counting

1 Introduction

Triangle counting is a fundamental problem in network analysis and triangle structure is widely used in many applications such as spam detection [10], hidden thematic layers uncovery, [11] and community detection [25]. There is a rich related work on enumerating all triangles [12–15]. However, exhaustive counting is not scalable, since it has to explore every triangle. Even with state-of-art algorithm, enumeration of all triangles has prohibitive cost for real-world large networks.

One alternative way is to use sampling algorithm to speed up the counting process with acceptable error [8,16], but both of these methods need to know the complete data of the network. For example, in [8], the authors proposed to count triangles by wedge sampling. In their method, we have to know the topology of the network in advance, but most OSN's service providers are unwilling to share the complete data for public use.

As noted in [2], for a typical online social network (OSN), the underlying social network may be available only through a public interface, in the form of an application programming interface (API) which may provide a list of a user's neighbors. As in [2] we assume that we can only have external access to the

© Springer International Publishing AG 2017
L. Chen et al. (Eds.): APWeb-WAIM 2017, Part I, LNCS 10366, pp. 346–361, 2017.
DOI: 10.1007/978-3-319-63579-8_27

social network via its public interface, and only a fraction of the users/nodes can be sampled. Graph sampling through random walk is widely used in this scenario to estimate graph properties such as degree distribution [4,5,9], clustering coefficient [2], graph size [1,2] and graphlet statistics [6,17]. However, to our knowledge, no one has considered how to get the counting of triangle or even larger graphlets in OSNs without knowing the complete data.

In this paper, we propose a random walk-based method for triangle counting in OSN (online social network), where we have no assumption on the graph, the complete data of the OSNs is not available, and the number of vertices and edges in the graph is not known. Specifically, our method runs a random walk based on a *subgraph relationship graph* (SR graph) of the original graph by using API calls during which some nodes of the SR graph are sampled and some states of an *expanded Markov chain* that could be built on the random walk process are also sampled. Then, based on the sampled nodes, we estimate the number of edges/links in the SR graph (which could be instantiated as the number of edges in the original graph in some case), based on which and also the sampled states, we estimate the number of triangles in the original graph.

Our contributions are summarized as follows:

- We propose a novel random walk based algorithm to estimate the number of edges and triangles in networks with restricted accesses. To the best of our knowledge, we are the first to estimate the number of edges and triangles in such a scenario.
- We derive a theoretical bound on the sample size for achieving an (ϵ, δ)-approximation.
- Experiments are conducted on publicly available real-world networks which verified the effectiveness of our algorithms.

The rest of the paper is organized as follows. Section 2 introduces the related work. Section 3 introduces our problem, background and sampling algorithm. Section 4 introduces our estimators, and Sect. 5 gives some implementation details and cost analysis. Section 6 reports our experiment results, and Sect. 7 concludes the paper.

2 Related Work

In graphs with full access, edge counting is a trivial task, but triangle counting has been a hot topic in graph analysis for a long time and a number of algorithms have been proposed to solve this problem. Most of the algorithms can be classified into two categories: exact counting and estimation by sampling.

Exact Counting. Enumeration of all triangle counting has a rich history [12–15]. However, as the graphs become much larger, exhaustive counting is not scalable, since it has to explore every triangle. Recently, many MapReduce based algorithms have been proposed to solving exact triangle counting problem [18–20]. In addition, using external memory is also considered as a solution to this problem [21–23].

Estimation by Sampling. Sampling methods are more related to our work. *Doulion* [16] samples a subgraph from the original graph by keeping each edge with probability p and then estimating triangle counting based on the sampled graph. Another important algorithm is wedge sampling [24]. It samples several wedges from the network as the sample set, and then estimating the number of triangles based on how many triangles exists in the sampled wedge set. However, these sampling methods are based on the facts that the topology of the network is known and we can access the whole network, which are not the assumptions in our scenario.

Most previous works on networks with restricted access are based on random walk methods, such as estimating degree distribution [4,5,9], clustering coefficient [2] and graph size [1,2]. Recently, Chen et al. [17] proposed the state-of-art algorithms for graphlet statistics. To the best of our knowledge, edge or triangle counting has not been studied before in this setting.

3   Problem, Background, and Sampling Algorithm

An OSN can be captured by a simple graph $G = (V, E)$ which is accessible via APIs: given a query vertex u, the API returns all the neighbors of u. We are interested in the problem of estimating two properties of the social network, namely (1) the number of edges in G, $|E|$, and (2) the number of triangles in G, denoted by N. To the best of our knowledge, this is the first attempt for these problems in an online context.

Our solution is based on a random walk on a *subgraph relationship graph* [6,7], from which we build an *expanded Markov chain* [17], and then construct the count estimator based on some sampled states of the expanded Markov chain and the sampled nodes of the random walk.

Our experiments show that our algorithm works best on the subgraph relationship graph when it reduces to the original graph, but since presenting the algorithm based on the original graph requires the same complexity as the general SR graph, we present our algorithm based on a general subgraph relationship graphs in this paper

In the following, we first give some background knowledge about a subgraph relationship graph and an expanded Markov chain process, and then introduce our sampling algorithm.

3.1   Subgraph Relationship Graphs

Given a graph $G = (V, E)$, and an integer $d \geq 1$, the d-node subgraph relationship graph of G (SR graph in short) [6,7], denoted by $G^{(d)} = (H^{(d)}, R^{(d)})$, is defined as follows. First, each node in $H^{(d)}$ corresponds to a connected and induced d-node subgraph in G. For example, each node in $H^{(1)}$ corresponds to a vertex in graph G and each node in $H^{(2)}$ corresponds to an edge in graph G. Second, two nodes in $H^{(d)}$ are connected by an edge in $G^{(d)}$ if and only if they share $d-1$ common vertices of G (exceptionally for $d = 1$, two nodes are connected by an edge if there exists an edge between their corresponding vertices in G), and all these edges constitute $R^{(d)}$. Note that $G^{(1)} = G$. Given a node y in $G^{(d)}$, $N(y)$ denotes the set of nodes adjacent to y (i.e., the neighbors of y) and d_y denotes the degree of y, i.e., $d_y = |N(y)|$. Given two nodes y and y', $\theta(y, y')$ denotes the number of common neighbors of y and y', i.e., $\theta(y, y') = |N(y) \cap N(y')|$.

Example 1. For illustration, consider Fig. 1, where a graph G and its corresponding 2-node SR graph $G^{(2)}$ and 3-node SR graph $G^{(3)}$ are shown. Here, each dashed circle corresponds to a node in a SR graph. In $G^{(2)}$, consider the node y containing vertices 1 and 5 and another node y' containing vertices 2 and 5, d_y is 3, $d_{y'}$ is 3, and $\theta(y, y')$ is 1.

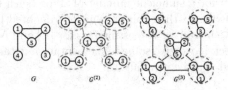

Fig. 1. $G^{(1)}, G^{(2)}$ and $G^{(3)}$

In this paper, we use the subgraph relationship graph $G^{(d)}$ with three settings of d, namely 1, 2, and 3 ($G^{(d)}$ with larger d's is unnecessarily complicated for estimating the counts of edges and triangles since each of them has at most 3 vertices).

3.2 Expanded Markov Chain

A simple random walk process on a graph is to start at a random node, and then randomly pick a neighbor of the current node, go to that node and then repeat the step. It can be regarded as a Markov chain where each node is a state. An *expanded Markov chain* on $G^{(d)}$ [17] based on a random walk remembers *the last l nodes* as the current state, i.e., each possible l consecutive steps/nodes in the random walk process, denoted by $x^{(l)} = (y_1, y_2, \dots, y_l)$, is considered as a state of the expanded Markov chain. We use $M^{(l)}$ to denote the state space of the expanded Markov chain. In this paper, depending on the parameter d of $G^{(d)}$, we use different l's. Specifically, for $G^{(1)}$, we use $l = 3$, for $G^{(2)}$, we use $l = 2$, and for $G^{(3)}$, we use $l = 1$. The reason is that each triangle in G is covered in $G^{(1)}$ by 3 nodes, or in $G^{(2)}$ by 2 nodes, or in $G^{(3)}$ by 1 node.

Existing studies show that the expanded Markov chain has a unique *stationary distribution*, denoted by π_M.

Theorem 1 [17]. *The stationary distribution π_M exists and is unique. For any $x^{(l)} = (y_1, y_2, \ldots, y_l) \in M^{(l)}$, we have*

$$\pi_M(x^{(l)}) = \begin{cases} \frac{d_{y_1}}{2|R^{(d)}|} & l = 1 \\ \frac{1}{2|R^{(d)}|} & l = 2 \\ \frac{1}{2|R^{(d)}|} \frac{1}{d_{y_2}} \cdots \frac{1}{d_{y_{l-1}}} & l > 2 \end{cases} \tag{1}$$

where $d_{y_i} (1 \leq i \leq l)$ is the number of neighbors of y_i.

3.3 Sampling Algorithm

The idea of our sampling algorithm is very simple, which is to run a random walk on the SR graph $G^{(d)}$ for a sufficient number of steps (such that the stationary distribution is attained) and then collect m nodes, denoted by $y_1, \ldots, y_m$.

Next, in Sect. 4, we introduce two estimators for $|E|$ and N based on the sampled nodes, and in Sect. 5, we give some implementation details of this algorithm and also the time and space cost analysis.

4 Estimators of $|E|$ and N

We introduce an estimator of the number of links/edges in $G^{(d)}$, i.e., $|R^{(d)}|$, in Sect. 4.1 (note that this estimator corresponds to one for the number of edges in G, i.e., $|E|$, when $d = 1$), and then based on this estimator, we design another estimator for the number of triangles N, in Sect. 4.2. We provide some accuracy results of these estimators in Sect. 4.3.

4.1 Estimator of $|R^{(d)}|$

Let Y and Y' be two *independent* nodes sampled from the stationary distribution. First, we define a random variable Φ as the degree of Y (or Y'), and a variable Ψ. That is,

$$\Phi = d_Y; \qquad \Psi = \frac{\theta(Y, Y')}{d_Y \cdot d_{Y'}} \tag{2}$$

Then, we have

$$E[\Phi] = E[d_Y] = \sum_{y \in H^{(d)}} d_y \cdot \frac{d_y}{2|R^{(d)}|} = \sum_{y \in H^{(d)}} \frac{d_y^2}{2|R^{(d)}|} \tag{3}$$

$$E[\Psi] = E\left[\frac{\theta(Y, Y')}{d_Y \cdot d_{Y'}}\right] = \sum_{y \in H^{(d)}} \sum_{y' \in H^{(d)}} \frac{d_y}{2|R^{(d)}|} \cdot \frac{d_{y'}}{2|R^{(d)}|} \cdot \frac{\theta(y, y')}{d_y \cdot d_{y'}} \tag{4}$$

$$= \sum_{y \in H^{(d)}} \frac{d_y^2}{4|R^{(d)}|^2} \tag{5}$$

The deduction within Eq. (4) is based on the assumption that Y and Y' are independent and the deduction from Eqs. (4) to (5) is based on the fact that each node y contributes d_y^2 times as a common neighbor for two nodes.

Eqs. (3) and (5) imply the following.

$$|R^{(d)}| = \frac{E[\Phi]}{2 \cdot E[\Psi]} \tag{6}$$

Thus, in order to get an estimator of $|R^{(d)}|$, we need estimators of $E[\Phi]$ and $E[\Psi]$. Since Φ and Ψ are based on one and two random nodes from $H^{(d)}$, respectively (see Eq. (2)), we design the estimators of $E[\Phi]$ and $E[\Psi]$, denoted by $\widehat{E[\Phi]}$ and $\widehat{E[\Psi]}$, as the average based on the m sampled nodes y_i $(1 \le i \le m)$ and the average based on a set Q of n pairs of sampled nodes (y_r, y_t) $(1 \le r < t \le m)$, respectively, where Q consists of those pairs of two sampled nodes y_r and y_t from the m sampled nodes such that y_r and y_t correspond to two nodes that are sampled at two steps far away from each other by a certain number f of steps in the random walk process, so we have $Q = \{(y_r, y_t)|t - r \ge f \wedge 1 \le r < t \le m\}$ and $|Q| = n$ (in this way, the two sampled nodes could be regarded as approximately independent sampled nodes according to the existing study [2], and in our experiments, following [2], we set f as 2.5%m).

$$\widehat{E[\Phi]} = \frac{1}{m} \cdot \sum_{i=1}^{m} d_{y_i}; \qquad \widehat{E[\Psi]} = \frac{1}{n} \cdot \sum_{(y_r, y_t) \in Q} \frac{\theta(y_r, y_t)}{d_{y_r} \cdot d_{y_t}} \tag{7}$$

Then, we design the estimator of $|R^{(d)}|$, denoted by $\widehat{|R^{(d)}|}$, based on Eq. (6) as follows.

$$\widehat{|R^{(d)}|} = \widehat{E[\Phi]}/(2 \cdot \widehat{E[\Psi]}) \tag{8}$$

Note that $\widehat{|R^{(1)}|}$ corresponds to an estimator of the number of edges in G.

4.2 Estimator of N

Let $x^{(l)}$ be a state in the expanded Markov Chain that involve three vertices in G. We define an indicator function $h(x^{(l)})$ such that $h(x^{(l)}) = 1$ if the three vertices form a triangle and $h(x^{(l)}) = 0$ otherwise.

We note that a certain number of states in $M^{(l)}$ based on $G^{(d)}$ correspond to the same triangle in G. For example, for the case of $G^{(1)}$ and $M^{(3)}$, for a triangle consisting of vertices v_1, v_2 and v_3, 6 states, namely $x_1^{(3)} = (v_1, v_2, v_3)$, $x_2^{(3)} = (v_1, v_3, v_2)$, $x_3^{(3)} = (v_2, v_1, v_3)$, $x_4^{(3)} = (v_2, v_3, v_1)$, $x_5^{(3)} = (v_3, v_1, v_2)$, and $x_6^{(3)} = (v_3, v_2, v_1)$, correspond to it. Similarly, it could be verified that 6 states for the case of $G^{(2)}$ and $M^{(2)}$ and 1 state for the case of $G^{(3)}$ and $M^{(1)}$ correspond to a triangle in G. Let α denote the number of states that correspond to the same triangle. We then know that the number of triangles is equal to the total number of states in which the vertices involved form a triangle divided by α.

$$N = \frac{1}{\alpha} \sum_{x^{(l)} \in M^{(l)}} h(x^{(l)}) \tag{9}$$

Let $X^{(l)}$ be random state with the stationary distribution π_M. Based on Eq. (9), we have the following.

$$N = \frac{1}{\alpha} \sum_{x^{(l)} \in M^{(l)}} h(x^{(l)}) = \frac{1}{\alpha} \sum_{x^{(l)} \in M^{(l)}} \frac{h(x^{(l)})}{\pi_M(x^{(l)})} \pi_M(x^{(l)}) \tag{10}$$

$$= \frac{1}{\alpha} E\left[\frac{h(X^{(l)})}{\pi_M(X^{(l)})}\right] = \frac{|R^{(d)}|}{\alpha} E\left[\frac{h(X^{(l)})}{\pi_M(X^{(l)}) \cdot |R^{(d)}|}\right] \tag{11}$$

Here, the deduction from Eqs. (10) to (11) is based on the fact that $X^{(l)}$ (and also $h(X^{(l)})$) follows the π_M distribution. We define a random variable T as follows.

$$T = \frac{h(X^{(l)})}{\pi_M(X^{(l)}) \cdot |R^{(d)}|} \tag{12}$$

Thus, we have

$$N = \frac{|R^{(d)}|}{\alpha} E\left[\frac{h(X^{(l)})}{\pi_M(X^{(l)}) \cdot |R^{(d)}|}\right] = \frac{|R^{(d)}|}{\alpha} E[T] \tag{13}$$

We note that based on a sampled state for $X^{(l)}$, T could be easily computed since $|R^{(d)}|$ would be canceled out in $\pi_M(X^{(l)}) \cdot |R^{(d)}|$ according to Eq. (1).

Based on Eq. 13, we know that in order to estimate N, we need an estimator of $E[T]$. Since T is based on one random state from $M^{(l)}$ with distribution π_M (See Eq. (12)), we design the estimator of $E[T]$, denoted by $\widehat{E[T]}$, as the average based on k states $x_i^{(l)}$ $(1 \leq i \leq k)$ of the expanded Markov chain built on the random walk process, where $k = m - l + 1$ and $x_i^{(l)} = (y_i, y_{i+1}, \ldots, y_{i+l-1})$.

$$\widehat{E[T]} = \frac{1}{k} \cdot \sum_{i=1}^{k} \frac{h(x_i^{(l)})}{\pi_M(x_i^{(l)}) \cdot |R^{(d)}|} \tag{14}$$

Then, we design the estimator of N, denoted by $\widehat{N}$, based on Eqs. (8) and (13) as follows.

$$\widehat{N} = \frac{\widehat{E[\Phi]} \widehat{E[T]}}{2\alpha \widehat{E[\Psi]}} \tag{15}$$

4.3   Accuracy Analysis

We did some analysis of the accuracies of $\widehat{|R^{(d)}|}$ and $\widehat{N}$ based on the settings of k, m and n, and the results are shown in the following theorem.

Theorem 2. *For any $\epsilon \leq 1/2$ and $\delta \leq 1$ we have*

$$Pr[N(1 - \epsilon) \leq \widehat{N} \leq N(1 + \epsilon)] \geq 1 - \delta \tag{16}$$

$$if \ k \geq \frac{48 \sum_{x^{(l)} \in M^{(l)}} \frac{h(x^{(l)})}{\pi_M(x^{(l)})} - \alpha^2 N^2}{\delta \epsilon^2 \alpha^2 N^2},$$

$$m \geq \frac{48[\sum_{y \in H^{(d)}} \frac{d_y^3}{2|R^{(d)}|} - (\sum_{y \in H^{(d)}} \frac{d_y^2}{2|R^{(d)}|})^2]}{\delta \epsilon^2 (\sum_{y \in H^{(d)}} \frac{d_y^2}{2|R^{(d)}|})^2}, \ and \qquad (17)$$

$$n \geq \max \left\{ \frac{384|R^{(d)}|^2}{\epsilon^2 \delta \sum_{y \in H^{(d)}} d_y^2}, \left[\frac{384\sqrt{2}|R^{(d)}| \cdot |H^{(d)}|}{\epsilon^2 \delta \sum_{y \in H^{(d)}} d_y^2}\right]^2 \right\}$$

Proof. Please see Appendix A. $\qquad \square$

The analysis of $\widehat{|R^{(d)}|}$ can be found in our technical report [26].

5   Implementation Details and Cost Analysis

With the algorithms introduce in the previous section, we now explain some details in our implementation, along with the time and space analysis.

First, we explain how to implement a random walk process based on $G^{(d)}$ during which m random nodes are collected. The core of the random walk process is the transition procedure which is to pick one neighbor of the current node $y = (v_1, \ldots, v_d)$ randomly. In the following, we explain how to compute the set of neighbors of y, based on which the transition procedure could be done easily. We consider three cases. Case 1, $d = 1$. In this case, the problem is easy and could be solved by invoking an API call based on y which corresponds to a vertex. Case 2, $d = 2$. In this case, $y = (v_1, v_2)$. The set of neighbors of y in $G^{(2)}$ corresponds to $\{(u,v) \mid u = v_1 \land v \in N(v_1) \backslash v_2\} \bigcup \{(u,v) \mid u = v_2 \land v \in N(v_2) \backslash v_1\}$, and thus it could be computed based on $N(v_1)$ and $N(v_2)$ which could be obtained by invoking two API calls, one with v_1 as input and the other with v_2 as input. Case 3, $d = 3$. In this case, $y = (v_1, v_2, v_3)$. We define $M(v_i, v_j) = N(v_i) \bigcup N(v_j) \backslash V(y)$, if $(v_i, v_j) \in E$ and $M(v_i, v_j) = N(v_i) \bigcap N(v_j) \backslash V(y)$, otherwise, where $V(y)$ denotes the set of vertices $\{v_1, v_2, \ldots, v_d\}$. Then the set of neighbors of y in $G^{(3)}$ corresponds to $\{(u, v, w) \mid u = v_1 \land v = v_2 \land w \in M(v_1, v_2)\} \bigcup \{(u, v, w) \mid u = v_2 \land v = v_3 \land w \in M(v_2, v_3)\} \bigcup \{(u, v, w) \mid u = v_1 \land v = v_3 \land w \in M(v_1, v_3)\}$, and thus it could be computed based on $N(v_1)$, $N(v_2)$ and $N(v_3)$ which could obtained by invoking three API calls.

Note that with the above implementation, each transition requires d API calls. Fortunately, since two nodes that are visited *consecutively* during the process are neighbors to each other (i.e., they share $d - 1$ vertices), the results of the previous API calls could be saved for the current node and thus exactly one new API call is needed. Therefore, each transition requires one API call on average.

We regard one API call as a unit of time and we exclude the time cost before the mixing time in the random walk process. Then the complexities of our algorithms are as follows, the proofs can be found in [26].

Theorem 3. *Let d_{max} be the maximum degree of $G^{(d)}$. The time complexity of the random walk process is $O(k)$ for $G^{(1)}$ and $O(k \cdot d_{max})$ for $G^{(2)}$ and $G^{(3)}$. The time complexities of computing $\widehat{R^{(d)}}$ and $\widehat{N}$ are $O(k \cdot d_{max})$. The space complexities for both processes are $O(k \cdot d_{max})$.*

6   Experiments

In this section, we report on our experimental findings. First we describe our experimental setup. We used 8 real datasets as shown in Table 1, which are used in the literature for estimating graphlet statistics and network size of OSNs [2, 17]. Following these studies, we simulate the scenario where we only have accesses to the datasets via APIs. In each network, we remove the directions of edges, self-loops and multi-edges. We used the largest connected component for each network (since the method could be similarly run on other connected components). The statistics of the largest connected components of networks are shown in Table 1 where N is the number of triangles and W is number of wedges.

Table 1. Statistics of Datasets

Network	$|V|$	$|E|$	N	W
Facebook [28]	6.34×10^4	8.17×10^5	3.5×10^6	7.1×10^7
Epinion [27]	7.6×10^4	4.06×10^5	1.6×10^6	7.4×10^7
Slashdot [27]	7.7×10^4	4.69×10^5	5.5×10^5	6.8×10^7
Gowalla [28]	1.97×10^5	9.5×10^5	2.3×10^6	2.9×10^8
Pokec [28]	1.6×10^6	2.23×10^7	3.26×10^7	2.08×10^9
Flickr [28]	2.2×10^6	2.28×10^7	8.38×10^8	2.33×10^{10}
Orkut [28]	3.08×10^6	1.17×10^8	6.28×10^8	4.56×10^{10}
Live Journal [28]	4.8×10^6	4.28×10^7	2.86×10^8	7.27×10^9

We adopt the following two metrics in our experiments.

- *Normalized root mean square error* (NRMSE): NRMSE is defined as

$$\text{NRMSE}(\widehat{N}) = \frac{\sqrt{\mathbb{E}[(\widehat{N} - N)^2]}}{N} = \frac{\sqrt{Var[\widehat{N}] + (N - \mathbb{E}[\widehat{N}])^2}}{N}, \quad (18)$$

Note that NRMSE captures both the variance and the bias of the estimator.
- *Confidence interval*: A $[p_1, p_2]$-confidence interval in our case corresponds to an interval $[L, U]$ such that $Pr[\widehat{N}/N \leq L] = p_1$ and $Pr[\widehat{N}/N \leq U] = p_2$.
  In our experiments, for each result pair, we run 200 simulations independently and get an estimate of the [5%, 95%]-confidence interval, $[L, U]$, such that L and U are the 5th and 95th percentile values, respectively.

We study the performance of three algorithms each corresponding to our sampling algorithm based on one of the following settings: (1) $d = 1, l = 3$, (2) $d = 2, l = 2$, and (3) $d = 3, l = 1$, the algorithms are denoted by SRWd, depending on the d value for the algorithm. Note that to the best of our knowledge, the problem of estimating the number of edges and triangles in an OSN has not been studied before, and thus we have no state-of-the-arts to compare with in our experiments.

All algorithms are implemented in C++, and we run experiments on a Linux machine with Intel 3.40 GHz CPU.

6.1 Performance Studies for Estimating N

Figure 2 shows the NRMSE results, where the x-axis is the number of random walk steps as a percentage of the set of nodes in the network, and the y-axis is the NRMSE. Each NRMSE value is calculated by averaging over 200 independent simulations. We summarize our results as follows.

- On all networks, SRW1 always achieves the lowest NRMSE among SRW1, SRW2 and SRW3, which means that SRW1 has the highest accuracy in estimating the number of triangles.
- SRW1, the best method in our framework, gives relatively accurate estimation. For example, when 2% of the nodes are sampled, the NRMSE of SRW1 is in the range [0.033, 0.23]. Besides, for most of the networks (6 out of 8 networks), the NRMSE of SRW1 is just around or less than 0.1.
- SRW3 is the worst method in terms of accuracy. It is slow and cannot finish running within a reasonable time on large networks, and as a result, its measurements on some datasets are not available.

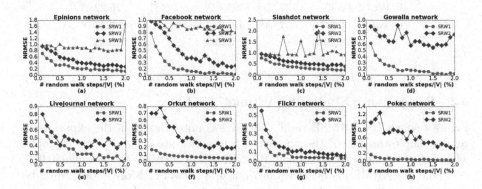

Fig. 2. NRMSE of triangle count estimation

Table 2 shows the [5%, 95%]-confidence interval of SRW1 and SRW2 when only 2% of the nodes are sampled. The confidence interval of SRW1 is much tighter than SRW2, which again demonstrates that SRW1 outperforms SRW2 significantly. This result is also consistent with the theoretical analysis in Sect. 4.3.

Table 2. [5%, 95%]-confidence interval of triangle count estimation

Network	SRW1	SRW2	Network	SRW1	SRW2
Facebook	[0.818, 1.180]	[0.612, 1.521]	Epinion	[0.764, 1.244]	[0.518, 1.438]
Slashdot	[0.629, 1.400]	[0.324, 1.788]	Gowalla	[0.823, 1.218]	[0.200, 2.187]
Pokec	[0.941, 1.066]	[0.669, 1.600]	Flickr	[0.944, 1.053]	[0.878, 1.119]
Orkut	[0.923, 1.083]	[0.754, 1.357]	Live Journal	[0.629, 1.353]	[0.496, 1.741]

Table 3 shows the running time of SRW1 and SRW2 when only 2% of the nodes are sampled. The running time of SRW1 is much smaller than SRW2, which demonstrates that SRW1 is much more efficient than SRW2. In addition, our algorithm SRW1 only takes quite short time to process these networks in experiments, which means our algorithm is very piratical. Figure 3 shows the running time of SRW1 and SRW2 with different sample size. We find that For both algorithms, the running time is nearly linear with the sample size.

Table 3. Running time when 2% of the total nodes are sampled

Network	SRW1	SRW2	Network	SRW1	SRW2
Facebook	0.0632 s	18.90 s	Epinion	0.1134 s	19.00 s
Slashdot	0.1152 s	17.77 s	Gowalla	0.44 s	66.90 s
Pokec	2.43 s	177.70 s	Flickr	39.81 s	226.22 s
Orkut	13.15 s	637.54 s	Live Journal	13.05 s	867.28 s

6.2   Performance Studies for Estimating $|E|$

Figure 4 shows the NRMSE results for edge counts. Based on the experiments results, our estimator of the number of edges is quite accurate. For example, when 2% of the nodes are sampled, the NRMSE of SRW1 is in the range [0.015, 0.14] and the NRMSE is below 0.1 in 7 out of 8 networks. More details could be found in the full version of the paper.

Remark: Our algorithm is designed for graphs with restricted access. To our knowledge, there is no published work on edge and triangle counting that we can compare with directly. A most relevant work is PSRW [6], which estimates certain graphlet statistics, but does not estimate the counts of edges or triangles. For

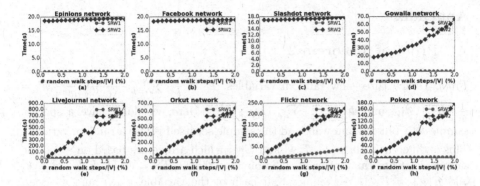

Fig. 3. Running time of triangle count estimation

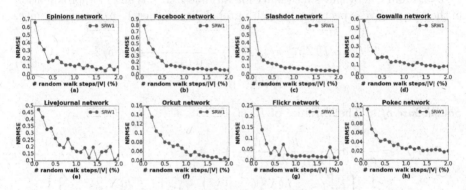

Fig. 4. NRMSE of edge count estimation

triangles, they consider only random walks on $G^{(2)}$. With our proposed method involving the estimation of the graph size of $|R^{(d)}|$, we can convert PSRW to compute the counts, in which case it becomes SRW2. However, we have shown that our new method SRW1 outperforms SRW2 significantly in Sect. 6.

7 Conclusion

In this paper, we present efficient random walk-based algorithms to estimate the number of triangles and the number of edges in OSNs with restricted access. We derive a theoretical bound on the number of samples needed for an accuracy guarantee. Our experiments on real-world OSNs showed that our algorithms provide accurate estimations.

Appendix

A   Proof of Theorem 2

Proof. Define three new random variables, $T' = \frac{1}{k} \cdot \sum_{i=1}^{k} \frac{h(X_i^{(l)})}{\pi_M(X_i^{(l)}) \cdot |R^{(d)}|}, \Psi' = \frac{1}{m} \cdot \sum_{i=1}^{m} d_{Y_i}$, and $\Phi' = \frac{1}{n} \cdot \sum_{(i,j) \in Q} \frac{\theta(Y_i, Y_j)}{d_{Y_i} \cdot d_{Y_j}}$. Here, we don't have a specific sample set from random walk and the sample set itself is also a random variable. This is different from $\widehat{E[T]}$, $\widehat{E[\Phi]}$ and $\widehat{E[\Psi]}$ which are values based on a specific sample set from random walk. It is obvious that $E[T'] = E[T]$, $E[\Psi'] = E[\Psi]$ and $E[\Phi'] = E[\Phi]$. We require that each of the variable T', Φ' and Ψ' is an $\epsilon/4$ approximation of their respective expected values with probability at least $1 - \delta/3$. From Chebyshev's inequality for variable Z, If Z is an $\epsilon/4$ approximation of their respective expected values with probability at least $1 - \delta/3$, then we have,

$$Pr(|Z - E[Z]| \geq \epsilon/4 \cdot E[Z]) \leq \frac{Var[Z]}{(\epsilon/4 \cdot E[Z])^2} \leq \delta/3$$

$$Var[Z] \leq \delta\epsilon^2(E[Z])^2/48 \tag{19}$$

Sample size of k. The Variance of T' is:

$$
\begin{aligned}
Var[T'] &= \frac{1}{k} \cdot Var\left[\frac{h(X^{(l)})}{\pi_M(X^{(l)}) \cdot |R^{(d)}|}\right] \\
&= \frac{1}{k} \cdot \left\{ E\left[\left(\frac{h(X^{(l)})}{\pi_M(X^{(l)}) \cdot |R^{(d)}|}\right)^2\right] - E^2\left[\frac{h(X^{(l)})}{\pi_M(X^{(l)}) \cdot |R^{(d)}|}\right]\right\} \\
&= \frac{1}{k} \cdot \sum_{x^{(l)} \in M^{(l)}} \frac{h(x^{(l)})}{\pi_M(x^{(l)}) \cdot |R^{(d)}|^2} - \frac{\alpha^2 N^2}{|R^{(d)}|^2}
\end{aligned}
\tag{20}
$$

With Eqs. (19) and (20), we can get the bound of sample size k:

$$k \geq \frac{48 \sum_{x^{(l)} \in M^{(l)}} \frac{h(x^{(l)})}{\pi_M(x^{(l)})} - \alpha^2 N^2}{\delta\epsilon^2\alpha^2 N^2} \tag{21}$$

Sample size of m.

$$
\begin{aligned}
Var[\Phi'] &= \frac{1}{m} \cdot Var[d_Y] = \frac{1}{m} \cdot (E[d_Y^2] - E^2[d_Y]) \\
&= \frac{1}{m} \cdot \left[\sum_{y \in H^{(d)}} \frac{d_y^3}{2|R^{(d)}|} - \left(\sum_{y \in H^{(d)}} \frac{d_y^2}{2|R^{(d)}|} \right)^2 \right]
\end{aligned}
\tag{22}
$$

so we have, $m \geq \dfrac{48[\sum_{y \in H^{(d)}} \frac{d_y^3}{2|R^{(d)}|} - (\sum_{y \in H^{(d)}} \frac{d_y^2}{2|R^{(d)}|})^2]}{\delta\epsilon^2(\sum_{y \in H^{(d)}} \frac{d_y^2}{2|R^{(d)}|})^2}$

Sample size of n

$$Var[\Psi'] = E[\Psi'^2] - E^2[\Psi'] \tag{23}$$

$E^2[\Psi']$ can be obtained from Eq. 5. Then, we discuss how to get $E[\Psi'^2]$.

$$
\begin{aligned}
E[\Psi'^2] &= \frac{1}{n^2} E[(\sum_{(i,j)\in Q} \theta(Y_i, Y_j) \cdot \frac{1}{d_{Y_i} \cdot d_{Y_j}})^2] \\
&= \frac{1}{n^2} \sum_{(i,j)\in Q} \sum_{(i',j')\in Q} E[\theta(Y_i, Y_j) \cdot \frac{1}{d_{Y_i} \cdot d_{Y_j}} \theta(Y_i', Y_j') \cdot \frac{1}{d_{Y_i'} \cdot d_{Y_j'}}]
\end{aligned} \tag{24}
$$

To calculate this summation easily, we divide it into three cases.
(a) The node pairs (Y_i, Y_j) and (Y_i', Y_j') are the same i.e. $i = i'$ and $j = j'$,

$$
\begin{aligned}
E[\theta(Y_i, Y_j) \cdot \frac{1}{d_{Y_i} \cdot d_{Y_j}} \theta Y_i', Y_j' \cdot \frac{1}{d_{Y_{i'}} \cdot d_{Y_{j'}}}] &= E[(\theta(Y_i, Y_j) \cdot \frac{1}{d_{Y_i} \cdot d_{Y_j}})^2] \\
&= \frac{1}{4|R^{(d)}|^2} \sum_{i\in H^{(d)}} \sum_{j\in H^{(d)}} \frac{(\theta(i,j))^2}{d_i \cdot d_j}
\end{aligned} \tag{25}
$$

Since for general OSN, the degree of every user should be no smaller than 2, we have $d_i \cdot d_j \geq d_i + d_j$. In addition, it is obvious that $d_i + d_j \geq \theta(i,j)$. Based on these results, we have, $\frac{(\theta(i,j))^2}{d_i \cdot d_j} \leq \frac{(\theta(i,j))^2}{d_i + d_j} \leq \theta(i,j)$. So, $E[(\theta(Y_i, Y_j) \cdot \frac{1}{d_{Y_i} \cdot d_{Y_j}})^2] \leq \frac{1}{4|R^{(d)}|^2} \sum_{i\in H^{(d)}} \sum_{j\in H^{(d)}} \theta(i,j) = \frac{1}{4|R^{(d)}|^2} \sum_{i\in H^{(d)}} d_i^2$.

Moveover, we have n such cases.
(b) The node pairs (Y_i, Y_j) and (Y_i', Y_j') share no common node.

$$
\begin{aligned}
&E[\theta(Y_i, Y_j) \cdot \frac{1}{d_{Y_i} \cdot d_{Y_j}} \theta(Y_i', Y_j') \cdot \frac{1}{d_{Y_i'} \cdot d_{Y_j'}}] \\
&= \frac{1}{16|R^{(d)}|^4} \sum_{i\in H^{(d)}} \sum_{j\in H^{(d)}} \theta(i,j) \cdot \sum_{i' H^{(d)}} \sum_{j'\in H^{(d)}} \theta(i',j') = \frac{1}{16|R^{(d)}|^4}(\sum_{i\in H^{(d)}} d_i^2)^2
\end{aligned}
$$

Hence, the number of such case is smaller than $n(n-1)$.
(c) The node pairs (Y_i, Y_j) and (Y_i', Y_j') share one common node.

$$
\begin{aligned}
&E[\theta(Y_i, Y_j) \cdot \frac{1}{d_{Y_i} \cdot d_{Y_j}} \theta(Y_i', Y_j') \cdot \frac{1}{d_{Y_i'} \cdot d_{Y_j'}}] \\
&= \frac{1}{8|R^{(d)}|^3} \sum_{i\in H^{(d)}} \sum_{j\in H^{(d)}} \sum_{i'\in H^{(d)}} \frac{\theta(i,j) \cdot \theta(j,i')}{d_j}
\end{aligned}
$$

Since $d_{Y_j} \geq \theta(Y_j, Y_i')$, we have,

$$
\begin{aligned}
&E[\theta(Y_i, Y_j) \cdot \frac{1}{d_{Y_i} \cdot d_{Y_j}} \theta(Y_i', Y_j') \cdot \frac{1}{d_{Y_i'} \cdot d_{Y_j'}}] \\
&\leq \frac{|H^{(d)}|}{8|R^{(d)}|^3} \sum_{i\in H^{(d)}} \sum_{j\in H^{(d)}} \theta(i,j) = \frac{|H^{(d)}|}{8|R^{(d)}|^3} \sum_{i\in H^{(d)}} d_i^2
\end{aligned} \tag{26}
$$

The number of such case is smaller than $2n\sqrt{2n}$. (there are n ways choosing the first pair, and the other pair shares one node with this pair, so there is two possible node to share with).

Then we can compute $Var[\Psi']$,

$$Var[\Psi'] = E[\Psi'^2] - E^2[\Psi'] \tag{27}$$

$$\leq \frac{1}{n}\frac{1}{4|R^{(d)}|^2}\sum_{i\in H^{(d)}} d_i^2 + \frac{1}{16|R^{(d)}|^4}(\sum_{i\in H^{(d)}} d_i^2)^2 \tag{28}$$

$$+\frac{2\sqrt{2n}}{n}\frac{|H^{(d)}|}{8|R^{(d)}|^3}\sum_{i\in H^{(d)}} d_i^2 - \left[\sum_{i\in H^{(d)}} \frac{d_i^2}{4|R^{(d)}|^2}\right]^2 \tag{29}$$

$$\leq \frac{1}{n}\frac{1}{4|R^{(d)}|^2}\sum_{i\in H^{(d)}} d_i^2 + \frac{2\sqrt{2n}}{n}\frac{|H^{(d)}|}{8|R^{(d)}|^3}\sum_{i\in H^{(d)}} d_i^2 \tag{30}$$

Substituting this into Eq. (19), and let both of the terms less than half of the right hand size in Eq. (19), we got the condition for n:

$$n \geq \max\{\frac{384|R^{(d)}|^2}{\epsilon^2\delta\sum_{i\in H^{(d)}} d_i^2}, \left[\frac{384\sqrt{2}|R^{(d)}|\cdot|H^{(d)}|}{\epsilon^2\delta\sum_{i\in H^{(d)}} d_i^2}\right]^2\} \tag{31}$$

This finishes the proof.   □

References

1. Katzir, L., Liberty, E., Somekh, O.: Estimating sizes of social networks via biased sampling. In: WWW, pp. 597–606 (2011)
2. Hardiman, S.J., Katzir, L.: Estimating clustering coefficients and size of social networks via random walk. In: WWW, pp. 539–550 (2013)
3. Jha, M., Seshadhri, C., Pinar, A.: Path sampling: a fast and provable method for estimating 4-vertex subgraph counts. In: WWW, pp. 495–505 (2015)
4. Lee, C.-H., Xu, X., Eun, D.Y.: Beyond random walk and metropolis-hastings samplers: why you should not backtrack for unbiased graph sampling. In: SIGMETRICS, pp. 319–330 (2012)
5. Li, R.-H., Yu, J., Qin, L., Mao, R., Jin, T.: On random walk based graph sampling. In: ICDE, pp. 927–938 (2015)
6. Wang, P., Lui, J.C.S., Ribeiro, B., Towsley, D., Zhao, J., Guan, X.: Efficiently estimating motif statistics of large networks. In: ICDE, pp. 8:1–8:27 (2014)
7. Bhuiyan, M., Rahman, M., Al Hasan, M.: Guise: uniform sampling of graphlets for large graph analysis. In: ICDM, pp. 91–100 (2012)
8. Seshadhri, C., Pinar, A., KoldaWedge, T.G.: Sampling for computing clustering coefficients and triangle counts on large graphs. Stat. Anal. Data Mining **7**(4), 233–235 (2014)
9. Gjoka, M., Kurant, M., Butts, C.T.: Walking in Facebook: a case study of unbiased sampling of OSNs. In: INFOCOM, pp. 1–9 (2010)
10. Becchetti, L., Boldi, P., Castillo, C., Gionis, A.: Efficient semi-streaming algorithms for local triangle counting in massive graphs. In: KDD, pp. 16–24 (2008)

11. Eckmann, J.-P., Moses, E.: Curvature of co-links uncovers hidden thematic layers in the World Wide Web. In: PNAS, pp. 5825–5829 (2002)
12. Chiba, N., Nishizeki, T.: Arboricity and subgraph listing algorithms. SIAM J. Comput. **14**, 210–223 (1985)
13. Berry, J.W., Fosvedt, L., Nordman, D., Phillips, C.A., Wilson, A.G.: Listing triangles in expected linear time on power law graphs with exponent at least 7/3. Technical report SAND 2010-4474C (2011)
14. Chu, S., Cheng, J.: Triangle listing in massive networks and its applications. In: KDD, pp. 672–680 (2011)
15. Latapy, M.: Main-memory triangle computations for very large (sparse (power-law)) graphs. Theoret. Comput. Sci. **407**, 458–473 (2008)
16. Tsourakakis, C.E., Kang, U., Miller, G.L., Faloutsos, C.: Doulion: counting triangles in massive graphs with a coin. In: KDD, pp. 837–846 (2009)
17. Chen, X., Li, Y., Wang, G.P., Lui, J.C.S.: A general framework for estimating graphlet statistics via random walk. CoRR abs/1603.07504 (2016)
18. Park, H.-M., Silvestri, F., Kang, U., Pagh, R.: MapReduce triangle enumeration with guarantees. In: ACM Conference Information Knowledge Manage, pp. 1739–1748 (2014)
19. Suri, S., Vassilvitskii, S.: Counting triangles and the curse of the last reducer. In: WWW, pp. 607–614 (2011)
20. Yoon, J.-H., Kim, S.-R.: Improved sampling for triangle counting with mapreduce. In: Lee, G., Howard, D., Ślęzak, D. (eds.) ICHIT 2011. LNCS, vol. 6935, pp. 685–689. Springer, Heidelberg (2011). doi:10.1007/978-3-642-24082-9_83
21. Hu, X., Tao, Y., Chung, C.-W.: Massive graph triangulation. In: SIGMOD, pp. 325–336 (2013)
22. Chu, S., Cheng, J.: Triangle listing in massive networks. ACM Trans. Knowl. Discov. Data **6**(4), 17 (2012)
23. Pagh, R., Silvestri, F.: The input/output complexity of triangle enumeration. In: ACM Symposium Principles Database System, pp. 224–233 (2014)
24. Seshadhri, C., Pinar, A., Kold, T.G.: Triadic measures on graphs: the power of wedge sampling. In: SDM (2013)
25. Berry, J.W., Hendrickson, B., LaViolette, R.A., Phillips, C.A.: Tolerating the community detection resolutionlimit with edge weighting. Phys. Rev. E **83**, 056119 (2011)
26. Counting Edges and Triangles in Online Social Networks via Random Walk. https://www.dropbox.com/sh/228dpiup7qgp6mw/AAA4ijK6jsVUosKNz2OHSlUba?dl=0
27. SNAP Datasets: Standford large network dataset collection. http://snap.standford.edu/data
28. KONECT Datasets: The koblenz network collection. http://konect.uni-koblenz.de

Fair Reviewer Assignment Considering Academic Social Network

Kaixia Li, Zhao Cao, and Dacheng Qu[✉]

Beijing Institute of Technology, Beijing, China
qudc@bit.edu.cn

Abstract. An important task in peer review is assigning papers to appropriate reviewers, which is known as Reviewer Assignment (RA) problem. Most of existing works mainly focus on the similarity between paper topics and reviewer expertise, few works consider multiple relationships between authors and reviewers on academic social network. However, these relationships could influence reviewer assessments on fairness. In this paper, we address RA problem considering academic social network to find a confident, fair and balanced assignment. We model papers and reviewers based on matching degree by combining collaboration distance and topic similarity, and propose Maximum Sum of Matching degree RA (MSMRA) problem. Two algorithms are designed for MSMRA problem: Simulated Annealing-based Stochastic Approximation, and Maximum Matching and Minimum Deviation. Experiments show that our methods achieved good performance both on overall effectiveness and fairness distribution within reasonable running time.

Keywords: Reviewer assignment · Academic social network · Fair

1 Introduction

One of the most challenging tasks in peer review is assigning papers to appropriate reviewers within a reasonable time, which is known as Reviewer Assignment (RA) problem. The opinions of reviewers determine whether papers should be accepted or not, so assigning papers to reviewers in a way that would maximize the quality of assessments, is the focus of RA problem. At present, most of conferences handle this task semi-automatically, which requires reviewers to bid on papers by providing preferences. Due to the large number of paper submissions, each reviewer would go through a great deal of papers (at least abstracts) to provide preferences. In some cases, reviewers choose preferences semi-randomly, which may lead to an inappropriate assignment. What is an appropriate reviewer assignment? We measure it by three important factors:

- **Confidence**: In order to ensure the quality of the assessment, reviewer should have adequate expertise on the topic of assigned paper.
- **Fairness**: For fairness, each reviewer should deal each paper in same manner, and each paper should be reviewed by a certain number of distinct reviewers.

© Springer International Publishing AG 2017
L. Chen et al. (Eds.): APWeb-WAIM 2017, Part I, LNCS 10366, pp. 362–376, 2017.
DOI: 10.1007/978-3-319-63579-8_28

- **Balance**: As the number of submitted papers is huge, making a balance workload for each reviewer is also a factor need to be considered.

To complete the reviewer assignment in the given assignment time window, which is usually a couple of days, **reasonable running time** is also an important criterion.

Automatic RA problem has already been studied in community like Dumais and Nielsen [1]. They modeled a paper as a query, attempted to retrieve reviewers with most relevant expertise based on content-reviewer expertise similarity, and expressed colleague relationship conflict-of-interest (COI) and reviewer workload as linear inequalities constraints. Long et al. [2] studied RA problem both on goodness and fairness. For goodness, they proposed to maximize the topic coverage between papers and reviewers. For fairness, they discussed COI as constraints. Most of the previous works mainly focus on the similarity between paper topics and reviewer expertise, and just add some constraints as linear inequalities for fairness. Few studies conduct an in-depth exploration of relationships between authors and reviewers on academic social network. The tendency of effects to spread from person to person, beyond an individual's direct social ties [3]. If we ignore these connections when assigning papers, these direct and indirect social ties are likely to affect reviewers' judgments, leading to partial review results.

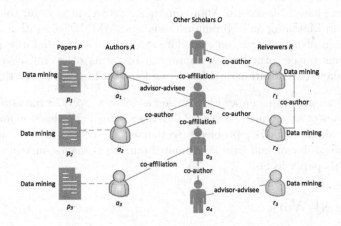

Fig. 1. An example of academic social network

For example, in Fig. 1, there are 3 papers about data mining, 4 scholars, and 3 reviewers. Each paper has one author. Every reviewer has expertise on data mining. Just according to the topic similarity, any paper in P can be assigned to any reviewer in R. However, paper p_1 cannot be assigned to reviewer r_1, because author a_1 has co-affiliation relationship with reviewer r_1. It is still not a fair assignment if we assign paper p_1 to reviewer r_2, since author a_1 is very likely to communicate with reviewer r_2 through scholar o_2. It does not mean that paper p_1 cannot be assigned to any reviewer in R. Our influence gradually dissipates and ceases to have a noticeable effect on people beyond the social frontier that lies at three degrees of separation [3]. As distance increase, influence decrease. The shortest path distance between author a_1 and reviewer r_3 is 4, the probability that author a_1 influences reviewer r_3 is much lower than reviewer r_1 and r_2. Comparing against reviewers r_1 and r_2, it is better to assign paper p_1 to reviewer r_3.

Motivated by above observation, we build an academic social network $G(V, E)$, which is used to mine multiple relationships between authors and reviewers, for fair assignment. We propose collaboration distance on academic social network for fairness, combining with topic similarity for confidence and some constraints for balance to find an appropriate reviewer assignment solution.

The main contributions in our work are summarized as follows:

1. To the best of our knowledge, this is the first work to address RA problem considering academic social network for fair assignment. We model papers and reviewers based on matching degree by combining collaboration distance on fairness with topic similarity on confidence, then we define Maximum Sum of Matching degree RA (MSMRA) problem subject to four constraints on balance: group size, relationship indicator, distinct assignment and balanced workload.
2. Two algorithms are proposed to solve MSMRA problem. One is Simulated Annealing-based Stochastic Approximation (SASA) algorithm, the other is Maximum Matching and Minimum Deviation (MMMD) algorithm.
3. Experimental evaluations on real datasets demonstrate that the proposed algorithms outperform baseline approach in terms of overall effectiveness, more balance distribution on fairness, and run within reasonable time.

The remaining of this paper is organized as follows: Sect. 2 summarizes relate work of reviewer assignment. Section 3 describes formal specification and formal definitions about MSMRA problem. Sections 4 and 5 present two algorithms to solve the problem, and Sect. 6 evaluates our experiments on real data. We conclude this paper in Sect. 7.

2   Related Work

A bulk of related works have studied the RA problem, we discuss these works from two aspects: how to compute confidence between papers and reviewers, and how to enhance fairness for each pair of paper and reviewer.

For **confidence** aspect, relevant studies can be divided into two methods. One is information retrieval method [1, 4–7]. Geller [4] implemented an intelligent knowledge-based reviewer selection program for IJCAI 1999. Basu et al. [5] mined reviewer abstracts from web and then use a TF-IDF weighted vector space model to rank reviewers for a given submitted paper. Hettich and Pazzani [6] measured the association of reviewers with applications by a TF-IDF weighted vector space model. Rodriguez and Bollen [7] identified related reviewers in a co-authorship network by using an energy distribution in a manner similar to the spreading activation techniques used for information retrieval. Mimno and Mccallum [8] proposed an author-persona-topic model for reviewer retrieving. The other is topic similarity method [2,9–17]. Ferilli et al. [9] adopted latent semantic indexing to identify paper topics and reviewer expertise. Xue et al. [10] introduced interval fuzzy ontology to compute the similarity of research subjects. Hartvigsen et al. [11] proposed that every paper should be sent to at least one reviewer whose expertise is greater than or equal to the threshold. Kou et al. [12] proposed a weighted-coverage group-based reviewer assignment problem. Li and Watanabe [13] assigned papers to reviewers by combining preference-based approach and topic-based approach. Kolasa and Krol [14] defined the quality degree based on the number of paper keywords covered by assigned reviewer keywords. Karimzadehgan and Zhai [15] studied matching of multiple aspects of expertise so that the assigned reviewers can cover all the aspects of a paper in a complementary manner. Kou et al. [16] demonstrated reviewer assignment system, which maximizes the topics coverage of each paper by the profiles of assigned reviewers. Charlin et al. [17] explored language model and collaborative filtering to learn confidence scores between papers and reviewers.

For **fairness** aspect, we mainly discuss avoiding conflict-of-interest (COI) methods. Karimzadehgan and Zhai [18] studied constrained multi-aspect expertise matching problem and modeled COI as one of inequality linear constraints. Kolasa and Król [19] used a Boolean variable bar to indicate COI. Benferhat and Lang [20] expressed geographical distribution as a relaxed constraint. Wang et al. [21] suggested that paper cannot be reviewed by friends or enemies of the authors, or by colleagues from the same institution. Conry et al. [22] modeled COI by hardwiring the desired entries of assignment matrix and took them out of play. Tang et al. [23] explored various domain-specific constraints in a constraint-based optimization framework. Garg et al. [24] modeled an edge-labelled bipartite graph of papers and reviewers, and connected an edge if two nodes have no COI. O'Dell et al. [25] represented problem as a weighted bipartite graph, and computing f-matching of the graph. Taylor [26] solved a variant of the bipartite matching problem to find an assignment between papers and reviewers that maximizes the total affinity subject to constraints on the number of reviewers that can be assigned to a paper and the load that can be assigned to any individual reviewer.

3 Problem Statements

3.1 Notation Specification

In this paper, the **data** of reviewer assignment problem consist of:

- A set of n reviewers, which is denoted as $R = \{r_1, r_2, \ldots, r_n\}$. Each reviewer $r \in R$ has four attributes: *id*, *name*, *topics*, and *affiliations*.
- A set of m papers, which is denoted as $P = \{p_1, p_2, \ldots, p_m\}$. Each paper $p \in P$ has three attributes: *id*, *topics*, and authors set $A(p)$. All authors of papers

form a set of m subsets, which is denoted as $A = A(p_1), A(p_2), \ldots, A(p_m)$. Each author a has three attributes: *id*, *name*, and *affiliations*.

- A set of s topics, which is denoted as $T = \{t_1, t_2, \ldots, t_s\}$. Each reviewer r is associated with a set of topics, which is denoted as $T(r)$, and $T(r) \subset T$. $t \in T(r)$ means reviewer r possess expertise t. Each paper p is associated with a set of topics, which is denoted as $T(p)$, and $T(p) \subset T$. $t \in T(p)$ means paper p covers subject t. Topics can be given by organizing committee as a topic keyword list, or learned from the abstract of papers by statistical topic models such as latent dirichlet allocation.
- A $s \times s$ similarity degree matrix S, which records the similarity degree of every pair of topics in T, and the value ranges in $[0, 1]$. Due to the cross-integration of knowledge domain, there are similarities between different topics, so a reviewer can also have other similar expertise. Topic similarity can be given or calculated by similarity measure such as cosine similarity.
- An integer k, which is the group size of required reviewers per paper.
- An undirected graph $G(V, E)$ of academic social network, which is used to represent multiple relationships between authors and reviewers. Node set V contains three types of nodes: reviewers in R, authors in A, and other scholars in an academic digital library, and $R \subset V$, $A \subset V$. Edge set E represents the relationship (co-author, co-affiliation, and advisor-advisee) between two nodes, and the weight of an edge is 1. Advisor-advisee relationship can be submitted by the authors, or extracted from personal homepage of reviewers. Co-affiliation relationship can be get from the attributes of authors and reviewers, and co-author relationship can be get from co-author dataset. $d(v_1, v_2)$ denotes the **distance** between two nodes v_1 and v_2, and the value is the shortest path between them in G. The value of $d(v_1, v_1)$ is 0.

The **output** is an assignment, which is a set with m two-tuple groups and is denoted as $\mathbb{R} = \{\langle p_1, R_{p_1} \rangle, \langle p_2, R_{p_2} \rangle, \ldots, \langle p_m, R_{p_m} \rangle\}$. R_p is a set with distinct reviewers being assigned for paper p, recorded as $\{r_p^1, r_p^2, \ldots, r_p^k\}$.

In remainder sections, R denotes a reviewer set, P denotes a papers set, T denotes a topic set, S denotes the similarity degree matrix of T, k denotes the group size, and G denotes the graph of academic social network.

3.2 Problem Definitions

In this section, we formally define the MSMRA problem. The goal of MSMRA is to find a reviewer assignment $\mathbb{R}$ such that the Sum of matching Degree (SM) of $\mathbb{R}$ is maximum subject to four constraints: group size, relationship indicator, distinct assignment and balance workload.

Definition 1 (Collaboration Distance). Given G, P and R, the collaboration distance between paper p and reviewer r is defined as:

$$D(p, r) = \min\left(d\left(a, r\right) | \forall a \in A(p)\right) \tag{1}$$

$MaxD$ denotes maximum collaboration distance between papers and reviewers. $MaxD$ can be set to 1 plus the maximum distance between all pairs of reviewers and authors, or a larger integer. If two nodes are not connected in G, the collaboration distance of them is set to $MaxD$.

Definition 2 (Topic Similarity). Given P, R, T and S, the topic similarity between paper p and reviewer r is defined as:

$$S(p,r) = \max\left(S\left[t_p\right]\left[t_r\right]| \forall t_p \in T(p), \forall t_r \in T(r)\right) \qquad (2)$$

Definition 3 (Matching Degree). Given $MaxD$, α, collaboration distance and topic similarity, the matching degree between paper p and reviewer r is defined as:

$$M(p,r) = \alpha \times \frac{D\left(p,r\right)}{MaxD} + (1-\alpha) \times S(p,r) \qquad (3)$$

According to formula (3), with longer collaboration distance and higher topic similarity, matching degree is higher. α is a weighting coefficient.

Definition 4 (Relationship Indicator). Given G, P and R, the relationship indicator between paper p and reviewer r is defined as:

$$B\left(p,r\right) = \begin{cases} 0 & \forall a \in A(p), d(a,r) > 1 \\ 1 & \text{others} \end{cases} \qquad (4)$$

Problem Definition (MSMRA problem). Given G, P, R, T, S and k, Maximum Sum of Matching degree Reviewer Assignment (MSMRA) problem is to find a reviewer assignment $\mathbb{R} = \{\langle p_1, R_{p_1}\rangle, \langle p_2, R_{p_2}\rangle, \ldots, \langle p_m, R_{p_m}\rangle\}$, such that:

$$\max\left(\sum_{p \in P} \sum_{r_p \in R_p} M(p, r_p)\right)$$

$$\begin{aligned} st.\ &|R_p| = k & &\text{(group size)} \\ &B(p, r_p) = 0 & &\forall p \in P, \forall r_p \in R_p \text{(relationship indicator)} \\ &r_p^i \neq r_p^j & &\forall r_p^i, r_p^j \in R_p, i \neq j \text{(distinct assignment)} \\ &r.count \leq \mu & &\forall r \in R \text{(balance workload)} \end{aligned} \qquad (5)$$

In formula (5), $r.count$ denotes the number of papers being assigned to reviewer r. μ denotes an upper bound of the number of average assigned papers per reviewer for balance workload, and can be calculated by formula (6):

$$\mu = \lceil k \times |P|/|R|\rceil \qquad (6)$$

According to formula (5), $\mathbb{R}$ satisfies that authors of paper p and reviewer r_p have no co-author, no co-affiliation, and no advisor-advisee relationship. Each paper is reviewed by k distinct reviewers, and the workload of reviewers is balanced. The Sum of collaboration Degree (SD), and the Sum of topic Similarity (SS) of $\mathbb{R}$ can also be maximized simultaneously.

In remainder sections, SM denotes sum of matching degree, SD denotes the sum of collaboration distance, SS denotes the sum of topic similarity, and $SM(\mathbb{R})$, $SD(\mathbb{R})$, and $SS(\mathbb{R})$ denote the SM, SD, and SS of $\mathbb{R}$ respectively.

4   SASA: Simulated Annealing-Based Stochastic Approximation Algorithm

We propose a stochastic approximation algorithm based on simulated annealing, which is denoted as SASA. Simulated Annealing algorithm was first used for combinatorial optimization problem by Kirkpatricket et al. [27], and has been proved to converge to an optimal solution in theory. The process of SASA consists of three steps:

Algorithm 1. Simulated Annealing-based Stochastic Approximation

---

input: $G, P, R, T, S, k, TEMP, MinT, RATE$
output: $\mathbb{R}, SM, SD,$ and SS of $\mathbb{R}$
 1: $temp \leftarrow TEMP$
 2: $\mathbb{R} \leftarrow \text{RandomAssignment}(G, P, R, k)$
 3: **while** $temp > MinT$ **do**
 4:      $p_1 \leftarrow \text{math.random}(1, |P|)$
 5:      $p_2 \leftarrow \text{math.random}(1, |P|)$
 6:      **if** $p_1 \neq p_2$ **then**
 7:          $r_1 \leftarrow \text{get minmum matching reviewer}(\mathbb{R}, p_1)$
 8:          $r_2 \leftarrow \text{get minmum matching reviewer}(\mathbb{R}, p_2)$
 9:          **if** $r_1 \neq r_2$ **and** $B(p_1, r_2) = 0$ **and** $B(p_2, r_1) = 0$ **then**
10:              $\Delta \leftarrow M(p_1, r_2) + M(p_2, r_1) - M(p_1, r_1) - M(p_2, r_2)$
11:              **if** $\Delta > 0$ **or** $\text{math.exp}(\Delta/temp) > \text{math.random}(0, 1)$ **then**
12:                  $\mathbb{R} \leftarrow \text{exchange reviewers}(p_1, r_1, p_2, r_2)$
13:      $temp \leftarrow temp \times RATE$
14: **return** $\mathbb{R}, SD(\mathbb{R}), SS(\mathbb{R}), SM(\mathbb{R})$

Step 1. **Initialization.** Generate a random assignment subject to four constraints of MSMRA, and $temp$ is set to initial temperature $TEMP$.

Step 2. **Simulated Annealing.** We use f_t to denote the solution at time t, and f_{t-1} denote the solution at time $t - 1$. Then loop following steps until $temp$ dropping to minimum temperature $MinT$:

Step 2.1. At time t, we randomly select two distinct papers p_1 and p_2, and get assigned reviewers with minimum matching degree of p_1 and p_2 respectively, which are denoted as r_1 and r_2. If $r_1 \neq r_2$ and relationship indicators of (p_1, r_2) and (p_2, r_1) are equal to 0, exchange r_1 and r_2, then a new solution f_t is obtained by formula (7):

$$f_t = f_{t-1} - \langle p_1, r_1 \rangle - \langle p_2, r_2 \rangle + \langle p_1, r_2 \rangle + \langle p_2, r_1 \rangle \tag{7}$$

Step 2.2. The deviation Δ of f_t and f_{t-1} is calculated by formula (8):

$$\begin{aligned}
\Delta &= SM(f_t) - SM(f_{t-1}) \\
&= M(p_1, r_2) + M(p_2, r_1) - M(p_1, r_1) - M(p_2, r_2)
\end{aligned} \tag{8}$$

If $\Delta > 0$, f_t is accepted; if $\Delta \leq 0$, f_t is accepted with a certain probability, and the probability decreases over time gradually. Accepted principle is defined as:

$$\mathbb{R} = \begin{cases} f_t & \Delta > 0 \ \textbf{or} \ \text{math.exp}(\Delta/temp) > \text{math.random}(0,1) \\ f_{t-1} & \text{others} \end{cases} \quad (9)$$

Step 2.3. Decrease $temp$ by multiplying a cooling rate $RATE$, which is a decimal number between 0 and 1. Then set $t = t + 1$.

Step 3. **Result**. Return the assignment $\mathbb{R}$, and the SD, SS, and SM of $\mathbb{R}$.

The pseudo code of SASA is presented in Algorithm 1.

The time complexity of RANDOMASSIGNMENT(G, P, R, k) is $O(k \times |P|)$. The outer loop of Algorithm 1 is $O(N)$. N depends on initial temperature $TEMP$, minimum temperature $MinT$, and cooling rate $RATE$. When $TEMP$ and $MinT$ are fixed, as $RATE$ is higher, N is larger. The time complexity of Algorithm 1 is $O(N \times k \times |P|)$. In order to achieve a good result, N is usually a large number, so the running time is mainly affected by N while parameters k and $|P|$ have little effect.

5   MMMD: Maximum Matching and Minimum Deviation Algorithm

Since the running time of SASA is easily influenced by the input parameters, we propose an exact polynomial algorithm for MSMRA problem, named Maximum Matching and Minimum Deviation algorithm, which is denoted as MMMD. A state matrix Q is used to record the assignment state, and Q could be adjusted according to minimum deviation principle. After initializing μ and state matrix Q, the process of MMMD mainly contains two steps:

Step 1. **Maximum matching degree assignment**. For each paper p, we compute Q of every reviewer and p by relationship indicator. If $B(p, r) = 0$, $Q[p][r] = 0$ means reviewer r is **assignable** for p; else $Q[p][r] = -1$ means reviewer r cannot be assigned to p.

We assign p to the assignable reviewer r with maximum matching degree, and set $Q[p][r] = 1$ for distinct assignment. $Q[p][r] = 1$ means reviewer r has been assigned to p. Then set $r.count = r.count + 1$.

If $r.count > \mu$, balanced workload is broken and needs to be adjusted, then jump to step 2; else perform step 1 until the number of reviewers assigned to paper p is equal to group size k.

Step 2. **Minimum deviation adjust**. Reviewers with less number of assigned papers than μ are **adjustable**, and can be adjusted in step 2.

For each paper $i \in [1, p]$ with $Q[i][r] = 1$, we first get reviewer t which is assignable, adjustable and with maximum matching degree for p. Then we calculate the deviation of $M(i, r)$ and $M(i, t)$, and use p_a· and r_a to denote the paper and the reviewer with minimum deviation $MinDev$. Finally, we set $Q[p_a][r] = -1$, $Q[p_a][r_a] = 1$, then return to step 1.

Algorithm 2. Maximum Matching and Minimum Deviation algorithm

---

input: G, P, R, T, S, k
output: $\mathbb{R}, SM, SD$,and SS of $\mathbb{R}$

1:   $\mu \leftarrow \lceil k \times |P|/|R| \rceil$
2:   $Q[|P|][|R|] \leftarrow \{0\}$
3:   **for each** $p \in P$ **do**
4:      $queue \leftarrow$ REVERSESORTREVIEWERS$(M(p, R))$
5:      **for** $i = 1 \rightarrow k$ **do**
6:        $r \leftarrow queue.\text{poll}()$
7:        **if** $B(p, r) = 1$ **then**
8:          $Q[p][r] \leftarrow -1$
9:        **else**
10:          $Q[p][r] \leftarrow 1$
11:          $r.count \leftarrow r.count + 1$
12:          **if** $r.count > \mu$ **then**
13:            $Q \leftarrow$ MINIMUMDEVIATIONADJUSTMENT(Q, r, p, μ)
14:        $i \leftarrow i + 1$
15: $\mathbb{R} \leftarrow$ get assignment (Q)
16: **return** $\mathbb{R}, SD(\mathbb{R}), SS(\mathbb{R}), SM(\mathbb{R})$

The pseudo code of MMMD is shown in Algorithm 2, and the pseudo code of MINIMUMDEVIATIONADJUSTMENT(Q, r, p, μ) is shown in Algorithm 3.

The time complexity of Algorithm 3 is $O(|P|(|P| + 1)/2)$. The running time of REVERSESORTREVIEWERS$(M(p, R))$ in Algorithm 2 is $O(|R| \times log|R|)$, so the time complexity of Algorithm 2 is $O(|P| \times (|R| \times log|R| + k \times (|P|(|P| + 1)/2)))$. Generally, $|R|$ is smaller than $|P|$, so the time complexity of Algorithm 2 is approximately equal to $O(k \times |P|^2)$. But not each assignment needs to be adjusted in practice, so the time complexity is much lower than the theoretical value, we will see it later in our experiments on real data sets.

Algorithm 3. Minimum Deviation Adjustment

1:   **function** MINIMUMDEVIATIONADJUSTMENT(Q, r, p, μ)
2:      $p_a, r_a \leftarrow 0$
3:      $MinDev \leftarrow \infty$
4:      **for** $i = 1 \rightarrow p$ **do**
5:        **if** $Q[i][r] = 1$ **then**
6:          $t \leftarrow$ get maximum, adjustable, and assignable reviewer (i)
7:          $\Delta \leftarrow M(i, s) - M(i, t)$
8:          **if** $\Delta < MinDev$ **then**
9:            $p_a \leftarrow i$
10:           $r_a \leftarrow t$
11:           $MinDev \leftarrow \Delta$
12:      $Q[p_a][r] \leftarrow -1$
13:      $Q[p_a][r_a] \leftarrow 1$
14:      **return** Q

---

6   Empirical Evaluation

In this section, we evaluate proposed algorithms for MSMRA problem. We implement all the algorithms in Java. The experiments are conducted on a computer with Intel (R) Core (TM) i5-2400 CPU (3. 10 GHZ) of 8G RAM, and Windows 10 64 bit.

6.1   Experimental Setup

Datasets: We collect accepted papers of SIGMOD 2014[1], 2015[2] and 2016[3] as the paper set P, and use the program committee of SIGMOD 2016 as the reviewer set R. The topic set T is collected from the topics of SIGMOD 2016. We randomly assign 1 to 3 topics as expertise of each reviewer. For each pair of topics, we manually assign a similarity degree range from 0 to 1 based on the similarity between research fields. Finally, there are 305 papers, 851 authors, 190 reviewers, 21 topics and a 21×21 similarity degree matrix S.

We use a snapshot of the DBLP dataset[4] taken on November 25, 2016. We get more than 1.81 million scholars, and collect co-author relationships between them in the DBLP dataset.

The academic social network G in our experiments contains more than 1,816,262 nodes and 8,021,719 edges. The running time of constructing G is 74 s.

Baseline Algorithms: We take two reasonable algorithms as baseline. One is Similarity-based Greedy (SimGreedy) algorithm subject to constraints of MSMRA, which is used in most academic conferences. Another is Random algorithm (Ran) subject to constraints of MSMRA.

BBFS for Shortest Path: We use Bidirectional Breadth First Search (BBFS) to compute the shortest path distance between authors and reviewers in G. The running time of BBFS for 851 authors and 190 reviewers is 63 s.

Parameters Setting: We vary $|P|$, $|R|$, k, and α and evaluate different results. In SASA algorithm, temperature $TEMP = 10^8$, minimum temperature $MinT = 10^{-8}$, and cooling rate $RATE = 0.9999$.

Optimality Ratio: A reasonable approach to evaluate the quality of assignment $\mathbb{R}$ is to compute its approximation ratio $SM(\mathbb{R})/SM(\mathbb{O})$ to the optimal assignment $\mathbb{O}$ [12]. But computing $\mathbb{O}$ will be very time consuming. We use an ideal assignment $\mathbb{I}$ to compute the optimality ratio $SM(\mathbb{R})/SM(\mathbb{I})$. To generate ideal assignment $\mathbb{I}$, we greedily select k maximum matching degree reviewers for each paper, and disregard other constraints in MSMRA. Therefore $SM(\mathbb{I}) > SM(\mathbb{O})$. So $SM(\mathbb{R})/SM(\mathbb{I})$ is a lower bound $SM(\mathbb{R})/SM(\mathbb{O})$.

---

[1] http://www.sigmod2014.org/.

[2] http://www.sigmod2015.org/.

[3] http://www.sigmod2016.org/.

[4] http://dblp.uni-trier.de/xml/.

6.2   Overall Effectiveness

In this section, we use 305 papers and 100 reviewers, and $k = 3$ to measure SM, SD, and SS of $\mathbb{R}$. It means that every paper needs 3 reviewers, and every reviewer needs to review no more than 10 papers. In order to obtain good experimental results, we vary α from 0 to 1, plus 0.1 per step. For showing optimality ratio in different cases, the values of k are 1, 3, 5, the values of α are 0.4, 0.5, 0.6. We use these measures to evaluate how well each algorithm maximize value of assignment in comparison to others.

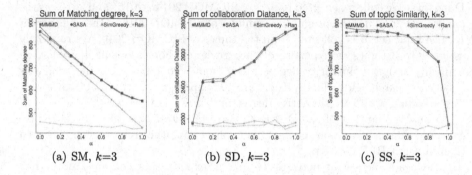

(a) SM, k=3   (b) SD, k=3   (c) SS, k=3

Fig. 2. Overall effectiveness: SM, SD, and SS

Sum of Matching Degree: In Fig. 2(a), when α is larger, SM is smaller On MMMD, SASA, and SimGreedy. This is because α is a weighting coefficient of collaboration distance, and normalized collaboration distance is smaller than topic similarity. MMMD and SASA are always outperform SimGreedy and Ran. While α is 0.4, 0.5 and 0.6, MMMD and SASA even have similar values, this phenomenon also appears in Fig. 2(b) and (c).

Sum of Collaboration Distance: In Fig. 2(b), MMMD and SASA are obviously superior to others. When α is larger, SD is smaller On MMMD, SASA. The reason is same as above. SimGreedy and Ran have no change when varying α, because they ignore the collaboration distance, which is an important factor when assigning reviewers.

Sum of Topic Similarity: In Fig. 2(c), when $\alpha \leq 0.6$, MMMD and SASA outperform SimGreedy and Ran. When $\alpha > 0.6$, the SS of MMMD and SASA decline rapidly, since the weight of topic similarity is $1-\alpha$. Considering similarity priority, SimGreedy always has a large value when varying α. The values of Ran are always low in Fig. 2(a), (b) and (c).

Optimality Ratio: In Fig. 3, the optimality ratios of MMMD and SASA are consistently greater than 0.9, and very close to 1. The optimality ratio of SimGreedy is also not less than 0.9. However, as shown in Fig. 2(b) and (c), although SS of SimGreedy is large, SD of SimGreedy is small. Ran is lower than 0.6 in most cases.

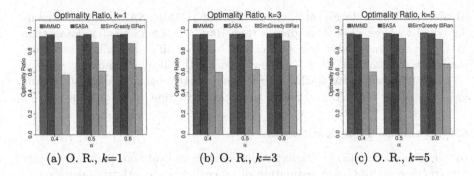

(a) O. R., k=1   (b) O. R., k=3   (c) O. R., k=5

Fig. 3. Overall effectiveness: optimality ratio

Given the superior performance in practice, both MMMD and SASA achieve much better Overall Effectiveness than SimGreedy and Ran.

6.3 Fairness Distribution Analysis

In this section, we use 305 papers and 100 reviewers totally with k=3. We vary α to evaluate assignment quality on individuals by distribution state. Subject to space restrictions, we only show the result of $\alpha = 0.6$.

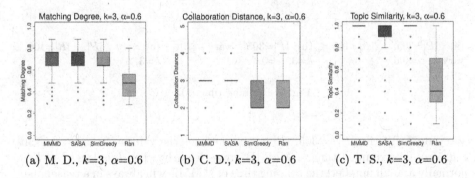

(a) M. D., k=3, α=0.6   (b) C. D., k=3, α=0.6   (c) T. S., k=3, α=0.6

Fig. 4. Fairness distribution analysis

Matching Degree: In Fig. 4(a). The medians of MMMD and SASA are larger than SimGreedy and Ran, which means most of papers are assigned to reviewers with matching degree greater than 0.7 by MMMD and SASA. The assignments of MMMD, SASA and SimGreedy are more convergent than Ran In Fig. 4(a), (b) and (c), which are fairer than Ran.

Collaboration Distance: In Fig. 4(b), there are more papers assigned to reviewers with collaboration distance 3 by MMMD and SASA. The medians of SimGreedy and Ran is only 2.

Topic Similarity: In Fig. 4(c), the medians of MMMD, SASA and SimGreedy are equal to 1, which are far greater than Ran. $Q1$ of SASA is 0.9, lower than MMMD and SimGreedy. As the similarity between different areas of expertise, the reviewer is also qualified to review papers with 0.9 topic similarity.

Since the results of MMMD and SASA have better balanced distribution and greater median, MMMD and SASA are superior to SimGreedy and Ran.

6.4   Running Time Evaluation

In this section, we use $\alpha = 0.6$, fix two parameters and vary one parameter. Since constructing graph G and computing distances between authors and reviewers have been preprocessed, the running time of them is not presented in Fig. 5, which is 137 s regarded as a reasonable and acceptable cost time. As shown in Fig. 5, all the algorithms can be completed in 2.5 s. The total running time of each algorithm is no more than 140 s.

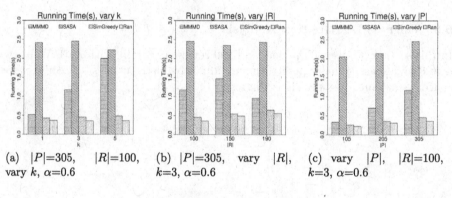

(a) $|P|$=305, $|R|$=100, vary k, α=0.6

(b) $|P|$=305, vary $|R|$, k=3, α=0.6

(c) vary $|P|$, $|R|$=100, k=3, α=0.6

Fig. 5. Running time (s)

Varying k: We fix $|P| = 305$, $|R| = 100$, and set the values of k as 1, 3, 5 in Fig. 5(a). The running time of MMMD grows linearly when k is lager. As k is normally a small number, the running time of MMMD will always in a reasonable range. The running time of SASA is not affected by k, and is mainly affected by the random initial solution and related parameters. SimGreedy and Ran have lower running time in Fig. 5(a), (b) and (c).

Varying $|R|$: We fix $|P| = 305$, $k = 3$, and set the values of $|R|$ as 100, 150, 190 in Fig. 5(b). When $|R|$ grows, the running time of MMMD firstly increase and then decrease. When $|R|$ is larger, the number of assignable reviewers for each paper is larger. There may be less adjustment, so the running time of MMMD will be lower. Therefore $|R|$ is not a major factor for the running time of MMMD. The running time of other algorithms are not obviously affected by $|R|$.

Varying $|P|$: We fix $|R| = 100$, $k = 3$, and set the values of $|P|$ as 105, 205, 305 in Fig. 5(c). The running time of MMMD is longer when $|P|$ is lager, but do not grow exponentially with $|P|$. This is much lower than the worst time complexity. The running times of other algorithms are not affected by $|P|$.

Based on above analyses, even in large conferences with thousands of papers, hundreds of reviewers, MMMD and SASA can also work within a reasonable time. Futhermore, MMMD always take less time to achieve a better result than SASA.

7 Conclusions

In this paper, we formulate reviewer assignment as a Maximum Sum of Matching degree Reviewer Assignment (MSMRA) problem subject to four constraints: group size, relationship indicator, distinct assignment and balanced workload. We introduce collaboration distance on academic social network to utilize multiple relationships between authors and reviewers for fair assignment. Combining collaboration distance and topic similarity, we model matching degree of papers and reviewers, aimed to find a confident, fair and balanced assignment. Two algorithms are proposed to solve this problem: Simulated Annealing-based Stochastic Approximation (SASA), and Maximum Matching and Minimum Deviation (MMMD). Our experimental evaluations show that, comparing against reasonable baseline approaches, our algorithms produced assignment with greater values on overall effectiveness and more balanced distribution on fairness, and also run within reasonable time.

Acknowledgments. The study has been supported by the National Natural Science Foundation of China under Grant No. 61370136.

References

1. Dumais, S.T., Nielsen, J.: Automating the assignment of submitted manuscripts to reviewers. In: International ACM SIGIR Conference on Research and Development in Information Retrieval, Copenhagen, Denmark, p. 233, June 1992
2. Long, C., Wong, C.W., Peng, Y., Ye, L.: On good and fair paper-reviewer assignment. In: IEEE International Conference on Data Mining, p. 1145 (2013)
3. Walker, S.K.: Connected: the surprising power of our social networks and how they shape our lives. J. Fam. Theory Rev. **3**, 220 (2012)
4. Geller, J.: Challenge: how IJCAI 1999 can prove the value of AI by using AI. In: International Joint Conference on Artifical Intelligence, p. 55 (1997)
5. Basu, C., Hirsh, H., Cohen, W.W., Nevill-Manning, C.: Recommending papers by mining the web. In: Proceedings of the IJCAI99 Workshop on Learning About Users, p. 1 (2000)
6. Hettich, S., Pazzani, M.J.: Mining for proposal reviewers: lessons learned at the national science foundation. In: Twelfth ACM SIGKDD International Conference on Knowledge Discovery and Data Mining, Philadelphia, PA, USA, p. 862, August 2006
7. Rodriguez, M.A., Bollen, J.: An algorithm to determine peer-reviewers, p. 319 (2006)

8. Mimno, D., Mccallum, A.: Expertise modeling for matching papers with reviewers. In: ACM SIGKDD International Conference on Knowledge Discovery and Data Mining, San Jose, California, USA, p. 500, August 2007

9. Ferilli, S., Di Mauro, N., Basile, T.M.A., Esposito, F., Biba, M.: Automatic topics identification for reviewer assignment. In: Ali, M., Dapoigny, R. (eds.) IEA/AIE 2006. LNCS, vol. 4031, pp. 721–730. Springer, Heidelberg (2006). doi:10.1007/11779568_78

10. Xue, N., Hao, J.X., Jia, S.L., Wang, Q.: An interval fuzzy ontology based peer review assignment method (2012)

11. Hartvigsen, D., Wei, J.C., Czuchlewski, R.: The conference paper-reviewer assignment problem. Decis. Sci. **30**, 865 (1999)

12. Kou N.M., Hou, U.L., Mamoulis N., Gong Z.: Weighted coverage based reviewer assignment. In: ACM SIGMOD International Conference, p. 2031 (2015)

13. Li, X., Watanabe, T.: Automatic paper-to-reviewer assignment, based on the matching degree of the reviewers. Procedia Comput. Sci. **22**, 633 (2013)

14. Kolasa, T., Krol, D.: A survey of algorithms for paper-reviewer assignment problem. IETE Tech. Rev. **28**, 123 (2014)

15. Karimzadehgan, M., Zhai, C.X., Belford, G.: Multi-aspect expertise matching for review assignment. In: ACM Conference on Information and Knowledge Management, p. 1113 (2008)

16. Kou, N.M., Hou, U.L., Mamoulis, N., Li, Y., Li, Y., Gong, Z.: A topic-based reviewer assignment system. Proc. VLDB Endow. **8**, 1852 (2015)

17. Charlin, L., Zemel, R.S., Boutilier, C.: A framework for optimizing paper matching. In: Proceedings of the Twenty-Seventh Conference on Uncertainty in Artificial Intelligence, UAI 2011, Barcelona, Spain, p. 86, July 2012

18. Karimzadehgan, M., Zhai, C.X.: Constrained multi-aspect expertise matching for committee review assignment. In: ACM Conference on Information and Knowledge Management, CIKM 2009, Hong Kong, China, p. 1697, November 2009

19. Kolasa, T., Król, D.: ACO-GA approach to paper-reviewer assignment problem in CMS. In: Jędrzejowicz, P., Nguyen, N.T., Howlet, R.J., Jain, L.C. (eds.) KES-AMSTA 2010. LNCS, vol. 6071, pp. 360–369. Springer, Heidelberg (2010). doi:10.1007/978-3-642-13541-5_37

20. Benferhat, S., Lang, J.: Conference paper assignment. Int. J. Intell. Syst. **16**, 1183 (2001)

21. Wang, F., Zhou, S., Shi, N.: Group-to-group reviewer assignment problem. Comput. Oper. Res. **40**, 1351 (2013)

22. Conry, D., Koren, Y., Ramakrishnan, N.: Recommender systems for the conference paper assignment problem. In: ACM Conference on Recommender Systems, p. 357 (2009)

23. Tang, W., Tang, J., Tan, C.: Expertise matching via constraint-based optimization. In: Main Conference Proceedings on IEEE/WIC/ACM International Conference on Web Intelligence, WI 2010, Toronto, Canada, August 31–September 3 2010, p. 34 (2010)

24. Garg, N., Kavitha, T., Kumar, A., Mehlhorn, K., Mestre, J.: Assigning papers to referees. Algorithmica **58**, 119 (2010)

25. O'Dell, R., Wattenhofer, M., Wattenhofer, R.: The paper assignment problem (2005)

26. Taylor, C.J.: On the optimal assignment of conference papers to reviewers (2012)

27. Kirkpatrick S., Gelatt Jr. C., Vecchi M.P.: Optimization by simulated annealing. In: Readings in Computer Vision, vol. 220, p. 606 (1987)

Viral Marketing for Digital Goods in Social Networks

Yu Qiao, Jun Wu, Lei Zhang, and Chongjun Wang[✉]

State Key Laboratory for Novel Software Technology, Nanjing University,
Nanjing 210023, China
yuqiao@smail.nju.edu.cn, {wujun,zhangl,chjwang}@nju.edu.cn

Abstract. Influence maximization is a problem of finding a small set of
highly influential individuals in social networks to maximize the spread
of influence. However, the distinction between the spread of influence and
profit is neglected. The problem of profit maximization in social network
extends the influence maximization problems to a realistic setting aiming
to gain maximum profit in social networks. In this paper, we consider
how to sell the digital goods (near zero marginal cost) by viral marketing
in social network. The question can be modeled as a profit maximization
problem. We show the problem is an unconstrained submodular maximization and adopt two efficient algorithms from two approaches. One
is a famous algorithm from theoretical computer science and that can
achieve a tight linear time (1/2) approximation. The second is to propose a profit discount heuristic which improves the efficiency. Through
our extensive experiments, we demonstrate the efficiency and quality
of the algorithms we applied. Based on results of our research, we also
provide some advice for practical viral marketing.

Keywords: Influence maximization · Viral marketing · Social network ·
Profit maximization · Digital goods

1 Introduction

A combination of recent economic and computational trends, such as negligible
cost of duplicating digital goods and, more importantly, the emergence of the
online social network as one of the most important arenas for marketing, has
created a number of new pricing and marketing problems [2,3]. In 2001, Domingos and Richardson first introduced the concept of viral marketing to the social
network [6]. With the in-depth research and development of social network such
as Facebook and Google+, successful online social network (OSN) is creating a
media environment for viral marketing. The spread of awareness about a specific
product in social network through "word of mouth" is a trend of advertising.

Influence maximization problem is a hot topic for social network analysis
which is first defined as an optimization problem by Kempe et al. [1] : A social
network is modeled as an undirected graph $G = (V, E)$, with vertices in V
modeling the users in the network and weighted edges E reflecting the influence

© Springer International Publishing AG 2017
L. Chen et al. (Eds.): APWeb-WAIM 2017, Part I, LNCS 10366, pp. 377–390, 2017.
DOI: 10.1007/978-3-319-63579-8_29

between users, and a positive number k. The goal is to find a seed set A, including k nodes, such that with a given propagation model, the information propagation range of A is maximized. Influence is propagated in the network according to a stochastic cascade model. Three cascade models, namely the Independent Cascade (IC) model , the Weight Cascade (WC) model, and the Linear Threshold (LT) model are extracted from mathematical sociology and considered in [1]. While computer scientists are trying to use mathematic theories and computing devices to understand the diffusion process in online social network (OSN), numerous researchers concern themselves with the efficient approximation algorithms and applications of influence maximization problem [1,4,5] (see Sect. 2).

In a real-world scenario, developers and application makers will give away a paid software for free temporarily in *App Store*. Then, if the user has a good experience, they will recommend it to their friends, and by that time the application will likely come back to its regular price, which should help bring in some profit. This is a typical case of viral marketing on the Internet. Digital goods such as softwares and songs are a suitable target for marketing online, since their duplicating and delivery costs are minimal. Digital goods is a popular topic of theoretical computer science [2,7].

Nowadays most researchers who study viral market are focusing on the spread of influence rather than the original goal of viral marketing: profit. To address the aforementioned limitation, we introduce the concept of free samples and profit achieved by activated users to the influence maximization problem and propose the profit maximization problem for digital goods (PMDG) in social networks. The study is focused on selecting a set of users and giving them free samples to promote the product in order to gain the maximum profit by the diffusion process under the LT model and IC model. The problem is a non-negative, non-monotone and submodular optimization problem without knapsack, which is more complex than the problem of spread of influence. More sophisticated algorithms are needed to solve the problem. Thus, we adopt two efficient algorithms to obtain the maximum profit. One is a famous algorithm from theoretical computer science which can achieve a tight linear time $(1/2)$-approximation. The other is a highly efficient profit discount heuristic. Through our extensive experiments, we show both algorithms are effective and efficient. Based on results of our research, we also provide some advice for practical viral marketing.

Roadmap. The rest of the paper is organized as follows. Section 2 discusses some related work. Section 3 formalizes the viral marketing problem and discusses properties of our proposed problem. Section 4 presents the efficient algorithms. Data sets and experimental results are discussed in Sect. 5. The last section provides conclusion, possible future work and some practical advice.

2   Related Work

Influence Maximization: Kempe et al. [1] are the first to propose a greedy algorithm for influence maximization. Due to a nice property called submodularity, the greedy algorithm produces near-optimal solution with a theoretical guarantee. However, it suffers from scalability, since the large number of Monte-Carlo simulations are necessary for accuracy. The typical algorithm Cost-Effective Lazy Forward (CELF) utilizes the submodularity to reduce the number of necessary influence estimations with the same performance as the original greedy algorithm [5]. Unfortunately, these improved greedy algorithm is still too inefficient because of the heavy Monte-Carlo simulation. The degree discount heuristics [8] are likely to the scalable solutions to the influence maximization for large scale real-life social networks with their influence spread getting close to that of the greedy algorithm and their extremely fast speed. Recently, Lu et al. [4] proposed a method which scales well to the large scale networks by breaking down the big network to small subgraphs.

Digital Goods: In the domain of marketing for digital goods, there has also been lots of work by theoretical computer scientists who study the pricing strategies for digital goods. For instance, [3] and [2] studied the competitive auction for digital goods aiming to maximize revenue. Hartline et al. [9] studied the marketing strategies for digital goods and proposed the influence-and-exploit(IE) framework. The paper mainly made a study of optimal pricing strategy for digital goods in social networks and the valuation of a user is determined by the buyers who own the product.

Profit Maximization: Lu and Lakshmanan [10] addressed the difference between influence and profit in their Linear Threshold model with user valuations (LT-V). The LT-V model is focusing on the pricing strategies and the additional adopting process. In contrast, our work is focused on "unbudgeted" seeds selection in social networks. The "unbudgeted" greedy (U-Greedy) algorithm is proposed for seeds selection in the profit maximization, starting with an empty set A, and then add the node v with maximum marginal profit in each round, i.e., $v = argmax_{v \in V \setminus A} R(A \cup \{v\}) - R(A)$, into A until the profit starting decreasing. $R(A)$ is the profit (see Sect. 3). The U-Greedy algorithm can not give a tight bound and its performance is unstable in some conditions. It also suffers the scalability due to the heavy Monte-Carlo computation in large scale networks.

Submodular Maximization: As we mentioned in Sect. 1, in our setting, the problem for promoting digital goods is an unconstrained submodular maximization, which is different from the influence maximization problem. From a pure algorithm perspective, without a cardinality constraint, maximizing a submodular function is well known to be NP-Hard and an (2/5) approximation can be achieved by local search [11]. Moreover, a linear time algorithm has been proposed for unconstrained submodular maximization, which can achieve an (1/2) approximation [12].

3   Preliminaries

In Sect. 3, we discuss the proposed model for viral marketing problem and define the objective function. Then we will discuss the properties of the problem. Table 1 gives the important notations used in this paper.

Table 1. Notations

Notations	Descriptions
$G = (V, E)$	An undirected network with nodes set V and edge set E
M	Number of the edges in G
N	Number of the nodes in G
P_v	Expected profit earned from node v
A	Seed set
W_{vu}	Influence weight of node v to u in LT model
Q_{vu}	Probability that node v will influence u in IC model
$R(A)$	Total expected profit achieved by seed set A
$\gamma(A)$	Set of active nodes at the end of propagation

3.1   Problem Statement

We consider a marketer who wants to sell the digital goods to a group of users in communities or social networks. The goal of marketer is to maximize the revenue rather than spread of awareness the products. Since the delivery cost and manufacturing cost of the digital goods are near zero, the supply is unlimited. When the supply of goods is unlimited, the marketer could give lots of free samples to the selected users aiming to promoting the product. In other words, the free samples are given to the seed set, the selected users will help to influence their neighbors to buy the product. We formalize the adopting mechanism in the setting of LT model [1,16] as an example. The mechanism for IC model can be reached analogously, so we omit here. Given a graph $G = (V, E)$, each edge $(v, u) \in E$ is associated with an influence weight for each $W_{v,u}$. $Nei_a(v)$ means the set of v's active neighbors. The diffusion process begins with an initial set A of active users who are given free samples at round $t = 0$. At each round i, an inactive (who do not own the product) node v will be activated if the sum of influence weight $\sum_{u \in Nei_a(v)} W_{u,v}$ from its active neighbor $Nei_a(v)$ is larger than the influence threshold θ_v. Once the user is activated, he will buy the product and stay active. The propagation terminates when no more nodes can be activated.

Definition 1. *The profit maximization problem for digital goods (PMDG) is defined as: given a graph $G = (V, E)$, the goal is to find out an unconstrained* set A, $A \subseteq V$, such that profit function $R(A)$ is maximizing. The profit function $R(A)$, is defined as:

$$R(A) = \sum_{v \in \gamma(A)} P_v - \sum_{v \in A} P_v, \qquad (1)$$

where $\gamma(A)$ is the set of active nodes at the end of propagation and P_v means the profit earned by single node v. The goal is to maximizing $R(A)$.

The first term in the above definition is the sum of profit achieved by activated user, the second term is the sum of users' profit who are given free samples for marketing.

3.2 Properties

In the Definition 1, $\sum_{v \in \gamma(A)} P_v$ is a typical influence maximization problem of weighted nodes. However, PMDG remains the submodularity which is a natural "diminishing returns" property [19]: the marginal gain from adding an element to a set A is at least as high as the marginal gain form adding the same element to a superset of A. Formally, a submodular function satisfies:

$$f(A \cup \{v\}) - f(A) \geq f(T \cup \{v\}) - f(T)$$

for all elements v and all pairs of sets $A \subseteq T$. However, unlike influence maximization, PMDG is not monotone. In view of the above discussion, PMDG will be a consequence of the follows:

Theorem 1. *The profit function $R(A)$ is a non-negative, non-monotone and submodular function.*

Proof. (Non-negativity): In (1), the first term of $R(A)$ is always larger than the second one, since $A \subseteq \gamma(A)$. Thus, $R(A)$ will be a positive number.

(Submodularity): Observe that $\sum_{v \in \gamma(A)} P_v$ is a influence function and its submodularity has been proven in [1]. $(-\sum_{v \in A} P_v)$ is a linear function and it's a special submodular function. $R(A)$ can be written as $\sum_{i \in \gamma(A)} P_v + (-\sum_{v \in A} P_v)$ and it's a sum of two submodular function. A positive linear combination of submodular functions is still a submodular function [19]. Therefore, $R(A)$ remains submodularity.

(Non-monotonicity): Obviously, $R(\emptyset)$ and $R(V)$ are both zero. Without loss of generality, we assume there exists a node v who can influence at least one of her neighbors. For the set $\{v\}$ which only have one element, $\sum_{v \in \gamma(\{v\})} P_v$ is larger than $\sum_{v \in \{v\}} P_v$ and $R(\{v\})$ is positive. When adding the elements to the set A from null set to full set, $R(A)$ will be from zero to be positive and then from a positive number to zero again. Thus, Non-monotonicity holds.

This completes the proof.   □

Next, we show the hardness of PMDG.

Theorem 2. *The problem of profit maximization for digital goods is NP-Hard for both Independent Cascade model and Linear Threshold model.*

Proof (Independent Cascade model). Firstly, we give the proof in the condition of IC model. Consider an instance of the NP-complete *Set Cover* problem, defined by a collection of subset $\{S_1, S_2 \ldots S_k\}$ of a ground set $V = \{v_1, v_2 \ldots v_n\}$; the problem is to identify the smallest of sub-collection of S whose union equals the universe. This can be viewed as a special case of PMDG.

Given an arbitrary instance of Set Cover problem, we can model the problem as maximizing $\{|N| - |S|\}$, where $|N|$ is the cardinality of the ground set V and m is the cardinality of feasible solution. In [1], Kempe et al. propose the snapshot simulation based on "flip coins". The propagation of each node is determined in advance and the result can be viewed as a subgraph. So the subgraphs for each node are equivalent to sub-collection S in Set Cover problem. The reduced profit for free sample can viewed as the $|S|$ the cardinality of feasible solution in Set Cover problem. Thus we can reduce NP-Hard Set Cover problem to our problem.

(Linear Threshold model). In an instance of *Minimum Vertex Cover* problem, defined by an undirected n-node graph $G = (V, E)$, there exists a vertex set V' which is subset of V such that $uv \in E \Rightarrow u \in V' \vee v \in V'$. It means that the set will cover the edges of G. Next we will show that Minimum Vertex Cover problem can be viewed as a special case of our problem.

Given an instance of *Minimum Vertex Cover* problem and a graph G. Minimum Vertex Cover problem also can be modeled as optimization problem, max $\{|M| - |V'|\}$. $|M|$ is the number of edges in the graph. The special case of profit maximization problem can be regarded as using fewest free samples to influence all users. If we want to activated all user, all edges need to be activated. Thus, the problem can be modeled as $max \{|M| - |A|\}$. This matches the optimization objective of minimum vertex cover problem.

Combining the proofs of two settings gives Theorem 2.   □

4   Algorithms for PMDG

In this section, we attempt to find good strategies for our problem. What set A of seeds, should we initially give the samples for free so the expected profit is maximizing? It means that we want to find a set A which maximizes $R(A)$. As aforementioned, we have shown that our problem is *NP-Hard* and it's computational intractable, but PTAS is available. Though we do not compute the optimal set A, we compute an A that gives a good approximation. Unlike for influence maximization, the function is non-monotone and "unbudgeted", thus standard greedy hill-climbing strategy [19] is not applicable here. In [10], U-Greedy is proposed for approximation and it's a standard greedy algorithm with terminating condition. It achieves an approximation better than $(1 - 1/e) \cdot R(A^*) - \Theta(max\{|A_g|, |A^*|\})$, where A_g is set returned by U-Greedy and A^* is optimal solution. As for digital goods setting, since 10%–15% of nodes may be selected as seeds, the performance of U-Greedy is unstable and computational complexity is quite high. Motivated by this, we make use of the recently developed non-monotone submodular maximization technic in [12].

4.1   Approximation Algorithm

There are two algorithms in [12] to maximize the non-monotone submodular function, *Deterministic Algorithm* and *Randomized Algorithm*. The deterministic algorithm is a linear time method that can compute the seed set A and can achieve at least a $(1/3)$-fraction of optimal profit. Furthermore, the randomized algorithm can achieve an $(1/2)$ approximation and the result can also be achieved in a linear time. Now we will show how to apply those technics to our problem.

Algorithm 1. Deterministic Algorithm

Input: R, $G=(V,E)$
1: $i=0$
2: $X_0 \leftarrow \emptyset$ and $Y_0 \leftarrow V$
3: sort V in an arbitrary order $v_1, v_2, .., v_N$
4: **while** $i < |V|$ **do**
5:     $a_i \leftarrow R(X_{i-1} \cup \{v_i\}) - R(X_{i-1})$
6:     $b_i \leftarrow R(Y_{i-1} \backslash \{v_i\}) - R(Y_{i-1})$
7:     **if** $a_i \geq b_i$ **then**
8:         $X_i \leftarrow X_{i-1} \cup \{v_i\}$ and $Y_i \leftarrow \{Y_{i-1}\}$
9:     **else**
10:         $X_i \leftarrow \{X_{i-1}\}$ and $Y_i \leftarrow Y_{i-1} \backslash \{v_i\}$
11:     **end if**
12: **end while**
Output: X_n or Y_n

The deterministic algorithm is presented in Algorithm 1. Two sets are used in the algorithm. X is initially an empty set and Y equals the node set V. The node set V need to be sorted in an arbitrary order, $v_1, v_2, .., v_N$. The algorithm will proceed in N rounds, where N is the number of nodes in the graph. In each round i, the node v_i will be added to X_{i-1} or removed from Y_{i-1} by comparing the marginal contribution of node v_i for X_{i-1} and Y_{i-1}. At the end of the iterations, X_n equals Y_n and the two set are both solutions for our problem. Obviously, Algorithm 1 runs in a linear time.

The Algorithm 2 implements the randomized algorithm. The initial setting of Algorithm 2 is the same as Algorithm 1. The adding or removing decision of v_i for Algorithm 1 is deterministic, but Algorithm 2 makes a decision with uncertainty, which is also based on the a_i and b_i. However, it provides a guarantee of $(1/2)$ approximation to our problem.

Complexity Analysis. The time complexity of standard greedy algorithm for influence maximization is $O(N^2 M)$, where N is the number of nodes in the graph and M is the number of edges. In contrast, the complexities of both deterministic algorithm and randomized algorithm are $O(NM)$. In addition, both methods provide a tight approximation guarantee [12].

Algorithm 2. Randomized Algorithm

Input: R, $G=(V,E)$
1: $i=0$
2: $X_0 \leftarrow \emptyset$ and $Y_0 \leftarrow V$
3: sort V in an arbitrary order $v_1, v_2, .., v_N$
4: **while** $i < |V|$ **do**
5:     $a_i \leftarrow R(X_{i-1} \cup \{v_i\}) - R(X_{i-1})$
6:     $b_i \leftarrow R(Y_{i-1} \backslash \{v_i\}) - R(Y_{i-1})$
7:     $a_i' \leftarrow max\{a_i, 0\}$
8:     $b_i' \leftarrow max\{b_i, 0\}$
9:     with probability $a_i'/(a_i' + b_i')$ **do:**
10:     $X_i \leftarrow X_{i-1} \cup \{v_i\}$ and $Y_i \leftarrow \{Y_{i-1}\}$
11:     else with compliment probability **do:**
12:     $X_i \leftarrow \{X_{i-1}\}$ and $Y_i \leftarrow Y_{i-1} \backslash \{v_i\}$
13: **end while**
Output: X_n or Y_n

4.2   Profit Discount

Inspired by [8], we propose a heuristic named *Profit Discount* for IC model. Although the algorithms discussed in Sect. 4.1 improve the efficiency dramatically, they're still suffered from heavy Monte-Carlo simulation. Thus, we are motivated to propose a fast heuristic based on degree. In sociology literature, degree and other centrality-based heuristics are commonly used to estimate the influence nodes in social networks [8,20]. In the setting of our problem, we regard the degree as the expected profit gained by single point. The expected profit achieved by single point v is measured by $\sum_{u \in Nei(v)} Q_{vu} \cdot P_u$, where $Nei(v)$ means neighbors of v. P_u and Q_{vu} mean the expected profit from u and influencing probability in IC model.

Algorithm 3. Profit Discount

Input: $A=\emptyset, G=(V,E), R$
1: $i = 0$
2: **while** *true* **do**
3:     **while** *each node v in $V \backslash A$* **do**
4:         compute *expected profit $D(v)$ of v*
5:     **end while**
6:     $u_i = argmax_{v \in V \backslash A} D(v)$
7:     **if** $R(A) - R(A \cup \{u_i\}) \geq 0$ **then**
8:         $A = A \cup \{u_i\}$
9:     **else**
10:         *break*
11:     **end if**
12: **end while**
Output: A

We adopt the thought of "discount" to our problem, which implies that the influence of v's seed neighbor should be taken into consideration when deciding whether to select v as a seed. Firstly, we assume the probability of propagation is small, thus the indirect influence of v will be neglected. Because v is a neighbor of seed u, with probability at least Q_{uv}, v will influenced by u. Thus, at probability Q_{uv}, we do not need to take v into the initial set. Furthermore, selecting the node v will reduce the expected profit of its neighbor in the seed set. The heuristic will select the node with maximum expected profit in each round until the total expected profit begins decreasing. This is the main idea of the profit discount heuristic.

Let $Nei_a(v)$ be the set of v's seed neighbors. The expected profit earned by node v, donated as $D(v)$, in IC model is defined as:

$$D(v) = \prod_{t \in Nei_a(v)} (1 - Q_{tv}) \cdot \sum_{u \in Nei(v) \backslash Nei_a(v)} (P_u \cdot Q_{vu}) - \sum_{t \in Nei_a(v)} (P_v \cdot Q_{tv}) \quad (2)$$

where $\prod_{t \in Nei_a(v)} (1 - Q_{tv})$ means that probability that the node v will not be influenced by any of the already selected seeds. $\sum_{u \in Nei(v) \backslash Nei_a(v)} (P_u \cdot Q_{vu})$ are the profit earned by v from its neighbors which is not seeds. The last term in (2) presents the reduced expected profit when free sample is provided to v. The selecting procedure for seeds will terminate when the total profit begin to decrease.

The profit discount heuristic is formulized in Algorithm 3. The speed of the algorithm is much faster than U-Greedy, deterministic algorithm and randomized algorithm, since mass Monte-Carlo simulations are unnecessary. In Sect. 5, we will show that the performance of profit discount heuristic approaches to the results achieved by U-Greedy.

5   Empirical Evaluations

We conduct the experiments on three real-life networks to evaluate the proposed model and algorithms. All algorithms are implemented in C++ and experiments are conducted on a server with 1.8 GHz eight-core Intel E5-2428 CPU, 32 GB for memory, and running Windows Server 2010 operating system. CELF algorithm is applied to accelerating the U-Greedy. We run 10,000 simulations and take the average of profit achieved. In order to reduce the uncertainty of propagation and obtain accuracy in IC model, the technic of snapshot is adopted in simulations.

5.1   Datasets

We use collaboration network datasets from real-life for empirical valuation. Three networks are undirected graphs. The number of vertices, edges and the other detail information are summarized in Table 2. Detailed description of datasets is as follows.

- **ca-HepPh.** This data set is from the e-print Arxiv High Energy Physics category. If the author v and u published a paper together, the network contains an edge connecting nodes v to u.

- **ca-CondMat.** The network is from the e-print Arxiv and covers scientific collaborations between authors' papers submitted to Condense Matter category. If v and u co-authored a paper, the edge of v and u will be built.
- **ca-AstroPh.** This data set is a collaboration network of Arxiv Astro Physics category. If v and u are co-author of a paper, the network will contain an edge connecting node v to u.

Table 2. Statistic of network data

Name	Node	Edge	Average degree	Maximum degree
ca-HepPh	15233	58831	7.7	314
ca-CondMat	23133	93497	8.1	560
ca-AstroPh	18772	198110	21.1	504

5.2   Parameters Sets

- **Independent Cascade Model.** The probability of each edge is selected uniformly at random from {0.01, 0.03, 0.05}.
- **Linear Threshold Model.** The threshold for every node is uniformly and randomly chosen from 0 to 1. For trivalency, the edge weight is selected uniformly at random from {0.03, 0.06, 0.09}.
- **Profit Distribution.** Following [10], we set the profit subject to normal distribution $N(\mu, \sigma^2)$ with $\mu = 0.53$ and $\sigma = 0.13$.

5.3   Baseline Methods

We compare our applied algorithms with three algorithms in both IC and LT models on three networks.

- **U-Greedy.** This is a modified standard greedy algorithm. In each round, the maximum of marginal contributor will be selected. When the total profit begins decreasing, the algorithm will terminate.
- **Random.** As a baseline comparison, simply select a random user in the graph. Terminate when the profit achieved is less than last round.
- **Degree.** As a comparison, a simple heuristic that selects a node with the maximum degree will terminate, when the total profit begin to decrease.

5.4   Experiments Results

We will show the decreasing curve of profit achieved by the U-Greedy and profit discount algorithm when number of seeds grows larger, though both algorithms will terminate when the profit starts decreasing. Deterministic algorithm and randomized algorithm just return the seed set and the selecting procedure is not progressive thus two straights are used to present the results of deterministic algorithm and randomized algorithm.

Discussion on Results. The results are shown in Figs. 1, 2, 3, 4, 5 and 6. From the figures, we can observe that our adopted algorithms perform well in all settings. The profit achieved by the heuristic approaches the performance of U-Greedy. From Table 2, we could know that the connectivity of ca-AtroPh is stronger than the others. Thus, the relative size of solutions for ca-AtroPh is smaller than the two others'. The relative size of the optimal solution set A is inversely proportional to the connectivity.

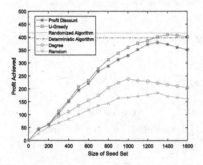

Fig. 1. Profit on ca-HepPh(IC)

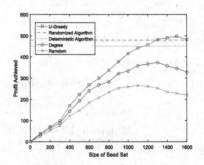

Fig. 2. Profit on ca-HepPh(LT)

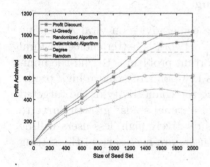

Fig. 3. Profit on ca-CondMat(IC)

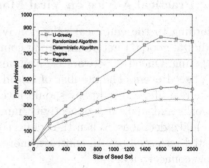

Fig. 4. Profit on ca-CondMat(LT)

Running Time. The Table 2 shows the running time of each algorithm in the network datasets. The speed of profit discount heuristic in IC model is impressive. It takes just few minutes and achieves a reasonable results. The two approximation algorithms are also faster than the modified standard greedy algorithm. Both algorithms finish in a linear time (Table 3).

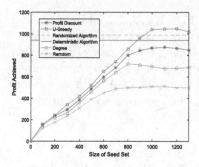

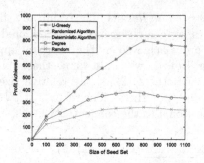

Fig. 5. Profit on ca-AstroPh(IC)   **Fig. 6.** Profit on ca-AstroPh(LT)

Table 3. Running time in hours

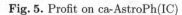

	ca-HepPh		ca-CondMat		ca-AstroPh	
	IC model	LT model	IC model	LT model	IC model	LT model
Profit discount	0.1	-	0.1	-	0.1	-
Deterministic	5.4	6.1	8.4	9.2	7.8	8.7
Randomized	6.1	4.7	7.3	9.0	6.4	7.9
U-Greedy	18.8	22.7	28.2	37.5	26.7	36.7

5.5 Practical Advice on Viral Marketing

According to the empirical valuation, we give some advice on viral marketing for digital goods. First of all, strong connected network is better choice for viral marketing. Once the network is given, an important problem is to find the number of seed users. The number of initial set is key point to achieve high profit. In the real life, the company should investigate the connectivity of the targeted networks and make a "connectivity/number of optimal seeds" growth curve to help make decisions. When the connectivity is relatively high, less users will be selected into seed set. In contrast, more seed users are necessary in the setting of low connectivity.

6 Conclusion

In this paper, we make a study of the profit maximization problem for digital goods in social networks and extend the influence maximization problem to PMDG in both IC and LT model. We also show the properties of PMDG, such as submodularity and non-monotonicity. In addition, famous approximation algorithms are applied to PMDG. We also propose a highly efficient heuristic. Our experimental results show that our applied algorithms outperform the baseline in both efficiency and effectiveness.

As for future work, *mechanism design* is one popular topic for both fields of theoretical computer science and economics [17,18]. How to let agents tell the truth is the focus of the research. In [15], the technic of algorithmic game theory is applied to the problem of social network analysis. It is interesting to research whether the algorithmic game theory can be applied to our setting for self-interested users.

Influence maximization in heterogeneous social networks is also an interesting topic in social network analysis [13,14]. Another future direction is for this research. We plan to introduce the concept of money into heterogeneous social networks and study the viral marketing in multi-platform setting.

Acknowledgments. This paper is supported by the National Key Research and Development Program of China (Grant No. 2016YFB1001102) and the National Natural Science Foundation of China (Grant Nos. 61375069, 61403156, 61502227), this research is supported by the Collaborative Innovation Center of Novel Software Technology and Industrialization, Nanjing University.

References

1. Kempe, D., Kleinberg, J., Tardos, E.: Maximizing the spread of influence through a social network. In: International Conference on Knowledge Discovery and Data Mining, pp. 137–146 (2003)
2. Goldberg, A.V., Hartline, J.D., Wright, A.: Competitive auctions and digital goods. In: Twelfth ACM-SIAM Symposium on Discrete Algorithms. pp. 735–744 (2001)
3. Chen, N., Gravin, N., Lu, P.: Optimal competitive auctions. In: ACM Symposium on Theory of Computing, pp. 253–262 (2014)
4. Lu, W.X., Zhang, P., Zhou, C., Liu, C., Gao, L.: Influence maximization in big networks: an incremental algorithm for streaming subgraph influence spread estimation. In: International Joint Conference on Artificial Intelligence, pp. 2076–2082 (2015)
5. Leskovec, J., Krause, A., Guestrin, C., Faloutsos, C., Vanbriesen, J.M., Glance, N.: Cost-effective outbreak detection in networks. In: Knowledge Discovery and Data Mining, pp. 420–429 (2007)
6. Domingos, P., Richardson, M.: Mining the network value of customers. In: Knowledge Discovery and Data Mining, pp. 57–66 (2001)
7. Alaei, S., Malekian, A., Srinivasan, A.: On random sampling auctions for digital goods. Electronic Commerce, pp. 187–196 (2009)
8. Chen, W., Wang, Y., Yang, S.: Efficient influence maximization in social networks. In: Knowledge Discovery and Data Mining, pp. 199–208 (2009)
9. Hartline, J.D., Mirrokni, V., Sundararajan, M.: Optimal marketing strategies over social networks. In: International World Wide Web Conferences, pp. 189–298 (2008)
10. Lu, W., Lakshmanan, L.V.: Profit maximization over social networks. In: International Conference on Data Mining, pp. 189–298 (2008)
11. Feige, U., Mirrokni, V., Vondrak, J.: Maximizing non-monotone submodular functions. In: Foundations of Computer Science, pp. 461–471 (2007)
12. Buchbinder, N., Feldman, M., Naor, J., Schwartz, R.: A tight linear time (1/2)-approximation for unconstrained submodular maximization. In: Foundations of Computer Science, pp. 649–658 (2012)

13. Wang, Y., Huang, H., Feng, C., Yang, X.: A co-ranking framework to select optimal seed set for influence maximization in heterogeneous network. In: Cheng, R., Cui, B., Zhang, Z., Cai, R., Xu, J. (eds.) APWeb 2015. LNCS, vol. 9313, pp. 141–153. Springer, Cham (2015). doi:10.1007/978-3-319-25255-1_12
14. Zhan, Q., Yang, H., Wang, C., Xie, J. A solution to influence maximization problem under cost control. In: International Conference on Tools with Artificial Intelligence, pp. 849–856 (2013)
15. Singer, Y.: How to win friends and influence people, truthfully: influence maximization mechanisms for social networks. In: Web Search and Data Mining, pp. 733–742 (2012)
16. Granovetter, M.: Threshold models of collective behavior. Am. J. Sociol. **83**, 1420–1443 (2015)
17. Zhang, L., Chen, H., Wu, J., Wang, C., Xie, J.: False-name-proof mechanisms for path auctions in social networks. In: European Conference on Artificial Intelligence, pp. 323–332 (2016)
18. Nisan, N., Roughgarden, T., Tardos, E., Vazirani, V.V.: Algorithmic Game Theory. Cambridge University Press, Cambridge (2007)
19. Nemhauser, G., Wolsey, L.A., Fisher, M.L.: An analysis of approximations for maximizing submodular set functions—I. Math. Program. **14**, 265–294 (1978)
20. Wasserman, S., Faust, K.: Social Network Analysis : Methods and Applications. Cambridge University Press, Cambridge (1994)

Change Detection from Media Sharing Community

Naoki Kito, Xiangmin Zhou$^{(\boxtimes)}$, Dong Qin, Yongli Ren, Xiuzhen Zhang,
and James Thom

School of Science, RMIT University, Melbourne, Australia
naokikito1@gmail.com,
{Xiangmin.zhou,Dong.qin,Yongli.ren,Xiuzhen.zhang,James.Thom}@rmit.edu.au

Abstract. This paper investigates how social images and image change detection techniques can be applied to identify the damages caused by natural disasters for disaster assessment. We propose a framework that takes advantages of near duplicate image detection and robust boundary matching for the change detection in disasters. First we perform the near duplicate detection by local interest point-based matching over image pairs. Then, we propose a novel boundary representation model called *relative position annulus* (RPA), which is robust to boundary rotation, location shift and editing operations. A new RPA matching method is proposed by extending *dynamic time wrapping* (DTW) from time to position annulus. We have done extensive experiments to evaluate the high effectiveness and efficiency of our approach.

Keywords: Change detection · Social images · Disaster assessment

1 Introduction

Before, during and after the natural disasters, information about the damages and the current situations are vital for people to make decisions for their next actions. For example, the Nepal earthquake in 2015 did a huge damage to everything, including buildings, roads, infrastructure, and people, which resulted in 8,019 people died and 17,866 people injured; in a Tokyo earthquake, people are still able to walk back home since the earthquake damage the infrastructure, but not the buildings and paths. A recent study [1] found that people would like to know earthquake size and epicentre. Moreover, knowing these information could prevent the secondary and tertiary disasters. It is also found that in Tokyo, only 67.8% of people managed to get back their house on the day of an earthquake, and the rest 32.2% had to become *"homeless"*, among them 2% failed to going back home only because they could not find the safe path. People want the information about earthquakes, but there is always the question "How could the people get the information of earthquakes?".

We focus on the problem of *change detection* in sharing communities. In image processing, *change* is defined as the difference between two pixels or the objects in different images. The *difference* varies in different situations. In this

© Springer International Publishing AG 2017
L. Chen et al. (Eds.): APWeb-WAIM 2017, Part I, LNCS 10366, pp. 391–398, 2017.
DOI: 10.1007/978-3-319-63579-8_30

paper, the *difference* is limited to the damages to the roads, the buildings or the infrastructures, which are caused by natural disasters. For instance, when a bridge breaks down during an earthquake, the damage to the bridge will be the *change* in this situation. *Change detection* is a process of finding the damages which have been caused by certain natural disasters such as earthquake and flood. Recently, many researchers had approached to detect the damages caused by natural disasters by using the change detection techniques with the aerial images. However, the aerial images consume longer times to retrieve and harder to get compare to the other images. For another, social media is pervasive, and updates very quickly especially on large events, e.g. natural disasters, by millions of people all the time. Thus, we investigate the problem of change detection from social images, so as to let the public aware of the latest situations of natural disasters on the spot. One of the challenges here is the large-scale of social images, which makes current change detection techniques infeasible if not impossible. One of the limitation of using the large-scale images with current change detection techniques is time cost. As existing techniques detect changes based on pixel level comparison [2] without index support or query optimization or both, the time cost for image comparison is high. The second challenge is the unavailability of some special features, like building shady, used in traditional change detection [3]. In shared communities, most social images do not have shady, thus the shady-based matching cannot be conducted. Forcing the existing techniques on the social images will cause low detection quality. Finally, traditional change detection for aerial images suppose the image pairs are known to the same location points, which is not true in media shared communities.

To address these issues, we propose a framework for change detections in sharing communities. First, we conduct PCA-SIFT based matching, which determines if two images are really referring to the same source. Then, we propose a novel boundary representation called *relative position annulus* (RPA), which is robust to the view point rotation and other global transformations of the same objects in images. RPA-based boundary matching between images is performed by extending DTW from series to annulus, which decides if a change has happened after disaster. The rest of the paper is structured as follows: Sect. 2 reviews the related work followed by the proposed change detection method; Sect. 4 includes the experiment evaluation; Sect. 5 concludes the paper.

2   Related Work

We review the related research, including the image copy detection for same source media [18] and change detection. Image copy detection is done by first extracting the descriptors of key points in each image, and counting the number of matched pairs between two compared ones. Examples on descriptors include SIFT [4], PCA-SIFT [5], SURF [6], GLOH [7], and Eff^2 [8] etc. In [4], the SIFT is extracted by scale-space extrema selection, keypoint localization, orientation assignment, and descriptor computation. A 128-d vector is computed for each local point. SIFT is invariant to the image variations. However, the matching over it can be expensive due to its high dimensionality. To improve the efficiency, SIFT variants was proposed [5–8]. PCA-SIFT applies principal component analysis to the normalized gradient patches, and is obtained by projecting the gradient image vector computed for each patch into a 36-d space [5]. In [6], SURF was proposed based on the Hessian matrix to approximate the previous descriptor as a vector of length 64. GLOH extended the SIFT by varying the location grid and using PCA to reduce the size [7]. In [8], Eff^2 was proposed by detecting the key points using Difference of Gaussian over different scales, and extracting the information of 8 orientation buckets over each of 3 grid cells around the point. This generates a 72-d vector for each key point. As PCA-SIFT has the stable performance in all situations and has lowest dimensionality [9], we select it in our image detection. To match the descriptor sets, there are mainly two methods: one-many matching and one to one symmetric matching (OOS). In [4], the similarity of two sets is measured by identifying the nearest neighbor of each descriptor based on L_2 distance, and calculating the number of their matched key point pairs. Using this approach, matches over noise key points can be introduced. In [10], Zhao et al. proposed OOS based on a cosine distance-based partial similarity. One key point in a query can only be matched with a single key point of an image data, so the noise matches can be excluded. Considering the superiority of OOS, we choose it for our descriptor similarity.

Object-based change detection (OBCD) algorithms were proposed for satellites, remote sensing and the SAR images to detect the damage and geographical changes from them. In [11], Murakami et al. identified changes by subtracting the digital surface model (DSM) from another DSM. In [3], Turker et al. used the watershed segmentation to create the segmented building vectors and calculate the shadow area of the segmented building to detect the damages. In [12], Matikainen et al. use object-based GIS model data with the overlap analysis algorithm change detection. Gong et al. [13] used VHR Terra SAR-X for finding the changes during earthquake. In [14], Tu et al. use the 3D GIS model image to detect the building damages during Beichuan earthquake. The 3D GIS model extracts the vectors of building images, and the height of a building is estimated using the shadow detection. Using the satellite, remote and SAR images with damage detection algorithms can achieve high accuracy. Though these images can cover the urban area, there is an issue of retrieving pre-event data sets. Taking all the imagery data for whole country or land is always hard to achieve. Media sharing communities provide great sources for capturing disasters during crisis. Thus, it is demanded to conduct the single pre- and post-event image detection in social communities for disaster management. However, existing methods use typical features that can not be obtained in images from public uploading.

3   Change Detection over Large Data Sets

We present the change detection over large-scale social images, including how to identify the image pairs from the same source and the changes in images.

3.1   Image Copy Detection

We deploy a PCA-SIFT-based matching over images in dataset, which will be used to decide if two images refer to the same objects by finding the similarity between objects. The main objects are the buildings, statues and other objects which are not moving. Also, the unrelated images selected by the PCA-SIFT algorithm will be removed. The kept image pairs are passed to change detection stage for assessing if any damages had happened in a disaster. We apply the OOS to the similarity between the local descriptor sets of images [15,16]. Given two descriptors, the similarity of them is measured by the *Cosine* similarity between these two vectors. For two key points from two images, they are match pair candidate if their similarity is bigger than a threshold value. OOS further check if any one of these two points is the nearest neighbor of the other among all the descriptors of its image. If they are nearest neighbors of each other, they are a real matched pair. The final similarity between two images is calculated by the average similarity of all their matched pairs. To improve the key points matching, we use the LIP-IS index [15] as well. All these ensure that effective and efficient PCA-SIFT-based matching is performed.

3.2   Change Detection

Change detection assesses the damages caused during disasters. Intuitively, two pictures to the same location point contains same buildings or other objects, each is described as a boundary. If there is no damage happened at this place, the object boundaries in two images match. Otherwise, if there exists boundary missing or unmatched from the before image to the after one, the damages could have been caused in the disaster. This section proposes a new image object boundary modelling together with a novel boundary matching, which are robust to image transformations, rotations or editing, for effective change detection.

Model Image Boundary: Existing works model building shadow area boundaries [3], boundary shapes [12] using the coordinates of each pixel falling on the boundary of objects in an image. However, these boundary modelling incurs low effectiveness of detection for social images due to the possible object changes with respect to the viewpoints, rotations, space shift etc. To address this issue, we propose a robust boundary representation, called *relative position annulus* (RPA), which describe each boundary as an annulus of the difference of neighboring edge lengths. Specifically, we exploit sober edge detector to detect a number of boundaries in an image, since it can detect the emphasising edges while reduce the effect of noise edges. Given a boundary consisting of m vertexes $\{v_1, \ldots, v_m\}$, we represent the boundary as $\{d(v_1, v_2) - d(v_2, v_3), \ldots, d(v_{m-2}, v_{m-1}) - d(v_{m-1} - v_m), d(v_{m-1}, v_m) - d(v_m, v_1)\}$, where any element can be the start point while other points are ordered clockwise. As such, the RPA will be robust to object rotation, viewpoint change and space shift in social images.

Matching Boundaries: As an image may contain multiple objects, each image is described as a set of relative position annulus of multiple boundaries. To assess the damages in a disaster, we need to do two steps matching: (1) the measure between two RPAs; (2) the measure between two images. To further reduce the influence of small noise objects, we only use the top T biggest boundaries for boundary comparison between two images. Given two RPAs, $Q :< q_1, \ldots, q_m >$ and $\mathcal{D} :< v_1, \ldots, v_n >$, we measure their similarity by extending the DTW [19] for RPAs. We consider Q as a series, and D as a set of n series, where each series in the set takes v_i $(i = 1, \ldots, n)$ as the start point of the boundary and the remaining ones are ordered clockwise. Denote the series to v_i as $\mathcal{D}_i$, and its elements as $< v_1^i, \ldots v_n^i >$, where $v_j^i = v_{((i+j-1) \pmod n)}$. Then the similarity between Q and $\mathcal{D}_i$ is measured by:

$$SRPA_i(Q, \mathcal{D}_i) = \begin{cases} 0 & m = m_1 - 1 \; or \; n = n_1 - 1 \\ max\{SRPA_i(Q_{m-1}, \mathcal{D}_{n-1}) + Sim(q_m, v_n^i), \\ \quad SRPA_i(Q_m, \mathcal{D}_{n-1}), SRPA_i(\mathcal{D}_{m-1}, \mathcal{D}_n)\} & otherwise \end{cases}$$
(1)

where Sim is the similarity between q_m and v_n^i computed based on L_1 distance: $Sim(q_m, v_n^i) = 1/(1 + |q_m - v_n^i|)$. The final boundary distance is the maximal DTW between Q and $\mathcal{D}_i$

$$SRPA = \max_{i=1}^{n} SRPA_i.$$
(2)

4   Experiment

This section examines the effectiveness and efficiency of the proposed method.

4.1   Experimental Setup

We use the dataset collected from Flickr by focusing on images relevant to the *Nepal earthquake*, which is also known as *Gorkha earthquake*, in 2015. 100,000 images are collected, which include images *before* and *after* the earthquake. We label the first set of ground-truth image pairs, denoted G_1, for near-duplicate image pair detection (first component of our system), and the second set of ground-truth image pairs, denoted G_2, for change detection (second component of our system). The ground-truth is manually identified by the authors of this paper via careful comparison of all *before* and *after* images. Specifically, for a *ground-truth* in G_1, both *before* and *after* images have to contain at least one same building, but the corresponding image contents, angles, resolutions, colours and light effects could be different. For a *ground-truth* in G_2, there is at least one damaged building found in *after* image comparing with *before* image. We conduct the effectiveness evaluation on 10,000 after and before the earthquake images from the whole dataset, and the efficiency tests on the whole dataset.

To evaluate the effectiveness of algorithms, we used two metrics in [17], the probability of miss detection and false alarm (*Pmiss* and *Pfa*). Specifically, the *missed detection* means that the algorithm fails to detect the ground-truth, and the *false alarm* means the detection of non-target pairs. *Pmiss* and *Pfa* are defined as follows: $Pmiss = \frac{number\ of\ missed\ detections}{number\ of\ ground\ truth}$, $Pfa = \frac{number\ of\ false\ alarms}{number\ of\ non-targets}$.

We evaluate the efficiency of our approach in terms of the overall time cost of near duplicate image detection, and that of change detection. We compare the time cost of different change detection approaches. Experiments are conducted on Window 7 with Intel (R) Core (TM) i7-4770S CPU (3.4 GHz)and 8 GB RAM.

4.2   Experimental Evaluation

We first test the effect of parameters in change detection, the boundary similarity threshold τ in SRPA and the boundary number threshold T. We then compare the proposed approach with the shape-based change detection ($SBCD$) [12].

Effect of τ: We test the effect of τ in our SRPA measure by varying it from 0.1 to 1. For each τ, we test the $Pmiss$ and Pfa values by setting the number of boundaries considered in each image T to 20, 30, 40, 50. Figure 1(a) and (b) show the results. It is observed that the $Pmiss$ increases slowly with the increase of τ upto 0.4, then increases dramatically with the further increase of τ after 0.4. On the other hand, the Pfa drops sharply with τ increasing upto 0.4, then decreases slightly after the further increasing of τ. This is because a bigger τ will force a strict constraint on change determination, thus more changes are missed while less false alarms are introduced. Considering a good balance between $Pmiss$ and Pfa, we set the default value of τ to 0.4.

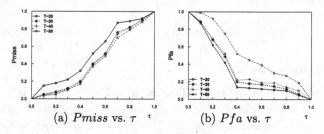

(a) $Pmiss$ vs. τ (b) Pfa vs. τ

Fig. 1. (a) $Pmiss$ vs. τ (b) Pfa vs. τ

Effect of T: We set τ to it default value and test the effect of T on the effectiveness of change detection by varying T from 5 to 50. Figure 2 (a) and (b) report the results. Clearly, the $Pmiss$ of the detection drops with T increasing to 25, and keeps steady after that. Meanwhile, its Pfa keeps steady with the increase of T to 25, and increases after 25. To balance the $Pmiss$ and Pfa, we set the default value of T to 25.

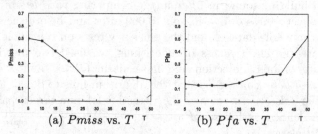

(a) $Pmiss$ vs. T (b) Pfa vs. T

Fig. 2. (a) $Pmiss$ vs. T (b) Pfa vs. T

Comparison of Change Detection Techniques: We compare the effectiveness of our SRPA approach with existing shape-based change detection ($SBCD$) [12]. We conduct change detection over the results returned in the first step with the $k = 50$ to avoid missed near duplicate image pairs. We follow all the parameter settings for shape-based matching in [12], and set the parameters τ and T in our SRPA matching to their default values. We test the $Pmiss$ and Pfa values of two approaches. Table 1 reports the results. Clearly our SRPA achieves much lower $Pmiss$ and Pfa, which indicates much better effectiveness, because our RPA is more robust to object variations than the shape representation.

Table 1. Comparison

	SRPA	SBCD
$Pmiss$	0.14	0.22
Pfa	0.11	0.52

We compare the time cost of our SRPA matching and existing SBCD matching over the near duplicates detected by varying the data size from 10,000 to 100,000. As shown in Fig. 3, our SRPA takes lower time cost than SBCD. This is because SBCD adopts RAP that is a one dimensional annulus. SBCD uses the coordinates of each pixel on the boundary, each is a 2-digital pair, thus more complex than RAP. Compared with SBCD, our SRPA achieves high effectiveness and efficiency, which has proved the superiority of our approach.

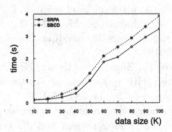

Fig. 3. Comparison

5   Conclusion

In this paper, we study the problem of change detection from media sharing communities for damage assessment in natural disasters. First, we exploit near duplicate detection technique to find each image pair from the same source. Then, we propose a robust boundary representation model together with the matching over it for effective change detection for damage assessment. Finally, we have conducted extensive experiments to evaluate the effectiveness and efficiency of our proposed change detection framework. The experimental results have proved the superiority of our approach over the state-of-the-art method.

References

1. Naoya, H.U.S., Ryota, N., Shuntaro, W., Hidenori, H.: Questionnaire survey concerning stranded commuters in metropolitan area in the east Japan great earthquake. J. Soc. Saf. Sci. 343–353 (2011)
2. Ilsever, M., Unsalan, C.: Two-Dimensional Change Detection Methods: Remote Sensing Applications. SpringerBriefs in Computer Science. Springer, London (2012)
3. Turker, M., Sumer, E.: Building-based damage detection due to earthquake using the watershed segmentation of the post-event aerial images. Int. J. Remote Sens. 29(11), 3073–3089 (2008)
4. Lowe, D.G.: Object recognition from local scale-invariant features. In: ICCV , vol. 2, pp. 1150–1157 (1999)
5. Ke, Y., Sukthankar, R.: PCA-SIFT: a more distinctive representation for local image descriptors. In: CVPR, vol. 2, pp. II-506-II-513 (2004)
6. Bay, H., Tuytelaars, T., Gool, L.: SURF: speeded up robust features. In: Leonardis, A., Bischof, H., Pinz, A. (eds.) ECCV 2006 Part I. LNCS, vol. 3951, pp. 404–417. Springer, Heidelberg (2006). doi:10.1007/11744023_32
7. Mikolajczyk, K., Schmid, C.: A performance evaluation of local descriptors. IEEE Trans. PAMI 27(10), 1615–1630 (2005)
8. Lejsek, H., Ásmundsson, F.H., Jónsson, B.T., Amsaleg, L.: Scalability of local image descriptors: a comparative study. In: MM, pp. 589–598 (2006)
9. Juan, L., Gwun, O.: A comparison of SIFT, PCA-SIFT and SURF. IJIP 3(4), 143–152 (2009)
10. Zhao, W.-L., Ngo, C.-W., Tan, H.-K., Wu, X.: Near-duplicate keyframe identification with interest point matching and pattern learning. IEEE Trans. MM 9, 1037–1048 (2007)
11. Murakami, H., Nakagawa, K., Hasegawa, H., Shibata, T., Iwanami, E.: Change detection of buildings using an airborne laser scanner. ISPRS 54(2), 148–152 (1999)
12. Matikainen, L., Hyypp, J., Ahokas, E., Markelin, L., Kaartinen, H.: Automatic detection of buildings and changes in buildings for updating of maps. Remote Sens. 2(5), 1217–1248 (2010)
13. Gong, L., Wang, C., Wu, F., Zhang, J., Zhang, H., Li, Q.: Earthquake-induced building damage detection with post-event sub-meter VHR TerraSAR-X staring spotlight imagery. Remote Sens. 8(11), 887 (2016)
14. Tu, J., Sui, H., Feng, W., Song, Z.: Automatic building damage detection method using high-resolution remote sensing images and 3D GIS model. ISPRS Ann. 3, 43–50 (2016)
15. Zhao, W., Ngo, C., Tan, H., Wu, X.: Near-duplicate keyframe identification with interest point matching and pattern learning. IEEE Trans. MM 9(5), 1037–1048 (2007)
16. Zhou, X., Zhou, X., Chen, L., Bouguettaya, A., Xiao, N., Taylor, J.A.: An efficient near-duplicate video shot detection method using shot-based interest points. IEEE Trans. MM 11(5), 879–891 (2009)
17. Zhou, X., Chen, L.: Event detection over Twitter social media streams. VLDB J. 23(3), 381–400 (2014)
18. Zhou, X., Zhou, X., Chen, L., Shu, Y., Bouguettaya, A., Taylor, J.A.: Adaptive subspace symbolization for content-based video detection. IEEE Trans. Knowl. Data Eng. 22(10), 1372–1387 (2010)
19. Berndt, D.J., Clifford, J.: Using dynamic time warping to find patterns in time series. In: SIGKDD, AAAIWS 1994, pp. 359–370. AAAI Press (1994)

Measuring the Similarity of Nodes in Signed Social Networks with Positive and Negative Links

Tianchen Zhu, Zhaohui Peng[(⊠)], Xinghua Wang,
and Xiaoguang Hong

School of Computer Science and Technology,
Shandong University, Jinan, China
ztc@mail.sdu.edu.cn, {pzh,hxg}@sdu.edu.cn,
wang.xingh@foxmail.com

Abstract. Similarity measure in non-signed social networks has been extensively studied for decades. However, how to measure the similarity of two nodes in signed social networks remains an open problem. It is challenging to incorporate both positive and negative relationships simultaneously in signed social networks due to the opposite opinions implied by them. In this paper, we study the similarity measure problem in signed social networks. We propose a basic node similarity measure that can utilize both positive and negative relations in signed social networks by comparing the immediate neighbors of two objects. Moreover, we exploit the propagation of similarity in networks. Finally, we perform extensive experimental comparison of the proposed method against existing algorithms on real data set. Our experimental results show that our method outperforms other approaches.

Keywords: Similarity measure · Signed networks · Positive and negative links

1 Introduction

Measuring the similarity of nodes in signed social networks [1–3] is an important but still not fully explored problem due to the following challenges. First, the existence of negative links in signed networks challenges many concepts for unsigned networks. In addition, it is challenging to incorporate both positive and negative relationships simultaneously in signed networks. As shown in Fig. 1, users can express trust or distrust on others in Epinions.com in which "+" indicates trust and "−" indicates distrust.

We can see from Fig. 1, user v_1 and user v_3 have two common neighbors, user v_1 and user v_6 have two common neighbors as well. If we ignore the polarity of this social network, we can conclude that user v_3 and user v_6 are both similar to user v_1. However, when we take into account the polarities of the links, things go differently. From the Fig. 1, we can see that user v_1 and user v_3 have different evaluations for the same persons, while user v_1 and user v_6 have identical evaluations for the same persons. So we can deduce that user v_1 is more similar to user v_6, and user v_1 is more dissimilar to user v_3.

© Springer International Publishing AG 2017
L. Chen et al. (Eds.): APWeb-WAIM 2017, Part I, LNCS 10366, pp. 399–407, 2017.
DOI: 10.1007/978-3-319-63579-8_31

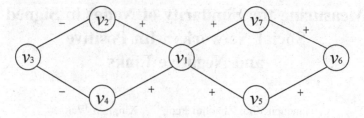

Fig. 1. Example of signed social network.

In this paper, we study the problem of measuring the similarity of nodes in signed social networks. Compared to existing approaches, our method considers the influence of negative relationships in addition to positive relationships. In particular, we define a basic similarity measure that captures effectively the proximity of two nodes in signed social networks. Besides, we also exploit the propagation of similarity in social networks. For example, considering the user v_3 and user v_6 in Fig. 1, these two users have no immediate neighbors in the signed social network, however, we can infer that user v_3 is dissimilar to user v_6 because user v_3 is dissimilar to user v_1 and user v_1 is similar to user v_6.

The major contributions of this paper are summarized as below.

- It investigates the similarity measure problem in signed social networks, a new but increasingly important issue due to the continuous development of social networks.
- It proposes a basic similarity measure method in signed social networks, which can utilize both positive and negative relationships.

The rest of this paper is organized as follows. Section 2 summarizes the related work, whereas Sect. 3 denotes a node similarity measure in signed social networks. Experimental results are given in Sect. 4. Finally, Sect. 5 concludes this paper.

2 Related Work

Unsigned social networks have been studied for decades [4–6]. There are many similarity measure method based on unsigned social networks. Traditional approaches to measure the similarity of nodes are mainly based on their feature values, such as Jaccard coefficient, Cosine similarity and Euclidean distance. However, these similarity measures are mainly focus on the nodes features, they do not consider link structures among objects.

The link based approaches measure the similarity of nodes defined on networks, such as shortest path algorithm, RWR (Random Walk with Restart) [7], SimRank [8], Personalized PageRank [9] and etc. Liben and Kleinberg [10] claimed that the identification of the shortest path between any pair of nodes in a graph can be used for friend recommendation. RWR iteratively explores the global structure of the network to estimate the proximity between two nodes. Starting at a node, the walker faces two choices at each step: either moving to a randomly chosen neighbor, or jumping back to the starting node. SimRank is a symmetric similarity measure that says "two objects are considered to be similar if they are referenced by similar objects". Personalized PageRank is an asymmetrical similarity measure that evaluates the probability starting from source nodes to target nodes by RWR. However, all above these methods do not consider the negative links, i.e., these studies are built on a fundamental assumption that all links in networks are positive.

FriendTNS [11, 12] discussed the similarity measure problem in signed social networks based on status theory. FriendTNS evaluates the similarity of two nodes in terms of their positive in-degree, negative out-degree, positive out-degree and negative in-degree. However, FriendTNS considers the degree of nodes only, it does not consider the relationship between two nodes. Besides, FriendTNS cannot apply in undirected signed social networks.

3   Node Similarity Measure

In this section, we define a basic node similarity measure to determine the proximity between a pair of nodes in signed social networks. For node v_i and node v_j, we define a specific function $sim(v_i, v_j)$ to express their corresponding similarity.

3.1   Basic Idea

To capture proximity between two nodes, we consider the example shown in Fig. 2. Figure 2(a) means user u_1 and user u_2 have many common friends, and Fig. 2(b) means that user u_1 and user u_2 have many common enemies. We can infer that u_1 and u_2 are likely to be similar, because they have the same evaluations on the same people. In contrast, Fig. 2(c) and (d) show that the evaluations of user u_1 and user u_2 on the same people are diametrically opposite. So, we can infer that u_1 and u_2 are most likely not similar.

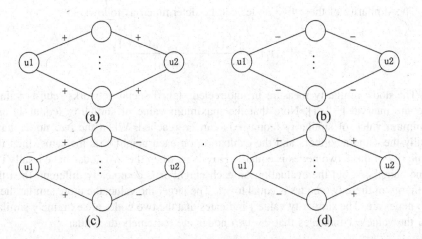

Fig. 2. Example of two neighbor connected users (nodes) in a signed social network.

This inspires us that if a pair of users will have high similarity in a signed social network, they should be satisfied the following condition:

- They have many common friends.
- They have many common enemies.

3.2 Similarity Measure in Undirected Signed Social Networks

Based on the above intuition, in an undirected signed social network, given a pair of nodes v_i and v_j, we can determine their similarity by considering their immediate neighbor sets. The followings are the immediate neighbor nodes, positive immediate neighbor nodes and negative immediate neighbor nodes of node v_i.

$$N_i = \{v_k|(v_i, v_k) \in \mathcal{E}\} + \{v_i\} \tag{1}$$

$$N_i^+ = \{v_k|(v_i, v_k) \in \mathcal{E}^+\} + \{v_i\} \tag{2}$$

$$N_i^- = \{v_k|(v_i, v_k) \in \mathcal{E}^-\} \tag{3}$$

We have discussed above that the more same evaluations on common neighbors two nodes have, the more similar they are. In contrast, the more contrary evaluations on common neighbors two nodes have, the more dissimilar they are.

Given a pair of nodes v_i and v_j, we can get the sets of nodes that they hold same evaluations and different evaluations.

$$C_s(v_i, v_j) = \left\{v_k \middle| \left(v_k \in N_i^+ \wedge v_k \in N_j^+\right) \vee (v_k \in N_i^- \wedge v_k \in N_j^-)\right\} \tag{4}$$

$$C_d(v_i, v_j) = \left\{v_k \middle| \left(v_k \in N_i^+ \wedge v_k \in N_j^-\right) \vee (v_k \in N_i^- \wedge v_k \in N_j^+)\right\} \tag{5}$$

The similarity of these two nodes can be determined as follows.

$$sim(v_i, v_j) = \frac{\left|C_s(v_i, v_j)\right| - \left|C_d(v_i, v_j)\right|}{|N_i \cup N_j|} \tag{6}$$

The node similarity measure in undirected signed social networks returns values into the interval $[-1, 1]$. Note that the maximum value of similarity (equal 1) and minimum value of similarity (equal -1) can be reached. When the two nodes have totally the same neighbors and the evaluation on each neighbor is the same, then the similarity of these two nodes is equal 1. However, when the two nodes have totally the same neighbors, but the evaluation on each neighbor is completely different, then the similarity of these two nodes is equal to -1. The larger the value, the more similar these two nodes are. The similarity value 1 indicates that the two nodes are extremely similar, and the value -1 indicates that the two nodes are extremely dissimilar.

Now, let us calculate some similarity values on the graph of Fig. 1 using Eq. 6. Considering the node v_1 and node v_3, $N_1 = \{v_1, v_2, v_4, v_5, v_7\}$, $N_3 = \{v_2, v_3, v_4\}$, $N_1^+ = N_1$, $N_1^- = \emptyset$, $N_3^+ = \{v_3\}$, $N_3^- = \{v_2, v_4\}$, $C_s(v_1, v_3) = \emptyset$, $C_d(v_1, v_3) = \{v_2, v_4\}$, so $sim(v_1, v_3) = \frac{0-2}{6} = -\frac{1}{3}$. Similarly, the similarity between node v_1 and v_6 is: $sim(v_1, v_6) = \frac{1}{3}$. Thus, the similarity score between nodes v_1, v_3 is less than that of v_1, v_6, and the result is also consistent with our intuitive experience.

3.3 Similarity Measure in Directed Signed Social Networks

In this section, we present how to measure similarity of a pair of nodes in a directed signed social network. In a directed social network, a user can express his attitudes to others, but can also receive the evaluations by other people. So, we separate neighbor set into four parts: (1) positive input neighbor set; (2) negative input neighbor set; (3) positive output neighbor set; (4) negative output neighbor set. Taking Fig. 3 as an example, we consider the similarity between node u_1 and node u_2.

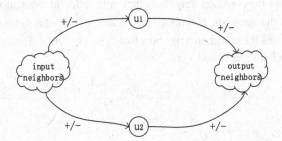

Fig. 3. Illustration of the intuition that measure similarity in a directed signed social network.

Based on the above intuition, u_1 and u_2 may be similar if:

- The evaluations on the two users that given by a lot of people are the same.
- The evaluations on a group of persons that given by the two users are the same.

Besides, if the two users hold different evaluations on many same persons and many people have different opinions about the two users, they may be dissimilar. It is an extension of the intuition we described in Sect. 3.1.

The followings are the positive input neighbor set, negative input neighbor set, positive output neighbor set and negative output neighbor set of node v_i in a directed signed social network.

$$I_i^+ = \{v_k | (v_k, v_i) \in \mathcal{E}^+\} + \{v_i\} \tag{7}$$

$$I_i^- = \{v_k | (v_k, v_i) \in \mathcal{E}^-\} \tag{8}$$

$$O_i^+ = \left\{ v_k | (v_i, v_k) \in \mathcal{E}^+ \right\} \tag{9}$$

$$O_i^- = \left\{ v_k | (v_i, v_k) \in \mathcal{E}^- \right\} \tag{10}$$

So, given a pair of nodes v_i and v_j in a directed signed social network,

$$C_s^I(v_i, v_j) = \left\{ v_k | \left(v_k \in I_i^+ \wedge v_k \in I_j^+ \right) \vee (v_k \in I_i^- \wedge v_k \in I_j^-) \right\} \tag{11}$$

$$C_d^I(v_i, v_j) = \left\{ v_k | \left(v_k \in I_i^+ \wedge v_k \in I_j^- \right) \vee (v_k \in I_i^- \wedge v_k \in I_j^+) \right\} \tag{12}$$

$$C_s^O(v_i, v_j) = \left\{ v_k | \left(v_k \in O_i^+ \wedge v_k \in O_j^+ \right) \vee (v_k \in O_i^- \wedge v_k \in O_j^-) \right\} \tag{13}$$

$$C_d^O(v_i, v_j) = \left\{ v_k | \left(v_k \in O_i^+ \wedge v_k \in O_j^- \right) \vee (v_k \in O_i^- \wedge v_k \in O_j^+) \right\} \tag{14}$$

where $C_s^I(v_i, v_j)$ means the common neighbors that hold the same evaluations on v_i and v_j, $C_d^I(v_i, v_j)$ means the common neighbors that hold different evaluations on v_i and v_j, $C_s^O(v_i, v_j)$ means the common neighbors that v_i and v_j hold the same evaluations on, and $C_d^O(v_i, v_j)$ means the common neighbors that v_i and v_j hold different evaluations on. So, the similarity between v_i and v_j is:

$$sim(v_i, v_j) = \frac{|C_s^I(v_i, v_j)| - |C_d^I(v_i, v_j)| + |C_s^O(v_i, v_j)| - |C_d^O(v_i, v_j)|}{|N_i \cup N_j|} \tag{15}$$

where $N_i = I_i \cup O_i$, $N_j = I_j \cup O_j$.

The basic similarity measure defined in directed signed social networks returns values into the interval $[-1, 1]$ as well. Interval $(0, 1]$ indicates two nodes are similar and interval $[-1, 0)$ indicates two nodes are dissimilar. Specifically, 0 is a special value that denotes neither similar nor dissimilar. The larger the similarity value, the more similar they are.

3.4   Propagation of Node Similarity

Based on the above discussion, the similarity values between all non-neighbor nodes in a graph G are zero. It is unreasonable. By propagating the similarity in the network, we can solve this problem. Notice that, we only propagate the similarity in the network in two hops according to the principle of balance theory [13, 14] that says "*the friend of my friend is my friend*", "*the enemy of my enemy is my friend*". That is to say, if two users have neither immediate common neighbors nor common related users (i.e. the similarity value determined by the basic similarity measure is not zero), the similarity between them is zero eventually. We define a extended similarity between two nodes v_i and v_j, denoted as $exsim(v_i, v_j)$.

$$exsim(v_i, v_j) = \begin{cases} sim(v_i, v_j), \text{if } v_i, v_j \text{ have immediate common neighbors} \\ \frac{1}{|\mathcal{R}(v_i, v_j)|} \sum_{v_k \in \mathcal{R}} sim(v_i, v_k) \cdot sim(v_k, v_j), otherwise \end{cases} \quad (16)$$

where $\mathcal{R}(v_i, v_j)$ is the set of common related users between user v_i and user v_j, $\mathcal{R}(v_i, v_j) = \{v_k | sim(v_i, v_k) \neq 0 \wedge sim(v_k, v_j) \neq 0\}$.

4 Experiments

In this section, we compare experimentally our approach with existing similarity measure algorithms.

4.1 Data Sets

We use the Epinions [15] data set, which is a signed social network. We compute the in-degree and out-degree distributions of Epinions graph, treating both the positive and negative edges alike, as shown in Fig. 4.

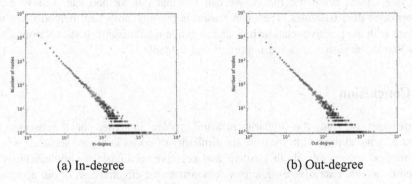

(a) In-degree   (b) Out-degree

Fig. 4. Degree distributions of Epinions data set.

4.2 Experimental Setup and Evaluation

To evaluate the proposed approach, we recommend a list of k friends (top-k list) to a target user. Our evaluation considers the division of neighbors of each target user into two sets: (1) the training set $\mathcal{E}_T$ is treated as known information and, (2) the verification set $\mathcal{E}_V$ is used for verification. We use *precision* and *accuracy* metric as performance measures for recommendations. In order to validate the effectiveness of our approach in signed social networks, we compare our model with FriendTNS algorithm and the Shortest Path algorithm, denoted as F-TNS and Shortest Path, respectively.

4.3   Experimental Results

In this section, we quantitatively compare our proposed method with baselines. Figure 5 shows the results conducted in Epinions data set.

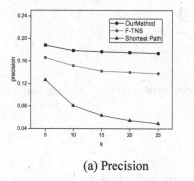   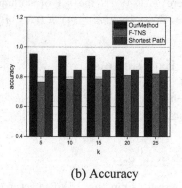

(a) Precision   (b) Accuracy

Fig. 5. Comparison of proposed method with baselines in Epinions.

The horizontal axis of Fig. 5 represents the number of friends that we recommend to a target user. From the Fig. 5, we can see that our method can achieve better performance than baselines. The main reason is that our proposed method takes into account both the positive relationships and negative relationships. It shows from a side view that the negative links are important and valuable.

5   Conclusion

In this paper, we study the similarity measure problem in signed social networks. We proposed a novel method to measure the similarity of nodes in signed social networks. Our method incorporates both positive and negative relationships simultaneously in signed networks. Extensive evaluations demonstrate the effectiveness of our approach.

Acknowledgement. This work is supported by NSF of China (No. 61602237), 973 Program (No. 2015CB352501), NSF of Shandong, China (No. ZR2013FQ009), the Science and Technology Development Plan of Shandong, China (Nos. 2014GGX101047, 2014GGX101019).

References

1. Kunegis, J., Lommatzsch, A., Bauckhage, C.: The Slashdot Zoo: mining a social network with negative edges. In: WWW 2009, pp. 741–750 (2009)
2. Leskovec, J., Huttenlocher, D.P., Kleinberg, J.M.: Predicting positive and negative links in online social networks. In: WWW 2010, pp. 641–650 (2010)
3. Tang, J., Chang, Y., Aggarwal, C., Liu, H.: A survey of signed network mining in social media. ACM Comput. Surv. **49**(3), 1–37 (2016)

4. Wang, S., Hu, X., Yu, P.S., Li, Z.: MMRate: inferring multi-aspect diffusion networks with multi-pattern cascades. In: KDD 2014, pp. 1246–1255 (2014)

5. Wang, S., Yan, Z., Hu, X., Yu, P.S., Li, Z.: Burst time prediction in cascades. In: AAAI 2015, pp. 325–331 (2015)

6. Shi, C., Li, Y., Zhang, J., Sun, Y., Yu, P.S.: A survey of heterogeneous information network analysis. IEEE Trans. Knowl. Data Eng. **29**(1), 17–37 (2017)

7. Pan, J., Yang, H., Faloutsos, C., Duygulu, P.: Automatic multimedia cross-modal correlation discovery. In: KDD 2004, pp. 653–658 (2004)

8. Jeh, G., Widom, J.: SimRank: a measure of structural-context similarity. In: KDD 2002, pp. 538–543 (2002)

9. Jeh, G., Widom, J.: Scaling personalized web search. In: WWW 2003, pp. 271–279 (2003)

10. Liben-Nowell, D., Kleinberg, J.M.: The link prediction problem for social networks. In: CIKM 2003, pp. 556–559 (2003)

11. Symeonidis, P., Tiakas, E., Manolopoulos, Y.: Transitive node similarity for link prediction in social networks with positive and negative links. In: RecSys 2010, pp. 183–190 (2010)

12. Symeonidis, P., Tiakas, E.: Transitive node similarity: predicting and recommending links in signed social networks. World Wide Web **17**(4), 743–776 (2014)

13. Heider, F.: Attitudes and cognitive organization. J. Psychol. **21**, 107–112 (1946)

14. Cartwright, D., Harary, F.: structure balance: a generalization of Heider's theory. Psychol. Rev. **63**(5), 277–293 (1956)

15. Leskovec, J., Huttenlocher, D.P., Kleinberg, J.M.: Signed networks in social media. In: CHI 2010, pp. 1361–1370 (2010)

Data Mining and Data Streams

Elastic Resource Provisioning for Batched Stream Processing System in Container Cloud

Song Wu[1]($\boxtimes$), Xingjun Wang[1], Hai Jin[1], and Haibao Chen[2]

[1] Services Computing Technology and System Lab,
Cluster and Grid Computing Lab, School of Computer Science and Technology,
Huazhong University of Science and Technology, Wuhan 430074, China
{wusong,wangxingjun,hjin}@hust.edu.cn
[2] School of Computer and Information Engineering, Chuzhou University,
Chuzhou 239000, China
chb@chzu.edu.cn

Abstract. Batched stream processing systems achieve higher throughput than traditional stream processing systems while providing low latency guarantee. Recently, batched stream processing systems tend to be deployed in cloud due to their requirement of elasticity and cost efficiency. However, the performance of batched stream processing systems are hardly guaranteed in cloud because static resource provisioning for such systems does not fit for stream fluctuation and uneven workload distribution. In this paper, we propose *EStream*: an elastic batched stream processing system based on Spark Streaming, which transparently adjusts available resource to handle workload fluctuation and uneven distribution in container cloud. Specifically, *EStream* can automatically scale cluster when resource insufficiency or over-provisioning is detected under the situation of workload fluctuation. On the other hand, it conducts resource scheduling in cluster according to the workload distribution. Experimental results show that *EStream* is able to handle workload fluctuation and uneven distribution transparently and enhance resource efficiency, compared to original Spark Streaming.

Keywords: Elastic · Resource provisioning · Stream processing · Container

1 Introduction

Due to data explosion in recent years, massive stream data that require to be processed in real time are increasingly common in business community and academia. Parallel stream processing systems play a critical role in such scenarios and provide low latency and high throughput. Traditional parallel stream processing systems, such as S4 [10] and Storm [15], are mostly designed based on operator model. In these systems, an application is represented with a topology of operators. Each operator implements a particular processing logic and is usually accelerated with a number of instances. Records in stream are processed by

© Springer International Publishing AG 2017
L. Chen et al. (Eds.): APWeb-WAIM 2017, Part I, LNCS 10366, pp. 411–426, 2017.
DOI: 10.1007/978-3-319-63579-8_32

the operators within the topology. In that way, records are handled one by one with latency in tens of milliseconds. Batched stream processing model, which combines the features of batch processing with stream processing, is proposed to meet demands of higher throughput and fault tolerance in recent years [20].

With the prevalence of cloud computing, cloud data center is increasingly being explored to deploy big data applications, such as parallel stream processing application, instead of building a local cluster. Recently, container technology is on the road to be a lightweight alternative to virtual machine (VM). Compared to VM, container imposes less overhead for resource virtualization since they share the OS kernel of physical machine. Until now, most researches discuss the elasticity of stream processing system in VM-based cloud. For example, Stela [18] is presented to adjust parallelism of operators in Storm with VMs. Unfortunately, the long startup time (usually in seconds) prevents VM from providing real-time resource provisioning. Besides, resource adjustment in VM is also complex.

Since stream workload fluctuation is common in production and it is difficult to make accurate estimation of resource requirements to handle the workload in advance or adjust resource configuration on the fly. As a result, there will be extra payments for the waste of cloud resource for customers. Many efforts have been paid to estimate reasonable resource provisioning to optimize performance for parallel distributed applications in cloud [6,14]. Besides, in batched stream processing systems, sometimes blocks in batches are in different sizes, which leads to uneven execution time of tasks. As a result, the parallelism of stream processing is influenced. Similar problem exists in Storm due to different traffics on operators and adjusting the number of operators or slot reallocation is common in existing researches [3,7]. Some researches also pay attention to handle this problem with dynamic operator placement or parallelism adjustments [17,18]. However, few researches focus on batched stream processing system, thus, an elastic resource provisioning method for bathed stream processing systems in cloud is promising and still in demand.

In this paper, we solve the above problems and propose *EStream*, an elastic distributed batched stream processing system which bridges the gap between stream processing applications and resource management in cloud.

In summary, we make the following contributions in this paper:

- We build an optimization model of batched stream processing system and determine when it needs resource adjustment.
- We introduce cluster scaling and local resource scheduling with historical execution information, targeting to satisfy the requirement of system stability and higher resource efficiency under fluctuant workload. Then we design and implement *EStream* based on Spark Streaming and container technology (i.e., Docker).
- We conduct a series of experiments to evaluate the performance of *EStream*, the results show that *EStream* is able to adapt to workload fluctuation and uneven distribution and the system resource efficiency is also enhanced.

The rest of this paper is organized as follows: In Sect. 2, we review and discuss some related work. Section 3 describes background of our design. We introduce an optimization mathematical model which guides our design in Sect. 4. Section 5 shows the overall system design of *EStream* and explains how it works. Performance evaluation results are presented in Sect. 6. Finally, we conclude the paper in Sect. 7.

2   Related Work

There are a great deal of researches that concentrate on resource managements of parallel stream or batch processing systems. Many researches on stream processing systems mainly focus on operator-based systems. For instance, Fu et al. [3] propose a resource scheduler which dynamically allocates processors to operators to ensure real-time response. They model the relationship between resources and response time and use optimization models to get processor allocation strategy. Xu et al. [18] present a VM operator instance scaling technique according to throughput of operators. Their approach does not work on batched stream processing system, which is not operator-based. What is more, the time and monetary cost of VM scaling is expensive. Cervino et al. [1] propose an adaptive approach for provisioning VMs for stream processing system based on input rates but latency guarantees are ignored. In addition, Madsen and Zhou [9] integrate fault-tolerance and dynamic resource managements to avoid waste of excessive processing delay. Wu and Tan [16] introduce ChronoStream, which provides vertical and horizontal elasticity. In a word, there are few researches on resource management of containerized batched stream processing system.

As for batch processing systems, a flexible slot management scheme to accelerate task execution for off-line MapReduce jobs is shown by Guo et al. [4]. Their work partly inspires our design. Ruan et al. [14] present an approach based on monetary cost model for cloud resource provisioning for Spark with sample running, which inspires us to leverage the historical information of previous batches in batched streaming system to make resource allocation decision in our design. Park et al. [12] dynamically increase or decrease the computing capability of each node to enhance locality-aware task scheduling with dynamic virtual CPU number configurations.

There are some recent studies that aim to improve performance of batched stream processing system. For example, Das et al. [2] ensure system stability and low latency with dynamic batch sizing, which changes task parallelism in runtime. However, resource efficiency is ignored in their work.

Our paper is different from previous work in several aspects. First, we model batched stream processing system with considering the features of both batch and stream and get the scheduling target which guides our design. Second, we design our elastic resource provisioning approach with execution information of previous processed batches. The approach automatically scales containerized cluster and carries out resource scheduling with the help of container technology. Finally, we implement *EStream* which takes full advantages of containerized environments.

3   Background

3.1   Batched Stream Processing

In this subsection, we take Spark Streaming as an example to describe batched stream processing.

Data Model. *Discretized stream* (DStream) is the abstraction of data stream in Spark Streaming which is a series of *resilient distributed datasets* (RDDs)[19]. RDD is introduced in Spark framework to persist data in memory or disk. Hot data can be cached in memory to avoid frequent IO operations to improve performance of batch jobs. As shown in Fig. 1, each RDD in DStream is the received stream data in a batch interval and stored with a number of blocks. The block size is determined by block interval in application.

Execution Model. Figure 1 also demonstrates the execution model of Spark Streaming. In Spark Streaming, user's application gives the definition of transformations on RDDs. When a batch is received and persisted as an RDD, a job is accordingly generated and submitted to the job scheduler. Each job waits in a scheduling queue to be executed if there are former jobs that are not completed. Jobs will finally be transformed into tasks and each task is executed on a certain executor in cluster.

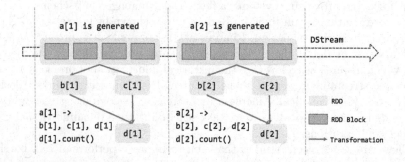

Fig. 1. Batched stream processing model with a sample RDD transformation definition in Spark Streaming. In the example, RDD a[i] is generated in batch[i]. RDD b[i], c[i], d[i] are transformed by different APIs from a[i]

The characteristics of batched stream processing system are: (1) Jobs are tiny and latency sensitive. Due to this feature, resource provisioning method should be proactive, simple, and efficient; (2) Jobs are recurrent [5,13], this nature inspires us to use runtime information of former processed batches to make workload prediction for further resource provisioning decision.

3.2 Container Technology

Container is an OS-level virtualization technology and gaining great much attention in recent years. One of the core technologies in container is CGroup, which is a mechanism for grouping and limiting resource of tasks, and all their future children, into hierarchical groups with specialized behavior in Linux kernel. With container, users can build a more flexible and lightweight platform for applications. There is a trend for container technology to replace traditional VM-based application deployment mode due to its lightweight, fast startup time, and near-native performance. Since container provides near-native performance and millisecond-level startup time, it is suitable for elastic stream processing system, which requires fast and efficient resource provisioning. In this paper, we use Docker[1] in our design to provide elasticity because it is one of the most popular container technologies and is widely used for application in production system.

4 Modeling

Before modeling the batched stream processing, some basic concepts are described as follows.

- *Batch interval*: the time period used to divide real-time data stream into batches, denoted by I.
- *Batch processing time*: the time consumption for processing a batch, denoted by p.
- *Scheduling delay*: the waiting time for a batch from when it is received to the time when it begins to be processed, denoted by s.
- *Latency*: the time cost for a batch from when it is received and prepared to be processed to the time when it is completely processed, denoted by l. It is obvious that $l = p + s$.

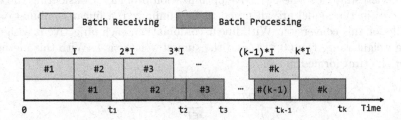

Fig. 2. A scenario of batched stream processing. Batched stream processing application is started at time $t_0 = 0$, and $batch_k$ is received and ready to be processed at time $k * I$, then it is completely processed at time t_k

---

[1] http://docker.com/.

In this section, we discuss a scenario of a batched stream processing system which starts an application at time $t_0 = 0$ and n batches are handled in the system until now. The i_{th} batch $batch_i$ is completely finished at time t_k. Processing time of $batch_i$ is p_k and its scheduling delay is s_k. Figure 2 displays the scenario. In order to simplify our discussion, we make the assumption that the system in our discussion always runs normally and no crash or other faults exist.

If $t_{k-1} > k * I$, it means when $batch_k$ is received and ready for processing, the system is busy in processing $batch_{k-1}$, in other words, $batch_k$ is blocked in the queue, and its block time is $BT_k = t_{k-1} - k * I$. When $t_{k-1} < k * I$, $batch_{k-1}$ is processed at once when it is ready and system is idle before processing $batch_k$, and the system idle time is $ST_k = k * I - t_{k-1}$. Things are quite simple when $t_{k-1} = k * I$: $batch_k$ begins to be processed at once when it is ready. Therefore, there is:

$$t_k = max\{t_{k-1}, k * I\} + p_k \tag{1}$$

In order to obtain better resource utilization, the system idle time should be as less as possible. On the other hand, the system should process batches as soon as possible when they are received. Therefore, the goals of our system design are to minimize the sum of batch blocked time (BT) and the sum of system idle time (ST). This can be modeled as an optimization problem:

$$\min_{p_i} \quad \sum_{i \in B} BT_i, \sum_{i \in D} ST_i$$

$$s.t. \begin{cases} t_k = max\{t_{k-1}, k * I\} + p_k, & 2 \leq k \leq n \\ t_1 = I + p_1 \end{cases} \tag{2}$$

where B is the set of batches that are blocked before processing and D is the set of batches that begin to be processed at once when they are received. There is $U = B \cup D$, where U is the set of all n batches in the scenario.

Since this is a multiple-objective and non-linear optimization model and cannot be solved directly, it is common to convert the original problem with multiple objectives into a single-objective optimization problem. Considering the two objectives in the model are all urged to be minimized, a linear combination is suitable for this conversion. With linear combination, each objective is assigned with a weight range from 0 to 1, and the sum of weights is 1. With this method, we get the transformed model:

$$\min_{p_i} \quad \alpha \cdot \sum_{i \in B} BT_i + \beta \cdot \sum_{i \in D} ST_i \quad (\alpha + \beta = 1)$$

$$s.t. \begin{cases} t_k = max\{t_{k-1}, k * I\} + p_k, & 2 \leq k \leq n \\ t_1 = I + p_1 \end{cases} \tag{3}$$

To simplify the problem, we give the same importance to the two objectives and choose $\alpha = 0.5$ and $\beta = 0.5$. Note that BT_i and ST_i can be unified to be demonstrated with $|t_{i-1} - i * I|$. We finally transform the above model to:

$$\min_{p_i} \quad \sum_{i=1}^{n} |t_{i-1} - i * I|/2$$

$$s.t. \begin{cases} t_k = max\{t_{k-1}, k * I\} + p_k, \quad 2 \le k \le n \\ t_1 = I + p_1 \end{cases} \tag{4}$$

By solving this model, we can get the solution:

$$p_1 = p_2 = p_3 = \cdots = p_n = I \tag{5}$$

The result indicates that if processing time of batches is equal to batch interval, there will be $l_k = t_k - k * I = p_k = I$ and $s_k = 0$. In that case, batches in data stream are processed in time and the system is never idle. This conclusion means that our resource provisioning mechanism should target to make processing time of batches reach as closer to batch interval as possible. Therefore, when we find $p \ne I$, we should adjust resource provisioning of the system.

5   System Design

5.1   Overview

EStream aims to make proactive resource provisioning by collecting and analyzing the execution information of processed batches. The overview of *EStream* is shown in Fig. 3. Two main components that run in Spark driver are: *Execution Information Collector* (EIC) and *Elasticity Manager* (EM). Executors of batched stream processing system run in Docker containers and they are managed by NodeManager and ContainerDaemon on the host server. NodeManager is in charge of starting or shutting down containers according to the requests from ClusterResourceManager. ContainerDaemon receives resource scheduling request and adjusts resource allocation of containers. There are mainly three steps to elastically provision resource:

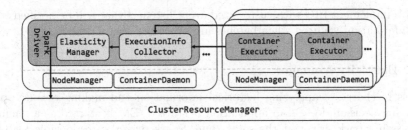

Fig. 3. Overview of *EStream*

1. *Execution information collection*: EIC gathers historical execution information about the workload handled on each container executor and execution time in previous processed batches. Processing time of last batch is also recorded by EIC. Then EIC sends the information to EM.
2. *Workload prediction and execution analysis*: After receiving execution information from EIC, EM firstly makes a workload prediction for each executors. Then it analyzes execution of the last batch and determines a resource provisioning plan. The resource provisioning plan is sent to ClusterResourceManager.
3. *Resource allocation*: When ClusterResourceManager receives the resource provisioning plan from EM, it will send this request to each NodeManager and ContainerDaemon. Then NodeManagers start or shutdown containers in response to the request from ClusterResourceManager. ContainerDaemon performs resource scheduling among containers.

In summary, EIC collects historical execution information and transmits it to EM. EM analyzes the historical information and determines whether a resource adjustment is needed. If so, EM triggers elastic resource provisioning to prevent the system from performance degradation in the future. In the next, we are going to explain more details of EIC and EM.

5.2   Execution Information Collector

EIC is designed to perform collection of historical execution information for later analysis. It is activated every time when a batch or a task is completed to obtain runtime information. The EIC collects the following information:

1. Workload handled by each executor in the last N batches, a window of size N is maintained in EIC. N is set to a default value of 5 in *EStream* according to our experimental results.
2. Total execution time on each executor in the last processed batch. It is calculated with $theLastTaskFinishTime - theFirstTaskStartTime$.
3. Processing time of last batch.

We implement EIC with customized event listeners that will be notified with the information we need when batches are processed.

5.3   Elasticity Manager

EM is in charge of analyzing historical information and making resource provisioning decision. When an application is started, EM is started as a daemon process in Spark driver at the same time. Spark driver is responsible for task scheduling of applications in Spark Streaming.

First, EM makes a workload prediction. For each executor in the cluster, EM predicts its workload in next batch with *Exponential Weighted Moving Average* (EWMA), which is a simple but efficient prediction algorithm. As shown in

Eq. 6, workload on an executor in i_{th} batch w_i is used to predict w_{i+1}. w' is the difference of workload handled by this executor between the last two batches. λ indicates the weights of w_i in prediction of w_{i+1}, ranging from 0 to 1. We set $\lambda = 0.8$ in our design and experimental results show it works well.

$$w_{i+1} = \lambda * w_i + (1 - \lambda) * w' \tag{6}$$

Second, EM conducts execution analysis on last processed batch. As discussed in Sect. 4, our target is to make batch processing time as closer to batch interval I as possible. Since it is infeasible to guarantee that batch processing time is equal to I all the time and we should detect the resource problem in advance to make a proactive resource adjustment in order to avoid future performance degradation. A threshold $\mu(0 < \mu < 1)$ is given to address this problem. When processing time of the last batch exceeds $\mu * I$, EM will try to prevent the risk of performance degradation through elastic resource provisioning. We set 0.9 as the default value of μ in our design according to lots of experiments. We introduce an *Elastic Resource Provisioning* (ERP) algorithm in EM which aims to handle two situations as follows:

Workload Fluctuation. When facing workload fluctuation, the system may suffer a performance degradation from resource insufficiency or it is under low resource utilization. ERP algorithm recognizes these two situations with analysis on execution time on executors and last batch processing time. We introduce a cluster scaling strategy in ERP to ensure performance and resource utilization of the system. If resource over-provisioning occurs, we scale in the cluster to save resource usage. On the contrary, if resource is insufficient, we scale out the cluster in response to increasing workload. The details of two situations are as follows:

1. *Scaling out*: If processing time of last completed batch is larger than $\mu * I$, executors whose execution time is larger than $\mu * I$ are regarded as slow executor. If the number of slow executors exceeds $\delta * clusterSize$, where δ is a threshold with default value of 0.4 according to our experimental results. ERP scales out the cluster to handle the increasing workload.
2. *Scaling in*: If processing time of last completed batch is smaller than $\mu * I$, executors whose execution time is smaller than $\gamma * I$ are marked as fast executor, where γ is a threshold with default value of 0.5 to identify fast executor. If the number of fast executors exceeds $\delta * clusterSize$, ERP scales in the cluster to save resource.

When cluster scaling is determined, ERP firstly calculates the average processing speed ($avgSpeed$) of each executor with:

$$avgSpeed = \frac{totalWorkloadInLastBatch}{clusterSize * processingTimeOfLastBatch} \tag{7}$$

Then the new cluster size is estimated with:

$$clusteSize' = \lceil \frac{workload_{predict}}{\mu * I * avgSpeed} \rceil \tag{8}$$

where $workload_{predict}$ is predicted workload of next batch.

Uneven Workload Distribution. When RDD blocks of batches are in different sizes, execution time on executors in the cluster is influenced. Some executors may suffer heavy workload which results in long task execution time. As a result, total execution time is increased on these executors and batch processing time is increased. In such situation, a common idea is to make load balancing across executors in the cluster, but this cannot be completed in real time and will lead to extra overhead, such as network bandwidth. Fortunately, container technology provides another approach to this problem. It is possible to adjust computing resource among executors in containers (Docker). ERP leverages this mechanism to make resource scheduling to handle uneven workload distribution.

Specifically, when last batch processing time is larger than $\mu * I$, ERP algorithm calculates the number of slow executors. If the number of slow executor is smaller than $\delta * clusterSize$, i.e., there exists an uneven workload distribution, ERP will conduct resource scheduling among executors. In this paper, our ERP only focuses on CPU resource allocation, because it is considered as the main bottleneck of Spark framework [11]. Docker supports CPU resource limitation with the help of CGroup and it is easy to carry out an adjustment with RESTful APIs. In our design, we introduce a proportional allocation strategy for CPU resource scheduling. ERP allocates CPU usage among container executors with:

$$cpuRatio_i = \frac{workload_i}{\sum workload_i} \cdot totalCpuRatio \qquad (9)$$

The pseudo-code of ERP algorithm is shown in Algorithm 1. The key idea is to scale cluster when over-provisioning or resource insufficiency exists and adjust resource allocation to handle uneven workload distribution. More specifically, if the last batch processing time exceeds the threshold $\mu * I$, executors whose execution time is larger than $\mu * I$ is regarded as slow executors (line 3–6). If the number of slow executor is larger than $\delta * clusterSize$, EM decides to scale out the cluster (line 7–10). Otherwise, *EStream* will carry out a CPU resource reallocation among container executors (line 11–16). The details of cluster scaling are as follows: (1) estimating average handling speed on executor (line 8); (2) calculating the new cluster size according to predicted total workload and the average handling speed (line 9). If the batch processing time is less than the threshold $\mu * I$, EM counts the number of fast executors (line 18–20). If the number of fast executors is larger than $\delta * clusterSize$, EM will scale in the cluster (line 21–24).

The Cost Analysis. Suppose the cluster size is m when ERP algorithm is called. The space complexity of ERP algorithm is $O(m)$ and the time complexity is $O(m)$. This is because the amount of information of each executor gathered by EIC is fixed and the amount of space needed only depends on m. When the cluster needs to be scaled, the average handling speed of executor is calculated in $O(m)$ time. Then ERP calculates the desired cluster size in $O(1)$ time. As for CPU resource scheduling, ERP firstly gets total CPU ratio in $O(m)$ time. Then new CPU ratio of each executor is calculated in $O(1)$ time.

Algorithm 1. Elastic resource provisioning algorithm

Input: (1) Batch processing time in last batch: pt; (2) Batch interval: I; (3) Workload handled on executors in last N batches: $W[N][excutorId : workload]$; (4) CPU usage of executors: $cpuRatio[executorId : percentage]$; (5) Execution time on executors in last batch: $d[executorId : executionTime]$;

Output: New cluster size or CPU resource allocation plan.

1: $slowExecutorNum \leftarrow 0$, $fastExecutorNum \leftarrow 0$, $clusterSize \leftarrow executorIds.size$, $hostServers \leftarrow$ host servers of container executors, $w \leftarrow W[N-1]$;

2: Predict workload on each executor: $w'[executorId : workload] \leftarrow EWMA(W)$

3: **if** $pt > \mu * I$ **then**

4:     **for** each $id \in executorIds$ **do**

5:         **if** $d[id] > \mu * I$ **then**

6:             $slowExecutorNum + +$

7:     **if** $slowExecutorNum > \delta * clusterSize$ **then**

8:         $avgSpeed = \frac{\sum w[i]}{clusterSize*pt}$

9:         $clusterSize' = \lceil \frac{\sum w'[i]}{\mu*I*avgSpeed} \rceil$

10:     **return** $clusterSize'$

11:     **if** $0 < slowExecutorNum < \delta * clusterSize$ **then**

12:         **for** each $host \in hostServers$ **do**

13:             $executors \leftarrow$ executors on $host$.

14:             **for** each $id \in executors$ **do**

15:                 $cpuRatio'[id] = \frac{w'[id]}{\sum w'[id]} * \sum cpuRatio[id]$

16:     **return** $cpuRatio'[]$

17: **else**

18:     **for** each $id \in executorIds$ **do**

19:         **if** $d[id] < \gamma * I$ **then**

20:             $fastExecutorNum + +$

21:     **if** $fastExecutorNum > \delta * clusterSize$ **then**

22:         $avgSpeed = \frac{\sum w[i]}{clusterSize*pt}$

23:         $clusterSize' = \lceil \frac{\sum w'[i]}{\mu*I*avgSpeed} \rceil$

24:     **return** $clusterSize'$

---

6 Evaluation

6.1 Experiments Setup and Method

Our experiments are conducted on a private container cloud. Each physical host is comprised of two quad-core 2.4 GHz Intel CPU, 24 GB memory, 1 TB disk and

1 Gbps Ethernet card. We implement *EStream* on Spark-1.6.0 and the version of Docker is 1.12.3. Each container executor is configured with 2 GB memory, 2 cores and 5% CPU usage limitation when it is started. We use WordCount, a widely used benchmark application [8,20] which counts the number of words in a fixed time interval in our experiments. The batch interval of WordCount application is 2 s.

6.2   Experiment with Workload Fluctuation

In this experiment, we simulate a data stream with a sinusoidal input rate ranging from 1 to 3 MB/s. We generate records from a set of words with the same length as the seed of input stream. The cluster in this experiment is initialized with 6 container executors. We run *EStream* with simulated data stream for 180 s. Figure 4(a) shows that, with the fluctuation of input rate, the cluster size of *EStream* changes accordingly, the maximum and minimal cluster size are 9 and 6, respectively. Figure 4(b) shows batch processing time in *EStream*. As shown in the figure, when input rate changes, batch processing time in *EStream* is relatively stable.

As for Spark Streaming, we deploy a cluster of 9 container executors, which is the maximum cluster size needed to handle peak traffic in *EStream* test. Figure 5(a) shows the input rate and cluster size in this test. Figure 5(b) shows batch processing time in original Spark Streaming test. Batch processing time in Spark Streaming suffers a fluctuation with the variation of input rate because of resource reservation configuration of Spark Streaming and it cannot elastically scale its container cluster. As we know, cloud payment is calculated by the number of instances and the runtime. We build a function $f(t)$ to denote the relationship between cluster size and time in our test, thus, resource utilization in this test can be expressed with $\int_0^{180} f(t)dt$. With this formulation, we can figure out that *EStream* saves a resource usage of about 30%, compared to Spark Streaming.

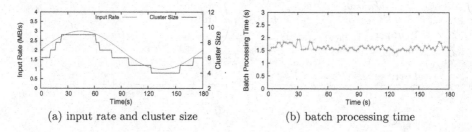

(a) input rate and cluster size   (b) batch processing time

Fig. 4. Performance of *EStream* with sinusoidal input rate

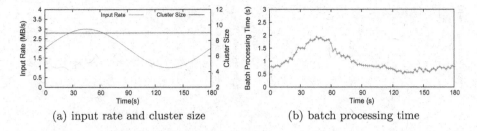

(a) input rate and cluster size   (b) batch processing time

Fig. 5. Performance of Spark Streaming with sinusoidal input rate

6.3 Experiment with Uneven Workload Distribution

To evaluate the adaptability of *EStream* to uneven workload distribution, we simulate a data stream with constant input rate of 3 MB/s. We generate two files with the same size of 3 MB as the seed of our simulated data stream. One file $file_1$ is a set of words with same length, while the other one $file_2$ is a set of words with random length. This means the number of words in these two files are different. The simulated data stream changes its seed in every 10 s. In each round, the data source sends the data from $file_1$ for 10 s, then $file_2$ is selected as the seed for the next 10 s. We setup a cluster of 9 container executors to carry out this experiment. We run WordCount in both *EStream* and Spark Streaming for 60 s. As mentioned above, uneven workload distribution leads to the difference of execution time on executors. We collect and normalize the execution time on each executor. We use the *variance of normalized execution time* (VNET) to measure the impact on parallelism caused by uneven workload distribution. Figure 6 shows the results of *EStream* and Spark Streaming under the simulated data stream in this experiment.

As we can observe in Fig. 6(a), when workload changes every 10 s, the VNET suddenly increases, due to the uneven workload distribution, as well as batch processing time. However, the two metrics decreases soon. This is because *EStream* makes resource adjustment among executors to deal with the uneven workload distribution. Note that when the seed of data stream is changed to $file_1$, the two metrics increase because the CPU allocation in the previous 10 s does not match the current workload distribution. But *EStream* will soon solve this problem.

Figure 6(b) shows the performance of Spark Streaming under the same condition. It is obvious that after the changing of data stream seed to $file_2$, both of the VNET and batch processing time no longer decrease in the next 10 s. By comparing the batch processing time when the system is under uneven workload distribution in this test, *EStream* reduces batch processing time by 19% on average.

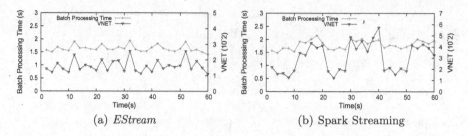

Fig. 6. Batch processing time and *variance of normalized execution time* (VNET) in *EStream* and Spark Streaming with uneven workload distribution

7 Conclusion

In this paper, we concentrate on common problems in stream data processing: workload fluctuation and uneven distribution. In order to achieve higher resource efficiency while provide performance guarantees in cloud, we design and implement *EStream*, which providing elastic resource provisioning for batched stream processing applications. We exploit container technology to realize fast cluster scaling and efficient resource reallocation in runtime to handle workload fluctuation and uneven workload distribution. In contrast to the original Spark Streaming, experimental results show *EStream* is able to adapt to varying workload and save a resource usage of about 30%. Besides, *EStream* can transparently schedule CPU resource among executors to enhance parallelism and reduce batch processing time.

Acknowledgments. This research is supported by National Key Research and Development Program under grant 2016YFB1000501, 863 Hi-Tech Research and Development Program under grant No. 2015AA01A203, and National Science Foundation of China under grants No. 61232008. This work is also supported by the Anhui Natural Science Foundation of China under grant (No. 1608085QF147), and Key Project of Support Program for Excellent Youth Scholars in Colleges and Universities of Anhui Province (No. gxyqZD2016332).

References

1. Cervino, J., Kalyvianaki, E., Salvachua, J., Pietzuch, P.: Adaptive provisioning of stream processing systems in the cloud. In: Proceedings of 2012 IEEE 28th International Conference on Data Engineering Workshops (ICDEW), pp. 295–301. IEEE (2012)
2. Das, T., Zhong, Y., Stoica, I., Shenker, S.: Adaptive stream processing using dynamic batch sizing. In: Proceedings of the ACM Symposium on Cloud Computing (SoCC), pp. 1–13. ACM (2014)
3. Fu, T.Z., Ding, J., Ma, R.T., Winslett, M., Yang, Y., Zhang, Z.: Drs: dynamic resource scheduling for real-time analytics over fast streams. In: Proceedings of 2015 IEEE 35th International Conference on Distributed Computing Systems (ICDCS), pp. 411–420. IEEE (2015)

4. Guo, Y., Rao, J., Jiang, C., Zhou, X.: Flexslot: moving hadoop into the cloud with flexible slot management. In: Proceedings of International Conference for High Performance Computing, Networking, Storage and Analysis (SC), pp. 959–969. IEEE (2014)
5. Jyothi, S.A., Curino, C., Menache, I., Narayanamurthy, S.M., Tumanov, A., Yaniv, J., Goiri, Í., Krishnan, S., Kulkarni, J., Rao, S.: Morpheus: towards automated slos for enterprise clusters. In: Proceedings of the 12th USENIX Symposium on Operating Systems Design and Implementation (OSDI), p. 117. USENIX (2016)
6. Kambatla, K., Pathak, A., Pucha, H.: Towards optimizing hadoop provisioning in the cloud. In: Proceedings of USENIX Workshop on Hot Topics in Cloud Computing (HotCloud), vol. 9, p. 12. USENIX (2009)
7. Kumbhare, A., Frincu, M., Simmhan, Y., Prasanna, V.K.: Fault-tolerant and elastic streaming mapreduce with decentralized coordination. In: Proceedings of 2015 IEEE 35th International Conference on Distributed Computing Systems (ICDCS), pp. 328–338. IEEE (2015)
8. Lin, W., Qian, Z., Xu, J., Yang, S., Zhou, J., Zhou, L.: Streamscope: continuous reliable distributed processing of big data streams. In: Proceedings of USENIX Symposium on Networked System Design and Implementation (NSDI), pp. 439–454. USENIX (2016)
9. Madsen, K.G.S., Zhou, Y.: Dynamic resource management in a massively parallel stream processing engine. In: Proceedings of the 24th ACM International Conference on Information and Knowledge Management (CIKM), pp. 13–22. ACM (2015)
10. Neumeyer, L., Robbins, B., Nair, A., Kesari, A.: S4: distributed stream computing platform. In: Proceedings of 2010 IEEE International Conference on Data Mining Workshops (ICDMW), pp. 170–177. IEEE (2010)
11. Ousterhout, K., Rasti, R., Ratnasamy, S., Shenker, S., Chun, B.G.: Making sense of performance in data analytics frameworks. In: Proceedings of USENIX Symposium on Networked Systems Design and Implementation (NSDI), pp. 293–307. USENIX (2015)
12. Park, J., Lee, D., Kim, B., Huh, J., Maeng, S.: Locality-aware dynamic vm reconfiguration on mapreduce clouds. In: Proceedings of the 21st International Symposium on High-Performance Parallel and Distributed Computing (HPDC), pp. 27–36. ACM (2012)
13. Rasley, J., Karanasos, K., Kandula, S., Fonseca, R., Vojnovic, M., Rao, S.: Efficient queue management for cluster scheduling. In: Proceedings of the 11th European Conference on Computer Systems (EuroSys), p. 36. ACM (2016)
14. Ruan, J., Zheng, Q., Dong, B.: Optimal resource provisioning approach based on cost modeling for spark applications in public clouds. In: Proceedings of the Doctoral Symposium of the 16th International Middleware Conference, p. 6. ACM (2015)
15. Toshniwal, A., Taneja, S., Shukla, A., Ramasamy, K., Patel, J.M., Kulkarni, S., Jackson, J., Gade, K., Fu, M., Donham, J., Bhagat, N., Mittal, S., Ryaboy, D.: Storm@ twitter. In: Proceedings of the 2014 ACM SIGMOD International Conference on Management of Data, pp. 147–156. ACM (2014)
16. Wu, Y., Tan, K.L.: Chronostream: elastic stateful stream computation in the cloud. In: Proceedings of 2015 IEEE 31st International Conference on Data Engineering (ICDE), pp. 723–734. IEEE (2015)
17. Xing, Y., Zdonik, S., Hwang, J.H.: Dynamic load distribution in the borealis stream processor. In: Proceedings of 2005 21st International Conference on Data Engineering (ICDE), pp. 791–802. IEEE (2005)

18. Xu, L., Peng, B., Gupta, I.: Stela: enabling stream processing systems to scale-in and scale-out on-demand. In: Proceedings of IEEE International Conference on Cloud Engineering (IC2E), pp. 22–31. IEEE (2016)
19. Zaharia, M., Chowdhury, M., Das, T., Dave, A., Ma, J., McCauley, M., Franklin, M.J., Shenker, S., Stoica, I.: Resilient distributed datasets: a fault-tolerant abstraction for in-memory cluster computing. In: Proceedings of USENIX Symposium on Networked Systems Design and Implementation (NSDI), p. 2. USENIX (2012)
20. Zaharia, M., Das, T., Li, H., Hunter, T., Shenker, S., Stoica, I.: Discretized streams: fault-tolerant streaming computation at scale. In: Proceedings of the 24th ACM Symposium on Operating Systems Principles (SOSP), pp. 423–438. ACM (2013)

An Adaptive Framework for RDF Stream Processing

Qiong Li[1,3], Xiaowang Zhang[1,3($\boxtimes$)], and Zhiyong Feng[2,3]

[1] School of Computer Science and Technology, Tianjin University,
Tianjin 300350, People's Republic of China
{liqiong,xiaowangzhang}@tju.edu.cn
[2] School of Computer Software, Tianjin University,
Tianjin 300350, People's Republic of China
zyfeng@tju.edu.cn
[3] Tianjin Key Laboratory of Cognitive Computing and Application,
Tianjin 300350, People's Republic of China

Abstract. In this paper, we propose a novel framework for RDF stream processing named PRSP. Within this framework, the evaluation of C-SPARQL queries on RDF streams can be reduced to the evaluation of SPARQL queries on RDF graphs. We prove that the reduction is sound and complete. With PRSP, we implement several engines to support C-SPARQL queries by employing current SPARQL query engines such as Jena, gStore, and RDF-3X. The experiments show that PRSP can still maintain the high performance by applying those engines in RDF stream processing, although there are some slight differences among them. Moreover, taking advantage of PRSP, we can process large-scale RDF streams in a distributed context via distributed SPARQL engines, such as gStoreD and TriAD. Besides, we can evaluate the performance and correctness of existing SPARQL query engines in processing RDF streams in a unified way, which amends the evaluation of them ranging from static RDF data to dynamic RDF data.

Keywords: RDF stream · RSP · SPARQL · C-SPARQL

1 Introduction

RDF stream, as a new type of dataset, can model real-time and continuous information in a wide range of applications, e.g. environmental monitoring and smart cities [15,16]. RDF streams have played an increasingly important role in many application domains such as sensors, feeds, and click streams [6]. SPARQL is recommended by W3C as the standard query language for RDF data. Though SPARQL engines are capable of querying integrated repositories and collecting data from multiple sources [7], the large knowledge bases now accessible via SPARQL are static, and knowledge evolution is not adequately supported.

© Springer International Publishing AG 2017
L. Chen et al. (Eds.): APWeb-WAIM 2017, Part I, LNCS 10366, pp. 427–443, 2017.
DOI: 10.1007/978-3-319-63579-8_33

For RDF stream processing (RSP), there are many works by extending SPARQL to support queryies over RDF streams such as C-SPARQL [4], EP-SPARQL [2], CQELS [12], SPARQLstream [6] etc. Continuous SPARQL (C-SPARQL) [4] extends SPARQL by adding window operators to manage RDF streams. Event Processing SPARQL (EP-SPARQL) [2] extends SPARQL by introducing ETALIS (a rule-based event language) to reason with events. Continuous Query Evaluation over Linked Streams (CQELS) [12] introduces three operators, namely, *window operator*, *relational operator*, and *streaming operator* to manage both RDF streams and RDF graphs. SPARQLstream [6] is an extension of SPARQL by introducing *window-to-stream operators* which are used to convert a stream of windows into a stream of tuples to process RDF streams. Besides, SPARQLstream can also provide the ontology-based streaming data access service since sources link their data content to ontologies through S_2O mappings.

Though those existing languages can represent many expressive continuous queries for RDF streams, there are a few prototype implementations for processing RDF streams [11] (e.g., S4 [9], CQELS-Cloud [13], and WAVES [10]) due to the complicacy of implementations and the requirement of highly efficiency in query evaluation [16]. On the other hand, there are many popular and efficient SPARQL query engines only supporting static RDF graphs such as Jena [7], RDF-3X [17], gStore [20]. Since those continuous query languages extend SPARQL by adding some extra operators to manage streams, it becomes an interesting problem to employ current SPARQL query engines to evaluate continuous queries. Barbieri et al. [5] employs Apache Jena [7] (a SPARQL query engine) to evaluate C-SPARQL queries in their implementation. Especially, those popular distributed SPARQL query engines (e.g., TriAD [8], gStoreD [19]) for large-scale RDF data could become very helpful to process large-scale RDF streams, which is a big challenge so far [11,16].

In this work, we propose a novel framework for RDF stream processing named *PRSP*, which is briefly introduced in [14], where the evaluation of continuous queries on RDF streams can be reduced to the evaluation of SPARQL queries on RDF graphs. For conveying our idea simply, we mainly discuss C-SPARQL queries in this paper. We argue that our framework could also support most of continuous query languages extending SPARQL, such as EP-SPARQL and CQELS. Our major contributions are summarised as follows:

- We formalize a C-SPARQL query as a 5-tuple to ensure the soundness and completeness of our reduction from the evaluation of C-SPARQL queries over RDF streams to the evaluation of SPARQL queries over RDF data. Besides, we present the semantics of C-SPARQL in a refined way.
- We develop an adaptive framework named PRSP for processing RDF stream by constructing four new modules, namely, *query parser* (to obtain parameters of windows and core patterns), *trigger* (to call queries and capture windows periodically), *data transformer* (to transform RDF graphs from RDF streams), and *SPARQL API* (to support various SPARQL endpoints to process RDF streams).

– We implement PRSP and evaluate on the YABench [11] by applying three centralized SPARQL endpoints, namely, Jena, gStore, and RDF-3X and two distributed SPARQL endpoints, namely, gStoreD and TriAD. The experiments show that PRSP can still maintain the high performance of those engines in RDF stream processing although there are some slight differences among them. Besides, we investigate that, given a C-SPARQL query, the correctness of results is sensitive to its window size and step among those query engines.

The remainder of this paper is structured as follows: the next section recalls RDF stream and C-SPARQL. Section 3 defines our sound and complete formalization of C-SPARQL, and Sect. 4 designs our framework PRSP. Section 5 presents experiments and evaluations. Finally, we summarize our work in the last section.

2   RDF Stream and C-SPARQL

In this section, we briefly recall RDF stream and C-SPARQL.

2.1   RDF Stream

Let I, B, and L be infinite sets of *IRIs*, *blank nodes* and *literals*, respectively.

A triple $(s, p, o) \in (I \cup B) \times I \times (I \cup B \cup L)$ is called an *RDF triple*. An *RDF graph* is a finite set of RDF triples.

An RDF stream S is defined as an (possibly infinite) ordered sequence of pairs which are quadruples, and each pair is made of an RDF triple (s_i, p_i, o_i) and a timestamp τ_i ($i \in \mathbb{Z}$, i.e., an integer) as follows:

$$S(t) = \{ \langle (s_i, p_i, o_i), \tau_i \rangle \mid t \leq \tau_i \leq \tau_{i+1} \}. \tag{1}$$

Note that $S(t)$ is the prefix of S ending at t.

Example 1. Let us consider an RDF stream S_{Sensor} generated in the YABench Benchmark [11]. It is associated with the temperature values from the environmental monitoring sensors. Table 1 shows the pairs of S_{Sensor}.

In Table 1, every sensor is identified via its id, e.g., A1; every temperature value is taken as an object; and the timestamp is represented as a 13-bit integer.

2.2   C-SPARQL

C-SPARQL (Continuous SPARQL) [4] extends SPARQL by adding new operators, namely, *registration* and *windows*, to support processing RDF streams. For simplification, we mainly introduce the basic aspects of new operators. We follow the formalization of C-SPARQL [4].

Table 1. S_{Sensor}: an RDF stream of sensors

Subject (sub)	Predicate (pre)	Object (obj)	Timestamp
...	...	...	...
observation:A2-1	om-owl:observedProperty	weather:AirTemperature	1483850233586
observation:A2-1	om-owl:procedure	sensor:A2	1483850233586
observation:A2-1	om-owl:result	measure:A2-1	1483850233586
measure:A2-1	om-owl:floatValue	73.0^^xsd:float	1483850233586
observation:A1-2	om-owl:observedProperty	weather:AirTemperature	1483850237596
observation:A1-2	om-owl:procedure	sensor:A1	1483850237596
observation:A1-2	om-owl:result	measure:A1-2	1483850237596
measure:A1-2	om-owl:floatValue	69.0^^xsd:float	1483850237596
...	...	...	...

Query registration C-SPARQL queries should be continuously registered to provide continuous querying services. It can be indicated in the following REGISTRATION QUERY clause.

Registration → 'REGISTRATION QUERY' QueryName
['COMPUTED EVERY' *Number TimeUnit*] 'AS' Query
TimeUnit → 'ms' | 's' | 'm' | 'h' | 'd'

The optional COMPUTED EVERY clause identifies the frequency of the update of the query. If it's not been assigned, it depends on the system the frequency of the query automatically computes. Every registered C-SPARQL query yields continuous results whose type and form are similar to standard SPARQL query. Apart from this, the output of C-SPARQL allows new RDF streams through the following REGISTRATION STREAM clause.

Registration → 'REGISTRATION STREAM' QueryName
['COMPUTED EVERY' *Number TimeUnit*] 'AS' Query

Window C-SPARQL does not process a whole RDF stream but its snapshots every time. For this reason, C-SPARQL introduces the notion of *window*, storing snapshots of RDF streams. *Window* can be defined via the FROM clause.

FromStrClause → 'FROM' ['NAMED']'STREAM' *StreamIRI* '[RANGE' Window']'
Window → *LogicalWindow* | *PhysicalWindow*
LogicalWindow → *Number TimeUnit WindowOverlap*
WindowOverlap → 'STEP' *Number TimeUnit* | 'TUMBLING'
PhysicalWindow → 'TRIPLES' *Number*

Note that windows depend upon two parameters, namely, the window size (RANGE: the maximal time-interval of RDF stream quadruples) and the window step (STEP: the frequency of updates of windows).

Example 2. Let Q_{Sensor} denote a C-SPARQL query *SensorQuery* shown in the table:

REGISTER QUERY *SensorQuery* AS
PREFIX om-owl: ⟨*http://knoesis.wright.edu/ssw/ont/sensor-observation. owl#*⟩
PREFIX weather: ⟨*http://knoesis.wright.edu/ssw/ont/weather.owl#*⟩
SELECT ?sensor ?obs ?value
FROM STREAM *SensorStreams* [RANGE 5s   STEP 4s]
WHERE { ?obs observedProperty AirTemperature ;
    om-owl:procedure ?sensor ;
    om-owl:result [om-owl:floatValue ?value] . }

Thus *SensorQuery* is registered. And the window size of Q_{Sensor} is 5 s and the window step of Q_{Sensor} is 4 s. Besides, the WHERE clause together with the SELECT clause is the core part of *SensorQuery*.

Analogously, the semantics of C-SPARQL is defined in terms of sets of so-called *mappings* which are simply total functions $\mu: V \to U$ on some finite set V' of variables. Formally, let Q be a C-SPARQL query, S an RDF stream, and t a time (taken as the initial time). We can use $[\![Q]\!]_{S(t)}$ to denote the semantics of Q over S at t (where $S(t)$: the prefix of S ending at t) defined via both its indicated window [4] and the core patterns as SPARQL patterns [18].

Furthermore, let k be a natural number. We use $[\![Q]\!]_{S(t,k)}$ to denote the semantics of Q over the k-th window of S starting at t. Intuitively, $[\![Q]\!]_{S(t,k)}$ is a local semantics restricted on a window. In other words, $[\![Q]\!]_{S(t,k)}$ is a subset of $[\![Q]\!]_{S(t)}$. In this sense, $[\![Q]\!]_{S(t,0)}$ denotes the semantics of Q over the *initial window* starting at t.

In Examples 1 and 2, consider an initial time $t = 1483850232556$. We have $[\![Q_{\text{Sensor}}]\!]_{S_{\text{Sensor}}(t,0)} = \{\mu_0\}$ and $[\![Q_{\text{Sensor}}]\!]_{S_{\text{Sensor}}(t,1)} = \{\mu_1\}$, where μ_0 and μ_1 are shown in Table 2.

Table 2. R_{Sensor}: The results of the example.

No	?sensor	?obs	?value
μ_0	sensor:A2	observation:A2-1	$73.0^{\wedge\wedge}$xsd:float
μ_1	sensor:A1	observation:A1-2	$69.0^{\wedge\wedge}$xsd:float

3   From C-SPARQL to SPARQL

In this section, we present a formal specification of C-SPARQL query and then reduce the evaluation of C-SPARQL query to the evaluation of SPARQL query equivalently.

3.1   Formal Specification of C-SPARQL Query

In Sect. 2.2, it can be seen that through adding new production, i.e., windows (including the window size, i.e., RANGE, and the frequency of updates of windows, i.e., STEP), registration and so on, to the standard grammar of SPARQL to extend into C-SPARQL in order to process RDF streams.

Definition 1. *Formally, a C-SPARQL query Q can be taken as a 5-tuple of the form:*

$$Q = [\text{Req}, S, \text{w}, \text{s}, \rho(Q)] \tag{2}$$

where

- Req: *the registration;*
- S: *the RDF stream registered;*
- w: *RANGE, i.e., the window size;*
- s: *STEP, i.e., the updating time of windows;*
- $\rho(Q)$: *a SPARQL query.*

For convenience, let Q be a C-SPARQL query. We use $\text{Req}(Q)$, $S(Q)$, $\text{w}(Q)$, $\text{s}(Q)$, and $\rho(Q)$ to denote the registration, the RDF stream registered, RANGE, STEP, and the SPARQL query of Q, respectively. Now, we consider an example:

Example 3. Consider the *SensorQuery* Q_{Sensor} in Example 2. Based on Eq. (2), we can find:

- $\text{Req}(Q_{\text{Sensor}}) = $ **REGISTER QUERY** *SensorQuery* AS;
- $S(Q_{\text{Sensor}}) = $ *SensorStreams;*
- $\text{w}(Q_{\text{Sensor}}) = 5s$;
- $\text{s}(Q_{\text{Sensor}}) = 4s$;
- $\rho(Q_{\text{Sensor}})$ is a SPARQL query as follows:

PREFIX   om-owl: ⟨*http://knoesis.wright.edu/ssw/ont/sensor-observation. owl#*⟩
PREFIX   weather: ⟨*http://knoesis.wright.edu/ssw/ont/weather.owl#*⟩
SELECT ?sensor ?obs ?value
WHERE { ?obs observedProperty AirTemperature ;
                 om-owl:procedure ?sensor ;
                 om-owl:result [om-owl:floatValue ?value] . }

By comparing Example 3 with Example 2, we have the following observations:

Remark 1. Comparing SPARQL, C-SPARQL has some slight differences.

- The registration of C-SPARQL query ensures the continuous recall of C-SPARQL query periodically. However, SPARQL query does not support such a continuous mechanism for recalling query. So we need to design an extra trigger to provide the mechanism (discussed later, see the next section).
- C-SPARQL can support RDF streams but SPARQL cannot due to the timestamp of tuples of RDF streams. We note that the timestamp of quadruples in a window of an RDF stream can be ignored. In this sense, SPARQL can characterize the core pattern of C-SPARQL in a window of an RDF stream.
- C-SPARQL query consists of RDF streams which are to be processed. Indeed, C-SPARQL query processes periodical windows of RDF streams and windows can be stored in the present RDF dataset which can be evaluated by SPARQL queries.

Based on discussions above, *window* is an important notion in transforming C-SPARQL to SPARQL.

Let k be a natural number. A k-th (logical) *window*, denoted by $W(S, \mathrm{w}, \mathrm{s}, t, k)$, for an RDF stream S, a window size (RANGE) w, an updating time of windows (STEP) s, and an initial time t are a collection of quadruples defined as follows:

$$W(S, \mathrm{w}, \mathrm{s}, t, k) = \{(\langle s, p, o \rangle, \tau) \in S \mid t + k\mathrm{s} - \mathrm{w} \leq \tau \leq t + k\mathrm{s}\}. \qquad (3)$$

Intuitively, a window is a snapshot of a stream. Accordingly, when $k = 0$, it is the initial window.

For instance, in Example 3, the initial window $W(S_{\mathrm{Sensor}}, 5s, 4s, 148385\,0232556, 0)$ is shown in Table 3 and the first window $W(S_{\mathrm{Sensor}}, 5s, 4s, 148385023\,2556, 1)$ is shown in Table 4:

Table 3. W_0: The initial window of S_{Sensor} starting at t

Subject	Predicate	Object	Timestamp
observation:A2-1	om-owl:observedProperty	weather:AirTemperature	1483850233586
observation:A2-1	om-owl:procedure	sensor:A2	1483850233586
observation:A2-1	om-owl:result	measure:A2-1	1483850233586
measure:A2-1	om-owl:floatValue	73.0^^xsd:float	1483850233586

The C-SPARQL query evaluation on a stream can be reduced to the evaluation on windows. Then we can have the following:

Lemma 1. *Let Q be a C-SPARQL query. For any RDF stream S and any initial time t, if S is registered in Q then for any natural number k, $[\![Q]\!]_{S(t,k)} = [\![Q]\!]_{W_k}$, where $W_k = W(S, \mathrm{w}, \mathrm{s}, t, k)$.*

Table 4. W_1: The 1st window of S_{Sensor} starting at t

Subject	Predicate	Object	Timestamp
observation:A1-2	om-owl:observedProperty	weather:AirTemperature	1483850237596
observation:A1-2	om-owl:procedure	sensor:A1	1483850237596
observation:A1-2	om-owl:result	measure:A1-2	1483850237596
measure:A1-2	om-owl:floatValue	69.0^^xsd:float	1483850237596

Proof (Skecth). We mainly discuss the function of RANGE. This equation holds since the evaluation of C-SPARQL queries is only restricted in those tuples of RDF streams within RANGE from those queries, which are already in the window W by definition.

Let W be a window. We use G(W) to denote the collection of RDF triples in W. In this sense, G(W) is an RDF graph by removing the timestamp of quadruples. As a result, we can reduce the evaluation of C-SPARQL queries on a window to the evaluation of SPARQL queries on an RDF graph generated from that window by removing timestamp.

The following property shows that the reduction from one of window from an RDF stream to its RDF graph is equivalent.

Lemma 2. *Let Q be a C-SPARQL query. For any RDF stream S and any initial time t, if S is registered in Q then for any natural number k, $[\![Q]\!]_{W_k} = [\![\rho(Q)]\!]_{G(t)}$, where $W_k = \mathrm{W}(S, w, s, t, k)$ and $G(t, k) = \mathrm{G}(W_k)$.*

By Lemmas 1 and 2, we can conclude the main result of this paper:

Theorem 1. *Let Q be a C-SPARQL query. For any RDF stream S and any initial time t, if S is registered in Q then for any $k = 0, 1, 2, \ldots,$ $[\![Q]\!]_{S(t,k)} = [\![\rho(Q)]\!]_{G(t,k)}$, where $G(t, k) = \mathrm{G}(W_k)$.*

In Examples 1 and 3, considering an initial time $t = 1483850232556$, we have that $[\![\rho(Q_{\text{Sensor}})]\!]_{G(W_0)} = \{\mu_0\}$ and $[\![\rho(Q_{\text{Sensor}})]\!]_{G(W_1)} = \{\mu_1\}$, where μ_0 and μ_1 are already stated in Table 2. Clearly, $[\![Q_{\text{Sensor}}]\!]_{W_k} = [\![\rho(Q_{\text{Sensor}})]\!]_{G(t,k)}$ $(k = 0, 1)$. That is, $[\![Q_{\text{Sensor}}]\!]_{S_{\text{Sensor}}(t,k)} = [\![\rho(Q_{\text{Sensor}})]\!]_{G(t,k)}$ $(k = 0, 1)$.

Theorem 1 ensures that the evaluation problem of C-SPARQL queries over RDF streams can be equivalent to the evaluation problem of SPARQL queries over RDF graphs. Moreover, Theorem 1 can show that the evaluation problem of C-SPARQL has the same computational complexity as SPARQL [4].

4   A Framework for RDF Stream Processing

In this section, we introduce PRSP, an adaptive framework for processing RDF stream.

The framework of PRSP is shown in Fig. 1, which contains four main modules: query parser, trigger, data transformer, and SPARQL API. We give a

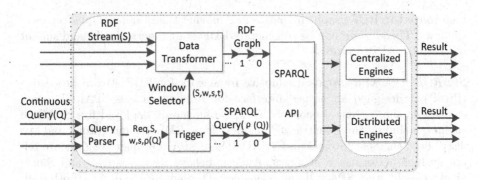

Fig. 1. The framework of PRSP

detailed description about the framework below. Both C-SPARQL queries and RDF streams as the input of PRSP, they are transformed by query parser module and data transformer module, respectively. And the output of query parser can be of immediate use or can be processed by trigger module, which is to call SPARQL queries and produce window selector periodically. After that, data transformer module generates RDF graphs periodically via the input of RDF streams and window selector. Meanwhile, SPARQL API module is capable of getting RDF graphs and SPARQL queries obtained from the former modules as inputs and producing the final results. PRSP is an adaptive framework for RDF stream processing since it can apply various SPARQL query engines to process RDF stream. And the right box, consisting of any SPARQL query engine, is used as a black box to evaluate RDF graphs.

Query Parser. The query parser module replies on the information captured by Denotational Graph (i.e., D-Graph) [5] which is defined as a view on the O-Graph [18], to obtain parameters of windows and core patterns from the input, i.e., continuous query (Q). The output of this module is the 5-tuple (i.e., Req, S, w, s, $\rho(Q)$) of a Q defined in Sect. 3, and they can be addressed in trigger module.

Trigger. The 5-tuple (i.e., Req, S, w, s, $\rho(Q)$) of a C-SPARQL (Q) is as the input of the trigger module which is to call SPARQL queries ($\rho(Q)$) and produce window selector (S, w, s, t) periodically. Let t_0 be an initial time. In Sect. 3, for any $k = 0, 1, \ldots$, we have $t = t_0 + ks$. And let k denote the k-th window of RDF stream S starting at t_0 over Q and the update frequency of SPARQL queries $\rho(Q)$ parsing from query parser module. The window selector (S, w, s, t) is captured by data transformer module, and SPARQL query is pushed into SPARQL API to be executed in a query engine.

Data Transformer. Via Esper or another DSMS, the data transformer module transforms RDF streams S specified in continuous query Q to capture snapshots based on the window selector (S, w, s, t) obtained from query parser module, and then convert to RDF graphs by means of removing timestamps of quadruples in windows. Therefore, it can be to generate RDF graphs periodically, and submit to SPARQL API to process.

SPARQL API. Our proposed adaptive framework for RDF stream processing (PRSP) is designed an unified interface for running various SPARQL query engines. In the current version of PRSP, we have only deployed five SPARQL engines, including three centralized engines, Jena, RDF-3X and gStore, and two distributed engines, gStoreD and TriAD. Those SPARQL query engines are relatively high-performance or newer. And we believe that it is easy and convenient to apply other SPARQL query engines. Through SPARQL API, both RDF graphs $G(t, k)$ and SPARQL query $\rho(Q)$ as the input of PRSP, it will output the continuous and real-time query results that users need. Because the most systems are considered scientific prototypes and work in progress, there is no doubt that they can't support all capabilities and querying services. For instance, gStoreD merely provide query model of BGP (basic graph pattern), and can't support the operators of filter and so on.

5   Experiments and Evaluations

5.1   Experimental Setup

Implementations and Running Environment. All centralized experiments, including exploiting Jena, RDF-3X and gStore to process RDF streams, were carried out on a machine running Ubuntu 14.04.5 LTS, which has 4 CPUs with 6 cores (E5-4607) and 64 GB memory. And a cluster of 5 machines (1 master and 4 workers) with the same performance as the former were used for distributed experiments. All query systems, including three centralized engines (Jena, RDF-3X, and gStore) and two distributed systems (TriAD and gStoreD), are SPARQL query engines for subgraph matching.

Dataset and Continuous Queries. For evaluation, we utilized YABench RSP benchmark [11], which provided a real world dataset describing different water temperatures captured by sensors spread throughout the underground water pipeline systems. In our experiments, we perform tumbling windows with a 5-seconds-window which slides every 5 s, and sliding windows with a 5-seconds-window which slides every 4 s. The experiments were carried out under five different input loads (i.e., $s = 500/1000/1500/2000/2500$ *sensors*) for windows using the query template Q_{Sensor}. The complexity of the scenarios was in the ascending order, from the least complex configuration ($s = 500$ *sensors*) that loaded roughly 42,000 triples to the most complex configuration ($s = 2500$ *sensors*) that injected more than 210,000 triples. For comparison, consider that different SPARQL query engines have different capabilities, and some systems can not support complex queries (e.g., filter operator, aggregation operator), so the experiments chose a BGP query template Q_{Sensor} with four triple patterns.

5.2 Performance Analysis

When processing RDF streams, it can be considered as three procedures: triples load time (TLT), query response time(QRT), and engine execution time (EET). TLT indicates the total time of RDF graphs obtained from data transformer module loading to SPARQL query engines. QRT denotes the registered query response time. EET means the total execution time of a SPARQL engine while processing all windows, containing TLT and QRT. We evaluate performance in terms of average time of the three procedures from the whole streams whose duration set at 30 s, respectively.

Figures 2, 3 and 4 compare the different systems within PRSP under the five scenarios, and corresponding the three processes (TLT, QRT, and EET) are depicted in Fig. 5(a)–(d), respectively. For most of the cases, all processes increase varying degrees with the increase in the amount of RDF dataset in windows, but there are slightly different. In addition to gStore, the centralized SPARQL query engines outperform the distributed engines by orders of magnitude under the lower load owing to the unnecessary for centralized systems to communicate with other nodes. Moreover, the gaps among the three procedures between gStoreD and TriAD decrease slightly along with the increase of the dataset, because they are designed to handle large datasets. It also reveals when the input load ranges from $s = 500$ sensors to $s = 1000$ sensors (i.e., the lower load), RDF-3X has a better performance than Jena and gStore, whereas Jena outperforms both RDF-3X and gStore under the higher load (i.e., $s = 2500$ sensors). TriAD is superior to gStoreD under the five scenarios. Besides that, the TLT of both gStore and gStoreD occupy a large rate of EET, resulting in their lower efficiency for processing RDF streams.

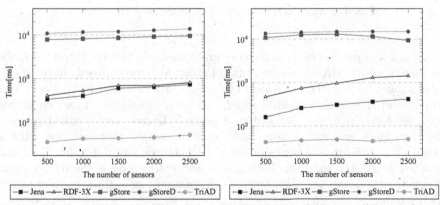

(a) The triples load time over tumbling window (b) The triples load time over sliding window

Fig. 2. Triples load time in different scenarios within PRSP

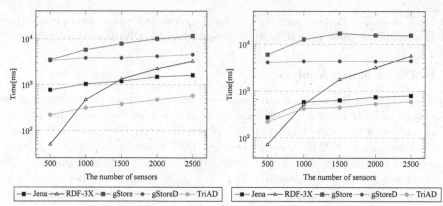

(a) Query response time over tumbling window   (b) Query response time over sliding window

Fig. 3. Query response time in different scenarios within PRSP

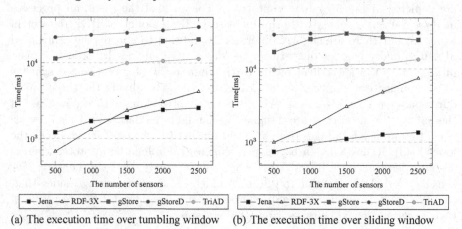

(a) The execution time over tumbling window   (b) The execution time over sliding window

Fig. 4. Execution time in different scenarios within PRSP

Jena stores its data in main memory and its schema takes the strategy space for time. Resource URIs and simple literal values are stored directly in the statement table in Jean. RDF-3X stores everything in a clustered $B^+ - Tree$. And triples, sorted in lexicographical order, can be compressed well, which makes them efficiently scan and fast lookup if prefix is known. TriAD combines join-ahead pruning by using a novel form of RDF graph summarization with a locality-based, horizontal partitioning of RDF triples into a grid-like, i.e., distributed index structure. But both gStore and gStoreD parse the RDF graphs to construct indexes, i.e., $VS*tree$, which consumes more time, in order to get results faster.

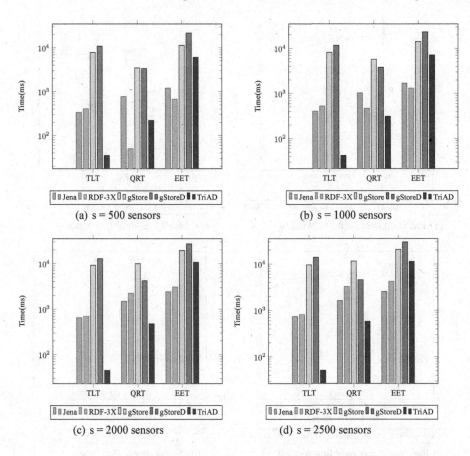

Fig. 5. RDF stream for processing time within PRSP

To assess the scalability performances between the two distributed engines (i.e., gStoreD and TriAD), the experimentation is based on four different set-ups by increasing the number of nodes: 2 nodes, 3 nodes, 4 nodes, and 5 nodes. The graph in Fig. 6(d) indicates the engines execution time for the windows over RDF streams for different number of nodes under $s = 1000$ sensors. Owing to the lack of available and ready-to-use distributed RSP engines, we just compare the two distributed systems. It is noticeable that until the two engines with 5 nodes in distribution model, the *EET* reduces along with the increase of the nodes. It implies the communication of the master with slaves with fewer nodes occupies a large rate of *EET*.

5.3 Correctness Analysis

In our experiments, we validated the correctness of the output from our framework PRSP over the five SPARQL query engines, by means of oracle metrics from YABench RSP benchmark. The oracle tests and verifies the effectiveness

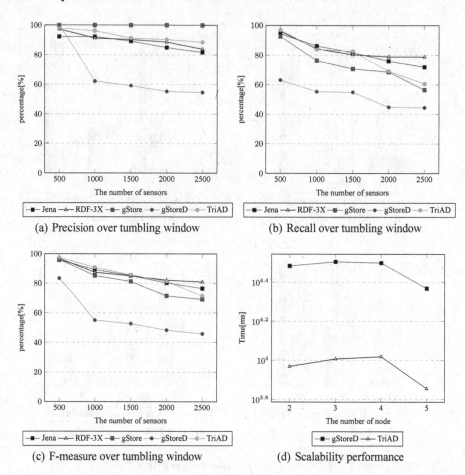

(a) Precision over tumbling window

(b) Recall over tumbling window

(c) F-measure over tumbling window

(d) Scalability performance

Fig. 6. (a–c), the correctness for PRSP, and (d) the scalability performance between gStoreD and TriAD under s = 1000 sensors

(i.e., correct and incorrect) of the output from SPARQL query engines by measuring the precision, recall, and f-measure score (i.e., the weighted harmonic mean of precision and recall) which reflecting the overall index. And the correctness analysis help us to find out which system is better to process RDF streams. Table 5 and Fig. 6(a)–(c) illustrate the results of precision, recall and f-measure scores from the experiments under the five load scenarios within PRSP. Along with more input loads for windows, most of them enjoy lower recalls with relatively high accuracy. We can observe that gStore succeeds to maintain 100% precision even though under higher load (i.e., s = 2500 sensors), but achieves lower recall. Generally, as we can see that recall scores are lower than precisions for all the five SPARQL query engines, and the values drop down dramatically when the engines are put under larger loads. Since every SPARQL query engine only can process a certain amount of data at a given time, which leads to lower recall scores under higher load.

In summary, our experiments show that different query engines in processing RDF streams exhibit different performance, and their correctness in answering is severely sensitive to the window sizes and steps to be selected.

6 Conclusions

In this paper, we present a novel framework for processing RDF streams by reducing RDF streams to RDF data. Taking advantage of the reduction, our framework can ideally support the most SPARQL endpoints including those under construction. Besides, our framework can be used to evaluate the performance and correctness of existing SPARQL query engines in processing RDF streams in a unified way, which amends the evaluation of them ranging from static RDF graphs to dynamic RDF streams. We also find that the efficiency of evaluations over RDF streams could influence the correctness of querying slightly different to RDF data. In the future work, we will adapt more query engines for RDF data, such as a redesign of database architecture [1] in order to take advantage of modern hardware and a compressed bit-matrix structure BitMat [3] for storing huge RDF graphs.

Table 5. Precision/recall/F-measure results

		Jena	RDF-3X	gStore	gStoreD	TriAD
Sensor = 500	Precision	92.3%	97.2%	100%	100%	97.9%
	Recall	94.4%	96.1%	92.6%	63.1%	96.9%
	F-Measure	95.8%	96.7%	96.1%	83.4%	97.4%
Sensor = 1000	Precision	92.2%	91.2%	100%	62.1%	96.2%
	Recall	86.2%	84.3%	76.3%	55.3%	83.9%
	F-Measure	89.2%	87.4%	85.1%	55.1%	90.8%
Sensor = 1500	Precision	89.2%	90.3%	100%	59.1%	91.3%
	Recall	81.7%	80.3%	70.7%	54.9%	82.5%
	F-Measure	85.2%	85.0%	81.2%	52.8%	85.6%
Sensor = 2000	Precision	84.9%	88.6%	100%	55.2%	90.3%
	Recall	76.0%	78.9%	68.6%	44.8%	69.1%
	F-Measure	80.2%	82.1%	71.5%	48.4%	81.3%
Sensor = 2500	Precision	81.7%	83.9%	100%	54.6%	88.6%
	Recall	71.9%	78.8%	56.5%	44.5%	60.6%
	F-Measure	76.5%	80.9%	69.2%	45.9%	71.5%

Acknowledgments. This work is supported by the programs of the National Natural Science Foundation of China (61672377), the National Key Research and Development Program of China (2016YFB1000603), and the Key Technology Research and Development Program of Tianjin (16YFZCGX00210).

References

1. Boncz, P.A., Kersten, M.L., Manegold, S.: Breaking the memory wall in MonetDB. Commun. ACM **51**(12), 77–85 (2008)
2. Anicic, D., Fodor, P., Rudolph, S., Stojanovic, N.: EP-SPARQL: a unified language for event processing and stream reasoning. In: Proceedings of WWW 2011, pp. 635–644 (2011)
3. Atre, A., Chaoji, V., Zaki, M.J., Hendler, J.A.: Matrix "Bit" loaded: a scalable lightweight join query processor for RDF data. In: Proceedings of WWW 2010, pp. 41–50 (2010)
4. Barbieri, D.F., Braga, D., Ceri, S., Della Valle, E., Grossniklaus, M.: Querying RDF streams with C-SPARQL. SIGMOD Rec. **39**(1), 20–26 (2010)
5. Barbieri, D.F., Braga, D., Ceri, S., Grossniklaus, M.: An execution environment for C-SPARQL queries. In: Proceedings of EDBT 2010, pp. 441–452 (2010)
6. Calbimonte, J.-P., Corcho, O., Gray, A.J.G.: Enabling ontology-based access to streaming data sources. In: Patel-Schneider, P.F., Pan, Y., Hitzler, P., Mika, P., Zhang, L., Pan, J.Z., Horrocks, I., Glimm, B. (eds.) ISWC 2010. LNCS, vol. 6496, pp. 96–111. Springer, Heidelberg (2010). doi:10.1007/978-3-642-17746-0_7
7. Carroll, J.J., Dickinson, I., Dollin, C., Reynolds, D., Seaborne, A., Wilkinson, K.: Jena: implementing the semantic web recommendations. In: Proceedings of WWW 2004 (Alternate Track Papers & Posters), pp. 74–83 (2004)
8. Gurajada, S., Seufert, S., Miliaraki, I., Theobald, M.X.: TriAD: a distributed shared-nothing RDF engine based on asynchronous message passing. In: Proceedings of SIGMOD 2014, pp. 289–300 (2014)
9. Hoeksema, J., Kotoulas, S.: High-performance distributed stream reasoning using s4. In: Proceedings of Ordring Workshop at ISWC 2011 (2011)
10. Khrouf, H., Belabbess, B., Bihanic, L., Kepeklian, G., Curé, O.: WAVES: big data platform for real-time RDF stream processing. In: Proceedings of SR+SWIT@ISWC 2016, pp. 37–48 (2016)
11. Kolchin, M., Wetz, P., Kiesling, E., Tjoa, A.M.: YABench: a comprehensive framework for RDF stream processor correctness and performance assessment. In: Bozzon, A., Cudre-Maroux, P., Pautasso, C. (eds.) ICWE 2016. LNCS, vol. 9671, pp. 280–298. Springer, Cham (2016). doi:10.1007/978-3-319-38791-8_16
12. Le-Phuoc, D., Dao-Tran, M., Xavier Parreira, J., Hauswirth, M.: A native and adaptive approach for unified processing of linked streams and linked data. In: Aroyo, L., Welty, C., Alani, H., Taylor, J., Bernstein, A., Kagal, L., Noy, N., Blomqvist, E. (eds.) ISWC 2011. LNCS, vol. 7031, pp. 370–388. Springer, Heidelberg (2011). doi:10.1007/978-3-642-25073-6_24
13. Le-Phuoc, D., Nguyen Mau Quoc, H., Le Van, C., Hauswirth, M.: Elastic and scalable processing of linked stream data in the cloud. In: Alani, H., et al. (eds.) ISWC 2013. LNCS, vol. 8218, pp. 280–297. Springer, Heidelberg (2013). doi:10.1007/978-3-642-41335-3_18
14. Li, Q., Zhang, X., Feng, Z.: PRSP: a plugin-based framework for RDF stream processing. In: Proceedings of WWW 2017, poster, pp. 815–816 (2017)

15. Margara, A., Cugola, G.: Processing flows of information: from data stream to complex event processing. In: Proceedings of DEBS 2011, pp. 359–360 (2011)
16. Margara, A., Urbani, J., Van Harmelen, F., Bal, H.: Streaming the web: reasoning over dynamic data. J. Web Semant. **25**(1), 24–44 (2014)
17. Neumann, T., Weikum, G.: The RDF-3X engine for scalable management of RDF data. VLDB J. **19**(1), 91–113 (2010)
18. Pérez, J., Arenas, M., Gutierrez, C.: Semantics and complexity of SPARQL. In: Cruz, I., Decker, S., Allemang, D., Preist, C., Schwabe, D., Mika, P., Uschold, M., Aroyo, L.M. (eds.) ISWC 2006. LNCS, vol. 4273, pp. 30–43. Springer, Heidelberg (2006). doi:10.1007/11926078_3
19. Peng, P., Zou, L., Özsu, M.T., Chen, L., Zhao, D.: Processing SPARQL queries over distributed RDF graphs. VLDB J. **25**(2), 243–268 (2016)
20. Zou, L., Özsu, M.T., Chen, L., Shen, X., Huang, R., Zhao, D.: gStore: a graph-based SPARQL query engine. VLDB J. **23**(4), 565–590 (2014)

Investigating Microstructure Patterns of Enterprise Network in Perspective of Ego Network

Xiutao Shi[1], Liqiang Wang[1], Shijun Liu[1,2(✉)], Yafang Wang[1], Li Pan[1,2], and Lei Wu[1,2]

[1] School of Computer Science and Technology, Shandong University, Jinan 250101, China
{shixtwy,wanglq1989}@163.com
[2] Engineering Research Center of Digital Media Technology, Ministry of Education, Jinan 250101, China
{lsj,yafang.wang,panli,i_lily}@sdu.edu.cn

Abstract. In social networks the behavior of individuals can be researched through the evolution of the microstructure. As we know, triad is the basic atom shape to build the whole social network. However we find that quad plays the basic role rather than triad in Enterprise Network (EN). In particular, we focus on four typical microstructure patterns including triad, 4-cycle, 4-chordalcycle and 4-clique in EN. We propose algorithms to mine these microstructure patterns and compute the frequencies of each type of microstructure patterns in an efficient parallel way. We also analyze the structural features of these microstructure patterns in a perspective of ego network. Additionally we present the evolutionary rules between these microstructure patterns based on the statistical analysis. Finally we combine the features into traditional methods to solve the link prediction problem. The results show that these features and our combination methods are effective to predict links between enterprises in EN.

Keywords: Microstructure patterns · Enterprise network · Ego network · Triad · Quad · Evolutionary rules · Link prediction

1 Introduction

Motivation. Nowadays, the links between enterprises are becoming tighter and the inter-enterprise relations are getting more and more complex. These links and relations are producing huge amounts of data, e.g., the information of the enterprise, the business interactions between enterprises and the unstructured data on social business. One of the most valuable data is the large enterprise network (EN).

In the social sciences, social structure is the patterned social arrangements in society [1]. These structures are emergent from the actions of the individuals,

© Springer International Publishing AG 2017
L. Chen et al. (Eds.): APWeb-WAIM 2017, Part I, LNCS 10366, pp. 444–459, 2017.
DOI: 10.1007/978-3-319-63579-8_34

thus we can analyze the features of individual actions through the social structures. The microstructures in EN can not only reflect the behavior of a single enterprise, but also reflect the way and characteristics of interactions between enterprises. It is possible to study the interaction patterns among enterprises and make more accurate recommendations and predictions based on these features. Researchers have studied enterprise networks at multiple levels of structures, including the dyad [2], the ego network [3], and the overall network [15]. However, less attention has been paid to triads and quads yet.

Contributions. We propose a parallel algorithm to search both triad and quad graphlets [5] (inc. 4-cycle, 4-chordalcycle and 4-clique) in EN, which reduced time complexity compared to other approaches. We conduct experiments to study and evaluate triad and quad graphlets to analyze the behaviors of different enterprises. We also apply triad and quad features for link prediction, and the results demonstrate that the methods with quad features performs better than the other features in EN.

2   Related Work

Microstructures Analysis. While analysing microstructures of 3-nodes and 4-nodes, Ahmed et al. [4] proposed a fast, efficient and parallel algorithm for counting microstructures of size k=3, 4-nodes. Trpevski et al. [17] found that vertex signature vector and microstructure correlation matrix are powerful tools for network analysis. Yanardag and Vishwanathan [20] presented a framework to learn latent representations of sub-structures for graphs. It leveraged the dependency information between sub-structures to improve classification accuracy. Madhavan et al. [15] found a characteristic of enterprise networks: enterprises tend to form triads defined by geography. This is a situation similar to network closure. Biswas and Biswas [7] focused on ego centric community detection in network data, mainly emphasizing on structural aspects, i.e., reachability and isolability. Dunbar et al. [8] studied the internal structure of these networks to determine whether they have the same kind of layered structure as offline face-to-face networks. Toral et al. [16] analyzed the social network structures in both macro and micro view. Li and Daie [13] proposed a hierarchical clustering method for configuring assembly supply chains to evaluate the coupling according to the product variety information. Girard et al. [10] studied how students social networks emerge by documenting systematic patterns in the process of friendship formation of incoming students. The shape of local and global network structures resulting from this process. Gordon and McCann [11] distinguished three ideal-typical models of processes which may underlie spatial concentrations of related activities. Survey data is used for the London conurbation to explore the relations between concentration and different forms of linkage. These works are mostly focus on analyzing features and evolution of triad. In this paper, we investigated quads in EN to analyze the behaviors between enterprises.

Link Prediction. Lou et al. [14] employed graphical model to predict reciprocity and triadic closure. Huang et al. [12] studied the triadic closure information in dynamic social network and proposed a probabilistic factor model for modeling and prediction. Dong et al. [9] predicted links in heterogeneous networks by a ranking factor graph model. Bhuiyan et al. [6] proposed GUISE, which uses a Markov Chain Monte Carlo (MCMC) sampling method for constructing the approximate Graphlet Frequency Distribution (GFD) of a large network. Wasserman and Pattison [19] described a series of models for investigating the structures in social networks, including several generalized stochastic block models. This paper evaluated the effectiveness of microstructure patterns for link prediction in EN.

3   Mining Microstructure Pattern in EN

3.1   Definitions

Enterprise Network (EN). As defined, EN = (ENT, E). ENT is the enterprise node collection, $i \in$ ENT, i is an enterprise and E is the edge collection. $e \in E$, $e = (i, j)$, i is a supplier and j is a manufacturer, the edge is from i to j [18] (Fig. 1).

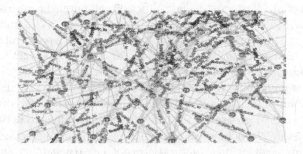

Fig. 1. The visualization of a part of EN.

Ego Network. Researchers usually examine the business patterns in the ego network with a central enterprise, for example, manufacturers concern more about their direct-connected suppliers, so we focus on the microstructures in the perspective of ego network.

Ego network is a well-known social phenomenon which deals with individual interest and the relationship [7]. Ego networks consist of a central node (ego) and the nodes to whom ego is directly connected (these are called alters). We can choose an arbitrary node to be the ego and different egos lead to different types of ego networks. For example, take a manufacturer as the ego, the manufacturer and its suppliers form an ego network. This ego network has different characteristics with the ego network that has a supplier ego (Fig. 2).

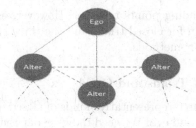

Fig. 2. Example of ego network. Any arbitrary central node (Ego) which connects other nodes (Alters) in the network (solid lines) and connectivity among alters (dashed lines). Connectivity with rest of the network is represented with dashed-dotted lines.

3.2   Category of Microstructures

Among the multiple levels of microstructure, triad—subsets of three network nodes and the possible ties among them—are the important topic of early research on social networks. However triads are not so valuable in EN. In a real business scenario, two enterprises A and B usually interact with each other directly rather than through other enterprises. Quad contains more information than triad including both indirect and direct relationships between two connected enterprises. Therefore, quads are more important than triads in EN. As EN is directional, there are more kinds of triads and quads compared with the undirected graph. Additionally, there is no two-way relationship between enterprises in this directed network, so there are only 8 possible different closed triads.

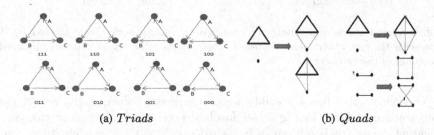

(a) *Triads*   (b) *Quads*

Fig. 3. Triad code representation and the formation of quads.

As shown in Fig. 3(a), we encode every triad with a 3-length binary number. A, B and C represent the enterprise. We defined that $A \rightarrow B$, $B \rightarrow C$ and $C \rightarrow A$ is positive direction represented by 1. The relationship AB represents that enterprise A supplies products to enterprise B. Meanwhile, 0 means negative direction. For example, 111 represents one type of triad that contains $A \rightarrow B$, $B \rightarrow C$ and $C \rightarrow A$. Figure 3(b) shows the forming of several critical microstructures, which works as the basic rule of our closed quads search algorithm. A closed triad and a point associated with it will form a 4-clique or 4-chordalcycle [4]. Two completely different relationships form a 4-cycle [4] by linking the corresponding points together. However, as the EN is directed, a structural classification for closed triads and the three classical quads are the main features used in our methods.

3.3   The Patterns of Triads and Quads

The closed triads and three representative kinds of closed quads are classified by relations between alters and ego. We need to figure out the specific classification and the mentioned structures in each category. In this way, we can not only calculate the frequency of different types of graphlets that each enterprise has, but also generalize the isomorphic patterns. We show the classifications in Fig. 4.

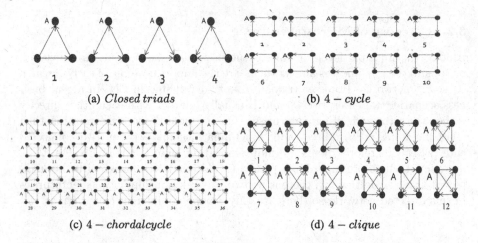

(a) *Closed triads*   (b) 4 − *cycle*

(c) 4 − *chordalcycle*   (d) 4 − *clique*

Fig. 4. Closed triads and quads are classified by ego network. Node labeled by A represents the ego, other nodes represent alters. These triads and quads are classified according to the ego network of A.

Obviously, triad has 4 possible 3-nodes variants. Subsequently, we get the data statistics (shown in Fig. 5) for involved enterprises in terms of this classification. Some triads (shown in Fig. 4(a)) can be categorized into these four variants by rotating around A. For example, the triad 110 becomes triad 100 by rotation and both of them belong to triad No. 3 shown in Fig. 4(a). Similarly, some 4-nodes graphlets can also be converted into the same type shown. For example, 4-cycle 0000 and 4-cycle 1111 belong to 4-cycle No. 1 shown in Fig. 4(b). The 4-clique 111111, 110001, 101111, 110000, 001010 and 000010 all belong to 4-clique No. 1 shown in Fig. 4(d), and 4-chordalcycle 000012, 111112 belong to 4-chordalcycle No. 1 shown in Fig. 4(c).

3.4   Mining Algorithms

In this section, we describe our approach for querying and recording the microstructure patterns. This algorithm takes only a fraction of the time compared with the prevailing methods. Given the enterprise graph $G = (V, E)$. Firstly, our proposed algorithm traverses all the edges when querying the triad. We define that $N(u)$ and $N(v)$ are the set of neighbors of u and v respectively. Given a single edge $e = (u, v) \in E$, if a node $w \in N(u) \cap N(v)$, we record u, v and w with their information and relationship between them, then we get a triad. Based on these triads, we start to search 4-nodes microstructures (Quads). As shown in Fig. 3(b), the searching principle is that the 4-chordalcycle and 4-clique can be decomposed into 3-nodes triads and the 4-cycle can be decomposed into two edges. We summarize this procedure in the following steps:

STEP 1: For each edge $e = (u, v)$, find all the common neighborhood nodes of u and v to form closed triad, and record their information and relationships between them.

STEP 2: For each closed triad, find all possible nodes which can form the 4-chordalcycle or 4-clique, and record the information of nodes and relationships.

STEP 3: For each edge $e1 = (u1, v1)$, find another edge $e2 = (u2, v2)$ to form the 4-cycle, and record the information of this 4-cycle.

STEP 4: For all closed triad and 4-nodes quads, divide them into different types and analyze them.

Algorithm 1. Mining Triad Graphlets

Input: $EN\ G = (V, E)$
Output: $TRIAD$.
1: **for** $edge\ e = (u, v) \in E$ **do**
2:     **if** u==v **then**
3:         then continue;
4:     $Union = N(u) \cap N(v)$;
5:     **for** $w \in Union$ **do**
6:         $recordToTriad(u, v, w, GetNodeRelation(u, v, w))$;

Lines 1–5 of Algorithm 1 show how to find and record triads in EN. Only when node w has relationships with both u and v in edge $e = (u, v)$, the pattern (u, v, w) could be a closed triad. However, the additional relevant functions, such as the approach to record the microstructures or remove the duplicates, are beyond our discussions, as they should be designed specifically for a concrete requirement.

Lines 1–9 of Algorithm 2 show how to find and record 4-clique and 4-chordalcycle. For any node w, w could form a 4-clique or 4-chordalcycle with triad t, if and only if $|N(w)|$ is not less than 2. We define $Linkpoints$ as the set of overlapping nodes in the neighborhoods of w and t, and define $TRIAD$ as the set of all triads in EN and $QUAD$ as the set of all quads. If t is completely in the ego network of w, it illustrates that every node has links with each other and these four nodes could form a 4-clique. And if the number of $Linkpoints$ is 2,

Algorithm 2. Mining Quad Graphlets.

Input: $EN, G = (V, E), TRIAD$
Output: $QUAD$.
   //Find 4-Clique And 4-Chordalcycle
1: **for** $node\ w \in V$ **do**
2:   **if** $|N(w)| < 2$ **then**
3:     then continue;
4:   **for** $triad\ t \in TRIAD$ **do**
5:     **if** $w \in t$ **then**
6:       then continue;
7:     **if** $t \in N(w)$ **then**
8:       $recordTo4clique(w, t, GetRelation(w, t))$
9:     **else**
10:      $Linkpoints = N(w) \cap t$
11:     **if** $|Linkpoints| == 2$ **then**
12:      $recordTo4chordalcycle(w, Linkpoints, GetRelation(w, Linkpoints)$
   // Find 4-Cycle
13: **for** $edge\ e1 = (u1, v1) \in E$ **do**
14:   **if** $u1 == v1$ **then**
15:     then continue;
16:   **for** $edge\ e2 = (u2, v2) \in E$ **do**
17:     **if** $e1 == e2\ or\ AnyPointSame(e1, e2)$ **then**
18:     then continue;
19:     **if** $u1 \in N(u2)\ and\ v1 \in N(v2) and\ u1 \notin N(v2)\ and\ v1 \notin N(u2)$ **then**
20:      $recordTo4cycle(e1, e2, GetEdgeRelation(e1, e2))$
21:     **if** $u1 \in N(v1)\ and\ v1 \in N(u2)\ and\ u1 \notin N(u2)\ and\ v1 \notin N(v2)$ **then**
22:      $recordTo4chordalcycle(w, Linkpoints, GetRelation(w, Linkpoints)$

---

it means that two nodes in t have relationships with w and these four nodes could be a 4-chordalcycle. Then we just need to select and record the information of these nodes and relationships, such as ID, direction of relationships. Lines 10–17 describe the process of finding and recording the 4-cycle. For every edge $e1 = (u1, v1)$, we try to find a related $e2 = (u2, v2)$ to form a 4-cycle. A feasible $e2$ requires that $e3 = (u1, u2)$ and $e4 = (v1, v2)$ exist while $u1, v1$ are not the neighbor of $v2, u2$ respectively, or similarly $e3 = (u1, v2)$ and $e4 = (u2, v1)$exist while $u1, v1$ are not the neighbor of $u2, v2$ respectively. Of course, if any two nodes are the same in $e1$ and $e2$, they cant form a 4-cycle.

4 Statistic Analysis of Ego Microstructures on the Entire Network

4.1 Ego Triads and Quads Characteristics in EN

Firstly, in Fig. 5(a), we can see that the frequency of type 2 is 0 and the frequency of other types is the same. Because there is no transitive supply in enterprise network, that is to say, there is no three enterprises A,B,C existing where A supplies B, B supplies C and C supplies A. The frequencies of other types are the same. The reason is that two different related suppliers supply to the same manufacturer. Secondly, Fig. 5(b) shows that type 7 and type 10 are the most commonly occurring 4-cycle. Both 4-cycle No. 7 and No. 10 are characterized by the distinct division resulting in two manufacturers and two suppliers, which share the manufacturers. As we can see, 4-cycle, 4-chordalcycle and 4-clique have structural dependence with each other. Therefore, by investigating the structures of highest frequency among the 4-chordalcycle, shown in Fig. 5(c), we discover the most important feature of these types of 4-chordalcycle compared with 4-cycle: there is a relationship between two manufacturers. Finally, Fig. 5(d) shows the number of different types of 4-clique. There is no circular supply chain, and 4-cliques whose frequency are 124 are isomorphic, which means that they share the same structure. In this structure it has a top manufacturer, two related tier suppliers, and a bottom supplier, where the bottom supplier not only indirectly supply to the top manufacturer through two tier suppliers but also directly supply to the top manufacturer.

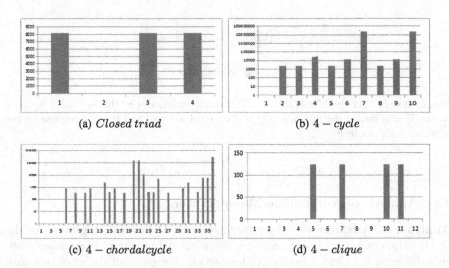

(a) *Closed triad*

(b) *4 − cycle*

(c) *4 − chordalcycle*

(d) *4 − clique*

Fig. 5. Number on the X-axis means different types of triad or quad that corresponding to Fig. 4. The frequencies of these types are represented by number on the Y-axis.

4.2   Ego Triads and Quads Characteristics Between Different Enterprise Types

As mentioned above, 4-cycle No. 7 and No. 10 are isomorphic and their frequencies are equal, while the distribution of the frequencies of 4-cycle No. 7 and No. 10 that every enterprise contains varies. Figure 6(b) shows that these enterprises have roughly equal frequency of 4-cycle No. 10. As for the distribution of 4-cycle

No. 7 in Fig. 6(a), the frequencies in several enterprises are typically high while
the rest are relatively small, even as low as 0. It also reflects the fact that the
number of suppliers is much larger than the number of manufacturers, and the
number of suppliers that supplies to large manufacturers is huge compared to
the number of manufacturers that large suppliers supply.

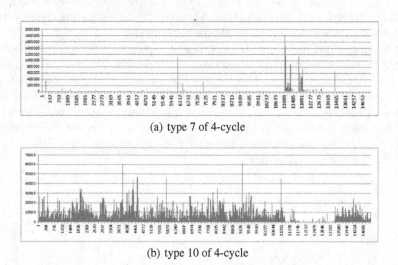

(a) type 7 of 4-cycle

(b) type 10 of 4-cycle

Fig. 6. Number on the X-axis means ID of these enterprises in this network. Number
on the Y-axis means the number of type 7 or 10 of 4-cycle, and these enterprises serve
as point A shown in Fig. 5(b).

4.3   Analysis of Automobile Manufacturers

Analysis of Automobile Manufacturers in Different Regions. As the
architectures of automobile manufacturing industry in different regions are probably different, we select a number of famous enterprises from the whole network
to study the differences in microstructures. Through the result, we can discover
the latent features underlying the enterprise network. Figure 7 shows the average
quantitative distribution of these enterprises in different microstructures.

From the Fig. 7(a), we can see that the enterprises from Korea, America and
Europe only appear as manufacturers. Most German enterprises are manufacturers and few German enterprises are intermediaries. Enterprises in China and
Japan are not only manufacturers but also intermediaries. From the other figures,
most enterprises in different regions play the role of first-tier manufacturer that
suppliers directly in microstructures of 4-nodes, and some of these enterprises
also serve as the second-tier manufacturers that suppliers indirectly by supplying
other enterprises. At the same time, some Chinese and Japanese enterprises are
intermediaries, through which suppliers supply manufacturers indirectly.

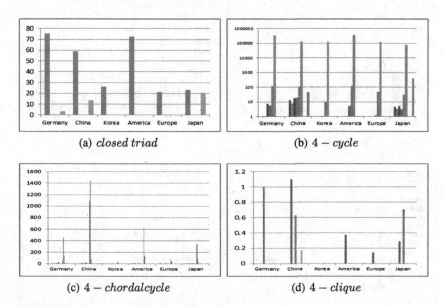

(a) *closed triad*

(b) 4 − *cycle*

(c) 4 − *chordalcycle*

(d) 4 − *clique*

Fig. 7. X-axis: different countries and regions. The number on the Y-axis means the average frequencies of different types of triad or quad. It should be noted that when we count the frequency of triads or quads about one enterprise, this enterprise must be the point A shown in Fig. 5.

Analysis of Automobile Manufacturers Producing Different Cars. The supply network of automobile enterprises that produce different kinds of cars may be different. Figure 8 shows the average distributions of enterprise that produce passenger cars, trucks, buses and commercial vehicles.

Form Fig. 8(a), we can find that these enterprises are mainly manufacturers except bus manufacturers which also appear as intermediaries in network. In addition, few passenger car manufacturers serve as suppliers in enterprise supply network. From the other figures, most of all these enterprises play the role of first tier manufacturer that suppliers directly supply in microstructures of 4-nodes and some of these enterprises also serve as the second tier manufacturers. Moreover, the passenger car manufacturers and truck manufacturers sometimes supply other enterprises directly. However, there is no enterprise that supplies an enterprise indirectly by another enterprise.

5   Microstructure Evolution and Link Predictions

5.1   Evolution Between Microstructure Patterns

In this section, we would focus on network evolution among four involved structures, analyzing four processes: 4-cycle evolves into 4-chordalcycle, 4-chordacycle evolves into 4-clique, triad evolves into 4-chordalcycle and triad evolves into 4-clique.

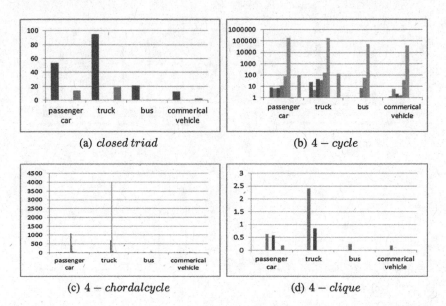

(a) *closed triad*

(b) *4 − cycle*

(c) *4 − chordalcycle*

(d) *4 − clique*

Fig. 8. X-axis: different motorcycle type for different purposes. The number on the Y-axis means the average frequencies of different types of triad or quad. It should be noted that when we count the frequencies of triads or quads about one enterprise, this enterprise must be the point A shown in Fig. 5.

In fact, type 1 to 4 of 4-chordalcycle could be constructed by attaching a bevel edge on type 1 of 4-cycle, which means that there exists a circular supply chain and a relationship where a supplier A supply to C indirectly through B or D. Because the frequency of 4-cycle 1 is 0, the frequencies of 4-chordalcycle No. 1–No. 4 are also 0 shown in Fig. 5. However, frequencies of some types in 4-chordalcycle are 0 except 4-chordalcycle No. 1–No. 4. The reason is that these types of 4-chordalcycle contain triad No. 2. The most frequently appearing 4-chordalcycle structures: type 36, 20 and 21, could be regarded as 4-cycle No. 7 or No. 10 added with a supply relationship between the manufacturers. The types 22, 34 and 36 similarly evolves from 4-cycle No. 7 or 4-cycle No. 10 but added with a supply relationship between the suppliers, rarely appear in Fig. 5(c). It demonstrates that two manufacturers are more likely to build a supply relationship between them than suppliers in enterprise network. At the same time, it explains why the frequency of triad is less than 9000 while the frequencies of 4-cycle No. 7 and No. 10 are more than twenty million: it is rare for two suppliers supplying to the same enterprise to build a supply relationship. However, there is a good chance for two enterprises with the same supplier to establish a relationship in enterprise network. We draw the inferences by observing and analyzing the type 36, 34 and 35 of 4-chordalcycle.

As we can see, there are only types 5, 7, 10 and 11 of 4-clique whose frequencies are not 0. In fact, these four types based on 4-chordalcycle No. 20 or No. 21 are isomorphic and they all mean that two manufacturers share two suppliers.

In addition, a supply relationship exists in two suppliers (or two manufacturers). When a supply relationship is generated between enterprise B and D, this 4-chordalcycle become a 4-clique.

It is not difficult to find that every 4-chordalcycle contains two triads, thus 4-chordalcycle has all the features of triad. If any type of 4-chordalcycle contains the type 2 of triad, the frequency of it must be 0, as shown in Fig. 5(c). Similar to 4-chordalcycle, every 4-clique can be viewed as a structure combined with four triads and thus the frequencies of types contained triad No. 2 in 4-clique is 0, shown in Fig. 5. There are not too many 4-cliques in enterprise network because there is no relationship between two suppliers that supply directly to the same enterprise.

5.2   Link Predictions Model Based on Microstructure Patterns

We use closed triad, 4-cycle, 4-chordalcycle and 4-clique as features to improve the prediction accuracy of different algorithms (PMF [22], GPLVM [23], SIM-RANK [24]). The goals of these algorithms are predicting whether a relationship will appear between two enterprises. *EE, ET, EF* and *ETF* are used to help prediction because they contain the relationship between enterprises and different kind of microstructures. We carried out a set of experiments to evaluate the performance with *EE, ET, EF* and *ETF*. The definition of *EE, ET, EF* and *ETF* are as follow.

(1) *EE* represents algorithms predicting the relationship by using only the enterprise-enterprise relationship network (Matrix **EE**) without those microstructures,

(2) *ET* means that algorithm uses the relationship between enterprise and closed triads No. 1–No. 4 (shown in Fig. 4) for predictions.

(3) *EF* means that algorithm uses the relationship between enterprise and 4-nodes microstructures (quads in Fig. 4), including 4-cycle, 4-chordalcycle and 4-clique for predictions.

(4) *ETF* represents using the relationship between enterprise and all microstructures (both closed triads and quads in Fig. 4).

For example, we fuse information from the enterprise-enterprise link matrix **EE** and enterprise-patterns matrix **EP** for probabilistic matrix factorization(PMF) and to predict links between enterprises. In PMF, the user-item matrix $R \in R^{M \times N}$ will be factorized into matrix $U \in R^{M \times D}$ and matrix $V \in R^{D \times N}$. U and V are latent user and item feature matrices. We use matrix **EE** and matrix **EP** to form matrix **R** shown in Fig. 10. These matrices that contain links between enterprises and these microstructure patterns are shown in Fig. 9. We try to add the matrix **EP** to the matrix **EE** to increase the accuracy of the prediction. We let $E_{1..n}$ be the ID of enterprises in EN and $P_{1..m}$ be the microstructure patterns shown in Fig. 4. The relationship between enterprises and microstructure patterns will be represented in matrix **EP**. If we only use triad as feature to help prediction(ET), the value of m is 4 and $P_{1..4}$ are

No. 1–No. 4 of triad shown in Fig. 4(a). Similarly, when using EF and ETF, the value of m are 58 and 62 respectively. Here, k_{ij} which value is 0 or 1 represents whether there is a relationship between enterprise i and enterprise j. Meanwhile, c_{ij} which value is 0 or 1 represents whether enterprise i exists in graphlets j or not.

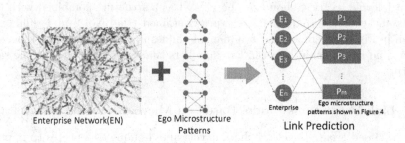

Enterprise Network(EN)   Ego Microstructure Patterns   Link Prediction

Fig. 9. Link prediction by using these microstructure patterns.

$$
E \quad \begin{array}{c} E \\ \hline \begin{array}{ccccccccc} & E_1 & E_2 & \cdots & E_n & P_1 & P_2 & P_3 & \cdots & P_m \\ E_1 & k_{11} & k_{12} & \cdots & k_{1n} & c_{11} & c_{12} & c_{13} & \cdots & c_{1m} \\ E_2 & k_{21} & k_{22} & \cdots & k_{2n} & c_{21} & c_{22} & c_{23} & \cdots & c_{2m} \\ \cdots & \cdots & \cdots & \cdots & \cdots & \cdots & \cdots & \cdots & \cdots & \cdots \\ E_n & k_{n1} & k_{n2} & \cdots & k_{nn} & c_{n1} & c_{n2} & c_{n3} & \cdots & c_{nm} \end{array} \end{array}
$$

Fig. 10. Matrix

5.3   Experiments

Dataset. To obtain the information of Enterprise Network(EN), we we produce a dataset by crawling several open websites, including the Autohome service [1] and an open enterprise network website[2]. The data is stored in Neo4j graph database. Then we use Algorithms 1 and 2 to record links between enterprises and those graphlets, such as triads, 4-cycle, 4-chordalcycle and 4-clique. The matrix described in Fig. 10 is constructed according to these information. There are 62,368 unique enterprises and 106,877 links between them in our dataset. These links represent the supply relationship between auto suppliers and manufacturers.

---

[1] http://www.autohome.com.cn.
[2] http://www.chinaautosupplier.com.

Evaluation Setting. We randomly split links between enterprises in matrix EE into training and test sets with a 4:1 ratio, where we use the training set for training. We also use a validation set from the training set to find the optimal hyperparameters for different algorithms with different patterns. Another part EP is complete in experiment. To evaluate the accuracy of link prediction on the enterprise network, we use the Area under Curve (AUC) measure, which is considered robust in the presence of imbalance [21]. Higher AUC value indicates a better performance.

5.4 Result

The results of different algorithms are shown in Table 1. It can be seen that the results of prediction achieve improvements when using the characteristics of microstructure patterns. In using different microstructure patterns, we got different AUC rates reported in Table 1. As shown in Table 1, $ETFs$ are more effective than other patterns (EE, ETs, EFs) for link prediction. The baseline method algorithm with EE, did not achieve good performance because it is difficult to predict the link between enterprises without any auxiliary information. These microstructure patterns (triad, 4-cycle, 4-chordalcycle and 4-clique) are the basic forms of the network, and can reflect the underlying structural information of EN. The results of AUC reported in Table 1 show $ETF > EF > ET > EE$. For the similar reason, 4-nodes microstructure(or graphlet) have more useful information in these methods than 3-nodes microstructure. Because we only consider closed triad in 3-nodes microstructure and 4-nodes microstructure contains more various relationship between one nodes and other three nodes. And using closed triads and 4-nodes microstructure together (ETF) get highest AUC rate in all the three algorithms.

Table 1. Predictive performance

ALGORITHM	SIMRANK	PMF	GPLVM
EE	0.7579	0.8283	0.8150
ET	0.7980	0.8380	0.8532
EF	0.8373	0.8573	0.8934
ETF	0.8434	0.8575	0.9047

In addition, the microstructure patterns of corresponding enterprise will change after the prediction. For example, if predicting hypotenuse existing in 4-cycle, then a 4-cycle will become a 4-chordalcycle. Similarly, these microstructures patterns will make evolution after the link prediction.

6 Conclusion

In this paper, we proposed a fast and efficient algorithm for querying and recording closed quads from enterprise network by using the discovered triads.

Furthermore, we classified the microstructure, including closed triad, 4-cycle, 4-chordalcycle and 4-clique, according to the ego network of node A. Then a statistical study on each enterprise leads us to find some interesting and meaningful rules in enterprise network. Finally, we use different state-of-the-art algorithms combining with these microstructure patterns to predict potential links on the real automobile supplying network dataset. The experimental results demonstrated the effectiveness of these microstructure patterns for link prediction.

In our future work, we plan to utilize our discoveries in predicting the possible neighbors of nodes in network, which would be well applied on recommendation systems. Furthermore, as our statistical results shown, some enterprises distinguish themselves from others on micro-structure distribution. So we would check and research these enterprises deeply.

Acknowledgment. The authors would like to acknowledge the support provided by the National Natural Science Foundation of China (61402263, 91546203), the National Key Research and Development Program of China (2016YFB0201405), the Fundamental Research Funds of Shandong University (2016JC011), the Natural Science Foundation of Shandong Province (ZR2014FQ031), the Shandong Provincial Science and Technology Development Program (2016GGX101008, 2016ZDJS01A09), the special funds of Taishan scholar construction project and the Taishan Industrial Experts Programme of Shandong Province No. tscy20150305.

References

1. Social structure. https://en.wikipedia.org/wiki/Social_structure
2. Dyad. https://en.wikipedia.org/wiki/Dyad_(sociology)
3. Ego. https://en.wikipedia.org/wiki/Ego
4. Ahmed, N.K., Neville, J., Rossi, R.A., Duffield, N.G., Willke, T.L.: Graphlet decomposition: framework, algorithms, and applications. Knowl. Inf. Syst. **50**(3), 689–722 (2017)
5. Graphlets. https://en.wikipedia.org/wiki/Graphlets
6. Bhuiyan, M.A., Rahman, M., Rahman, M., Al Hasan, M.: Guise: uniform sampling of graphlets for large graph analysis. In: 2012 IEEE 12th International Conference on Data Mining, pp. 91–100. IEEE (2012)
7. Biswas, A., Biswas, B.: Investigating community structure in perspective of ego network. Expert Syst. Appl. **42**(20), 6913–6934 (2015)
8. Dunbar, R., Arnaboldi, V., Conti, M., Passarella, A.: The structure of online social networks mirrors those in the offline world. Soc. Netw. **43**, 39–47 (2015)
9. Dong, Y., Tang, J., Wu, S., Tian, J., Chawla, N.V., Rao, J., Cao, H.: Link prediction and recommendation across heterogeneous social networks. In: 2012 IEEE 12th International Conference on Data Mining, pp. 181–190. IEEE (2012)
10. Girard, Y., Hett, F., Schunk, D.: How individual characteristics shape the structure of social networks. J. Econ. Behav. Organ. **115**, 197–216 (2015)
11. Gordon, I.R., McCann, P.: Industrial clusters: complexes, agglomeration and/or social networks? Urban stud. **37**(3), 513–532 (2000)
12. Huang, H., Tang, J., Wu, S., Liu, L., et al.: Mining triadic closure patterns in social networks. In: Proceedings of the 23rd International Conference on World Wide Web, pp. 499–504. ACM (2014)

13. Li, S., Daie, P.: Configuration of assembly supply chain using hierarchical cluster analysis. Procedia CIRP **17**, 622–627 (2014)
14. Lou, T., Tang, J., Hopcroft, J., Fang, Z., Ding, X.: Learning to predict reciprocity and triadic closure in social networks. ACM Trans. Knowl. Disc. Data (TKDD) **7**(2), 5 (2013)
15. Madhavan, R., Gnyawali, D.R., He, J.: Two's company, three's a crowd? Triads in cooperative-competitive networks. Acad. Manag. J. **47**(6), 918–927 (2004)
16. Toral, S., Martínez-Torres, M.D.R., Barrero, F.: Analysis of virtual communities supporting OSS projects using social network analysis. Inf. Softw. Technol. **52**(3), 296–303 (2010)
17. Trpevski, I., Dimitrova, T., Boshkovski, T., Kocarev, L.: Graphlet characteristics in directed networks. arXiv preprint arXiv:1603.05843 (2016)
18. Wang, L., Liu, S., Pan, L., Wu, L., Meng, X.: Enterprise relationship network: build foundation for social business. In: 2014 IEEE International Congress on Big Data, pp. 347–354. IEEE (2014)
19. Wasserman, S., Pattison, P.: Logit models and logistic regressions for social networks: I. An introduction to markov graphs andp. Psychometrika **61**(3), 401–425 (1996)
20. Yanardag, P., Vishwanathan, S.: Deep graph kernels. In: Proceedings of the 21th ACM SIGKDD International Conference on Knowledge Discovery and Data Mining, pp. 1365–1374. ACM (2015)
21. Stager, M., Lukowicz, P., Troster, G.: Dealing with class skew in context recognition. In: 26th IEEE International Conference on Distributed Computing Systems Workshops (ICDCSW 2006), pp. 58–58. IEEE (2006)
22. Salakhutdinov, R., Mnih, A.: Probabilistic matrix factorization. In: NIPS, vol. 20, pp. 1–8 (2011)
23. Lawrence, N.D., Urtasun, R.: Non-linear matrix factorization with gaussian processes. In: Proceedings of the 26th Annual International Conference on Machine Learning, pp. 601–608. ACM (2009)
24. Jeh, G., Widom, J.: SimRank: a measure of structural-context similarity. In: Proceedings of the Eighth ACM SIGKDD International Conference on Knowledge Discovery and Data Mining, pp. 538–543. ACM (2002)

Neural Architecture for Negative Opinion Expressions Extraction

Hui Wen[1,2(✉)], Minglan Li[1,2], and Zhili Ye[1,2]

[1] Institute of Software Chinese Academy of Sciences, Beijing 100190, China
{wenhui2015,minglan2015,zhili2015}@iscas.ac.cn
[2] University of Chinese Academy of Sciences, Beijing 100049, China

Abstract. Opinion expressions extraction is one of the main frameworks in opinion mining. Extracting negative opinions is more difficult than positive opinions because of indirect expressions. Especially, in the domain of consumer reviews, consumers are easier to be influenced by negative reviews when making decision. In this paper, we focus on the extraction of negative opinion expressions of consumer reviews. State-of-art methods heavily depend on task specific knowledge in the form of handcrafted features and data pre-processing. In this paper, we use a neural architecture by combining word embeddings, Bi-LSTM and CRF. We add a conditional random fields (CRF) layer to bidirectional long-short term memory (Bi-LSTM) recurrent neural network language model, which provides sentence level tag information and improves the result of experiment. Our model requires no feature engineering and outperforms feature dependent methods when experimenting on real-world reviews from Amazon.com.

Keywords: Neural architecture · Opinion extraction

1 Introduction

Opinion expressions extraction is one of the main frameworks in opinion mining. Existing works mainly focus on subjective expressions extraction and opinion target extraction [10]. Subjective expressions extraction is to distinguish the sentence or phrase is subjective or objective and extract subjective ones out. Opinion target extraction focus on finding the target terms implying sentiment in sentences, for example, "software" is the target term of the review "Updating with the latest software didn't help". However, what is important to us is the opinion implied by the target term not the term itself. Opinion expression is consisted by not only the target term but also the description of the term ("software didn't help" in the last example). Therefore, the task of extracting the whole opinion expressions or determining the opinion phrase boundary is important for further opinion mining tasks such as polarity classification and sentiment summarizations.

In the domain of consumer reviews opinion mining, precisely extracting opinion expressions is useful for further sentiment classification and summarization,

© Springer International Publishing AG 2017
L. Chen et al. (Eds.): APWeb-WAIM 2017, Part I, LNCS 10366, pp. 460–474, 2017.
DOI: 10.1007/978-3-319-63579-8_35

which is very important for helping consumers to make decisions and helping merchant to improve the quality of their product. Also, reviews that imply negative sentiment is always much more difficult to analysis than positive ones because the obscure or indirect expression. In this work, we focus on the negative opinion expressions extraction in the domain of customer reviews from Amazon.com provided by Mukherjee [14].

Most previous works on opinion target extraction have used parsing, sematic, syntactic features, language patterns and task specific knowledge. The huge amount of features make the task complex and time-consuming. Conditional random fields (CRF) [8] and other improved methods based on CRF are the most popular methods for this task. However, CRF is a linear model and heavily depend on handcrafted features. Therefore, the effectiveness of previous work is limited by the selection of features and the grammatical accuracy of sentences in the dataset.

In recent years, deep learning methods have been widely studied on natural language processing tasks. Mikolov et al. [13] presents word embeddings training with neural network which becomes effective representations of words and phrases. Recurrent neural network [12] becomes a new effective way for building language model compared with probabilistic methods. Also, long short-term memory [5] cell makes it possible for recurrent neural network to learn long sequence.

For more complex natural language processing tasks, such as named entity recognition and part-of-speech tagging, deep learning models have also outperformed feature-based methods. Neural architecture is famous for its non-linear character and can automatically learn the semantic and syntactic information of the sentences, which remedy the limitation of CRF.

To this end, we use a neural architecture by combining pre-trained word embeddings, Bi-LSTM and CRF. Pre-trained word embeddings provide word presentations learnt from huge amount of data, which also contains the context information. Bidirectional long-short term memory (Bi-LSTM) recurrent neural network language model provides non-linear character in modeling sequential data. Linear conditional random fields provides sentence level tag information.

We take the task as a sequence labeling task and use deep architecture to solve the task of opinion expressions extraction (phrase boundary detection). The main contribution of our work are summarized as follows:

- We solve the negative opinion expressions extraction task with a neural architecture with no need of feature engineering and outperform feature-based methods in real-word consumer review data.
- We prove that adding a linear CRF layer which provide sentence level tag information to neural network language model can significantly improve the result of the task of opinion expressions extraction.

The rest of this paper is organized as follows: Sect. 2 formulates the problem and describes the model. Section 3 discusses the experimental setup, training and evaluation of the network. Section 4 reviews the related work. Section 5 concludes the paper and presents our future work.

2   Neural Network Architecture

In this Section we formulate the problem of opinion expressions extraction, describe our neural network architecture components from bottom to up and describe the algorithm to apply the neural architecture to negative opinion expressions extraction.

2.1   Problem Statement

The opinion expressions extraction task can be formally described as follows: Given a sentence, $s = (w_1, w_2, ..., w_n)$, find sub-sequence of the sentence which contains target of opinion and opinion expression, $o = (w_p, ..., w_i, ..., w_q)$ ($p \geq 1, q \leq n$). The aim is to correctly decide the phrase boundary of the sentence.

We take the task as a sequence labeling task. We represents the sentence into the BIO format (Beginning of opinion expressions, Inside of opinion expressions, End of opinion expressions). For a word in sentence, we label the word B if the word is the first word in opinion expressions, we label the word I if the word is inside but not in the first position of the opinion expression, and for other words which is not in the opinion expression, we label it O. For example, in the sentence below, the opinion expression is inside [[]], we label the sentence as follow:

Updating(O) with(O) the(O) latest(O) [[software(B) didn't(I) help(I)]].

The task is that given a sentence , finding the right place for the position of BIO, which means correctly predicting every word in the sentence to be B, I or O.

2.2   Word Embeddings

Word embeddings map words to vectors using methods including probabilistic models, neural networks and so on. Word embeddings have exhibit remarkable boost in many natural language processing (NLP) tasks with the property of being able to encode the analogy between words. Since Bengio et al. [1] propose neural language and Collobert et al. [3] train word embeddings on a large dataset and show effectiveness of word embddings. Since that, word embeddings have been used in NLP tasks in wide range.

In NLP task using neural architecture, word embedding can be used in two ways: co-trained with the task and pre-trained based on larger dataset. In our work, we use Glove pre-trained 200-dimensional word embeddings trained on 2 billions twitters Twitter. [17] Also, in Sect. 3.6 we discuss the impact of word embeddings on our task using the Stanford Glove [17] two different pre-trained word embeddings trained Wikipedia and Twitter respectively with different dimensions and the 300-dimensional word embeddings proposed by Mikolov et al. [13] trained on 100 billion words from Google News.

2.3   Bi-LSTM RNN Language Model

Recurrent Neural Network (RNN) is a neural architecture feasible for dealing with sequence data, because it makes use of information and performs the same task for every element of a sequence. RNN language model has been proved to be more effective and less computational complexity compared with traditional language model by Mikolov et al. [12].

However, RNN suffers vanishing/exploding gradients when learning long sequence. Long Short-term Memory (LSTM) [5] solves the problem of RNN with the addition of a memory cell and uses input gate i_t, output gate o_t, forget gate f_t to control the proportion of the input and the previous state to the memory. The LSTM outputs memory cell c_t, hidden state h_t at each time step is shown as follows:

$$
\begin{aligned}
i_t &= \sigma(W^i x_t + U^i h_{t-1} + b^i) \\
f_t &= \sigma(W^f x_t + U^f h_{t-1} + b^f) \\
o_t &= \sigma(W^o x_t + U^o h_{t-1} + b^o) \\
g_t &= tanh(W^g x_t + U^g h_{t-1} + b^g) \\
c_t &= f_t \odot c_{t-1} + i_t \odot g_t \\
h_t &= o_t \odot tanh(c_t)
\end{aligned}
\tag{1}
$$

where σ is the element-wise sigmoid and hyperbolic tangent functions, $\odot$ is the element-wise multiplication operator.

Bi-directional LSTM, [4] composed of a forward LSTM and a backward LSTM, can capture both the past and future information respectively. The feature of Bi-LSTM makes it effective in many NLP tasks, because it takes the advantage of contextual information.

2.4   CRF

Conditional random fields (CRF) [8] is used in wide range for natural language processing sequence labeling tasks such as part-of-speech tagging, name entity recognition. CRF takes context into account, using tagging information at sentence level to model tagging decisions jointly

For a sequence labeling task, CRF gives a score for labeling result of a sentence as follows:

$$
score(y|x) = \sum_{j=1}^{m} \sum_{i=1}^{n} \lambda_j f_j(y_i, y_{i-1}, x)
\tag{2}
$$

where f_j is feature function and λ_j is its weight. Each feature function is composed with the label of current word y_i and the label of the previous word y_{i-1}. That means, feature functions in CRF depend on current and previous label rather than arbitrary word. $\Lambda = \{\lambda_k\}$ is the weights of feature functions. In CRF, normalize the score into probability $p(y|x, \Lambda)$ with the normalization as follows:

$$Z(x, \Lambda) = \sum_{y} (exp(\sum_{j=1}^{m} \sum_{i=1}^{n} \lambda_j f_j(y_i, y_{i-1}, x))) \tag{3}$$

$$p(y|x, \Lambda) = \frac{exp(\sum_{j=1}^{m} \sum_{i=1}^{n} \lambda_j f_j(y_i, y_{i-1}, x))}{Z(x, \Lambda)} \tag{4}$$

During training, we minimize the negative log-probability of the correct label.

$$\Lambda = argmin_\Lambda(-\sum_{j} log(p(y_i|x_i, \Lambda))) \tag{5}$$

where is (x_i, y_i) is in the training examples. Once we have trained a CRF model, given a new sentence, we can get its labeling sequence using dynamic programming, such as Viterbi Decoding.

2.5   Bi-LSTM-CRF

In our model, we combine Bi-LSTM with CRF similar with architecture in [9,11]. First, we feed word embeddings of sentence into bidirectional long-short term memory (Bi-LSTM) recurrent neural network language model. Then the output is fed into the conditional random fields (CRF) layer which take into account neighboring tags, yielding the final predictions for every word. The combination of no-linear and linear model improves the result of our experiment, which will be discussed in Sect. 3.5. The architecture is shown in Fig. 1.

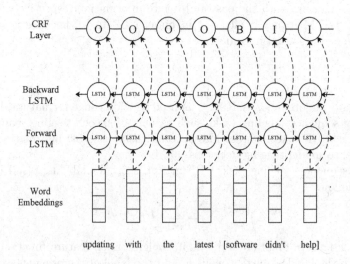

Fig. 1. The architecture of Bi-LSTM-CRF

2.6 Apply the Neural Architecture to Negative Opinion Expressions Extraction

In this part, we will discuss the technological process to apply the neural architecture to negative opinion expressions extraction. Given a set of real-world negative reviews from Amazon.com. The only pre-processing for the data is simply tokenize the words with blank character and padding the sentences as discussed in Sect. 3.3. Then we present the word with pre-trained word embeddings, and feed them into our neural architecture. Then we apply backpropagation algorithm to minimize the loss function and upgrade the parameters. The flow of our algorithm is as follows (Table 1):

Table 1. The flow for opinion expressions extraction

Input: A set of consumer reviews $S = \{s_1, s_2, ... s_n\}$
Output: The parameter of Bi-LSTM-CRF
Parameters Initialization: Intialize the matrix and bias with a normal distribution with mean equals 0 and standard deviation equals 0.1. Padding and initial reviews using pre-trained word embeddings. **for** each review $s_i \in S$ **do:**

1 Feed x_i, the vector prestantion of s_i to Bi-LSTM-CRF and generate pridicted sequence label y_i.
    2 Compute the loss of y_i.
    3 Use backpropagation algorithm to update the parameters of the network.

end for

3 Experimental Setup

In this section, we describe the experimental setup of our paper, including the way we train our neural architecture and the efficiency of our model in detail.

3.1 Datasets

Most of the datasets in domain of opinion mining focus on the sentiment classification. For the task of opinion expressions extraction, the sentiment analysis dataset developed by Qiu et al. [22] and the SemEval2014 dataset [18] focus on the opinion target term extraction. The dataset provided by Wang et al. [26] expand the SemEval2014 dataset by adding manually labeled opinion terms. The dataset developed by Mukherjee [14] is for the task of opinion expressions boundary detection.

We use the dataset developed by Mukherjee [14], which provides consumer reviews from Amazon.com. The dataset contains six domains data, the number of negative and total sentences are as below: Router (1284/5063), GPS (632/2075), Keyboard (667/1446), Mouse (494/2488), MP3-Player (174/352), Earphone (359/678).

In our model, we require the recurrent network language model with long-short term memory to learn the long term context and dependencies between words in the input sentences. We need to deal with the variation of the length of sentences, we need to find a feasible max sentence length so that many of the sentences would share the same length and the length won't be too long to make the model too complex. We will describe the distribution of the sentence lengths of our dataset and the way of padding in detail in the Sect. 3.3.

3.2   Comparative Approaches

We compare our model with four feature based methods described in [14].

UHB: A rule based method by finding the constitution around the opinion term, such as a positive sentiment and a negator. And decide the window to find sentiment negator by experiment.

CRF: The traditional Markov linear-chain conditional random fields (CRF). The features contains two classes: Pivot Features (POS Tags, Phrase Chunk Tags, Prefixes, Suffixes, Word Sentiment Polarity), Latent Semantics.

CRF-L2R: The traditional Markov linear-chain conditional random fields (CRF) with regularization on the feature weights. During training, the log-likelihood is modified as follows:

$$\Lambda = argmin_\Lambda(-\sum_j log(p(y_i|x_i, \Lambda))) + \sum_k \lambda_k^2 \qquad (6)$$

CRF-PSC: A constrained model which summing over only the valid phrases produced by CRF. In traditional CRF, we $p(y|x, \Lambda)$ with the normalization $Z(x, \Lambda)$, which summing over all possible sequence labeling. In CRF-PSC , the normalization sums over only the valid sequence labeling.

3.3   Network Training

In this part, we will discuss the details about training the neural network in our work.

We implement Tensorflow library in our experiment. The details contain about parameter initialization, the sequence length for recurrent neural network language model, loss function with padding in sequence, optimization algorithm and ways to avoid overfitting.

Parameter Initialization: To verify the effectiveness of our neural network in simple pre-processing of data, the only data pre-processing is basically split the sentence into a list of words, without removing punctuations or stop words. We use Glove pre-trained 200-dimensional word embeddings trained on 2 billions twitters [17]. For the words which are not in the pre-trained words set, we random them uniformly distributed over the half-open interval $[-0.1, 0.1)$ so that any value within the given interval is equally likely to be drawn.

Matrix parameters and bias is used in two circumstances: (1) The softmax layer between Bi-LSTM and the output of the sequence labeling model in the contrast method of no adding a CRF layer, (2) The activation function between Bi-LSTM and the CRF layer. Matrix parameters and bias are randomly set both with a normal distribution with mean equals 0 and standard deviation equals 0.1.

Sequence Length for RNN Language Model: During the training of recurrent neural network, we have to feed the network the same length data in a training batch. However, the sentences in our dataset is not in the same length. Padding is the most popular ways to deal with this problem. We will pad the sentences with zero vectors to fill up the remaining part, so that in a training batch all the sentences share the same length.

The max length of padded data is one of the most important hyperparameters in RNN language model. Too long sequence length would not help due to gradient vanish, even LSTM suffer from this problem. We analysis the distribution of the sentence lengths in our data. As we want to find the opinion expressions, we must keep them in our data. So we analysis the length of sentence after truncating at the end of opinion expressions as Fig. 2. We choose the max sequence length as 50, so that many of our sentences would share the same length. We randomly keep part of sentences longer than that with no break of opinion expressions and do padding for shorter ones.

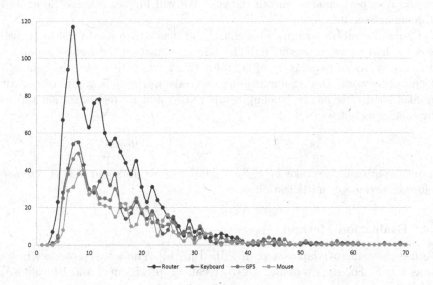

Fig. 2. The sentence length distribution of our dataset

Loss Function with Padding in Sequence: The loss function need to be calculate accordingly after the data being padded. We mask the sequence with the actual length when training. So that during training, the padded part would not been considered into loss function and not been trained.

Optimization Algorithm: We choose the mini-batch stochastic gradient descent (SGD) with momentum 0.9. Momentum [21] is a method to help avoid stuck into local minima when applying SGD. As the size of dataset is not large enough, we finally set the batch size at 1 after experiment on different batch size, which becomes standard SGD, but the time for training is still acceptable and the result is the best. We also use Adam [7] optimization algorithm when choosing the hidden size of LSTM preliminarily, because Adam has a fast convergence velocity which helps us tune our network more quickly. The hidden size for each domain is Router(64), Keyboard(30), GPS(40), Mouse(24).

We set the learning rate at 0.001 and early stop [2] based on performance on validation sets, all the results are based on 4-fold cross validation. To avoid gradient exploding, we also use gradient clipping method [16]. Also, we shuffle the data randomly before training.

Ways to Avoid Overfitting: In our work, we use two ways to avoid overfitting: dropout and L2-regulazition on matrix weights and bias. Dropout [23] is an effective method to deal with overfitting by randomly drop units and their connections to avoid co-adapting of units. When some neurons are randomly dropped out during training, the co-adapting between them is avoided. This makes the network have better generalization performance. We use the dropout at the input of the Bi-LSTM and put a dropout wrapper on both forward and backward cell of Bi-LSTM. We set the dropout rate at 0.6, which helps us to get the best performance on our dataset. We will further discuss the effect of dropout in Sect. 3.6.

L2-regularization term to the weights and bias is also one of the frequently-used method to avoid overfitting. The L2-regularization term gives penalty on the large values of parameters of neural network which increases the generality of neural network. During our training, we take our loss as a sum of negative log-likelihood of sequence labeling result (NLL) and L2-regularization term of parameters as follows:

$$Loss = NLL + \beta_1 R(weights) + \beta_2 R(bias) \tag{7}$$

In our experiment, we take $\beta_1 = \beta_2 = 0.001$, the weights and bias is discussed before in parameter initialization.

3.4   Evaluation Method

We use the same overlap matrix described in [14]. The set of sentences in test dataset is S. For each sentence $s \in S$, recall(r), precision(p) and F1 value(F_1) as follows:

$$p = avg_{s \in S, |s_c| \neq 0}\left(\frac{|s_c| \bigcap |s_p|}{|s_c|}\right)$$

$$r = avg_{s \in S, |s_p| \neq 0}\left(\frac{|s_c| \bigcap |s_p|}{|s_p|}\right) \tag{8}$$

$$f_1 = \frac{2pr}{p + r}$$

where s_c, s_p denote the correct and predicted opinion expressions spans in sentence.

3.5 Model Effectiveness

In this part, we present the performance of our model. We verify the effectiveness of our model compared with feature-based methods described in Sect. 3.2.

Compared with feature-based methods [14]: Table 2 shows the precision, recall, f1-value on the dataset described in Sect. 3.1 with four domains (Router, Keyboard, GPS, Mouse). The recall and f1-value has been improved using our neural architecture. Noting we used the same architecture and no feature engineering to get the improvement on different domains of data. In the keyboard domain, however, we don't get result good enough, by analysis the raw data, we find there are too many numbers and html links. To show the advantage of our model, we have not done preprocessing to these circumstances.

Table 2. Performance of four types datasets

Dataset	Method	P	R	F1	Dataset	Method	P	R	F1
Router	UHB	67.0	73.7	70.2	Keyboard	UHB	64.4	68.1	66.0
	CRF	87.9	76.9	82.0		CRF	86.8	74.8	80.3
	CRF-L2R	88.7	77.1	82.5		CRF-L2R	87.6	75.7	81.2
	CRF-PSC	92.0	80.1	85.7		CRF-PSC	90.4	78.3	83.9
	Bi-LSTM	87.0	81.3	84.0		Bi-LSTM	76.3	74.0	75.1
	Bi-LSTM+CRF	87.6	87.8	**87.7**		Bi-LSTM+CRF	80.1	82.6	81.3
GPS	UHB	67.5	67.7	67.6	Mouse	UHB	60.7	59.1	59.8
	CRF	84.1	71.9	77.5		CRF	83.8	61.6	70.9
	CRF-L2R	84.8	72.7	78.3		CRF-L2R	84.5	61.9	71.4
	CRF-PSC	86.7	73.5	79.5		CRF-PSC	86.1	63.6	73.1
	Bi-LSTM	81.2	77.6	79.4		Bi-LSTM	84.0	75.9	79.7
	Bi-LSTM+CRF	85.0	84.3	**84.7**		Bi-LSTM+CRF	80.1	81.2	**80.6**

Also, In Table 2, we show the result of adding a CRF layer. Compared with feed the output of Bi-LSTM to Softmax, the CRF layer improve the result significantly. Adding a CRF layer has improved the recall a lot and thus improve the F1-value.

3.6 Impact of Dropout and Word Embeddings

In this part, we discuss about the impact of dropout and pre-trained word embeddings on our experiment.

Dropout: Deep neural networks always suffer from overfitting because of large number of parameters, especially when dataset for training is not large enough. Dropout is a powerful method solving this problem [6]. In our experiment, the dataset is kind of small of neural network, so dropout is extremely necessary for the performance. We use dropout on the input and Bi-LSTM layer when training, the effect of Dropout is shown as Table 3, the parameter is the same as Sect. 3.5. As we can see from the figure, dropout is effective in improving the performance of our neural architecture model in different domains of data.

Table 3. Impact of dropout on four types datasets

	Router			Keyboard			Gps			Mouse		
	P	R	F1	P	R	F1	P	R	F1	P	R	F1
Dropout	87.6	87.9	87.7	80.1	82.6	81.3	85.0	84.3	84.7	80.1	81.2	80.6
No dropout	84.1	83.3	83.7	70.1	73.0	71.5	75.3	78.0	76.6	77.2	79.4	78.3

Word Embeddings: In order to find the effect of pre-trained word embeddings on our model, we performed experiments with three different pre-trained word embeddings on the dataset in Router domain, the F1-value using different pre-trained word embeddings are shown in Table 4.

(1) Glove, trained on wikipedia 2014 and Gigaword 5 (6 billions tokens, 400K vocabulary) with dimensions of 50, 100, 200, 300 [17].
(2) Glove Twitter, trained on Twitter (2 billions tweets, 12M vocabulary) with dimensions of 25, 50, 100, 200 [17].
(3) Word2Vec, trained on 100 billion words from Google News with dimensions of 300 produced by Mikolov et al. [13].

As we can see from the figure, different pre-trained word embeddings effect the performance of our model. The choice of pre-trained word embeddings is important when the training dataset is not large enough to co-train the word embeddings while solving the task. Using word embeddings pre-trained on larger dataset is helpful to get better performance.

4 Related Work

Opinion expressions extraction is a major task in opinion mining. Hu and Liu [6] take the opinion expressions extraction as a frame consisting of: (1) Opinion Holder: the person making the evaluation (2) Target: a named entity belonging

Table 4. Different pre-trained word embeddings

	25D	50D	100D	200D	300D
Glove	-	80.1	83.1	81.8	83.2
Glove Twitter	83.6	86.1	84.6	87.7	-
Word2Vec	-	-	-	-	87.7

to a class of interest (e.g., iPhone) (3) Aspect: a part, member or related object, or attribute of the Subject (Target) (e.g., size, cost) (4) Evaluation: a phrase expressing an evaluation or the opinion holder's mental/emotional attitude (e.g., too bulky). Opinion extraction task means filling these slots for each evaluation expressed in text.

Opinion target extraction is first studied by Hu and Liu [6], they classify the opinion target into two kinds: explicit and implicit, but only deal with the explicit target with rule-based method. Popescu and Etzioni [19] asume the class of the product is already known and use the mutual information between words and the product class to find the target words. Zhiqiang and Wenting [27] use CRF as a sequence labeler with features such as POS tags and WordNet taxonomies. In SemEval 2014 [18] task 4, a dataset of two domains (Laptop and Restaurant) is provided for target extraction and polarity classification, many methods based on task-specific knowledge, sematic and syntactic structure have been proposed. Also, there are many methods based on topic model [15]. Neural architecture in recent years has shown improvement in this task, Poria et al. [20] use a deep CNN combining with rules and get the state-of-art result on the dataset of SemEval 2014.

Recent years, the co-extraction of opinion target and opinion evaluation terms has been a promising task. Qiu et al. [22] use rules and relations between them. Wang et al. [26] use recursive neural network combining conditional random fields and proposed an extended dataset for this task based on the dataset of SemEval 2014.

The task in our work is find the phrase boundaries of opinion expressions, which is similar to opinion target extraction but different from it from getting the full opinion expressions of aspect and evaluation. Mukherjee [14] is the first to focus on this task, and proposes the dataset for the task of opinion expressions extraction.

In recent years, neural architecture has shown significant improvement on natural language processing tasks. Mikolov et al. [13] presents the distributed representations of words and phrases with neural network, which is known as word embeddings. Pennington et al. [17] make the training process of word embedding possible on large dataset. And the recurrent neural network [12] has shown effectiveness in modeling natural language with long short-term memory [5] cell to solve the problem of learning long sequence.

On the sequence information labeling and analysis tasks, methods with improvement based on Hidden Markov Model (HMM) and LDA haven been proved effective [24, 25]. Lample et al. [9] established two neural networks, one is a bidirectional LSTM with a sequential conditional random layer above it, the other is a new model that a new model that constructs and labels chunks of input sentences using an algorithm inspired by transition-based parsing with states represented by stack LSTMs. The model presents state-of-art name entity recognition task in different languages. Ma and Hovy [11] propose a neural architecture with the combination of CNN character level word embeddings, LSTM and CRF on the end-to-end sequence labeling tasks. They experiment on the two major sequence labeling tasks: part-of-speech tagging and name entity recognition, on both of tasks the model proves effectiveness over traditional feature-based methods.

Different from the target terms extraction discussed above, we focus on the task of phrase boundaries detection. And different from the model [9, 11] discussed above, we don't use the CNN character level word embeddings or stack LSTM. Our model outperforms the previous feature-based methods such as feature-based CRF or rule-based methods.

5   Conclusion and Future Work

In this paper, we use a neural architecture by combining word embeddings, bidirectional long-short term memory (Bi-LSTM) recurrent neural network language model and conditional random fields (CRF) to solve the task of opinion expressions extraction. Our model has no need to make handcrafted features and outperforms the feature-based methods on real-world negative consumer reviews. Also, our work shows that adding a CRF layer to Bi-LSTM can significantly improve the result as import the sentence level tagging information. We also study about the methods of prevent overfitting when the dataset is not large enough and the influence of the size and corpus of the pre-trained word embeddings.

There are many directions for our future work: First, we can bring in finer granularity information to the model such as character level information of the words in sentence. Another direction is to combine our model with further task of opinion summarization and make a joint model for opinion extraction and summarization. Also, as our model has no need of task specific knowledge or handcrafted features, we can use it in other domain, such as find viewpoint in news, standpoint in academic papers and so on.

References

1. Bengio, Y., Ducharme, R., Vincent, P., Jauvin, C.: A neural probabilistic language model. J. Mach. Learn. Res. **3**(Feb), 1137–1155 (2003)
2. Caruana, R., Lawrence, S., Giles, L.: Overfitting in neural nets: backpropagation, conjugate gradient, and early stopping. In: NIPS (2000)
3. Collobert, R., Weston, J., Bottou, L., Karlen, M., Kavukcuoglu, K., Kuksa, P.: Natural language processing (almost) from scratch. J. Mach. Learn. Res. **12**(Aug), 2493–2537 (2011)

4. Graves, A., Schmidhuber, J.: Framewise phoneme classification with bidirectional LSTM and other neural network architectures. Neural Netw. **18**(5), 602–610 (2005)
5. Hochreiter, S., Schmidhuber, J.: Long short-term memory. Neural Comput. **9**(8), 1735–1780 (1997)
6. Hu, M., Liu, B.: Mining and summarizing customer reviews. In: Proceedings of the Tenth ACM SIGKDD International Conference on Knowledge Discovery and Data Mining. ACM (2004)
7. Kingma, D., Ba, J.: Adam: a method for stochastic optimization. arXiv preprint arXiv:1412.6980 (2014)
8. Lafferty, J., McCallum, A., Pereira, F., et al.: Conditional random fields: probabilistic models for segmenting and labeling sequence data. In: Proceedings of the Eighteenth International Conference on Machine Learning, ICML, vol. 1 (2001)
9. Lample, G., Ballesteros, M., Subramanian, S., Kawakami, K., Dyer, C.: Neural architectures for named entity recognition. arXiv preprint arXiv:1603.01360 (2016)
10. Liu, B., Zhang, L.: A survey of opinion mining and sentiment analysis. In: Aggarwal, C., Zhai, C. (eds.) Mining Text Data. Springer, Boston (2012). doi:10.1007/978-1-4614-3223-4_13
11. Ma, X., Hovy, E.: End-to-end sequence labeling via bi-directional lstm-cnns-crf. arXiv preprint arXiv:1603.01354 (2016)
12. Mikolov, T., Karafiát, M., Burget, L., Cernocký, J., Khudanpur, S.: Recurrent neural network based language model. In: Interspeech, vol. 2, p. 3 (2010)
13. Mikolov, T., Sutskever, I., Chen, K., Corrado, G.S., Dean, J.: Distributed representations of words and phrases and their compositionality. In: Advances in Neural Information Processing Systems (2013)
14. Mukherjee, A.: Extracting aspect specific sentiment expres-sions implying negative opinions. In: Proceedings of the 17th International Conference on Intelligent Text Processing and Computational Linguistics (2016)
15. Mukherjee, A., Liu, B.: Aspect extraction through semi-supervised modeling. In: Proceedings of the 50th Annual Meeting of the Association for Computational Linguistics: Long Papers, vol. 1. Association for Computational Linguistics (2012)
16. Pascanu, R., Mikolov, T., Bengio, Y.: On the difficulty of training recurrent neural networks. In: ICML, vol. 28, no. 3, pp. 1310–1318 (2013)
17. Pennington, J., Socher, R., Manning, C.D.: Glove: global vectors for word representation. In: EMNLP, vol. 14 (2014)
18. Pontiki, M., Galanis, D., Pavlopoulos, J., Papageorgiou, H., Androutsopoulos, I., Manandhar, S.: Semeval-2014 task 4: aspect based sentiment analysis. In: Proceedings of SemEval (2014)
19. Popescu, A.M., Etzioni, O.: Extracting product features and opinions from reviews. In: Kao, A., Poteet, S.R. (eds.) Natural Language Processing and Text Mining. Springer, London (2007). doi:10.1007/978-1-84628-754-1_2
20. Poria, S., Cambria, E., Gelbukh, A.: Aspect extraction for opinion mining with a deep convolutional neural network. Knowl.-Based Syst. **108**, 42–49 (2016)
21. Qian, N.: On the momentum term in gradient descent learning algorithms. Neural Netw. **12**(1), 145–151 (1999)
22. Qiu, G., Liu, B., Bu, J., Chen, C.: Opinion word expansion and target extraction through double propagation. Comput. Linguist. **37**(1), 9–27 (2011)
23. Srivastava, N., Hinton, G.E., Krizhevsky, A., Sutskever, I., Salakhutdinov, R.: Dropout: a simple way to prevent neural networks from overfitting. J. Mach. Learn. Res. **15**(1), 1929–1958 (2014)
24. Su, B., Ding, X.: Linear sequence discriminant analysis: a model-based dimensionality reduction method for vector sequences. In: ICCV (2013)

25. Su, B., Ding, X., Liu, C., Wu, Y.: Heteroscedastic max-min distance analysis. In: Proceedings of the IEEE Conference on Computer Vision and Pattern Recognition (2015)
26. Wang, W., Pan, S.J., Dahlmeier, D., Xiao, X.: Recursive neural conditional random fields for aspect-based sentiment analysis. arXiv preprint arXiv:1603.06679 (2016)
27. Zhiqiang, T., Wenting, W.: Dlirec: aspect term extraction and term polarity classification system (2014)

Identifying the Academic Rising Stars via Pairwise Citation Increment Ranking

Chuxu Zhang[1,2], Chuang Liu[1(✉)], Lu Yu[3], Zi-Ke Zhang[1], and Tao Zhou[4]

[1] Alibaba Research Center for Complexity Sciences,
Hangzhou Normal University, Hangzhou, China
liuchuang@hznu.edu.cn
[2] Department of Computer Science and Engineering,
University of Notre Dame, Notre Dame, USA
[3] King Abdullah University of Science and Technology, Thuwal, Saudi Arabia
[4] Big Data Research Center, University of Electronic Science
and Technology of China, Chengdu, China

Abstract. Predicting the fast-rising young researchers (the Academic Rising Stars) in the future provides useful guidance to the research community, e.g., offering competitive candidates to university for young faculty hiring as they are expected to have success academic careers. In this work, given a set of young researchers who have published the first first-author paper recently, we solve the problem of how to effectively predict the top $k\%$ researchers who achieve the highest citation increment in Δt years. We explore a series of factors that can drive an author to be fast-rising and design a novel pairwise citation increment ranking (PCIR) method that leverages those factors to predict the academic rising stars. Experimental results on the large ArnetMiner dataset with over 1.7 million authors demonstrate the effectiveness of PCIR. Specifically, it outperforms all given benchmark methods, with over 8% average improvement. Further analysis demonstrates that temporal features are the best indicators for rising stars prediction, while venue features are less relevant.

Keywords: Scientific impact prediction · Bayesian personalized ranking · Data engineering

1 Introduction

The growing scientific activities lead to the expanding body of literature, as well as the increasing academic population. Figure 1a and b report the rapid increase of publication volume each year and the accumulative number of authors from 1960 in the ArnetMiner [1] academic dataset of Computer Science [9]. Despite the large number of researchers, their scientific impacts are different. Generally, the citation count is used as a measurement of an author's scientific impact. Figure 1c depicts the distribution of individual researcher' contribution in terms

---

[1] https://aminer.org/AMinerNetwork.

© Springer International Publishing AG 2017
L. Chen et al. (Eds.): APWeb-WAIM 2017, Part I, LNCS 10366, pp. 475–483, 2017.
DOI: 10.1007/978-3-319-63579-8_36

of the citation count, showing that less than 7% of researchers have more than 20 citations. Meanwhile, the citation increasing trends of different authors are different. Figure 1d reports the citation increment distribution of all authors from year 2008 to 2012, showing that the distribution is power-law like and less than 10% of authors have the fast-rising trend (each has the increment of citations being larger than 20) of scientific impact.

Many previous works [3,8,10–13] have been conducted to predict the citation values of authors or papers while the forecast of different authors' or papers' rising trends w.r.t. citation count remains to be solved. An intuitive meaningful question arises that can we effectively predict the fast-rising ones (namely the Academic Rising Stars - ARSes) among a set of young scholars who start academic research recently, as early identification of the ARSes may offer useful guidance to the research community like young faculty recruiting of university. Accordingly, we define the fast-rising researchers as the authors reach relatively large citation increments in a given time period and our contributions are three-fold: (1) We formalize the problem of identifying the ARSes as a citation increment ranking task. (2) We introduce a series of factors that are correlated with authors' future citation increments and design a novel pairwise citation increment ranking method to identify ARSes. (3) We conduct experiments to evaluate the performances of our method on a large academic dataset. The result shows that our method outperforms all given baseline methods, with over 8% average improvement.

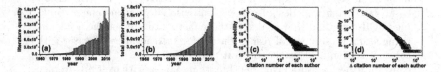

Fig. 1. (a) The volume of research literatures in each year. (b) The accumulative number of authors in each year from year 1960. (c) Distribution of all researchers' citation counts till year 2012. (d) Distribution of all researchers' citation increments from year 2008 to year 2012.

2   Data and Problem

2.1   Notations

For ease of representation, letter A and L represent the author set and the paper set respectively. Letter c denotes author's citation and Δc_a is researcher a's citation increment in the given time period. The true and predicted impact increment scores for a are symbolized by s_a and $\hat{s}_a$ respectively. Let A^* represents the set of young researchers who publish first firs-author paper at recent timestamp t_{1st} and A_r^* is the subset of A^* belongs to research topic r. In addition, we use $A_{r,k}^*$ and $\hat{A}_{r,k}^*$ to denote the set of true ARSes who achieve top $k\%$ citation increments and the set of predicted ARSes who have top $k\%$ predicted citation increment scores for topic r, respectively. In this work, the true impact increment score of author a is quantified by citation increment value, i.e., $s_a = \Delta c_a$.

2.2 Data

In this paper, we use a large real-world dataset from ArnetMiner which is a well known online service for academic search and analysis. The dataset contains 1,712,433 authors and 2,092,356 papers from major computer science venues for more than 50 years (from 1960 to 2014). Each paper contains content information on the title, authorship, abstract, publication time, publication venue and references. In total, we extract 4,258,615 collaboration relationships among authors and 8,024,869 citation relationships among papers from the dataset.

2.3 Researchers Division

We should take it as an independent prediction task for each topic due to different influences and audiences. As a widely used method for topic modeling, we use Latent Dirichlet Allocation[2] (LDA) to categorize the corpus into R different topics. We run a R-topics LDA on the title and abstract content of L_{A^*} and it returns the probability distribution $p(r|l)$ over topic r for each paper $l \in L_a$ of each researcher $a \in A^*$. We define the correlation between a and topic r as accumulative probability $C_{r|a}$ over $p(r|l)$ for each $l \in L_a$: $C_{r|a} = \sum_{l \in L_a} p(r|l)$. Note that most of researchers' works cover several different topics, we divide all researchers of A^* into R groups and each $a \in A^*$ belongs to m ($m = 3$) different groups according to the top m values of $C_{r|a}$ for all topics.

2.4 Problem Definition

After categorizing all researchers of A^* into different groups, we define the problem as: *Top $k\%$ Academic Rising Stars Prediction - Given the publication corpus L_t before the current year t and a set of young researchers A^* who publish the first first-author paper at the recent year t_{1st}, the task is to predict the fast-rising scholars $A^*_{r,k}$ who rank in top $k\%$ in A^*_r for each topic r according to the citation increments (or impact increment scores) after Δt years.* The schematic diagram of this work is illustrated in Fig. 2.

3 Method

3.1 Pairwise Citation Increment Ranking

Inspired by Bayesian preference ranking [7] technique, we design a Pairwise Citation Increment Ranking (PCIR) method for ARSes prediction. Let $(a_i, a_j) \in T_r$ denote an author pair of topic r. In order to capture impact increment rankings of different authors categorized in the same topic, we maximize the posterior probability over parameter ω: $p(\omega| >_r) \propto p(>_r |\omega) \cdot p(\omega)$, where notation

---

[2] http://radimrehurek.com/gensim/.

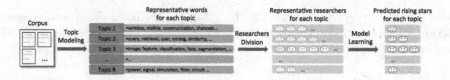

Fig. 2. Schematic diagram of this work. We divide all young researchers into R groups then extract a series features of each author and design a ranking algorithm to predict ARSes of each research topic.

$>_r = \{a_i >_r a_j : ((a_i, a_j) \in T_r) \cap (s_{a_i} > s_{a_j})\}$ represents the pairwise structure of topic r and $p(\omega)$ is the prior probability. We take citation increment as the true impact increment score, i.e., $s_{a_i} = \Delta c_{a_i}$. Let set $T_{>_r}$ represent all instances in $>_r$ and set $T_{\leq_r}$ consists of the remaining cases not included in $T_{>_r}$. In general, we assume that each case in $>_r$ is independent and the likelihood function can be written as a product of single density for all cases in $>_r$:

$$p(>_r | \omega) = \prod_{(a_i, a_j) \in T_r} p(a_i >_r a_j | \omega)^{\delta((a_i, a_j) \in T_{>_r})} \cdot (1 - p(a_i >_r a_j | \omega))^{\delta((a_i, a_j) \in T_{\leq_r})}$$

With antisymmetric nature of $>_a$, the log form of the objective becomes:

$$ln\, p(\omega | >_r) = ln \prod_{(a_i, a_j) \in T_{>_r}} p(a_i >_r a_j | \omega) \cdot p(\omega) = \sum_{(a_i, a_j) \in T_{>_r}} ln\, p(a_i >_r a_j | \omega) + ln\, p(\omega)$$

Let $p(a_i >_r a_j | \omega) = \sigma(d_{(a_i, a_j) \in T_{>_r}}(\omega))$, where σ is the logistic function: $\sigma(x) = \frac{1}{1 + e^{-x}}$, and $d_{(a_i, a_j) \in T_{>_r}}(\omega)$ measures the difference of impact increments between a_i and a_j. The predicted impact increment score $\hat{s}_{a_i}$ of a_i based on academic features $\boldsymbol{f}$ (described in **Feature Selection** subsection) extracted from the corpus is formulated as $\hat{s}_{a_i} = \sum_{k=1}^{K} \omega_k \cdot f_{ik}$, where f_{ik} represents the k-th factor of a_i and K is the number of all factors. Intuitively, we define $d_{(a_i, a_j) \in T_{>_r}}(\omega) = \hat{s}_{a_i} - \hat{s}_{a_j}$ and parameter prior distribution as $\omega \sim \mathcal{N}(\boldsymbol{0}, \lambda_\omega \boldsymbol{I})$. Therefore the objective of PCIR becomes:

$$PCIR_{obj} \equiv \sum_{(a_i, a_j) \in T_{>_r}} ln\, \sigma(\hat{s}_{a_i} - \hat{s}_{a_j}) - \lambda_\omega \cdot \|\omega\|^2,$$

where λ_ω is the model specific regularization parameter. The above objective function is differentiable thus we use stochastic gradient descent for learning.

3.2 Feature Selection

In PCIR, we formalize the predicted impact increment score $\hat{s}_a$ of author a as the combined influence of various factors. It includes *author*, *social*, *venue*, *content* and *temporal* features of each author, as listed in Table 1. We give the description and discussion of each selected feature in the follows.

– **Author features.** The author's impact increment in the future is naturally correlated with their current attributes [6,11]. We extract three author features

Table 1. Feature definition. All of the feature values are obtained before the year t (here we set $t = 2008$). Each PR score is rescaled by multiplying 10^6.

Group	Feature	Definition and description
Author	A-#-L (F_1)	The number (#) of the author's previous publications
	A-c (F_2)	The author's current citation #
	A-c-ave (F_3)	The author's current average (ave.) citation # of previous papers
Social	S-#-coa (F_4)	The author's previous co-authors #
	S-c-ave-coa (F_5)	The ave. value of the author's co-authors' citation #
	S-PR_{ACN} (F_6)	The author's PR score on ACN
	S-PR_{ACCN} (F_7)	The author's PR score on ACCN
	S-PR_{ACN}-ave-coa (F_8)	The ave. value of co-authors' PR scores on ACN
	S-PR_{ACCN}-ave-coa (F_9)	The ave. value of co-authors' PR scores on ACCN
Venue	V-ave-c (F_{10})	The ave. value of venues' citations of the author's previous papers
	V-ave-c-two (F_{11})	The ave. value of venues' citations of the author's papers of last two years
	V-PR_{VCCN} (F_{12})	The ave. PR score of venues of the author's previous papers on VCCN
Content	C-diversity$_{LDA}$ (F_{13})	The author's diversity value
	C-authority$_{LDA}$ (F_{14})	The author's authority score
Temporal	T-one-Δc (F_{15})	The citation increment of the author in one year
	T-two-Δc (F_{16})	The ave. citation increment of the author in two years
	T-one-Δl (F_{17})	The paper increment of the author in one year
	T-two-Δl (F_{18})	The ave. paper increment of the author in two years

related with paper number and citation count, namely (1) the number of the author's previous papers, (2) the author's current citation number and (3) the author's current average citation value of previous papers.

- **Social features.** Social interactions among different researchers may influence an author's citation increment [1,4]. To explore such effect, we extract the weighted collaboration network (ACN) among all authors. In ACN, each edge represents a collaboration relationship between two authors and the weight of an edge is defined as the corresponding collaboration frequency. Besides, we assume a widely cited author is an authority researcher who has large impact, and construct authors' citing-cited network (ACCN). Unlike ACN, the ACCN is a weighted directed network and each link denotes a citing-cited relationship between two authors. The PageRank (PR) [5] is used to quantify the authority value. We introduce 6 social attributes of each author including: (1) the number of co-authors, (2) the average value of co-authors' citation counts, (3) the author's PR score on ACN, (4) the author's PR score on ACCN, (5) the average value of co-authors' PR scores on ACN and (6) the average value of co-authors' PR scores on ACCN.

– **Venue features.** Different venues have different impacts due to various reputations and audiences. To quantify the venue's influence, we compute the average citation value of all papers in a venue and name it as a venue's citation. Besides, we construct the weighted directed venues' citing-cited networks (VCCN) of all venues to measure the authority of each venue. Similar to ACCN, the PR score is used to quantify the venue's authority value. Three venue features of each author are extracted: (1) the average value of venue's citations of the author's previous publications, (2) the average value of venue's citations of the author's previous publications in the last two years and (3) the average PR score of the venues of the author's previous publications on VCCN.

– **Content features.** As a popular method for content analysis, topic modeling is useful for predicting paper's impact [2,11]. In **Researchers Division** subsection, we generate the topic distribution of each paper $l \in L_{A^*}$ by LDA. In general, papers with various topics attract attentions from various research fields. We define topic diversity of author a as the average Shannon entropy [2] over their papers' topic distribution: $diversity(a) = \frac{\sum_{l \in L_a} \sum_r -p(r|l) \cdot log p(r|l)}{|L_a|}$, where $p(r|l)$ is the probability distribution over topic r for each paper l. Besides the author's diversity, we further define the author's authority over topics as: $authority(a) = \frac{\sum_r \sum_{l \in L_a} p(r|l) \cdot c_l}{R}$.

– **Temporal features.** Temporal features of publications have been applied to model paper's scientific impact [2,12]. Similarly, the previous increment of the author's paper number or citation can be good indication for the author's future citation increment. Thus we extract 4 temporal features of each author including: (1) the author's citation addition in previous one year, (2) the author's citation addition in previous two years, (3) the author's paper number increment in previous one year and (4) the author's paper number increment in previous two years.

4   Experiment

4.1   Comparison Methods

For comparison, we use three categories of baseline methods. *Category I*: A series of regression learning methods[3] which predict the rising stars according to the predicted impact increment score. It contains Logistic Regression, Naive Bayesian (NB), Random Forest, Support Vector Machine. We only report the results of NB because it performs best. *Category II*: Various ranking algorithms which predict the rising stars according to the predicted rankings of

---

[3] http://scikit-learn.org/.

impact increment scores. It includes RankNet, RankBoost (RankB), AdaRank and Coordinate Ascent. RankB achieves the best performance and we only report it. *Category III*: Naive method Base-1 which predicts the rising stars according to the author's current citation number and naive method Base-2 which ranks the authors by using their average citation increments in the previous two years.

4.2   Evaluation Metric

As a general machine learning task, we divide A_r^* for each topic r into two parts, one for training set $A_{r,train}^*$ and the other for test set $A_{r,test}^*$. The model is trained on $A_{r,train}^*$ and we evaluate its performance on $A_{r,test}^*$. We take the recall value of the top $k\%$ fastest-rising authors in $A_{r,test}^*$ as the evaluation metric: $Recall@k\% = \frac{|\hat{A}_{r,k_{test}}^* \cap A_{r,k_{test}}^*|}{|A_{r,k_{test}}^*|}$.

4.3   Performances Comparison

We construct $A_{r,train}^*$ with 50% instances of A_r^* and use it for model training. The performance is evaluated on the remaining 50% instances in $A_{r,test}^*$. In this work, the parameters of the ARSes prediction problem are set as: $R = 10$, $t = 2008$, $t_{1st} = 2006$ and $\Delta t = 4$. And the parameters in PCIR algorithm are fixed as: $\alpha = 0.01$ and $\lambda_\omega = 0.01$. We adjust feature value f of each author via log-transform: $f = ln(f + 1)$. Figure 3 reports the performances of different methods for different topics when $k = (10, 20)$. Overall, NB, RankB and PCIR perform much better than Base-1 and Base-2. It indicates that the selected features are effective in predicting ARSes. PCIR reaches the best performance, i.e., (0.51, 0.58) average accuracy for all topics, with over 8% average improvement than the other baselines. PCIR performs best for most topics because: (a) It transforms the regression task to a pairwise classification problem and can achieve good prediction accuracy for nonlinear (power-law like) author's citation increment distribution. (b) The posterior probability of Bayesian ranking model well captures the uncertainty of the future ranking orders of authors' citation increment with the knowledge of current ranking. We therefore choose PCIR as the primary predicator for further analysis.

4.4   Feature Contribution Analysis

We examine the contributions of features by two ways: (1) *+Feature.* Keep only one group of features for model training. (2) *−Feature.* Remove the selected group of features and use the remaining features for model training. The *Recall@k%* ($k = 10$) of PCIR with different feature settings are reported in Table 2. According to the result, the temporal features group is the best indicator for the rising stars prediction. The PCIR keeps over 90% accuracy with only temporal features present and drops over 6% if temporal features are removed. The temporal features capture the rising-trend of each researcher in the previous few years, which is strong correlated with future rising trend. Author and social features also have strong influence on prediction result. Meanwhile, the venue

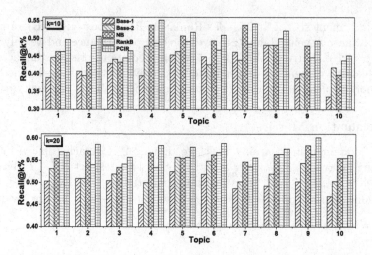

Fig. 3. Performance comparison of different methods ($k = 10, 20$). NB, RankB and PCIR perform much better than Base-1 and Base-2. Indeed, PCIR reaches the best performance for most topics.

Table 2. Contribution analysis of features in different groups on different topics. Temporal features are most influential. Author and social features also have strong effect. Venue features are shown to be least significant.

+/−	+					−					All
Feature	Author	Social	Venue	Content	Temp.	Author	Social	Venue	Content	Temp.	
Topic 1	0.417	0.394	0.246	0.314	0.463	0.469	0.440	0.480	0.463	0.429	0.497
Topic 2	0.420	0.500	0.346	0.370	0.420	0.469	0.444	0.469	0.469	0.490	0.506
Topic 3	0.433	0.429	0.290	0.349	0.441	0.454	0.466	0.450	0.454	0.454	0.466
Topic 4	0.421	0.421	0.252	0.336	0.487	0.529	0.486	0.486	0.481	0.479	0.542
Topic 5	0.464	0.448	0.328	0.344	0.486	0.497	0.492	0.486	0.486	0.481	0.505
Topic 6	0.449	0.463	0.323	0.354	0.449	0.468	0.471	0.483	0.490	0.498	0.504
Topic 7	0.470	0.462	0.311	0.318	0.487	0.515	0.500	0.523	0.530	0.515	0.542
Topic 8	0.482	0.438	0.313	0.348	0.482	0.509	0.491	0.509	0.509	0.482	0.522
Topic 9	0.395	0.447	0.217	0.289	0.434	0.461	0.434	0.454	0.461	0.461	0.489
Topic 10	0.342	0.383	0.336	0.308	0.425	0.445	0.425	0.445	0.438	0.397	0.449
Average	0.429	0.442	0.296	0.333	0.457	0.482	0.465	0.480	0.479	0.469	0.503

features are confirmed to be the least significant. Removing them results in small loss and using only those features will lead to poor performance. It indicates that the papers' impacts of rising stars are highly depend on the author's attributes and papers' content quality.

5   Conclusion

In this work, we formalize a new problem for predicting the top $k\%$ fastest rising young researchers w.r.t. citation values. To solve this problem, we explore a series of factors that can drive a young researcher to be ARS and design a pairwise citation increment ranking method to effectively predict the ARSes. The experimental result on a real dataset shows that our method outperforms the other counterparts. We further find that the temporal features are the best indicators for rising stars prediction. In addition, the author and social features also have strong effects on the prediction while the venue features are little relevant. Overall, our work identifies ARSes in advance and may offer useful guidance for research community like young faculty hiring of university.

Acknowledgements. This work was partially supported by Natural Science Foundation of China (Grant Nos. 61673151, 61503110 and 61433014), Zhejiang Provincial Natural Science Foundation of China (Grant Nos. LY14A050001 and LQ16F030006).

References

1. Bethard, S., Jurafsky, D.: Who should i cite: learning literature search models from citation behavior. In: CIKM (2010)
2. Dong, Y., Johnson, R.A., Chawla, N.V.: Will this paper increase your h-index?: scientific impact prediction. In: WSDM (2015)
3. Li, L., Tong, H., Tang, J., Fan, W.: ipath: forecasting the pathway to impact. In: SDM (2016)
4. Martin, T., Ball, B., Karrer, B., Newman, M.E.J.: Coauthorship and citation patterns in the physical review. Phys. Rev. E **88**(1), 012814 (2013)
5. Page, L., Brin, S., Motwani, R., Winograd, T.: The PageRank citation ranking: bringing order to the web. Technical report 1999-66, Stanford InfoLab (1999)
6. Petersen, A.M., Fortunato, S., Pan, R.K., Kaski, K., Penner, O., Rungi, A., Riccaboni, M., Stanley, H.E., Pammolli, F.: Reputation and impact in academic careers. Proc. Natl. Acad. Sci. U.S.A. **111**(43), 15316–15321 (2014)
7. Rendle, S., Freudenthaler, C., Gantner, Z., Schmidt-Thieme, L.: BPR: Bayesian personalized ranking from implicit feedback. In: UAI (2009)
8. Shen, H.-W., Wang, D., Song, C., Barabási, A.-L.: Modeling and predicting popularity dynamics via reinforced poisson processes. In: AAAI (2014)
9. Tang, J., Zhang, J., Yao, L., Li, J., Zhang, L., Arnetminer, Z.: Extraction and mining of academic social networks. In: KDD (2008)
10. Wang, D., Song, C., Barabási, A.-L.: Quantifying long-term scientific impact. Science **342**(6154), 127–132 (2013)
11. Yan, R., Huang, C., Tang, J., Zhang, Y., Li, X.: To better stand on the shoulder of giants. In: JCDL (2012)
12. Yan, R., Tang, J., Liu, X., Shan, D., Li, X.: Citation count prediction: learning to estimate future citations for literature. In: CIKM (2011)
13. Zhang, C., Yu, L., Lu, J., Zhou, T., Zhang, Z.-K.: AdaWIRL: a novel bayesian ranking approach for personal big-hit paper prediction. In: Cui, B., Zhang, N., Xu, J., Lian, X., Liu, D. (eds.) WAIM 2016. LNCS, vol. 9659, pp. 342–355. Springer, Cham (2016). doi:10.1007/978-3-319-39958-4_27

Fuzzy Rough Incremental Attribute Reduction Applying Dependency Measures

Yangming Liu[1], Suyun Zhao[1,2(✉)], Hong Chen[1], Cuiping Li[1], and Yanmin Lu[3]

[1] School of Information, Renmin University of China, Beijing 100872, China
zhaosuyun@ruc.edu.cn
[2] Key Laboratory of Data Engineering and Knowledge Engineering,
Renmin University of China, MOE, Beijing, China
[3] Building 6, 1388 Zhangdong Road Pudong New Area, Shanghai 201203,
People's Republic of China

Abstract. Since data increases with time and space, many incremental rough based reduction techniques have been proposed. In these techniques, some focus on knowledge representation on the increasing data, some focus on inducing rules from the increasing data. Whereas there is less work on incremental feature selection (i.e., attribute reduction) from the increasing data, especially the increasing real valued data. And fuzzy rough sets is then applied in this incremental method because fuzzy rough set can effectively reduce attributes from the real valued data. By analyzing the basic concepts, such as lower approximation and positive region, of fuzzy rough sets on incremental datasets, the incremental mechanisms of these concepts are then proposed. An incremental algorithm is then designed. Finally, some numerical experiments demonstrate that the incremental algorithm is effective and efficient compared to non-incremental attribute reduction algorithms, especially on the datasets with large number of attributes.

Keywords: Feature selection · Incremental learning · Fuzzy rough set · Dependency function

1 Introduction

Recently, rough theory has already attracted much attention. And then many techniques based on rough set theory and its generalizations have been developed [1–4]. Most of these approaches, however, are proposed based on the assumption that the data are static. Once the data are updated or dynamically increase with time or space, these techniques have to be re-computed on the updated database, which are very costly or even intractable [5]. Incremental learning is a promising approach to refreshing data mining results, because it utilizes previously saved results or data structures to avoid the expense of re-computation [6–8]. To deal with a dynamically increasing dataset, there already exist a lot of rough based researches in an incremental manner and they have successfully been used to data analysis in the real time applications and the applications with limited memory and computability [4–13].

© Springer International Publishing AG 2017
L. Chen et al. (Eds.): APWeb-WAIM 2017, Part I, LNCS 10366, pp. 484–492, 2017.
DOI: 10.1007/978-3-319-63579-8_37

Considering their applications, the rough based incremental approaches are split into the incremental approaches for knowledge representation, feature selection and rule induction. In rough philosophy, feature selection is also called attribute reduction. In the case of attribute reduction, some researchers constructed an incremental attribute reduction algorithm when new objects are added into a decision information system one by one [13]. Furthermore, some researchers proposed some incremental ways when objects are updated in groups [12], some other researchers conducted attribute reduction methods when an attribute set are updated [14].

By the above analysis, it is easy to see that incremental attribute reduction approaches received relatively less attention. The key reason is that it is hard to design an incremental mechanism for attribute reduction algorithms because the relation of granularity in the updated knowledge space is hard to measure before and after attribute reduction, especially for the fuzzy granularity.

This motivates the investigation of incremental attribute reduction in dynamic real valued data. Considering fuzzy rough set technique is a useful tool to handle real values, this paper develops a fuzzy rough based incremental attribute reduction algorithm in real valued dynamic datasets. Here we focus on the dynamic datasets with increasing objects, in the near future we would work on dynamic system with increasing attributes. Our main contributions in this paper include:

(1) Incremental mechanisms for fuzzy positive region and dependency function are proposed by analyzing their differences before and after some new objects are added.
(2) What is more, the incremental mechanism for reduction is proposed based on strict theoretical reasoning.
(3) Then, we propose an incremental attribute reduction algorithm, which is effective and efficient, especially on the datasets with a large number of attributes.

The remainder of this paper is organized as follows. In Sect. 2, we briefly review FRS. Section 3 proposes some incremental mechanisms. Section 4 designs an incremental attribute reduction algorithm. And then in Sect. 5, we give some numerical experiments. Section 6 concludes this paper.

2 Preliminaries

2.1 Some Notations of Fuzzy Rough Set

The data set can be described as one decision table, denoted by a triple $DT = <U, A, D>$. Assumed that $B(B \subseteq A)$ is one subset attributes.

Definition 1. Based on the subset B, the similarity of x_i and x_j is defined as $r_B(x_i, x_j)$, the distance of x_i and x_j is defined as $d_B(x_i, x_j)$.

(1) $d_B(x_i, x_j) = max\{|x_{B_i} - x_{B_j}|\}, 0 < i, j < |B|$;
(2) $r_B(x_i, x_j = 1 - d_B(x_i, x_j)$.

Proposition 1. In Definition 3, the positive region can be simplified as follows:

$$POS_B^U(x) = \min_{\{u \in U, u \notin D_x\}} \{d_B(x,u)\}, x \in D_x$$

Proposition 1 shows the relation between fuzzy positive region and lower approximation. Then, the dependency degree of subset B is defined as follows.

Definition 2. The dependency degree of $B \subseteq A$ corresponding to the decision attributes D is as follows.

$$\gamma_B^U = \frac{\sum_{x \in U} POS_B^U(x)}{|U|}$$

Theorem 1. If $B_1 \subseteq B_2 \subseteq A$, then $\gamma_{B_1}^U \leq \gamma_{B_2}^U \leq \gamma_A^U$.

Definition 3. Given a fuzzy decision table $DT = <U, A, D>$ and attribute subset $B \subseteq A$, when B satisfies the following two conditions, we say that B is a reduction,

(1) $\forall a \in B, \gamma_{B-a}^U < \gamma_B^U$;
(2) $\gamma_B^U = \gamma_A^U$;

Then the non-incremental attribute reduction algorithm, shortened by *NonIAR*, is described in Algorithm 1 [1, 2].

Algorithm 1: NonIAR
INPUT: $DT = <U, A, D>$
OUTPUT : *redu*

---

1. $B = \phi$
2. $Lef = A$
3. Calculate γ_A^U
4. While $\gamma_B^U < \gamma_A^U$ do
5. $\gamma_{B+a}^U = max\{\gamma_{B+a_i}^U\}, a_i \in lef$
6. $B = B \cup a$
7. **End while**
8: $i \leftarrow 0$
9: **While** $i < |B|$ do
10:     **if** $\gamma_{B-B_i}^U < \gamma_A^U$ **then** $red = red \cup B_i$;
12:     **end if**
13:     $i++$;
14: **End while**
15: **Output** *redu*

---

3 Fuzzy Rough Based Incremental Mechanisms

3.1 Incremental Mechanism for Positive Region

Theorem 2. Given a fuzzy decision table $DT = <U, A, D>$ and an attribute subset $B \subseteq A$, If some new objects $\Delta U = \{x_{n+1}, x_{n+2}, \ldots, x_{n+s}\}$ are added, then

If $x \in U$, then $POS_B^{U \cup \Delta U} = POS_B^U(x) - \Delta POS_B^U(x); \Delta POS_B^U(x) = POS_B^U(x) - \min\limits_{u \in \Delta U, U \notin D_x}$ $d_B(x, u), POS_B^U(x) > \min\limits_{u \in \Delta U, U \notin D_x}\{d_B(x, u)\}$; $\Delta POS_B^U(x) = 0, otherwise.$ If $x \in \Delta U$, then $POS_B^U(x) = \min\limits_{u \in U \cup \Delta U, u \notin D_x}\{d_B(x, u)\}$.

Proof.
By Proposition 1, in the case of $x \in \Delta U$, it is apparent. In the case of $x \in U$,
$POS_B^{U \cup \Delta U}(x) = \min\limits_{u \in U \cup \Delta U, u \notin D_x}\{d_B(x, u)\} = \min\{\min\limits_{u \in U \cup \Delta U, u \notin D_x}\{d_B(x, u)\}, \min\limits_{u \in \Delta U, u \notin D_x}$
$\{d_B(x, u)\}\} = \min\limits_{u \in U, u \notin D_x}\{d_B(x, u)\} - \Delta POS_B^U(x) = POS_B^U(x) - \Delta POS_B^U(x)$ ∎

3.2 Incremental Mechanism for Fuzzy Dependency

Theorem 3. Given a fuzzy decision table $DT = <U, A, D>$ and an attribute subset $B \subseteq A$, If some new objects ΔU are added in, then

$$\gamma_B^{U \cup \Delta U} = \frac{|U|\gamma_B^U - \sum_{x \in U} \Delta POS_B^U(x) + \sum_{x \in \Delta U} \Delta POS_B^{U \cup \Delta U}(x)}{|U| + |\Delta U|}$$

Proof.

$$\sum_{x \in U \cup \Delta U} POS_B^U(x) = \sum_{x \in U} POS_B^{U \cup \Delta U}(x) + \sum_{x \in \Delta U} POS_B^{U \cup \Delta U}(x)$$
$$= \sum_{x \in U}(POS_B^U(x) - \Delta POS_B^U(x)) + \sum_{x \in \Delta U} POS_B^{U \cup \Delta U}(x) = \sum_{x \in U} POS_B^U(x) - \sum_{x \in U} \Delta POS_B^U(x) + \sum_{x \in U} POS_B^{U \cup \Delta U}(x)$$
$$\gamma_B^{U \cup \Delta U} = \frac{\sum_{x \in U \cup \Delta U} POS_B^{U \cup \Delta U}(x)}{|U| + |\Delta U|} = \frac{\sum_{x \in U} POS_B^U(x) - \sum_{x \in U} \Delta POS_B^U(x) + \sum_{x \in \Delta U} \Delta POS_B^{U \cup \Delta U}(x)}{|U| + |\Delta U|}$$
$$= \frac{|U|\gamma_B^U - \sum_{x \in U} \Delta POS_B^U(x) + \sum_{x \in \Delta U} POS_B^{U \cup \Delta U}(x)}{|U| + |\Delta U|}$$ ∎

3.3 Incremental Mechanism for Reduction

Theorem 4. Given a fuzzy decision table $DT = <U, A, D>$, and an attribute subset $B \subseteq A$, when one attribute a is added into B, let $\Delta S = \{x \in U \wedge POS_B^U(x) < POS_C^U(x)\}$, then

$$\gamma_{B+a}^{U} = \frac{|U|\gamma_B^U - \sum_{x \in \Delta S} POS_B^U(x) + \sum_{x \in \Delta S} POS_{B+a}^U(x)}{|U|}$$

Proof.

If $B \subseteq A = \{A_1, A_2, \ldots, A_n\}, x \in U \Rightarrow \forall x \in U, POS_B^U(x) \leq POS_A^U(x)$. If $x \notin \Delta S$, then $POS_B^U(x) = POS_A^U(x)$, which means $POS_B^U(x)$ would not grow any more.

As a result, when adding one attribute, we just need to consider

$$\gamma_{B+a}^{U} = \frac{|U|\gamma_B^U - \sum_{x \in \Delta S} POS_B^U(x) + \sum_{x \in \Delta S} POS_{B+a}^U(x)}{|U|}, x \in \Delta S.$$

∎

4   Incremental Algorithm for Reduction

Algorithm 2:   **FIAR** (Dependency Function based Incremental Reduction Algorithm)
INPUT: $DT = <U, A, D>, \Delta U, redu, \gamma_{redu}^U, \bigcup_{x \in U} POS_{redu}^U, \gamma_C^U, \bigcup_{x \in U} POS_c^U$
OUTPUT: $newredu$

1: **Let** $B = redu$;
2:        **Calculate** $\gamma_B^{U \cup \Delta U}$;
3:        **Calculate** $\gamma_C^{U \cup \Delta U}$
4: **Let** $lef = C - redu$;
5: **While** $\gamma_B^{U \cup \Delta U} < \gamma_C^{U \cup \Delta U}$ **do**
6:        $besta = \{a \mid \gamma_{besta}^{U \cup \Delta U} = max\{\gamma_{a_i}^U\}, a_i \in lef$;
7:        $B = B \cup besta$;
8: **End while**
9: $i \leftarrow 0$
10: **While** $i < |B|$ **do**
11:        **If** $notredundancy(B_i)$ **then** $red = red \cup B_i$;
12:        **End if**
13:        $i++$;
14: **End while**
15: **Output** red

Both NonIAR and FIAR are heuristic. The complexity of FIAR is O(M $*||lef||\Delta S||U +\Delta U|$), whereas NonIAR is O(K $*N |U + \Delta U|^2$). When N is far larger than $|U|$, FIAR is significantly faster than NonIAR.

5   Numerical Experiments

In this section, we conduct some numerical experiments on a series of UCI datasets [15]. The dataset detailed are summarized in Table 1. All the experiments have been carried out on Ubuntu release 16.0, i7-4790 CPU @ 3.60 GHz with 8 GB and C++.

Table 1. The description of the selected datasets

Datasets	Attribute no.	Object no.	Classes
Waveform	22	5000	3
Letter	17	20000	26
Shuttle	10	58000	2
Credit	25	30000	2
Gene12	9182	174	5
Gene14	3312	203	5

5.1 Execution Time

New Data are Added in One Time

In this part, every dataset is equally split into two parts. We use FIAR-2 represents FIAR run on such new data added in one time.

Table 2 demonstrates that the ratios of FIAR/NonIAR are always smaller than 1. This shows that the incremental algorithm accelerates the reduction.

Table 2. The execution time when new data are added in one time

Datasets	NonIAR (CPU seconds)	FIAR (CPU seconds)	Ratio (FIAR/NonIAR)
Waveform	209.67	135	0.64
Letter	2521	1479	0.58
Shuttle	3380	1141	0.33
Credit	3633	1979	0.54
Gene12	4240	497	0.117
Gene14	876	55	0.062

Table 2 also shows that FIAR works dramatically better than NonIAR on the datasets with high dimension. For example, FIAR is dramatically faster than NonIAR on Gene9, Gene12 and Gene14. This is because the time complexity of FIAR is $O(M * |lef||\Delta S||U + \Delta U|)$, whereas NonIAR is $O(K * N|U + \Delta U|^2)$. When N is much larger, it needs more loops for NonIAR's $\gamma_B^{U \cup \Delta U}$ approaching to $\gamma_C^{U \cup \Delta U}$, which means $M < < K$. Thus, FIAR works significantly faster than NonIAR on the datasets with high dimension.

New Data are Added Successively

In this part, data is split into six parts. One part is seen as the old data, the other parts are seen as the successive added-in data. We use FIAR-6 represents it.

In Fig. 1, the trends of *NonIAR* go upward dramatically with the successively increasing data. Comparatively, the trends of *FIAR* go upward just obviously with the successively increasing data. This demonstrates that *FIAR* works significantly faster

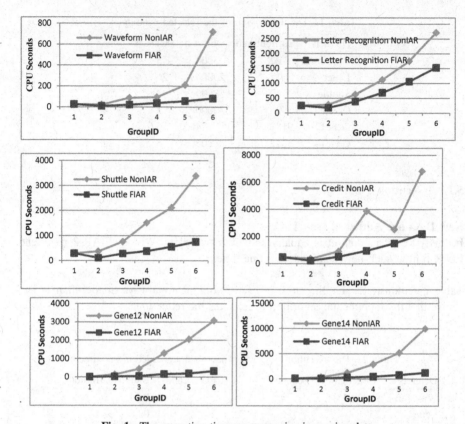

Fig. 1. The execution time on successive increasing data

than *NonIAR* with the data successively increasing. This is because *NonIAR* need calculate all the data, whereas *FIAR* just need consider the new-added data. The faster the data dynamically update, the better FIAR works than NonIAR.

5.2   Reduction Ratio

In this part, we compare the effectiveness of reduction between *NonIAR*, *FIAR-2* and *FIAR-6*. The comparison results are summarized in Table 3.

Table 3.   Reduction ratio

Data sets with low dimension				Dataset with high dimension				
Datasets	NonIAR	FIAR-2	FIAR-6	Baseline	Datasets	NonIAR	FIAR-2	FIAR-6	Baseline
Waveform	14	14	13	21					
Letter	10	9	9	17	Gene12	38	38	39	9182
Shuttle	4	5	5	10	Gene14	20	20	21	3312
Credit	8	11	10	25					
Ratio	0.49	0.53	0.50	–	Ratio	0.004	0.005	0.0049	–

In Table 3, the reduction ratios of these three algorithms are similar. They are 0.49, 0.53, 0.50 on the dataset with high dimension, and 0.004, 0.005, 0.0049 on the datasets with high dimension. This shows that the proposed incremental algorithm FIAR has the similar reduction results with the non-incremental algorithm NonIAR. Or say, the proposed incremental algorithm FIAR is an effective attribute reduction algorithm on dynamic increasing data.

5.3   Classification Performance

In this part, we apply KNN to check the classification quality of the reductions obtained by different algorithms. The comparison results are summarized in Table 4.

Table 4. The classification performance of the reductions

Datasets	NonIAR	FIAR-2	FIAR-6	Original
Waveform	66.56	66.56	66.49	67.71
Letter	77.44	77.34	77.34	77.89
Shuttle	29.16	30.02	29.56	30.82
Credit	27.68	27.13	27.71	32.94
Gene12	80.34	80.26	80.71	81.85
Gene14	70.11	70.01	70.55	72.91

In Table 4, it is easy to see that these algorithms have the similar classification accuracy. Comparing to the Baseline, i.e., the whole datasets without reduction, the accuracy lost is limited.

Table 4 also shows that the reductions of FIAR-2 and FIAR-6 have the similar classification performance. This shows that it does not affect the classification performance of reduction no matter the new data are added in one time or successively.

6   Conclusion

In this paper, we provides new views on dealing with attribute reduction on real valued datasets with the following characters:

(1)   Incremental mechanisms of some notations of fuzzy rough sets are proposed.
(2)   The incremental algorithm is designed based on fuzzy rough sets on the real valued datasets.
(3)   The numerical experiments demonstrate that the proposed incremental algorithm is effective and efficient, especially on the datasets with a large number of attributes.

References

1. Pawlak, Z.: Rough Sets: Theoretical Aspects of Reasoning about Data, vol. 9. Springer Science & Business Media, Heidelberg (2012)
2. Pawlak, Z., Skowron, A.: Rough sets: some extensions. Inf. Sci. **177**(1), 28–40 (2007)
3. Düntsch, I., Gediga, G.: Uncertainty measures of rough set prediction. Artif. Intell. **106**(1), 109–137 (1998)
4. Qian, Y., Liang, J., Pedrycz, W., Dang, C.: Positive approximation: an accelerator for attribute reduction in rough set theory. Artif. Intell. **174**(9–10), 597–618 (2010)
5. Chen, H., Li, T., Luo, C., Horng, S., Wang, G.: A decision-theoretic rough set approach for dynamic data mining. IEEE Trans. Fuzzy Syst. **23**(6), 1958–1970 (2015)
6. Guyon, I., Elisseeff, A.: An introduction to variable and feature selection. J. Mach. Learn. Res. **3**(Mar), 1157–1182 (2003)
7. Ye, J., Li, Q., Xiong, H., Park, H., Janardan, R., Kumar, V.: IDR/QR: an incremental dimension reduction algorithm via qr decomposition. IEEE Trans. Knowl. Data Eng. **17**(9), 1208–1222 (2005)
8. Zhang, Y., Chen, S., Wang, Q., Yu, G.: MapReduce: incremental MapReduce for mining evolving big data. IEEE Trans. Knowl. Data Eng. **27**(7), 1906–1919 (2015)
9. Zhang, J., Li, T., Ruan, D., Liu, D.: Rough sets based matrix approaches with dynamic attribute variation in set-valued information systems. Int. J. Approx. Reason. **53**(4), 620–635 (2012)
10. Zeng, A., Li, T., Hu, J., Chen, H., Luo, C.: Dynamical updating fuzzy rough approximations for hybrid data under the variation of attribute values. Inf. Sci. **378**, 363–388 (2017)
11. Zhang, J., Wong, J., Pan, Y., Li, T.: A parallel matrix-based method for computing approximations in incomplete information systems. IEEE Trans. Knowl. Data Eng. **27**(2), 326–339 (2015)
12. Liang, J., Wang, F., Dang, C., Qian, Y.: A group incremental approach to feature selection applying rough set technique. IEEE Trans. Knowl. Data Eng. **26**(2), 294–308 (2014)
13. Hu, F., Wang, G., Huang, H., Wu, Y.: Incremental attribute reduction based on elementary sets. In: Ślęzak, D., Wang, G., Szczuka, M., Düntsch, I., Yao, Y. (eds.) RSFDGrC 2005. LNCS, vol. 3641, pp. 185–193. Springer, Heidelberg (2005). doi:10.1007/11548669_20
14. Zeng, A., Li, T., Liu, D., Zhang, J., Chen, H.: A fuzzy rough set approach for incremental feature selection on hybrid information systems. Fuzzy Sets Syst. **258**, 39–60 (2015)
15. https://archive.ics.uci.edu/ml/datasets.html/

Query Processing

SET: Secure and Efficient Top-k Query in Two-Tiered Wireless Sensor Networks

Xiaoying Zhang[1,2], Hui Peng[3], Lei Dong[1,2], Hong Chen[1,2], and Hui Sun[1,2(✉)]

[1] School of Information, Renmin University of China, Beijing 100872, China
`sun_h@ruc.edu.cn`
[2] Key Laboratory of Data Engineering and Knowledge Engineering
of Ministry of Education, Beijing, China
[3] The Fifth Electronic Research Institute of MIIT, Guangzhou 510000, China

Abstract. Top-k query is one of important queries in wireless sensor networks (WSNs). It provides the k highest or lowest data collected in the entire network. Due to abundant resources and high efficiency, many future large-scale WSNs are expected to follow a two-tiered architecture with resource-rich master nodes. However, the sensor network is unattended and insecure. Since master nodes store data collected by sensor nodes and respond to queries issued by users, they are attractive to adversaries. It is challenging to process top-k query while protecting sensitive data from adversaries. To address this problem, we propose SET, a framework for secure and efficient top-k query in two-tiered WSNs. A renormalized arithmetic coding is designed to enable each master node to obtain exact top-k result without knowing actual data. Besides, a verification scheme is presented to allow the sink to detect compromised master nodes. Finally, theoretical analysis and experimental results confirm the efficiency, accuracy and security of our proposal.

Keywords: Top-k query · Wireless sensor networks · Privacy preservation · Integrity verification

1 Introduction

As the indispensable building block for Internet of Things, wireless sensor networks (WSNs) have been widely used in many applications, including smart home, e-health and environment monitoring. In these applications, top-k query is an important query which interests users. It is to seek for the k highest or lowest data in the sensor network, which is meaningful for monitoring extreme conditions. However, due to the openness and non-supervision of WSNs, severe privacy problems have been exposed in practical applications. For instance, in smart homes, sensors and actuators are distributed in houses to collect information and make our lives more comfortable. The power provider wants to know

This work is supported by National Science Foundation of China (Grant No. 61532021), National Basic Research Program of China (973 Program) (No. 2014CB340403) and National High Technology Research and Development Program of China (863 Program) (No. 2014AA015204).

L. Chen et al. (Eds.): APWeb-WAIM 2017, Part I, LNCS 10366, pp. 495–510, 2017.
DOI: 10.1007/978-3-319-63579-8_38

which are the k highest electricity consumption of a community in the peak hours. During the process of data transmission and aggregation, household information may be overheard by an adversary such that privacy of every resident leaks. Therefore, privacy-preserving top-k query processing is greatly urgent.

On account of resource savings, rapid response and high scalability [16], the two-tiered architecture [10] will be adopted in future large-scale WSNs. As shown in Fig. 1, a two-tiered WSN consists of a great many sensor nodes in the lower tier and a few master nodes, also called storage nodes, in the upper tier. Sensor nodes gather data while master nodes store data and return results to the sink. In spite of its advantages, this tiered network also brings difficulties in performing privacy-preserving top-k query. On the one hand, since master nodes store all data collected by sensor nodes, adversaries are apt to compromise them instead of sensor nodes. Once compromising a master node, the adversary can obtain all sensitive data, which violates **data privacy**. Moreover, the adversary can manipulate the compromised master node to submit fake or incomplete result, which breaches **result integrity**. On the other hand, master nodes are required to process queries efficiently and correctly, which is hindered if data are encrypted. As a result, it is challenging to preserve data privacy and result integrity as well as achieving efficient performance and correct result.

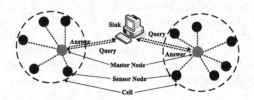

Fig. 1. Architecture of two-tiered wireless sensor networks

To address the above problem, this paper investigates secure top-k query in tiered WSNs with the following contributions:

- We first design a renormalized arithmetic coding scheme (RAC) suitable for WSNs. The RAC scheme encodes sensitive data of sensor nodes to tuples of codes which hold order-preserving property.
- We then propose SET, a framework for **S**ecure and **E**fficient **T**op-k query in two-tiered WSNs. SET relies on the RAC scheme to preserve data privacy and enable master nodes to perform top-k queries efficiently and correctly without knowing actual data.
- We further present an integrity verification scheme to examine result integrity. Global correlation between data helps the sink to detect the incorrect top-k result.
- Theoretical analysis and experimental results indicate that our proposal can reduce more communication cost and save more energy while accomplishing accurate result, data privacy and result integrity.

The rest of paper is organized as follows. Section 2 summarizes related work. Section 3 describes models and requirements. Our SET framework is elaborated in Sect. 4. We analyze and evaluate the proposed SET in Sects. 5 and 6, respectively. Finally, this paper is concluded in Sect. 7.

2   Related Work

Privacy-preserving query in WSNs has drawn more concerns. Existing work on privacy-preserving query mainly focuses on aggregation query, range query and top-k query. Our work focuses on top-k query. Privacy-preserving top-k query in two-tiered WSNs has been explored in [3,5–8,12–15,17].

Zhang *et al.* [14] for the first time present a verifiable fine-grained top-k query algorithm in tiered WSNs. Three schemes are proposed to verify the authenticity and completeness of result. These schemes rely on the basic idea that sensor nodes embed relationships among the data. The above work is further extended in [15] by proposing the random probing scheme, the query conversion scheme, and the random witness scheme. VSFTQ [8] sorts all data and constructs order relationships between data collected by each node. The final top-k result is verified using order relationships. Since work [8,14,15] only pays attention to authenticity and completeness of top-k result, serious privacy issues still exist.

SafeTQ [3] is a verifiable privacy-preserving top-k query algorithm in tiered WSNs. The whole network is separated into different cells. In each cell, the head node collaborates with the aided computing node to determine the k-th maximum (minimum) by using secure computation [11]. Probabilistic neighbor checking is to examine result integrity. Since its privacy preservation is ensured, the selection criterion of head nodes and aided computing nodes is not given.

Order-preserving encryption scheme is adopted in [7,12]. PriSecTopk [7] is a series of privacy-preserving top-k query mechanisms on the time slot data set in two-tier WSNs. The basic PriSecTopk use order-preserving symmetric encryption to hide original data and calculate results. To protect order-relation and distance-relation privacy, the basic PriSecTopk is improved by secret perturbation. One obvious drawback is that the result is imprecise. The sink verifies result integrity through sample-based hypothesis testing method combined with computing commitment information, and cannot detect the incomplete result completely. Work proposed in [12] transforms the input distribution to target distribution such that the final result may be also imprecise.

SVTQ [17] answers top-k queries in WSNs without leaking sensitive information. It computes two sets: the prefix family and the prefix range for each datum, and then performs prime aggregation to the numericalized prefixes. Comparison between two data is converted to checking whether their greatest common divisor equals to 1. Since there is no known efficient formula for primes, prime aggregation scheme requires sensor nodes to prestore enough prime numbers and their sequences. For instance, if the data domain is $[0, 127]$, each sensor node should prestore 512 prime numbers. Due to the limited storage space of sensor nodes, SVTQ is only suited to extremely limited domain.

ADVQ [13] can safely compute top-k result by data anonymization, but lots of extra information of neighboring nodes increases communication overhead of both sensor nodes and master nodes, and introduces false positives. In addition, the randomized and distributed ordering preserving encryption also delays result response. ETQFD [5] supports secure top-k queries based on filters. Nevertheless, maintenance of filters in sensor nodes is costly and result integrity cannot be verified. Jonsson *et al.* [6] present a prototype system for secure distributed top-k aggregation, which is unsuitable for the tiered WSNs.

3   Models and Requirements

3.1   Network Model

The architecture of a two-tiered WSN is displayed in Fig. 1. It is composed of three types of nodes: sensor nodes, master nodes and a sink. Sensor nodes are restricted in various resources, e.g., energy, storage and bandwidth. In contrast, master nodes have more but not inexhaustible resources. Only the sink has unlimited resources and powerful capabilities. Sensor nodes can communicate with their neighbors and their master nodes. Master nodes can also communicate with each other and the sink. In fact, the entire network is separated into several non-overlapping cells and each cell comprises one master node and some sensor nodes.

Sensor nodes collect data from surroundings with a specified frequency, and then periodically send data to their master nodes. The message that a sensor node s_i sends to its master node M is

$$s_i \rightarrow M : i, t, \{d_1, d_2, ..., d_\lambda\},$$

where i denotes the unique identifier of s_i, t denotes an epoch, λ is the number of sensed data and d_j $(1 \leq j \leq \lambda)$ is the data collected by s_i at epoch t. Suppose all sensors are synchronized so that they keep consistent in epoches.

The sink directly issues the top-k query $Q = \langle k, T, C \rangle$ to master nodes, where C is the set of interested cells, and k means that the user inquires the k highest data generated in area C at the time slot T. After receiving this query, master nodes in area C immediately search for proper data and submit the result to the sink. The result should contain the k highest data and the identifiers of corresponding sensor nodes. Unless otherwise stated, "top" in top-k query is defined as "highest" in this paper.

3.2   Adversary Model

As assumed in [3,5–8,12–15,17], the sink is reliable. During query processing, adversaries attempt to eavesdrop on sensitive data through the wireless link. It is unsafe to transmit data in the form of plaintext. Moreover, adversaries are more inclined to compromise master nodes than compromise sensor nodes. There are two reasons. First, master nodes store large quantity of sensitive data. Once compromising a master node, it enables the adversary to obtain all data stored in this master node. Second, the adversary may manipulate the compromised master node to submit wrong result to the sink. It is obvious that the compromised master node leads to more harmful threats than the compromised sensor node. Therefore, as in state of the art work [3, 6–8, 12–15, 17], our paper focuses on the problem of compromised master nodes.

3.3   Design Requirements

A good secure top-k query algorithm in tiered WSNs should fulfill the following requirements:

- Data privacy. Since all sensory data are private and sensitive, they should not be obtained by master nodes. Furthermore, data should only be known by its owner and the sink.
- Result integrity. Three conditions should be satisfied: (1) The number of data in the result should be equal to k; (2) All data in the result should be authentic, i.e., forged data are not permitted; (3) Any datum outside the result should not be greater than all data in the result. In other words, the result should include only data satisfying the query and exclude data dissatisfying the query. At least, the algorithm should guarantee that the sink is able to detect unauthentic or incomplete result.
- Efficiency. Communication cost is one important metric for energy consumption. It is generated during data submission from sensor nodes to master nodes and result submission from master nodes to the sink. The less communication cost, the higher efficiency.

4   Framework for Secure and Efficient Top-k Query

In this section, we propose a **S**ecure and **E**fficient **T**op-k query framework (SET) for two-tiered WSNs. To prevent adversaries from knowing sensitive data, the naïve but effective method is to encrypt data using secret keys. However, it is impossible for master nodes to process the top-k query just based on encrypted data. To solve this problem, we design a renormalized arithmetic coding scheme (RAC) to encode data and preserve privacy while processing queries. In the following, we first present the RAC scheme, and then elaborate four phases of SET framework, including system initialization, data submission, result response and integrity verification. For clear description, Table 1 summarizes the notation used in this paper.

4.1   Arithmetic Coding

Arithmetic coding is one of entropy encoding for lossless data compression. It is first introduced in [9]. The key idea of Arithmetic coding is to represent the entire message by a number between 0 and 1. Figure 2 shows an example of arithmetic

Table 1. Notation

Symbol	Meaning
k	The parameter for a top-k query
t	An epoch
T	A time slot
M	A master node
s_i	A sensor node with the unique identifier i
$k_{i,t}$	A secret key of s_i at epoch t
d	The sensory data
$E()$	A function for encryption
$\mathbb{B}()$	A function for binary conversion
$\mathbb{R}()$	A function for RAC scheme
$\mathbb{R}^0()$	The number of 0 bits before the first 1 bit in the RAC code
$\mathbb{R}^1()$	The 0–1 code beginning with the first 1 bit in the RAC code
$P(m)$	The probability of the source symbol m
$C(m)$	The cumulative distribution of the source symbol m

Symbol(m)	P(m)	C(m)
A	0.5	0
B	0.3	0.5
C	0.2	0.8

(a) Symbol Distribution Table   (b) Encoding

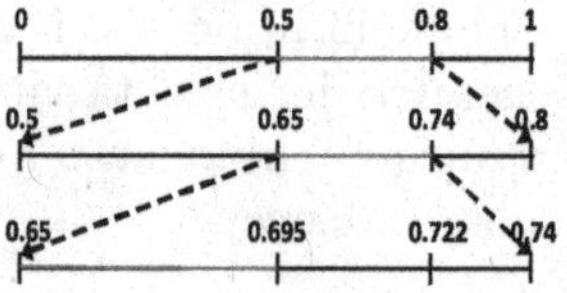

Fig. 2. An example of arithmetic coding

coding. To simplify description, assume any message is one combination of three source symbols: A, B, C. Given a symbol m, $P(m)$ denotes its probability and $C(m)$ denotes its cumulative distribution with $C(m) = \sum_{i=1}^{m-1} P(i)$. The process of encoding a message "BBA" is detailed in Fig. 2(b). Initial interval is set to $[0, 1]$. For the first symbol "B", the interval is updated to $[0.5, 0.8]$ highlighted by red line. Then divide the new interval $[0.5, 0.8]$ into sub-intervals according to the probability and cumulative distribution. For the second symbol "B", the interval is further updated to $[0.65, 0.74]$ and divided. For the last symbol "A", the interval is finally updated to $[0.65, 0.695]$. Therefore, the message "BBA" can be represented by any number lying within the interval $[0.65, 0.695]$, e.g. 0.68.

However, the final interval may collapse into a single point when the message is too long, which reduces the precision. Although renormalization is used to overcome this problem in [4], these algorithms are too complex to be applied

---

Scheme 1 Renormalized Arithmetic Coding

Input:

  M : message

  n : length of message

  I: size of initial interval

  $P(m)$: probability of symbol m

  $C(m)$: cumulative distribution of symbol m

Output:

  $code$: code of message

1: $l \leftarrow 0$

2: $u \leftarrow I - 1$

3: $code \leftarrow null$

4: $count \leftarrow 0$

5: **for** read a symbol m of M from 1 to n **do**

6:   $r \leftarrow u - l + 1$

7:   $l \leftarrow l + r * C(m)$

8:   $u \leftarrow l + r * P(m) - 1$

9:   $r \leftarrow u - l + 1$

10:   **while** $r < I/4$ **do**

11:     **if** $l >= I/2$ **then**

12:       $code \leftarrow code + "1"$

13:       $u \leftarrow 2 * u - I + 1$

14:       $l \leftarrow 2 * l - I$

15:       **for** from 0 to $count$ **do**

16:         $code \leftarrow code + "0"$

17:       **end for**

18:       $count \leftarrow 0$

19:     **else**

20:       **if** $u < I/2$ **then**

21:         $code \leftarrow code + "0"$

22:         $u \leftarrow 2 * u + 1$

23:         $l \leftarrow 2 * l$

24:         **for** from 0 to $count$ **do**

25:           $code \leftarrow code + "1"$

26:         **end for**

27:         $count \leftarrow 0$

28:       **else**

29:         $u \leftarrow 2 * u - I/2 + 1$

30:         $l \leftarrow 2 * l - I/2$

31:         $count \leftarrow count + 1$

32:       **end if**

33:     **end if**

34:     $r \leftarrow u - l + 1$

35:   **end while**

36: **end for**

37: **if** $l >= I/2$ **then**

38:   $code \leftarrow code + "1"$

39:   **for** from 0 to $count$ **do**

40:     $code \leftarrow code + "0"$

41:   **end for**

42:   $code \leftarrow code + "0"$

43: **else**

44:   $code \leftarrow code + "0"$

45:   **for** from 0 to $count$ **do**

46:     $code \leftarrow code + "1"$

47:   **end for**

48:   $code \leftarrow code + "1"$

49: **end if**

50: return $code$;

---

to WSNs. Here we design a simple but efficient renormalized arithmetic coding (RAC) specialized for WSNs, as detailed in Scheme 1. Assume the initial interval is set to $[0, I-1](I > 1)$, the main idea of RAC scheme is to keep the interval size not smaller than $I/4$ by enlarging the interval once it becomes too small. Given a message, read each symbol m successively and update both the interval $[l, u]$ and interval size s (line 6–9). If the size of current interval is smaller than $I/4$, renormalization is required (line 10–35). There are three cases to be considered: (1) If the interval completely locates within the top half of $[0, I-1]$, which means the next interval gets closer to $I - 1$ than 0, expand the interval and output a 1 followed by $count$ 0s (line 11–18), where $count$ is determined by the third case; (2) If the interval completely locates within the bottom half of $[0, I - 1]$, which means the next interval gets closer to 0 than $I - 1$, expand the interval and output a 0 followed by $count$ 1s (line 19–27); (3) If the lower bound of the interval falls in the bottom half and the upper bound falls in the top half, the trend of next interval cannot be determined. Keep track the expansion by increasing *count* and expand the interval (line 28–31). After the final interval is calculated, choose a number of the interval (line 37–49). For example, the message "*BBA*" is encoded to 10101. RAC scheme will be used to encode data in our SET. Now we begin to elaborate the SET framework.

4.2 System Initialization

The network is initialized before executing a task. Without loss of generality, let $[d_{min}, d_{max}]$ be the domain of data which are positive integers. Each sensor node s_i shares a seed key $k_{i,0}$ only with the sink. s_i encrypts its data by $k_{i,t}$ which is the secret key of s_i at epoch t. Let $k_{i,t} = hash(k_{i,t-1})$ and erase $k_{i,t-1}$ at epoch t. Let $\{00, 01, 10, 11\}$ be the set of source symbols. The sink constructs a symbol distribution table (SDT). SDT is a set of tuples $\langle m, P(m), C(m) \rangle$, where $P(m)$ and $C(m)$ respectively denote the probability and cumulative distribution of the symbol m. $k_{i,t}$ and SDT are preloaded in every sensor node. To avoid a brute force attack, all sensor nodes change SDT simultaneously and periodically based on the initial SDT and an agreed rule.

4.3 Data Submission

In data submission, we describe how a sensor node processes its data before sending them to the closest master node. Assume $d_1, d_2, ..., d_\lambda$ are the data collected by sensor node s_i at epoch t. s_i processes data in the following five steps:

- Step1. Sort $d_1, d_2, ..., d_\lambda$ in descending order with $d_1 > d_2 > ... > d_\lambda$.
- Step2. Convert each d_i to a binary form. For example, the integer 21 is converted to 10101. Assume the domain is $[0, 255]$, at least 8-bits (always even) is required to represent any value in this domain. In order to use the proposed RAC scheme, insert 000 to 10101 and shift 10101 to the right, i.e., 00010101. Let $\mathbb{B}()$ denote the binary conversion, and d_i is converted to $\mathbb{B}(d_i)$.
- Step3. Encode each $\mathbb{B}(d_i)$ using the RAC scheme (presented in Sect. 4.1) and the symbol distribution table SDT. For instance, assume $I = 100$ and 0.4, 0.3, 0.2, 0.1 respectively denote the probability of source symbols 00, 01, 10, 11, 00010101 is encoded to 0011101. Let $\mathbb{R}()$ denote the RAC scheme, and $\mathbb{B}(d_i)$ is encoded to $\mathbb{R}(\mathbb{B}(d_i))$.
- Step4. Represent each $\mathbb{R}(\mathbb{B}(d_i))$ by two parts: $\mathbb{R}^0(\mathbb{B}(d_i))$ and $\mathbb{R}^1(\mathbb{B}(d_i))$, where $\mathbb{R}^0(\mathbb{B}(d_i))$ denotes the number of 0 bits before the first 1 bit in $\mathbb{R}(\mathbb{B}(d_i))$ and $\mathbb{R}^1(\mathbb{B}(d_i))$ denotes the rest code of $\mathbb{R}(\mathbb{B}(d_i))$ excluding $\mathbb{R}^0(\mathbb{B}(d_i))$. For example, $\langle 2, 11101 \rangle$ represents 0011101. Therefore $\mathbb{R}(\mathbb{B}(d_i))$ is represented by $\langle \mathbb{R}^0(\mathbb{B}(d_i)), \mathbb{R}^1(\mathbb{B}(d_i)) \rangle$.
- Step5. Construct a structure $\langle f_i | d_i | b_i \rangle$ for each d_i, where $f_i = d_{i-1} - d_i$ ($f_1 = d_{max}$ for d_1) and $b_i = d_i - d_{i+1}$ ($b_\lambda = d_\lambda + 1$ for d_λ), which are useful for integrity verification. Then encrypt $\langle f_i | d_i | b_i \rangle$ using the secret key $k_{i,t}$. Let $\mathbb{E}()$ denote the encryption function, i.e., $\mathbb{E}(f_i | d_i | b_i)$.

After these five steps, the message that sensor node s_i submits to its master node M is

$$s_i \rightarrow M : i, t, \{\mathbb{E}(f_1|d_1|b_1), \mathbb{E}(f_2|d_2|b_2), ..., \mathbb{E}(f_\lambda|d_\lambda|b_\lambda)\},$$

$$\{\langle \mathbb{R}^0(\mathbb{B}(d_1)), \mathbb{R}^1(\mathbb{B}(d_1))\rangle, \langle \mathbb{R}^0(\mathbb{B}(d_2)), \mathbb{R}^1(\mathbb{B}(d_2))\rangle,$$

$$..., \langle \mathbb{R}^0(\mathbb{B}(d_\lambda)), \mathbb{R}^1(\mathbb{B}(d_\lambda))\rangle\}.$$

Without knowing correct SDT, RAC scheme and secret key $k_{i,t}$, even given $\mathbb{E}(f_i|d_i|b_i)$ and $\langle \mathbb{R}^0(\mathbb{B}(d_i)), \mathbb{R}^1(\mathbb{B}(d_i))\rangle$, the master node cannot obtain actual data.

4.4 Result Response

In result response, we concern how a master node searches for data required by a top-k query without knowing about actual data. Receiving a top-k query denoted as $Q = \langle k, T, C\rangle$, the master node M in C begins to process this query on all data transmitted from its affiliated sensor nodes at epoch $t \in T$. Given two positive integers d_i and d_j, let $\langle \mathbb{R}^0(\mathbb{B}(d_i)), \mathbb{R}^1(\mathbb{B}(d_i))\rangle$ and $\langle \mathbb{R}^0(\mathbb{B}(d_j)), \mathbb{R}^1(\mathbb{B}(d_j))\rangle$ be calculated according to the steps of data submission (in Sect. 4.3). If d_i and d_j are collected by the same sensor node at the same epoch, it is easily to determine which is the greater one according to the descending order of sorted data. If d_i and d_j are collected by different sensor nodes or at different epoches, there are three cases of comparison between d_i and d_j based on $\langle \mathbb{R}^0(\mathbb{B}(d_i)), \mathbb{R}^1(\mathbb{B}(d_i))\rangle$ and $\langle \mathbb{R}^0(\mathbb{B}(d_j)), \mathbb{R}^1(\mathbb{B}(d_j))\rangle$.

(1) $\mathbb{R}^0(\mathbb{B}(d_i)) \neq \mathbb{R}^0(\mathbb{B}(d_j))$. If $\mathbb{R}^0(\mathbb{B}(d_i)) > \mathbb{R}^0(\mathbb{B}(d_j))$, there is $d_i < d_j$. For instance, integers 21 and 25 are encoded to $\langle 2, 11101\rangle$ and $\langle 1, 1000001\rangle$, respectively. Because $\mathbb{R}^0(\mathbb{B}(21)) = 2 > \mathbb{R}^0(\mathbb{B}(25)) = 1$, the master node knows encrypted 21 is less than encrypted 25.

(2) $\mathbb{R}^0(\mathbb{B}(d_i)) = \mathbb{R}^0(\mathbb{B}(d_j))$ and $|\mathbb{R}^1(\mathbb{B}(d_i))| = |\mathbb{R}^1(\mathbb{B}(d_j))|$. If $\mathbb{R}^1(\mathbb{B}(d_i)) > \mathbb{R}^1(\mathbb{B}(d_j))$, there is $d_i > d_j$. For example, integers 21 and 20 are encoded to $\langle 2, 11101\rangle$ and $\langle 2, 11011\rangle$, respectively. There is $\mathbb{R}^0(\mathbb{B}(21)) = \mathbb{R}^0(\mathbb{B}(20)) = 2$ and $|\mathbb{R}^1(\mathbb{B}(21))| = |\mathbb{R}^1(\mathbb{B}(20))| = 5$. Because $\mathbb{R}^1(\mathbb{B}(21)) = 11101 > \mathbb{R}^1(\mathbb{B}(20)) = 11011$, the master node knows encrypted 21 is greater than encrypted 20.

(3) $\mathbb{R}^0(\mathbb{B}(d_i)) = \mathbb{R}^0(\mathbb{B}(d_j))$ and $|\mathbb{R}^1(\mathbb{B}(d_i))| \neq |\mathbb{R}^1(\mathbb{B}(d_j))|$. Without loss of generality, assume $|\mathbb{R}^1(\mathbb{B}(d_i))| > |\mathbb{R}^1(\mathbb{B}(d_j))|$, insert $|\mathbb{R}^1(\mathbb{B}(d_i))| - |\mathbb{R}^1(\mathbb{B}(d_j))|$ 0 bits to $\mathbb{R}^1(\mathbb{B}(d_j))$ and shift $\mathbb{R}^1(\mathbb{B}(d_j))$ to the left. Now this case is equivalent to the second one. For instance, integers 21 and 22 are encoded to $\langle 2, 11101\rangle$ and $\langle 2, 111011\rangle$, respectively. There is $\mathbb{R}^0(\mathbb{B}(21)) = \mathbb{R}^0(\mathbb{B}(22)) = 2$ and $|\mathbb{R}^1(\mathbb{B}(21))| = 5 < |\mathbb{R}^1(\mathbb{B}(22))| = 6$. Convert $\mathbb{R}^1(\mathbb{B}(21))$ to 111010. As $111010 < 111011$, the master node knows encrypted 21 is less than encrypted 22.

Adopting the above scheme, master node M finds the highest k data $d'_1, d'_2, ..., d'_k$ respectively collected by sensor nodes $s'_1, s'_2, ..., s'_k$, and then responds to the sink with the message as follows:

$$M \to \ the \ sink : \{\langle s_i', \mathbb{E}(f_i'|d_i'|b_i') \rangle \, |d_i' \in \mathbb{D}, s_i' \in \mathbb{S}\},$$

$$\{\langle s_j, \mathbb{E}(f_1|d_1|b_1) \rangle \, |s_j \notin \mathbb{S}\},$$

where $\mathbb{D} = \{d_1', d_2', ..., d_k'\}$, $\mathbb{S} = \{s_1', s_2', ..., s_k'\}$, $\{\langle s_i', \mathbb{E}(f_i'|d_i'|b_i') \rangle \, |d_i' \in \mathbb{D}, , s_i' \in \mathbb{S}\}$ denotes the local top-k result set, and $\{\langle s_j, \mathbb{E}(f_1|d_1|b_1) \rangle \, |s_j \notin \mathbb{S}\}$ denotes the verification set. For the sensor node $s_j \notin \mathbb{S}$, M still needs to submit the first encrypted part $\mathbb{E}(f_1|d_1|b_1)$ of s_j for result verification. It can be seen that although the master nodes know nothing about actual data, they still get the correct result.

4.5   Integrity Verification

If each master node in area C submits its local top-k result honestly, the sink can calculate the exact final result on the basis of all local results. However, master nodes are easily to become the target of attacks. A compromised master node may return fake or incomplete result. Therefore, it is necessary for the sink to verify result integrity.

Suppose the sink receives a message:

$\{\langle s_i', \mathbb{E}(f_i'|d_i'|b_i') \rangle \, |d_i' \in \mathbb{D}, s_i' \in \mathbb{S}\}, \{\langle s_j, \mathbb{E}(f_1|d_1|b_1) \rangle \, |s_j \notin \mathbb{S}\}$. To simplify explanation, we let

$RS = \{\langle s_i', \mathbb{E}(f_i'|d_i'|b_i') \rangle \, |d_i' \in \mathbb{D}, s_i' \in \mathbb{S}\}$ and $VS = \{\langle s_j, \mathbb{E}(f_1|d_1|b_1) \rangle \, |s_j \notin \mathbb{S}\}$.
The result is considered to be correct if and only if the following five conditions are all satisfied: (1) The sink can decrypt any encrypted parts $\mathbb{E}(f|d|b)$ correctly; (2) $|RS| = k$; (3) $\forall d(d \in RS)$ and $d'(d' \in VS)$, there must be $d \geq d'$; (4) $\forall d(d \in RS)$ with its corresponding f, there must be $d + f \in RS$; (5) $\forall d(d \in RS)$ with its corresponding b, $\exists d' \in RS$, if $d - b \geq d'$, there must be $d - b \in RS$. Condition (1) verifies the authenticity while other conditions verify the completeness. More analysis will be detailed in Sect. 5. If all local results are valid, the sink further calculates the final top-k result.

5   Analysis

5.1   Privacy Analysis

In this paper, we only focus on the case where master nodes are compromised, because master nodes are more attractive to adversaries and it is difficult to protect sensor nodes from being compromised unless the hardware progresses.

If a master node M is compromised, M will attempt to obtain the real data stored in it through either the cyphertexts or the corresponding RAC codes. To decrypt the cyphertext, M has to get the correct key. Assume the length of the key is l_k, the probability that M guesses the correct key is 2^{-l_k}. The probability is negligible when l_k is large enough. To determine the mapping relationship between the real data and the RAC codes, M has to obtain the exact SDT. Assume SDT contains Y source symbols and P_i denotes the probability of the i-th symbol. There are infinite combinations of SDT satisfying $\sum_{i=1}^{Y} P_i = 1$, so the probability that M guesses the exact SDT is extremely tiny. Besides, keys and SDT vary constantly, which make the inference more difficult.

Therefore, even though master nodes are compromised, data privacy is well protected by our proposal.

5.2 Integrity Analysis

The local top-k result of a master node should satisfy the five conditions mentioned in Sect. 4.5. Now we begin to give the detailed integrity analysis.

We also assume the sink receives a message $\{\langle s_i', \mathbb{E}(f_i'|d_i'|b_i')\rangle |d_i' \in \mathbb{D}, s_i' \in \mathbb{S}\}, \{\langle s_j, \mathbb{E}(f_1|d_1|b_1)\rangle |s_j \notin \mathbb{S}\}$ from a master node M, let $RS = \{\langle s_i', \mathbb{E}(f_i'|d_i'|b_i')\rangle |d_i' \in \mathbb{D}, s_i' \in \mathbb{S}\}$ and $VS = \{\langle s_j, \mathbb{E}(f_1|d_1|b_1)\rangle |s_j \notin \mathbb{S}\}$. The sink is able to detect the unauthentic and incomplete result as follows:

(1) If $\mathbb{E}(f_i'|d_i'|b_i')$ of sensor node s_i' cannot be decrypted by the secret key $k_{i,t}$ shared with the sink, the sink can know the data is unauthentic. Because only data encrypted by the correct key is able to be decrypted by the same key.
(2) If $|RS| \neq k$, the sink can detect this obvious error. Because k is much less than the number of sensor nodes such that exact k data must be included in RS.
(3) If $\exists d_1$ collected by sensor node $s_j \notin \mathbb{S}$ and $d_i' \in RS$ satisfying: $d_1 > d_i'$, the sink is able to know the local result is incorrect for the reason that d_1 should be contained in the local top-k result as long as d_i' ($d_i' < d_1$) exists in this result.
(4) If $\exists d_i' \in RS$ with its corresponding $f_i'(f_i' \neq d_{max})$ satisfying: $(d_i' + f_i') \notin RS$, the sink can also find this error. Similar to the second case, $d_i' + f_i'$ ($d_i' + f_i' > d_i'$) should belong to the local top-k result.
(5) If $\exists d_i' \in RS$ with its corresponding $b_i'(b_i' \neq d_i' + 1)$ and $d_j' \in RS$ satisfying: $d_i' - b_i' > d_j'$ and $(d_i' - b_i') \notin RS$, the sink can still detect this error. Similar to the second case, $d_i' - b_i'$ should exist in the local top-k result.

Therefore, our integrity verification scheme is effective to detect the forged and incomplete result.

6 Performance Evaluation

In this section, we thoroughly evaluate the performance of the proposed SET by comparing with the state-of-the-art work—ADVQ [13], SVTQ [17] and PriSecTopk [7] in terms of communication cost and result accuracy.

6.1 Experiment Setup

SET, ADVQ, SVTQ and PriSecTopk are implemented on the simulator OMNet++4.1 [2]. The area of sensor network is set to $400\,\text{m} \times 400\,\text{m}$. 200 more sensor nodes are uniformly deployed in the network. Assume the network is separated into four identical cells and a master node is placed at the center of each cell. The transmission radius of each sensor node is 50 meters. We build the dataset by randomly selecting different data from a real dataset LUCE [1].

For SET, ADVQ, SVTQ and PriSecTopk, we adopt the typical symmetrical encryption technique 64-bit DES to encrypt private data. For ADVQ and PriSecTopk, we suppose the size of both $HMAC$s and MACs is 64 bits. In our SET, the parameters of SDT for RAC scheme are set as follows: the size of initial interval $I = 1000$ and the set of source symbols is $\{00, 01, 10, 11\}$ with probability $\{0.4, 0.3, 0.2, 0.1\}$ and cumulative distribution $\{0, 0.4, 0.7, 0.9\}$.

In the experiments, there are two parameters, that is network size and submission period. Network size is defined as the number of sensor nodes in the whole network, ranging from 200 to 600. Assume each sensor node samples a datum every 10 s, then submission period is defined as the interval time between two successive submissions, ranging from 10 to 50 s.

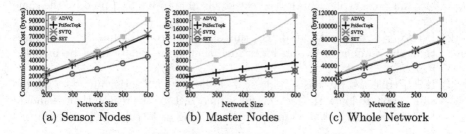

(a) Sensor Nodes   (b) Master Nodes   (c) Whole Network

Fig. 3. Impact of network size on communication cost

6.2   Communication Cost

In WSNs, communication is the dominant factor to consume energy which affects network lifetime. The lower communication cost, the better efficiency.

Figure 3 displays the impact of the network size on the communication cost when the submission period is 40 s and $k = 10$. As network size grows, more data are transmitted from sensor nodes to master nodes. The communication cost of sensor nodes in SET is much less than that in other three algorithms and increases more slowly. The reason is that: In ADVQ, sensor nodes send too much $HMAC$s, virtual line segments, and neighborhood information, and the number of neighbors rises rapidly as network density increases; In SVTQ, sensor nodes send the encrypted data and two big integers representing prime aggregation results; In PriSecTopk, sensor nodes send the encrypted data and the corresponding MACs; In our SET, sensor nodes only send the encrypted data and two simple codes generated upon the RAC scheme. With the growth of network size, more data are transmitted from master nodes to the sink, and the communication cost of master nodes in SET is still less than that in ADVQ and PriSecTopk. That is because master nodes in SET only submit k encrypted data as the top-k result and k' ($N - k \leq k' \leq N - 1$ and N denotes the network size) encrypted data for verification, whereas master nodes in ADVQ submit all data included in the $\eta - 1 + k$ highest virtual line segments and their neighbors' information, where η ($\eta > 1$) is a system parameter, and master nodes in PriSecTopk submit k $MACs$, the computing commitment information combined several sensing records for sample-based hypothesis testing method except k encrypted data. It is observed that the communication cost of master nodes in SVTQ is equal to that in SET, because master nodes in SVTQ return the same message to the sink as SET does. Finally, the communication cost of whole network in SET is the least of four algorithms.

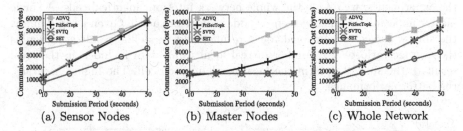

(a) Sensor Nodes   (b) Master Nodes   (c) Whole Network

Fig. 4. Impact of submission period on communication cost

Figure 4 demonstrates the impact of the submission period on communication cost when the network size is 400 and $k = 10$. As the submission period grows, the number of data sent by each sensor node increases. We can see that the communication cost of sensor nodes in SET is still much less than that in other algorithms. The reason is similar to that of Fig. 3(a). When a sensor node samples a datum, different algorithms will takes different action as follows. ADVQ will construct the $HMAC$ for the datum, update the data counter and two extra $HMACs$, and may enlarge the virtual line segment. Except the encrypted part, SVTQ, PriSecTopk and our SET will respectively generate two prime aggregation results, a MAC and two smalll codes. The submission period has neglectable effect on the communication cost of master nodes in SET and SVTQ, because their communication cost of master nodes is nearly related to the network size for a given top-k query. In ADVQ, the number of data in a virtual line segment is proportional to the submission period. If the submission period becomes greater, the $\eta - 1 + k$ highest virtual line segments will cover more data so that master nodes will produce more communication. In PriSecTopk, we assume $s = k + 5$, which means master nodes select s sensing data in every selection of the query processing. As mentioned in work [7], master nodes will generate at most $\lceil \frac{\overline{N}-k}{s-k} \rceil$ computing commitment information for result verification, where $\overline{N}$ denotes the number of sensing data collected in one submission period. When the submission period increases by 10 s, $\overline{N}$ increases by the network size, and consequently, master nodes need to transmit more computing commitment information. The communication cost of whole network is mainly determined by that of sensor nodes. Therefore, SET saves the most communication cost for the whole network.

6.3   Result Accuracy

In the ideal condition, the sink can directly receive the top-k result including all data satisfying the query and excluding data dissatisfying the query. In fact, SET, ADVQ, PriSecTopk and SVTQ all adopt the two-tiered network model. Due to this special network architecture, each master node first submits its local top-k result to the sink, and then the sink calculates the final result based on these local results, so it is crucial to find the accurate local top-k results. In the experiments, we divide the whole network into four identical cells. A single master node and its affiliated sensor nodes constitute a cell. Given a top-k query, the sink will receive four local top-k results, and we define result accuracy as the average ratio of the number of real local top-k data included in the local result to the sum of data in the local result in each cell. The higher accuracy, the better performance.

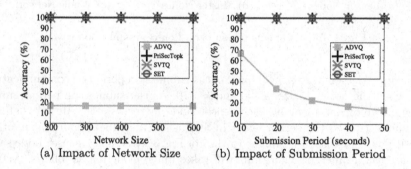

(a) Impact of Network Size   (b) Impact of Submission Period

Fig. 5. Result accuracy

Figure 5 displays the impact of different parameters on result accuracy. It can be seen that no matter how these parameters change, master nodes in SET, PriSecTopk and SVTQ always compute and return the correct local top-k data to the sink, i.e., the result accuracy of SET, PriSecTopk and SVTQ is 100%. In ADVQ, the network size has no impact on the accuracy as illustrated in Fig. 5(a) (submission period is 40 s and $k = 10$), because its accuracy is only related to both k and the number of data included in the $\eta - 1 + k$ highest virtual line segments. In Fig. 5(b) (network size is 400 and $k = 10$), the accuracy of ADVQ decreases sharply as the submission period grows. The reason is that lager submission period makes sensor nodes generate more data in each virtual line segment. As a result, master nodes need to send more unsatisfactory data to the sink.

In summary, experimental results demonstrate that our SET not only gains the accurate top-k result but also achieves low communication cost, which is more suitable for the practical applications of WSNs.

7 Conclusion

Privacy-preserving top-k query in WSNs is significant and challenging. In this paper, we present a secure and efficient top-k query framework—SET in two-tiered WSNs. In the proposed SET, data privacy is protected while the top-k result is correctly calculated by using the RAC scheme. Besides, the sink is able to verify result integrity through a series checking. Theoretical analysis and simulation results confirm the high efficiency, accuracy and security of SET.

References

1. Luce Deployment. http://lcav.epfl.ch/cms/lang/en/pid/86035
2. Omnet++ 4.1. http://www.omnetpp.org
3. Fan, Y., Chen, H.: A secure top-k query protocol in two-tiered sensor networks. Chin. J. Comput. **49**(3), 1947–1958 (2012). (in chinese)
4. Hong, D., Eleftheriadis, A.: Memory-efficient semi-quasi renormalization for arithmetic coding. IEEE Trans. Circuits Syst. Video Technol. **17**(1), 106–110 (2007)
5. Huang, H., Juan, F., Wang, R., Qin, X.: An exact top-k query algorithm with privacy protection in wireless sensor networks. Int. J. Distrib. Sens. Netw. (IJDSN) **2014**, 1–10 (2014)
6. Jnsson K.V., Palmskog, K., Vigfusson, Y.: Secure distributed top-k aggregation. In: ICC, Ottawa, ON, Canada, pp. 804–809, June 2012
7. Liao, X., Li, J.: Privacy-preserving and secure top-k query in two-tier wireless sensor network. In: IEEE Global Communications Conference (GLOBECOM), Anaheim, CA, USA, December 2012
8. Ma, X., Song, H., Wang, J., Gao, J., Min, G.: A novel verification scheme for fine-grained top-k queries in two-tiered sensor networks. Wirel. Pers. Commun. **75**(3), 1809–1826 (2014)
9. Martin, G.N.N.: Range encoding: an algorithm for removing redundancy from a digitised message. In: Video and Data Recording Conference (1979)
10. Paek, J., Greenstein, B., Gnawali, O., Jang, K.Y., Joki, A., Vieira, M.A.M., Hicks, J., Estrin, D., Govindan, R., Kohler, E.: The tenet architecture for tiered sensor networks. ACM Trans. Sens. Netw. (TOSN) **6**(4), 1–42 (2010)
11. Vaidya, J., Clifton, C.W.: Privacy-preserving kth element score over vertically partitioned data. IEEE Trans. Knowl. Data Eng. (TKDE) **21**(2), 253–258 (2009)
12. Yao, Y., Ma, L., Liu, J.: Privacy-preserving top-k query in two-tiered wireless sensor networks. Int. J. Adv. Comput. Technol. (IJACT) **4**(6), 226–235 (2012)
13. Yu, C.M., Ni, G.K., Chen, I.Y., Gelenbe, E., Kuo, S.Y.: Top-k query result completeness verification in tiered sensor networks. IEEE Trans. Inf. Forensics Secur. (TIFS) **9**(1), 109–124 (2014)
14. Zhang, R., Shi, J., Liu, Y., Zhang, Y.: Verifiable fine-grained top-k queries in tiered sensor networks. In: IEEE Conference on Computer Communications (INFOCOM), San Diego, CA, USA, pp. 1199–1207, March 2010

15. Zhang, R., Shi, J., Zhang, Y., Huang, X.: Secure top-k query processing in unattended tiered sensor networks. IEEE Trans. Veh. Technol. (TVT) **63**(9), 4681–4693 (2014)
16. Zhang, X., Dong, L., Peng, H., Chen, H., Li, D., Li, C.: Achieving efficient and secure range query in two-tiered wireless sensor networks. In: IEEE/ACM International Symposium on Quality of Service (2014)
17. Zhou, T., Lin, Y., Zhang, W., Xiao, S., Li, J.: Secure and verifiable top-k query in two-tiered sensor networks. In: Zia, T., Zomaya, A., Varadharajan, V., Mao, M. (eds.) SecureComm 2013. LNICSSITE, vol. 127, pp. 19–34. Springer, Cham (2013). doi:10.1007/978-3-319-04283-1_2

Top-k Pattern Matching Using an Information-Theoretic Criterion over Probabilistic Data Streams

Kento Sugiura[1]([⊠]) and Yoshiharu Ishikawa[2]

[1] Graduate School of Information Science, Nagoya University, Nagoya, Japan
sugiura@db.ss.is.nagoya-u.ac.jp
[2] Graduate School of Informatics, Nagoya University, Nagoya, Japan
ishikawa@i.nagoya-u.ac.jp

Abstract. As the development of data mining technologies for sensor data streams, more sophisticated methods for complex event processing are demanded. In the case of event recognition, since event recognition results may contain errors, we need to deal with the uncertainty of events. We therefore consider *probabilistic event data streams* with occurrence probabilities of events, and develop a *pattern matching method based on regular expressions*. In this paper, we first analyze the semantics of pattern matching over non-probabilistic data streams, and then propose the problem of top-k pattern matching over probabilistic data streams. We introduce the use of *an information-theoretic criterion* to select appropriate matches as the result of pattern matching. Then, we present an efficient algorithm to detect top-k matches, and evaluate the effectiveness of our approach using real and synthetic datasets.

Keywords: Complex event processing · Probabilistic data streams · Pattern matching · Regular expressions · Information-theoretic criterion

1 Introduction

It has become popular to obtain and analyze sensor data streams, which are the results of human activities monitored by smartphones and wearable devices [5, 14,25]. In the analyses, we recognize human activities based on machine learning techniques and detect primitive events, such as "walk" and "jog." These events, however, are too primitive to understand human activities because such an event shows only an action at a certain time step. Thus, it is required to detect more complex events, such as "a user is walking for a while," using *complex event processing* [8].

When we apply complex event processing to the results of machine learning techniques, we need to consider the uncertainty of classification. The accuracy of classification easily decreases because of the noises of sensors and the ambiguity between classes. For example, consider the recognition of human activities, such as "walk," "stUp (stair up)," and "stDown (stair down)," by using the

© Springer International Publishing AG 2017
L. Chen et al. (Eds.): APWeb-WAIM 2017, Part I, LNCS 10366, pp. 511–526, 2017.
DOI: 10.1007/978-3-319-63579-8_39

event symbol	time step						
	1	2	3	4	5	6	7
stay (s)	0.15	0.25	0.80	0.05	0.05	0.05	0.05
walk (w)	0.20	0.20	0.05	0.40	0.40	0.40	0.40
jog (j)	0.25	0.15	0.05	0.05	0.05	0.05	0.05
stUp (u)	0.20	0.20	0.05	0.25	0.25	0.25	0.25
stDown (d)	0.20	0.20	0.05	0.25	0.25	0.25	0.25

Fig. 1. A probabilistic data stream of human activity

acceleration sensors of smartphones. Since these activities are very similar, it is difficult to classify them correctly using only the sensor information. In the real world, classification becomes more difficult because some accidents may disturb classifiers. Thus, we should deal with *probabilistic data streams* that have the occurrence probabilities of events. Figure 1 shows a probabilistic data stream of the above example. In the probabilistic data stream, we can understand that "walk" event is most likely at time step 4, but we cannot determine which events occur in the real world.

In this paper, we apply *pattern matching* to probabilistic data streams, and detect *matches* as complex events. Existing methods apply variety of techniques to probabilistic data streams, such as database-like query processing [6,22], top-k query processing [12], frequent item detection [27], and clustering [1]. However, it is difficult for these methods to describe complex event queries and detect them efficiently. Thus, we use *regular expressions* to specify complex events, and propose an efficient algorithm for their detection.

When we apply pattern matching to probabilistic data streams, we have to consider how to measure the appropriateness of matches. We may detect a lot of matches as the result of pattern matching because of the uncertainty of data streams. Since it is difficult to check such numerous matches, we need to select appropriate matches using some criterion. The existing method [15] uses the occurrence probabilities of matches, but it has two problems. We explain it using the following two matches that are detected by applying pattern $p = \langle \mathtt{w}^+ \rangle$ to the data stream in Fig. 1. Note that "$\mathtt{w}$" is the abbreviation of "walk" and each subscript indicates a time step.

$$m_1 = \langle \mathtt{w}_1, \mathtt{w}_2 \rangle, \qquad\qquad P(m_1) = 4.0 \times 10^{-2}$$
$$m_2 = \langle \mathtt{w}_4, \mathtt{w}_5, \mathtt{w}_6, \mathtt{w}_7 \rangle, \qquad\qquad P(m_2) \simeq 2.6 \times 10^{-2}$$

First, if patterns contain Kleene closures (* and $^+$), the occurrence probabilities may mislead users. m_1 is superior than m_2 according to their occurrence probabilities. However, it is inappropriate to naïvely consider m_2 as an inferior match because m_2 indicates a more complex event than m_1. As we consider the intention of Kleene closures, m_2 is probably a more appropriate match. Second, occurrence probabilities cannot consider the accuracy of classification. Let $[t_s : t_e]$ be a time interval from t_s to t_e. In Fig. 1, the classification results in the time interval $[1 : 2]$ are uncertain. Thus, we should detect m_2 from the time interval $[4 : 7]$, but occurrence probabilities are not useful for the decision.

In this paper, we propose the use of *deviation of amount of information*, an information-theoretic criterion. The reason of the above problems is to compare matches using the absolute values of their occurrence probabilities. We therefore consider comparing amount of information, which is a basic concept in information theory [7]. Let us continue the above example with m_1 and m_2. First, suppose that we select m_1 instead of $\langle s_1, s_2 \rangle$ in the time interval $[1 : 2]$. The amounts of information are $-\log_2 P(m_1) \simeq 4.6$ and $-\log_2 P(\langle s_1, s_2 \rangle) = -\log_2 (0.15 \times 0.25) \simeq 4.7$. This means that we should select m_1 instead of $\langle s1, s2 \rangle$ – smaller amount of information corresponds to a more frequent match. This example shows that we can compare a match and other sequences in the same time interval, but we cannot compare matches in different time intervals (e.g., m_1 and m_2) because a long time interval usually results in small probability. Thus, to compare matches in different time intervals, we propose to use the deviation of amount of information between a match and other sequences. For example, let S_{m_1} be the universal set of sequences in the same time interval with m_1. The deviation of amount of information of m_1 is calculated as follows:

$$\frac{1}{|S_{m_1}|} \sum_{s \in S_{m_1}} (-\log_2 P(s)) - (-\log_2 P(m_1)) \simeq 4.68 - 4.64 = 0.04.$$

The first term of the expression is the average of information of the sequences in the time interval $[1 : 2]$, and the second term is the amount of information of match m_1. Since match m_1 has slightly high probability than the average, its amount of information is slightly smaller than the average. Therefore, the result is 0.04, a positive value. If we use the same formula for match m_2 in the time interval $[4 : 7]$, we get the result 5.9. It means that the probability of match m_2 is significantly larger than the average. Comparing two results, we select m_2 as a more significant match.

In this paper, we describe top-k pattern matching based on the deviation of amount of information over probabilistic data streams. The remainder of the paper is organized as follows. We introduce the related work in Sect. 2, and explain preliminaries, such as the definition of patterns and deviation of amount of information, in Sect. 3. Then, we define top-k pattern matching in Sect. 4, and propose an efficient matching algorithm in Sect. 5. Section 6 evaluates our approach by experiments, and Sect. 7 concludes the paper.

2   Related Work

2.1   Probabilistic Data Streams

Ré et al. proposed pattern matching methods that consider the temporal correlation of event occurrence [19]. For example, when a user is going up the stairs at a certain time step, it is rare to suddenly go down the stairs at the next time step. Such temporal correlation is represented by the Markov property. They, however, only consider whether matches occur or not at every time step, and do not detect matches as concrete event sequences. Therefore, we cannot apply their methods in our context. Besides, since they calculate the absolute values of occurrence probabilities, their methods have the problem described in Sect. 1.

Li et al. proposed top-k pattern matching methods with sliding windows [15]. However, they use occurrence probabilities as evaluation metrics, and assume that events are skipped in pattern matching. For example, suppose that we apply pattern $p = \langle j^+ \; w^+ \rangle$ to the stream in Fig. 1. Their methods detect $\langle j_1, w_4 \rangle$ and $\langle j_1, w_5 \rangle$ with the top-1 occurrence probability because they skip events with small probabilities, such as "j_2." However, we should detect $\langle j_1, w_2 \rangle$ because continual event sequences are more appropriate in this case.

2.2 Non-probabilistic Data Streams

Many methods for pattern matching were proposed in the literature of non-probabilistic data streams. The SASE project lets a user specify detailed conditions for pattern matching [24,26]. [3,4,16] treat data streams with special features such as disordered and/or distributed streams. [17] considers occurrence frequencies of events and [23] uses field-programmable gate arrays to increase the throughput. [20] considers the decidability of pattern matching when ambiguous patterns and infinite data streams are given.

Pattern matching on strings has been well studied. The Thompson construction method [21] is particularly useful. It generates a non-deterministic finite automaton (NFA) from a regular expression pattern. Besides, we can easily construct an NFA that implements the Knuth–Morris–Pratt algorithm [13] or the Aho–Corasick algorithm [2] by using the Thompson construction method [18]. In this paper, we also use the Thompson construction method to detect matches efficiently.

3 Preliminaries

In this section, we define a probabilistic data stream, a query patten, and deviation of amount of information of a match.

3.1 Probabilistic Data Streams

We define a *probabilistic event* as an entry of a probabilistic data stream.

Definition 1. *Let α be the type of an event e with Σ, the universal set of event symbols. A* probabilistic event e_t *is an event with its occurrence probability $P(e_t = \alpha)$ for each $\alpha \in \Sigma$ at time step t. $P(e_t = \alpha)$ satisfies the following properties:*

$$\forall \alpha \in \Sigma, \; 0 \leq P(e_t = \alpha) \leq 1 \tag{1}$$

$$\forall \alpha, \beta \in \Sigma, \; \alpha \neq \beta \rightarrow P(e_t = \alpha \wedge e_t = \beta) = 0 \tag{2}$$

$$P(\bigvee_{\alpha \in \Sigma} e_t = \alpha) = \sum_{\alpha \in \Sigma} P(e_t = \alpha) = 1 \tag{3}$$

In the following, we use the symbol α_t as $e_t = \alpha$. Our approach can be extended for probabilistic events with the Markov property as in [19]; we omit the details for the sake of brevity.

We define a *probabilistic data stream* in terms of probabilistic events.

Definition 2. *A probabilistic data stream $PDS = \langle e_i, e_{i+1}, ..., e_j, ... \rangle$ is a sequence of probabilistic events.*

For example, Fig. 1 shows a probabilistic data stream $\langle e_1, e_2, e_3, e_4, e_5, e_6, e_7 \rangle$ with the symbols $\Sigma = \{s, w, j, u, d\}$.

3.2 Query Patterns and Matches

We define the grammar for query patterns.

Definition 3. *Let α be an event symbol in Σ, and let ϵ be the empty symbol. An input pattern is generated by the following grammar:*

$$p ::= \alpha \mid \epsilon \mid p\,p \mid p \vee p \mid p^* \mid p^+ \mid (p) \tag{4}$$

In other words, we assume that patterns are specified as regular expressions.

We introduce a *match* and its occurrence probability.

Definition 4. *A match m is a sequence of symbols that fits a specified pattern. Let $m.t_s$ and $m.t_e$ be the start and end time steps of m, respectively. An occurrence probability of m is calculated by the following equation.*

$$P(m) = \prod_{t=m.t_s}^{m.t_e} P(\alpha_t) \tag{5}$$

3.3 Deviation of Amount of Information

We propose deviation of amount of information to measure the appropriateness of matches.

Definition 5. *Let S_m be the universal set of sequences in the same time interval with a match m. The deviation of amount of information of m is calculated as follows:*

$$\mathcal{D}_I(m) = \frac{1}{|S_m|} \sum_{s \in S_m} (-\log_2 P(s)) - (-\log_2 P(m)) \tag{6}$$

$\mathcal{D}_I(m)$ takes a positive value when the amount of information of m is lower than the average of information over all sequences. Intuitively, since the amount of information takes a smaller value with a higher probability, $\mathcal{D}_I(m)$ becomes a large value when the occurrence of m is more likely. For example, $\mathcal{D}_I(\langle w_4, w_5, w_6, w_7 \rangle) \simeq 5.9$ is larger than that of other sequences such as $\mathcal{D}_I(\langle s_4, s_5, s_6, s_7 \rangle) \simeq -6.1$, and $\langle w_4, w_5, w_6, w_7 \rangle$ is the most likely sequence in the time interval $[4:7]$.

We can calculate deviation of amount of information $\mathcal{D}_I(m)$ efficiently by modifying (6). The calculation based on (6) is inefficient because the number of sequences in S_m increases exponentially. Thus, we first modify (6) as follows:

$$\mathcal{D}_I(m) = \log_2 \frac{P(m)}{\left(\prod_{s \in S_m} P(s)\right)^{\frac{1}{|S_m|}}} \tag{7}$$

Let E_t be the set of event symbols $\alpha \in \Sigma$ with $P(\alpha_t) > 0$. Since we assume that the occurrence of events is independent, the denominator of (7) is calculated by the following equation:

$$\left(\prod_{s \in S_m} P(s)\right)^{\frac{1}{|S_m|}} = \prod_{t=m.t_s}^{m.t_e} \left(\prod_{\alpha \in E_t} P(\alpha)\right)^{\frac{1}{|E_t|}}. \tag{8}$$

We can derive the following equation by substituting Eqs. (5) and (8) in (7):

$$\mathcal{D}_I(m) = \sum_{t=m.t_s}^{m.t_e} \log_2 \frac{P(\alpha_t)}{\left(\prod_{\alpha \in E_t} P(\alpha)\right)^{\frac{1}{|E_t|}}}. \tag{9}$$

That is, we can calculate $\mathcal{D}_I(m)$ by summarizing the deviation of amount of information of an event at each time step. For example, $\mathcal{D}_I(\langle \mathtt{w_1}, \mathtt{w_2} \rangle)$ with Fig. 1 is calculated as follows:

$$\begin{aligned}
IG(\langle \mathtt{w_1}, \mathtt{w_2} \rangle) &= \log_2 \frac{P(\mathtt{w_1})}{\left(\prod_{\alpha \in E_1} P(\alpha)\right)^{\frac{1}{|E_1|}}} + \log_2 \frac{P(\mathtt{w_2})}{\left(\prod_{\alpha \in E_2} P(\alpha)\right)^{\frac{1}{|E_2|}}} \\
&\simeq 1.86 \times 10^{-2} + 1.86 \times 10^{-2} \\
&\simeq 3.7 \times 10^{-2}
\end{aligned}$$

Since the number of events in E_t is constant, calculation based on (9) is efficient.

4   Top-k Pattern Matching Problem

In this section, we consider the semantics for pattern matching over probabilistic data streams, and then describe the problem definition.

4.1   Semantics for Pattern Matching

If we simply select top-k matches from all possible matches, inappropriate matches may be detected. For example, suppose that we apply pattern $p = \langle \mathtt{w^+} \rangle$ to the stream in Fig. 1, and detect top-2 matches based on $\mathcal{D}_I$ values in Sect. 3. Figure 2 shows some detected matches. In this case, m_2 is the top-1 match and m_3 is the top-2 match. However, since m_3 is a subsequence of m_2, they have an overlap of event occurrence. That is, when we consider the occurrence of pattern events in the real world, m_2 and m_3 correspond to the same event "a user was

match	time step							$\mathcal{D}_I(m)$
	1	2	3	4	5	6	7	
m_1	w	w						0.04
m_2				w	w	w	w	5.88
m_3				w	w	w		4.41

Fig. 2. Matches of $p = \langle w^+ \rangle$ and their $\mathcal{D}_I(m)$ scores

sequence	time step							matches
	1	2	3	4	5	6	7	
s_1	.	.	.	.	.	.	.	$\emptyset$
s_2	.	.	.	w	w	w	w	$\{m_2\}$
s_3	.	.	.	w	w	w	.	$\{m_3\}$
s_4	w	w	.	.	.	.	.	$\{m_1\}$
s_5	w	w	.	w	w	w	w	$\{m_1, m_2\}$
s_6	w	w	.	w	w	w	.	$\{m_1, m_3\}$

Fig. 3. No-overlapping matches in Fig. 2 and their corresponding sequences

walking over the time interval [4 : 7]." We therefore need to detect matches in different time intervals by defining semantics of pattern matching.

We consider appropriate semantics based on the semantics in the literature of non-probabilistic data streams. The existing method [20] enumerates three semantics for regular expression pattern matching as follows:

1. all-path semantics
2. no-overlap semantics
3. use-and-throw semantics

All-path semantics detects all matches over probabilistic data streams. However, as we explain in the above example, this semantics is inappropriate because redundant matches may be detected.

No-overlap semantics detects sets of no-overlapping matches. For example, in Fig. 2, no-overlap semantics can detect the set $\{m_1, m_2\}$, but cannot detect $\{m_2, m_3\}$. Let "." be any event symbols. When we use no-overlap semantics, the sets of matches can be expressed by the regular expression $\langle (. \vee p)^* \rangle$. That is, we can express the sets of matches by corresponding sequences as shown in Fig. 3. Since we can construct a finite automaton from the regular expression $\langle (. \vee p)^* \rangle$, we can apply existing efficient algorithms, such as the Viterbi algorithm [10]. No-overlap semantics, however, cannot detect some desired sets of matches. For example, suppose that pattern $p = \langle u^+ \vee d^+ \rangle$ is given. As "u" and "d" are different event symbols, users may want to detect $\langle u_1, u_2 \rangle$ and $\langle d_1, d_2 \rangle$ at the same time. However, since these matches overlap with each other, no-overlap semantics cannot detect $\{\langle u_1, u_2 \rangle, \langle d_1, d_2 \rangle\}$.

Use-and-throw semantics is proposed to solve the problems of no-overlap semantics. It uses each event only once in pattern matching. For example, suppose that we apply pattern $p = \langle u^+ \vee d^+ \rangle$ to the stream in Fig. 1. The matches

$\{\langle u_1, u_2 \rangle, \langle d_2, d_3 \rangle\}$ can be output because every event $\{u_1, u_2, d_2, d_3\}$ is used only once. In contrast, $\{\langle u_1, u_2 \rangle, \langle u_2, u_3 \rangle\}$ cannot be output because "u_2" is used twice. Use-and-throw semantics detects matches flexibly, but cannot express the results of pattern matching using regular expressions. That is, it is difficult to detect matches efficiently using a finite automaton.Besides, since the conditions of the semantics are complex, it is also difficult to simultaneously apply multiple queries such as in YFilter [9].

In this paper, we use no-overlap semantics for pattern matching. No-overlap semantics cannot detect some desired matches, such as $\{\langle u_1, u_2 \rangle, \langle d_1, d_2 \rangle\}$ for $p = \langle u^+ \vee d^+ \rangle$, but we can solve this problem by dividing the query into $p_1 = \langle u^+ \rangle$ and $p_2 = \langle d^+ \rangle$. Since we can detect no-overlapping matches efficiently by using a finite automaton of $\langle (. \vee p)^* \rangle$, we can apply multiple queries at the same time. Note that concrete algorithms for multiple queries are our future work, and we describe algorithms for only one query in the following.

4.2   Problem Definition

We consider top-k pattern matching based on no-overlap semantics. Since the combination of matches is restricted by no-overlap semantics, it is inappropriate to naïvely select matches in descending order of $\mathcal{D}_I$ values. To solve this problem, we consider the sum of $\mathcal{D}_I$ values. As usual top-k queries select answers in descending order of a specified score, the sum of scores is maximized. Thus, we maximize the sum of $\mathcal{D}_I$ values where the number of matches is k or less. We allow that the number of matches is less than k because large k value may divide matches inappropriately. For example, suppose that we apply top-3 pattern matching to the stream in Fig. 1 with $p = \langle w^+ \rangle$. The set of matches $\{\langle w_1, w_2 \rangle, \langle w_4, w_5, w_6, w_7 \rangle\}$ maximizes the sum of $\mathcal{D}_I$ values with $k = 2$. However, since the $\mathcal{D}_I$ values of other matches, such as $\langle w_3 \rangle$, is less than zero, the sum of $\mathcal{D}_I$ values with $k = 3$ is maximized by dividing $\langle w_4, w_5, w_6, w_7 \rangle$ into $\langle w_4, w_5 \rangle$ and $\langle w_6, w_7 \rangle$. It means that the best $\mathcal{D}_I$ score for $k = 3$ is worse than that for $k = 2$. We therefore use the condition that the number of matches is k or less, and detect $\{\langle w_1, w_2 \rangle, \langle w_4, w_5, w_6, w_7 \rangle\}$ with $k = 3$.

Now we define our top-k pattern matching problem.

Definition 6. *Let* ts_overlap(m_1, m_2) *be a predicate that indicates m_1 overlaps with m_2. When a probabilistic data stream PDS, a query pattern p, and the maximum number of matches k are given, we detect the set of matches M according to the following equation:*

$$\text{maximize} \quad \sum_{m \in M} \mathcal{D}_I(m) \qquad (10)$$

$$\text{subject to} \quad |M| \le k$$

$$\forall m_1, m_2 \in M, \neg\text{ts\_overlap}(m_1, m_2)$$

5   Pattern Matching Algorithm

We can obtain an optimal solution by enumerating all the combinations of matches, but this naïve method is obviously inefficient. Since the number of matches increases polynomially with pattens containing Kleene closures, it takes time to detect all matches. Besides, it is difficult to enumerate all combinations of matches because they increase exponentially.

Therefore, we transform the problem into an equivalent one, which can be processed efficiently. Remember that a set of matches can be represented as a sequence under the no-overlap semantics as described in Sect. 4.1. We transform the problem of finding M, the sets of matches with the largest $\mathcal{D}_I$ value, into an equivalent problem of finding a sequence S for the pattern $\langle(.\vee p)^*\rangle$ with the largest $\mathcal{D}_I$ value, where we define $\mathcal{D}_I(S)$ of a sequence S as the sum of $\mathcal{D}_I$'s of matches in S. In the process, we first detect the sequence with the maximum $\mathcal{D}_I$ value and then extract matches from the sequence. We let the amount of information of an arbitrary event "." be zero.

5.1   Algorithm Using Dynamic Programming

In our algorithm, we use an ϵ-NFA for regular expression $\langle(.\vee p)^*\rangle$. The construction of an ϵ-NFA consists of five steps as follows:

1. generate an ϵ-NFA for pattern p by using the Thompson construction [11]
2. convert the ϵ-NFA to a corresponding DFA (deterministic finite automaton) and minimizing the DFA [11]
3. add ϵ-transitions from all the final states to the initial state, and convert the initial state to the final state
4. convert the redundant final states to normal states
5. add a transition with "." from the initial state to itself

Steps 1 and 2 generate a DFA that becomes the base of the resulting ϵ-NFA. Step 3 adds ϵ-transitions to detect multiple matches repeatedly. Note that the initial state is also the final state after Step 3 because the initial state is reachable from all the final states with ϵ-transitions. Step 4 convert the redundant final states, which are the final states in the original DFA, to normal states. Since we can accept all the sequences of $\langle(.\vee p)^*\rangle$ at the initial state, which is converted to the final state at Step 3, we do not need the other final states. Step 5 adds a transition with "." to allow sequences containing "." among matches. For example, Fig. 4 shows the ϵ-NFA for pattern $p = \langle w^+\rangle$.

Fig. 4. ϵ-NFA of $\langle(.\vee w^+)^*\rangle$

We propose a pattern matching algorithm based on dynamic programming. We first illustrate an important property. Consider the sequences in Fig. 3. At time step 3, there are two sequences $\langle .1, .2, .3 \rangle$ and $\langle w_1, w_2, .3 \rangle$. $\mathcal{D}_I(\langle .1, .2, .3 \rangle)$ is zero and $\mathcal{D}_I(\langle w_1, w_2, .3 \rangle)$ is 0.04. As both sequences reach the initial state in Fig. 4, the same events are added after time step 4. That is, the additional amount of information to these sequences will be the same. We therefore remove $\langle .1, .2, .3 \rangle$ at time step 3, and process only $\langle w_1, w_2, .3 \rangle$ after time step 4. We can detect an optimal sequence by retaining a sequence with the maximum $\mathcal{D}_I$ value at every state over a data stream.

However, when we restrict the number of matches to k or less, as shown in Definition 6, we cannot detect an optimal solution using the above method. Suppose that $k = 1$ is given in the above example. $s_2 = \langle .1, .2, .3, w_4, w_5, w_6, w_7 \rangle$ is an optimal solution with $k = 1$. However, we cannot detect s_2 because $\langle .1, .2, .3 \rangle$ is rejected at time step 3.

We therefore use a layered state transition diagram to distinguish sequences with the number of matches. Figure 5 shows such a diagram for the ϵ-NFA in Fig. 4. In this state transition diagram, sequences shift to the next layer when the number of matches increases. That is, when sequences use the ϵ-transition in Fig. 4, they shift to the next layer. We can detect an optimal solution according to k by using this layered state transition diagram. For example, at time step 3, we retain $\langle .1, .2, .3 \rangle$ at state 0 in layer 0 and $\langle w_1, w_2, .3 \rangle$ at state 0 in layer 1. Thus, we can detect $s_2 = \langle .1, .2, .3, w_4, w_5, w_6, w_7 \rangle$ at state 0 in layer 1 with $k = 1$ and $s_5 = \langle w_1, w_2, .3, w_4, w_5, w_6, w_7 \rangle$ at state 0 in layer 2 with $k = 2$.

We explain our algorithm by using an example with the probabilistic data stream in Fig. 1, pattern $p = \langle w^+ \rangle$, and $k = 2$. Figure 6 shows the generation process of sequences until time step 2. Note that the underline in each state indicates an optimal sequence, and we distinguish matches by square brackets. We initialize the initial state in layer 0 to have an empty sequence. At time step 1, we extend the empty sequence to reachable states. In this example, the extended sequences reach state 0 in layer 0, state 1 in layer 0, and state 0 in

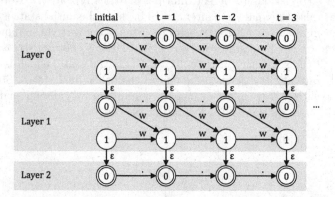

Fig. 5. State transition diagram layered by the number of matches

t	layer	state	
		0	1
initial	0	$\{\langle\,\rangle\}$	$\emptyset$
	1	$\emptyset$	$\emptyset$
	2	$\emptyset$	
1	0	$\{\langle._1\rangle\}$	$\{\langle w_1\rangle\}$
	1	$\{\langle[w_1]\rangle\}$	$\emptyset$
	2	$\emptyset$	
2	0	$\{\langle._1, ._2\rangle\}$	$\{\langle._1, w_2\rangle, \langle w_1, w_2\rangle\}$
	1	$\{\langle._1, [w_2]\rangle, \langle[w_1, w_2]\rangle, \langle[w_1], ._2\rangle\}$	$\{\langle[w_1], w_2\rangle\}$
	2	$\{\langle[w_1], [w_2]\rangle\}$	

Fig. 6. Generation process of sequences based on Fig. 5

layer 1. Since each state has only one sequence, all the sequences are retained at each state. At time step 2, we extend these sequences similarly to reachable states. As multiple sequences reach state 1 in layer 0 and state 0 in layer 1, we retain only $\langle w_1, w_2\rangle$ and $\langle[w_1, w_2]\rangle$ at each state because these sequences have the maximum $\mathcal{D}_I$ values. This process continues until time step 7. Finally, we select $\langle[w_1, w_2], .3, [w_4, w_5, w_6, w_7]\rangle$ because this sequence reaches the final state in layer 2 with the maximum $\mathcal{D}_I$ values. We then output matches $\{\langle w_1, w_2\rangle, \langle w_4, w_5, w_6, w_7\rangle\}$ in the sequence as pattern matching results.

6   Experiments

We use real and synthetic datasets in the experiments. We evaluate the effectiveness of our approach using the real dataset provided by the Lahar project [19]. The dataset represents an indoor space as a graph, as in Fig. 7, and records the estimated locations of a person walking around the space. For every second, the probability that a person is located at the node is estimated based on the sensor data. The dataset also provides the ground truth, a non-probabilistic data stream of user locations. There are nine room entry/exit events in the dataset. In addition, we use a synthetic dataset. We generate a probabilistic data stream that comprises ten thousand events, using the Markov model in Fig. 8 with $\Sigma = \{a, b, \ldots, z\}$. The model shows the cyclic event occurrence from "a" to "z". The generation starts from the initial state and shifts to the next state according to the probability of each edge. Each state corresponds to each event symbol such as state 1 and "a", and outputs a probabilistic event that contains five symbols with their probabilities. The probability of each symbol is maximized at the corresponding state, and decreases as the current state moves away from it.

6.1   Evaluation of Detection Quality

Evaluation Metrics. We use precision, recall, and F-measure to analyze detection quality. Let M be the set of matches and C be the set of correct sequences.

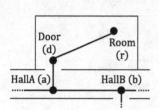

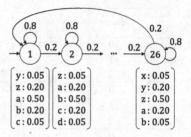

Fig. 7. Graph structure in the real dataset

Fig. 8. Markov model for the synthetic dataset

Since matches are rarely identical with correct sequences, we define these metrics by using the following overlap ratios:

$$\text{overlap}(m, c) = \frac{\min(m.t_e, c.t_e) - \max(m.t_s, c.t_s) + 1}{\max(m.t_e, c.t_e) - \min(m.t_s, c.t_s) + 1}$$

Precision indicates the accuracy of detection, and is usually calculated by $|M \cap C|/|M|$. We regard the average of overlap ratios of M as the precision score:

$$\text{precision}(M, C) = \frac{1}{|M|} \sum_{m \in M} \max_{c \in C}(\text{overlap}(m, c)) \tag{11}$$

Recall indicates the completeness of detection, and is usually calculated by $|G \cap C|/|C|$. In this paper, we regard the average of the overlap ratios of C as recall:

$$\text{recall}(M, C) = \frac{1}{|C|} \sum_{c \in C} \max_{m \in M}(\text{overlap}(m, c)) \tag{12}$$

F-measure indicates the overall performance of detection, and calculated by the harmonic mean of precision and recall:

$$\text{F}(M, C) = \frac{2 \cdot \text{precision}(M, C) \cdot \text{recall}(M, C)}{\text{precision}(M, C) + \text{recall}(M, C)} \tag{13}$$

Comparison with the Occurrence Probability-Based Approach. We compare our approach with top-k pattern matching based on occurrence probabilities. That is, the competitor uses the following objective function in (10).

$$\text{maximize} \sum_{m \in M} P(m) \tag{14}$$

Note that we can detect results of the competitor by extending the proposed algorithm in Sect. 5.

Compared with the competitor, the proposed method achieves greater precision, recall, and F-measure scores. In the experiments, we apply pattern

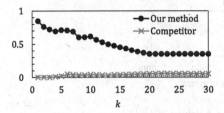

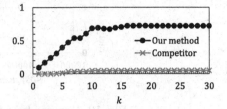

Fig. 9. Precision for different values of k **Fig. 10.** Recall for different values of k

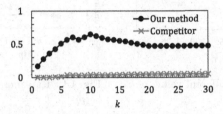

Fig. 11. F-measure for different values of k

$p = \langle \text{Door}^+ \ \text{Room}^+ \ \text{Door}^+ \rangle$ to the real dataset with the different values of k. Figures 9, 10, and 11 show precision, recall, and F-measure scores, respectively. Every metrics value of the competitor is less than 0.1 because it is difficult for the competitor to detect correct sequences. Each correct sequence occurs over a time interval from 20 to 40 s, and their occurrence probabilities are less than 1.0×10^{-3}. Thus, we cannot detect correct sequences by selecting matches in descending order of probabilities. On the other hand, our method compares the probabilities of matches with those of other sequences in the same time interval. In other words, we detect matches in time intervals that the probabilities of "Door" and "Room" are larger than those of "HallA" and "HallB." Unless there is no significant error in the classification, such matches overlap greatly with the correct sequences. Thus, every metrics of the proposed method is larger than that of the competitor.

In addition, the experimental results show the effectiveness of our approach. The precision scores decrease as we increase k. That is, the proposed method preferentially selects matches that overlap greatly with correct sequences. On the other hand, the recall scores increase until $k = 10$, and then maintain that value. Since the number of correct sequences is 9, this result indicates that the proposed method outputs appropriate matches with almost no waste. F-measure scores also indicate the effectiveness of the proposed method because they are maximized around $k = 9$, the number of correct sequences.

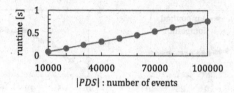

Fig. 12. Runtime for different values of $|PDS|$

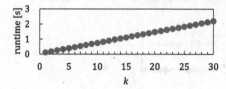

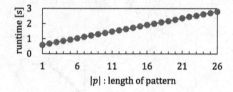

Fig. 13. Runtime for different values of k

Fig. 14. Runtime for different values of $|p|$

6.2   Evaluation of Efficiency

We measure runtime as we modify the number of events $|PDS|$, k, and pattern p. In the experiments, we use the synthetic dataset, and set $|PDS| = 100,000$, $k = 10$, and $p = \langle \mathsf{a}^+ \ \mathsf{b}^+ \ \mathsf{c}^+ \rangle$ as the default values.

   The experimental results show the efficiency of our approach. Figures 12, 13, and 14 show that runtime increases linearly as we increase each parameter. As shown in Sect. 5, the proposed method repeats the expansion of sequences over a probabilistic data stream. Since $|PDS|$ affects only the number of the expansion, the effect of $|PDS|$ on runtime is linear. Similarly, as k and p affect only the number of extended sequences, their effect is also linear. Note that we modify patterns naïvely, such as $p = \langle \mathsf{a}^+ \rangle$ as $|p| = 1$ and $p = \langle \mathsf{a}^+ \ \mathsf{b}^+ \rangle$ as $|p| = 2$, in this experiment. Thus, if we use more complex patterns, the runtime may increase faster.

7   Conclusion

We proposed top-k pattern matching based on an information-theoretic criterion over probabilistic data streams. We defined deviation of amount of information to measure the appropriateness of matches. Besides, we defined the problem definition of top-k pattern matching by using no-overlap semantics [20]. To detect top-k matches efficiently, we proposed the use of a state transition diagram layered by the number of matches. We evaluated the effectiveness and efficiency of our approach by the experiments using the real and synthetic datasets. Future work includes the expansion to real time processing and the simultaneous processing of multiple queries.

Acknowledgment. This research was partially supported by the Center of Innovation Program from Japan Science and Technology Agency (JST) and KAKENHI (16H01722, 26540043).

References

1. Aggarwal, C.C., Yu, P.S.: A framework for clustering uncertain data streams. In: 2008 IEEE 24th ICDE, pp. 150–159 (2008)
2. Aho, A.V., Corasick, M.J.: Efficient string matching: an aid to bibliographic search. Commun. ACM **18**(6), 333–340 (1975)
3. Akdere, M., Çetintemel, U., Tatbul, N.: Plan-based complex event detection across distributed sources. Proc. VLDB Endow. **1**(1), 66–77 (2008)
4. Chandramouli, B., Goldstein, J., Maier, D.: High-performance dynamic pattern matching over disordered streams. Proc. VLDB Endow. **3**(1–2), 220–231 (2010)
5. Chen, L., Nugent, C., Wang, H.: A knowledge-driven approach to activity recognition in smart homes. IEEE TKDE **24**(6), 961–974 (2012)
6. Cormode, G., Garofalakis, M.: Sketching probabilistic data streams. In: Proceedings of 2007 ACM SIGMOD, pp. 281–292 (2007)
7. Cover, T.M., Thomas, J.A.: Elements of Information Theory. Wiley, Hoboken (2012)
8. Cugola, G., Margara, A.: Processing flows of information: from data stream to complex event processing. ACM Comput. Surv. **44**(3), 15:1–15:62 (2012)
9. Diao, Y., Fischer, P., Franklin, M.J., To, R.: YFilter: efficient and scalable filtering of XML documents. In: Proceedings of 18th ICDE, pp. 341–342 (2002)
10. Forney Jr., G.D.: The Viterbi algorithm. Proc. IEEE **61**(3), 268–278 (1973)
11. Hopcroft, J.E., Motwani, R., Ullman, J.D.: Introduction to Automata Theory, Languages, and Computation. Addison Wesley, Boston (2000)
12. Jin, C., Yi, K., Chen, L., Yu, J.X., Lin, X.: Sliding-window top-k queries on uncertain streams. Proc. VLDB Endow. **1**(1), 301–312 (2008)
13. Knuth, D.E., Morris Jr., J.H., Pratt, V.R.: Fast pattern matching in strings. SIAM J. Comput. **6**(2), 323–350 (1977)
14. Lara, O.D., Labrador, M.A.: A survey on human activity recognition using wearable sensors. IEEE Commun. Surv. Tutor. **15**(3), 1192–1209 (2013)
15. Li, Z., Ge, T., Chen, C.X.: ε-matching: event processing over noisy sequences in real time. In: Proceedings of 2013 ACM SIGMOD, pp. 601–612 (2013)
16. Liu, M., Golovnya, D., Rundensteiner, E.A., Claypool, K.T.: Sequence pattern query processing over out-of-order event streams. In: 2009 IEEE 25th ICDE, pp. 784–795 (2009)
17. Mei, Y., Madden, S.: ZStream: a cost-based query processor for adaptively detecting composite events. In: Proceedings of 2009 ACM SIGMOD, pp. 193–206 (2009)
18. Nakata, I.: Generation of pattern-matching algorithms by extended regular expressions. Japan Soc. Softw. Sci. Tech. **5**, 1–9 (1993)
19. Ré, C., Letchner, J., Balazinska, M., Suciu, D.: Event queries on correlated probabilistic streams. In: Proceedings of 2008 ACM SIGMOD, pp. 715–728 (2008)
20. Santini, S.: Querying streams using regular expressions: some semantics, decidability, and efficiency issues. VLDB J. **24**(6), 801–821 (2015)
21. Thompson, K.: Programming techniques: regular expression search algorithm. Commun. ACM **11**(6), 419–422 (1968)
22. Tran, T.T.L., Peng, L., Diao, Y., McGregor, A., Liu, A.: CLARO: modeling and processing uncertain data streams. VLDB J. **21**(5), 651–676 (2012)

23. Woods, L., Teubner, J., Alonso, G.: Complex event detection at wire speed with FPGAs. Proc. VLDB Endow. **3**(1–2), 660–669 (2010)
24. Wu, E., Diao, Y., Rizvi, S.: High-performance complex event processing over streams. In: Proceedings of 2006 ACM SIGMOD, pp. 407–418 (2006)
25. Yin, J., Yang, Q., Pan, J.J.: Sensor-based abnormal human-activity detection. IEEE TKDE **20**(8), 1082–1090 (2008)
26. Zhang, H., Diao, Y., Immerman, N.: On complexity and optimization of expensive queries in complex event processing. In: Proceedings of 2014 ACM SIGMOD, pp. 217–228 (2014)
27. Zhang, Q., Li, F., Yi, K.: Finding frequent items in probabilistic data. In: Proceedings of 2008 ACM SIGMOD, pp. 819–832 (2008)

Sliding Window Top-K Monitoring over Distributed Data Streams

Zhijin Lv[1], Ben Chen[1], and Xiaohui Yu[1,2(✉)]

[1] School of Computer Science and Technology, Shandong University, Jinan 250101,
Shandong, China
allen3jin@163.com, CBStubborn@163.com, xyu@sdu.edu.cn
[2] School of Information Technology, York University, Toronto, ON M3J 1P3, Canada

Abstract. The problem of distributed monitoring has been intensively investigated recently. This paper studies monitoring the top k data objects with the largest aggregate numeric values from distributed data streams within a fixed-size monitoring window W, while minimizing communication cost across the network. We propose a novel algorithm, which reallocates numeric values of data objects among distributed monitoring nodes by assigning revision factors when local constraints are violated, and keeps the local top-k result at distributed nodes in line with the global top-k result. Extensive experiments are conducted on top of Apache Storm to demonstrate the efficiency and scalability of our algorithm.

Keywords: Data stream · Distributed monitoring · Top-k query · Stream processing

1 Introduction

The prior studies for distributed top-k query [5,7] focus on providing results to one-time top-k queries in distributed settings. These studies are not suitable to continuously query top-k result over distributed data streams. In this paper, we study a new problem of distributed top-k monitoring, which is continuously querying the top k data objects with the largest aggregate numeric values over distributed data streams within a fixed-size monitoring window. Each data stream contains a sequence of data objects associated with numeric values, and the aggregate numeric value of each data object is calculated from distributed data streams. The continuous top-k query we studied is restricted to the most recent portion of the data stream, and the numeric values of data objects are changed correspondingly as the monitoring window slides.

The study of distributed top-k monitoring is significant in a variety of application scenarios, such as network monitoring, sensor data analysis, web usage logs, and market surveillance. The purpose of many applications tends to track the exceptionally large (or small) numeric values relative to the major numeric values of data objects. For example, in the field of traffic flow monitoring, it is

© Springer International Publishing AG 2017
L. Chen et al. (Eds.): APWeb-WAIM 2017, Part I, LNCS 10366, pp. 527–540, 2017.
DOI: 10.1007/978-3-319-63579-8_40

necessary to continuously monitor the top k largest number of road traffic within the last 15 min in order to monitor the traffic jams in time. Another example, consider a system that monitors a large network for distributed denial of service (DDoS) attacks. The DDoS attacks may issue an unusually large number of Domain Name Service (DNS) lookup requests to distributed DNS servers from a single IP address. Hence, it is necessary to monitor the DNS lookup requests with potentially suspicious behavior. In this case, the monitoring infrastructure continuously reports the top k IP addresses with the largest number of requests at distributed servers in recent time. Since requests are frequent and rapid at distributed DNS servers, the solution of forwarding all requests to a central location and processing them is infeasible, incurring a huge communication overhead.

The major challenge of our top-k monitoring problem is numeric values of data objects varying independently at distributed nodes. Tracking the top k data objects with the largest aggregate numeric values from distributed nodes results in huge communication overhead, because the global top-k result is affected by local changes of data objects at distributed nodes. It is imperative to find solutions that can effectively monitor the global top-k result, while minimizing communication cost across the network.

Existing algorithms for distributed top-k query such as the Threshold Algorithm [7] focus on efficiently providing results to one-time top-k queries. Though distributed top-k monitoring could be implemented by repeatedly executing one-time query alogrithms, it is useless to execute query if the top-k result remains unchanged. These algorithms do not include mechanisms for detecting the changes of top-k result, incurring unnecessary communication overhead. Most of solutions proposed for sliding window top-k monitoring [14, 17] are inappropriate to our monitoring problem, because their ranking function is based on the dominance relationship of data objects, rather than the aggregate numeric values from distributed data streams. Babcock and Olston present an original algorithm for distributed top-k monitoring [3], which maintains arithmetic constraints at distributed data sites to ensure that the provided top-k answer remains valid. Their algorithms assume that a single node may violate constraints each time, which is unrealistic. Moreover, it is not suitable to the case of sliding window, which focuses on the impact of recent data objects.

For continuous data monitoring, we adopt a time-based sliding window model [14], where the data objects generated within W time-stamps from the current time-stamp are target for monitoring. In this paper, we consider a model in which there is one coordinator node $\mathcal{C}$ and a set of m distributed nodes N connected to the coordinator node as shown in Fig. 1.

The coordinator node tracks the global top-k result and assigns constraints to each monitoring node, at which local top-k result should be in alignment with the global top-k result. Each monitoring node receives data objects from an input stream and detects the potential violations of local constraints whenever the window slides. When local constraints are violated at some monitoring nodes, it is necessary to send the violated data objects and their numeric values to the coordinator node. Then, the coordinator node tries to resolve the violations,

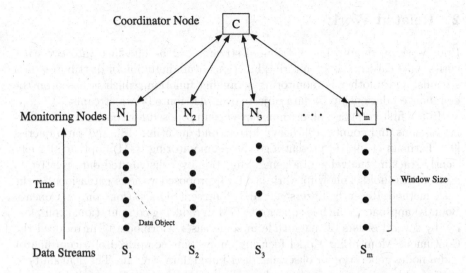

Fig. 1. Sliding window distributed monitoring architecture

called *partial resolution*. If the global constraint is satisfied by assigning new local constraints to the violated nodes, then the global top-k result remains valid. Otherwise, the coordinator node requests all distributed nodes for current numeric values of violated objects to determine whether the global constraint is still satisfied. We refer to this process as *global resolution*, which does not always occur.

We implement our distributed top-k algorithm on top of Apache Storm [1], an open-source distributed stream processing platform, on which we conduct extensive experiments to evaluate the performance of our solutions.

Our main contributions can be summarized as follows.

- We investigate the problem of sliding window top-k monitoring over distributed data streams. To the best of our knowledge, there is no prior work regarding this.
- We propose a novel algorithm for top-k monitoring over distributed data streams, which achieves a significant reduction in communication cost.
- We implement our algorithms on top of Apache Storm, and conduct extensive experiments to evaluate the performance of our algorithms with real world data, which have demonstrated the efficiency and scalability of our algorithms.

The rest of the paper is organized as follows. Section 2 reviews previous work on monitoring over distributed data streams. Section 3 formally defines the top-k monitoring problem studied in this paper. We describe our top-k monitoring algorithm in detail in Sect. 4. Section 5 experimentally evaluates the performance of our algorithms. Finally, we conclude the paper in Sect. 6.

2   Related Work

Prior work on monitoring distributed streams can be classified into two categories. One category is monitoring functions over the union of distributed data streams, and another is monitoring a ranking function, which is based on the dominance relationship of data objects over distributed data streams.

In the first category, algorithms have been proposed for continuous monitoring of sums and counts [10], heavy hitters and quantiles [18], and ratio queries [9]. Sharfman et al. [16] present a geometric monitoring (GM) approach for efficiently tracking the value of a general function over distributed data relative to a given threshold. Followup work [8,11,13] proposed various extensions to the basic method. Recently, Lazerson et al. [12] presented a CB (for Convex/Concave Bounds) approach, which is superior to GM in reducing computational complexity, by several orders of magnitude in some cases. Cormode [6] introduced the Continuous Monitoring Model focusing on systems comprised of a coordinator and n nodes generating or observing distributed data streams. The goal shifts to continuously compute a function depending on the information available across all n data streams and a dedicated coordinator.

There are also plenty of works in the second category. These works study the monitoring problem with essentially different semantics compared to the first category. Mouratidis et al. [14] proposed an efficient technique to compute top-k queries over sliding windows. They make an interesting observation that a top-k query can be answered from a small subset of the objects called k-skyband [15]. Existing top-k processing solutions are mainly based on the dominance property between data stream. The dominance property states that object O_a dominates object O_b iff O_a has a higher score than O_b. Amagata et al. [2] presented algorithms for distributed continuous top-k dominating query processing, which reduces both communication and computation costs. Unfortunately, their algorithms are inappropriate for the top-k monitoring problem we studied.

Further problems related to our distributed top-k monitoring are distributed one-time top-k queries [4,5,7]. Fagin et al. [7] examined the Threshold Algorithm (TA) and considered both exact answers and approximate answers with relative error tolerance. TA goes down the sorted lists in parallel, one position at a time, and calculates the sum of the values at that position across all the lists. The sum of the values is called "threshold". TA stops when it finds k objects whose values are higher than the "threshold" value. Cao and Wang [5] proposed an efficient algorithm called "Three-Phase Uniform Threshold" (TPUT), which reduces network bandwidth consumption by pruning away ineligible objects, and terminates in three round-trips regardless of data input. However, these studies are interested in algorithms that can obtain the initial top-k result efficiently and provide top-k result to one-time queries. Our study focuses on monitoring whether the top-k result have changed after an initial answer has been obtained.

3 Problem Definitions

Now we more formally define the problem studied in this paper. As described above, there is one coordinator node $\mathcal{C}$ and m distributed monitoring nodes $N_1, N_2, ..., N_m$. Each monitoring node N_j continuously receives data records from an input stream. Collectively, the monitoring nodes track a set $\mathcal{O}$ of n logical data objects $\mathcal{O} = \{O_1, O_2, ..., O_n\}$. Each data object is associated a numeric value within the current monitoring window. The numeric value of each data object is updated at distributed nodes as the monitoring window slides. For each monitoring node N_j, we define partial numeric values $C_{1,j}(t), C_{2,j}(t), ..., C_{n,j}(t)$ representing node N_j's view of the data stream S_j within monitoring window W at time t, where

$$C_{i,j}(t) = |\{O_i^{t'} \in S_j \mid t - t' \leq W\}|. \tag{1}$$

The aggregate numeric value of each object O_i from distributed monitoring nodes is defined to be $C_i(t) = \sum_{1 \leq j \leq m} C_{i,j}(t)$. Tracking $C_i(t)$ exactly requires alerting the coordinator node every time data object O_i arrives or expires, so the goal is to track $C_i(t)$ approximately within an ϵ-error. The coordinator node is responsible for tracking the top k data objects within a bounded error tolerance. We define the approximate top-k set maintained by the coordinator node as $\mathcal{T}$, which is considered valid if and only if:

$$\forall O_a \in \mathcal{T}, \forall O_b \in \mathcal{O} - \mathcal{T} : C_a(t) + \epsilon \geq C_b(t) \tag{2}$$

where $\epsilon \geq 0$ is an user-specified approximation parameter. If $\epsilon = 0$, then the top-k set is exact, otherwise a corresponding degree of error is permitted in the top-k set. The goal of our approach is to provide an approximate top-k set that is valid within an ϵ-error in the case of sliding window, while minimizing the overall communication cost to the monitoring infrastructure.

3.1 Revision Factors and Slack

We realize that the global top-k set is valid, if distributed monitoring nodes have the same top-k set locally. Since the actual local top-k set at distributed monitoring nodes may be differ from the global top-k set, we use *revision factors*, labeled $\delta_{i,j}$, to reallocate the numeric values of data object O_i to monitoring node N_j to satisfy the following local constraint:

$$\forall O_a \in \mathcal{T}, \forall O_b \in \mathcal{O} - \mathcal{T} : C_{a,j}(t) + \delta_{a,j} \geq C_{b,j}(t) + \delta_{b,j} \tag{3}$$

In addition, the coordinator node maintains partial revision factors of data objects as global slack, labeled $\delta_{i,0}$. To ensure correctness, the sum of revision factors for each data object O_i should satisfy: $\sum_{0 \leq j \leq m} \delta_{i,j} = 0$.

In order to reallocate numeric values of data objects among nodes, it is necessary to compute additional slacks of data objects at each node. We define resolution set which contains data objects from global top-k set $\mathcal{T}$ and all violated objects as $\mathcal{R}$. Our algorithm selects the maximum values $\mathcal{P}_j$ of data object not in the resolution set $\mathcal{R}$ as a baseline for computing additional slacks of data objects at each node N_j:

$$\mathcal{P}_j = \max_{O_i \in \mathcal{O} - \mathcal{R}} (C_{i,j}(t) + \delta_{i,j}) \tag{4}$$

Thus, the overall slack $\mathcal{S}_i$ for each data object O_i from the resolution set $\mathcal{R}$ is given by:

$$\forall O_i \in \mathcal{R} : \mathcal{S}_i = \sum_{1 \leq j \leq m} (C_{i,j}(t) - \mathcal{P}_j) = C_i(t) - \sum_{1 \leq j \leq m} \mathcal{P}_j \tag{5}$$

It is important to reallocate the overall slack of data objects among the coordinator node and distributed monitoring nodes. If the slack is tight at monitoring node, the violation of local constraints would be frequent. However, smaller slack at coordinator node results in higher probability of violation of global constraints. The optimum slack allocation polices balances these two costs.

3.2   Sliding Window Unit

In the sliding window scenario, distributed monitoring nodes track numeric values of data objects within the monitoring window W. Based on the arrival order of each object in W, the data objects in window W could be partitioned into several small window unit $s_0, s_1, ..., s_{l-1}$ ($l = \frac{W}{w}$). The size of sliding window unit w is specified according to the actual application scenario. Small window unit is more suitable for near real-time applications, but incurring more communication and computation costs.

As shown in Fig. 2, the monitoring window W slides, whenever a new sliding window unit s_{new} has been created and the expired window unit s_{exp} is removed. Thus, the partial numeric values $C_{i,j}(t)$ of each data object O_i at monitoring node N_j updates within the new monitoring window W' ($W' = W + s_{new} - s_{exp}$). Obviously, the changes of data objects may violate the current local constraints.

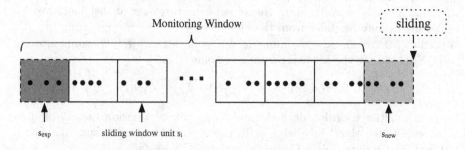

Fig. 2. Sliding window unit structure

4 Top-K Monitoring Algorithm

We now describe our algorithm in detail for sliding window top-k monitoring over distributed data streams. At the outset, the coordinator node initializes the global top-k set by running an efficient algorithm for one-time top-k queries, e.g. TPUT algorithm [5]. Once the global top-k set T has been initialized, the coordinator node C sends a message containing T and initial revision factors $\delta_{i,j} = 0$ corresponding to each monitoring node N_j. Upon receiving this message, the monitoring node N_j creates local constraint (3) from T and revision factors to detect potential violations due to local changes of data objects.

If one or more local constraints are violated, the global top-k set T may have become invalid. We use a distributed process called *resolution algorithm* to determine whether current top-k set is still valid and resolve the violations if not.

4.1 Resolution Algorithm

Resolution algorithm is initiated if one or more local constraints are violated at some monitoring nodes $N_{\mathcal{R}}$. Our resolution algorithm consists of three phases, and the third phase does not always occur.

Algorithm 1. Partial Resolution Algorithm

Input: $T, \delta_{i,j}, \{\mathcal{R}_j\}, \{C_{i,j}(t)\}$
Output: succeed or failed

1: **for** $\forall N_j \in N_{\mathcal{R}}$ **do**
2:     $bound_j \leftarrow \max(C_{i,j}(t) + \delta_{i,j}), O_i \in \mathcal{R}_j - T$ //find the max value of violated data object at each violated node.
3: /* try to reallocate the numeric value of $O_i \in T$ among violated nodes to resolve violation of constraints */
4: **for** $\forall O_i \in T$ **do**
5:     $slack_i \leftarrow \delta_{i,0}$
6:     **for** $\forall N_j \in N_{\mathcal{R}}$ **do**
7:         $used_i \leftarrow bound_j - (C_{i,j}(t) + \delta_{i,j})$
8:         $slack_i \leftarrow slack_i - used_i$
9: $bound_0 \leftarrow \max(\delta_{i,0}), O_i \notin T$
10: /* determine whether the coordinator node satisfies the Constraint (3) */
11: **for** $\forall O_i \in T$ **do**
12:     **if** $slack_i + \epsilon < bound_0$ **then**
13:         **return** failed
14: **return** succeed

- **Local Alert Phase.** The monitoring node N_j at which violated constraints have been detected sends a message containing a local resolution set $\mathcal{R}_j$ (containing data objects from global top-k set T and current local top-k set) and a set of partial numeric values $C_{i,j}(t)$ of data object O_i in the local resolution set to coordinator node.

- **Partial Resolution Phase.** The coordinator node determines whether all violations can be solved based on the messages from violated nodes $N_\mathcal{R}$ and itself alone according to the Algorithm 1. If the coordinator node resolves all violations successfully by assigning updated revision factors to the violated nodes, the global top-k set remains unchanged and resolution process terminates. Otherwise, the coordinator node is unable to rule out all violations during this phase, the third phase is required.
- **Global Resolution Phase.** The coordinator node requests the current partial values $C_{i,j}(t)$ of data objects O_i in overall resolution set $\mathcal{R} = \cup_{N_j \in N_\mathcal{R}} \mathcal{R}_j$ as well as the baseline value $\mathcal{P}_j$ from all monitoring nodes. Once the coordinator node receives responses from all monitoring nodes, it computes a new top-k set and new revision factors of data objects in the resolution set $\mathcal{R}$, and notifies all monitoring nodes of a new top-k set $\mathcal{T}'$ and their new revision factors. Our algorithm adopts even policy to divide the overall slack $\mathcal{S}_i$ of data object O_i among monitoring nodes and coordinator node. The revision factors allocation algorithm is described in Algorithm 2.

For notational convenience, we extend our notation for partial numeric values and baseline value to the coordinator node by defining $C_{i,0}(t) = 0$ for all data object O_i and $\mathcal{P}_0 = \max_{O_i \in \mathcal{O}-\mathcal{R}} \delta_{i,0}$. We also define nodes set $\mathcal{A}$ as all nodes involved in the resolution process. For each object O_i, $C_{i,\mathcal{A}}(t) = \sum_{0 \le j \le m}(C_{i,j}(t) + \delta_{i,j})$. Similarly, we define the sum of the baseline values from the nodes set $\mathcal{A}$, $\mathcal{P}_\mathcal{A} = \sum_{0 \le j \le m} \mathcal{P}_j$.

Algorithm 2. Revision Factors Allocation Algorithm

Input: $\mathcal{T}', \mathcal{R}, \{\mathcal{P}_j\}, \{C_{i,j}(t)\}$
Output: $\{\delta_{i,j}\}$

1: /* compute the overall slack of $O_i \in \mathcal{R}$ */
2: **for** $\forall O_i \in \mathcal{R}$ **do**
3:     **if** $O_i \in \mathcal{T}'$ **then**
4:         $\mathcal{S}_i = C_{i,\mathcal{A}}(t) - \mathcal{P}_\mathcal{A} + \epsilon$
5:     **else**
6:         $\mathcal{S}_i = C_{i,\mathcal{A}}(t) - \mathcal{P}_\mathcal{A}$
7: /* compute new revision factors $\{\delta_{i,j}\}$ using even policy */
8: **for** $\forall O_i \in \mathcal{R}$ **do**
9:     $ps \leftarrow \mathcal{S}_i/(1 + m)$
10:     **for** $j = 0 \to m$ **do**
11:         **if** $O_i \in \mathcal{T}'$ and $j = 0$ **then**
12:             $\delta_{i,j} = \mathcal{P}_j - C_{i,j}(t) + ps - \epsilon$
13:         **else**
14:             $\delta_{i,j} = \mathcal{P}_j - C_{i,j}(t) + ps$

4.2   Correctness and Cost Analysis

The goal of our algorithm is to keep the local top-k set at each node in line with the global top-k set. If the global constraint is satisfied, the global top-k remains valid. When local constraints are violated at distributed nodes, our algorithm reallocates the numeric values of violated data objects by assigning revision factors to distributed nodes.

Example 1. Consider a simple scenario with two monitoring nodes N_1 and N_2 and three data objects O_1, O_2 and O_3, and current revision factors are zero. At time t, the current data values at N_1 are $C_{1,1}(t) = 4, C_{2,1}(t) = 6$ and $C_{3,1}(t) = 10$ and at N_2 are $C_{1,2}(t) = 3, C_{2,2}(t) = 4$ and $C_{3,2}(t) = 3$. Let $k = 1, \epsilon = 0$, the current top-k set $\mathcal{T} = \{O_3\}$. However, the local top-k set at N_2 is $\{O_2\}$, which violates the constraints. Our algorithms find that partial resolution phases are failed to resolve the violations, due to slack at coordinator node are zero. The global resolution phase computes the new revision factors assigned to the monitoring nodes, at coordinator node are $\delta_{2,0} = 1, \delta_{3,0} = 2$ and at N_1 are $\delta_{2,1} = -1, \delta_{3,1} = -4$ and at N_2 are $\delta_{2,2} = 0, \delta_{3,2} = 2$. Then, the local constraints at distributed node are satisfied and the global top-k set $\mathcal{T} = \{O_3\}$ is valid.

Data objects not in the resolution set $\mathcal{R}$ can not be candidates for new top-k set $\mathcal{T}'$, because their numeric values satisfy the current local constraints. Therefore, the sum of all baseline values $\mathcal{P}_A$ should be less than the minimum numeric values $C_l(t)$ of data object O_l in the previous top-k set $\mathcal{T}$. Furthermore, each data object O_i in the new top-k set $\mathcal{T}'$ satisfies: $C_i(t) \geq C_l(t) \geq \mathcal{P}_A$. And, the overall slacks of data objects in resolution set $\mathcal{R}$ satisfy the following inequation:

$$\forall O_a \in \mathcal{T}', \forall O_b \in \mathcal{R} - \mathcal{T}' : \mathcal{S}_a \geq \mathcal{S}_b \ and \ \mathcal{S}_a \geq 0 \tag{6}$$

As described in Algorithm 2, we evenly allocate the overall slack $\mathcal{S}_i$ of each object O_i in the resolution set $\mathcal{R}$ to all nodes. As a result, the new local top-k set computed by new revision factors at distributed nodes must be in line with the new global top-k set, and local constraints (3) at distributed nodes are satisfied.

Our resolution algorithm maintains global slack at the coordinator node, which is significant at partial resolution phase. If the partial resolution phase resolves the violations successfully, the third phase does not require. Thus, the communication cost at this phase is just assigning updated revision factors to the violated node, and number of $2 * |N_\mathcal{R}|$ messages are exchanged altogether. If all three phases are required, the total of $|N_\mathcal{R}| + 3m$ messages are necessary to perform complete resolution.

If the global slack retained at coordinator node is tight, the probability of failure at partial resolution phase becomes high, incurring more communication cost at global resolution phase. However, tight slacks at distributed monitoring nodes result in frequent violations of local constraints. Our even policy for allocating additional slacks balances these two costs well.

5   Experiments

In this section, we provide an experimental evaluation on the communication cost of our resolution algorithm. We implement two different top-k monitoring algorithms as baseline algorithms. One algorithm retains zero slack at coordinator node (LSA), and another algorithm retains zero slack at distributed monitoring nodes (GSA).

5.1   Setup

The experiments are conducted on a cluster of 16 Dell R210 servers with Gigabit Ethernet interconnect. Each server has a 2.4 GHz Intel processor and 8 GB RAM. As shown in Fig. 1, one server works as the coordinator node and the remaining nodes work as the monitoring nodes. The monitoring node can exchange messages with the coordinator node, but can not communicate with each other. Additionally, the coordinator node can send broadcast messages received by all monitoring nodes.

We implement our algorithm on top of Apache Storm, a free and open source distributed realtime computation system, which makes it easy to reliably process unbounded streams of data. All of nodes are implemented as *Bolt* components within the Storm system, and receive data objects continuously from a *Spout* component, which is a source of data streams. They constitute a *Topology* run on the Storm system. The version of Apache Storm we used is 1.0.2 in our experiments.

We evaluated the efficiency of our algorithm against sliding window unit size w, number of monitoring nodes m, approximation parameter ϵ respectively. The default values of the parameters are listed in Table 1. Parameters are varied as follows:

- number of monitoring nodes m: 3, 5, 8, 10, 15
- sliding window unit size w: 5s, 10s, 15s, 20s
- approximation parameter ϵ: 0, 25, 50, 75, 100

Table 1. Experimental parameters

Notation	Definition (Default value)
k	Number of objects to track in top-k set (10)
m	Number of monitoring nodes (10)
ϵ	Approximation parameter (0)
W	Monitoring window size (15 min)
w	Sliding window unit size (10s)

5.2 Data and Queries

We conducted our experiments on both synthetic dataset and real dataset. The datasets are described in detail as follows:

- **Synthetic Dataset**: The synthetic dataset includes random data records, which follow Zipf distribution [19]. The distribution parameter we used is 2. Each data record contains ID of data object and the time of generation. The goal of experiment is continuous querying the top k data objects with the largest number of occurrences.
- **Real Dataset**: The real dataset consists of a portion of real vehicle passage records from the traffic surveillance system of a major city. The dataset contains 5,762,391 passage records, which are generated within 6 h (about 267 passage records per second), and involves about 1000 detecting locations on the main roads. Our experiments continuously monitor the top k detecting locations with the largest number of vehicle passage records.

Our experiments continuously monitor the top k data objects over distributed data streams within last 15 min, and the total communication cost is the number of messages exchanged for processing 100 sliding windows.

5.3 Evaluation

As shown in Figs. 3 and 4, we vary the number of monitoring nodes m with diverse window unit size w to demonstrate the efficiency and scalability of our resolution algorithm using synthetic dataset and real dataset.

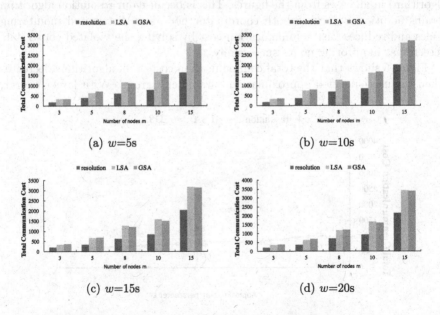

(a) w=5s   (b) w=10s

(c) w=15s   (d) w=20s

Fig. 3. Varying number of nodes m using synthetic dataset

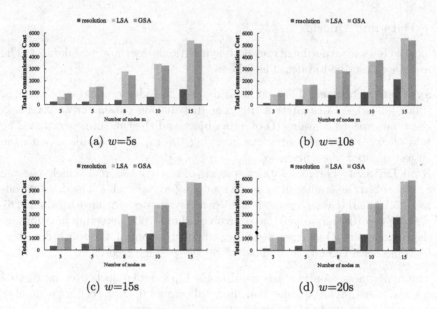

(a) w=5s   (b) w=10s

(c) w=15s   (d) w=20s

Fig. 4. Varying number of nodes m using real dataset

Normally, as the number of monitoring nodes m increases, the overall communication cost of monitoring infrastructure is increased correspondingly. Because the global resolution phase in our resolution algorithm needs to request informations from all distributed nodes to resolve violations of local constraints. Our resolution algorithm outperforms baseline algorithms (LSA algorithm and GSA algorithm) in all cases from the figures. This is because our resolution algorithm retains additional slack at both coordinator node and distributed monitoring nodes and reduces vast communication cost by solving the violated constraints detected at monitoring nodes successfully.

Figure 5 shows that the total communication cost of all algorithms decreases when the user-specified approximation parameter ϵ grows. With larger ϵ-error,

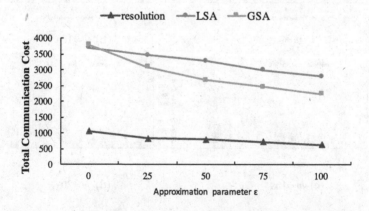

Fig. 5. Varying approximation parameter ϵ using real dataset, $w = 10$s, $m = 10$

there are less violations of local constraints at distributed nodes, resulting in lower communication overhead. However, the global top-k result is not accurate, and the error tolerance lies on the various application scenarios.

In all cases, our algorithm achieves a significant reduction in communication cost compared to baseline algorithms. Moreover, with the increase of monitoring nodes, the gap between our resolution algorithm and baseline algorithms becomes wider. These experiments results demonstrate the efficiency and scalability of our algorithm.

6   Conclusions

In this paper, we studied the problem of top-k monitoring over distributed data streams in sliding window case. We propose a novel algorithm, which reallocates numeric values of data objects among distributed monitoring nodes by assigning revision factors to deal with distributed top-k monitoring problem and implement our algorithm on top of Apache Storm, on which extensive experiments are conducted to demonstrate the efficiency and scalability of our algorithm. Future work will concentrate on monitoring other functions over distributed data streams.

Acknowledgment. This work was supported in part by the National Basic Research 973 Program of China under Grant No. 2015CB352502, the National Natural Science Foundation of China under Grant Nos. 61272092 and 61572289, the Natural Science Foundation of Shandong Province of China under Grant Nos. ZR2012FZ004 and ZR2015FM002, the Science and Technology Development Program of Shandong Province of China under Grant No. 2014GGE27178, and the NSERC Discovery Grants.

References

1. Twitter storm. http://storm.apache.org/
2. Amagata, D., Hara, T., Nishio, S.: Sliding window top-k dominating query processing over distributed data streams. Distrib. Parallel Databases **34**(4), 535–566 (2016). http://dx.doi.org/10.1007/s10619-015-7187-9
3. Babcock, B., Olston, C.: Distributed top-k monitoring. In: Proceedings of the 2003 ACM SIGMOD International Conference on Management of Data, San Diego, California, USA, 9–12 June 2003, pp. 28–39 (2003). http://doi.acm.org/10.1145/872757.872764
4. Bruno, N., Gravano, L., Marian, A.: Evaluating top-k queries over web-accessible databases. In: Proceedings of the 18th International Conference on Data Engineering, San Jose, CA, USA, 26 February–1 March 2002, pp. 369–380 (2002). http://dx.doi.org/10.1109/ICDE.2002.994751
5. Cao, P., Wang, Z.: Efficient top-k query calculation in distributed networks. In: Proceedings of the Twenty-Third Annual ACM Symposium on Principles of Distributed Computing, PODC 2004, St. John's, Newfoundland, Canada, 25–28 July 2004, pp. 206–215 (2004). http://doi.acm.org/10.1145/1011767.1011798
6. Cormode, G.: The continuous distributed monitoring model. SIGMOD Rec. **42**(1), 5–14 (2013). http://doi.acm.org/10.1145/2481528.2481530

7. Fagin, R., Lotem, A., Naor, M.: Optimal aggregation algorithms for middleware. In: Proceedings of the Twentieth ACM SIGACT-SIGMOD-SIGART Symposium on Principles of Database Systems, Santa Barbara, California, USA, 21–23 May 2001 (2001). http://doi.acm.org/10.1145/375551.375567

8. Giatrakos, N., Deligiannakis, A., Garofalakis, M.N., Sharfman, I., Schuster, A.: Prediction-based geometric monitoring over distributed data streams. In: Proceedings of the ACM SIGMOD International Conference on Management of Data, SIGMOD 2012, Scottsdale, AZ, USA, 20–24 May 2012, pp. 265–276 (2012). http://doi.acm.org/10.1145/2213836.2213867

9. Gupta, R., Ramamritham, K., Mohania, M.K.: Ratio threshold queries over distributed data sources. In: Proceedings of the 26th International Conference on Data Engineering, ICDE 2010, Long Beach, California, USA, 1–6 March 2010, pp. 581–584 (2010). http://dx.doi.org/10.1109/ICDE.2010.5447920

10. Kashyap, S.R., Ramamirtham, J., Rastogi, R., Shukla, P.: Efficient constraint monitoring using adaptive thresholds. In: Proceedings of the 24th International Conference on Data Engineering, ICDE 2008, 7–12 April 2008, Cancún, México, pp. 526–535 (2008). http://dx.doi.org/10.1109/ICDE.2008.4497461

11. Keren, D., Sharfman, I., Schuster, A., Livne, A.: Shape sensitive geometric monitoring. IEEE Trans. Knowl. Data Eng. 24(8), 1520–1535 (2012). http://dx.doi.org/10.1109/TKDE.2011.102

12. Lazerson, A., Keren, D., Schuster, A.: Lightweight monitoring of distributed streams. In: Proceedings of the 22nd ACM SIGKDD International Conference on Knowledge Discovery and Data Mining, San Francisco, CA, USA, 13–17 August 2016, pp. 1685–1694 (2016). http://doi.acm.org/10.1145/2939672.2939820

13. Lazerson, A., Sharfman, I., Keren, D., Schuster, A., Garofalakis, M.N., Samoladas, V.: Monitoring distributed streams using convex decompositions. PVLDB 8(5), 545–556 (2015). http://www.vldb.org/pvldb/vol8/p545-lazerson.pdf

14. Mouratidis, K., Bakiras, S., Papadias, D.: Continuous monitoring of top-k queries over sliding windows. In: Proceedings of the ACM SIGMOD International Conference on Management of Data, Chicago, Illinois, USA, 27–29 June 2006, pp. 635–646 (2006). http://doi.acm.org/10.1145/1142473.1142544

15. Papadias, D., Tao, Y., Fu, G., Seeger, B.: Progressive skyline computation in database systems. ACM Trans. Database Syst. 30(1), 41–82 (2005). http://doi.acm.org/10.1145/1061318.1061320

16. Sharfman, I., Schuster, A., Keren, D.: A geometric approach to monitoring threshold functions over distributed data streams. In: Proceedings of the ACM SIGMOD International Conference on Management of Data, Chicago, Illinois, USA, 27–29 June 2006, pp. 301–312 (2006). http://doi.acm.org/10.1145/1142473.1142508

17. Yang, D., Shastri, A., Rundensteiner, E.A., Ward, M.O.: An optimal strategy for monitoring top-k queries in streaming windows. In: Proceedings of 14th International Conference on Extending Database Technology, EDBT 2011, Uppsala, Sweden, 21–24 March 2011, pp. 57–68 (2011). http://doi.acm.org/10.1145/1951365.1951375

18. Yi, K., Zhang, Q.: Optimal tracking of distributed heavy hitters and quantiles. In: Proceedings of the Twenty-Eigth ACM SIGMOD-SIGACT-SIGART Symposium on Principles of Database Systems, PODS 2009, Providence, Rhode Island, USA, 19 June–1 July 2009, pp. 167–174 (2009). http://doi.acm.org/10.1145/1559795.1559820

19. Zipf, G.K.: Selected studies of the principle of relative frequency in language. Language 9(1), 89–92 (1932)

Diversified Top-*k* Keyword Query Interpretation on Knowledge Graphs

Ying Wang, Ming Zhong[✉], Yuanyuan Zhu, Xuhui Li,
and Tieyun Qian

State Key Laboratory of Software Engineering,
Wuhan University, Wuhan 430072, China
{wysklse,clock,yyzhu,lixuhui,qty}@whu.edu.cn

Abstract. Exploring a knowledge graph through keyword queries to discover meaningful patterns has been studied in many scenarios recently. From the perspective of query understanding, it aims to find a number of specific interpretations for ambiguous keyword queries. With the assistance of interpretation, the users can actively reduce the search space and get more relevant results.

In this paper, we propose a novel diversified top-*k* keyword query interpretation approach on knowledge graphs. Our approach focuses on reducing the redundancy of returned results, namely, enriching the semantics covered by the results. In detail, we (1) formulate a diversified top-*k* search problem on a schema graph of knowledge graph for keyword query interpretation; (2) define an effective similarity measure to evaluate the semantic similarity between search results; (3) present an efficient search algorithm that guarantees to return the exact top-*k* results and minimize the calculation of similarity, and (4) propose effective pruning strategies to optimize the search algorithm. The experimental results show that our approach improves the diversity of top-*k* results significantly from the perspectives of both statistics and human cognition. Furthermore, with very limited loss of result precision, our optimization methods can improve the search efficiency greatly.

Keywords: Diversification · Keyword query interpretation · Top-k search · Knowledge graph

1 Introduction

1.1 Motivation

Recently, keyword search is well recognized as a popular and effective approach to acquire knowledge from the large-scale knowledge graphs, such as DBPedia [3], Yago [8], Freebase [4], Probase [11], etc. However, keyword search suffers from a trade-off between expressiveness and ease-of-use, which results in the ambiguities of users' information needs. Therefore, keyword query interpretation is proposed for predicating the most relevant query semantics to users' information needs. As a result, the users can still issue keyword queries initially, and then, they will formulate more expressive and relevant queries from the returned intermediate results, thereby narrowing the search space and improving the quality of final results.

© Springer International Publishing AG 2017
L. Chen et al. (Eds.): APWeb-WAIM 2017, Part I, LNCS 10366, pp. 541–555, 2017.
DOI: 10.1007/978-3-319-63579-8_41

The existing keyword query interpretation approaches [12–15] mainly focus on improving the relevance of results and the efficiency. There is still a lack of discussion of an important property of interpretation results, namely, diversity. Unfortunately, according to our observation on real-world knowledge graphs, only considering the relevance of results often leads to lots of similar results which are redundant to the user and also may not reach the different user's intention. That is because the most relevant results mostly have the same nodes, edges and even structures that are preferred by scoring functions. Let us consider the following example.

Example 1. Given a keyword query "London, Paris" on DBPedia, the top-4 relevant results with and without diversification with respect to some specific scoring function and similarity function are shown in Fig. 1. Each tree is an interpreted result, and can be seen as a graph query actually. Their leaf nodes contain the two keywords respectively. Moreover, the nodes in different colors represent different classes in DBPedia. Intuitively, the four trees on the top (without diversification) are very similar to each other. They are all rooted at the same class node, and almost share the same classes. In contrast, the four trees on the bottom (with diversification) are quite different from each other, and demonstrate various relationships between "London" and "Paris", such as biological relationship between two plants, soccer player who served in two clubs, or geographic relationship between two locations.

In order to make the interpretations of a keyword query meet the various information needs of different users, we need to diversify the results.

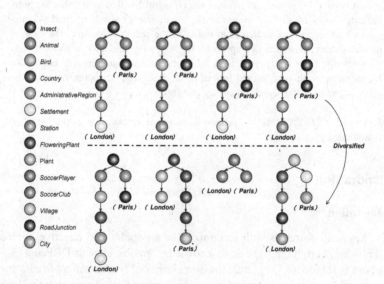

Fig. 1. An example of top-k interpretations with and without diversification.

1.2 Related Work

Keyword Query Interpretation. It is a popular research topic in the communities of semantic web, information retrieval and database. There are generally two kinds of methodologies. The first is to map the keyword query to semantic patterns precomputed from the graph. For example, Pound et al. [5, 6] generates a list of possible elements in the knowledge graph for each keyword phrase in the query, sorts the elements by syntactic similarity, and lastly processes the sorted lists by using a variation of Threshold Algorithm. The second is to search the relationships between keywords in the schema graph and compose patterns. For example, Tran et al. [9, 10] models the interpretation results as subgraphs of the schema graph that connect all keywords in the query. The top-k results ranked with respect to relevance are returned. Overall, the current research works try to find the best individual results but not the best result set.

Diversified Top-k Search. It has been studied in many applications recently, such as [16–18]. The most straightforward solutions (e.g., [1, 2]) assume the rankings of all search results are known in advance, and diversified search algorithms are given to output k results with respect to score and similarity. However, graph search usually returns a very large number of results, so that it is not feasible to rank all results in the context of this paper. Then, Qin et al. [7] propose general frameworks to handle the diversified top-k search problem, which can stop early without exploring all search results. But it cannot be directly applied to solve our keyword query interpretation problem due to the different problem definition.

1.3 Our Contributions

In this paper, we propose a novel diversified top-k keyword query interpretation approach to address the redundancy problem of interpreted results, so that the top-k results could cover richer semantics and satisfy different users. Our approach follows the line of [10] to search for semantic patterns on a schema graph of knowledge graph for a given keyword query, and tries to return k patterns that have the most possible high scores and meanwhile are not similar to each other.

Our main contributions are as follows.

- We reasonably formulate a diversified top-k search problem on a schema graph of knowledge graph for keyword query interpretation.
- We define an effective similarity measure to evaluate the semantic similarity between search results.
- We present an efficient top-*k* algorithm to address the problem. The algorithm can guarantee to return the exact top-*k* results, and minimize the calculation of similarity. Moreover, two heuristic pruning strategies are proposed to improve the search efficiency with possible losses of exact top-*k* results.
- We perform experiments on DBPedia. The results show that our approach improves the diversity of top-*k* results effectively from the perspective of both statistics and human cognition. Meanwhile, with a small loss (on average 3%) of result precision as trade-off, the pruning strategies reduce the response time dramatically.

The rest of this paper is organized as follows. Section 2 introduces the background and problem definition. We present our similarity measure and search algorithm in Sects. 3 and 4 respectively. Section 5 shows the experimental results. Lastly, we conclude in Sect. 6.

2 Preliminaries

2.1 Data and Query Models

Knowledge Graph. Without loss of generality, we simply model a knowledge graph as $G = (V, E, H, K)$ where V is a set of nodes, $E = V \times V$ is a set of edges between the nodes, H is a set of classes of nodes, and K is a set of keywords contained by the nodes. For each node $v \in V$, we denote by $class(v) \in H$ and $keyword(v) \subset K$ its class and set of keywords, respectively.

Schema Graph. Given a data graph G, let its schema graph $G_s = (V_s, E_s)$, where V_s is a set of class nodes, one for each class in H, and $E_s = V_s \times V_s$ is a set of edges between the class nodes. For a node $v \in V$ and a class node $v_s \in V_s$, $v \in v_s$ if and only if $class(v) \subset class(v_s)$, so that V_s is actually a partition of V with respect to the classes of nodes. Moreover, we denote by $keyword(v_s) \subset K$ the union of keywords contained by all nodes of v_s, namely, $keyword(v_s) = \cup_{v \in v_s} keyword(v)$. Lastly, an edge $(v_s, u_s) \in E_s$ if and only if there exist nodes $v \in v_s$ and $u \in u_s$ such that $(v, u) \in E$.

Keyword Query. A keyword query $Q \subseteq K$ is simply a set of keywords. In particular, we call a class node $v_s \in V_s$ keyword node if and only if there exists a keyword $q \in Q$ such that $q \in keyword(v_s)$.

Pattern Tree. Given a schema graph G_s, we interpret a keyword query Q to a set of pattern trees, where each node is an embedding of a class node on the schema graph and the leaf nodes are the keyword nodes of each keyword in Q, the root node is the same class node in the end of each path. Formally, for a pattern tree $T = (V_t, E_t)$, there is a mapping $f : V_t \mapsto V_s$. For each edge $(v_t, u_t) \in E_t$ we have $(f(v_t), f(u_t)) \in E_s$. Moreover, a pattern tree is comprised of $|Q|$ keyword-to-root paths called *search paths*. For a search path $P_q = v_{t1}/\ldots/v_{tn}$ outgoing from the keyword $q \in Q$, we have $q \in keyword(f(v_{t1}))$.

2.2 Scoring Metrics

Interpreting a keyword query by searching on a schema graph usually returns a huge number of results due to the explosive combinations of nodes and edges, most of which are irrelevant. To address the problem, existing keyword query interpretation works have considered a variety of scoring metrics in order to evaluate how well an interpreted result matches the user's information needs. For example, some widely-used metrics are introduced as follows.

Compactness. In the context of keyword search over graphs, a basic assumption is that more tightly-connected nodes comprise a more meaningful answer. For example, the answer trees with less edges or levels are ranked higher. Thus, the pattern trees comprised of shorter search paths are preferred.

Popularity. For each node in the schema graph, we can compute a popularity score by means like PageRank. Like web page ranking, pattern trees that contain more popular nodes should be ranked higher.

Relevance. As a common measure in IR, TF/IDF can also be used to evaluate the relevance of pattern trees to the given keywords. For a keyword node, we can compute an initial relevance score.

For a search path P, we denote by $score(P)$ its score that incorporates path length, popularity of nodes on the path, relevance of keyword node, and even other metrics. The scoring function is featured by (1) the higher the score, the better the search path, and (2) for a search path $P' = P/.../v$ extended from another search path P, we have $score(P') < score(P)$. The details of scoring function are omitted because it is not the focus of this paper.

Based on the scoring function of search path, we evaluate the score of a pattern tree as follows.

$$score(T) = \sum_{P \in T} score(P) \tag{1}$$

Obviously, the scoring function of pattern tree is monotonic in the context of search algorithms, thereby facilitating efficient top-k search algorithm and effective search path pruning during the search (see Sect. 4).

2.3 Problem

Consider a set of results $S = \{T_1, T_2, \ldots\}$. For each $T_i \in S$, the score of T_i is denoted as $score(T_i)$. Given two results $T_i, T_j \in S$, the similarity between them is denoted as $sim(T_i, T_j)$ with $0 \leq sim(T_i, T_j) \leq 1$. The parameter τ is defined as the threshold to determine whether two pattern trees are similar. When $\tau \leq sim(T_i, T_j) \leq 1$, T_i is similar to T_j, and vice versa. The definition of the diversified top-k results is as follows.

Definition 1 (Diversified Top-k Results). Given an integer k with $1 \leq k \leq |S|$, the diversified top-k results of S is S_k such that

1. $S_k \subseteq S$ and $|S_k| \leq k$;
2. for any two results $T_i, T_j \in S_k$ with $T_i \neq T_j$, we have $sim(T_i, T_j) < \tau$, namely, they are not similar to each other;
3. for each result $T_i \in S_k$, if $T_j \in S$ and $score(T_j) > score(T_i)$, we have either $T_j \in S_k$ or $\exists T_l \in S_k$ such that $score(T_l) > score(T_j)$ and $\tau \leq sim(T_l, T_j) \leq 1$.

Example 2. Figure 2 shows an example result set $S = \{T_1, T_2, \ldots, T_7\}$ and the corresponding S_k with $k = 4$. Each node in the figure represents a pattern tree, and the edges indicate that the two connected pattern trees are similar. Moreover, the labels of nodes are their scores. The diversified top-k results S_k includes four nodes T_3, T_1, T_6 and T_7, which are sorted by their scores. For T_2 and T_5, their scores are just too low. For T_4, although its score is higher than T_6 and T_7, it is not qualified for top-k because it is similar to T_3 and T_1. Thus, we diversify the top-k results by abandoning T_4 and importing T_6 and T_7.

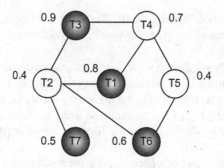

Fig. 2. An example of diversified top-k results.

With Definition 1, the problem addressed in this paper is defined as follows.

Problem. Given a schema graph G_s and a keyword query Q, let S be the set of pattern trees interpreted from Q on G_s, compute the diversified top-k results S_k of S without generating the whole S.

3   Similarity Measure

In order to diversify the pattern trees, we need to measure the similarity between them.

Given a keyword query Q, each pattern tree returned will have $|Q|$ search paths outgoing from each keyword in Q. We firstly abstract each search path as a feature vector. Each feature represents the frequency of a class in the path. Given a schema graph G_s, the dimensionality of feature vector is the number of classes, namely, $|V_s|$. Let $F(P) = (I_1, I_2, \ldots, I_{|V_s|})$ be the feature vector of a search path P. We calculate I_i as follows.

$$I_i = \frac{n_i}{|P|} \tag{2}$$

where $|P|$ is the total number of nodes on the path P and n_i is the number of nodes that belong to the i-th class node of V_s on the path.

Let P_{q1} and P_{q2} be two search paths outgoing from a keyword q in two different pattern trees. We use the angle cosine formula to calculate the similarity of the two paths.

$$sim(P_{q1}, P_{q2}) = \frac{F(P_{q1}) \cdot F(P_{q2})}{|F(P_{q1})| * |F(P_{q2})|} \tag{3}$$

Lastly, the similarity of two pattern trees is the mean of the similarity of all their corresponding paths.

$$sim(T_1, T_2) = \frac{\sum_{P_{q1} \in T_1, P_{q2} \in T_2} sim(P_{q1}, P_{q2})}{|Q|} \tag{4}$$

Intuitively, our similarity function measures how redundant the classes in two pattern trees are. We do not consider the structural similarity measures like tree edit distance because of the high computational complexity and low additional profit.

4 Diversified Top-*k* Search

As mentioned above, we interpret a keyword query by searching its pattern trees on the schema graph. In this section, we present the top-*k* search algorithm and optimization techniques.

4.1 Search Algorithm

The main goal of our algorithm is to avoid the unnecessary similarity comparison, which is relatively expensive and could be used very frequently during the search. It is due to the fact that lots of pattern trees with low scores generated during the search are unlikely to become the top-*k* results and thereby are not needed for similarity comparison. Thus, our algorithm calculates the similarity for a pattern tree only when its score meets a specific condition.

The pseudo codes are given in Algorithm 1. Let C be the candidate set, which is a priority queue of generated pattern trees in descending order of score, and S_k be the diversified top-*k* result set, which is also sorted in descending order of score. We denote by $\overline{unseen}$ the upper bound of score for the unseen results. For simplicity, the function $search()$ is used to traverse the schema graph and generate a set of pattern trees in each iteration. We do not discuss how to schedule the graph traversals and how to generate pattern trees here, which have been well studied by existing research works like [10]. Obviously, the value of $\overline{unseen}$ will decrease gradually, and meanwhile, there will emerge pattern trees with higher scores than $\overline{unseen}$. Then we compute the similarity between the first emerged pattern tree and results in S_k. If it is not similar to any result in S_k, it will be put into S_k. Otherwise, it will be abandoned directly. The algorithm terminates as soon as there are k results in S_k.

It is easy to prove that, each candidate popped up from C is a pattern tree with the highest score in all remaining results, so that it is certainly the next top-*k* result if it is not similar to any existing top-*k* result. Thus, the correctness of Algorithm 1 is guaranteed.

Then we prove that we reduce the cost of similarity comparison to the minimum. The pattern trees with a score higher than the last result in S_k have to be compared with the results in S_k for identifying whether it is qualified for becoming one of top-k. While, the similarity comparison of the rest pattern trees is totally unnecessary to find the top-k results. Our algorithm only calculates the similarity while the score of the pattern is higher than $\overline{unseen}$, which is always no less than the score of the last result in S_k. So our algorithm minimizes the calculation of similarity.

Algorithm 1. DivSA

Input: a schema graph G_s, a keyword query Q
Output: S_k
1:   $C \leftarrow \emptyset$, $S_k \leftarrow \emptyset$, $\overline{unseen} \leftarrow 1$;
2:   **while** $|S_k| < k$ **do**
3:      $temp\_set \leftarrow search()$; //see prune details in Section 4.2
4:      **for each** $T \in temp\_set$
5:         put T into the candidate set C;
6:      **end for**
7:      update $\overline{unseen}$;
8:      **while** $C.peek() > \overline{unseen}$ **and** $|S_k| < k$ **do**
9:         $T \leftarrow$ the first candidate in C;
10:        **if** T is not similar to any results in S_k **do**
11:           put T into S_k;
12:        **end if**
13:     **end while**
14: **end while**

Execution Example. Consider the example in Fig. 2. Table 1 describes the search procedure of DivSA. Initially, the value of $\overline{unseen}$ is 1, and candidate set C and diversified top-k results set S_k are both empty. The pattern trees are generated by calling

Table 1. An example search procedure of DivSA.

iteration #	new results	C	S_k	$\overline{unseen}$
0	$\emptyset$	$\emptyset$	$\emptyset$	1
1	$\{T_1, T_2\}$	$\{T_1, T_2\}$	$\emptyset$	0.9
2	$\{T_3, T_4, T_5\}$	$\{T_3, T_1, T_4, T_2, T_5\}$	$\{T_3\}$	0.85
3	$\{T_6\}$	$\{T_1, T_4, T_6, T_2, T_5\}$	$\{T_3, T_1\}$	0.75
4	$\{T_7\}$	$\{\cancel{T_4}, T_6, T_7, T_2, T_5\}$	$\{T_3, T_1, T_6, T_7\}$	0.45

search() iteratively. After two iterations, T_1, T_2, T_3, T_4 and T_5 have been generated, and $\overline{unseen}$ decreases to 0.85. Then, the pattern tree T_3 ($score(T_3) = 0.9 > 0.85$) is moved to S_k because it is certainly the best of all possible results. In the next iteration, T_6 is generated and $\overline{unseen}$ decreases to 0.75. Then, T_1 ($score(T_1) = 0.8 > 0.75$) becomes the best of rest results, and meanwhile it is not similar to T_3, thereby being moved from C to S_k. When $\overline{unseen}$ decreases to 0.45, there are five pattern trees in C, i.e., $\{T_4, T_6, T_7, T_2, T_5\}$, where T_4, T_6 and T_7 have higher score than $\overline{unseen}$. So they will be compared with the results in S_k one by one. Lastly, T_6 and T_7 are moved to S_k, and T_4 is removed from C because it is similar to T_3.

4.2 Optimization

In order to improve the search efficiency, we propose two rules to prune the search path that is traversed by *search*() in each iteration.

Rule 1. Let m_q be the maximum score of search paths outgoing from the keyword $q \in Q$, and $score_{max}$ be the highest score of currently generated pattern trees. For a new path $P_{q'}$ outgoing from a keyword $q' \in Q$, we prune $P_{q'}$ if $score(P_{q'}) + \sum_{q \in Q, q \neq q'} m_q < \mu * score_{max}$.

The rationality behind Rule 1 is as follows. Firstly, we prove that the left side of the inequality is the upper bound of score of pattern trees that contain the path $P_{q'}$. As indicated in Sect. 2.2, the score of a search path composed of another path and a new edge is certainly less than the original path. So the possibly best pattern tree that contains the path $P_{q'}$ is composed of the current best paths outgoing from each keyword. According to Eq. (1), $score(P_{q'}) + \sum_{q \in Q, q \neq q'} m_q$ is the upper bound.

The right side of the inequality is the estimated lower bound of score of top-k results. The lower bound is to prevent the similarity computation for results with very low scores. The empirical parameter μ is used to balance the prune effect and result completeness. It is possible to return fewer than k results when the value of μ is relatively large. But in practice the returned results can be guaranteed to be complete by carefully tuning the value of μ.

Rule 2. For two search paths P_q and P'_q outgoing from a same keyword q with $score(P_q) > score(P'_q)$, if $sim\left(P_q, P'_q\right) = 1$, we can prune the search path P'_q safely, and if $1 > sim\left(P_q, P'_q\right) > \tau$, we will prune the search path P'_q with a probability $\delta = 1 - \cos^{-1} sim\left(P_q, P'_q\right) / \cos^{-1} \tau$.

First, assume that the pattern trees T_i and T_j contain P_q and P'_q respectively. And the other search paths of T_i and T_j are same. If $score(P_q) > score(P'_q)$ and $sim\left(P_q, P'_q\right) = 1$, then $score(T_i) > score(T_j)$ and $sim(T_i, T_j) = 1$. The similarity between two paths is one do not mean that the two paths are exactly the same. Because the feature vector of the path does not describe all details of the path. The similarity between pattern trees is calculated entirely based on the similarity between paths. So we can know $sim(T_i, T_j) = 1$. The score of the pattern tree is the sum of the score of all its paths. So we can know $score(T_i) > score(T_j)$.

In one case, when T_i is in the S_k, because $score(T_i) > score(T_j)$ and $sim(T_i, T_j) = 1$, T_j can't be a result in the S_k. In another case, when T_i is not a result in the S_k, there must be a pattern tree T_x in S_k having higher score than T_i and $1 > sim(T_i, T_x) > \tau$. Since the feature vector of all paths of T_i and T_j are the same, we can know that $1 > sim(T_j, T_x) > \tau$. And obviously, $score(T_x) > score(T_j)$. So T_j can't be a result in the S_k.

In summary, T_j is not needed in the final results. And that also means that P_q' is not needed during the generation of pattern trees. Because if there is any pattern tree containing P_q', there is always another pattern tree containing P_q making the previous one useless. So we can prune the search path P_q'.

When $1 > sim(P_q, P_q') > \tau$, δ is a good simulation of the possibility of $1 > sim(T_j, T_x) > \tau$ (According to the preceding proof, it is easy to understand that it is also the possibility to prune the search path P_q'). In fact, when P_q is more similar to P_q', P_q' is more likely to be pruned. δ simulates this trend with very good effect. So we use it as a probability to prune the path P_q'.

5 Experiments

5.1 Setup

Dataset. We perform the experiments on DBPedia, a popular real-world RDF dataset which contains over two million entities and nearly ten million relationships. From DBPedia, we extract a schema graph with 272 class nodes, such as "Aircraft", "BaseballPlayer", "ChemicalCompound", etc. and nearly 20 K edges. Our keyword query interpretation approach is to search for pattern trees on the schema graph and generate diverse patterns covering various classes.

Metrics. To evaluate the effectiveness of our approach, we introduce the following two metrics

- *Coverage.* Each top-k pattern tree could be a representative of many other similar patterns. Thus, we try to count how many patterns are covered by (namely, similar to) the top-k pattern trees, which is a reasonable measure of their diversity. Certainly, the more pattern trees covered by the top-k results, the more diverse the results. Formally, we define the coverage of diversified top-k results S_k as

$$coverage(S_k) = \frac{|\{T \in S, LT \in S_k | score(T) \geq score(LT)\}|}{|S_k|} \tag{5}$$

where S is the result set including all the pattern trees once generated and LT is the last pattern tree in S_k.

- *Precision*. Since our probabilistic pruning method could result in losses of exact top-k results, we evaluate the effectiveness of pruning method by using the precision of top-k results, namely, the percentage of "correct" top-k results that are also returned by the original algorithm. Formally, we define the precision of top-k results S_k as

$$precision(S_k) = \frac{|\{P \in S_k | P \in S_k'\}|}{|S_k|} \qquad (6)$$

where S_k' is the set of exact top-k results corresponding to S_k.

5.2 Effectiveness

Test 1. We firstly evaluate the effectiveness in the sense of coverage. We test 50 random keyword queries that have two or three keywords respectively, with $k = 10$ and varying values of τ. As shown in Fig. 3, the value of coverage increases significantly with the decrease of τ. For example, when $\tau = 0.8$, there are averagely 4655 pattern trees represented by our diversified top-10 results of two-keyword queries. Moreover, if the queries have more keywords, the coverage of diversified top-k results is even higher, because there are more redundant pattern trees combined from similar paths.

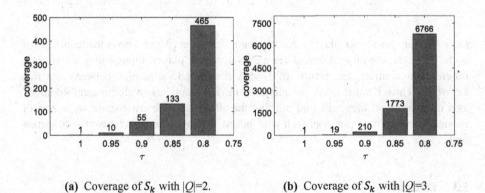

(a) Coverage of S_k with $|Q|$=2.   (b) Coverage of S_k with $|Q|$=3.

Fig. 3. Coverage of S_k.

Test 2. We also evaluate the effectiveness by conducting a user case study. Given a query "Beckham, Ronaldo", Table 2 shows the top-5 results without diversification and Table 3 shows the top-5 results of DivSA with $\tau = 0.9$. We can see that, the top-1 result without diversification is indeed the most desired semantic relationship between

Table 2. A user case study: the top-5 results without diversification.

Root	Paths
Soccer player	Soccer club—Soccer player (Beckham)
	Soccer club—Soccer player (Ronaldo)
Soccer player	Soccer club—Soccer player—Person (Beckham)
	Soccer club—Soccer player (Ronaldo)
Soccer player	Soccer club—Soccer player—Album (Beckham)
	Soccer club—Soccer player (Ronaldo)
Soccer player	Soccer club—Soccer player (Beckham)
	Soccer club—Stadium (Ronaldo)
Soccer player	Soccer club—Soccer player—Person (Beckham)
	Soccer club–Stadium (Ronaldo)

Table 3. A user case study: the top-5 results of DivSA.

Root	Paths
Soccer player	Soccer club—Soccer player (Beckham)
	Soccer club—Soccer player (Ronaldo)
Broadcast network	Country—Administrative region—Settlement (Beckham)
	Country—Administrative region—School (Ronaldo)
Music genre	Album (Beckham)
	Musical artist (Ronaldo)
Book	Language—Book (Beckham)
	Country—Administrative region—School (Ronaldo)
Record LABEL	Album (Beckham)
	Country—Administrative region—School (Ronaldo)

these two famous soccer players, namely, another soccer player who is the teammate of both. However, the other 4 results are all about soccer player, though they are slightly different. In contrast, the results of DivSA reveal rich semantics between the two keywords. Thus, if some users are interested in "Beckham" as an album and "Ronaldo" as a musical artist, they will find out that the album and the artist share some sort of musical genre. Thus, our approach can indeed improve the search results in human sense.

5.3 Efficiency

We compare the efficiency of two algorithms: DivSA1 and DivSA2. DivSA1 is our diversified top-k search algorithm DivSA using Rule 1 for pruning. DivSA2 is the DivSA using both Rule 1 and 2 for pruning.

We test 50 random keyword queries with two or three keywords respectively by using each algorithm with varying values of parameters. The followings are our observations.

1. Figure 4 depict the average response time of all algorithms with $\tau = 0.9$, $\mu = 0.5$ and $k = 20, 40, \ldots, 100$ for queries with two and three keywords respectively. In most of the time, DivSA1 can find the top-k results within tens of seconds, since it reduces the overheads of calculating similarity significantly. Moreover, the optimized DivSA2 is more efficient than DivSA1. Specifically, DivSA2 is averagely 2.43 and 2.92 times faster than DivSA1 when the keyword number is 2 and 3 respectively. It verifies the effectiveness of our pruning strategy.

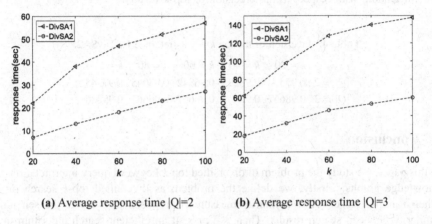

(a) Average response time $|Q|=2$ (b) Average response time $|Q|=3$

Fig. 4. Average response time.

With the increase of the value of k, the response time of both DivSA1 and DivSA2 increases sub-linearly.

With the increase of the keyword number in query, the response time of both DivSA1 and DivSA2 increases rapidly, due to the explosive combinations of paths. Thus, keyword query cleaning is useful when there are many keywords.

2. Figure 5 demonstrate the effectiveness of pruning. Generally, we can see that the number of search paths is reduced significantly by pruning. For the queries with two

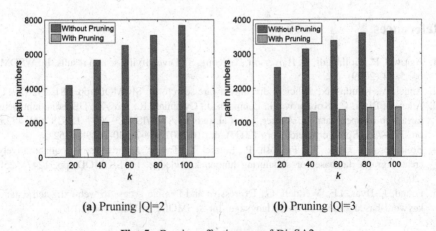

(a) Pruning $|Q|=2$ (b) Pruning $|Q|=3$

Fig. 5. Pruning effectiveness of DivSA2.

keywords, 66% of search paths are pruned. For the queries with three keywords, 59% of search paths are pruned. Meanwhile, although the pruning is heuristic, the precision is testified to be quite high.
3. Since our heuristic pruning is not guaranteed to be safe, we need to testify the precision of returned results. Table 4 shows the average precision of top-k results of DivSA2. We can see that the average precision is generally higher than 95%. Thus, although DivSA2 improves the efficiency dramatically by using unsafe pruning, it is still reliable with respect to the precision of top-k results.

Table 4. Precision of the top-k results returned by DivSA2.

	$k = 20$	$k = 40$	$k = 60$	$k = 80$	$k = 100$
$|Q| = 2$	0.975	0.98295	0.96842	0.97045	0.97545
$|Q| = 3$	0.98667	0.965	0.96667	0.96667	0.95286

6 Conclusion

In this paper, we study the problem of diversified top-k keyword query interpretation on knowledge graphs. Firstly, we define the problem as diversified top-k search on a schema graph. An effective similarity measure is proposed to evaluate the semantic similarity between search results. Then, we present an efficient search algorithm that guarantees to return the exact top-k results and minimize the calculation of similarity. In order to further optimize the algorithm, we propose two heuristic pruning strategies. Lastly, we perform experiments on a real-world knowledge graph to verify the effectiveness and efficiency of our approach.

Acknowledgments. This work was supported by National Natural Science Foundation of China under contracts 61202036, 61572376, 61502349, and 61272110, and by Wuhan Morning Light Plan of Youth Science and Technology under contract 20140727040111250.

References

1. Agrawal, R., Gollapudi, S., Halverson, A., Ieong, S.: Diversifying search results. In: WSDM, pp. 5–14 (2009)
2. Angel, A., Koudas, N.: Efficient diversity-aware search. In: SIGMOD, pp. 781–792 (2011)
3. Auer, S., Bizer, C., Kobilarov, G., Lehmann, J., Cyganiak, R., Ives, Z.: DBpedia: a nucleus for a web of open data. In: Aberer, K., et al. (eds.) ASWC/ISWC -2007. LNCS, vol. 4825, pp. 722–735. Springer, Heidelberg (2007). doi:10.1007/978-3-540-76298-0_52
4. Bollacker, K., Evans, C., Paritosh, P., Sturge, T., Taylor, J.: Freebase: a collaboratively created graph database for structuring human knowledge. In: SIGMOD, pp. 1247–1250 (2008)
5. Pound, J., Ilyas, I.F., Weddell, G.: Expressive and flexible access to web-extracted data: a keyword-based structured query language. In: SIGMOD, pp. 423–434 (2010)

6. Pound, J., Hudek, A.K., Ilyas, I.F., Weddell, G.: Interpreting keyword queries over web knowledge bases. In: CIKM, pp. 305–314 (2012)
7. Qin, L., Yu, J.X., Chang, L.: Diversifying top-k results. In: VLDB, pp. 1124–1135 (2012)
8. Suchanek, F.M., Kasneci, G., Weikum, G.: Yago: a core of semantic knowledge unifying wordnet and wikipedia. In: WWW, pp. 697–706 (2007)
9. Tran, T., Cimiano, P., Rudolph, S., Studer, R.: Ontology-based interpretation of keywords for semantic search. In: Aberer, K., et al. (eds.) ASWC/ISWC -2007. LNCS, vol. 4825, pp. 523–536. Springer, Heidelberg (2007). doi:10.1007/978-3-540-76298-0_38
10. Tran, T., Wang, H., Rudolph, S., Cimiano, P.: Top-k exploration of query candidates for efficient keyword search on graph-shaped (RDF) data. In: ICDE, pp. 405–419 (2009)
11. Wu, W., Li, H., Wang, H., Zhu, K.: Probase: a probabilistic taxonomy for text understanding. In: SIGMOD, pp. 481–492 (2012)
12. Wu, Y., Yang, S., Srivatsa, M., Iyengar, A., Yan, X.: Summarizing answer graphs induced by keyword queries. In: VLDB, pp. 1774–1785 (2013)
13. Zeng, Z., Bao, Z., Le, T.N., Lee, M.L., Ling, W.T.: ExpressQ: identifying keyword context and search target in relational keyword queries. In: CIKM, pp. 31–40 (2014)
14. Zhao, F., Zhang, X., Tung, A.K.H., Chen, G.: BROAD: Diversified keyword search in databases. In: VLDB, pp. 1355–1358 (2011)
15. Zhou, Q., Wang, C., Xiong, M., Wang, H., Yu, Y.: SPARK: adapting keyword query to semantic search. In: Aberer, K., et al. (eds.) ASWC/ISWC -2007. LNCS, vol. 4825, pp. 694–707. Springer, Heidelberg (2007). doi:10.1007/978-3-540-76298-0_50
16. Garbonell, J.G., Goldstein, J.: The use of MMR, diversity-based reranking for reordering documents and producing summaries. In: SIGIR, pp. 335–336 (1998)
17. Demidova, E., Fankhauser, P., Zhou, X., Nejdl, W.: DivQ: diversification for keyword search over structured databases. In: SIGIR, pp. 331–338 (2010)
18. Golenberg, K., Kimelfeld, B., Sagiv, Y.: Keyword proximity search in complex data graphs. In: SIGMOD, pp. 927–940 (2008)

Group Preference Queries for Location-Based Social Networks

Yuan Tian[1], Peiquan Jin[1,2(✉)], Shouhong Wan[1,2], and Lihua Yue[1,2]

[1] School of Computer Science and Technology,
University of Science and Technology of China, Hefei 230027, China
jpq@ustc.edu.cn
[2] Key Laboratory of Electromagnetic Space Information,
Chinese Academy of Sciences, Hefei 230027, China

Abstract. Location-based social networks involve a great number of POIs (points of interest) as well as users' check-in information and their ratings on POIs. We note that users have their own preferences for POI categories. In addition, they have their own network of friends. Therefore, it is necessary to provide for a group of users (circle of friends) a new kind of POI-finding service that considers not only POI preferences of each user but also other aspects of location-based social networks such as users' locations and POI ratings. Aiming to solve this problem, in this paper we present a new type of query called *Spatial Group Preference* (*SGP*) query. For a group of users, an SGP query returns top-k POIs that are most likely to satisfy the needs of users. Specially, we propose a new evaluation model that considers user preferences for user preferences for POI categories, POI properties including locations and ratings, and the mutual influence between POIs. Based on this model, we develop algorithms based on R-tree to evaluate SGP queries. We conduct experiments on a simulation dataset and the results suggest the efficiency of our proposal.

Keywords: Location-based social network · Group preference · Spatial query

1 Introduction

Recently, with the popularity of GPS-enabled smart phones, location-based social network becomes a hot topic. Location-based social networks allow a group of users (circle of friends) to share their location information each other. Together with other information about POIs (points of interest) such as POI locations, POI category labels (i.e., restaurant, hotel, café, etc.), and POI ratings, we can provide a variety of services for people [1].

In this paper, we study an interesting type of query called *Spatial Group Preference* (SGP) query. Given a set of POIs and a group of users, an SGP query retrieves k POIs that are expected to satisfy the overall need of the group of users. We assume that each user in a group has his or her current location and a preference list for POI categories, e.g., restaurant, theater, and shopping mall.

One basic solution to the SGP query is using distance to select the best candidates, which has been widely studied in previous spatial databases. This approach neglects the

© Springer International Publishing AG 2017
L. Chen et al. (Eds.): APWeb-WAIM 2017, Part I, LNCS 10366, pp. 556–564, 2017.
DOI: 10.1007/978-3-319-63579-8_42

users' preferences for POI categories as well as the POI properties (e.g., POI ratings). Many previous studies [11] have proposed to integrate POI properties into POI recommendation. However, these works are not user-aware because they do not consider the preferences of users. To the best of our knowledge, there are very few works that can be directly used to effectively answer SGP queries. As a result, it lacks of effective evaluation models that can evaluate the quality of POIs according to the needs of a group of users.

In this paper, we first give the formal definition on the SGP query. Then, we propose a new evaluation model for SGP queries. Our model integrates different metrics, including distance, POI ratings, user preferences, and influence between POIs, to provide a comprehensive ranking for POI candidates. Based on the model, we propose an R-tree-based algorithm for evaluating SGP queries. Briefly, we make the following contributions in this paper:

- We define a new kind of query called SGP query for location-based social networks (Sect. 3).
- We propose a new evaluation model that considers user preferences for user preferences for POI categories, POI properties including locations and ratings, and the mutual influence between POIs. Compared with existing approaches, this model is user-aware and can provide more reasonable ranking for POIs (Sect. 4).
- We conduct experiments on a simulation dataset as well as on a real dataset with respect to various configurations. The results suggest the efficiency of our proposal (Sect. 5).

2 Related Work

In 2004, Papadias et al. first proposed the concept of group nearest neighbor (GNN) query [2] in spatial databases, which is to find a suitable gathering place for a group of users scattered throughout the Euclidean distance space. They established the R-tree index to organize candidate gathering points and further extended this approach to Aggregate Nearest Neighbor (ANN) query [3]. Both of them studied the group nearest neighbor query in Euclidean distance space, which is not suitable for road networks [4, 5] due to the differences between road networks and the Euclidean distance space. There are other researches [6–10] paying attention to improve the efficiency and to reduce the I/O cost of queries in Euclidean distance space or road network. However, all the above works do not consider group preferences.

In 2007, Yiu et al. proposed a new kind of query called top-k spatial preference query [11]. This query returns top-k candidate objects whose rankings are defined by the quality of other objects around them. Further, the work in [12] proposed an improved scoring method based on textual similarity for getting candidate objects. The idea of considering the influence of near objects was also applied to road networks [13–16]. For example, in [16] the authors mapped distances and the scores of other kinds of objects surrounding the candidate object into a two-dimensional space based on distance and score, and then used the dynamic skyline method to reduce the number of candidate objects. However, existing works on spatial preference queries are not user-aware, meaning that they do not consider the user preferences for POIs.

In 2016, Li et al. studied location-aware group preference queries [17] that are a combination of GNN query and group preference. They return top-k sites that consist of those sites having the minimum distance to the locations of scattered uses in a group and the sites matching the group preference. However, this approach only uses distance to evaluate the group preference queries. It is not suitable for location-based social networks where POI ratings need to be considered in query evaluation.

3 Problem Statement

We assume that there are m categories of POIs, which are represented as $C = \{c_1, c_2, c_3, c_4, \ldots, c_m\}$. Each $c_i (1 \le i \le m)$ represents a category. A set of POIs is represented by $P = \{p_1, p_2, p_3, p_4, \ldots, p_l\}$. Each $p_i (1 \le i \le l)$ is represented by a triple $\langle p_i.loc, p_i.t, s(p_i) \rangle$. Here, $p_i.loc$ is the location of p_i; $p_i.t$ is the category associated with p_i and $s(p_i)$ is the rating of p_i, which is a value between 0 and 1 and can be obtained from social network platforms.

Definition 1 (SGP Query). *Given a set of POIs P, a group of querying users* $Q = \{u_1, u_2, u_3, \ldots, u_n\}$, *a targeted category c, and an integer k, a Spatial Group Preference (SGP) query retrieves a set* $S \subseteq P$ *that consists of k POIs, such that:*

(1) $(\forall x)(x \in S) \rightarrow x.t = c$
(2) $(\forall x)(\forall y)(x \in S \wedge y \in P - S \wedge y.t = c) \rightarrow sd(x, Q) \ge sd(y, Q)$
Here, sd(x, Q) returns the satisfaction degree between x and Q.   □

$Q = \{u_1, u_2, u_3, \ldots, u_n\}$ represents a set of users and each user u_i is denoted as a tuple $\langle u_i.loc, u_i.CW \rangle$, where $u_i.loc$ is the current location of user u_i and $u_i.CW$ represents the preferences for POI categories of the user. To quantify the preferences, we assign a weight to each preferred POI category for each user. Consequently, we have $u_i.CW = \{\langle u_i.c_1, u_i.w_1 \rangle, \langle u_i.c_2, u_i.w_2 \rangle, \ldots, \langle u_i.c_{m'}, u_i.w_{m'} \rangle\}$. Each tuple in $u_i.CW$, e.g., $\langle u_i.c_1, u_i.w_1 \rangle$, means that the category c_1 has a weight of w_1. We assume that $0 \le u_i.w_j \le 1$, yielding $\sum_{j=1}^{m'} u_i.w_j = 1$. In addition, we use C_Q^r to represent all the categories contained in Q, i.e., $C_Q^r = \bigcup_{i=1}^{n} \bigcup_{j=1}^{m'} u_i.c_j$.

The key issue of answering an SGP query is how to define the satisfaction degree model $sd(x, Q)$. In location-based social networks, there are mainly two factors that impact the satisfaction degree of POIs: (1) the distances between the locations of users in the group and candidate POIs labeled with the category given in the query; (2) the influence of surrounding candidate POIs that have categories in C_Q^r. We define two metrics, namely *distance relevance* and *preference relevance*, to formally reflect these two factors in computing satisfaction degree.

Definition 2 (Distance Relevance). *Given a candidate POI p and an SGP query Q, the distance relevance between p and Q is defined by* (1) [17]:

$$\delta(p,Q) = 1 - \frac{d(Q,p)}{D_{max} \cdot |Q|} \tag{1}$$

Here, $d(Q,p)$ represents the sum of the distances of each user to p. D_{max} is the diagonal length of the MBR covered by all the involved POIs. □

Definition 2 (Preference Relevance). *Given a candidate POI p and an SGP query Q, the preference relevance between p and Q is defined by (2):*

$$\tau(p,Q) = \sum_{c_i \in c_Q^r} \frac{1}{|Q|} \cdot \frac{W(c_i) \cdot r_{c_i}^r(p)}{1 + \frac{d(p,p_{c_i}^{\prime r}(p))}{r}} \tag{2}$$

Here, $p_{c_i}^{\prime r}(p)$ is one POI, labeled with c_i, within the range r of candidate POI p. And $r_{c_i}^r(p)$ indicates the comprehensive score of $p_{c_i}^{\prime r}(p)$, which is obtained from social network. We assume there is $0 \le r_{c_i}^r(p) \le 1$. A greater value represents a greater comprehensive score. And $W(c_i)$ indicates the degree of attention of all users in group to category c_i. The calculation of it has been mentioned above. There $d\left(p, p_{c_i}^{\prime r}(p)\right)$ shows the distance between p and p' in Euclidean distance space. Note that gathering category c is the special case of C_Q^r, and $d(p, p_c^{\prime r}(p)) = 0$ this time. The similar form of this formula is used to balance the relations between the textual relevance and network proximity of two objects. Here, we use the score of candidate POI affected by other POIs with group preference around it instead of textual relevance and definition in the denominator is similar to original definition in that paper. □

Definition 3 (Satisfaction Degree). *Given a set of POIs P, a group of querying users $Q = \{u_1, u_2, u_3, \ldots, u_n\}$, the satisfaction degree of p as a gathering point of Q is defined by (3), where α is a smoothing parameter.*

$$sd(p,Q) = \alpha \cdot \delta(p) + (1 - \alpha) \cdot \tau(p)(1) \tag{3}$$

When $\alpha = 1$, the SGP query is similar to the GNN query. If $\alpha = 0$, the SGP query becomes a variant of the top-k spatial preference query. □

4 Algorithms for SGP Queries

The SGP query is a new query that has not been mentioned before, so there is no ready-made method to solve the problem. But there are some basic approaches to it.

4.1 Baseline Algorithm (BA)

In the baseline algorithm, we build Rtrees for POIs in P labeled with each category, respectively. Rtree is one of the most popular and widely used data structure for indexing spatial objects. First of all, traversal the R_c (Rtree of category c) starting from the root node. If it is a non-leaf node, we execute a recursive call to all of its child nodes; else for each POI p in the leaf node, we traversal each R_{c_i} with $c_i \in C_Q^r$ from top to bottom and compute the value according to the model of satisfaction. At the end of the algorithm, it returns top-k POIs labeled with category c, and with the highest satisfaction.

4.2 Pruning Algorithm (PA)

In the BA algorithm, all candidate POIs as well as POIs around them labeled with group preference categories are retrieved. Thus, we present the pruning algorithm to reduce the number of POIs are retrieved. In this method, aRtrees [11] instead of Rtree are established for all POIs in P, according with different categories. The aRtree is similar to Rtree, but is added some aggregate information to each node in Rtree. In this paper, aggregate information tagged to a node is max comprehensive score of POIs in the child nodes of this node. Algorithm 1 shows the details of the pruning process.

Algorithm 1. $f_{pruning}(node\ N)$

1   **If** $sd^+(N) > kth\_bestvalue$ // *pruning strategy 1*
2   **If** N is a non-leaf node
3   execute $f_{pruning}()$ recursively for each child node of N
4   **Else**
5   **For** all entities $e \in N$
6   **For** each $c_i \in C_Q^r \wedge c_i \neq c$ do
7            $Score\_comput(e, N, R_{c_i}.Root)$
8   Compute the value of $sd(e)$ and Update H and $kth\_bestvalue$
    $Score\_comput(Point\ p, Node\ N_1, Node\ N_2)$
9   **If** $sd^+(N_1, N_2) > kth\_bestvalue$
10   **If** N_2 is a non-leaf node then
11   execute $Score\_comput()$ recursively for each child node of N
12   **Else**
13   **For** each entry $e' \in N_2$ and $dist(p, e') \leq r$ do
14   Compute and return the actual value of $\dfrac{1}{|Q|} \cdot \dfrac{w(c_i) \cdot r_{c_i}^r(p)}{1 + \frac{d(p,e')}{r}}$

End $f_{pruning}$

Here, $sd^+(N)$ and $sd^+(N, N')$ are upper bounds of $sd(p, Q)$, where p is in node N. Only if these upper bounds are greater than $kth\_bestvalue$ (a global variable indicating the kth highest satisfaction degree), the algorithm will continue. Firstly, we invoke function $f_{pruning}()$ at the root node of aR_c (indexing all the POIs labeled with category c) and traverse it from top to bottom. Then, *kth_bestvalue* is compared with the upper bound of each node that is visited in the traversal route. This procedure is recursively performed on a non-leaf node until a leaf node is encountered. The *Score_comput()* method is invoked to compute the $\frac{1}{|Q|} \cdot \frac{w(c_i) \cdot r_{c_i}^r(p)}{1 + \frac{d\left(p, p_{c_i}^{r}(p)\right)}{r}}$ in aR_{c_i} for each POI p in the leaf node of aR_c. Finally, H (one global variable that maintains the results to be returned) and *kth_bestvalue* are updated according to $sd(p, Q)$ of p. The implementation of the *Score_comput()* method is similar to that of $f_{pruning}()$. Line 10–11 describes the recursive calls on non-leaf nodes of aR_{c_i} and Line 13–14 indicates how to obtain the value of $\frac{1}{|Q|} \cdot \frac{w(c_i) \cdot r_{c_i}^r(p)}{1 + \frac{d\left(p, p_{c_i}^{r}(p)\right)}{r}}$ for p in aR_c and $p_{c_i}^{r}(p)$ in aR_{c_i}.

4.3 Optimized Pruning Algorithm (OPA)

In fact, same POIs with group preference may be within the ranges r of the adjacent POIs associated with category c. In other words, the satisfaction degree of respective candidate POIs in one leaf node of aR_c may be affected by the same POIs with group preference. Therefore, we take a set of all the POIs in one leaf node N_1 of aR_c, instead of just one POI, within the range r of N_1 one time and compute the component score on aR_{c_i}. We describe this improved pruning scheme in Algorithm 2.

Algorithm 2. *OPA(Node N_1, Node N_2)*

1    $P = \{p | p \in N_1\}$
2   **If** $sd^+(N_1, N_2) > kth\_bestvalue$
3   **If** N_2 is a non-leaf node then
4   execute *Score_comput()* recursively for each child node of N_2
5   **Else**
6   **For** each entry $e \in P$ do
7   **For** each entry $e' \in N_2$ and $dist(e, e') \leq r$ do
8   Compute and return the actual value of $\frac{1}{|Q|} \cdot \frac{w(c_i) \cdot r_{c_i}^r(e)}{1 + \frac{d(e, e')}{rad}}$

End *OPA*

Algorithm 2 is an optimized version of Algorithm 1. Nodes N_1 and N_2 are two parameters of the function, where N_1 is visited leaf node in aR_c and N_2 is root node of a$R_{c_i}(c_i \in C_Q^r \wedge c_i \neq c)$. When a leaf node N_1 is visited when traversing aR_c, its all candidate POIs are obtained one time and stored in a set P. Then, the component scores $\frac{1}{|Q|} \cdot \frac{w(c_i) \cdot r_{c_i}^r(e)}{1 + \frac{d\left(p, p_{c_i}^{r}(e)\right)}{r}}$ of them are computed concurrently at a single traversal of the aR_{c_i}. Obviously, the candidate POI set corresponding to one leaf node in aR_c access less tree nodes in aR_{c_i}, which also speeds up the query.

5 Experiments

We conduct experiments on a simulated data set that contains a set of POIs. Each POI is tagged with a geographical location and a category description. We randomly assign a value between 0 and 1 for each POI to represent the rating in social networks. All algorithms are implemented in Java, and an Intel® Core™ i3-4210 M CPU @2.60 GHz with 8 GB RAM is used for the experiments. The index structure is memory resident, and the maximum number entries of a node is set to 100. In all experiments, we run 100 queries and report the average costs of the queries. The default settings of the parameters are: $C_Q^r = 8$, $n = 8$, $r = 100$, $a = 0.5$, and $k = 8$.

We first evaluate the impact of the number of POIs. The runtime and pruning rates are shown in Tables 1 and 2, respectively. The runtime increases in all algorithms when the number of POIs ranges from 100 k to 2 M. The runtime of BA increases rapidly, because it employs no pruning strategies. OPA gets the best runtime due to its optimization on PA. The pruning rate also increases in all algorithms with the increase of the number of POIs.

Table 1. Runtime (ms) of varying the number of POIs (l)

Algorithm	#POI (l)				
	100000	500000	1000000	1500000	2000000
BA	888.7	8328.24	21188.5	37399.36	56305.26
PA	284.92	876.38	1606.84	2295.94	3151.42
OPA	170	619.2	1173.54	1693.74	2425.86

Table 2. Pruning rate (%) of varying the number of POIs (l)

Algorithm	#POI (l)				
	100000	500000	1000000	1500000	2000000
BA	0	0	0	0	0
PA	16.15	55.14	65.98	71.85	75.35
OPA	16.78	55.79	66.61	72.46	75.96

Next, we study the performance of our algorithms for different numbers of categories. As shown in Table 3, the runtime of all algorithms decreases with the increase of m. In the case of a constant number of POIs that are indexed, the larger the number

Table 3. Runtime (ms) of varying the number of categories (m)

Algorithm	#Categories (m)			
	4	8	16	32
BA	72082.85	25236.1	9400.4	3808.6
PA	3051.68	1595	866.24	501.34
OPA	2281.16	1156.04	631.56	356.54

of POIs is, the smaller average number of POIs labeled with each category is. This leads to the fact that a smaller number of POIs labeled with group-preference categories are retrieved.

6 Conclusion

In this paper we study a new kind of spatial queries named spatial group preference queries in location-based social networks. We define a satisfaction degree model to measure whether a candidate POI meets the group's needs. Then, we propose two algorithms based on R-tree-based pruning strategies. Our preliminary experimental results over a simulation dataset show the efficiency of the algorithms. Our future work will focus on devising new efficient indexes to accelerate SGP queries. We will also evaluate our proposal on real datasets.

Acknowledgements. This work is supported by the National Science Foundation of China (61379037 and 61672479). Peiquan Jin is the corresponding author.

References

1. Jin, P., Cui, T., Wang, Q., Jensen, C.S.: Effective similarity search on indoor moving-object trajectories. In: Navathe, S.B., Wu, W., Shekhar, S., Du, X., Wang, X.S., Xiong, H. (eds.) DASFAA 2016. LNCS, vol. 9643, pp. 181–197. Springer, Cham (2016). doi:10.1007/978-3-319-32049-6_12
2. Papadias, D., Shen, Q., Tao, Y., et al.: Group nearest neighbor queries. In: 20th International Conference on Data Engineering, Boston, MA, USA, pp. 301–312. IEEE (2004)
3. Papadias, D., Tao, Y., Mouratidis, K., et al.: Aggregate nearest neighbor queries in spatial databases. ACM Trans. Database Syst. **30**(2), 529–576 (2005)
4. Yiu, M.L., Mamoulis, N., Papadias, D.: Aggregate nearest neighbor queries in road networks. IEEE Trans. Knowl. Data Eng. **17**(6), 820–833 (2005)
5. Ioup, E., Shaw, K., Sample, J., et al.: Efficient AKNN spatial network queries using the M-tree. In: 15th International Symposium on Advances in Geographic Information Systems, Article 46, Seattle, Washington, USA (2007)
6. Xie, X., Jin, P., Yiu, M., et al.: Enabling scalable geographic service sharing with weighted imprecise voronoi cells. IEEE Trans. Knowl. Data Eng. **28**(2), 439–453 (2016)
7. Li, H., Lu, H., Huang, B., et al.: Two ellipse-based pruning methods for group nearest neighbor queries. In: 13th Annual ACM International Symposium on Advances in Geographic Information Systems, Bremen, Germany, pp. 192–199 (2005)
8. Li, F., Yao, B., Kumar, P.: Group enclosing queries. IEEE Trans. Knowl. Data Eng. **23**(10), 1526–1540 (2011)
9. Li, Y., Li, F., Yi, K., et al.: Flexible aggregate similarity search. In: 2011 International Conference on Management of Data, Athens, Greece, pp. 1009–1020. ACM (2011)
10. Yan, D., Zhao, Z., Ng, W.: Efficient processing of optimal meeting point queries in Euclidean space and road networks. Knowl. Inf. Syst. **42**(2), 319–351 (2015)

11. Yiu, M.L., Dai, X., Mamoulis, N., et al.: Top-k spatial preference queries. In: 23rd International Conference on Data Engineering, Istanbul, Turkey, pp. 1076–1085. IEEE (2007)
12. Gao, Y., Wang, Y., Yi, S.: Preference-aware top-k spatio-textual queries. In: Song, S., Tong, Y. (eds.) WAIM 2016. LNCS, vol. 9998, pp. 186–197. Springer, Cham (2016). doi:10.1007/978-3-319-47121-1_16
13. Rocha-Junior, J.B., Gkorgkas, O., Jonassen, S., Nørvåg, K.: Efficient processing of top-k spatial keyword queries. In: Pfoser, D., Tao, Y., Mouratidis, K., Nascimento, Mario A., Mokbel, M., Shekhar, S., Huang, Y. (eds.) SSTD 2011. LNCS, vol. 6849, pp. 205–222. Springer, Heidelberg (2011). doi:10.1007/978-3-642-22922-0_13
14. Cho, H., Kwon, S., Chung, T.: ALPS: An efficient algorithm for top-k spatial preference search in road networks. Knowl. Inf. Syst. **42**(3), 599–631 (2015)
15. Attique, M., Cho, H.J., Jin, R., et al.: Top-k spatial preference queries in directed road networks. ISPRS J. Geo-Inf. **5**(10), 170 (2016)
16. Rocha-Junior, J.B., Vlachou, A., Doulkeridis, C., Nørvåg, K.: Efficient processing of top-k spatial preference queries. Proc. VLDB Endow. **4**(2), 93–104 (2010)
17. Li, M., Chen, L., Cong, G., et al.: Efficient processing of location-aware group preference queries. In: 25th International on Conference on Information and Knowledge Management, Indianapolis, Indiana, USA, pp. 559–568. ACM (2016)

A Formal Product Search Model with Ensembled Proximity

Zepeng Fang[1], Chen Lin[1,2]([✉]), and Yun Liang[3]

[1] Department of Computer Science, Xiamen University, Xiamen 361005, China
zpfang@stu.xmu.edu.cn
[2] Shenzhen Research Institute of Xiamen University, Shenzhen 518057, China
chenlin@xmu.edu.cn
[3] College of Mathematics and Informatics, South China Agricultural University,
Guangzhou 510642, China
sdliangyun@163.com

Abstract. In this paper we study the problem of product search, where products are retrieved and ranked based on how their reviews match the query. Current product search systems suffer from the incapability to measure correspondence between a product feature and its desired property. A proximity language model is presented to embed textual adjacency in the frequency based estimation framework. To tailor for product search problem, we explore strategies for distinguishing product feature and desired property, quantifying pair-wise proximity based on conditional probability, and aggregating review opinions at product level. Experiments on a real data set demonstrate good performances of our model.

Keywords: Information Retrieval · Proximity model · Product search

1 Introduction

Recently, there has been increasing interest in the problem of product search, due to the abundance of online reviews. Product review sites, such as TripAdvisor[1], Yelp[2], have attracted numerous users, and thus have generated an incredible volume of comments. Unfortunately, it is impossible for users to absorb all the information for every candidate product. Product search is then considered to be a prominent tool to explore online reviews and to make smart consumptions.

Product search queries usually consists of consumption preferences on multiple product features. The goal of product search engine is to locate the right product reviews, and rank them based on how they meet people's demands. For example, the query in Fig. 1 seeks for a restaurant with nice decor that serves hot pot. Product B is relevant, since the second review is a supporting evidence which explicitly states that the two features of restaurant satisfy the query need.

---

[1] http://www.tripadvisor.com/
[2] http://www.yelp.com/

© Springer International Publishing AG 2017
L. Chen et al. (Eds.): APWeb-WAIM 2017, Part I, LNCS 10366, pp. 565–572, 2017.
DOI: 10.1007/978-3-319-63579-8_43

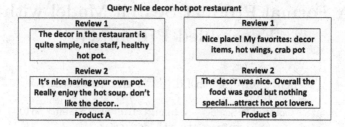

Fig. 1. An illustration of a product search problem

To retrieve preferred products from online reviews, several approaches [1,2] have been proposed, most of which are based upon a probabilistic model that measures query-review relevance. They consider a review as supporting evidence, if all the query keywords appear in the review. This type of models is limited, as it treats the query as a plain, unstructured "bag of words", and does not distinguish the pair-wise correspondence between preferences and features. As illustrated in Fig. 1, the first review of product A contains all query keywords. Nevertheless, it is not a supporting evidence. Because the preferred opinion "nice" does not correspond to "decor", instead it is used to describe the feature "staff".

The key issue in product search systems is to quantify the relevance between the desired property and the corresponding feature in the reviews. One may easily recognize that relevance is reflected by textual adjacency. Information Retrieval (IR) models incorporating term proximity like PLM [3] and BM25P [4] have been shown to significantly enhance the performance of IR systems. However, directly applying them for product search is problematic. On one hand, these models hold a constraint of closeness to all query terms, which might be too strict to harm the accuracy of product ranking. For example, in Fig. 1, the average proximity scores for all query terms in review 2 for product B is in fact the largest of four reviews. But since both the preferred opinion keywords are close to the feature keywords ("nice" to "decor", "hot" to "pot"), this review is a supporting evidence. On the other hand, given a pair of product feature and preference, traditional IR models are only capable of capturing their association at document level, while product search is implemented at entity level. Therefore, it is necessary to quantify the overall degree of association.

In this paper, we present a formal model to address the above two problems. Following the classic framework of language modeling, we compute the likelihood of observing the query for each product. The query likelihood is factorized to conditional probability of preferred opinion given the product feature. We study the estimation of conditional probability and present several strategies to embed proximity in estimation. Furthermore, we study the effect of aggregated proximity from the review corpus.

2   Related Work

The great potential of entity search [5] has been acknowledged in recent years. Typical entity search paradigms include expert search [6], query driven product retrieval [1,2,7]. Product retrieval are either built upon a keyword search framework [2], or a probabilistic language model framework [1]. Sources for product search are mainly product profiles and online reviews. When retrieving product from reviews, a critical property is that reviews are opinionated on different product aspects, and thus demands special treatment. In [1], relevance is evaluated by aggregating over all query aspects and achieve a good effect. To measure relevance of product aspect and opinion, a new indexing unit, Maximal Coherent Semantic Unit is defined and employed in the ranking process [7].

Language modeling approaches have been extensively studied in IR community, e.g. query likelihood, divergence and relevance and so on [8]. Beyond general frameworks for computing unigram relevance, one may also want to reward documents in which query terms appear close to each other. To exploit such proximity heuristics, some researchers attempt to capture word dependency by utilizing a larger matching unit, e.g. bigram [9]. To avoid making the indexing space too sparse, Markov Random Field model [10] is presented to collectively score unigram, bigram and textual unit within a certain window. Other researchers incorporate query term proximity into an existing retrieval model either directly or indirectly. Directly applying term proximity usually involves defining a combination of relevance score from existing retrieval models and adjusting scores from proximity heuristics [11]. Indirect methods embed proximity measures and term frequencies in a unified model. In [3], a language model for each position of a document that takes into account propagation of word count from other places. In [4], it extends the well-established BM25 [12] model by taking a linear combination of Ngram proximity based BM25 models for different N. This line of researches also include CRTER [13], which introduces a pseudo term that is the combination of the individual terms, and is weighted by the intersection of impact propagated from each individual at different positions.

3   Model

Let $D = \{d_1, d_2, \cdots, d_N\}$ be the product universe and $C = \{R_d\}$ is the collection of all reviews, where R_d is the set of review documents for product d. The query consists of several preference phrases on multiple product features, $q = \{(o, f)\}$ in which o denotes the preferred opinion terms and f represents the corresponding feature keywords. Our goal is to estimate the likelihood of generating query from the hidden product model.

$$p(q|d) = \Pi_{(f,o)}p(f,o|d) = \Pi_{(f,o)}p(f|d)p(o|f,d) \tag{1}$$

The first part $p(f|d)$ is the probability of selecting feature f from the product, which can be estimated by Dirichlet Prior smoothing with parameter μ.

The second part $p(o|f,d)$ defines the relevance between an opinion o and a feature f in d's reviews. In the next subsection, we will elaborate how to incorporate term proximity between opinion and feature keywords into its estimation.

3.1 Conditional Probability Estimation

Proximity Parameterized The first strategy **PP** is to represent $p(o|f,d)$ as the probability density function with respect to term proximity $d(o,f,R_d)$. We assume that given the feature f, the author will select opinion o according to a Gaussian distribution $p(o|f,d) \sim N(0,\sigma^2)$. Note that the probability for a Gaussian achieves its maximum at its mean, and decreases as the value is distant from the the mean. Therefore, if the distance between opinion term o and the feature word f is smallest, then we will get the maximum $p(o|f,d)$. The above observations lead to the following functional form

$$p(o|f,d) = \frac{1}{\sqrt{2\pi}\sigma} \exp(-\frac{d(o,f,R_d)^2}{2\sigma^2}) \tag{2}$$

Proximity Adjusted. Another strategy **PA** is to first compute the probability $p(o|f,d)$, then modify it by the proximity. With Jelinek-Merccer smoothing, we have $p(o|f,d) = (1-\lambda)\frac{c(o,f,R_d)}{c(f,R_d)} + \lambda\frac{c(o,f,C)}{c(f,C)}$, where λ is the parameter.

However, the above definition depends only on the co-occurrences, thus ignores the impact of proximity. We employ an exponential weighting scheme to simulate the negative correlation between terms proximity and conditional probability, so that the confidence of a frequency-based estimation $c(o,f,R_d)$ decreases as the absolute distance increases. In order to guarantee the adjusted function is a probability, i.e. $\Sigma_o p(o|f,d) = 1$, the dominator $c(f,R_d)$ should be regularized accordingly. Note that, $\int_{-\infty}^{+\infty} \exp{-x^2}dx = \sqrt{\pi}$. Therefore, we have

$$p(o|f,d) = (1-\lambda)\frac{c(o,f,R_d)\exp\left(-d(o,f,R_d)^2\right)}{c(f,R_d)\sqrt{\pi}} + \lambda\frac{c(o,f,C)}{c(f,C)} \tag{3}$$

Proximity Censored. Finally we consider probability estimation by directly manipulating the event space. As in the **PA** strategy, $p(o|f,d)$ is proportional to the co-occurrence frequency of the opinion term o and feature keyword f. But such a relatedness is actually invalid if the two words are far away. Therefore we define the event of observing terms o,f in a text window of size ϵ as $c_\epsilon(o,f,R_d)$. The probability, according to strategy **PC**, is defined as

$$p(o|f,d) = \frac{c_\epsilon(o,f,R_d)}{c(f,R_d)} \tag{4}$$

where $c_\epsilon(o,f,R_d) = |\{d(o,f,R_d) < \epsilon\}|$, and obviously $\Sigma_o p(o|f,d) = 1$

3.2 Proximity Aggregation

For a document, the proximity $d(o,f)$ is the absolute difference of the positions of terms o and f. Because both terms o,f might appear at multiple positions in the product reviews R_d, we need to study the aggregation of term proximity.

Min strategy returns the minimal proximity between the opinion and feature in a product. Intuitively, the **Min** strategy suggests that the most significant evidence is adopted, i.e. if only one consumer gives positive feedback, the product will be regarded as relevant.

Avg strategy measures average proximity of two terms in the product reviews. This strategy assumes that all reviews matter, i.e. the product is deemed relevant when the overall feedbacks are good.

Max strategy calculates the maximal proximity between an opinion and its neighboring feature in the product specific reviews. Intuitively, the **Max** strategy only considers the weakest evidence, i.e. the product is relevant if the most critical consumer speaks highly of it.

The above three heuristics are simple and intuitive. The **Min** strategy and **Max** strategy are based on single evidence instead of the collective opinions. The **Avg** strategy is a global measurement, but it is still sensitive to outliers. A typical problem for mining social documents is that it usually involves diverse social behaviors. In the setting of product search, reviews for a specific product generally contain quite distinct or even opposite opinions. As a result, we may need a more accurate measurement to reflect the collected opinions on the required product features. Thus, we present a **ClusterMin** strategy as follows.

First we represent each review $r \in R_d$ as multiple $V-$dim vectors r^o, each for a product feature in the query $o \in q$. The $v-$th element of the feature specific review vector is the minimal distance $r_v^o = \min_{v \in r} d(o, v)$ between the given product feature o and the $v-$th word of the opinion lexicon in the review. We adopt K-means algorithm to cluster all reviews $r \in R_d$ for product d. The centroid of each cluster is represented by multiple feature specific vectors. In the assigning step, calculate the distance between a review and a centroid s^k in cluster k by the combination of feature specific Euclidean distance $\|r - s^k\|_2 = \Sigma_o \|r^o - s^{o,k}\|_2$. As opinions are naturally divided into three categories: positive, neutral and negative, we set $K = 3$ for clustering. When the clustering converges, we choose the cluster centroid which has the nearest query specific feature-opinion distance $s^i = s^k | \min_{k=1 \to 3} \Sigma_{<o,f> \in q} s_f^{o,k}$, and set the aggregated distance as the corresponding element defined by the cluster centroid $d(o, f) = s_f^{o,i}$.

4   Experiment

We evaluate our model with the open benchmark [1]. The data set consists of reviews for hotels in different cities and cars of various years. For the purpose of this study, only queries which contain both opinions and associated product features are remained. Statistics of experimental data set is shown as Table 1.

The proposed retrieval model is implemented on the Terrier [14] platform. To enhance efficiency, reviews for a single product are merged to a unified document unit. Term distance across reviews is assigned a large value 400. In preprocessing, stop words are not removed, and porter stemmer is adopted.

We adopt NDCG@10 (Normalized Discounted Cumulative Gain) as evaluation metric in the following experiments.

Table 1. Statistics of data set

Hotels		Cars	
No. cities	5	No. years	3
Avg. no. hotels	143.2	Avg. no. cars	199.3
Avg. no. reviews per hotel	60	Avg. no. reviews per car	67.7
Avg. document length	1219.4	Avg. document length	1097.3
Avg. no. queries	5	Avg. no. queries	5

4.1   Conditional Probability Strategy

We first tune the smoothing parameter μ for each strategy. The proximity aggregation strategy is fixed to be **Min**. As shown in Fig. 2(a), (b) and (c), the smoothing parameter μ for good performance tends to be large. A larger μ indicates that the feature probability relies more on the global feature probability $p(f|C)$. It is reasonable since the product specific reviews are associated with a limited number of features. Also, we observe that the effect of decaying confidence in the **PA** significantly shrinks the smoothing parameter μ as shown in Fig. 2(b).

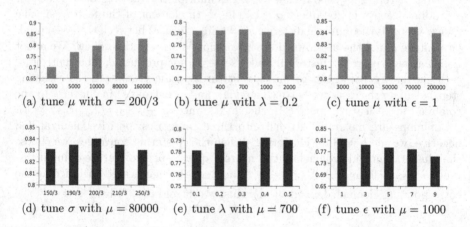

(a) tune μ with $\sigma = 200/3$ (b) tune μ with $\lambda = 0.2$ (c) tune μ with $\epsilon = 1$

(d) tune σ with $\mu = 80000$ (e) tune λ with $\mu = 700$ (f) tune ϵ with $\mu = 1000$

Fig. 2. Parameter tuning for **PP**, **PA** and **PC** strategy

Next we report the effect of strategy specific parameters in Fig. 2(d), (e) and (f). Best performance for **PP** is achieved when σ is around $200/3$. For a Gaussian distribution with zero mean, 99.7% of the data points are in the range of $[-3\sigma, +3\sigma]$, which is in line with our assumption that each product feature should at least appear in the same passage with the corresponding opinion (passage length $= 400$ as mentioned above). Best performance for **PA** is obtained when $\lambda = 0.4$, which suggests the confidence plays a more significant effect than the global statistic. For **PC** strategy, $\epsilon = 1$ performs best, which means when an opinion is directly adjacent to it associating feature, it is the most relevant.

4.2 Proximity Aggregation Strategy

We next study the performance of four aggregation strategies. The comparison is carried in the hotel data set, Beijing category. Parameters $\mu = 80000, \sigma = 200/3$ for **PP**, $\mu = 700, \lambda = 0.2$ for **PA**. As shown in Table 2, **Min** and **ClusterMin** both achieve satisfying results, which verifies the most significant evidence in the reviews play a great role in the calculation of relevance.

Table 2. The performance of four aggregation strategies

	Min	Avg	Max	ClusterMin
PP	0.8400	0.5201	0.5285	0.9086
PA	0.7871	0.4803	0.5274	0.7514

4.3 Comparative Study

We finally analyze the performances of our work, compared with state-of-the-art systems including (1) traditional IR models BM25 [12], PL2 [14]; (2) positional models BM25P [4], CRTER [13], MRF [10] and PLM [3]. These models are not dedicated to entity search scenarios, thus we first rank the reviews, and then re-rank the products by counting the number of reviews for each product in the top 100 search results; (3) product search framework, i.e. OpinionRank [1]. Parameters $\mu = 80000, \sigma = 200/3$ for **PP**, $\mu = 700, \lambda = 0.4$ for **PA**, $\mu = 50000, \epsilon = 1$ for **PC**. And we use **Min** aggregation for all strategies. From Table 3, we have the following conclusions. (1) In general, positional models outperform traditional IR models, which highlight the importance of proximity constraints in IR system. (2) A unified model designed for product search performs significantly better than the re-ranking scheme based retrieval. (3) Our models are comparable to OpinionRank. Among the three paradigms, **PP** is most stable and generally obtains best results, which verifies the competency of our contribution.

Table 3. Performance of various product search systems

Models	BM25	PL2	rrBM25P	rrCRTER	rrMRF	rrPLM	OpinionRank	PP	PA	PC
Hotels										
Beijing	0.5179	0.5234	0.7753	0.7620	0.7685	0.8276	**0.8521**	0.8346	0.7927	0.8472
Dubai	0.6160	0.6228	0.6450	0.7066	0.6400	0.8106	0.8401	**0.8579**	0.7149	0.8246
New-Delhi	0.4323	0.4360	0.6319	0.6532	0.6576	0.7462	**0.8130**	0.8045	0.6820	0.7345
San-Francisco	0.4551	0.4619	0.7839	0.6532	0.7603	0.8227	0.8130	**0.8702**	0.8328	0.8274
Shanghai	0.5097	0.5206	0.7249	0.6865	0.7603	0.8178	0.8239	**0.8276**	0.7460	0.7849
Average	0.5062	0.5129	0.7122	0.6923	0.7173	0.8050	0.8284	**0.8389**	0.7537	0.8037
Cars										
2007	0.8908	0.8900	0.9133	0.9259	0.9152	0.9349	**0.9458**	0.9443	0.9369	0.9198
2008	0.8781	0.8788	0.9174	0.9167	0.9257	0.9308	0.9347	**0.9376**	0.9248	0.9179
2009	0.9176	0.9163	0.9129	0.9256	0.9257	0.9186	0.9494	**0.9526**	0.9430	0.9320
Average	0.8955	0.8950	0.9145	0.9227	0.9222	0.9281	0.9429	**0.9453**	0.9349	0.9233

5   Conclusion

In this paper we present a positional language model for product search problems. Our contributions are two-fold: (1) we incorporate pairwise proximity into the estimation of conditional probability of generating an opinion given a product feature; (2) we explore the aggregation strategies to ensemble review evidences to evaluate the relevance of a product. Experiments on real data set verify the competence of the presented framework. In the future, we plan to extend the model to tolerate noisy query segmentation. Also, clustering on the fly is a potential direction to speed up the computation for **ClusterMin** aggregation.

Acknowledgment. Chen Lin is partially supported by China Natural Science Foundation under Grant No. NSFC61472335, Base Research Project of Shenzhen Bureau of Science, Technology, and Information under Grand No. JCYJ20120618155655087, Baidu Open Research Grant No. Z153283. Yun Liang is supported by the Science and Technology Planning Project of Guangdong Province (2016A050502050), and the NSFC-Guangdong Joint Fund (U1301253).

References

1. Ganesan, K., Zhai, C.: Opinion-based entity ranking. Inf. Retrieval **15**(2), 116–150 (2012)
2. Duan, H., Zhai, C., Cheng, J., et al.: Supporting keyword search in product database: a probabilistic approach. Proc. VLDB Endow. **6**(14), 1786–1797 (2013)
3. Lv, Y., Zhai, C.: Positional language models for information retrieval. In: SIGIR, pp. 299–306 (2009)
4. He, B., Huang, J.X., Zhou, X.: Modeling term proximity for probabilistic information retrieval models. Inf. Sci. **8**(14), 3017–3031 (2011)
5. Yao, T., Liu, Y., Ngo, C.W., et al.: Unified entity search in social media community. In: WWW, pp. 1457–1466 (2013)
6. Balog, K., Azzopardi, L., de Rijke, M.: Formal models for expert finding in enterprise corpora. In: SIGIR, pp. 43–50 (2006)
7. Choi, J., Kim, D., Kim, S., et al.: CONSENTO: a new framework for opinion based entity search and summarization. In: CIKM, pp. 1935–1939 (2012)
8. Zhai, C.: Statistical language models for information retrieval a critical review. Found. Trends Inf. Retrieval **2**(3), 137–213 (2008)
9. Srikanth, M., Srihari, R.: Incorporating query term dependencies in language models for document retrieval. In: SIGIR, pp. 405–406 (2003)
10. Metzler, D., Croft, W.B.: A Markov random field model for term dependencies. In: SIGIR, pp. 472–479 (2005)
11. Büttcher, S., Clarke, C.L.A., Lushman, B.: Term proximity scoring for ad-hoc retrieval on very large text collections. In: SIGIR, pp. 621–622 (2006)
12. Robertson, S.E., Walker, S., Hancock-Beaulieu, M., et al.: Okapi at TREC-4. NIST SPECIAL PUBLICATION Sp. 73–96 (1996)
13. Zhao, J., Huang, J.X., Ye, Z.: Modeling term associations for probabilistic information retrieval. ACM Trans. Inf. Syst. **32**(2), 7:1–7:47 (2014)
14. Ounis, I., Amati, G., Plachouras, V., et al.: Terrier: a high performance and scalable information retrieval platform. In: SIGIR Workshop on Open Source Information Retrieval (2006)

Topic Modeling

Incorporating User Preferences Across Multiple Topics into Collaborative Filtering for Personalized Merchant Recommendation

Yunfeng Chen[1], Lei Zhang[2], Xin Li[3], Yu Zong[4,5], Guiquan Liu[1(⊠)],
and Enhong Chen[1]

[1] University of Science and Technology of China, Hefei, China
cyunfeng@mail.ustc.edu.cn, {gqliu,cheneh}@ustc.edu.cn
[2] Anhui University, Hefei, China
zl@ahu.edu.cn
[3] IFlyTek Research, Hefei, China
xinli2@iflytek.com
[4] West Anhui University, Lu'an, China
Nick.zongy@gmail.com
[5] Texas A&M University, College Station, USA

Abstract. Merchant recommendation, namely recommending personalized merchants to a specific customer, has become increasingly important during the past few years especially with the prevalence of Location Based Social Networks (LBSNs). Although many existing methods attempt to address this task, most of them focus on applying the conventional recommendation algorithm (e.g. Collaborative Filtering) for merchant recommendation while ignoring harnessing the hidden information buried in the users' reviews. In fact, the information of user real preferences on various topics hidden in the reviews is very useful for personalized merchant recommendation. To this end, in this paper, we propose a graphical model by incorporating user real preferences on various topics from user reviews into collaborative filtering technique for personalized merchant recommendation. Then, we develop an optimization algorithm based on a Gaussian model to train our merchant recommendation approach. Finally, we conduct extensive experiments on two real-world datasets to demonstrate the efficiency and effectiveness of our model. The experimental results clearly show that our proposed model outperforms the state-of-the-art benchmark approaches.

1 Introduction

With the prevalence of Location Based Social Networks (LBSNs), personalized merchant recommendation has become very popular and attracted much attention from industry and academia. Personalized merchant recommendation not only satisfies users personalized preferences for visiting new merchants, such as restaurants, stores and theatres, but also increases merchants revenues. As customers can easily make their evaluation by rating and express their opinions

© Springer International Publishing AG 2017
L. Chen et al. (Eds.): APWeb-WAIM 2017, Part I, LNCS 10366, pp. 575–590, 2017.
DOI: 10.1007/978-3-319-63579-8_44

POI

Meskerem
Category:Restaurant

601 South Kings Drive
Charlotte, NC 28204

Rating & Review for User 1

⊙⊙⊙⊙○
What a nice **treat**! The bread was **fresh** and seasoned with **Italian** herbs, and **served** with olive oil & red pepper flakes, yummy! The **shrimp** and **crab** risotto was **delicious**, creamy & buttery, a true risotto. We also ordered the Pollo Alla Griglia, a seasoned grilled chicken, **served** with mashed potatoes & the vegetable , which I substituted for the grilled asparagus, **delish**! The **service** was great and the **wine** selection perfect!

Rating & Review for User 2

⊙⊙⊙⊙○
I have been here twice recently. I love the **patio** here and enjoy the overall feel of it. I have had the **spaghetti** with meat sauce and the ravioli with meat sauce. The food is awesome and the **service** is good as well. I noticed they have a **pizza** oven as a center piece so I would love to try their **pizza** next time. But overall a great find for a really **decent price**.

Fig. 1. Example of Yelp rating and review.

freely by writing their reviews on merchants, mining user real preferences on various topics from user reviews will be useful for personalized merchant recommendation. To solve the data sparsity and the cold start problem in real recommendation system, some methods further incorporate contextual information, such as social networks, temporal information and geographical location to alleviate the cold start problem [3, 10, 24].

Though current techniques achieve great progress to alleviate these problems, it is still very difficult to learn user preferences across various topics from user reviews. As we know, review texts may consist of many topics. From all the reviews written by a specific user, we can infer what topics the specific user gives more attention to. However, users with same rating on same merchant may have different preferences on different topics. Let us consider a real restaurant example shown in Fig. 1 where two users have same rating on same restaurant. From the perspective of the user, we can infer that user 1 pays more attention to the *flavour of the food* and the *service*, while user 2 pays more attention to the *price* and the *service*. From the perspective of the merchant, we can conclude that this restaurant may be an Italian restaurant, which offers seafood with a good *service* and charges a decent *price*. So we can come to the conclusion that the rating reflects user general evaluation on the merchant, while the review reflects user preferences across many topics. Therefore, both the rating and the review should be integrated for merchant recommendation.

In this paper, we propose an integrated model by fusing user ratings and user real preferences on various topics from user reviews in a unified framework, named "Collaborative Filtering meets Reviews" (CFMR), which combines Collaborative Filtering technique and Topic Modeling method based on a graphical model. We assume that the overall rating consists of two parts. One part is the collaborative rating, which is based on the rating behavior of users in the past.

The other is the latent rating, which is based on the current user's preference. The experimental results on two real-world datasets show that our proposed model outperforms the state-of-the-art benchmark approaches in terms of effectiveness and efficiency.

The structure of the remainder of this paper is as follows. In Sect. 2 we will give an overview of the related work. In Sect. 3 we will describe the problem definition. Then, in Sect. 4, we will detail our new model in the form of graphical model and the concrete algorithm for merchant recommendation based on our model. In Sect. 5 we will present and discuss the experimental result on two datasets. The settings of the parameters is also included in this section. Finally, we will conclude our main contributions and discuss the future work in Sect. 6.

2 Related Work

In this section, we will review a number of existing works on recommender systems.

Recommender systems can be generally classified into two groups: Content-based Filtering [20] and Collaborative Filtering [8]. The latent factor model based on matrix factorization is one of the typical models in Collaborative Filtering, such as basic Matrix Factorization [14] and Probabilistic Matrix Factorization [21]. In the recent years, some additional information, such as item taxonomy information [23] and social networks [17], have been incorporated to improve the prediction quality. In [6], the authors presented the performance of matrix factorization algorithms and memory based models on the task of recommending long tail items.

With the prevalence of mobile devices and Web 2.0 technologies, location-based social networks (LBSNs) allow users to share their experiences and opinions on merchants. There are a lot of applications, such as user behavior study [25] and online retail store placement [12] based on Points-of-Interest (POI) recommendation. In recent years, many additional information have been explored to improve the recommendation quality for POI recommendation in LBSNs, such as geographical location [4] and social connections [5]. Our approach differs from previous methods in that we try to learn user preferences by mining the reviews.

As review texts contain very useful information to learn user preferences, there are a few efforts to combine this information to make predictions for ratings. In the work [9], the authors created bag of words from the top frequent words in all reviews and proposed the combination of Linear Regression model. The authors of [11] proposed the decomposition of hidden topics over all reviews and predicted rating over every hidden topic. In [19], the authors proposed a generic rating prediction model by incorporating user preferences in sentiment rating prediction to improve the correlation with ground ratings. The most related study to our work is the work of [16] where the boot-strapping aspect segmentation algorithm proposed in [22] is directly used to define the meaningful aspects. However, the aspect representation based on the boot-strapping aspect segmentation algorithm proposed in [22] needs to provide pre-specified aspect keywords by users. The major difference between our work and these review based methods is that, we proposed a graphical model by fusing Collaborative Filtering technique and Topic Modeling approach to learn user preferences without users' specification of aspect keywords.

3   Notation and Definition

In this section, we introduce the relevant notations and the formal definitions. The input is a set of ratings of some merchants, where each rating has a review. All the notations we used in this paper are shown in Table 1.

Formally, let $D = \{\mu_1, \ldots, \mu_N\}$ be the set of users. Let $P = \{v_1, \ldots, v_M\}$ be the set of merchants. A rating R_{ij} user μ_i gives to merchant v_j indicates the preference the user μ_i shows for the merchant v_j, where high rating means the high preference. Here, ratings are integers ranging from 1 (star) indicating few interest to 5 (star) indicating strong interest. The rating R_{ij}^c indicates the collaborative rating which is achieved by general Collaborative Filtering method. The rating R_{ij}^d indicates the latent rating which is mined in the reviews. The notation d_j^i represents the review written by user μ_i for merchant v_j, which implicitly indicates the user preferences and the merchant quality. We aggregate all the reviews written for the merchant v_j, i.e., $\mathbf{d}_j = \{d_j^1, \ldots, d_j^i, \ldots, d_j^N\}$, to analyze the quality of the merchant v_j. In order to understand our proposed model, we introduce the following definitions.

Table 1. Mathematical notations.

Symbol	Size	Description
D	N	All the users
P	M	All the merchants
R	$N \times M$	Overall rating matrix
R^c	$N \times M$	Collaborative rating matrix
R^d	$N \times M$	Latent rating matrix
U	$N \times L$	User latent factor matrix
V	$M \times L$	Merchant latent factor matrix
d_j^i	–	Review on merchant j written by user i
T	$(.)$	Multiple topics from user reviews
K	–	The number of keywords in each topic representation
W_j	$K \times |T|$	Word frequency matrix for merchant j
Z_j	$1 \times |T|$	Topic ratings vector for merchant j
γ	$K \times |T|$	Word sentiment polarity matrix
β_i	$1 \times |T|$	Preference vector for user i
$\sigma, \sigma_U, \sigma_V, \sigma_\beta$	R	Variance of the priors

Definition (Topics). The review is made up of many topics. Users may pay more attention to some topics. The topics are denoted as $\mathbf{T} = \{T^1, \ldots, T^i, \ldots, T^{|T|}\}$. The topics can be *price*, *food* and *service* in the case of a restaurant.

Definition (Topic Representation). Every topic $\mathbf{T}^i$ consists of K most representative keywords, i.e., $\mathbf{T}^i = \{w_1^i, w_2^i, \ldots, w_K^i\}$. Each topic keyword is selected from the review corpus.

Definition (Topic Ratings). Topic ratings is a $|T|$ dimensional vector of score over all topics, denoted as $\mathbf{Z}_j = \{Z_j^1, \ldots, Z_j^k, \ldots, Z_j^{|T|}\}$. Topic ratings indicate the quality offered by the merchant on topics. A high topic rating means the merchant does well on this topic.

Definition (User Preferences). User preferences is a $|T|$ dimensional vector of weights over all topics, where the k-th dimension indicates the degree of user preference toward the topic T^k, denoted as $\boldsymbol{\beta}_i = \{\beta_i^1, \ldots, \beta_i^k, \ldots, \beta_i^{|T|}\}$. A high weight means more attention is paid to the corresponding topic.

4   Model Specification

In this section, we first introduce the topic representation and the calculation of the topic ratings. Then, we introduce the proposed model, titled "Collaborative Filtering meets Reviews" (CFMR).

4.1   Topic Representation and Calculation of the Topic Ratings

A major challenge in our work is defining meaningful topics. Because we do not have the supervision information about how many topics we should define and what each topic is made up of. In this paper, we apply Latent Dirichlet Allocation (LDA) [2] model to find the underlying topics and components of the corresponding topic. We train the LDA model on the review corpus made up of all the reviews from all users. For each topic, there exists a word distribution. Every word is associated with a probability value which represents the percentage that this word makes up of the corresponding topic. We first sort the words by the probability value descendingly for each topic. Then we select K most representative words with high probability as our topic keywords.

The calculation of the topic ratings is as follows. Suppose that we have $|T|$ topics and each topic is represented by K keywords. We get a $K \times |T|$ keyword matrix. We aggregate all the reviews written for merchant v_j, i.e., $\mathbf{d}_j = \{d_j^1, \ldots, d_j^i, \ldots, d_j^N\}$. Based on the keyword matrix, we can map the aggregated review $\mathbf{d}_j$ to a word frequency matrix W_j in which each column corresponds to a keyword frequency vector in terms of this topic representation. Each element in the matrix W_j represents the frequency of the word in the aggregated review $\mathbf{d}_j$. In natural language processing, each word is associated with a negative, neutral or positive sentiment which shows the level of reviewer preferences.

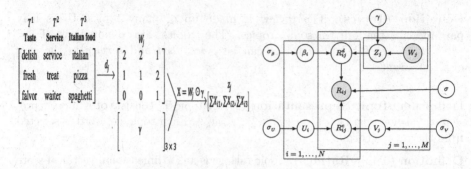

Fig. 2. Left panel: topic representation and calculation of the topic ratings. Right panel: the graphical model for the CFMR model.

So we introduce a sentiment polarity parameter matrix γ which can be learned by minimizing the objective function given by Eq. 10. Each topic rating is the dot product of the keyword frequency vector in terms of this topic representation and the corresponding sentiment polarity vector. The Hadamard product of matrix W_j and matrix γ is denoted as matrix X, i.e.

$$X = W_j \odot \gamma, \tag{1}$$

where $\odot$ is the Hadamard product. Then we can calculate the topic ratings Z_j by doing the column sum on matrix X, i.e.

$$Z_j = \left[\sum X_{i1}, \sum X_{i2}, \ldots, \sum X_{i|T|} \right]. \tag{2}$$

Figure 2 (left panel) shows a simple example where we assume the $\mathbf{d}_j$ consists of two reviews shown in Fig. 1. We only present three topics and show three keywords in each topic due to space limitation.

4.2 Collaborative Filtering Meets Reviews

In this section, we apply an integrated model, titled "Collaborative Filtering meets Reviews" (CFMR), which combines the collaborative rating and the latent rating to approximate the overall rating.

The Generation Assumption. We assume that the overall rating consists of two parts. One part is the collaborative rating, which is based on the rating behavior of users in the past. The other is the latent rating, which is based on the current user's preference. We further assume that the latent rating is a weight sum of all the topic ratings, where the weights show the relative preference the user placed on each topic. The graphical model for the CFMR model is shown in Fig. 2 (right panel).

The CFMR Model. As aforementioned, the overall rating consists of the collaborative rating and the latent rating. We adopt a probabilistic model with Gaussian observation noise (see Fig. 2 (right panel)). We define the condition distribution over the observed ratings as

$$p(R|R^c, R^d, \sigma^2) = \prod_{i=1}^{N} \prod_{j=1}^{M} \left[\mathcal{N}(R_{ij}|R_{ij}^c + R_{ij}^d, \sigma^2) \right]^{I_{ij}}, \tag{3}$$

where $\mathcal{N}(x|\mu, \sigma^2)$ represents the probability density function of the Gaussian distribution with mean μ and variance σ^2, and I_{ij} is the indicator function that is equal to 1 if user i rated merchant j and equal to 0 otherwise.

The collaborative rating R_{ij}^c can be calculated as follows based on matrix factorization [14]:

$$R_{ij}^c = U_i V_j^T, \tag{4}$$

where the row vector U_i and V_j represent the user feature vector and the merchant feature vector respectively. We place zero-mean spherical Gaussian priors [1,7] on user and merchant feature vectors as follows:

$$p(U|\sigma_U^2) = \prod_{i=1}^{N} \mathcal{N}(U_i|0, \sigma_U^2 I), \qquad p(V|\sigma_V^2) = \prod_{j=1}^{M} \mathcal{N}(V_j|0, \sigma_V^2 I). \tag{5}$$

As we assume that the latent rating is a weight sum of all topic ratings, where the weights show the relative preference the user placed on each topic, so we define the latent rating R_{ij}^d as

$$R_{ij}^d = \beta_i Z_j^T, \tag{6}$$

where the row vector β_i and Z_j represent the preference vector of user i and the topic ratings vector of merchant j respectively.

Note that the parameter matrix γ characterizes the sentiment polarity of all the keywords. In order to simplify the discussion, we place a uniform distribution on parameter matrix γ. As we show in Sect. 4.1, topic ratings vector Z_j depends on the parameter γ and the word frequency matrix W_j. So we do not place any priors on topic ratings vector Z_j. We just place zero-mean spherical Gaussian priors [1,7] on user preference vectors as follows:

$$p(\beta|\sigma_\beta^2) = \prod_{i=1}^{N} \mathcal{N}(\beta_i|0, \sigma_\beta^2 I). \tag{7}$$

From Eqs. 3, 4, 6, we can get the following equation, i.e.

$$p(R|U, V, \beta, \sigma^2) = \prod_{i=1}^{N} \prod_{j=1}^{M} \left[\mathcal{N}(R_{ij}|U_i V_j^T + \beta_i Z_j^T, \sigma^2) \right]^{I_{ij}}. \tag{8}$$

For the simplification of the equation, let $\Omega = \{U, V, \beta, \gamma\}$ and $\Theta = \{\sigma, \sigma_U, \sigma_V, \sigma_\beta\}$. The log of the posterior distribution over user feature, merchant feature and user preference vectors is given by

$$
\begin{aligned}
\ln p(\Omega|R, \Theta) = {} & \ln p(R|U, V, \beta, \gamma, \sigma^2) + \ln p(U|\sigma_U^2) + \ln p(V|\sigma_V^2) + \\
& \ln p(\beta|\sigma_\beta^2) + \ln p(\gamma) - \ln p(R) \\
= {} & -\frac{1}{2\sigma^2} \sum_{i=1}^{N} \sum_{j=1}^{M} \boldsymbol{I}_{ij} (R_{ij} - U_i V_j^T - \beta_i Z_j^T)^2 - \frac{1}{2\sigma_U^2} \sum_{i=1}^{N} U_i U_i^T - \\
& \frac{1}{2\sigma_V^2} \sum_{j=1}^{M} V_j V_j^T - \frac{1}{2\sigma_\beta^2} \sum_{i=1}^{N} \beta_i \beta_i^T - \frac{1}{2} \left(\sum_{i=1}^{N} \sum_{j=1}^{M} \boldsymbol{I}_{ij} \right) \ln \sigma^2 - \\
& \frac{1}{2} \left(LN \ln \sigma_U^2 + LM \ln \sigma_V^2 + |T| N \ln \sigma_\beta^2 \right) + C,
\end{aligned}
\tag{9}
$$

where L represents the length of the user feature vector and C represents a constant that does not depend on the parameters. Maximizing the log-posterior in Eq. 9 with hyperparameters $(\sigma^2, \sigma_U^2, \sigma_V^2, \sigma_\beta^2)$ kept fixed is equivalent to minimizing the sum-of-squared-errors objective function with quadratic regularization terms:

$$
\begin{aligned}
L = {} & \frac{1}{2} \sum_{i=1}^{N} \sum_{j=1}^{M} \boldsymbol{I}_{ij} (R_{ij} - U_i V_j^T - \beta_i Z_j^T)^2 + \frac{\lambda_U}{2} \sum_{i=1}^{N} \|U_i\|_{Fro}^2 + \\
& \frac{\lambda_V}{2} \sum_{j=1}^{M} \|V_j\|_{Fro}^2 + \frac{\lambda_\beta}{2} \sum_{i=1}^{N} \|\beta_j\|_{Fro}^2,
\end{aligned}
\tag{10}
$$

where $\lambda_U = \sigma^2/\sigma_U^2$, $\lambda_V = \sigma^2/\sigma_V^2$, $\lambda_\beta = \sigma^2/\sigma_\beta^2$, and $\|.\|_{Fro}^2$ denotes the Frobenius Norm. Note that topic ratings vector Z_j depends on the parameter matrix γ. A local minimum of the objective function given by Eq. 10 can be calculated by performing the gradient descent in U, V, β, γ by moving in the opposite direction of the gradient. The gradient of the U, V, β, γ is as followed:

$$
\frac{\partial L}{\partial U_i} = \sum_{j} \boldsymbol{I}_{ij} [(U_i V_j^T + \beta_i Z_j^T) - R_{ij}] V_j + \lambda_U U_i
$$

$$
\frac{\partial L}{\partial V_i} = \sum_{i} \boldsymbol{I}_{ij} [(U_i V_j^T + \beta_i Z_j^T) - R_{ij}] U_i + \lambda_V V_j
$$

$$
\frac{\partial L}{\partial \beta_i} = \sum_{j} \boldsymbol{I}_{ij} [(U_i V_j^T + \beta_i Z_j^T) - R_{ij}] V_j + \lambda_\beta \beta_i
$$

$$
\frac{\partial L}{\partial \gamma_k} = \sum_{i,j} \boldsymbol{I}_{ij} [(U_i V_j^T + \beta_i Z_j^T) - R_{ij}] \beta_i^k W_j^{;,k}
\tag{11}
$$

where γ_k is a word sentiment polarity vector for k-th topic.

Algorithm 1. Merchant recommendation.

Input : Reviews D , Rating Matrix R
Output: $\Omega = \{U, V, \beta, \gamma\}$

1 *Random initialize Ω*;
2 **for** *step* = 1 **to** *MAXSTEP* **do**
3   **for** $i \leftarrow 0$ **to** N **do**
4          $U_i \leftarrow U_i - \eta \nabla_{U_i}$;
5          $\beta_i \leftarrow \beta_i - \eta \nabla_{\beta_i}$;
6   **for** $j \leftarrow 0$ **to** M **do**
7          $V_j \leftarrow V_j - \eta \nabla_{V_j}$;
8          $\gamma_j \leftarrow \gamma_j - \eta \nabla_{\gamma_j}$;
9   **if** *converge* **then**
10   stop iteration;
11   *change the learning rate η as step increases*;
12 **for** *each user i* **do**
13   **for** *each merchant j* **do**
14   calculate the rating R_{ij};
15   recommend merchants with high rating for user i;

Merchant Recommendation by CFMR Model. Merchant recommendation can be implemented by selecting merchants with high ratings. The algorithm of merchant recommendation is detailed in Algorithm 1.

Complexity Analysis. The review preprocessing and the training process of LDA model only cost a constant time, which is not relevant to complexity of our algorithm. The main complexity of our algorithm is evaluating the objective function in Eq. 10 and its gradient against variables in Eq. 11. Owing to the spartity of matrix R, the computational complexity of the objective function is $O(\rho(F + |T|))$, where ρ is the number of the nonzero entries in matrix R, F is the dimension of the row vector U_i and the $|T|$ is the dimension of the row vector β_i. The computaional complexity of gradients is also $O(\rho(F + |T|))$. Therefore, the computational time of our algorithm is linear with repect to the number of observations in the matrix R.

5   Experimental Results

In this section, we compare our CFMR model with some baseline methods in terms of effectiveness and efficiency.

5.1 Data Set and Setting

We perform our experiments on two real-world datasets, Yelp dataset[1] and TripAdvisor dataset[2]. The two datasets offer the ratings and the corresponding reviews, making them ideal datasets for our model. For Yelp dataset, we choose the restaurant category which covers more than 1/3 of the total reviews and filter out users and merchants with less than 20 records. For TripAdvisor dataset, we filter out users and merchants with less than 10 records. The final Yelp dataset consists of 6335 users, 4096 merchants and 220454 reviews. The final TripAdvisor dataset consists of 13930 users, 4489 merchants and 216929 reviews.

We perform some necessary preprocessing on the reviews before Topic Modeling: (i) transform the words to lower cases; (ii) filter out the stop words; (iii) stem each word in the review corpus to its root.

In the experiments, we try various values for the learning rate and experiment with various values of L (the length of the user feature vector U_i), finally we chose to use a learning rate of 0.005 and the value of L is set to 30. We also investigate the impact of the number keywords in each topic representation and the number of topics, the number keywords in each topic representation is fixed to 10 (i.e., $K = 10$) and the number of topics is fixted to 20 (i.e., $|T| = 15$).

5.2 Baseline Methods

We compare our method with the following four baseline models:

- **PMF**: Probabilistic Matrix Factorization is proposed in [21] by modeling user preference matrix as a product of two lower-rank user and merchant matrices.
- **LDAMF**: This method is proposed in [18] which utilizes the information buried in the review texts by fitting an LDA model on the review corpus and then treating the learned topic distribution on merchants (or users) as the latent factors in Matrix Factorization.
- **SVD++**: This is the state-of-the-art method [13] whose author had won the Netflix Prize. This approach is based on a matrix factorization model by incorporating implicit feedback information.
- **GM-L2**: This method is proposed in [16] which follows the boot-strapping aspect segmentation algorithm proposed in [22] to achieve the aspect representation.

5.3 Accuracy Prediction

Evaluation Metrics. We use two typical accuracy metrics, Mean Absolute Error (MAE) and Root Mean Square Error (RMSE), to measure the performance by comparing the observed rating against the predicted rating. The definitions are as follows:

$$MAE = \frac{1}{|L|} \sum_{i,j} |R_{i,j} - \widehat{R}_{i,j}|, \quad RMSE = \sqrt{\frac{1}{|L|} \sum_{i,j} (R_{i,j} - \widehat{R}_{i,j})^2},$$

---

[1] https://www.yelp.com/dataset_challenge.

[2] http://times.cs.uiuc.edu/~wang296/Data/.

where $R_{i,j}$ and $\widehat{R}_{i,j}$ indicate the observed rating and the predicted rating respectively, and $|L|$ is the number of all test cases.

Experimental Results. For Yelp dataset, the regularization parameters are set to $\lambda_U = \lambda_V = \lambda_\beta = 0.003$. For TripAdvisor dataset, the regularization parameters are set to $\lambda_U = \lambda_V = \lambda_\beta = 0.001$. We use different ratio of training data to test all algorithms and the results are shown in Table 2. Compared with the baseline methods, our model achieves better performance on accuracy prediction. The reason is that our model incorporates the review information and takes user preferences across multiple topics into consideration. Our approach also performs better than the recently proposed method GM-L2 [16] because the number of topics we consider is more than the number of aspects used in GM-L2 [16]. We also find that the more data used to training, the more accurate prediction results we got. The comparison results on TripAdvisor dataset are similar to the results on Yelp dataset.

5.4   Cold Start Problem

Experimental Configuration. We randomly split the total dataset into two parts: training set and testing set according to the split ratio of 20 : 80. For any test user, we randomly select 20% of user total rating records as the observed merchant ratings and the remaining as the held-out merchant ratings. Our main

Table 2. Comparison on accuracy prediction.

Datasets	Ratio	Metric	LDAMF	PMF	SVD++	GM-L2	CFMR
Yelp	90%	MAE	0.8211	0.7957	0.7733	0.7544	0.7464
		Improved	10.01%	6.61%	3.60%	1.07%	
		RMSE	1.0592	1.0224	1.0019	0.9787	0.9684
		Improved	9.38%	6.19%	3.46%	1.06%	
	70%	MAE	0.8254	0.7938	0.7774	0.7573	0.7497
		Improved	10.10%	5.88%	3.69%	1.01%	
		RMSE	1.0662	1.0266	1.0071	0.9823	0.9727
		Improved	9.61%	5.54%	3.54%	0.99%	
TripAdvisor	90%	MAE	0.7946	0.7629	0.7431	0.7259	0.7154
		Improved	11.07%	6.64%	3.87%	1.47%	
		RMSE	1.0255	0.9821	0.9562	0.9349	0.9198
		improved	11.49%	6.77%	3.96%	1.64%	
	70%	MAE	0.8003	0.7707	0.7460	0.7308	0.7192
		Improved	11.27%	7.16%	3.73%	1.61%	
		RMSE	1.0332	0.9928	0.9600	0.9431	0.9232
		Improved	11.92%	7.54%	3.99%	2.16%	

objective is to use the ratings of the observed merchants to predict the ratings of the held-out merchants. In order to measure the performance of our model on the cold start problem, we divide the test users into groups according to the number of their observed ratings in the training set, i.e., 1–3, 4–5, 6–7, 8–9, 10–11, 12–13 on Yelp dataset.

Experimental Results. For Yelp dataset, the regularization parameters are set to $\lambda_U = \lambda_V = \lambda_\beta = 0.005$. For TripAdvisor dataset, the regularization parameters are set to $\lambda_U = \lambda_V = \lambda_\beta = 0.002$. Figure 3 (left panel) shows the distribution of the observed ratings in the training set according to the split ratio of 20:80 on Yelp dataset. Figure 3 (left panel) and Fig. 3 (right panel) report the MAE and RMSE results on Yelp dataset respectively. We notice that the group where users have more ratings achieves better accuracy than other groups. We can also observe that our approach outperforms the other baseline methods because our model can learn user preferences across multiple topics. LDAMF method behaves worst among these methods, since it only incorporates the review information. We also present the performance of all algorithms on TripAdvisor dataset in Fig. 4. From Fig. 4, we can see that the experimental results of all algorithms on TripAdvisor dataset are not as well as the performance on Yelp dataset because TripAdvisor dataset is more sparser than Yelp dataset.

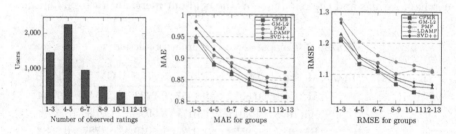

Fig. 3. Yelp: Left panel: distribution of users in the training dataset. Middle panel: MAE for groups. Right panel: RMSE for groups.

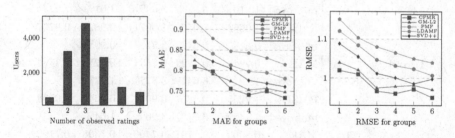

Fig. 4. TripAdvisor: Left panel: distribution of users in the training dataset. Middle panel: MAE for groups. Right panel: RMSE for groups.

5.5   Long Tail Effect

Evaluation Metrics. We utilize the recall measurement to evaluate the performance of our model to discover long tail merchant (less popular merchant). This testing methodology adopted in this paper is applied in [11], which has been widely used to evaluate the recommender system. The detailed procedure about how this experiment is conducted is as followed. For each dataset, known ratings are split into two subsets: training set and testing set. The testing set contains only 5-stars ratings. For each user, we randomly select one long tail rating with 5-stars which will be added into the testing set. The remaining ratings are as the training set. As expected, the testing set is not used for training. In order to calculate recall, we first train the model on the training set. Then for each long tail merchant i rated 5-stars by user u in the testing set: (i) randomly select 800 additional merchants unrated by user u. (ii) predict the ratings for the merchant i as well as the additional 800 merchants. (iii) form a ranked list by order in all the 801 merchants based on their ratings. (iv) form a top-N recommendation list by selecting the top-N ranked merchants from the ranked list. If the merchant i is in the top-N recommendation list, we have a hit ($hit = 1$). Otherwise we have a miss ($hit = 0$).

The recall is defined by averaging over all test cases:

$$Recall = \frac{\sum hit}{|Test|},$$

where $|Test|$ is the number of all test cases.

Experimental Results. For Yelp dataset, the regularization parameters are set to $\lambda_U = \lambda_V = \lambda_\beta = 0.008$. For TripAdvisor dataset, the regularization parameters are set to $\lambda_U = \lambda_V = \lambda_\beta = 0.001$. We present the performance on top-N in the range from 0 to 50, since the larger N is meaningless in the typical top-N recommendation task. Figure 5 (left panel) reports the performance of recommendation algorithms on Yelp dataset. Clearly, the models achieve different

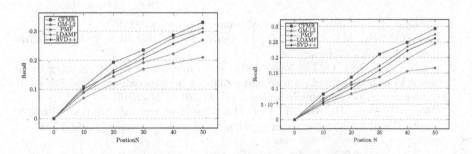

Fig. 5. Left panel: recall on Yelp. Right panel: recall on TripAdvisor.

performance in terms of top-N recommendation. The recall of CFMR at $N = 30$ is 0.225, i.e., the model can place a long tail merchant in the top-30 with the probability of 22.5%. From Fig. 5 (left panel), we can see that our model is better than the model GM-L2 [16] because the number of topics we considered is more than the number of aspects in GM-L2 [16]. We can also find that the model of SVD++ presents better performance than the method of PMF because the SVD++ model incorporates implicit feedback information. Figure 5 (right panel) reports the performance of all algorithms on TripAdvisor dataset. It is apparent that the trend of the comparison result is similar to that of Fig. 5 (left panel).

6   Conclusion

In this paper, we propose an integrated graphical model named CFMR by incorporating user real preferences on various topics from user reviews into Collaborative Filtering technique for personalized merchant recommendation. We assume that the overall rating consists of two parts. One part is the collaborative rating, which is based on the rating behavior of users in the past. The other is the latent rating, which is based on the current user's preference. Our work mainly focuses on the topic representation and the calculation of the latent rating. We conducted extensive experiments on two real-world datasets and the experimental results demonstrate that our model outperforms other methods in terms of effectiveness and efficiency. In future, we would like to consider the impact of the additional information, such as geographic information [4] and social information [15], to further improve the performance of merchant recommendation.

Acknowledgments. This research was partially supported by grants from the National Natural Science Foundation of China (NSFC, Grant No. U1605251), the National Science Foundation for Distinguished Young Scholars of China (Grant No. 61325010), and the NSFC Major research program (Grant No. 91546103).

References

1. Bishop, C.M.: Probabilistic principal component analysis (1997)
2. Blei, D.M., Ng, A.Y., Jordan, M.I.: Latent Dirichlet allocation. J. Mach. Learn. Res. **3**, 993–1022 (2003)
3. Chaney, A.J., Blei, D.M., Eliassi-Rad, T.: A probabilistic model for using social networks in personalized item recommendation. In: Proceedings of the 9th ACM Conference on Recommender Systems, pp. 43–50. ACM (2015)
4. Cheng, C., Yang, H., King, I., Lyu, M.R.: Fused matrix factorization with geographical and social influence in location-based social networks. AAAI **12**, 17–23 (2012)
5. Cho, E., Myers, S.A., Leskovec, J.: Friendship and mobility: user movement in location-based social networks. In: Proceedings of the 17th ACM SIGKDD International Conference on Knowledge Discovery and Data Mining, pp. 1082–1090. ACM (2011)

6. Cremonesi, P., Koren, Y., Turrin, R.: Performance of recommender algorithms on top-n recommendation tasks. In: Proceedings of the Fourth ACM Conference on Recommender Systems, pp. 39–46. ACM (2010)

7. Dueck, D., Frey, B., Dueck, D., Frey, B.J.: Probabilistic sparse matrix factorization. University of Toronto technical report PSI-2004-23 (2004)

8. Ekstrand, M.D., Riedl, J.T., Konstan, J.A.: Collaborative filtering recommender systems. Found. Trends Hum.-Comput. Interact. 4(2), 81–173 (2011)

9. Fan, M., Khademi, M.: Predicting a business star in yelp from its reviews text alone. arXiv preprint arXiv:1401.0864 (2014)

10. Gao, H., Tang, J., Hu, X., Liu, H.: Exploring temporal effects for location recommendation on location-based social networks. In Proceedings of the 7th ACM Conference on Recommender Systems, pp. 93–100. ACM (2013)

11. Huang, J., Rogers, S., Joo, E.: Improving restaurants by extracting subtopics from yelp reviews. In: iConference 2014 (Social Media Expo) (2014)

12. Karamshuk, D., Noulas, A., Scellato, S., Nicosia, V., Mascolo, C.: Geo-spotting: mining online location-based services for optimal retail store placement. In: Proceedings of the 19th ACM SIGKDD International Conference on Knowledge Discovery and Data Mining, pp. 793–801. ACM (2013)

13. Koren, Y.: Factorization meets the neighborhood: a multifaceted collaborative filtering model. In Proceedings of the 14th ACM SIGKDD International Conference on Knowledge Discovery and Data Mining, pp. 426–434. ACM (2008)

14. Koren, Y., Bell, R., Volinsky, C., et al.: Matrix factorization techniques for recommender systems. Computer 42(8), 30–37 (2009)

15. Li, H., Ge, Y., Zhu, H.: Point-of-interest recommendations: learning potential check-ins from friends. In: Proceedings of the 22th ACM SIGKDD International Conference on Knowledge Discovery and Data Mining. ACM (2016)

16. Li, X., Xu, G., Chen, E., Li, L.: Learning user preferences across multiple aspects for merchant recommendation. In: 2015 IEEE International Conference on Data Mining (ICDM), pp. 865–870. IEEE (2015)

17. Ma, H., King, I., Lyu, M.R.: Learning to recommend with social trust ensemble. In: Proceedings of the 32nd International ACM SIGIR Conference on Research and Development in Information Retrieval, pp. 203–210. ACM (2009)

18. McAuley, J., Leskovec, J.: Hidden factors and hidden topics: understanding rating dimensions with review text. In Proceedings of the 7th ACM Conference on Recommender Systems, pp. 165–172. ACM (2013)

19. Mukherjee, S., Basu, G., Joshi, S.: Incorporating author preference in sentiment rating prediction of reviews. In: Proceedings of the 22nd International Conference on World Wide Web, pp. 47–48. ACM (2013)

20. Pazzani, M.J., Billsus, D.: Content-based recommendation systems. In: Brusilovsky, P., Kobsa, A., Nejdl, W. (eds.) The Adaptive Web. LNCS, vol. 4321, pp. 325–341. Springer, Heidelberg (2007). doi:10.1007/978-3-540-72079-9_10

21. Salakhutdinov, R., Mnih, A.: Probabilistic matrix factorization. In: NIPS 20, pp. 1–8 (2011)

22. Wang, H., Lu, Y., Zhai, C.: Latent aspect rating analysis on review text data: a rating regression approach. In: Proceedings of the 16th ACM SIGKDD International Conference on Knowledge Discovery and Data Mining, pp. 783–792. ACM (2010)

23. Weng, L.-T., Xu, Y., Li, Y., Nayak, R.: Exploiting item taxonomy for solving cold-start problem in recommendation making. In: 2008 20th IEEE International Conference on Tools with Artificial Intelligence, vol. 2, pp. 113–120. IEEE (2008)

24. Yuan, Q., Cong, G., Sun, A.: Graph-based point-of-interest recommendation with geographical and temporal influences. In: Proceedings of the 23rd ACM International Conference on Conference on Information and Knowledge Management, pp. 659–668. ACM (2014)
25. Zheng, V.W., Cao, B., Zheng, Y., Xie, X., Yang, Q.: Collaborative filtering meets mobile recommendation: a user-centered approach. AAAI **10**, 236–241 (2010)

Joint Factorizational Topic Models for Cross-City Recommendation

Lin Xiao[1]([✉]), Zhang Min[2], and Zhang Yongfeng[3]

[1] Institute of Interdisciplinary Information Sciences,
Tsinghua University, Beijing, China
jackielinxiao@gmail.com
[2] Tsinghua National Laboratory for Information Science and Technology,
Department of Computer Science and Technology, Tsinghua University,
Beijing 100084, China
z-m@tsinghua.edu.cn
[3] College of Information and Computer Science,
University of Massachusetts Amherst, Amherst, MA 01003, USA
yongfeng@cs.umass.edu

Abstract. The research of personalized recommendation techniques today has mostly parted into two mainstream directions, namely, the factorization-based approaches and topic models. Practically, they aim to benefit from the numerical ratings and textual reviews, correspondingly, which compose two major information sources in various real-world systems, including Amazon, Yelp, eBay, Netflix, and many others.

However, although the two approaches are supposed to be correlated for their same goal of accurate recommendation, there still lacks a clear theoretical understanding of how their objective functions can be mathematically bridged to leverage the numerical ratings and textual reviews collectively, and why such a bridge is intuitively reasonable to match up their learning procedures for the rating prediction and top-N recommendation tasks, respectively.

In this work, we exposit with mathematical analysis that, the vector-level randomization functions to harmonize the optimization objectives of factorizational and topic models unfortunately do not exist at all, although they are usually pre-assumed and intuitively designed in the literature.

Fortunately, we also point out that one can simply avoid the seeking of such a randomization function by optimizing a Joint Factorizational Topic (JFT) model directly. We further apply our JFT model to the cross-city Point of Interest (POI) recommendation tasks for performance validation, which is an extremely difficult task for its inherent cold-start nature. Experimental results on real-world datasets verified the appealing performance of our approach against previous methods with pre-assumed randomization functions in terms of both rating prediction and top-N recommendation tasks.

© Springer International Publishing AG 2017
L. Chen et al. (Eds.): APWeb-WAIM 2017, Part I, LNCS 10366, pp. 591–609, 2017.
DOI: 10.1007/978-3-319-63579-8_45

1   Introduction

The vast amount of items in various web-based applications has made it an essential task to construct reliable Personalized Recommender Systems (PRS) [25]. With the ability to leverage the wisdom of crowds, the Collaborative Filtering (CF)-based [16,29] approaches have achieved significant success and wide application, especially for those Latent Factor Models (LFM) [12] based on Matrix Factorization (MF) [30] techniques, which attempt to model the preferences of users and items collectively through multivariate hidden factors, so as to make recommendations based on numerical star rating predictions.

Recently, researchers have been putting attention on another important information source in many online systems, namely, the textual user reviews. Usually, the ratings and reviews come in pairs in many typical applications, *e.g.*, Amazon and Yelp. While the ratings act as integrated indicators of user attitudes towards products, the reviews serve as more detailed explanations of what aspects users care about and why the corresponding rating is made [33,34].

As such, the application of Topic Models [4] has gained attention to leverage the textual reviews for personalized recommendation, especially the frequently used Latent Dirichlet Allocation (LDA) [5] technique and its variants, for their ability to extract latent topics/aspects from reviews, which represent the inherently actual factors that users care about when making numerical ratings [20,21]. This further leads to the recent research direction to bridge the LFM and LDA models, which makes use of the ratings and reviews collectively for personalized recommendation [2,20,21,23,32].

However, without a clear mathematical understanding of how the objective functions of LFM and LDA interact with each other when bridged for unified model learning, current approaches have to base themselves on unvalidated and pre-assumed designations to bridge the inherently heterogenous objective functions. For example, McAuley et al. [21] transform the latent factors in LFM to topic distributions in LDA through a manually designed randomization function based on logistic normalization, while Ling et al. [20] let the factors and topics be the same by assuming them to be sampled from mixture Gaussian distributions.

In this work, we investigate the mathematical relations between the probability of recommending an item to a user and the estimated user-item correlations by LFM or LDA models. Based on this, we prove that a multiplicatively monotonic randomization function that transforms latent factors in LFM to topic distributions in LDA actually does not exist at all. As a result, although some normalization-based transformations seem to be intuitial in previous work [21], they actually make the objective functions of LFM and LDA conflict with each other during optimization procedure, where a higher value of log-likelihood in the LDA component may force a lower rating prediction in the LFM component, which is not favoured when bridging the two models.

Fortunately, we further find that instead of transforming a latent factor to a topic distribution separately, we can simply transform the product of latent factors in LFM to the corresponding product of topic distributions in LDA as a whole, so as to avoid the seeking of a theoretically nonexistent randomization function. This is because what we really care about in practice is the final product of the user/item latent factors (in LFM) or topic distributions (in LDA), where the former accounts for the predicted user-item ratings, and the latter affects the log-likelihood of the observed reviews. Based on these findings, we propose the Joint Factorizational Topic (JFT) model to bridge LFM and LDA, so as to adopt the numerical ratings and textual reviews collectively, and at the same time guarantee the inner-model consistency between the LFM and LDA components.

2 Related Work

With the continuous growth of various online items across a vast range of the Web, Personalized Recommender Systems (PRS) [25] have set their missions to save users from information overload [14], and they have been widely integrated into various online applications in the forms of, for example, product recommendation in e-commerce [19], friend recommendation in social networks [3], news article recommendation in web portals [7], and video recommendation in video sharing websites [8], etc.

Early systems of personalized recommendations rely on content-based approaches [22], which construct the user/item content profiles and make recommendation by paring users with the contently similar items. Content-based approaches usually gain good accuracy but functionally lack the ability of providing recommendations with novelty, serendipity, and flexibility. Besides, they usually require a large amount of expensive human annotations [25]. This further leads to the prospering of Collaborative Filtering (CF)-based recommendation algorithms [16,29] that leverage the wisdom of the crowds. Typically, they construct the partially observed user-item rating matrix and conduct missing rating prediction based on the historical records of a user, as well as those of the others.

With widely recognized performance in rating prediction, scalability, and computational efficiency, the Latent Factor Models (LFM) [12] based on Matrix Factorization (MF) [30] techniques for CF have been extensively investigated by the research community, and widely applied in practical systems. Perhaps the most early and representative formalization of LFM for recommendation dates back to Koren et al. [15], and other variants for personalization include Non-negative Matrix Factorization (NMF) [17], Probabilistic Matrix Factorization (PMF) [26,27], and Maximum Margin Matrix Factorization (MMMF) [28], etc. Despite the important success in rating prediction, the CF approaches based solely on the numerical ratings suffer from the problems of explainability [34], cold-start [18], and the difficulty to provide more specific recommendations that meet targeted item aspects [13]. Besides, related research results show that the performance on numerical rating prediction does not necessarily relate to the performance on practical top-N recommendations [6], and that the numerical star ratings may not always be a reliable indicator of users' attitudes towards items [35].

To alleviate these problems, researchers have been investigating the incorporation of textual reviews for recommendation, which is another important information source beyond the star ratings in many systems [31]. Early approaches rely on manually extracted item aspects from reviews for more informed recommendation [1,13] and rating prediction [9,11], which improved the performance but also required extensive human participations. As a results, researchers recently have begun to investigate the possibility of integrating the automatic topic modeling techniques on textual reviews and the latent factor modeling approach on numerical ratings for boosted recommendation, and have achieved appealing results [2,21,32].

However, without a clear mathematical exposition of the relationships between latent factor models and topic modeling, current approaches have to base themselves on manually designed randomization functions or probabilistic distributions. In this work, however, we attempt to make an exposition on the relationships between the two types of objective functions, and further bridge the inherently heterogenous models in a harmonious way for recommendation with the power of both numerical ratings and textual reviews.

3 Preliminaries and Definitions

3.1 Latent Factor Models (LFM)

Latent Factor Models (LFM) [16] attempt to encode user and item preferences in a latent factor space so as to estimate the user-item relations for rating prediction, which account for many of the frequently used Matrix Factorization (MF) [30] techniques. Among those, a 'standard' and representative formalization [15] predicts the user-item ratings $r_{u,i}$ with user/item biases and latent factors by,

$$rate(u,i) = \alpha + \beta_u + \beta_i + \gamma_u \cdot \gamma_i \tag{1}$$

where α is the global offset, β_u and β_i are user and item biases, γ_u and γ_i are the K-dimensional latent factors of user u and item i, respectively, and "$\cdot$" denotes vector multiplication. Intuitively, γ_u can be interpreted as the preference of user u to some latent factors, while γ_i is the property embedding of item i on those latent factors. Based on a set of observed training records $\mathcal{R}$, the model is typically targeted with the goal of providing accurate rating predictions, where we determine the parameter set $\Theta = \{\alpha, \beta_u, \beta_i, \gamma_u, \gamma_i\}$ with the following minimization problem,

$$\Theta = \underset{\Theta}{argmin} \sum_{r_{u,i} \in \mathcal{R}} \left(rate(u,i) - r_{u,i}\right)^2 + \lambda \Omega(\Theta) \tag{2}$$

and $\Omega(\Theta)$ is a regularization term. A variety of methods exist to minimize Eq. (2), for example, Stochastic Gradient Descent (SGD) or Alternating Least Squares (ALS) [15]. However, this model merely takes into account the numerical ratings and leaves out the textual reviews, which is information-rich and may well help to provide better recommendations.

3.2 Latent Dirichlet Allocation (LDA)

Different from LFM, the LDA model attempts to learn a number of K latent topics from documents (textual reviews in this work), where each word w is assigned to a topic z_w, and each topic z is associated with a word distribution ϕ_z. Based on this, each document $d \in \mathcal{D}$ is represented with a K-dimensional topic distribution θ_d, where the j-th word $w_{d,j}$ in document d discusses its corresponding topic $z_{d,j}$ with probability $\theta_{d,z_{d,j}}$. It is usually convenient to also define the word distribution $\phi_{z,w}$, which is the probability that word w is used for topic z in the whole corpus $\mathcal{D}$. The final model conducts parameter learning by maximizing the likelihood of observing the whole $\mathcal{D}$:

$$P(\mathcal{D}|\theta, \phi, z) = \prod_{d \in \mathcal{D}} \prod_{j=1}^{L_d} \theta_{d,z_{d,j}} \phi_{z_{d,j}, w_{d,j}} \tag{3}$$

where L_d is the length (number of words) of document d. Intuitively, we are multiplying the probability of seeing a particular topic in θ_d with the likelihood of seeing a particular word given the topic to estimate the likelihood of seeing the whole corpus.

3.3 Randomization Function

Let $\gamma \in \mathbb{R}^K$ be an arbitrary vector and $\theta \in [0,1]^K$ be a stochastic vector, where their dimensions are the same K as the latent factors γ_u, γ_i and latent topics θ_d in the previous subsections. According to the definition, we have $0 \le \theta_k \le 1$ and $\|\theta\|_1 = \sum_{k=1}^K \theta_k = 1$. The target of a randomization function $f : \mathbb{R}^K \to \mathbb{R}^K$ is to convert an arbitrary vector γ to a probabilistic distribution $\theta = f(\gamma)$. The inherent nature of a randomization function is the key component to bridge the gap between LFM and LDA models, which links the latent factors γ in LFM to the topic distributions θ in LDA, and thus makes it possible to model the numerical ratings and textual reviews in a joint manner.

In the background of personalized recommendation, a desired randomization function is expected to be *monotonic* in the sense that it preserves the orderings, so that the largest value of γ should also correspond to the largest value in θ, thus the dimensions of the LFM model and the LDA model are inherently aligned during the model learning process to express the user-item relations in a shared feature space. As a result, the basic properties of a randomization function $f(\cdot)$ can be summarized as follows:

$$\begin{cases} 0 \le f(\gamma)_i \le 1, \|f(\gamma)\|_1 = 1 \\ \gamma_i < \gamma_j \to f(\gamma)_i < f(\gamma)_j \end{cases}, \forall \gamma \in \mathbb{R}^K, 1 \le i,j \le K \tag{4}$$

For example, in [21] the authors designed a randomization function as:

$$\theta_k = f(\gamma)_k = \frac{\exp(\kappa \gamma_k)}{\sum_{k'} \exp(\kappa \gamma_{k'})} \tag{5}$$

which conducts logistic normalization on a latent factor. In the following, we investigate the relationship between the objective functions of LFM and LDA, and further point out the properties required on a randomization function to harmonize the models when bridging the two different functions.

4   Bridging Factors and Topics

4.1   Probability of Item Recommendation

In the Latent Factor Model (LFM), a recommendation list is constructed in descending order of the predicted ratings $rate(u, i)$ for a given user, which means that an item i with a higher rating prediction on user u also gains a higher probability of being recommended $P(i|u)$. As a result,

$$P_{\text{LFM}}(i|u) \propto rate(u, i) \propto \gamma_u \cdot \gamma_i \tag{6}$$

where $\propto$ denotes a positive correlation, and we leave out the parameters α, β_u and β_i because they are constants given a user and an item in the model learning process [15].

In Latent Dirichlet Allocation (LDA) for personalized recommendation, each user or item is represented by its corresponding set of textual reviews d_u or d_i, and the underlying intuition models the topical correlation between them by estimating the potential review $d_{u,i}$ that a user may write on an item, based on the topical distributions θ_{d_u} and θ_{d_i}. To simplify the notations, we use u, i, and k to denote the user or item document representations d_u, d_i and the k-th latent topic z_k interchangeably, and we thus have:

$$P(k|u) = \theta_{u,k} \text{ and } P(k|i) = \theta_{i,k} \tag{7}$$

The LDA model conducts likelihood maximization on each observed review $d_{u,i}$ given the corresponding user u and item i, and the embedded topical distribution represents the probability of observing each topic k, which is:

$$P(k|u, i) = \theta_{d,k} \tag{8}$$

LDA applies an indirect causal effect from users to items via latent topics [5], which means that user u and item i are conditionally independent given topic k, i.e., $P(u|i, k) = P(u|k)$, and this further gives us the following:

$$P(u, i, k) = \frac{P(u, k)P(i, k)}{P(k)} \tag{9}$$

By applying Eqs. (9) to (8), we decompose the topical distribution of a review into the topical representations of the corresponding user and item:

$$\theta_{d,k} = P(k|u, i) = P(k|u)P(k|i)\frac{P(u)P(i)}{P(k)P(u, i)} \propto P(k|u)P(k|i) = \theta_{u,k}\theta_{i,k} \tag{10}$$

where $P(u), P(i)$ and $P(u, i)$ are constants in the LDA procedure, and the latent topics z_k are identically independent from each other, giving us constant and equal valued $P(k)$'s over the K topics. As a result, we have the following conditional recommendation probability for LDA models:

$$
\begin{aligned}
P_{\text{LDA}}(i|u) &= \sum_{k=1}^{K} P(k|u)P(i|k) = \sum_{k=1}^{K} P(k|u)P(k|i)\frac{P(i)}{P(k)} \\
&\propto \sum_{k=1}^{K} P(k|u)P(k|i) = \sum_{k=1}^{K} \theta_{u,k}\theta_{i,k} = \theta_u \cdot \theta_i
\end{aligned}
\tag{11}
$$

and this result conforms with Eq. (10) in that, the probability of recommending an item given a user is positively correlated to the sum of topic probabilities that a user may textually review on an item.

4.2 Bridging the Objective Functions

According to the conditional item recommendation probabilities given a target user specified in Eqs. (6) and (11) for LFM and LDA models, respectively, a favoured approach to bridge the two models to leverage the power of both ratings and reviews should harmonize their objective functions, so that a higher value of $\gamma_u \cdot \gamma_i$ also corresponds to a higher value in $\theta_u \cdot \theta_i$. More precisely, except for the monotonic property defined in Eq. (4) on a vector itself, the randomization function $f(\cdot)$ from γ to θ is also required to be monotonic for vector multiplications:

$$
\gamma_1 \cdot \gamma_2 < \gamma_3 \cdot \gamma_4 \rightarrow f(\gamma_1) \cdot f(\gamma_2) < f(\gamma_3) \cdot f(\gamma_4), \ \forall \gamma_1, \gamma_2, \gamma_3, \gamma_4 \tag{12}
$$

In this way, the LFM and LDA components in a bridged objective function would not conflict with each other during the model learning process, because both of them increase/decease the recommendation probability $P(i|u)$ at the same time for each single iteration.

Previous work intuitively assumes that a randomization function satisfying the vector-level monotonic property in Eq. (4) will also be monotonic on product-level as Eq. (12). Frequently used examples are the normalization-based randomization functions, which normalize the elements of γ to construct θ so that they sum to one [2,20,21,32]. In [21] for example, a logistic normalization randomization function as in Eq. (5) is applied so as to minimize the following joint objective function to bridge the LFM and LDA models:

$$
\mathcal{O} = \sum_{r_{u,i} \in \mathcal{R}} \underbrace{\left(rate(u,i) - r_{u,i}\right)^2}_{\text{LFM component}} - \lambda \underbrace{\mathcal{L}\left(\mathcal{D}|\theta, \phi, z\right)}_{\text{LDA component}} \tag{13}
$$

where the LFM component still minimizes the error in predicted ratings, while the LDA component is the log-likelihood of the probability for the review corpus in Eq. (3).

Previous designations do seem intuitional and reasonable, and they indeed improve the performance of personalized recommendation in many cases. However, we would like to point out in this work that the vector-level monotonic property does not necessarily guarantee the monotonic property on a product-level. Actually, we prove that such a randomization function that satisfies both vector- and product-level monotonic properties does not exist at all, and the proof is omitted due to the page limit (details can be found in the appendix).

For this reason, forcing a vector-level randomization function on the latent factors to bridge the LFM and LDA models will result in a conflict between the two components during the procedure of objective optimization, i.e., while the LFM component gains a higher probability of item recommendation with a larger value of $\gamma_u \cdot \gamma_i$, the LDA component may reversely force a lower recommendation probability with $\theta_u \cdot \theta_i$ just because of the mathematical property of the randomization function, which is not favoured in model learning process. This further explains the observation that the prediction accuracy of Eq. (13) tends to fluctuate drastically during optimization, although the overall performance generally tends to increase along with the iterations.

4.3   Direct Product-Level Randomization

Despite that a randomization function with product-level monotonic property does not exist, we shall notice a simple fact that the de facto components that we need to consider so as to preserve the orderings of $P_{\mathrm{LFM}}(i|u)$ and $P_{\mathrm{LDA}}(i|u)$, are the final product of the latent factors or latent topics as a whole, i.e., $\gamma_u \cdot \gamma_i$ and $\theta_u \cdot \theta_i$, rather than each latent factor γ to a latent topic distribution θ separately.

More precisely, what we really need in the LDA model of Eq. (3) is the topic distribution of each document θ_d, where we have $\theta_{d,k} \propto \theta_{u,k}\theta_{i,k}$ by Eq. (10). As a result, we can apply a randomization function $f(\cdot)$ to the product of latent factors $\gamma_{u,k}\gamma_{i,k}$ directly, so as to obtain the product of latent topic distributions $\theta_{u,k}\theta_{i,k}$ as a whole, which is further positively correlated to $\theta_{d,k}$ that will finally be adopted by the LDA component for model learning.

A lot of normalization-based randomization functions guarantee the product-level monotonic property when applied to the product of latent factors directly. In this work, we adopt the logistic-normalization function to enforce $\theta_{u,k}\theta_{i,k}$ (and thus $\theta_{d,k}$) to be positive and sum to one:

$$\theta_{d,k} \propto \theta_{u,k}\theta_{i,k} = f(\gamma_{u,k}\gamma_{i,k}) \doteq \frac{\exp(\gamma_{u,k}\gamma_{i,k})}{\sum_{k'} \exp(\gamma_{u,k'}\gamma_{i,k'})} \tag{14}$$

which preserves the orderings of the dimensions from $\gamma_u \cdot \gamma_i$ to $\theta_u \cdot \theta_i$, and thus guarantees the positive correlation between $P_{\mathrm{LFM}}(i|u)$ and $P_{\mathrm{LDA}}(i|u)$ according to Eqs. (6) and (11). Based on this direct product-level randomization, we are fortunately able to bridge the LFM and LDA models to leverage the power of ratings and reviews collectively, and meanwhile make the two components harmonize with each other for model learning, which improves both the performance and stability of personalized recommendation.

In the following, we describe our Joint Factorizational Topic (JFT) model, as well as its application in the practical scenario of (cross-city) restaurant recommendation.

5 The Model

5.1 Joint Factorizational Topic Model (JFT)

The basic Joint Factorizational Topic (JFT) model bridges the LFM component as Eq. (1) and the LDA component in Eq. (3) according to the product-level randomization. Specifically, let $\theta_{d,k} = \frac{\exp(\gamma_{u,k}\gamma_{i,k})}{\sum_{k'}\exp(\gamma_{u,k'}\gamma_{i,k'})}$ in Eq. (3), the JFT model attempts to optimize the following objective function:

$$F(\Theta,\phi,z) = \sum_{r_{u,i}\in\mathcal{R}} \big(rate(u,i) - r_{u,i}\big)^2 - \lambda_l\mathcal{L}(\mathcal{D}|\theta,\phi,z) + \lambda_p\Omega(\Theta) \qquad (15)$$

where $\Theta = (\alpha,\beta_u,\beta_i,\gamma_u,\gamma_i)$ is the parameter set of the LFM component, $\mathcal{L}(\mathcal{D}|\theta,\phi,z)$ is the log-likelihood of the whole corpus whose document distributions θ come from the latent factors γ, and $\Omega(\Theta) = \sum_{u,i}(\beta_u^2+\beta_i^2+\|\gamma_u\|_2^2+\|\gamma_i\|_2^2)$ is the ℓ_2-norm regularizer for the latent parameters.

Intuitively, the LDA component $\mathcal{L}(\cdot)$ serves as another regularization term besides the traditional ℓ_2-norm regularizer $\Omega(\cdot)$ for numerical rating prediction, and we trade off between them two with λ_l and λ_p, respectively. In this way, the JFT model attempts to minimize the error in rating prediction, and meanwhile maximizes the likelihood of observing the corresponding textual reviews.

Most importantly, the LFM and LDA components are designed to be consistent with each other by product-level randomization in model learning, in that a smaller prediction error in the LFM component functionally invokes a larger likelihood of observing the corresponding textual review, which makes the two components collaborate rather than violate with each other when optimizing the objective function.

5.2 Incorporate City Factors and Novelty-Seeking

The problem of cross-city recommendation finds its fundamental importance in many Location-Based Services (LBS) like Foursquare and Yelp, where it is usual for users to expect personalized recommendations from the application when he is travelling outside the hometown in a new city.

However, previous approaches for point-of-interest recommendation (especially for LFM and its variants) encounter serious cold-start problems [18] in the application scenario of cross-city recommendation, where we may have only a few or even none historical rating records of a user who is traveling in a new city, although he/she may have made quite a number of ratings in his/her home city.

Our basic JFT model helps to alleviate the cross-city cold-start problem by bridging the textual reviews with numerical ratings, because the topic embeddings that we learn from the reviews of a new restaurant, may be similar to the embeddings that we can learn for the restaurants that a user previously liked in his/her hometown, which may help to provide personalized cross-city recommendations.

Nevertheless, the assumption of recommending similar items may not always be true in different scenarios, although it is one of the most basic assumptions that inherently drives the intuition of most personalization models. This is because the preferences of a user travailing in a new city may well diverge from his/her historical preferences in hometown. For example, some users may prefer to try new flavours of local features when in a new city, while others may still like to keep to their previous favourites.

As a result, we further introduce the city factors into the basic JFT model so as to learn the variant of user preferences for cross-city recommendation. To do so, we first model the user-item rating as:

$$rate'(u,i) = \alpha + \beta_u + \beta_i + (1 - \tau_u)(\gamma_u \cdot \gamma_i) + \tau_u(\gamma_i \cdot \gamma_c) \tag{16}$$

where $\gamma_u \cdot \gamma_i$ estimates the similarity between a user and a targeted item as with Eq. (1), while $\gamma_i \cdot \gamma_c$ models the similarity between the targeted item and its corresponding city, and the novel-seeking parameter $0 \leq \tau_u \leq 1$ indicates the degree that a user prefers to try local flavours.

The intuition here (which will later be verified in the experiments) lies in that, a user u with a high preference of novelty-seeking τ_u would put more interest on those restaurants whose factor representations γ_i are similar to that of the whole city γ_c (i.e., local flavour), while a user who prefers flavours she previously liked would be attracted more by those restaurants with similar factors of herself γ_u. We leave out the consideration of $\gamma_u \cdot \gamma_c$ because the factors γ_u and γ_c would be fixed parameters in model learning and when making recommendation given a user u and a targeted city c, as a result, this component would not make a difference in learning and recommendation procedures.

Correspondingly, we re-parameterize the topic distribution θ_d of each review document d from user u to item i by product-level randomization of the item-city factors $\gamma_{i,k}\gamma_{c,k}$:

$$\theta'_{d,k} = \frac{\exp(\gamma_{i,k}\gamma_{c,k})}{\sum_{k'} \exp(\gamma_{i,k'}\gamma_{c,k'})} \tag{17}$$

where each city c is similarly represented as the set of reviews d_c corresponding to the restaurants located therein. In this way, we reformulate the likelihood of the review corpus with weighted geometric mean of the user-item randomization θ_d in Eq. (14) and item-city randomization θ'_d in Eq. (17):

$$P'(\mathcal{D}|\theta, \theta', \phi, z) = \prod_{d \in \mathcal{D}} \prod_{j=1}^{L_d} \left(\theta_{d, z_{d,j}}\right)^{1 - \tau_u} \left(\theta'_{d, z_{d,j}}\right)^{\tau_u} \phi_{z_{d,j}, w_{d,j}} \tag{18}$$

Based on this, our JFT model for cross-city recommendation attempts to minimize the following objective function:

$$F'(\Theta', \phi, z) = \sum_{r_{u,i} \in \mathcal{R}} \left(rate'(u, i) - r_{u,i}\right)^2 - \lambda_l \mathcal{L}'(\mathcal{D}|\theta, \theta', \phi, z) + \lambda_p \Omega(\Theta')$$

(19)

Similar to the basic JFT model, $\Theta' = (\alpha, \beta_u, \beta_i, \gamma_u, \gamma_i, \gamma_c, \tau_u)$ is the parameter set of the LFM component, and $\mathcal{L}'(\mathcal{D}|\theta, \theta', \phi, z)$ is the log-likelihood of the corpus probability in Eq. (18).

5.3 Fitting the Model

We introduce the algorithm for model fitting in this subsection. For notational simplicity and also without loss of generality, we use $F(\Theta, \phi, z)$ to denote the objective function of both the basic and the cross-city JFT model, where the parameter set for the LFM component are $\Theta = (\alpha, \beta_u, \beta_i, \gamma_u, \gamma_i)$ and $\Theta = (\alpha, \beta_u, \beta_i, \gamma_u, \gamma_i, \gamma_c, \tau_u)$, respectively. The learning procedure for them are similar.

Typically, the LFM component (with ℓ_2-regularizer) can be easily fit with gradient descent, while the log-likelihood LDA component is usually optimized with Gibbs sampling. As our model jointly includes the two inherently heterogenous components, we construct a learning procedure that optimizes the two components alternatively:

$$Step1 : \{\Theta^t, \phi^t\} \leftarrow \underset{\Theta, \phi}{argmin} \, F(\Theta, \phi, z^{t-1}) \text{ by gradient descent}$$

$$Step2 : Logistic \ normalization \ on \ each \ topic \ vector \ \phi_k^t$$

$$Step3 : Sample \ z_{d,j}^t \ with \ probability \ P(z_{d,j}^t = k) = \theta_{d,k}^t \phi_{k,w_{d,j}}^t$$

In the first step, we fix the topic z of each word in each document, and further compute the gradient of each parameter in $\{\Theta, \phi\}$ while fixing the others. Based on these gradients, the parameters in $\{\Theta, \phi\}$ are updated one by one, where the step size for each parameter is determined by linear search.

Specifically, we should note that the parameter τ_u in the cross-city JFT model represents the probability that user u attempts to try new flavours different from his/her historical preferences. As a result, we only adopt those review records in $\mathcal{D}$ that user u visited a restaurant outside of his/her home city to construct document d_u and to update parameter τ_u (τ_u is kept stable if $d_u = \emptyset$), while the other parameters are updated with all their corresponding reviews.

However, the gradient descent procedure would not guarantee the word distribution ϕ of latent topics to be stochastic vectors. As a result, we conduct logistic normalization for each topic ϕ_k ($1 \leq k \leq K$) in the second step, where each dimension of ϕ_k is normalized as $\phi_{k,w} = \frac{\exp(\phi_{k,w})}{\sum_{w'} \exp(\phi_{k,w'})}$.

In the last step, we preserve the results from the previous steps, and update the topic assignment for each word in each document. Similar to LDA, which assigns each word to the k-th topic according to the likelihood of the word

---

Algorithm 1. TOP-N RECOMMENDATION

Input: $\mathcal{R}, \mathcal{D}, N$ Recommendation list of length N
1: $\mathcal{R}^+ \leftarrow \mathcal{R}$ with all ratings reset to be 1
2: Initialize model parameters $\alpha, \beta, \gamma, \phi, z$ randomly
3: Initialize $\tau_u \leftarrow 0.5$ for all users, $t \leftarrow 0$ **while** Not Convergence **or** $t < T$
4: $t \leftarrow t + 1, \mathcal{R}^- \leftarrow \emptyset$ **for** $(u, i) \in \mathcal{R}^+$
5: Sample item j from the same city of item i randomly, where $(u, j) \notin \mathcal{R}^+$
6: $\mathcal{R}^- \leftarrow \mathcal{R}^- \cup (u, j)$ with rating 0
7: Update model by $Step1 \sim 3$ with $\{\mathcal{R}^+ \cup \mathcal{R}^-, \mathcal{D}\}$
8: Rank items in descending order of rating prediction
9: Top-N Recommended items for each user

---

discussing topic k, we set $z_{d,j} = k$ with probability proportional to $\theta_{d,k}\phi_{k,w_{d,j}}$, where the indices pair $\{d, j\}$ denotes the j-th word of document d, θ_d is the the topic distribution of document d, and ϕ_k is the word distribution of topic k.

The major difference between LDA and the last step of our JFT model is that, the topic distributions θ_d are determined base on the product-level randomization from latent factors γ_u, γ_i and γ_c in our model, instead of sampling from a Dirichlet distribution in LDA. As a result, we only need to sample the topic assignments z in each iteration of our JFT model. The probabilistic interpretation of our approach and its inherent relationship with the LFM component have been exposited in the previous sections.

Finally, these steps are repeated iteratively until convergence, i.e., the ℓ_2-difference in Θ is sufficiently small between two consecutive iterations, or that an overfitting is observed in the validation set.

5.4 Top-N Recommendation

In this subsection, we further adapt our basic and cross-city JFT model to provide more practical personalized top-N recommendation lists beyond numerical rating predictions.

It is known that a good performance on rating prediction does not necessarily guarantee a satisfactory performance of top-N recommendation by ranking the items in descending order of the predicted ratings [6]. This is partly because of the contradiction between the goal of recommending items that users would potentially visit and the data (ratings) that we use for model training, i.e., users actually indeed visited the items in the dataset, no matter what numerical ratings they eventually made on them. Intuitively, a relatively low predicted rating does not necessarily mean that the user would not be attracted by the item at all, because of the many items with low ratings yet visited by the users.

As a result, we train our JFT model (and also the baseline approaches) for top-N recommendation in a different way from the task of rating prediction. Specifically, we feed the learning procedure with binary inputs, where the observed records in $\mathcal{R}$ are all treated as positive cases (rating=1), and the negative cases (rating=0) are sampled from the unobserved user-item pairs in a 1 : 1

negative sampling manner. For clarity, we exposit the sampling, learning, and recommendation produce in Algorithm 5.3.

6 Experiments

6.1 Experimental Setup

We collected user reviews from a major restaurant review website Dianping.com in China, including 253,749 reviews from 32,529 users towards 8,026 restaurants located in 194 cities, where each user made 20 or more reviews, including intro- and cross-city cases. We set the home city of a user according to the registration information in his/her profile.

Of the 253,749 reviews in the whole corpus, 233,802 records fall into intro-city reviews, and the remaining 19,947 records are cross-city reviews, where the ratio between intro-city and cross-city records is 11.72. Each review in the corpus consists of an integer rating ranging from 1 to 5 stars and a piece of textual comment, where the user expresses his/her opinions on the corresponding restaurant. The average length of textual comments is 41.5 words. To feed the LDA component with high quality textual inputs, we conduct part-of-speech tagging and stop word removing for each review with the widely used Stanford NLP toolkit[1].

We initialize $\tau_u = 0.5$ for the cross-city JFT model, and the eventual value of τ_u for each user is automatically determined by the model learning process. After careful tuning with grid search, we set the hyper-parameters $\lambda_l = 0.01$ and $\lambda_p = 0.001$, and five-fold cross-validation was conducted in performance evaluation for all methods.

6.2 Performance on Rating Prediction

In this section, we investigate the performance on rating prediction of our basic and cross-city JFT model, which are denoted as JFT and JFTC in the following. We also adopt the following baseline methods for performance comparison.

LFM: The basic LFM approach denoted in Eq. (2), which takes no advantage of the textual reviews.

EFM: The Explicit Factor Model presented in [34], which is the state-of-the-art recommendation approach based on textual reviews by phrase-level sentiment analysis.

HFT: The Hidden Factors and Topics model in [21], which also takes advantages of both LFM and LDA, but applies a vector-level randomization (Eq. (5)) on the latent factors.

We adopt Root Mean Square Error (RMSE) and Mean Absolute Error (MAE) for evaluation, and the results with the number of topics/factors $K = 10$

Table 1. RMSE and MAE when $K = 10$. Standard deviations for each method are ≤ 0.002.

Method	LFM	EFM	HFT	JFT	JFTC
RMSE	0.6688	0.6529	0.6532	0.6456	**0.6386**
MAE	0.5309	0.5283	0.5280	0.5213	**0.5128**

are shown in Table 1. The standard deviations in five-fold cross-validation for each method and metric are ≤ 0.002.

We find that all the other approaches gain better performance against LFM, which means that taking advantage of the textual reviews helps to make better rating predictions. Besides, our basic JFT model achieves better performance than both the EFM and HFT models. On considering that a major difference between our basic JFT model and HFT is the product- and vector-level randomization, this experimental result verifies our theoretical analysis to bridge the LFM and LDA components in Sect. 4. Finally, by incorporating user preferences in novelty-seeking in a cross-city scenario, our JFTC approach achieves the best performance.

To exhibit a clearer view of the performance on cross-city scenarios, we further take out the cross-city rating cases from the test set in each of the 5 folds, and conduct performance evaluation under different choices of the number of latent factors and topics K from 10 through 100. Results for RMSE and MAE are shown in Fig. 1. We see that our cross-city JFT approach outperforms all other baselines on all choices of topic numbers, which validates the superior performance when we consider the local features of a city and the user preference of novelty-seeking in cross-city scenarios, where it would be easy for other approaches to encounter the problem of cold-start.

6.3 Top-N Recommendation

In this subsection, we explore the performance of our approach in more practical top-N recommendation tasks. We adopt our JFTC for top-N recommendation algorithm with binary inputs and negative sampling described Sect. 5.4, and make comparison with the following baseline methods:

WRMF: Weighted Regularized Matrix Factorization described in [10], which is similar to LFM but applies weighted negative sampling to benefit top-N recommendations.

BPRMF: Bayesian Personalized Ranking (BPR) for MF presented in [24], which is the state-of-the-art algorithm for top-N recommendation based only on numerical ratings.

---

[1] http://nlp.stanford.edu/software/lex-parser.shtml.

HFT: The original HFT method achieves poor top-N performance in our settings. As a result, we optimize the HFT method with the same binary inputs and negative sampling approach as in our model for fair comparison.

To evaluate, we randomly hold out 5 records for each user, and provide top-5 recommendation list for each user, as with most practical applications. We adopt the measures of Precision@5 and NDCG@5, where the latter takes the positions of recommended items into consideration, and the results are shown in Table 2.

Table 2. Prec@5 and NDCG@5 with $K = 10$. Standard deviations for each method are ≤ 0.0006.

Method	WRMF	BPRMF	HFT	JFTC
Precision@5	0.0060	0.0057	0.0065	**0.0079**
NDCG@5	0.1616	0.1632	0.1653	**0.1790**

We see that both our JFTC approach and the HFT method (which make use of textual reviews) gain better performance than WRMF and BPRMF (which only make use of ratings). Further more, our JFTC method gains a 22% improvement against HFT in terms of precision, and 8.3% on NDCG, which is a superior achievement for practical applications.

Similar to the task of rating prediction, we also evaluate the top-N performance in cross-city settings. To do so, we select those user-city pairs that a user has at least 5 records in a city beyond his/her home city, and this results into 4,021 pairs corresponding to 1,739 users. We thus randomly hold out 5 records for a user in a corresponding city, and construct the recommendation list using the restaurants from that city, which gives us 4,021 lists in total for evaluation. We also conduct 5-fold cross-validation, and the standard deviations for both Precision and NDCG are ≤ 0.005. Figure 2 shows the results against the number of latent factors/topics K.

We see that the performance of our JFTC model is better than the baselines on nearly all choices of K, except that the NDCG of HFT is slightly better when $K = 50$, which means that our model sometimes may not rank the right items to the top, though with a much better precision. However, our approach still beats the baselines for nearly all the cases. Interestingly, we find that the overall cross-city performance is a magnitude better than that on the whole dataset. This means that user behaviours can be more predictable in cross-city settings, where users do visit local attractions beyond their historical preferences. This further validates the underlying intuition of our novelty-seeking component in the JFTC model for cross-city recommendations.

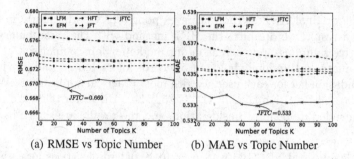

(a) RMSE vs Topic Number   (b) MAE vs Topic Number

Fig. 1. RMSE and MAE vs the number of topics or latent factors K in cross-city settings.

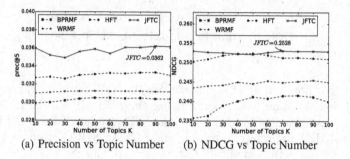

(a) Precision vs Topic Number   (b) NDCG vs Topic Number

Fig. 2. Precision@5 and NDCG@5 vs the number of topics or latent factors K in cross-city settings.

7   Conclusions

In this paper, we propose the Joint Factorizational Topic model that leverages the ratings and reviews in a collective manner for cross-city recommendation. For the first time, we examine the mathematical relationship between the LFM and LDA approaches for personalized recommendation, and we prove that vector-level randomization functions that are multiplicatively monotonic actually do not exist at all, although they are frequently used to bridge the LFM and LDA components in previous work. Fortunately, we also find that a direct product-level randomization approach can be used to bridge the two components and harmonize their behavior for model learning. Extensive experimental results on case studies, rating prediction, top-N recommendation, and inner-model analysis verified both the intuitional reasonability, theoretical basis, and the quantitative performance of our approach.

Acknowledgement. We thank the reviewers for their valuable suggestions. This work is supported by Natural Science Foundation of China (Grant Nos. 61532011, 61672311) and National Key Basic Research Program (2015CB358700).

Appendix

Let γ denote arbitrary vectors with length K, and $\mathcal{F}$ is the set of all randomization functions $f : \mathbb{R}^K \to \mathbb{R}^K$ satisfying:

$$\begin{cases} 0 \leq f(\gamma)_i \leq 1, \|f(\gamma)\|_1 = 1 \\ \gamma_i < \gamma_j \to f(\gamma)_i < f(\gamma)_j \end{cases}, \forall \gamma \in \mathbb{R}^K, 1 \leq i, j \leq K \qquad (20)$$

then there exists no randomization function $f \in \mathcal{F}$ with the product-level monotonic property of:

$$\gamma_1 \cdot \gamma_2 < \gamma_3 \cdot \gamma_4 \to f(\gamma_1) \cdot f(\gamma_2) < f(\gamma_3) \cdot f(\gamma_4), \ \forall \gamma_1, \gamma_2, \gamma_3, \gamma_4 \qquad (21)$$

PROOF: Suppose there exists a randomization function $f \in \mathcal{F}$ that meets Eq. (21). Let $t > 1$, and let α and β be vectors with $\alpha \cdot \beta > 0$, then we have $t\alpha \cdot \beta > \alpha \cdot \beta$. By applying the property of product-level monotonic in Eq. (21) we have:

$$f(t\alpha) \cdot f(\beta) > f(\alpha) \cdot f(\beta) \qquad (22)$$

and this can be equivalently written as:

$$\big(f(t\alpha) - f(\alpha)\big) \cdot f(\beta) > 0 \qquad (23)$$

Let $\Delta \doteq f(t\alpha) - f(\alpha)$, and according to the definition of randomization function in Eq. (20), we know that $\sum_k f(t\alpha)_k = \sum_k f(\alpha)_k = 1$, thus we have:

$$\sum_k \Delta_k = \sum_k \big(f(t\alpha) - f(\alpha)\big)_k = 0 \qquad (24)$$

According to Eq. (23) we know that $\Delta \neq \mathbf{0}$. Let $\mathcal{P}$ denote the indices of all positive elements in vector Δ, and $\mathcal{N}$ denote the indices of negative elements. We have:

$$\sum_{k \in \mathcal{P}} \Delta_k + \sum_{k \in \mathcal{N}} \Delta_k = 0 \qquad (25)$$

As Eq. (23) holds for any β with $\alpha \cdot \beta > 0$, without loss of generally, let β be a vector where $\beta_{k \in \mathcal{P}} = 0$ and $\beta_{k \in \mathcal{N}} = 1$. According to the vector-level monotonic property in Eq. (20) and the fact that $0 < 1$, we have $f(\beta)_{k \in \mathcal{P}} < f(\beta)_{k \in \mathcal{N}}$ and $0 \leq f(\beta)_{k \in \mathcal{P} \cup \mathcal{N}} \leq 1$. Combined with Eq. (25), we further obtain the following:

$$\Delta \cdot f(\beta) = \sum_{k \in \mathcal{P}} \Delta_k f(\beta)_k + \sum_{k \in \mathcal{N}} \Delta_k f(\beta)_k < 0 \qquad (26)$$

which is a direct contradiction with Eq. (23). As a result, there exists no randomization function $f \in \mathcal{F}$ that satisfies the product-level monotonic property in Eq. (21). $\qquad \square$

References

1. Aciar, S., Zhang, D., Simoff, S., Debenham, J.: Informed recommender: basing recommendations on consumer product reviews. Intell. Syst. **22**(3), 39–47 (2007)
2. Agarwal, D., Chen, B.C.: fLDA: matrix factorization through latent Dirichlet allocation. In: WSDM (2010)
3. Baatarjav, E.-A., Phithakkitnukoon, S., Dantu, R.: Group recommendation system for Facebook. In: Meersman, R., Tari, Z., Herrero, P. (eds.) OTM 2008. LNCS, vol. 5333, pp. 211–219. Springer, Heidelberg (2008). doi:10.1007/978-3-540-88875-8_41
4. Blei, D.M.: Probabilistic topic models. Commun. ACM **55**(4), 77–84 (2012)
5. Blei, D.M., Ng, A.Y., Jordan, M.I.: Latent Dirichlet allocation. JMLR **2003**(3), 993–1022 (2003)
6. Cremonesi, P., Koren, Y., Turrin, R.: Performance of recommender algorithms on top-N recommendation tasks. In: RecSys pp. 39–46 (2010)
7. Das, A., Datar, M., Garg, A., Rajaram, S.: Google news personalization: scalable online collaborative filtering. In: WWW pp. 271–280 (2007)
8. Davidson, J., Liebald, B., Liu, J., et al.: The YouTube video recommendation system. In: RecSys, pp. 293–296 (2010)
9. Ganu, G., Elhadad, N., Marian, A.: Beyond the stars: improving rating predictions using review text content. In: WebDB (2009)
10. Hu, Y., Koren, Y., Volinsky, C.: Collaborative filtering for implicit feedback datasets. In: ICDM (2008)
11. Jakob, N., Weber, S.H., Müller, M.C., et al.: Beyond the stars: exploiting free-text user reviews to improve the accuracy of movie recommendations. In: TSA (2009)
12. Knott, M., Bartholomew, D.: Latent Variable Models and Factor Analysis. Kendall's Library of Statistics 2 (1999)
13. Ko, M., Kim, H.W., Yi, M.Y., Song, J., Liu, Y.: MovieCommenter: aspect-based collaborative filtering by utilizing user comments. In: CollaborateCom (2011)
14. Konstan, J.A., Riedl, J.: Recommender systems: from algorithms to user experience. User Model. User-Adap. Interact. **22**(1–2), 101–123 (2012)
15. Koren, Y., Bell, R., Volinsky, C.: Matrix factorization techniques for recommender systems. Computer **42**, 8 (2009)
16. Koren, Y., Bell, R.: Advances in Collaborative Filtering. In: Ricci, F., Rokach, L., Shapira, B., Kantor, P.B. (eds.) Recommender Systems Handbook, pp. 145–186. Springer, Heidelberg (2011). doi:10.1007/978-0-387-85820-3_5
17. Lee, D.D., Seung, H.S.: Algorithms for non-negative matrix factorization. In: Proceedings of NIPS (2001)
18. Lika, B., Kolomvatsos, K., Hadjiefthymiades, S.: Facing the cold start problem in recommender systems. Expert Syst. Appl. **41**(4), 2065–2073 (2014)
19. Linden, G., Smith, B., York, J.: Amazon.com recommendations: item-to-item collaborative filtering. Internet Comput. **7**(1), 76–80 (2003)
20. Ling, G., Lyu, M.R., King, I.: Ratings meet reviews: a combined approach to recommend. In: RecSys (2014)
21. McAuley, J., Leskovec, J.: Hidden factors and hidden topics: understanding rating dimensions with review text. In: RecSys, pp. 165–172 (2013)
22. Pazzani, M.J., Billsus, D.: Content-based recommendation systems. In: Brusilovsky, P., Kobsa, A., Nejdl, W. (eds.) The Adaptive Web. LNCS, vol. 4321, pp. 325–341. Springer, Heidelberg (2007). doi:10.1007/978-3-540-72079-9_10
23. Purushotham, S., Liu, Y., Kuo, C.C.J.: collaborative topic regression with social matrix factorization for recommendation systems. In: ICML (2012)

24. Rendle, S., Freudenthaler, C., Gantner, Z., Thieme, L.S.: BPR: Bayesian Personalized Ranking from implicit feedback. In: UAI (2009)
25. Ricci, F., Rokach, L., Shapira, B.: Introduction to Recommender Systems Handbook. Springer, Heidelberg (2011)
26. Salakhutdinov, R., Mnih, A.: Bayesian probabilistic matrix factorization using Markov chain Monte Carlo. In: Proceedings of ICML (2008)
27. Salakhutdinov, R., Mnih, A.: Probabilistic matrix factorization. In: Proceedings of NIPS (2008)
28. Srebro, N., Rennie, J.D.M., Jaakkola, T.S.: Maximum-margin matrix factorization. In: NIPS (2005)
29. Su, X., Khoshgoftaar, T.M.: A survey of collaborative filtering techniques. In: Advances in AI, p. 4 (2009)
30. Takacs, G., Pilaszy, I., Nemeth, B., Tikk, D.: Investigation of various matrix factorization methods for large recommender systems. In: Proceedings of ICDM (2008)
31. Terzi, M., Ferrario, M.A., Whittle, J.: Free text in user reviews: their role in recommender systems. In: RecSys (2011)
32. Wang, C., Blei, D.M.: Collaborative topic modeling for recommending scientific articles. In: KDD (2011)
33. Xu, X., Datta, A., Dutta, K.: Using adjective features from user reviews to generate higher quality and explainable recommendations. IFIP Advances in Info. and Com. Tech. 389, 18–34 (2012)
34. Zhang, Y., Lai, G., Zhang, M., Zhang, Y., Liu, Y., Ma, S.: Explicit factor models for explainable recommendation based on phrase-level sentiment analysis. In: SIGIR (2014)
35. Zhang, Y., Zhang, H., Zhang, M., Liu, Y., et al.: Do users rate or review? boost phrase-level sentiment labeling with review-level sentiment classification. In: SIGIR (2014)

Aligning Gaussian-Topic with Embedding Network for Summarization Ranking

Linjing Wei[1,3], Heyan Huang[1,2], Yang Gao[1,2(✉)], Xiaochi Wei[1,3], and Chong Feng[1,2]

[1] Beijing Institute of Technology, Beijing, China
gyang@bit.edu.cn
[2] Beijing Engineering Research Center of High Volume Language Information Processing and Cloud Computing Applications, Beijing Institute of Technology, Beijing, China
[3] Beijing Advanced Innovation Center for Imaging Technology, Capital Normal University, Beijing 100048, People's Republic of China

Abstract. Query-oriented summarization addresses the problem of information overload and help people get the main ideas within a short time. Summaries are composed by sentences. So, the basic idea of composing a salient summary is to construct quality sentences both for user specific queries and multiple documents. Sentence embedding has been shown effective in summarization tasks. However, these methods lack of the latent topic structure of contents. Hence, the summary lies only on vector space can hardly capture multi-topical content. In this paper, our proposed model incorporates the topical aspects and continuous vector representations, which jointly learns semantic rich representations encoded by vectors. Then, leveraged by topic filtering and embedding ranking model, the summarization can select desirable salient sentences. Experiments demonstrate outstanding performance of our proposed model from the perspectives of prominent topics and semantic coherence.

Keywords: Query-oriented summarization · Embedding model · Topic

1 Introduction

Sentence ranking, a vital part of extractive summarization, has been extensively investigated. In typical query-oriented summarization ranking, it often stresses two key points, one is to select the coherent salient sentences, and the other is to capture the desired topic content from the query. To this end, most traditional ranking models [7,20,26] utilize features (e.g., term frequency or position) to generate final summarization. Feature engineering largely determines the final summarization performance. Although they have acceptable performance and efficiency advantage, they lack of deep understanding semantics mechanism, therefore are unable to extract salient summaries.

© Springer International Publishing AG 2017
L. Chen et al. (Eds.): APWeb-WAIM 2017, Part I, LNCS 10366, pp. 610–625, 2017.
DOI: 10.1007/978-3-319-63579-8_46

Delighted by the successful embedding models, such as word2vec [18,19] and PV [13], a great efforts [4,11,12,27] have been conducted to construct summarization model. These methods utilize embeddings directly to calculate relevant sentences. Moreover, the word embeddings are generated based on language models that depend on local context. However, they lack of structures that embrace topic characteristics at global level. Vector-based representation only including local context cannot well represent sentence semantics, therefore, is not able to capture desired topical content for summaries.

Some works have realised the importance of combining topic analysis and word embedding approaches [3,6,17]. Despite of great success of combining topics in word embedding model for many text mining and NLP tasks, at sentence level, less work focuses on representing sentences with semantic topics in summarization tasks to our knowledge. To incorporate sentence embedding and topical analysis for enhancing summarization effectiveness and accuracy, two interesting questions are arising: (1) how to represent the consistent space for documents, sentences and words, that can facilitate to discover coherent semantics; (2) how to extract sentences from data collection that are dependent on the focused topics from the user query.

In order to tackle these problems, we propose a novel query-oriented multidocument summarization approach, called **G**aussian-**T**opic with **E**mbeddings Network for Summarization **F**iltering and **R**anking (GTEFR). Our proposed model is designed to leverage the topical aspects for continuous vector representations to facilitate sentence modelling for summarization. Gaussian distributions capture a notion of centrality in space, hence semantically related sentences are clustered together in the space. In order to solve the first problem, we encodes a prior preference for semantic topics and learn the topic distribution by incorporating a gaussian distribution into embedding network. In this way, the trained sentence embeddings are semantically rich and coherent. As to the second problem, we extract topic related sentences according to the learnt topic distributions of both query and candidate sentences.

In this paper, we consider the learning process as a nonlinear embeddings from content in terms of Gaussian mixture model (GMM) and a neural network framework. The words, sentences and documents are represented by the GMM of vectors. Particularly for sentence embeddings, the semantically related sentences are localized in the space. Therefore, the topic categories can represent sentences in an abstraction way, which can assist sentence embedding in a high level of guidance. Besides, every word and document have their own topic assignment to assure the learning consistency. During training process, all the vector representations are learnt upon the neural network and gradually updated according to the word sequential context and topic assignments of sentences and documents. Once this process ends, the system outputs coherent encoded vectors of words, sentences, documents and mixture topic distributions of them. With the result for topic distributions of queries and sentences, irrelevant sentences are filtered out by utilising the consistent topics of queries and sentences, then query-oriented summarization has advantages reflected by content coherence and relevancy. We conduct experiments to verify the effectiveness of the proposed model on a benchmark dataset, and quantitatively demonstrate that our model outperforms those embedding models, the topic model, and the state-of-the-art topic-and-word embedding cooperation models.

The main contributions of our work include:

1. Aligning Gaussian-topic with embedding network is seamlessly integrated in the process of generating semantic sentences. Topics are jointly learnt in the GMM process as well as embedding learning process, which facilitate to aggregate topics accurately and capture intensive sentence semantics.
2. Representing queries and collection of sentences in terms of salient topics, in this way, desired topics are filtered especially based on the user specific needs. It enhances the topic coherence and relevance of summaries.
3. When referring to summarization tasks, we conduct experiments to demonstrate the effectiveness of our method for summarization ranking in real application.

The rest of this paper is organized as follows: Sect. 2 presents related work. We then proposed associated topics enhanced embedding model and summarization system in Sect. 3. Section 4 reports the experimental results and analysis. Finally, we conclude this paper in Sect. 5.

2   Related Work

Most existing extraction-based document summarization methods can be roughly divided into four categories, i.e., feature based method, deep learning based method, vector space based method and topic based method. Features such as term frequency [20,26], cue words [14] and topic theme [10] are used to measure the importance of words. There are also some methods utilizing deep neural network, such as, LSTM [8], RBM [16]. Because our model is related to continuous distributed vector representation and topic model, we focus on these two categories.

Since Mikolov et al. [18] proposed the efficient word embedding method, vector space model has attracted a growth of attention. But to the best of our knowledge, only [11,12] considered a direct summarization method using embeddings. Kågebäck et al. [11] proposed a summarization method, which maximized a submodular function defined by the summation of cosine similarity measure based on sentence embeddings. A summarization method based on document-level similarity [12] was proposed by Kobayashi et al., and they examined an objective function defined by a cosine similarity based on document embeddings instead of sentence embeddings.

Topic-based methods are widely applied in the summarization task. Parveen et al. [22] proposed an approach, which is based on a weighted graphical representation of documents obtained by topic modeling. Barzilay and Lee [1] used the Hidden Markov Model to learn a latent topic for each sentence. They chose the "important" topics that had the high probability of generating summary sentences. The work proposed by Gupta et al. [9] measured topic concentration in a direct manner: a sentence was considered relevant to the query if it contained at least one word from the query. While these work assume that documents related to the query only talk about one topic. Tang et al. [23] proposed a unified probabilistic approach to uncover query-oriented topics and four scoring methods to calculate the importance of each sentence in the document collection.

Although their success in the summarization task, semantic coherence and topic information are not encoded in these models. Sentence embedding with topic model has not previously been used in summarization tasks as far as we know, but there are same models incorporating vector representations and topics in other NLP tasks [3,6,17,25]. Das et al. [6] developed a variant of LDA that operated on continuous space embeddings of words rather than word types to impose a prior, which helped topics to be semantically coherent. Liu et al. [17] employed latent topic models to assign topics for each word in the text corpus, and learned topical word embeddings based on both words and their topics. Cao et al. [3] made use of the word embeddings available [18] and a neural network to explain the topic model on the embedding space. However, these methods learn the vector representations of words, rather than sentences, which is unfavourable for discovering comprehensive meaning on the summarization task. Our work is inspired by the recent neural probabilistic language models [25], which considered the ordering of words and the semantic meaning of sentences into topic modeling. While the work in [25] ignores the topic coherence among adjacent words. In our model, we add the sentence topic into context information which ensures topic coherence among adjacent words.

3   Proposed Model

In this section, we describe the details of the proposed method, which contains two stages, i.e. training embeddings and constructing summary based on the embeddings. Firstly, the embedding structure will be introduced, which seamlessly connects GMM and a embedding network. Then, we describe how to apply the proposed embedding method for summarization.

3.1   The Embedding Model GTE

The proposed model, aligning Gaussian-Topic with Embedding Network (GTE), is inspired by the recent work in learning vector representations of words and sentences using neural networks [13,17–19,25], we deploy a novel word prediction framework to exploit intensive sentence semantics. The model is built upon the GMM-based topic assumption of a continuous vector space. It is composed by two parts, the GMM centralised modelling and neural network learning framework, which is shown in Fig. 1. Given a document (sentence or word), the topical information is conveyed by its embedding vectors through the GMM. To be specific, the probability of words is calculated by the forward neural network and a

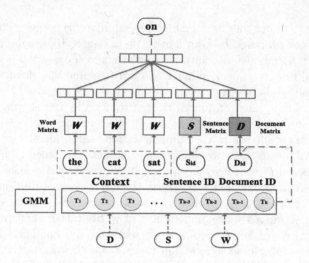

Fig. 1. The model architecture of the GTE model

nonlinear transformation function. In this section, we present the details of the proposed model GTE.

Embedding Process

Table 1 summarizes the notations. Given the collection of parameters of GMM, we use

$$P(x|\lambda) = \sum_{k=1}^{K} \pi_k N(x|\mu_k, \Sigma_k) \qquad (1)$$

to represent the probability distribution for sampling a vector x from the Gaussian mixture model.

Table 1. Notations

Symbol	Description	Symbol	Description
K	Number of topics	$vec(d)$	Vector of document d
W	Words collection	$vec(s)$	Vector of sentence s
S	Sentence collection	$vec(w)$	Vector of word w
D	Document collection	ϕ	All vector representation set
M	Number of documents	π	Mixture weights of GMM
d_i	The ith document	μ	Means of GMM
s_{ij}	The jth sentence in document d_i	Σ	Covariance matrices of GMM
$\mathbf{T}_{d_i}$	The topic vector assigned to d_i	λ	The parameters collection of GMM
$\mathbf{T}_{s_{ij}}$	The topic vector assigned to s_{ij}		

Given the model parameters λ and the vectors for documents, we can infer the posterior probability distribution of topics. For each document d_i, the posterior distribution of its topic is

$$q(z(d_i) = z) = \frac{\pi_z N(vec(d_i)|\mu_z, \Sigma_z)}{\sum_{k=1}^{K} \pi_k N(vec(d_i)|\mu_k, \Sigma_k)} \tag{2}$$

Based on the distribution, the topic of document d_i can be vectorized as $[q(z(d_i) = 1), q(z(d_i) = 2), \cdots, q(z(d_i) = K)]$.

Similarly, for each sentence s_{ij} in the document d_i, the topic of sentence s_{ij} can be vectorized as $[q(z(s_{ij}) = 1), q(z(s_{ij}) = 2), \cdots, q(z(s_{ij}) = K)]$.

The basic idea of GTE is that we model one word as a prediction task based on word sequential context and topic assignments of sentences and documents. Given the Gaussian mixture model λ, the predicted process is described as follows:

1. For each document d_i in corpus D
   (a) Choose a topic $T_{d_i} \sim \pi := (\pi_1, \pi_2, \cdots, \pi_T)$
   (b) Choose a vector representation $vec(d_i) \sim N(\mu_{T_{d_i}}, \Sigma_{T_{d_i}})$
   (c) For each sentence s_{ij} in document d_i
       i. Choose a topic $T_{s_{ij}} \sim \pi := (\pi_1, \pi_2, \cdots, \pi_T)$
       ii. Choose a vector representation $vec(s_{ij}) \sim N(\mu_{T_{s_{ij}}}, \Sigma_{T_{s_{ij}}})$
       iiii. For each word w_t in sentence s_{ij}
             A. Choose a topic $T_{w_t} \sim \pi := (\pi_1, \pi_2, \cdots, \pi_T)$
             B. Choose a vector representation $vec(w_t) \sim N(\mu_{T_{w_t}}, \Sigma_{T_{w_t}})$
2. For t-th word $vec(w_t)$ in sentence s_{ij}
   (a) Predict w_t according to the documents vector $vec(d_i)$ and topic $\mathbf{T}_{d_i}$, the current sentences vector $vec(s_{ij})$ and topic $\mathbf{T}_{s_{ij}}$, as well as at most m previous words in the same sentence.

Given the t-th location in j-th sentence, we represent its word realization by w_t, the objective of GTE is to maximize the probability

$$G_t = P(w_t|d_i, s_{ij}, w_{t-m}, \cdots, w_{t-1}) \tag{3}$$

As aforementioned, the assumption is that the word is predicted by the representation of the word's sentence and document as well as their assigned topics. Besides, as we all know, dot-product represents similarity of two vectors, so maximizing the similarity is utilized to guide vector embedding and topic mutually. Thus, we obtained the following objective function

$$G_t = \sigma(\mathbf{T}_{d_i} vec(d_i)^T + \mathbf{T}_{s_{ij}} vec(s_{ij})^T + \sum_{n=1}^{m} \mu_n^{w_t} vec(w_{t-n})^T) \tag{4}$$

where $\sigma(x) = 1/(1 + \exp(-x))$ and $\mu_n^{w_t} \in R^V$ is parameter of the model.

Combining the equations above, we use GMM as a prior and generate the next word by a sigmoid conditional distribution. So the log-likelihood of the generative model can be described as

$$J_t = log(P(\phi|\lambda)) + log(\sigma(\mathbf{T}_{d_i} vec(d_i)^T + \mathbf{T}_{s_{ij}} vec(s_{ij})^T + \sum_{n=1}^{m} \mu_n^{w_t} vec(w_{t-n})^T))$$

(5)

Estimating Model Parameters

The model parameters $\{\lambda, \mu_n^{w_t}, \phi\}$ are estimated by maximizing the likelihood of the generative model. A two-phase iteration process is conducted, as shown in Algorithm 1. Given $\{\mu_n^{w_t}, \phi\}$, the parameters of the Gaussian mixture model $\lambda = \{\pi_k, \mu_k, \Sigma_k\}$ are estimated by Expectation Maximization (EM) algorithm. Given λ, stochastic gradient descent (SGD) is adopted to find the optimized result.

Figure 2 shows a small example of the results after estimating model parameters and learning. We list several vector and topic representations of documents, sentences as well as words on document collection D301 of DUC2005. As shown in Fig. 2, the vector and topic representations of document d_1 and d_2, coming from document collection D301, can be learned since the GTE model is performed on multiple documents of the D301. Likewise, the proposed model also learn vector and topic representations of s_{11} and s_{21}, w_1 and w_2. What's more, w_1 and w_2 are selected from s_{11} and s_{21}, respectively.

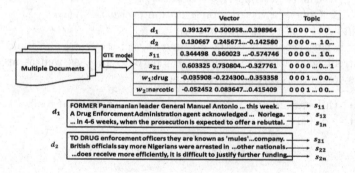

Fig. 2. The examples of vectors and topics

3.2   The Summarization Model GTEFR

We regard the informative results from the GTE model on different granularities of topical distributions, sentence and word embeddings as an exhaustive mining of the whole collection of documents. As a result, it will strongly support our summarization ranking system.

Figure 3 shows the graphical structure of the summarization model. In order to extract desired topic content and salient sentences, we construct two submodules i.e. sentence filtering and sentence ranking . Sentence filtering facilitates to

Algorithm 1

Input:
Documents D, sentences S, $|W|$ words contained in dictionary W, the learning rate α, the dimension of the vector V and the number of topics K.

Output:
Topic representations $\mathbf{T}_w$, $\mathbf{T}_s$ and $\mathbf{T}_d$. vector representations $vec(w)$, $vec(s)$ and $vec(d)$.

1: Randomly initialize parameters.
2: Fixing parameters $\mu_n^{w_t}$ and ϕ, run the EM algorithm:

   E-Step:

$$\gamma(i,k) = \frac{\pi_k N(x_i|\mu_k,\Sigma_k)}{\sum_{j=1}^{K} \pi_j N(x_i|\mu_j,\Sigma_j)}$$

   M-Step:

$$N_k = \sum_{i=1}^{N} \gamma(i,k)$$
$$\mu_k = \frac{1}{N_k} \sum_{i=1}^{N} \gamma(i,k)x_i$$
$$\Sigma_k = \frac{1}{N_k} \sum_{i=1}^{N} \gamma(i,k)(x_i - \mu_k)(x_i - \mu_k)^T$$
$$\pi_k = \frac{N_k}{N}$$

3: Fixing parameters λ, run the SGD algorithm:

   For each document d_i

      For the t-th word w_t of sentence s_{ij}

         Update the vector of document d_i, s_{ij} and w_t

         Update the factor of influence $\mu_n^{w_t}$

$$vec(d_i) \leftarrow vec(d_i) + \alpha \frac{\partial J_t(\mu_n^{w_t},\phi)}{\partial vec(d_i)}$$
$$vec(s_{ij}) \leftarrow vec(s_{ij}) + \alpha \frac{\partial J_t(\mu_n^{w_t},\phi)}{\partial vec(s_{ij})}$$
$$vec(w_t) \leftarrow vec(w_t) + \alpha \frac{\partial J_t(\mu_n^{w_t},\phi)}{\partial vec(w_t)}$$
$$vec(\mu_n^{w_t}) \leftarrow vec(\mu_n^{w_t}) + \alpha \frac{\partial J_t(\mu_n^{w_t},\phi)}{\partial vec(\mu_n^{w_t})}$$

      End For

   End For

---

filter out those irrelevant sentences to the user desired content from topic perspective. while for ranking, the system considers the performance of the summary at two different levels simultaneously i.e. word and sentence level.

Sentence Filtering

Sentence filtering in our summarization framework aims to learn the salient topics for updating a query-focused sentence collections and filter out those irrelevant sentences quickly and efficiently. Specifically, the GTE process outputs topic distributions of query and each sentence. With obtaining the topic distribution of one query, we can rank the topic of the query in descending order. Subsequently, the salient topics set $salient(T)$ of query can be defined as $salient(T) = \{T_k|k > p, k \in \{1,2,\cdots,K\}\}$, where T_k denotes the kth topic after ranking the topic distribution and p is a parameter in our model.

Similarly, with obtaining the topic distribution of one certain sentence, we select the top p sentence topics. If one of these topic is contained in the salient topics set $salient(T)$, the sentence will be appended to the new query-focused sentence collections.

Sentence Ranking

As obtaining word and sentence embedding after the GTE process, we use a weighted sum of all scores of the two different levels (word and sentence level) to measure the importance of sentences in the new query-focused sentence collections. Then, the sentence score is described as follows:

$$Score(s) = \beta \underbrace{\sum_{t=1}^{n_w} TF(w_t) + \gamma \sum_{t=1}^{n_w} Sim(vec(w_t), vec(Q_1))}_{word \quad level} + \underbrace{\delta Sim(vec(s), vec(Q_2))}_{sentence \ level} \qquad (6)$$

where n_w represents the number of words in sentence s. Q_1 is the word-based query[1], which is several key words and contributions to focus the user interests roughly. Q_2 represents the sentence-based query[2], which is a complete sentence and relates to a in-depth and comprehensive aspect of the subject. $Sim()$ represents the function of similarity, and we use cosine similarity in this paper. β, γ and δ are parameters in our summarization model.

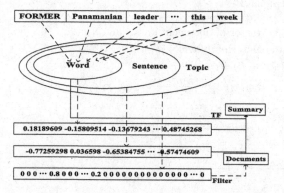

Fig. 3. The graphical structure of the summarization model

At word level, the importance of words is calculated by a weighted sum of term frequency and the relevant score between sentence s_{ij} and word-based query. Vector representation of a sentence retains the coherently semantic information. The most coherent and important sentences are selected by the cosine similarity which represents the similarity between each sentence and query.

4   Experiments

In this section, we present the evaluation of the proposed approach. The hypothesis is that our model is effective in two aspects as follows: (1) Incorporating vector

---

[1] In DUC, the word-based query is also called "title", such as "New hydroelectric projects".

[2] In DUC, the sentence-based query is also called "narrative", such as "What hydroelectric projects are planned or in progress and what problems are associated with them?".

representations of sentences and topics is contributed to capture the coherent semantic meaning and focused topical content, (2) our model is effective in summarization tasks.

4.1 Data Sets

In this study, we use the standard summarization benchmark DUC2005[3] for evaluation. DUC2005 contains 50 query-oriented summarization tasks. For each query, a relevant document cluster is assumed to be "retrieved", which contains 25–50 documents. Thus, the task is to generate a summary from the document cluster for answering the query[4]. The length of a result summary is limited by 250 tokens (whitespace delimited).

For the benchmark data sets, we preprocessed each document by(a) removing stopwords; (b) removing words that appear less than one time in the corpus; and (c) downcasing the obtained words.

4.2 Evaluation Measures

We conducted evaluations by ROUGE [15] metrics. The measure evaluates the quality of the summarization by counting the number of overlapping units, such as n-grams. Basically, ROUGE-N is an n-gram recall measure. Among the evaluation methods implemented in Rouge, Rouge-1 emphasises on the occurrence of the same words between candidate summary and reference summary, while Rouge-2 and Rouge-SU4 focus more on the readability of the candidate summary. We use these three metrics in the experiment.

4.3 Baseline Models

We compare the GTE model with several query-focused summarization methods.

1. **TF:** this model uses term frequency [20] for scoring words and sentences.
2. **Lead:** take the first sentences one by one from the document in the collection, where documents are ordered randomly. It is often used as an official baseline of DUC.
3. **Avg-DUC05:** average system-summarizer performance on DUC2005.
4. **LDA:** this method uses Latent Dirichlet Allocation [2] to learn the topic model. After learned the topic model, we give max score to the word of the same topic with query. The reader can refer to the paper [23] for the details.
5. **LLRSum:** This system [5] employs a log-likelihood ratio (LLR) test to select topic words. The sentence importance score is equal to the number of topic words divided by the number of words in the sentence.
6. **SNMF:** this system [24] is for topic-biased summarization. it used symmetric non-negative matrix factorization (SNMF) to cluster sentences into groups, then selected sentences from each group for summary generation.

---

[3] http://duc.nist.gov/data.html.
[4] In DUC, the query is also called "narrative" or "topic".

7. **Word2Vec:** the vector representations of words can be learned by Word2Vec [18,19] models. We use the vectors and calculate the three features, where the sentence-level representations is calculated by using a average of all word embeddings in the sentence.
8. **PV:** PV [13] learns sentence vectors based on Word2Vec Model. Thus, we use the same parameters as that in our approach to calculate the scores of sentences without word similarity.
9. **TWE:** TWE [17] learns topical word embeddings based on both words and their topics. The three features are calculate with the same parameters, where the sentence-level representations is calculated by using a average of all word embeddings in the sentence.
10. **GTER:** Comparing with GTEFR, we implement the summarization utilize sentence ranking submodule without sentence filtering.

4.4 Implementation Details

In the training stage, the learning rate α is set to 0.026 and gradually reduced to 0.0001. For each word, at most m = 6 previous words in the same sentence is used as the context. The word vector size is set to the same as the number of topics V = K = 100. These parameters (m, V, K) are empirically set in our experiment.

In the summarization stage, we use the MERT [21] to tune parameters for GTER and GTEFR, which is showed in Fig. 4(a) and (b). γ, δ and ρ are tuned from 1 to 10, with the step size of 0.1. What's more, the parameter p is tuned from 1 to 4, which is showed in Fig. 4(c), and finally we set $p = 2$. As mentioned above, a relevant document cluster is assigned to a query. Thus, documents related to the query talk about a main topic. Then our experiment show that a document cluster averagely talks about two topics. Similar result has been mentioned in Tang et al. [23], who think a document cluster talks about multiple topics.

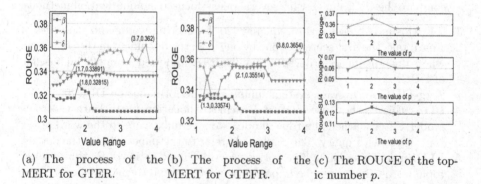

(a) The process of the MERT for GTER. (b) The process of the MERT for GTEFR. (c) The ROUGE of the topic number p.

Fig. 4. (a) the MERT for GTEFR and (b) the MERT for GTEFR. (c) ROUGE via tuning the parameter p

All experiments were carried out on a CentOS 7.2 Server with four Dual-Core Intel Xeon processors (2.6 GHz) and 8 GB memory. It took about 3 days for estimating the GTE model. There is a problem of high time complexity in training process. But we all know the training process is offline, and then the efficiency is not primary issue.

4.5   Experimental Results

In this subsection, we give the results of the experiments and the analysis.

Table 2. Results of the baseline methods and the GTE model on DUC2005

Method	Rouge-1	Rouge-2	Rouge-SU4
TF	0.3356	0.0580	0.1107
Lead	0.3354	0.0565	0.1102
Avg-DUC05	0.3434	0.0602	0.1148
LDA	0.3170	0.0533	0.1150
LLRSum	0.3363	0.0599	0.1166
SNMF	0.3501	0.0604	0.1229
Word2Vec	0.3459	0.0548	0.1154
PV	0.3541	0.0614	0.1186
TWE	0.3505	0.0606	0.1202
GTER	0.3600	0.0633	0.1231
GTEFR	**0.3620**	**0.0636**	**0.1244**

The results for the proposed model and baseline models are reported in Table 2. As shown in the table, the scores in bold are the highest ones in the column. It can be observed that our model gives the best summary compare to any other method in ROUGE measurements, which strongly testifies the effectiveness of the proposed summarization model. Meanwhile, our system significantly outperforms these methods based on features (i.e. TF and Lead). It is because that our method wisely considers deep semantics and desired topic content.

Impact by Topics. From Table 2, we can also obtain that, comparing with the embedding model like Word2Vec and PV, the GTER and GTEFR are always outperforming in all three ROUGE measurements. As aforementioned, Word2Vec and PV learn vector representations mainly based on the local context information, none of them take into consideration of global topic information that contributes to capture coherent semantics and obtain the desired topic content. This is main reason why our models are outperforming those state-of-the-arts high-quality embedding methods for summarization. Besides, the performance of the GTEFR model exceeds the GTER model, which also strongly demonstrates the effectiveness of sentence filtering based on topics for obtaining desired topics.

To reveal the reason why the topic-based approaches (e.g., GTEFR) can outperform the word-based approaches and the embedding-based approaches (e.g.,

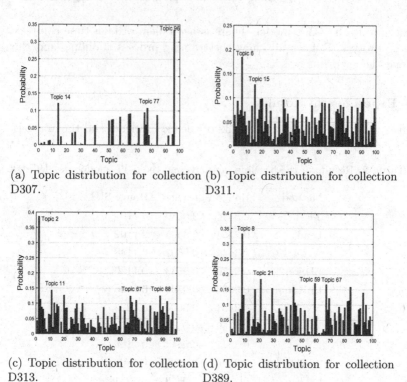

(a) Topic distribution for collection D307.

(b) Topic distribution for collection D311.

(c) Topic distribution for collection D313.

(d) Topic distribution for collection D389.

Fig. 5. Topic distribution for collections. The x axis denotes topics and the y axis denotes the occurrence probability of each topic in collections

TF, Word2Vec and PV), a topical analysis was performed on the document collections. As the Fig. 5 shows, we calculated the probability distribution of each topic, since for a certain sentence, its topic can be learned by the GTE model. From the Fig. 5(a), we see that there is a major topic (Topic 96, talking about new hydroelectric projects), but still same information is captured by the other topics (e.g., ♯14 and ♯77). While word-based approaches and embedding-based approaches can not learn the latent topic structure. It indicates that the proposed model can accelerate to focus the salient topics and further enhance summarization accuracy. Besides, topic coverage can be guaranteed, which facilitates to meet the summary need. We have also found that topics are unbalanced distributed in a collection. However, viewing the Fig. 5 integratedly, topics are covered by all collections, which denotes the data is balanced distributed rather than tend to a certain category. What's more, each document can be distinguished by topic distribution. Hence, aligning Gaussian-topic effectively boosts to extract topic information for summarizer.

Impact by Sentence Modelling. Our proposed model performs better than the TWE model, and the PV model performs better than the Word2Vec model. These results suggest that sentence-level methods can facilitate to discover more semantic contents than the word-level method. On the other hand, our model and the PV model perform better than the Word2Vec, especially in Rouge-2 and Rouge-SU4. As mentioned above, Rouge-2 and Rouge-SU4 focus more on the readability of the candidate summary. Because our model considers vector representation of sentence which retains the complete and coherent semantic content. What's more, our model is better than all the topic-based baseline models, such as LDA, LLRSum and SNMF. LDA aims to cover the main topics of documents and LLRSum pays more attention to find topic word. Although SNMF select sentences from each topic group, it ignores the importance of salient topics. More importantly, none of them take into consideration of sentence coherence. All of the aforementioned analysis of results prove that encoding sentences and topics into the same process can facilitate to discover coherent and intensive semantics, especially for summarization tasks.

In order to explore impact of the designed measure of word and sentence similarity deeply, we remove the part of features and keep the rest of feature parameters consistent. Table 3 presents the contribution analysis of each feature to the final summary generation. From the table, we can observe that sentence embedding achieves the best performance for summarization. What's is more, the performance of word embedding is comparable to general baseline models. we can imply that sentence similarity computation by our proposed sentence embedding plays the dominant impact for the summary and aligning Gaussian-topic with the neural network is effectively boosting to capture intensive sentence semantics.

Table 3. Contribution analysis of each feature to the final summary generation

Method			Rouge-1	Rouge-2	Rouge-SU4
TF-IDF	word_sim	sen_sim			
√	√		0.3228	0.0500	0.1107
	√	√	0.3308	0.0528	0.1146
√		√	0.3562	0.0641	0.1198
√	√	√	0.3654	0.0686	0.1254

5   Conclusion

A new model is proposed to leverage topics and vectors to capture complete sentence meaning. In the model, the topical embedding, document, sentence and word embeddings are jointly learnt through the seamlessly combinative GMM model and embedding network, and then supportively adapted to extract the more semantic relevant, coherent and topical focused summaries. Comparing to all baselines summarization approaches, our proposed model outperforms in all three ROUGE measurements. Besides, the proposed approach is quite flexible, which can also be applied to other areas, such as question answering and text classification.

624   L. Wei et al.

Acknowledgments. The work was supported by National Nature Science Foundation of China (Grant No. 61602036), National Basic Research Program of China (973 Program, Grant No. 2013CB329303), Beijing Advanced Innovation Center for Imaging Technology (BAICIT-2016007).

References

1. Barzilay, R., Lee, L.: Catching the drift: probabilistic content models, with applications to generation and summarization. Comput. Sci. 113–120 (2004)
2. Blei, D.M., Ng, A.Y., Jordan, M.I.: Latent dirichlet allocation. JMLR **3**(Jan), 993–1022 (2003)
3. Cao, Z., Li, S., Liu, Y., Li, W., Ji, H.: A novel neural topic model and its supervised extension. In: Proceedings of AAAI 2015 (2015)
4. Chen, K.Y., Liu, S.H., Wang, H.M., Chen, B., Chen, H.H.: Leveraging word embeddings for spoken document summarization. Comput. Sci. 1383–1387 (2015)
5. Conroy, J.M., Schlesinger, J.D., O'Leary, D.P.: Topic-focused multi-document summarization using an approximate oracle score. In: Proceedings of ACL 2006 (2006)
6. Das, R., Zaheer, M., Dyer, C.: Gaussian LDA for topic models with word embeddings. In: Proceedings of ACL 2015 (2015)
7. Galley, M.: A skip-chain conditional random field for ranking meeting utterances by importance. In: Proceedings of EMNLP 2007 (2006)
8. Ghosh, S., Vinyals, O., Strope, B., Roy, S., Dean, T., Heck, L.: Contextual LSTM (CLSTM) models for large scale NLP tasks (2016)
9. Gupta, S., Nenkova, A., Jurafsky, D.: Measuring Importance and Query Relevance in Topic-Focused Multi-document Summarization (2007)
10. Harabagiu, S., Lacatusu, F.: Topic themes for multi-document summarization. In: Proceedings of SIGIR 2005 (2005)
11. Kågebäck, M., Mogren, O., Tahmasebi, N., Dubhashi, D.: Extractive summarization using continuous vector space models. In: Proceedings of EACL 2014 (2014)
12. Kobayashi, H., Noguchi, M., Yatsuka, T.: Summarization based on embedding distributions. In: Proceedings of EMNLP 2015 (2015)
13. Le, Q.V., Mikolov, T.: Distributed representations of sentences and documents. Comput. Sci. 1188–1196 (2014)
14. Lin, C.Y., Hovy, E.: The automated acquisition of topic signatures for text summarization. In: Proceedings of COLING 2000 (2000)
15. Lin, C.Y., Hovy, E.: Automatic evaluation of summaries using n-gram co-occurrence statistics. In: Proceedings of ACL 2003 (2003)
16. Liu, Y.: Query-oriented multi-document summarization via unsupervised deep learning. In: Proceedings of AAAI 2012 (2012)
17. Liu, Y., Liu, Z., Chua, T.S., Sun, M.: Topical word embeddings. In: Proceedings of AAAI 2015 (2015)
18. Mikolov, T., Chen, K., Corrado, G., Dean, J.: Efficient estimation of word representations in vector space. CoRR abs/1301.3781 (2013)
19. Mikolov, T., Sutskever, I., Chen, K., Corrado, G., Dean, J.: Distributed Representations of Words and Phrases and Their Compositionality (2013)
20. Nenkova, A., Vanderwende, L., Mckeown, K.: A compositional context sensitive multi-document summarizer: exploring the factors that influence summarization. In: Proceedings of SIGIR 2006 (2006)
21. Och, F.J.: Minimum error rate training in statistical machine translation. In: Proceedings of ACL 2003 (2003)

22. Parveen, D., Ramsl, H., Strube, M.: Topical coherence for graph-based extractive summarization. In: Proceedings of EMNLP 2015 (2015)
23. Tang, J., Yao, L., Chen, D.: Multi-topic based query-oriented summarization. In: Proceedings of SDM 2009 (2009)
24. Wang, D., Li, T., Zhu, S., Ding, C.: Multi-document summarization via sentence-level semantic analysis and symmetric matrix factorization. In: Proceedings of SIGIR 2008 (2008)
25. Yang, M., Cui, T., Tu, W.: Ordering-sensitive and semantic-aware topic modeling. In: Proceedings of AAAI 2015 (2015)
26. Yih, W.T., Goodman, J., Vanderwende, L., Suzuki, H.: Multi-document summarization by maximizing informative content-words. In: Proceedings of IJCAI 2007 (2007)
27. Yin, W., Pei, Y.: Optimizing sentence modeling and selection for document summarization, In: Proceedings of IJCAI 2015 (2015)

Improving Document Clustering for Short Texts by Long Documents via a Dirichlet Multinomial Allocation Model

Yingying Yan[1,2], Ruizhang Huang[1,2,3(✉)], Can Ma[1,2], Liyang Xu[1,2], Zhiyuan Ding[1,2], Rui Wang[1,2], Ting Huang[1,2], and Bowei Liu[1,2]

[1] College of Computer Science and Technology, Guizhou University, Guiyang, Guizhou, China
yyingy0921@gmail.com, rzhuang@gzu.edu.cn, canma.love@gmail.com, lyxu90@gmail.com, subsontding@gmail.com, ruiwang1239@gmail.com, durant.huang@gmail.com, bwei.liu@gmail.com
[2] Guizhou Provincial Key Laboratory of Public Big Data, Guizhou University, Guiyang, Guizhou, China
[3] State Key Laboratory for Novel Software Technology, Nanjing University, Nanjing, People's Republic of China

Abstract. Document clustering for short texts has received considerable interest. Traditional document clustering approaches are designed for long documents and perform poorly for short texts due to the their sparseness representation. To better understand short texts, we observe that words that appear in long documents can enrich short text context and improve the clustering performance for short texts. In this paper, we propose a novel model, namely DDMA*fs*, which (1) improves the clustering performance of short texts by sharing structural knowledge of long documents to short texts; (2) automatically identifies the number of clusters; (3) separates discriminative words from irrelevant words for long documents to obtain high quality structural knowledge. Our experiments indicate that the DDMA*fs* model performs well on the synthetic dataset and real datasets. Comparisons between the DDMA*fs* model and state-of-the-art short text clustering approaches show that the DDMA*fs* model is effective.

Keywords: Short text clustering · Dirichlet multinomial allocation · Gibbs sampling algorithm

1 Introduction

With the rapid development of the Internet, huge amount of short texts are generated. Short text clustering is of great interest for many applications. For example, document clustering for twitter messages is of substantial usage for analyzing the public opinions and interests. However, directly applying traditional document clustering models to short texts is with poor performance. The main reason is that the representation of short texts is highly sparse. Short texts

© Springer International Publishing AG 2017
L. Chen et al. (Eds.): APWeb-WAIM 2017, Part I, LNCS 10366, pp. 626–641, 2017.
DOI: 10.1007/978-3-319-63579-8_47

are with a strict limit on the text length. For instance, twitter restricts the number of words in 140 characters for each message. As a result, discriminative terms are in short and the number of common terms shared by related short texts is small.

Compared with short texts, long documents are with rich content and a large amount of discriminative terms. There are a number of document clustering approaches that achieve promising performance for discovering latent structure for long documents. Besides, it is practical to collect long documents related to short texts in real usage. For example, related long documents of twitter messages can be found from various document sources, such as news websites and blogs sites of news analysis. Therefore, to deal with the sparse representation problem of short texts, it would be useful if the high quality structural knowledge discovered from long documents can be shared to short texts to improve the understanding of short texts.

In practice, not every word in long documents is useful. Long document is normally represented by a number of discriminative words and a large amount of non-discriminative words. Only discriminative words are useful for grouping documents. The involvement of irrelevant noise words confuses the clustering process and leads to poor clustering solution for long documents which limit the effect of sharing the structural knowledge of long documents to improve the document clustering performance for short texts. This situation aggravates when the number of clusters are unknown.

The second challenge for short text clustering is to determine the number of clusters. Traditional short text clustering approaches consider the number of cluster as a predefined parameter. However, given large-scale short texts, users have to scan the whole document collection with the purpose of estimating the number of clusters. Apparently, it is time-consuming. In addition, inappropriate estimations of the number of clusters misdirect the short text clustering process and lead to bad clustering results.

In this paper, we propose a novel model, namely Dual Dirichlet Multinomial Allocation with feature selection (DDMA*fs*) to (1) improve the discovery of document structure for short texts by sharing structural knowledge of long documents; (2) relieve the effect of poor quality of long document representation by separating discriminative words from non-discriminative words automatically; (3) automatically identify the number of clusters of both long documents and short texts simultaneously. DDMA*fs* model is developed based on the Dirichlet Multinomial Allocation model (DMA) [3] which shows promising performance on document clustering for both long documents [6,18] and short texts [17]. Long documents and short texts share the same set of latent clusters so that the structural knowledge can be transferred from long documents to short texts. Discriminative words are automatically separated from non-discriminative words for long documents. On the other hand, all terms in short texts are regarded as discriminative due to the sparse representation problem of short texts. Latent structure of short texts is further improved by only using the structural knowledge discovered from high quality discriminative words of long documents.

To determine the number of clusters, a Gibbs sampling algorithm is developed for the DDMAfs model. When a new data point arrives, it either rises from existing clusters or starts a new cluster. The number of clusters for short texts K_S and long documents K_L are discovered automatically along the Gibbs sampling algorithm. Noted that K_S and K_L are not necessarily the same in our development. It is more practical for users to collect a large amount of long documents without the needed to guarantee that every long document should be directly related to short texts.

We have conducted extensive experiments on our proposed model by using both synthetic and realistic datasets. We compared our approach with state-of-the-art document clustering approaches. Experimental results show that our proposed approach is effective.

2   Related Work

Existing works mainly focused on utilizing external resources to enrich the contexts of short texts. [14,19] aggregated the short texts into lengthy pseudo-documents for training topic model. Hong and Davison [4] presented several schemes to train a standard topic model with aggregated messages from Twitter. Hotho et al. [5] integrated Wordnet into the clustering process. In [10,11,16], Wikipedia is considered as a background base to enrich the knowledge of short texts.

There has been little work on sharing structural knowledge of long documents to short texts. In [9], Jin et al. proposed the Dual Latent Dirichlet Allocation (DLDA) model which enhances short text clustering by incorporating auxiliary long texts. However, DLDA is a probabilistic finite mixture model and the number of clusters is pre-defined. In reality, the number of clusters should be determined after the clustering process rather than got in advance.

Some methods have been introduced to find an estimation of the number of clusters K. The direct solution is to train the model with different value of K and pick the one with the highest likelihood on held-out dataset [12]. Another way is to assign a prior to K and then compute the posterior distribution of K to determine the most probable number of clusters [2]. In [17], Yin and Wang inferred the number of clusters by the GSDMM model. DPMFS model [18] and DPMFP model [6] was proposed to estimate the document collection structure by utilizing the Dirichlet Process model. However, none of these approaches considers to automatically infer number of clusters for long documents and short texts simultaneously.

3   DDMAfs Model

Formally, we define the following notations:

- A word w is the basic unit of discrete data, defined to be an item from a vocabulary indexed by $\{1, \cdots, V\}$.

- A document can be represented as V-dimensional vector $x_d = \{x_{d1}, \cdots, x_{dV}\}$, where x_{dj} is the number of appearance of the j-th word in the document x_d.
- A dataset is a collection of D documents which are composed of two parts. The first part is long document set which is a collection of L long documents denoted by $D_L = \{x_1, x_2, \cdots, x_L\}$. The other is short text set which is a collection of S short texts denoted by $D_S = \{x_1, x_2, \cdots, x_S\}$.

The DDMA*fs* model is a generative probability model for long documents and short texts. Following the feature partition model mentioned in [6], a latent binary vector γ is used to partition words of long documents to two groups, in particular, the discriminative words and non-discriminative words. A mixture of components is used to generate short texts and discriminative words of long documents, where each component corresponds to a latent cluster characterized by a distribution over words. Non-discriminative words for long documents are generated from a background cluster. The generative process of the DDMA*fs* model is as follows:

1. Choose $\gamma_j \mid \omega \sim B(1, \omega)$, where $j = 1, 2, \cdots, V$.
2. Choose $\phi_k \mid \beta \sim \text{Dirichlet}(\beta_1, \beta_2, \cdots, \beta_{V.})$, where $k = 1, 2, \cdots, K$.
3. Choose $\phi_0 \mid \lambda \sim \text{Dirichlet}(\lambda_1, \lambda_2, \cdots, \lambda_V)$.
4. Choose $\theta_S \mid \alpha \sim \text{Dirichlet}(\frac{\alpha}{K}, \frac{\alpha}{K}, \cdots, \frac{\alpha}{K})$;
   Choose $\theta_L \mid \alpha \sim \text{Dirichlet}(\frac{\alpha}{K}, \frac{\alpha}{K}, \cdots, \frac{\alpha}{K})$.
5. Choose $z_s \mid \theta_S \sim \text{Discrete}(\theta_{S1}, \theta_{S2}, \cdots, \theta_{SK})$, where $s = 1, 2, \cdots, S$;
   Choose $z_l \mid \theta_L \sim \text{Discrete}(\theta_{L1}, \theta_{L2}, \cdots, \theta_{LK})$, where $l = 1, 2, \cdots, L$.
6. Choose $x_s \mid \phi_{z_s} \sim \text{Multinomial}(\mid x_s \mid; \phi_{z_s})$, where $s = 1, 2, \cdots, S$;
   Choose $x_l \cdot \gamma \mid \phi_{z_l}, \gamma \sim \text{Multinomial}(\mid x_l \mid_\gamma; \phi_{z_l})$;
   Choose $x_l \cdot (1 - \gamma) \mid \phi_0, \gamma \sim \text{Multinomial}(\mid x_l \mid_{1-\gamma}; \phi_0)$, where $l = 1, 2, \cdots, L$.

where ω is the parameter of the Bernoulli distribution, which represents the probability of each word expected to be discriminative. $|x_s|$ and $|x_l|$ are the total appearance of words in a short text x_s or a long document x_l, respectively. ϕ_k is the multinomial parameter representing the cluster k. K is the overall total number clusters for both long documents and short texts. L and S are the number of long documents and short texts respectively. The K-dimensional parameter θ_S and θ_L are the mixture weights of clusters for short texts and long documents, respectively; z_s and z_l indicate the latent cluster assigned to short text x_s and long document x_l, respectively. The graphical representation of DDMA*fs* model is shown in Fig. 1.

The approximation of the probability density function of the dataset D_S and D_L given $\{z_1, z_2, \cdots, z_S\}$, $\{z_1, z_2, \cdots, z_L\}$, and γ can be represented as follows:

$$p(D_s | z_1, \cdots, z_S) \approx \prod_{s=1}^{S} \frac{|x_s|!}{\prod_{v=1}^{V} x_{sv}!} \cdot Q_\beta \tag{1}$$

$$p(D_L | z_1, \cdots, z_L, \gamma) \approx \prod_{l=1}^{L} \frac{|x_l|!}{\prod_{v=1}^{V} x_{lv}!} \cdot Q_{\beta,\lambda} \cdot Q_\beta \cdot Q_\lambda \tag{2}$$

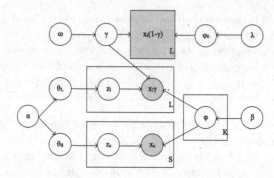

Fig. 1. Graphical representation of DDMA*fs* Model.

$$Q_{\beta,\lambda} = \left(\frac{\Gamma(\sum_{v=1}^{V} \beta_v)}{\prod_{v=1}^{V} \Gamma(\beta_v)} \right)^K \cdot \frac{\Gamma(\sum_{v=1}^{V} \lambda_v)}{\prod_{v=1}^{V} \Gamma(\lambda_v)} \tag{3}$$

$$Q_\beta = \prod_{k=1}^{K} \frac{\prod_{v=1}^{V} \Gamma(\beta_v + \sum_{\{s:z_s=k\}} x_{sv} + \sum_{\{l:z_l=k\}} x_{lv}\gamma_v)}{\Gamma(\sum_{v=1}^{V} \beta_v + \sum_{v=1}^{V}(\sum_{\{s:z_s=k\}} x_{sv} + \sum_{\{l:z_l=k\}} x_{lv}\gamma_v))} \tag{4}$$

$$Q_\lambda = \frac{\prod_{v=1}^{V} \Gamma(\lambda_v + \sum_{l=1}^{L} x_{lv}(1-\gamma_v))}{\Gamma(\sum_{v=1}^{V} \lambda_v + \sum_{v=1}^{V} \sum_{l=1}^{L} x_{lv}(1-\gamma_v))} \tag{5}$$

4   Algorithm

In this section, a blocked Gibbs sampling algorithm is designed to infer the latent clusters and select discriminative words for long documents simultaneously.

For the DDMA*fs* model, the state of Markov chain is $\bar{U} = \{\gamma, \theta_L, \theta_S, \phi, zs, zl\}$, where $\gamma = \{\gamma_1, \cdots, \gamma_V\}$, $zl = \{z_1, \cdots, z_L\}$, $zs = \{z_1, \cdots, z_S\}$, $\phi = \{\phi_0, \cdots, \phi_K\}$. After initializing latent variables $\{\gamma, zl, zs\}$ and parameters $\{\alpha, \beta, \lambda, \omega\}$, the blocked Gibbs sampling inference procedure is as follows:

(1) Update the latent discriminative words indicator γ by repeating the following Metropolis step R times: a new candidate γ_{new} which adds or deletes a discriminative word is generated by randomly picking one of the V indices in γ_{old} and changing its value. The new candidate is accepted with the probability:

$$min\{1, \frac{p(\gamma_{new} \mid D_L, zl)}{p(\gamma_{old} \mid D_L, zl)}\} \tag{6}$$

where $p(\gamma \mid D_L, zl) \propto p(D_L \mid \gamma, zl) \cdot p(\gamma)$ and $p(D_L \mid \gamma, zl)$ is provided by Eq. (4).

(2) Conditioned on the other latent variables, for $k = \{1, 2, \cdots, K\}$, if k is not in $\{z_1^*, z_2^*, \cdots, z_{K^*}^*\}$, draw ϕ_k from a dirichlet distribution with parameter

β. Otherwise, update ϕ_k by sampling a value from a dirichlet distribution with parameter:

$$\left\{\beta_1 + \sum_{x_l:z_l=k} x_{l1}\gamma_1 + \sum_{x_s:z_s=k} x_{s1},\cdots,\beta_V + \sum_{x_l:z_l=k} x_{lV}\gamma_V + \sum_{x_s:z_s=k} x_{sV}\right\} \quad (7)$$

(3) Update ϕ_0 by sampling a value from a dirichlet distribution with parameter:

$$\left\{\lambda_1 + \sum_{l=1}^{L} x_{l1}(1-\gamma_1),\cdots,\lambda_V + \sum_{l=1}^{L} x_{lV}(1-\gamma_V)\right\} \quad (8)$$

(4) Update θ_L by sampling a value from a dirichlet distribution with parameter:

$$\left\{\frac{\alpha}{K} + \sum_{l=1}^{L} I(z_l=1),\cdots,\frac{\alpha}{K} + \sum_{l=1}^{L} I(z_l=K)\right\} \quad (9)$$

where $I(z_l=k)$ is an indicator function which equals to 1 if $z_l=k$.

(5) Update θ_S by sampling a value from a dirichlet distribution with parameter:

$$\left\{\frac{\alpha}{K} + \sum_{s=1}^{S} I(z_s=1),\cdots,\frac{\alpha}{K} + \sum_{s=1}^{S} I(z_s=K)\right\} \quad (10)$$

where $I(z_s=k)$ is an indicator function which equals to 1 if $z_s=k$.

(6) Conditioned on the other latent variables, for $l=\{1,\cdots,L\}$, update z_l by sampling a value from a discrete distribution with parameter $\{p_{l1},\cdots,p_{lK}\}$ where

$$\sum_{k=1}^{K} p_{lk} = 1 \text{ and } p_{lk} \propto \theta_{Lk} p(x_l \mid \phi_k, \phi_0, \gamma) \quad (11)$$

(7) Conditioned on the other latent variables, for $s=\{1,2,\cdots,S\}$, update z_s by sampling a value from a discrete distribution with parameter $\{q_{s1},\cdots,q_{sK}\}$ where

$$\sum_{k=1}^{K} q_{sk} = 1 \text{ and } q_{sk} \propto \theta_{Sk} p(x_s \mid \phi_k) \quad (12)$$

Note that the inference procedure focus on three parameters, in particular z, ϕ and θ which are closely related to the allocation of documents to clusters, the cluster representative parameters, and the cluster weight partitions. The parameter z, K, and γ are needed to be initialized. Other parameters are sampled during the inference process without the necessity of initialization. In our inference process, z is simply initialized by selecting a random cluster from $\{1,2,\cdots,K\}$. The number of cluster K is initialized with a reasonably large value and can be automatically learned during the inference process. γ is initialized by randomly choosing one discriminative word in the dataset. All other words are initialized as non-discriminative with the value of γ equal to 0. The number of cluster estimated, denoted by K^*, is then determined by the size of $\{z_1^*, z_2^*, \cdots, z_{K^*}^*\}$ which is the set of non-empty clusters. K^* is much less than K with all the empty clusters eliminated. We can further divide K^* to K_S and K_L which are the number of clusters estimated for short texts and long texts. Note that K_S and K_L are not necessarily the same. Through experimental study, we found that K_S and K_L are close to the true value of the number of clusters.

After the Markov chain has reached its stationary distribution, we collect H samples of $\{\gamma_1, \gamma_2, \cdots, \gamma_V\}$. we regard a word w_j is discriminative when the average value of γ_j of the last H samples is bigger than a threshold σ. (We set σ as 0.7 in our experiments).

5   Experiment

We study the performance of our proposed DDMAfs model by two sets of experiments. For the first set of experiments, synthetic datasets are used. For the second set of experiments, the DDMAfs model is evaluated on real document datasets. For both set of experiments, we used a standard document evaluation metric, in particular, Normalized Mutual Information (NMI) to evaluate the clustering performance [20].

5.1   Synthetic Dataset Experiments

Experimental Datasets. We derived a synthetic dataset to evaluate the performance of our proposed DDMAfs model. The synthetic dataset consists of 3000 data points, in which 600 data points are used to represent short texts and 2400 data points are regarded as long documents. All data points were generated from 6000 features, in which 2000 features are regarded as discriminative features. The remaining 4000 features are regarded as non-discriminative features. For all data points, we derived one multinomial distribution to generate non-discriminative features for long documents. The parameter of the multinomial distribution, for non-discriminative features denoted as π_0, was generated randomly. Six multinomial distributions were used to represent latent clusters of discriminative features. Parameters of the six multinomial distributions, denoted as $\{\pi_1, \cdots, \pi_6\}$, were generated following the stick breaking approach of Dirichlet distribution [1]. In particular, one specific multinomial parameter $\pi_k = (u_1, \cdots, u_{2000})$ was generated as follows:

(1) For the first feature f_1, draw ι_1 from $Beta(\in_1, \sum_{j=2}^V \in_j)$ and then assign the probability of f_1 with ι_1, denoted as u_1.
(2) For feature f_i, where $2 \leq i < 2000$, draw ι_i from $Beta(\in_i, \sum_{j=i+1}^V \in_j)$ and then assign the probability of f_i, denoted as u_i, as follows:

$$u_i = \iota_i \pi_{j=0}^{i-1}(1 - \iota_j) \tag{13}$$

(3) u_{2000} is set with the remaining probability to ensure that $\sum_1^{2000} u_i = 1$.

In our experiment, we set $\epsilon_i = 0.5$, where $i = 1, 2, \cdots, V$.

Each short text data point consists of 15 features. All features are regarded as discriminative generated from a multinomial mixture model with five components $\{\pi_1^S, \cdots, \pi_5^S\}$, where $\{\pi_1^S, \cdots, \pi_5^S\}$ is a subset of $\{\pi_1, \cdots, \pi_6\}$ and was selected randomly. In particular, the generation process of a short text data point x_i is as follows:

(1) Randomly select a cluster $\pi_k^S \in \{\pi_1^S, \cdots, \pi_5^S\}$
(2) Draw $x_i \sim Multinomial(\pi_k^S, 15)$

Each long document data point consists of 2000 features. The probability of a feature in long document data points to be discriminative is set to 0.6. Discriminative features were derived from a multinomial mixture model with 6 components $\{\pi_1^L, \cdots, \pi_6^L\}$. Non-discriminative features were generated from a multinomial distribution with parameter π_0. In particular, the generation process of a long document data point $x = (f_1, \cdots, f_i, \cdots, f_{2000})$ is as follows: For each feature f_i of x:

(1) Randomly select a probability $p_i \sim U[0, 1]$
(2) If $p_i > 0.6$, then
   a) randomly select a cluster $\pi_k^L \in \{\pi_1^L, \cdots, \pi_6^L\}$
   b) draw $f_i \sim Multinomial(\pi_k^L, 1)$
(3) Otherwise draw $f_i \sim Multinomial(\pi_0, 1)$

For our proposed DDMA*fs* model, we set $K = 30$, $\alpha = 0.1$, $\beta = 0.1$, $\omega = 0.01$, and $\lambda = 0.1$. The Metropolis step R was set to be 200. We ran our proposed DDMA*fs* model 10 times. The performance is computed by taking the average of these 10 experiments. Each experiment was conducted with 2000 iterations in which the first 500 as burn-in.

Experimental Performances. We investigated the clustering performance for our proposed DDMA*fs* model by varying the number of long documents. Experimental results are depicted in Fig. 2. From experimental results, it shows that our proposed DDMA*fs* model is effective for improving the clustering performance for short texts by transferring high quality structural knowledge discovered from long documents. When the number of long documents is equal to 0, the DDMA*fs* model is reduced to the ordinary DMA model. Clustering performances can be obviously improved when the number of long documents is increased. The improvement of the clustering performances is significant with a relatively small number of long documents. When the number of long documents is reasonably large, the DDMA*fs* model identifies almost perfect cluster structure.

We also investigated the performances of the DDMA*fs* model in one typical run by varying the number of iterations. The number of long documents involved is set to 2400. From Fig. 3, it shows that the DDMA*fs* model reaches to a stable result within a few hundred iterations. Figures 4 and 5 demonstrate the number of discriminative features, the number of clusters for short texts estimated, and

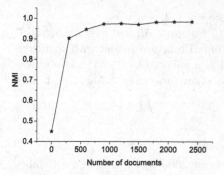

Fig. 2. Clustering performance of the DDMA*fs* model on the synthetic dataset with different number of long documents.

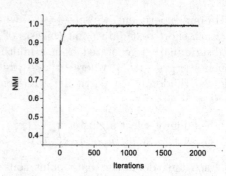

Fig. 3. Trace plot for NMI performance of the DDMA*fs* model for the synthetic dataset.

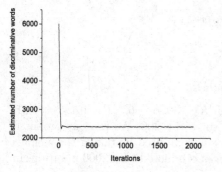

Fig. 4. Trace plot for the number of discriminative features for the synthetic datasets.

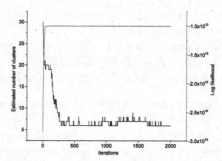

Fig. 5. Trace plot for the number of clusters for short text data points and log likelihood of the synthetic dataset.

the value of log likelihood with each iteration. The result shows that the number of discriminative features estimated is 2398 which is slightly larger than the real number. The feature selection process is faster to stabilize than the number of clusters estimated. The value of log likelihood increases obviously with decreasing number of clusters. DDMA*fs* model estimated 6 number of clusters for short texts and 14 clusters for long documents. The DDMA*fs* model is able to identify different numbers of clusters for short texts and long documents. Note that the number of clusters estimated are slightly larger than the real ones. However, after removing those extremely small clusters, the number of clusters for both short texts and long documents are exactly the same with the real ones. The cluster assignments for data points are depicted in Figs. 6 and 7. Apparently, short data

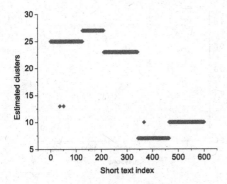

Fig. 6. Estimated cluster labels of the short text data points.

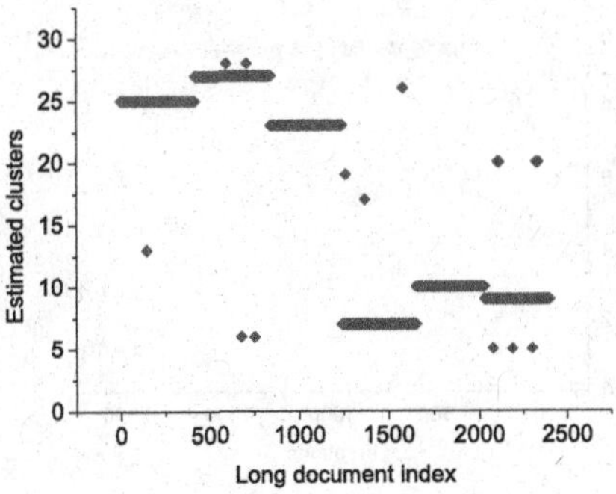

Fig. 7. Estimated cluster labels of the long document data points.

points are partitioned into 5 clusters and long data points are partitioned into 6 clusters which are the exact numbers of real clusters.

To evaluate the generality on the performance of the DDMA*fs* model, we derived another synthetic dataset with 450 short texts and 1200 long documents by using similar strategy discussed before. Short texts were randomly selected from 4 classes. Long documents are organized in 3 classes. We obtained similar experimental performances except two observations: (1) the converging process is slower; (2) the NMI value, which is 0.90, is slightly worse than the previous synthetic experiments. The main reason is that only part of short texts is improved by long documents because the number of classes of short texts is smaller than long documents.

Discussions. In this section, we investigated the sensitivity of the choices of hyper parameters α, β, λ, and ω for the DDMA*fs* model. Experiments were conducted on various values of these parameters. There are some other parameters, in particular, the initial number of clusters K and the Metropolis step parameter R. We discuss the setting of these parameters in detail in the following part of the section.

Choice of α, β, λ and ω: We investigated the sensitivity of the choice of hyper parameters α, β, λ, and ω with the performance of the DDMA*fs* model. α, β, and λ were set to 0.1, 1, and 10 which corresponds to a small, moderate, and large prior values. We also experimented with different values of ω where ω was set to be a small value 0.01, a moderate value 0.1, and a large value 1. For the different values of α, β, λ, and ω, we set other values of parameters with the same setting discussed before. Figures 8, 9, 10 and 11 show the clustering performance for short texts by varying the number of iterations. The DDMA*fs* model achieved almost perfect clustering structure in all these experiments. This indicates that the DDMA*fs* model is robust to the choice of hyper parameters.

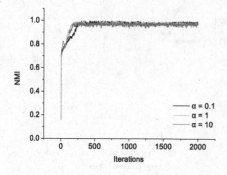

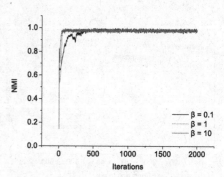

Fig. 8. The NMI result for short text data points when α gets different values.

Fig. 9. The NMI result for short text data points when β gets different values.

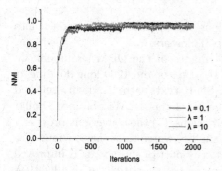

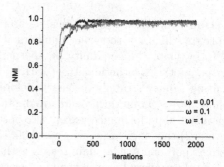

Fig. 10. The NMI result for short text data points when λ gets different values.

Fig. 11. The NMI result for short text data points when ω gets different values.

Choice of K and R: Theoretically, we should choose K to be the number of data points. In the process of experiment, we discovered that it is time-consuming. So we chose a relatively small K follow the advice of [7]. The number of Metropolis step R was set to be 200 in our algorithm because we found that lager value on R had little improvement on the clustering quality.

5.2 Real Dataset Experiments

Experimental Datasets. Two real-word text corpora were used to generate experimental datasets for conducting experiments. The first corpus is the AMiner-Paper collection [13]. Three different research areas were chosen, in particular, the "graphical image", the "computer network", and the "database", to form a subset of the AMiner-Paper collection. The first experimental dataset, namely, AMpaperSet, was then derived from the subset by extracting 600 paper titles for short texts and another 450 paper abstracts for long documents. The second dataset is called the TweetSet dataset. We crawled 79413 tweets from 3 hot topics on twitter from the hashtag "JeSuisParis", "RefugeesWelcome" and "PlutoFlyby". In these tweets, there are 6399 tweets containing URLs and accessible URLs are 5577, we therefore crawled the content of the accessible URLs to form the set of long documents. The TweetSet dataset was then derived by randomly selecting 2400 documents from the crawled long documents and 600 documents from tweets.

We pre-processed all the datasets by stop-word removal. The summary of these two text document datasets is shown in Table 1.

Table 1. Summary description of datasets.

Datasets	L	S	V	K
AMpaperSet	450	600	3586	3
TweetSet	2400	600	33462	3

(L: Long Text Sets, S: Short Text Sets, V: Vocabulary size, K: Number of clusters.)

Experimental Setup and Experimental Performances. For all set of experiments, we use the same parameter settings of α, β, ω, λ, and R of the synthetic dataset to real dataset experiment. For the number of initial clusters K, we set K to 30 for both AMpaperSet and TweetSet dataset.

For comparative study, we compare our models with four other approaches. We investigated the standard K-Means document clustering model taking the bag-of-word assumption as the first approach, labeled as KMEANS [8]. The KMEANS approach is used as the benchmark. The number of cluster is required as the input parameter in this model. The second and third approaches are state-of-art document clustering approaches for short texts, in particular, GSDMM [17] and STCC [15]. The GSDMM approach is designed based on the dirichlet multinomial mixture model. A collapsed gibbs sampling algorithm is employed to infer the number of clusters automatically. The STCC approach is designed with the help of convolutional neural networks, and takes the number of clusters as a pre-defined parameter. The fourth approach, labeled as DLDA model, is the most recent short text clustering model which transfers structural knowledge learned from auxiliary long documents to short texts [9]. Although DLDA model utilizes the long documents to cope with sparse problem, it can't infer the number of clusters automatically. The KMEANS, GSDMM and STCC approaches are not specially designed for studying how long documents improve the clustering performance of short texts. Therefore, the type of data points can't be identified in these two models. We evaluated the experimental performances with and without long document data points for the KMEANS and GSDMM approaches respectively. For experiments with long documents, we merged short text and long document data points to form a single dataset and evaluated the performance on short texts only. We studied the performance of KMEANS, DLDA, and STCC when right or wrong number of clusters are given. Each comparative experiment was run 10 times. The performance is computed by taking the average of the NMI results on short text data points of these 10 experiments. The clustering performance of long documents is not the focus of our paper.

Table 2. Cluster performance on short texts on the AMpaperSet and TweetSet datasets.

	AMpaperSet	TweetSet
DDMA*fs*	0.465	0.557
KMEANS (s)	0.136	0.126
KMEANS (K = 2)	0.341	0.091
KMEANS (K = 3)	0.428	0.075
KMEANS (K = 10)	0.260	0.293
GSDMM (s)	0.376	0.432
GSDMM	0.430	0.539
STCC (s)	0.16	0.18
DLDA (K = 2)	0.287	0.277
DLDA (K = 3)	0.342	0.311
DLDA (K = 10)	0.187	0.232

(KMEANS (s) (GSDMM (s), or STCC (s)) indicates the clustering performance of KMEANS (GSDMM, or STCC) approach without the aid of long documents.)

Table 2 depicts document clustering performances acquired by the DDMA*fs*, KMEANS, GSDMM, STCC, and DLDA models on the AMpaperSet and Tweet-Set dataset. From the experimental results, our proposed DDMA*fs* model apparently performs better compared with all other models. Therefore, the DDMA*fs* model is effective for discovering the latent document structure of short texts. Note that the DDMA*fs* model reduces to the ordinary DMA model, which shares the same document generation process with the GSDMM model, when no long documents are available. Compared the experimental performances of the DDMA*fs* and the GSDMM(s), it is obvious that long documents are able to help the clustering performance of short texts. The clustering performance can be greatly improved with the help of long documents. In all experiments with long documents, the DDMA*fs* model outperforms all other models. There are two main reasons. Firstly, long document data points are with a great number of non-discriminative features which deduce the quality of structural knowledge shared to short texts for all other models. Secondly, the KMEANS and GSDMM models are not able to identify long document and short text data points which results in losing the information on cluster partition of short texts and long documents.

Table 3. Number of clusters on short texts estimated on the AMpaperSet and Tweet-Set datasets.

	DDMA*fs*	GSDMM(s)	GSDMM
AMpaperSet	20	21	27
TweetSet	17	25	23

Table 3 shows the estimated number of clusters on two real dataset. Among the five methods, KMEANS, STCC and DLDA are given the true value of K. All estimation on the number of clusters are larger than the true one due to the reason of outlier documents. The DDMA*fs* model obtains a relatively accurate estimation compared with GSDMM on two real datasets.

Table 4. Top 8 words of three typical larger clusters discovered by DDMA*fs* model on the AMpaperSet dataset.

Cluster	Top words
1	Transact process optimization cost distribute system file memory
2	Sensor wireless traffic node rout protocol rate channel
3	Query relation model language object operation update semantic

Table 4 shows top 8 words of three typical larger clusters on the AMpaperSet dataset discovered by our proposed DDMA*fs* model. DDMA*fs* model captures meaningful words associated with three clusters, in particular, "graphical image", "computer network", and "database". For some general clusters, the DDMA*fs* model subdivides the cluster to more specific sub-clusters as the number of clusters is unknown. As a result, more clusters are discovered than the real ones. As shown in Table 4, cluster 1 and 3 are two related clusters under the general cluster of "database".

We explores the impact of long documents to short texts on the TweetSet dataset. Similar performance was obtained with synthetic dataset as shown in Fig. 2. We can find that the clustering performance is improved with increase of the number of long documents. It shows that the clustering effect of short texts will level up with the help of long documents in the real application.

6   Conclusion and Future Work

In this paper, we propose a DDMA*fs* model for the problem of short text clustering. Structural knowledge of long documents are shared to short texts so that the clustering performance of short texts can be greatly improved. A blocked Gibbs sampling technique is proposed to infer the cluster structure of short text set as well as the latent discriminative word subset of long document set. Our experiment shows that our approach achieves good performance with a reasonable set of long documents. The comparisons between DDMA*fs* and existing state-of-the-art models indicate that our approach is effective.

An interesting direction for future research is to study how to enhance the clustering of short texts via utilizing multi-source dataset rather than only long documents. Besides, we also concern to involve a small number of supervised information on long documents for improving the performance of short text clustering.

Acknowledgments. This work is supported by Nation Science Foundation of China (Nos. 61462011, 61202089), Introduced Talents Science Projects of Guizhou University (No. 2016050), Major Applied Basic Research Program of Guizhou Province (Grant No. JZ20142001) and the Graduate Innovated Foundation of Guizhou University Project (Nos. 2011015, 2016051).

References

1. Bela, A., Frigyik, A., Gupta, M.: Introduction to the dirichlet distribution and related processes. Department of Electrical Engineering, University of Washington (2010)
2. Cheeseman, P., Kelly, J., Self, M., Stutz, J., Taylor, W., Freeman, D.: Autoclass: a Bayesian classification system. In: Readings in Knowledge Acquisition and Learning, pp. 431–441. Morgan Kaufmann Publishers Inc., Burlington (1993)
3. Green, P.J., Richardson, S.: Modelling heterogeneity with and without the dirichlet process. Scand. J. Stat. **28**(2), 355–375 (2001)
4. Hong, L., Davison, B.D.: Empirical study of topic modeling in Twitter. In: Proceedings of the First Workshop on Social Media Analytics, pp. 80–88. ACM (2010)
5. Hotho, A., Staab, S., Stumme, G.: Wordnet improves text document clustering. In: Proceedings of the SIGIR 2003 Semantic Web Workshop, pp. 541–544 (2003)
6. Huang, R., Yu, G., Wang, Z., Zhang, J., Shi, L.: Dirichlet process mixture model for document clustering with feature partition. IEEE Trans. Knowl. Data Eng. **25**(8), 1748–1759 (2013)
7. Ishwaran, H., James, L.F.: Gibbs sampling methods for stick-breaking priors. J. Am. Stat. Assoc. **96**(453), 161–173 (2001)
8. Jain, A.K.: Data clustering: 50 years beyond k-means. Pattern Recogn. Lett. **31**(8), 651–666 (2010)
9. Jin, O., Liu, N.N., Zhao, K., Yu, Y., Yang, Q.: Transferring topical knowledge from auxiliary long texts for short text clustering. In: Proceedings of the 20th ACM International Conference on Information and Knowledge Management, pp. 775–784. ACM (2011)
10. Phan, X.H., Nguyen, C.T., Le, D.T., Nguyen, L.M., Horiguchi, S., Ha, Q.T.: A hidden topic-based framework toward building applications with short web documents. IEEE Trans. Knowl. Data Eng. **23**(7), 961–976 (2011)
11. Phan, X.H., Nguyen, L.M., Horiguchi, S.: Learning to classify short and sparse text & web with hidden topics from large-scale data collections. In: Proceedings of the 17th International Conference on World Wide Web, pp. 91–100. ACM (2008)
12. Smyth, P.: Model selection for probabilistic clustering using cross-validated likelihood. Stat. Comput. **10**(1), 63–72 (2000)

13. Tang, J., Zhang, J., Yao, L., Li, J., Zhang, L., Su, Z.: Arnetminer: extraction and mining of academic social networks. In: Proceedings of the 14th ACM SIGKDD International Conference on Knowledge Discovery and Data Mining, pp. 990–998. ACM (2008)

14. Weng, J., Lim, E.P., Jiang, J., He, Q.: TwitterRank: finding topic-sensitive influential twitterers. In: Proceedings of the Third ACM International Conference on Web Search and Data Mining, pp. 261–270. ACM (2010)

15. Xu, J., Wang, P., Tian, G., Xu, B., Zhao, J., Wang, F., Hao, H.: Short text clustering via convolutional neural networks. In: Proceedings of NAACL-HLT, pp. 62–69 (2015)

16. Yang, C.L., Benjamasutin, N., Chen-Burger, Y.H.: Mining hidden concepts: using short text clustering and wikipedia knowledge. In: 2014 28th International Conference on Advanced Information Networking and Applications Workshops (WAINA), pp. 675–680. IEEE (2014)

17. Yin, J., Wang, J.: A dirichlet multinomial mixture model-based approach for short text clustering. In: Proceedings of the 20th ACM SIGKDD International Conference on Knowledge Discovery and Data Mining, pp. 233–242. ACM (2014)

18. Yu, G., Huang, R., Wang, Z.: Document clustering via dirichlet process mixture model with feature selection. In: Proceedings of the 16th ACM SIGKDD International Conference on Knowledge Discovery and Data Mining, pp. 763–772. ACM (2010)

19. Zhao, W.X., Jiang, J., Weng, J., He, J., Lim, E.-P., Yan, H., Li, X.: Comparing Twitter and traditional media using topic models. In: Clough, P., Foley, C., Gurrin, C., Jones, G.J.F., Kraaij, W., Lee, H., Mudoch, V. (eds.) ECIR 2011. LNCS, vol. 6611, pp. 338–349. Springer, Heidelberg (2011). doi:10.1007/978-3-642-20161-5_34

20. Zhong, S.: Semi-supervised model-based document clustering: a comparative study. Mach. Learn. **65**(1), 3–29 (2006)

Intensity of Relationship Between Words: Using Word Triangles in Topic Discovery for Short Texts

Ming Xu, Yang Cai, Hesheng Wu, Chongjun Wang(✉), and Ning Li

National Laboratory for Novel Software Technology, Nanjing University,
Nanjing 210023, China
xuming0830@gmail.com, caiyang023@gmail.com, weierson@smail.nju.edu.cn,
{chjwang,ln}@nju.edu.cn

Abstract. Uncovering latent topics from given texts is an important task to help people understand excess heavy information. This has caused the hot study on topic model. However, the main texts available daily are short, thus traditional topic models may not perform well because of data sparsity. Popular models for short texts concentrate on word co-occurrence patterns in the corpus. However, they do not consider the intensity of relationship between words. So we propose the new way, called word-network triangle topic model (WTTM). In WTTM, we search for the word triangles to measure the relations between words. The results of experiments on real-world corpus show that our method performs better in several evaluation ways.

Keywords: Topic model · Short texts · Word co-occurrence network · Document classification

1 Introduction

Recent years, the technology of Internet is growing rapidly, people are more willing to get information from the Internet. Information surplus becomes a big problem. So topic distillation is crucial for many content analysis tasks. And a popular research field is topic model.

Topic model [4] is used to discover latent topics from given texts. It assumes that a document is generated from a mixture of topics, where a topic is a probabilistic distribution over words. Topic model is helpful in automatically extracting thematic information from large archive [1]. It's widely used in modeling text collections like news articles, research papers and blogs. Now the popular models are PLSA (also as PLSI) [5] and LDA [2].

However, with the great popularity of social networks and Q&A networks, short texts occupy the main space on the Web. Inferring latent topics based on these texts may help a lot in daily life. However, a bad news is that traditional topic models do not perform well in these situations. The main reason may

© Springer International Publishing AG 2017
L. Chen et al. (Eds.): APWeb-WAIM 2017, Part I, LNCS 10366, pp. 642–649, 2017.
DOI: 10.1007/978-3-319-63579-8_48

rely on that the parameters of topic model are calculated according to the co-occurrence of words in the texts, while short texts is short of such relations because of lack of words [6].

Under the circumstances, we need some improvements on topic model for short texts. One simple but useful idea is to bring together the similar short documents to structure longer documents. For example, combining the texts written by the same user in twitter [9] or combining the texts which own the same words [6]. Some others add extra information into short texts to enrich data density. However, methods above are highly data-dependent or task-dependent. Another popular approach now is to extend the documents based on the information from original texts, like BTM [10], RIBS-TM [7] and WNTM [11]. Though these models have good performance on the short texts, they ignore that the relations of the word co-occurrence patterns are not all so close. As we can see, not any two words in the documents share similarly close relation.

To solve the data sparsity on short texts and value relationship between words, we propose the method called $Word - network\ Triangle\ Topic\ Model$ (WTTM). The method came from the thoughts below. (1) Though topic model infers the parameters based on the word co-occurrence patterns, but not all patterns share the same intensity of relations. (2) How can we express the closer relationship among words as the word pairs do not show the intensity of relationship. Triangle is an important structure in graph theory. It represents strong relationship among the nodes in the graph. So we choose triangles as our basic structures. In WTTM, we build up the global word co-occurrence network of word pairs in the corpus, and find out the word triangles in the network. Then we make use of word triangles we find to train the models. Compared with the existing models, we use the word triangles to draw the close relationship among words. In this way, we reduce the influence of the weak-related patterns and strengthen the highly-related patterns.

Extensive experiments are conducted on real-world dataset to compare WTTM with the baseline models. Through the experiments on topic coherence and document semantic classification, the outstanding results prove that our method is a good choice of topic inferring for short texts.

The rest of the paper is organized as follows. We first give the detail analysis of our approach is followed in Sect. 2; Experimental results are explained in Sect. 3. At last, we'll conclude our present work in Sect. 4 and some measures to perfect the method are also pointed out.

2 Our Approach

Probabilistic topic models find the latent topics of the given documents based on word co-occurrence patterns in the documents, thus the results will be highly effected by data sparsity in short texts scenario. To overcome the difficulty, we came up with the idea called word-network triangle topic model (WTTM), which chooses the more related patterns and represent them with word triangles according to the word co-occurrence network from the corpus. The details will be described as follows.

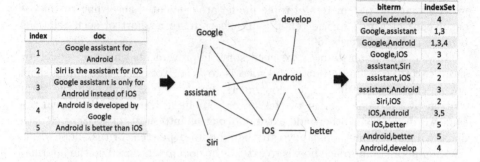

Fig. 1. A example of the word co-occurrence network from the input corpus

2.1 Word Co-occurrence Network

The word co-occurrence network (which is denoted as word network in the following text) is a global undirected graph of given texts. In the word network, nodes are the words appear in the corpus and an edge between two nodes shows that the two words appear in the same document. It's known that topic model is based on the bag of words model so it's possible for us to build up the global network of words in the corpus. Figure 1 shows an example of word network.

Definition 1. *Given a set of documents* $D = \{d_1, d_2, \ldots, d_n\}$, *and every document* d_k *is divided a group of words,* $d_k = \{w_1, w_2, \ldots, w_m\}$. *The word co-occurrence network is represented as* $N = <V, E, S>$.
where:
V *represents the nodes in the network* N, *namely the whole group of words in the given documents* D. $V = \{w_i | w_i \in d_k\}$.
E *represents the edges in the network* N, *namely the word co-occurrence patterns in the given documents* D. $E = \{e_{ij} = (w_i, w_j) | w_i, w_j \in d_k\}$.
$S_{e_{ij}}$ *represents the set of indexes of the edge* e_{ij}. *It expresses indexes of the documents where the words* w_i, w_j *can be found in* D. $S_{e_{ij}} = \{k | e_{ij} = (w_i, w_j) \wedge w_i, w_j \in d_k\}$.
S *is the group of all the index sets of every edge* $e_{ij} \in E$. $S = \{S_{e_{ij}} | e_{ij} \in E\}$.

In our idea, the co-occurrence of two words in the same context shows that the two words share a relationship. And we tag the edges with the indexes of the documents where the connected words co-occurred. It's obvious that the more tags of indexes one edge owns, the more triangles it may be chosen in. Based on the co-occurrence patterns, we build up the word network.

2.2 Word Triangle

With the help of the word network we built up above, we can find the triangle connections among the words in the corpus.

Traditional topic models take advantage of the word co-occurrence patterns by modeling word generation from the document level. However, patterns do not show the relations, but the intensity of relations of patterns should be taken into consideration and we need to choose a new structure to perform strong relations. Triangle is an important structure in graph theory. A triangle is a set of three nodes that are pairwise connected and is arguably one of the most important patterns in terms of understanding the inter-connectivity of nodes in real graphs [3]. And a triangle is the complete graph, that's to say nodes in a triangle share closer relationships than common graphs. So here we came up with the idea of valuing the relation of word co-occurrence patterns and using word triangles to express the closer relations.

Definition 2. *Given the word co-occurrence network $N = <V, E, S>$, for every three words $w_i, w_j, w_k \in V$, if:*

(1) $e_{ij} = (w_i, w_j), e_{ik} = (w_i, w_k), e_{jk} = (w_j, w_k) \wedge e_{ij}, e_{ik}, e_{jk} \in E$.
(2) $s_{e_{ij}}, s_{e_{ik}}, s_{e_{jk}} \in S, s_{e_{ij}} \neq s_{e_{ik}} \wedge s_{e_{ij}} \neq s_{e_{jk}} \wedge s_{e_{ik}} \neq s_{e_{jk}}$.

the three words w_i, w_j, w_k can structure the word triangle $t = (w_i, w_j, w_k)$.

We emphasize that the edges are labeled with different sets of tags to avoid the situation that any group of three words in the same context will be regarded as a word triangle. In our experiments we find that there are few words which are not included in any triangle, we tend to think they are unrelated in the corpus and ignore them.

2.3   Word-Network Triangle Topic Model

The key idea of BTM is to learn the latent topics over given short texts based on the aggregated word patterns in the whole corpus. However, relation of some patterns may not be so close. So we use triangles to evaluate patterns and exclude weak-related ones. In our model, each word triangle is drawn from a specific topic independently. And the probability of the triangle drawn from a specific topic is determined by the proportion that the three words in the triangle to be drawn from the topic. Suppose α and β are the Dirichlet priors. And the process of WTTM can be described as follows:

1. For each topic z
   (a) draw a topic-specific word distribution $\phi_z \sim Dir(\beta)$
2. Draw a topic distribution $\theta \sim Dir(\alpha)$ for the whole collection
3. For each triangle t in the triangle set T
   (a) draw a topic assignment $z \sim Multi(\theta)$
   (b) draw three words: $w_i, w_j, w_k \sim Multi(\phi_z)$

According to the procedure above, the probability of a word triangle $t = (w_i, w_j, w_k)$ can be calculated as:

$$P(t) = \sum_z P(z)P(w_i|z)P(w_j|z)P(w_k|z) = \sum_z \theta_z \phi_{i|z}\phi_{j|z}\phi_{k|z} \qquad (1)$$

And the possibility of the given texts is:

$$P(T) = \prod_{i,j,k} \sum_z \theta_z \phi_{i|z} \phi_{j|z} \phi_{k|z} \tag{2}$$

From the process we directly train the model using the word triangles as basic elements of the topics. We consider the close relationship of triangles to select the stronger relationships among words. What's more, the whole patterns in the corpus are aggregated to find the triangle connections. In consequence, we evaluate the patterns in the corpus to reveal the topics better. To train the parameters mentioned above, we adopt Gibbs sampling the same as BTM.

2.4   Inferring Topics for a Document

From the previous introduction of the procedure we can know that WTTM trains the model based on the whole corpus. Therefore, the training results do not give the topic distribution on the documents directly. We need some efforts to obtain the topics of every document and the distribution. The calculation is as follows:

We assume that the topic distribution of the document is the same as the expectation of the topic distribution of the group of word triangles containing the word patterns in the document:

$$P(z|d) = \sum_t P(z|t)P(t|d) \tag{3}$$

In the formula above, $P(z|t)$ can be calculated according to Bayes' formula based on the distribution of words in topics $P(w|z)$ and the distribution of topics in the corpus $P(z)$ got from WTTM:

$$P(z|t) = \frac{P(z)P(w_i|z)P(w_j|z)P(w_k|z)}{\sum_z P(z)P(w_i|z)P(w_j|z)P(w_k|z)} = \frac{\theta_z \phi_{i|z} \phi_{j|z} \phi_{k|z}}{\sum_z \theta_z \phi_{i|z} \phi_{j|z} \phi_{k|z}} \tag{4}$$

where $P(z) = \theta_z$ and $P(w_i|z) = \phi_{i|z}$.

And the other one to calculate is $P(t|d)$. We can get it easily based on the distribution of word triangles in the document:

$$P(t|d) = \frac{n_d(t)}{\sum_t n_d(t)} \tag{5}$$

where $n_d(t)$ is the frequency of the word triangle t of the group of word triangles in the document d.

And for the documents which structure no word triangle, we infer the topic distribution of such document using the expectation of the topic distributions of words in the documents. According to the formulas above, we can get the topic distribution for given documents.

3   Experiments

In this section, we conduct experiments in the real-world short texts collection of questions with labels from Zhihu to verify the improvements on the algorithm. We take two typical topic models, BTM and LDA, to perform our comparison. Besides, as BTM and WTTM do not model the generation process of documents so we cannot evaluate the models by calculating the perplexity. In consequence, we evaluate the performance of WTTM from two sides, topic coherence and semantic document classification.

3.1   Topic Coherence

In order to compare the methods with quantitative metrics, we introduce the topic coherence to measure the relations of words in the topic we get.

Topic coherence, also called UMass measure, is proposed by Mimno et al. [8] in 2011 for topic quality evaluation. Topic coherence weighs the semantic similarity degree of the words with high-frequency in the topics learned from the documents and then decides the quality of the model. The higher the topic coherence is, the better the topic quality is.

We compare WTTM with LDA and BTM on the short text collection. For all the models, we change the number of topics from 10 to 50. And the average results of topic coherence of the three models are listed in Table 1. Here the number of topics is 30 (P-value < 0.001 by T-test):

Table 1. Average results of topic coherence on the top K words in topics inferred by LDA, BTM, and WTTM on Zhihu questions.

T	5	10	20
LDA	-60.1 ± 4.5	-442.3 ± 18.9	-2752.2 ± 62.4
BTM	-58.0 ± 6.0	-365.7 ± 16.2	-2112.7 ± 30.1
WTTM	-37.4 ± 0.9	-200.0 ± 3.8	-1215.2 ± 17.7

T in Table 1 refers to the number of top words we choose in the topic, ranging from 5 to 20. From the table we find that the average topic coherence of WTTM is obviously better than BTM and LDA. In fact, LDA infers latent topics based on the limited word co-occurrence patterns in each document. So LDA cannot learn the topics of short texts accurately as lack of words. BTM outperforms LDA as it learns topics by modeling directly on the word co-occurrence patterns and overcomes the data sparsity. However, BTM doesn't evaluate the patterns and some weak-related patterns may result in bad influence. Our method exclude such patterns and uses word triangles to express the strong relations. That's the reason why WTTM performs better than BTM. Comparing with the two baseline methods, WTTM can find more related words in one topic.

3.2   Semantic Document Classification

To compare the topic distributions in documents which can express the short documents more accurately, we conduct the experiments to train the document classifiers on the question collection of Zhihu in this part. The topic distributions in each document can be used for dimensionality reduction, that's to say the documents can be represented with a set of topics which can express the contents well. Then we can use the topics and the proportion of topics as the feature vectors to express the documents. So it's convenient for us to value the accuracy of the topical representation of short documents for classification. In the experiments, splitspilt into training subset and testing subset with the ratio 9:1, then we classified them by using random forest in Weka. The number of trees is 50 and the max depth of each tree is 20. We show the accuracy on 10-fold cross validation in Fig. 2.

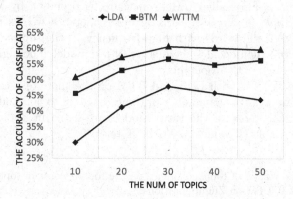

Fig. 2. The accuracy of document classification of WTTM, BTM, and LDA on Zhihu questions

From the results of Fig. 2, we find that WTTM outperforms BTM and LDA in classifying short texts. The results shows that the topics which WTTM infers can represent the documents better. We can figure out again that LDA is not good at topic inferring of short documents. And the reason is similar as what is said in experiments on topic coherence. The data sparsity in short texts affects the results of LDA. It shows that the topics which WTTM infers can represent the documents better. BTM performs better than LDA. WTTM performs better than BTM as it emphasizes the evaluation of the relations between words in the corpus. So we can conculde that WTTM can find better topics to express the texts.

4   Conclusion

Topic inference on short texts is becoming increasingly popular. In this paper, we proposed a new topic model for short texts, namely word-network triangle topic model (WTTM), which finds the word triangles in the word co-occurrence network of the corpus. By contrast, WTTM values the relationship of each pattern through the word triangles and excludes some weak related ones. We conducted experiments and the results proved that WTTM did better than other baseline models. Considering the outstanding performance of WTTM, We can say that WTTM is a good choice for topic inferring on short texts.

However, there are still a lot of improvements we can do to perfect the model. How to make use of the weight of triangles can be the next point to improve. Applying our method to real-world situations is also a good direction for us to explore.

Acknowledgments. This paper is supported by the National Key Research and Development Program of China (Grant No. 2016YFB1001102) and the National Natural Science Foundation of China (Grant No. 61375069, 61403156, 61502227), this research is supported by the Collaborative Innovation Center of Novel Software Technology and Industrialization, Nanjing University.

References

1. Blei, D.M.: Probabilistic topic models. Commun. ACM **55**(4), 77–84 (2012)
2. Blei, D.M., Ng, A.Y., Jordan, M.I.: Latent dirichlet allocation. J. Mach. Learn. Res. **3**(Jan), 993–1022 (2003)
3. Durak, N., Pinar, A., Kolda, T.G., Seshadhri, C.: Degree relations of triangles in real-world networks and graph models. In: Proceedings of the 21st ACM International Conference on Information and knowledge Management, pp. 1712–1716. ACM (2012)
4. Griffiths, T.L., Steyvers, M.: Finding scientific topics. Proc. Nat. Acad. Sci. **101**(suppl. 1), 5228–5235 (2004)
5. Hofmann, T.: Probabilistic latent semantic indexing. In: Proceedings of the 22nd Annual International ACM SIGIR Conference on Research and Development in Information Retrieval, pp. 50–57. ACM (1999)
6. Hong, L., Davison, B.D.: Empirical study of topic modeling in twitter. In: Proceedings of the First Workshop on Social Media Analytics, pp. 80–88. ACM (2010)
7. Lu, H.Y., Xie, L.Y., Kang, N., Wang, C.J., Xie, J.Y.: Don't forget the quantifiable relationship between words: using recurrent neural network for short text topic discovery. In: Thirty-First AAAI Conference on Artificial Intelligence (2017)
8. Mimno, D., Wallach, H.M., Talley, E., Leenders, M., McCallum, A.: Optimizing semantic coherence in topic models. In: Proceedings of the Conference on Empirical Methods in Natural Language Processing, pp. 262–272. Association for Computational Linguistics (2011)
9. Weng, J., Lim, E.P., Jiang, J., He, Q.: Twitterrank: finding topic-sensitive influential twitterers. In: Proceedings of the Third ACM International Conference on Web Search and Data Mining, pp. 261–270. ACM (2010)
10. Yan, X., Guo, J., Lan, Y., Cheng, X.: A biterm topic model for short texts. In: Proceedings of the 22nd International Conference on World Wide Web, pp. 1445–1456. ACM (2013)
11. Zuo, Y., Zhao, J., Xu, K.: Word network topic model: a simple but general solution for short and imbalanced texts. Knowl. Inf. Syst. **48**(2), 379–398 (2016)

Context-Aware Topic Modeling for Content Tracking in Social Media

Jinjing Zhang[1], Jing Wang[2], and Li Li[1(✉)]

[1] School of Computer and Information Science, Southwest University,
Chongqing, China
1476509610@qq.com, lily@swu.edu.cn
[2] Economy and Technology Developing District, Zhengzhou, Henan, China
382876766@qq.com

Abstract. Content in social media is difficult to analyse because of
its short and informal feature. Fortunately, some social media data like
tweets have rich hashtags information, which can help identify meaningful topic information. More importantly, hashtags can express the
context information of a tweet better. To enhance the significant effect
of hashtags via topic variables, this paper, we propose a context-aware
topic model to detect and track the evolution of content in social media
by integrating hashtag and time information named hashtag-supervised
Topic over Time (hsToT). In hsToT, a document is generated jointly by
the existing words and hashtags (the hashtags are treated as topic indicators of the tweet). Experiments on real data show that hsToT capture
hashtags distribution over topics and topic changes over time simultaneously. The model can detect the crucial information and track the
meaningful content and topics successfully.

Keywords: Topic model · Content evolution · Topic over time · Social
media

1 Introduction

In recent years, some conventional topic models such as LDA [1] and PLSA [2]
have been proposed successfully in mining topics for a diverse range of document
genres. However, for the data in tweets, they always fail to achieve high quality
underlying topics because of its short and informal feature. Likely, there are
several types of metadata could help identify the contents of tweets, such as the
associated short url, picture, and #hashtag [3,4]. Among these metadata types,
hashtags always play crucial roles in content analysis. Hashtags can not only
express the context information of a tweet to the fullest, but also act as weakly-
supervised information when sampling topics from certain tweets. Meanwhile,
hashtags enrich the expressiveness of topics.

Motivated by the above, we propose a context-aware topic model to identify
and track the evolution of the contents in social media named hashtag-supervised

© Springer International Publishing AG 2017
L. Chen et al. (Eds.): APWeb-WAIM 2017, Part I, LNCS 10366, pp. 650–658, 2017.
DOI: 10.1007/978-3-319-63579-8_49

Topic over Time(hsToT). This model extends the classical LDA [1] by integrating the hashtags and time information. In hsToT, the distribution of hashtags over topics directly affects the topic sampling for a document. A topic is defined as a set of words that is highly correlated. In addition, in order to capture topic evolution over time for our method, we model each topic with a multinomial distribution over timestamps and uses a beta distribution over a time span covering all the data.

The remainder of paper is organized as following. Section 2 review several representative works. Section 3 presents our approach in detail. We show the data preparation and discuss the experiment results in Sect. 4. The final section concludes the work.

2   Related Work

Topic Model in Social Media. As a powerful text mining tool, topic models have successfully applied to text analysis. Unfortunately, traditional topic models LDA [1] and PLSI [2] do not work well with the messy form of data in Twitter. To overcome the noise in tweets, [5] merges user's tweets as a document. However, they ignore the content detection from topics. Thus, [6] proposes a probabilistic model that a topic depends on not only the users preference but also the preference of users; TCAM [7,8] focuses on analyzing user behaviors combining users intrinsic interests and temporal context; Except the user features, mLDA [3] utilizes multiple contexts such as hashtags and time to discover consensus topics. While these works focus far more on user interests rather than content mining. Besides, some works take advantage semi-structured information, such as TWDA [9] and MA-LDA [10]. However, these methods ignore the dynamic nature of contents in social media.

Topic over Time. To capture the topics change over time, qualitative evolution and quantitative evolution are two main analysis patterns. Qualitative evolution focus on some aspects of a topic like word distribution, inter-topic correlation, vocabulary, etc. [11] uses state space model to model the time variation. DTM [12] and TTM [13] are two typical models. However, the time must be discretized and the length of time intervals must be determined. Quantitative evolution focuses on the amount of data related to some topic at some timestamp, and models the time variation as an attribute of topics. The pioneering works in the literature are TOT [14], COT [4] and [15] where each topic is associated with a beta distribution over time. In this paper, we prefer quantitative evolution and replace the beta distribution with Dirichlet distribution so that the parameters could be simply estimated by Gibbs sampling.

3   Modeling Content Evolution in Social Media

In this section, firstly we introduce some preliminaries, especially the parameters used in the method would be interpreted. Then we explain the details of our hashtag based topic modeling solutions, namely hsToT.

3.1 Preliminaries

As the most popular topic model, Latent Dirichlet Allocation (LDA) has achieved numerous successful extensions. hsToT is LDA-based model, and also a probabilistic generative model. While different from LDA, hsToT includes two additional variables, namely hashtags and timestamps. We can discover the content through a cluster of hashtags that frequently occur with a topic, then the content over time can be observed through topic distribution over timestamps.

Formally, we define a set of tweets as $\mathbf{W} = \{\mathbf{d}\}_{d=1}^{M}$. Each tweet is regarded as a document d and has a timestamp t. Suppose that document d is related to a word sequence $\mathbf{w}_d = \{w_1, w_2, \ldots, w_i, \ldots, w_N\}$ and a hashtag sequence $\mathbf{h}_d = \{h_1, h_2, \ldots, h_i, \ldots, h_L\}$, where N and H is the number of words and hashtags in document d. The rest notations used in this paper are listed in Table 1. For a corpus, T, N, and H are integer constants, while N and H are varying with different document. T is set manually.

Table 1. Notation in hsToT

M, K	Number of document and topics respectively
N, H, T	Number of words, hashtags and timestamps in a document respectively
z, w, h, t, d	Topic, word, hashtag, timestamps and document respectively
θ	Multinomial distribution over topics for a hashtag
φ	Multinomial distribution over words for a topic
ψ	Multinomial distribution over timestamps for a topic
α, β, μ	Dirichlet prior parameters for θ, φ and ψ respectively

3.2 Hashtag-Supervised Topic over Time

In this subsection, we describe hashtag-supervised Topic over Time (hsToT) to directly uncover the latent relationship among topic, hashtag, and time. In hsTor, hashtags act as the weakly-supervised information in topics sampling. Figure 1 shows the graphical models of the hsToT. In Fig. 1, each topic is typically represented by a distribution over words as φ with β as the Dirichlet prior. hsToT also includes two distributions over hashtags and timestamps with respect to topics. hsToT do not directly sample the distribution over topics for a document d. Instead, it sample a hashtag's distribution over topics from the $K \times H$ matrix as the topics distribution of the document. Furthermore, the time feature is first discretized and each tweet is annotated with a discrete timestamp label (e.g. day, month, year). Naturally, time modality is captured by the variable t, and consequently topic evolution over time is obtained using multinomial distribution. In particular, each hashtag is characterized by a distribution over topics as θ with α as the Dirichlet prior. We allocate a topic assignment z_i and a hashtag assignment h_i for each word w_i in the document d. In hsToT, each word is associated with a "hashtag-topic" assignment pair and a "topic-timestamp" pair. The generative process for hsToT is given as follows, shown in Fig. 1. We use variables z_i, h_i and t_i to represent a certain topic, hashtag, and timestamp associated with the word w_i respectively.

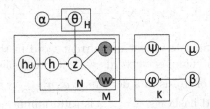

Fig. 1. Graphical model representation of hsToT.

1. For each hashtag $h = 1 : H$, draw the mixture of topics $\theta_h \sim Dir(\alpha)$
2. For each topic $z = 1 : K$, draw the mixture of words $\varphi_z \sim Dir(\beta)$
3. For each topic $z = 1 : K$, draw the mixture of timestamps $\psi_z \sim Dir(\mu)$
4. For each document $d = 1 : M$, draw its words length N, and give its hashtag set $\mathbf{h}_d$
   (a) For each word w_i, $i = 1 : N_d$
       i. Draw a hashtag $h_i \sim Uniform(\mathbf{h}_d)$
       ii. Draw a topic $z_i \sim Mult(\theta_{h_i})$
       iii. Draw a word $w_i \sim Mult(\varphi_{z_i})$
       iv. Draw a timestamp $t_i \sim Mult(\psi_{z_i})$

In hsToT, there are three posterior distributions: hashtag-topic distribution θ, topic-word distribution φ and topic-timestamp distribution ψ. We assume that "topic-word" distribution and "hashtag-topic" distribution are conditionally independent. To efficiently estimate posterior distribution, we employ Gibbs sampling [16]. Thus, the joint probability of words, topics, hashtags and timestamps is Eq. 1.

$$p(\mathbf{w}, \mathbf{h}, \mathbf{t}, \mathbf{z} | \alpha, \beta, \mu, \mathbf{h_d}) = \mathbf{p}(\mathbf{w}|\mathbf{z}, \beta) \cdot \mathbf{p}(\mathbf{h}|\mathbf{h_d}) \cdot \mathbf{p}(\mathbf{t}|\mathbf{z}, \mu) \cdot \mathbf{p}(\mathbf{z}|\mathbf{h_d}, \alpha) \qquad (1)$$

In order to infer the hidden variables, we compute the posterior distribution of the hidden variables. The likelihood of a document d is Eq. 2.

$$
\begin{aligned}
p(\mathbf{w}_d | \theta, \varphi, \psi, \mathbf{h}_d) &= \prod_{i=1}^{N} p(w_i | \theta, \varphi, \psi, \mathbf{h}_d) \\
&= \prod_{i=1}^{N_d} \sum_{j=1}^{L_d} \sum_{k=1}^{K} \sum_{s=1}^{T} p(w_i, z_i = k, h_i = j, t_i = s | \theta, \varphi, \psi, \mathbf{h}_d) \\
&= \prod_{i=1}^{N_d} \sum_{j=1}^{L_d} \sum_{k=1}^{K} \sum_{s=1}^{T} p(w_i | z_i = k, \varphi) p(z_i = k | h_i = j, \theta) p(t_i = s | z_i = k, \psi) p_{jh_i} \\
&= \prod_{i=1}^{N_d} \sum_{j=1}^{L_d} \sum_{k=1}^{K} \sum_{s=1}^{T} \varphi_{w_i, k} \theta_{k, j} \psi_{s, k} p_{jh_i}
\end{aligned}
\qquad (2)
$$

Where p_{jh_i} represents the probability of $h_i = j$ when sampling a hashtag h_i from $\mathbf{h}_d$. The generating probability of the corpus is:

$$p(\mathbf{W}|\theta, \varphi, \psi, \mathbf{h}) = \prod_{d=1}^{M} p(\mathbf{w}_d|\theta, \varphi, \psi, \mathbf{h}_d) \tag{3}$$

The estimation method of posterior distributions in hsToT is Eq. 4.

$$p(z_i = k, h_i = j|\mathbf{w}, \mathbf{h}^{-i}, \mathbf{t}, \mathbf{z}^{-i}) \propto \frac{n_{w_i,-i}^k + \beta}{n_{w',-i}^k + V\beta} \times \frac{n_{k,-i}^j + \alpha}{n_{k',-i}^j + K\alpha} \times \frac{n_{t_i,-i}^k + \mu}{n_{t',-i}^k + T\mu} \tag{4}$$

where $-i$ means assignments except for current word in the current document d. In Eq. 4, $n_{k,-i}^j$ is the number of words assigned to topic k in a document d, $n_{k',-i}^j$ is total number of words in a document d; $n_{w_i,-i}^j$ is the number of words assigned to topic k and hashtags j, $n_{k',-i}^j$ is total number of words assigned to topic k; $n_{t,-i}^j$ is the number of time words assigned to topic k and hashtags j, $n_{t',-i}^j$ is total number of words assigned to topic k and hashtags j. Finally, when the sampling process reaches the convergence, we can get the results of φ, θ, and ψ by:

$$\varphi_{w,k} = \frac{n_w^k + \beta}{n_{w'}^k + V\beta} \quad \theta_{k,j} = \frac{n_k^j + \alpha}{n_{k'}^j + K\alpha} \quad \psi_{s,k} = \frac{n_t^k + \mu}{n_{t'}^k + T\mu} \tag{5}$$

From the generative process of hsToT, time modality is involved in topic discovery. However, this may impact the homogeneity of topics because time modality is assumed having the same "weight" in word modality. In practice, it is not. To address this issue, we adopt the same strategy as in TOT [14] where a balancing hyperparameter is introduced in order to balance word and time contribution in topic discovery. Naturally, we set the hyperparameter as the inverse of the number of words n_d.

4   Experiments

Experiments are conducted based on evaluation on results of topic detection and topic evolution over time. We take three topic models TOT [14], COT [4] and hgToT, which is our another model as the baselines. (hgToT is generally similar to COT in the usage of hashtags, but whose time modeling method changes the beta distribution into the multinomial distribution.) TOT is extended from LDA by adding a Beta distribution over timestamps for topics. COT is extended from TOT by adding a multinomial distribution over hashtags for topics.

4.1   Data Preparation

The experiments are conducted on a twitter data set, named "TREC2011"[1]. The original data contains nearly 16 millions tweets posted from January 23rd to February 8th in 2011. Each tweet includes a user id and a timestamp. The

---

[1] http://trec.nist.gov/data/microblog2011.html.

process of the raw data is similar to the steps in [17]. The properties of the dataset are given in Table 2. To guarantee the convergence of Gibbs sampling, all results were obtained after 1000 iterations. Timestamps is divided by days. The hyperparameters in generative models (hsToT, hgToT, TOT) are set as $50/K$ for α and 0.04, 0.04, 0.01 for β, γ and μ respectively [11].

Table 2. Dataset properties

♯ tweets	♯ Unique words	♯ hashtags	♯ Average words	♯ Average hashtags
304,480	12,160	98,649	5.11	1.42

4.2 Evaluation on Topic Detection

To assess the effectiveness of topic models, a typical metric like perplexity on a held-out test set [17] has been widely used. The perplexity represents the performance of document modeling by comprehensively estimate the results of $p(z|d)$ and $p(w|z)$. Another automatic evaluation metrics *coherent score*[18] is proposed to measure the quality of topics from the perspective of topic visualization and semantic coherent. In this paper, we choose perplexity and coherent score as evaluation criteria.

Perplexity. Perplexity indicates the uncertainty in predicting a single word. The lower the perplexity score is, the higher the performance will be. We compute this metric according to the method [1]. To equilibrate the different usage of hashtags in these methods, the computation of $p(w_d)$ is different. For TOT, the computation method of $p(w_d)$ is same as [1]. While for hsToT, hgToT and COT, $p(w_d) = \sum_{N_d} p(w_i) + \sum_{L_d} p(h_i)$. In this step, we hold out 10% of data for test and train these methods on the remaining 90% data. Figure 2 states the perplexity results with topic number $k = 20, 30, 40, 50, 60$. From the Fig. 2, hsToT obviously outperforms other methods. This indicates that we can indeed improve the document modeling performance by taking advantage of hashtags especially regarding hashtags as weak-supervised information. In addition, For hsToT, the perplexity value reduces gradually with the increase of topic number, and then tends to stable when $k \geq 50$. While TOT is running into over-fitting. This means that our method is more stable. Based on this observation, the number of topic K is fixed at 50 in the remaining experiments.

Coherent Score. To intuitively investigate the quality of topics, we analyze the topics from visualization perspective. For each topic, we take top 5 words or hashtags ordered by $p(w|z)$ or $p(h|z)$) as their semantic representation. By observing the 50 topics, there are two major kind of topics. One is "common topics", which are related to users's daily lives. The other is "time-sensitive

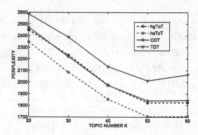

Fig. 2. Perplexity results with different topic number K.

topics" which may be some emergencies or hot news events. Overall, compared with TOT and hsToT can discover more meaningful hashtags and words highly related to a topic. Besides, hsToT can detect the content from a topic which TOT can not achieve. Table 3 lists an example of a common topic "EGYPT" learn by all hsToT and TOT methods.

Table 3. A sample of semantic representation of topic "EGYPT"

hgToT		hsToT		TOT
#egypt,	egypt	#egypt,	egypt	obama
#election,	obama	#mubarak,	people	mubarak
#25-Jan,	#egypt	#election,	obama	#mubarak
#tcot,	mubarak	#turbulence,	mubarak	turbulence
#sotu,	egyptian	#news,	egyptian	egypt

For TOT, both words and hashtags occur in the "topic-word" distribution. While in hgToT and hsToT, words and hashtags only occur in "topic-word" and "topic-hashtag" distribution respectively. In addition, topic "EGYPT" also shows that hsToT outperforms TOT in discovering meaningful hashtag i.e., the semantics of topics for a common topic. For example, TOT only discovers one hashtag "#mubarak", whereas hsToT discovered more meaningful hashtags.

In order to quantitatively evaluate the topic quality of all test methods, we further utilize the automated metric, namely *coherent score*. The coherence score is that words belonging to a single concept will tend to co-occur within the same documents [19]. A larger coherence score means the topics are more coherent. Given a topic z and its top n words $V^{(z)} = (v_1^{(z)}, \ldots, v_n^{(z)})$ ordered by $p(w|z)$, the coherent score can be defined as:

$$C(z; V^{(z)}) = \sum_{t=2}^{n} \sum_{l=1}^{t} \log \frac{D(v_t^{(z)}, v_l^{(z)}) + 1}{D(v_l^{(z)})} \tag{6}$$

where $D(v)$ is the document frequency of word v, $D(v, v')$ is the number of document in which words v and v' co-occurred. The final results are

$\frac{1}{K} \sum_k C(z_k; V^{(z_k)})$. For TOT, the average coherent score can be directly captured by this way. While for other three methods, a topic is jointly associated with a multinomial distribution over words and a multinomial distribution over hashtags. Therefore, we also need consider the top n hashtags when computing the coherent score of a given topic. The final average coherent score in these three methods is $\frac{1}{2K} \sum_k (C(z_k; V^{(z_k)}) + C(z_k; H^{(z_k)}))$.

The result is shown in Table 4, the number of top words ranges from 5 to 20. From Table 4, hsToT achieves the best performance(with p-value <0.01 by T-test), and hsToT receives the highest coherence score. hgToT and COT outperform TOT by 0.086 on average. It also shows hashtag co-occurrence frequency can help detect more coherent words, and improve the topic coherence greatly. Moreover, compared with the strategy of hashtags in COT, hsToT always receives better results. By observing the different results with the change of top n words (or hashtags), we find that the coherent score of all methods gradually reduces when the number of n increases.

Table 4. Average coherence score on the top n words (hashtags) in topics.

n	5	10	20
hgToT	−62.6	−239.3	−1043.5
hsToT	**−61.5**	**−236.5**	**−1029.2**
COT	−62.7	−239.4	−1042.9
TOT	−64.3	−243.6	−1066.4

5   Conclusions

In this paper, we introduced a new model (hsToT) of tracking topics over time in social media. By considering features such as hashtags, words and timestamps jointly, our model can successfully identify the meaningful topics precisely and track topic changes over time appropriately. Experiments on real dataset illustrate the effectiveness and efficiency of our methods.

Acknowledgments. The work was supported by the Fundamental Research Funds For the Central Universities (No. XDJK2017D059).

References

1. Blei, D.M., Ng, A.Y., Jordan, M.I.: Latent dirichlet allocation. J. Mach. Learn. Res. **3**, 993–1022 (2003)
2. Hofmann, T.: Probabilistic latent semantic indexing. In: 22nd ACM SIGIR, pp. 50–57 (1999)

3. Tang, J., Zhang, M., Mei, Q.: One theme in all views: modeling consensus topics in multiple contexts. In: Proceedings of the 19th ACM SIGKDD International Conference on Knowledge Discovery and Data Mining, pp. 5–13 (2013)
4. Alam, M.H., Ryu, W.J., Lee, S.: Context over time: modeling context evolution in social media. In: Proceedings of the 3rd Workshop on Data-Driven User Behavioral Modeling and Mining from Social Media, pp. 15–18 (2014)
5. Hong, L., Davison, B.D.: Empirical study of topic modeling in twitter. In: Proceedings of the First Workshop on Social Media Analytics, pp. 80–88 (2010)
6. Kim, Y., Shim, K.: TWITOBI: a recommendation system for twitter using probabilistic modeling. In: 2011 IEEE 11th International Conference on Data Mining (ICDM), pp. 340–349 (2011)
7. Yin, H., Cui, B., Chen, L., Hu, Z., Huang, Z.: A temporal context-aware model for user behavior modeling in social media systems. In: Association for Computing Machinery, Special Interest Group on Management of Data, pp. 1543–1554 (2014)
8. Yin, H., Cui, B., Chen, L., Hu, Z., Zhou, X.: Dynamic user modeling in social media systems. ACM Trans. Inf. Syst. **33**(3), 10 (2015)
9. Li, S., Huang, G., Tan, R., Pan, R.: Tag-weighted dirichlet allocation. In: 2013 IEEE 13th International Conference on Data Mining (ICDM), pp. 438–447 (2013)
10. Wang, J., Li, L., Tan, F., Zhu, Y., Feng, W.: Detecting hotspot information using multi-attribute based topic model. PLoS ONE **10**(10), e0140539 (2015)
11. Dermouche, M., Velcin, J., Khouas, L., Loudcher, S.: A joint model for topic-sentiment evolution over time. In: 2014 IEEE International Conference on Data mining (ICDM), pp. 773–778 (2014)
12. Blei, D.M., Lafferty, J.D.: Dynamic topic models. In: Proceedings of the 23rd International Conference on Machine learning, pp. 113–120 (2006)
13. Iwata, T., Watanabe, S., Yamada, T., Ueda, N.: Topic tracking model for analyzing consumer purchase behavior. In: IJCAI. Citeseer, vol. 9, pp. 1427–1432 (2009)
14. Wang, X., McCallum, A.: Topics over time: a non-markov continuous-time model of topical trends. In: Proceedings of the 12th ACM SIGKDD International Conference on Knowledge Discovery and Data Mining, pp. 424–433 (2006)
15. Yin, H., Cui, B., Lu, H., Huang, Y., Yao, J.: A unified model for stable and temporal topic detection from social media data. In: IEEE International Conference on Data Engineering, vol. 48, pp. 661–672 (2013)
16. Heinrich, G.: Parameter estimation for text analysis. Technical report (2005)
17. Cheng, X., Yan, X., Lan, Y., Guo, J.: BTM: topic modeling over short texts. IEEE Trans. Knowl. Data Eng. **26**(12), 2928–2941 (2014)
18. Mimno, D., Wallach, H.M., Talley, E., Leenders, M., McCallum, A.: Optimizing semantic coherence in topic models. In: Proceedings of the Conference on Empirical Methods in Natural Language Processing, pp. 262–272 (2011)
19. Yan, X., Guo, J., Lan, Y., Cheng, X.: A biterm topic model for short texts. In: Proceedings of the 22nd International Conference on World Wide Web, pp. 1445–1456 (2013)

Author Index